Automation, Communication and Cybernetics in Science and Engineering 2015/2016

Sabina Jeschke · Ingrid Isenhardt
Frank Hees · Klaus Henning
Editors

Automation, Communication and Cybernetics in Science and Engineering 2015/2016

Editors
Sabina Jeschke
Ingrid Isenhardt
Frank Hees
Klaus Henning
IMA/ZLW & IfU - RWTH Aachen University
Faculty of Mechanical Engineering
Aachen
Germany

Das Buch wurde gedruckt mit freundlicher Unterstützung der RWTH Aachen University

ISBN 978-3-319-82620-2   ISBN 978-3-319-42620-4   (eBook)
DOI 10.1007/978-3-319-42620-4

Mathematics Subject Classification (2010): 68-06, 68Q55, 68T30, 68T37, 68T40

CR Subject Classification: I.2.4, H.3.4

Cover figure: © Fotolia - red150770

Printed on acid-free paper

This Springer imprint is published by Springer Nature
The registered company is Springer International Publishing AG Switzerland

Foreword

Dear Reader,

Today we present the fourth instalment of our book series **Automation, Communication and Cybernetics**. Like its predecessors this book brings together our scientifically diverse and widespread publications from the period of July 2014 to June 2016. The peer-reviewed publications have been published in recognised journals, books or conference proceedings of the various disciplinary cultures. Below you find an up to date version of the organisational structure of our Cybernetics Lab. It is headed by Sabina Jeschke with Ingrid Isenhardt and Frank Hees as her Deputy and Vice Deputy. The former Head Klaus Henning still supports us as Senior Advisor. The Cybernetics Lab itself consists of three institutes: the **Institute of Information Management in Mechanical Engineering IMA**, the **Center for Learning and Knowledge Management ZLW** and the **Associated Institute for Management Cybernetics e.V. IfU**, which are managed by Tobias Meisen, Anja Richert and René Vossen respectively. Our research activities are arranged in nine different research groups whose activities are described further below.

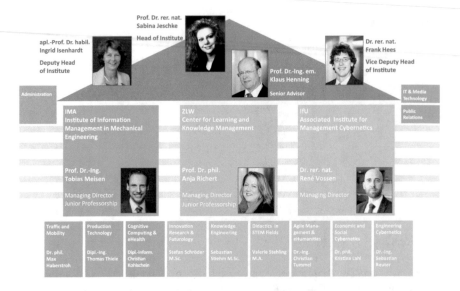

Although the structure itself has not changed the people or their statuses have. Our managing director of the IMA, Tobias Meisen, is now a Junior Professor and within the Cybernetics Lab we have several new research group leaders:

- At the IMA Thomas Thiele is now the leader of the group **Production Technology**, Christian Kohlschein of the group **Cognitive Computing & eHealth** and Max Haberstroh heads the group **Traffic and Mobility**.
- In the ZLW Christian Tummel now heads the research group **Agile Management & eHumanties**. Stefan Schröder took over the team **Innovation & Research Futurology**, Sebastian Stiehm the group **Knowledge Engineering** and Valerie Stehling the group **Didactics in STEM Fields**.
- At the IfU the Heads of the research groups are now Kristina Lahl and Sebastian Reuter for the groups **Economic and Social Cybernetics and Engineering Cybernetics** respectively.

The scientific core of the Institute of Information Management in Mechanical Engineering – IMA consists of three research groups:

- The scope of the research group **Production Technology** is to provide innovative research regarding information management for Industry 4.0. The group is specialised in methods and procedures from computer science to semantically integrate, to consolidate and to propagate data generated in these domains. In addition, their research focuses on visualisation and interaction techniques to enable the user to analyse the retrieved information in an explorative and interactive way. Thereby, their research covers a broad range of different areas especially virtual and automated production. Meeting the challenges of information management within these areas, the group studies information integration, descriptive and predictive analysis using a variety of techniques from artificial intelligence like regression, machine learning, natural language processing and data mining as well as visual analytics. Regarding the domain of virtual production, the group has shaped the concept of the Virtual Production Intelligence (VPI) to collaboratively and holistically support product-, factory- and machine planners. The work of the group provides essential basics to facilitate the realisation of cyber-physical production systems (CPPS) and therefore is a cornerstone of information management in Industry 4.0.

- The research group **Traffic and Mobility** is working on concepts for multi-modal freight transport and urban mobility, intelligent transport systems and on the design of user-friendly and barrier-free mobility solutions and human–machine-interaction. In its projects, the research group investigates concepts for autonomous vehicles, advanced driver assistant systems and the interactions and interdependencies between humans, organisation and technology. In order to develop holistic solutions, the interdisciplinary team combines skills and knowledge from engineering, computer science, sociology and economics. The applied methods of the research group range from simulator and real-life testing, over usage-centered design, empirical studies and acceptance or mental stress and strain analysis. One approach to reach the ideal of efficient freight traffic of the future is to use modular, worldwide usable loading units with appropriate transport carriers. All research is based upon the holistic consideration of the three recursion dimensions: human, organisation and technology. The activities of the research group include the research and development of new technologies as well as the development of methods and tools for the product development process in the above mentioned application fields.

- The research group **Cognitive Computing & eHealth** focuses on research considering information management supporting healthcare. The group is specialised to meet the challenges within the research fields predictive data analytics and visual analytics. The research group understands itself as an "integrator" within the eHealth domain: educating and providing experts in the mentioned research fields, but also understanding the importance of covering and dealing with problems of all phases (i.e. needs assessment, integration, evaluation, and deployment) of the information management cycle. Lately, the group focused on research topics occurring in scenarios of medical emergencies, thus developing an intelligent and reliable ad-hoc network structure streaming medical data in real-time from the case of incident to an expert. Predictive analytics is used to detect upcoming delays, future connection losses, or approaching quality reductions. The eHealth group coined the term "prescient profiling" which is used to describe an AI driven concept selecting relevant laypersons to nearby medical emergencies. To determine relevancy the solution considers for example traveling speed, known behavioural patterns (i.e. trajectories), current circumstances, and infrastructural limitations. Currently, the group works on an algorithm to predict the emotions of a driver interacting with a navigation system to adapt the systems behaviour accordingly. In the near future, the group will also use its expertise to establish a complex and highly

available information management system for rapidly changing ad-hoc infrastructures that are for example needed to ensure information availability in the
case of major incidents.

The Center for Learning and Knowledge Management – ZLW has four research
groups:

- The research group **Innovation Research and Futurology** focuses on two
  fields. Innovation research concentrates on a concept of innovation management, which not only comprises planning, realisation, and design of processes
  and structures to create innovation, but also stresses the innovative capability.
  The first research field focuses on innovation systems with various dimensions
  like regional and national innovation systems as well as their relevant subsystems, which are created and analysed from a cybernetic perspective. This is
  achieved by a holistic consideration of the system-intrinsic dimensions "human,
  organisation and technology", in order to produce innovative capability of the
  involved actors under competitive and sustainable conditions. The second field
  of the research group depicts futurology. Here, a monitoring approach is applied
  for different research, development and funding programs. Consequently, a
  range of future trends, scenarios and development strategies is derived for
  respective target groups. This expertise is supplied to experts in science,
  economy and policy.
- The research group **Knowledge Engineering** currently focuses on three topics:
  First, it supports and explores the development and steering of inter- and
  transdisciplinary networks and clusters of excellence with the aim to identify
  and promote synergies as a source of innovation. A continuous qualitative and
  quantitative evaluation of the research network as well as text mining of scientific publications with machine learning are realized. The use of interactive
  data visualizations for a feedback and exploration of the results is considered in
  both cases. Second, within the framework of demographic change in the labour
  world, the research group develops concepts for the evaluation and analysis of
  company's demographic alignment. Making use of a wide range of quantitative
  research methods, holistic demography management systems can be implemented, which also respect the perspectives of the various stakeholders
  involved. Third, the research group focuses on the identification of opportunities
  and potentials for (re-) integrating production sites into urban space with a
  holistic, transdisciplinary view. Realizing a socio-technical research-approach,
  the research group develops factors and scenarios of urban production by
  combining methods of empirical social research with data science.
- With an interdisciplinary team of communication scientists, engineers, psychologists, sociologists and computer scientists the research group **Didactics in
  STEM Fields** is dealing with challenges of didactics, especially those of the
  STEM Fields, including mathematics, computer sciences and engineering. To
  ensure the development of successful didactical concepts, the involvement of
  every actor actively participating in education is needed. Therefore groups of
  students, teaching staff, intermediate organisations and other experts on

university didactics are involved in our research activities. The user oriented approach of the research focuses on learning in virtual environments, learning with natural user interfaces and VR-technology, remote and virtual laboratories and other forms of computer and web based learning. Moreover, social aspects of learning in a higher education context are investigated. Here, the focus lies on mentoring concepts, students' mobility and service based learning methods. In all its activities, the research group considers the whole student life cycle, from pupils, bachelor and master students up to doctoral candidates.

- The research group **Agile Management & eHumanities** deals with the application of data analytics approaches in social sciences and humanities. The major effort is the examination of how computer-assisted processes and digital resources are systematically used in these disciplines while its main emphasis is put on the field of data analytics with special regards to social media. In order to manage the continuously increasing complexity and dynamics in organisational structures the field of Agile Management investigates the application and implementation of agile methods, techniques, principles, and values. As far as the application area of research on competencies is concerned, the analysis of the "digital footprints" from employers and staff is focused, which allows to draw conclusions on hidden profile characteristics. The identification of these hidden characteristics and their significance for tomorrow's job market are current research topics in this field. Furthermore, the research group conducts analyses on effects of these characteristics in order to enhance individual competencies by optimizing qualification processes and programmes in the context of academic teaching.

The Associated Institute for Management Cybernetics e.V. – IfU used the opportunity to extend its research focus once more:

- The research team **Economic and Social Cybernetics** deals with cybernetic methods and tools for industrial applications. The main research topics include the assessment of organisational culture and structure, business model innovation and development of decision support tools. In the context of evaluation and decision support enhanced economic assessment tools including uncertainty and soft aspects and sustainability assessment tools are generated. In interdisciplinary research projects cybernetic tools and solutions for complex problems in collaboration with industrial and research partners are developed. The employed methods include system dynamics, viable system model, organizational culture assessment instrument (OCAI) and business model canvas. Furthermore, cybernetic tools for the development of sustainable product strategies, design of efficient organisational structure, culture based implementation of quality management, and change processes are applied.
- The research team **Engineering Cybernetics** is a part of the Institute for Management Cybernetics at the RWTH Aachen University. Its research objectives are intelligent planning and control algorithms for technical systems. The focus is on mobile robotics within intralogistic applications as well as process planning and industrial robotics. Here the group addresses aspects of

human robot interaction and collaboration. The main goal is to endow the respective technical systems with autonomy and situational awareness in order to achieve more robust behaviour and an increased flexibility while at the same time simplifying the interaction with those systems. (Multi-)agent technologies, closed loop control systems and visual serving, and natural interface technologies play an important role. The research group also maintains the institute's school labs.

We would like to thank our scientific researchers who work hard and publish continuously and without whom we would not have accomplished the now fourth instalment of this diverse and comprehensive collection. Further, we would like to acknowledge the support of our administrative and technical staff who fight the battles with bureaucracy and IT-technologies for us to keep our minds focused on our research projects and the education of students. At last we would like to thank our Public Relations team and especially Miro Tommack for the unification of all these different articles.

Aachen, Germany   Sabina Jeschke
September 2016   Ingrid Isenhardt
                                                                         Frank Hees
                                                                      Klaus Henning

Contents

Contents

Contributors

Susanne Aghassi Fraunhofer Institute for Production Technology IPT, Aachen, Germany

Bahoz Abbas IMA/ZLW & IfU, RWTH Aachen University, Aachen, Germany

Anas Abdelrazeq IMA/ZLW & IfU, RWTH Aachen University, Aachen, Germany

Philipp Abel Institute of Textile Technology (ITA), RWTH Aachen University, Aachen, Germany

Toufik Al-Khawly IMA/ZLW & IfU, RWTH Aachen University, Aachen, Germany

Mohammad Alfraheed IMA/ZLW & IfU, RWTH Aachen University, Aachen, Germany

Ursula Bach IMA/ZLW & IfU, RWTH Aachen University, Aachen, Germany

Saskia Bakuhn IMA/ZLW & IfU, RWTH Aachen University, Aachen, Germany

Guido Becke artec,Forschungszentrum Nachhaltigkeit, Universität Bremen, Bremen, Germany

Wiebke Behrens IMA/ZLW & IfU, RWTH Aachen University, Aachen, Germany

Peter Bleses artec, Forschungszentrum Nachhaltigkeit, Universität Bremen, Bremen, Germany

Esther Borowski IMA/ZLW & IfU, RWTH Aachen University, Aachen, Germany

André Breakling Fraunhofer Institute for Production Technology IPT, Aachen, Germany

Jan Brinker Department of Mechanism Theory and Dynamics of Machines (IGM), RWTH Aachen University, Aachen, Germany

Andrea Buratti Werkzeugmaschinenlabor (WZL), RWTH Aachen University, Aachen, Germany

Meike Bücker IMA/ZLW & IfU, RWTH Aachen University, Aachen, Germany

Christian Büscher IMA/ZLW & IfU, RWTH Aachen University, Aachen, Germany

André Calero Valdez Human-Computer Interaction Center, RWTH Aachen University, Aachen, Germany

Oleg Cernavin BC Forschung GmbH, Wiesbaden, Germany

Kurt-Georg Ciesinger Gaus GmbH, Dortmund, Germany

Burkhard Corves Department of Mechanism Theory and Dynamics of Machines (IGM), RWTH Aachen University, Aachen, Germany

Johann Philipp von Cube Fraunhofer Institute for Production Technology IPT, Aachen, Germany

Michael Czaplik MedIT, RWTH Aachen University, Aachen, Germany; Department of Anaesthesiology, University Hospital Aachen, Aachen, Germany

Alicia Dröge IMA/ZLW & IfU, RWTH Aachen University, Aachen, Germany

Antje Ducki Fachbereich I: Wirtschafts- und Gesellschaftswissenschaften, Beuth Hochschule für Technik Berlin, Berlin, Germany

Julia Eich IMA/ZLW & IfU, RWTH Aachen University, Aachen, Germany

Jesko Elsner IMA/ZLW & IfU, RWTH Aachen University, Aachen, Germany

Philipp Ennen IMA/ZLW & IfU, RWTH Aachen University, Aachen, Germany

Urs Eppelt Lehrstuhl für Lasertechnik, RWTH Aachen University, Aachen, Germany

Daniel Ewert IMA/ZLW & IfU, RWTH Aachen University, Aachen, Germany

Alexia Fenollar Solvay IMA/ZLW & IfU, RWTH Aachen University, Aachen, Germany

Alexander Ferrein MASCOR Institute, Aachen University of Applied Sciences, Aachen, Germany

Tobias Fürtjes Werkzeugmaschinenlabor (WZL), RWTH Aachen University, Aachen, Germany

Thomas Gries ITA, RWTH Aachen University, Aachen, Germany

Michael Grosch Karlsruhe Institute of Technology, Karlsruhe, Germany

Kerstin Groß IMA/ZLW & IfU, RWTH Aachen University, Aachen, Germany

Julia Günther IMA/ZLW & IfU, RWTH Aachen University, Aachen, Germany

Max Haberstroh IMA/ZLW & IfU, RWTH Aachen University, Aachen, Germany

Marcel Haeske ITA, RWTH Aachen University, Aachen, Germany

Frank Hees IMA/ZLW & IfU, RWTH Aachen University, Aachen, Germany

Klaus Henning IMA/ZLW & IfU, RWTH Aachen University, Aachen, Germany

Frederik Hirsch Department of Anaesthesiology, University Hospital Aachen, Aachen, Germany

Max Hoffmann IMA/ZLW & IfU, RWTH Aachen University, Aachen, Germany

Lasse Härtel Fraunhofer Institute for Production Technology IPT, Aachen, Germany

Mathias Hüsing Department of Mechanism Theory and Dynamics of Machines (IGM), RWTH Aachen University, Aachen, Germany

Susanne Ihsen Gender Studies in Ingenieurwissenschaften, Technische Universität München, München, Germany

Ingrid Isenhardt IMA/ZLW & IfU, RWTH Aachen University, Aachen, Germany

Ulrich Jansen Chair for Nonlinear Dynamics of Laser Processing, RWTH Aachen University, Aachen, Germany

Daniela Janßen IMA/ZLW & IfU, RWTH Aachen University, Aachen, Germany

Yves Jeanrenaud Gender Studies in Ingenieurwissenschaften, Technische Universität München, München, Germany

Sabina Jeschke IMA/ZLW & IfU, RWTH Aachen University, Aachen, Germany

Claudia Jooß IMA/ZLW & IfU, RWTH Aachen University, Aachen, Germany

Sabine Kadlubek IMA/ZLW & IfU, RWTH Aachen University, Aachen, Germany

Achim Kampker WZL, RWTH Aachen University, Aachen, Germany

Diane Keng School of Engineering, Santa Clara University, Santa Clara, CA, USA

Walter Kimmelmann Werkzeugmaschinenlabor (WZL), RWTH Aachen University, Aachen, Germany

Rüdiger Klatt Forschungsinstitut für innovative Arbeitsgestaltung und Prävention e.V. (FIAP e.V.), Wissenschaftspark Gelsenkirchen, Gelsenkirchen, Germany

Philipp Kosse Werkzeugmaschinenlabor (WZL), RWTH Aachen University, Aachen, Germany

Markus Kowalski IMA/ZLW & IfU, RWTH Aachen University, Aachen, Germany

Kai Kreisköther WZL, RWTH Aachen University, Aachen, Germany

Helga Krieger Institute of Textile Technology (ITA), RWTH Aachen University, Aachen, Germany

Moritz Krunke Werkzeugmaschinenlabor (WZL), RWTH Aachen University, Aachen, Germany

Bernd Kujat AUDI AG, Ingolstadt, Germany

Larissa Köttgen IMA/ZLW & IfU, RWTH Aachen University, Aachen, Germany

Kristina Lahl IMA/ZLW & IfU, RWTH Aachen University, Aachen, Germany

Gerhard Lakemeyer Knowledge-based Systems Group, RWTH Aachen University, Aachen, Germany

Renaud De Landtsheer CETIC Research Centre, Charleroi, Belgium

Thomas Langhoff Prospektiv Gesellschaft für betriebliche Zukunftsgestaltungen mbH, Dortmund, Germany

Ingo Leisten IMA/ZLW & IfU, RWTH Aachen University, Aachen, Germany

Karsten Lensing Center of Higher Education, TU Dortmund University, Dortmund, Germany

Laura Lenz IMA/ZLW & IfU, RWTH Aachen University, Aachen, Germany

Riccardo Manzotti Department of Linguistics and Philosophy, Massachusetts Institute of Technology, Cambridge, MA, USA

Philippe Massonet CETIC Research Centre, Charleroi, Belgium

Dominik May Center of Higher Education, TU Dortmund University, Dortmund, Germany

Tobias Meisen IMA/ZLW & IfU, RWTH Aachen University, Aachen, Germany

Philipp Meisen IMA/ZLW & IfU, RWTH Aachen University, Aachen, Germany

Larissa Müller IMA/ZLW & IfU, RWTH Aachen University, Aachen, Germany

Sarah L. Müller IMA/ZLW & IfU, RWTH Aachen University, Aachen, Germany

Friedemann W. Nerdinger Institut für Betriebswirtschaftslehre, Universität Rostock, Rostock, Germany

Tim Niemueller Knowledge-based Systems Group, RWTH Aachen University, Aachen, Germany

Gustavo Ospina CETIC Research Centre, Charleroi, Belgium

Alexander Paulus IMA/ZLW & IfU, RWTH Aachen University, Aachen, Germany

Marcus Petermann Lehrstuhl für Feststoffverfahrenstechnik, Ruhr Universität Bochum, Bochum, Germany

Lana Plumanns IMA/ZLW & IfU, RWTH Aachen University, Aachen, Germany

Christophe Ponsard CETIC Research Centre, Charleroi, Belgium

Isabel Prause Department of Mechanism Theory and Dynamics of Machines (IGM), RWTH Aachen University, Aachen, Germany

Stephan Printz IMA/ZLW & IfU, RWTH Aachen University, Aachen, Germany

Marco Recchioni Airport Division, Inform GmbH Aachen, Aachen, Germany

Rudolf Reinhard IMA/ZLW & IfU, RWTH Aachen University, Aachen, Germany

Sebastian Reuter IMA/ZLW & IfU, RWTH Aachen University, Aachen, Germany

Anja Richert IMA/ZLW & IfU, RWTH Aachen University, Aachen, Germany

Michael Rix IMA/ZLW & IfU, RWTH Aachen University, Aachen, Germany

Anne Kathrin Schaar Human-Computer Interaction Center, RWTH Aachen University, Aachen, Germany

Daniel Schilberg IMA/ZLW & IfU, RWTH Aachen University, Aachen, Germany

Robert Schmitt Werkzeugmaschinenlabor (WZL), RWTH Aachen University, Aachen, Germany; Fraunhofer Institute for Production Technology IPT, Aachen, Germany

Stefan Schröder IMA/ZLW & IfU, RWTH Aachen University, Aachen, Germany

Günther Schuh Werkzeugmaschinenlabor (WZL), RWTH Aachen University, Aachen, Germany; Fraunhofer Institute for Production Technology IPT, Aachen, Germany

Stella Schulte-Cörne IMA/ZLW & IfU, RWTH Aachen University, Aachen, Germany

Wolfgang Schulz Chair for Nonlinear Dynamics of Laser Processing, RWTH Aachen University, Aachen, Germany; IMA/ZLW & IfU, RWTH Aachen University, Aachen, Germany

Katharina Schuster IMA/ZLW & IfU, RWTH Aachen University, Aachen, Germany

Tomas Sivicki IMA/ZLW & IfU, RWTH Aachen University, Aachen, Germany

Thorsten Sommer IMA/ZLW & IfU, RWTH Aachen University, Aachen, Germany

Valerie Stehling IMA/ZLW & IfU, RWTH Aachen University, Aachen, Germany

Sebastian Stiehm IMA/ZLW & IfU, RWTH Aachen University, Aachen, Germany

A. Erman Tekkaya Institute of Forming Technology and Lightweight Construction, TU Dortmund University, Dortmund, Germany

Sebastian Thelen IMA/ZLW & IfU, RWTH Aachen University, Aachen, Germany

Thomas Thiele IMA/ZLW & IfU, RWTH Aachen University, Aachen, Germany

Kerstin Thöing IMA/ZLW & IfU, RWTH Aachen University, Aachen, Germany

Christian Tummel IMA/ZLW & IfU, RWTH Aachen University, Aachen, Germany

Tobias Vaegs IMA/ZLW & IfU, RWTH Aachen University, Aachen, Germany

Hanno Voet Werkzeugmaschinenlabor (WZL), RWTH Aachen University, Aachen, Germany

René Vossen IMA/ZLW & IfU, RWTH Aachen University, Aachen, Germany

Florian Welter IMA/ZLW & IfU, RWTH Aachen University, Aachen, Germany

Freya Willicks IMA/ZLW & IfU, RWTH Aachen University, Aachen, Germany

Stephanie Winter IMA/ZLW & IfU, RWTH Aachen University, Aachen, Germany

Martina Ziefle Human-Computer Interaction Center, RWTH Aachen University, Aachen, Germany

Inna Zimmer IMA/ZLW & IfU, RWTH Aachen University, Aachen, Germany

Part I
Agile and Turbulence-Suitable Processes for Knowledge and Technology Intensive Organizations

Automated Heterogeneous Platoons in Unstructured Environment: Real Time Tracking of a Preceding Vehicle Using Video

Mohammad Alfraheed, Alicia Dröge, Daniel Schilberg
and Sabina Jeschke

Abstract In autonomous driving, object tracking is necessary to gather actual information about the object of interest. The longitudinal and lateral controls of automated highway systems need a target object not only to maintain the safety distance between vehicles but also to keep the following vehicle in the same track as the preceding vehicle. So far automated highway systems were only developed for urban and highway environment depending on lane markings. In future, their application should be extended to unstructured environments (e.g. desert) and be adapted for heterogeneous vehicles. In this paper an approach towards this is presented, where the back view of preceding vehicle is the target object. This solution is independent from the environmental structure as well as additional equipment like infrared emitters. In this paper, the tracking process of the back view is discussed using video streams recorded by a stereo vision system. For an accurate and fast tracking in unstructured environment and with heterogeneous platoons the proposed method is a supplement to the detection process. Therefore, the tracking process has to be (a) applicable under real time constraints and (b) adaptable in dynamic environments. Compared to other methods related to object detection and tracking, the proposed method reduces the running time for the tracking of the back view from reported 12–30 to 16–66 frame/s.

Keywords Automated Highway System · Unstructured Environment · Heterogeneous Platoon · Longitudinal and Lateral Control · Detection and Tracking Process · Stereo Vision System

M. Alfraheed (✉) · A. Dröge · D. Schilberg · S. Jeschke
IMA/ZLW & IfU, RWTH Aachen University, Dennewartstr. 27, 52068 Aachen, Germany
e-mail: mohammad.alfraheed@ima-zlw-ifu.rwth-aachen.de

A. Dröge
e-mail: alicia.droege@ima-zlw-ifu.rwth-aachen.de

D. Schilberg
e-mail: daniel.schilberg@ima-zlw-ifu.rwth-aachen.de

S. Jeschke
e-mail: sabina.jeschke@ima-zlw-ifu.rwth-aachen.de

Originally published in "5th International Conference on Information
and Communication Systems (ICICS) 2014", © IEEE 2014. Reprint
by Springer International Publishing Switzerland 2016,
DOI 10.1007/978-3-319-42620-4_1

1   Introduction

One application field of autonomous driving is an Automated Highway System (AHS). In AHS only the first vehicle is driven actively and the following vehicles automatically. These vehicles drive closely behind each other with just the necessary safety distance in order to optimize highway capacity [1]. Each vehicle (except the first vehicle) is thus able to drive with a low air resistance, which saves energy and fuel within a safety distance of about 10 m [2].

The longitudinal and lateral controls of AHS enable – with the help of other AHS components – vehicles to be coupled electronically and to form a semi-autonomous platoon. The longitudinal control's essential function is the measurement of the distance between the preceding and following vehicle to maintain the safety distance. For the latter, a relatively constant speed of the preceding vehicle is required [3]. The lateral control's essential function is to keep the following vehicle behind the preceding vehicle [4].

All techniques developed within several projects concerning AHS are based on structured environment like a highway. In future, their application range could be extended to unstructured environment like e.g. for unpaved roads. This would be especially beneficial for an application in third world countries who do not have well-developed infrastructures but suffer from traffic jams and congestions.

Currently, several of AHS projects have been established for highway environment. The first AHS project, the PATH project [1], ran from 1992 to 2003. In 2000, the DEMO 2000 project [5] enables efficiently the AHS to detect and recognize obstacles (i.e. small rocks) in structured environment.

Within the last years (from 1999 to 2008) three projects of the AHS have been developed in Germany. The first is CHAUFFEUR I project [4] which had been developed in the second project (CHAUFFEUR II) [6]. The third project is KONVOI coordinated by IMA/ZLW & IfU [7, 8] was carried out in collaboration with other institutes. In KONVOI, four homogeneous trucks were equipped and electronically coupled so that a platoon was successfully formed at highway speed on a highway with a distance of 10 m between the vehicles with only the leading vehicle being actively driven. However, those systems are not applicable in unstructured environments, because they use infrastructure based information like lane markings as references for the longitudinal and lateral controls.

The Energy ITS [2] (Intelligent Transport Systems) has been developed in Japan since 2008. Within this project V2V (Vehicle To Vehicle) communication is used to control both vehicle speed and inter-vehicle distance (longitudinal control).The vision camera, in turn, which is pointing onto the highway, calibrates the lateral control based on lane markings. Since the lane markings are used as a references point for the lateral control, the Energy ITS is not applicable in unstructured environment. Furthermore, a heterogeneous setup was not investigated.

To test the heterogeneous scenarios, the Ricardo UK Ltd. company started the SARTRE (Social Attitudes to Road Traffic Risks in Europe) [9] project in 2009. Here, the traffic efficiency and safety of the platoon vehicle is improved by considering

Human-Machine-Interaction. Several transportation solutions are designed, developed and integrated in order to enable a platoon to be driven on public motorways [10]. Despite being developed for heterogeneous scenarios this system is designed for highways and not for unstructured environments and thus most likely is not useable in unstructured environments.

Several challenges prevent current automated highway systems to be applied in unstructured environment. Some of these challenges relate to the unpaved environment, such as no lane markings. Other challenges arise due to winding roads or sharp turns, where the signal from the preceding vehicle gets lost and the platoon dissolves. To overcome the dependency of lane markings another reference point is required to keep the following vehicles behind the one in front. The method proposed employs a video-based stereo vision system (SVS) affixed on top of the vehicles, which then records the preceding vehicle on video. This video-based approach is less expensive than other vision systems [11]. The lateral and longitudinal controls use features of the back view of the preceding vehicle (BVPV) instead of fixing special tracker onto it or using lane markings as a reference points. The tracking process of the BVPV provides the AHS with the distance and deviation degree of the preceding vehicle. Further, the trajectory path of the preceding vehicle is calculated via SVS, which enables the system to track a disappearing preceding vehicle.

This paper discusses the tracking process of the BVPV using video streams, which show a preceding vehicle driven on an unstructured environment. This part-solution, that considers the detection and tracking process of the BVPV, enables the application of AHS in unstructured environment without motorways. Further, it offers the possibility to let different vehicles types join the platoon. Emphasis is on the development of a less technical intensive and less expensive solution. In addition, the platoon considered can also consist of heterogeneous (non-similar) vehicles.

Details about detection process, which runs prior to the tracking process, have already been published [12], but will be discussed shortly in Section 2, since the tracking process builds up on it. The developed method of the tracking process is presented in Section 3 and a real test of both processes is described in Section 4, concluding with a summary in Section 5.

2 Realization of the Back View Detection

In an earlier publication [12], a detection method applicable for AHS was presented based on a machine learning algorithm (called AdaBoost [13]), taking the BVPV as a reference point for the longitudinal and lateral controls. The detection method distinguished itself from other available detection methods (discussed in Section 3) through its ability to work under real time constraints. Moreover, the detection method can locate the BVPV even if it is not clearly visible due to environmental effects (i.e. reflection of the sunshine towards the camera). However, for AHS, which involve a lot of simultaneous processes (e.g. communication between vehicles, processing of various sensor data), an even shorter running time is required. Problems occurred for distances larger than 10 m between BVPV and camera, when semi-similar features (e.g. back views of cars parking at the side line) were closer to the camera as the BVPV and the latter was not detected in the captured frame.

3 Realization of the Back View Tracking

Although the successful results of the detection process are suitable to detect the BVPV, the unsuccessful results (semi-similar BVPV) prevent the following vehicle to trace the preceding vehicle. To eliminate these results, another single agent is required to track the BVPV based on the machine learning algorithm. Here, the tracking process supplements the detection process to follow the BVPV in the next video frames. In order to reduce the running time associated with detection process, the tracking process has to be run without having to check the whole next frame again. Therefore, the tracking process has to be also applicable under real time constraints. Additionally, the dynamic environment still represents a challenge for the tracking of the back view because the dynamic view often changes the appearance of BVPV. Therefore, the tracking process has to be adaptable in dynamic environment. Concerning of the latter, several of features (i.e. edges) have to be extracted for the BVPV at far distance (i.e. more than 10 m).

Several already developed methods have been published for object detection and tracking. Okuma et al. [14] extended particle filters to multi-target tracking for a changing background, in this case moving hockey players. Moreover, they proposed a probabilistic mixture model to incorporate information achieved from AdaBoost – a machine learning algorithm that is used for feature selection and classification [13] – and the dynamic model of the individual object. The latter measures the similarity of the color histogram based on statistical properties and estimates the tracker region of the interested object. The main drawback is that the dynamic model uses the color of the object of interest. Since this color might also be present in the environment the model might not be able to distinguish the object thus making it non-applicable for AHS.

Grabner et al. [15] therefore proposed a novel on-line AdaBoost feature selection method which is able to tackle a large variety of real time tasks (i.e. learning complex background model, visual tracking and object detection). Here the discrete AdaBoost algorithm [16] is used for feature selection and classification. To classify the foreground (object of interest) and background they partition the image into small blocks. The classifier associated with each block classifies it as foreground and background region. The latter is a statistical predictable block in the image. Despite of the successful results achieved by the AdaBoost algorithm, the mechanism used requires to classify the foreground and background in the frame is not applicable in the AHS because the appearance of the BVPV is changing while the vehicle is moving.

For real time applications, the SURF algorithm [17] was developed. Although it successfully detected the corresponding points between two semi-similar images,

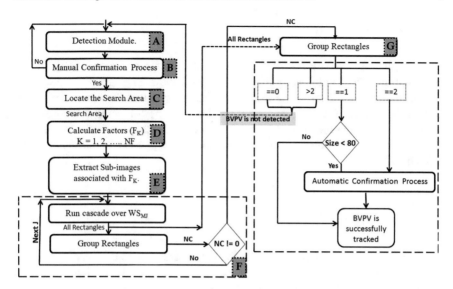

Figure 1 Flowchart of tracking method

it requires an original image of the object of interest to match the detected points. This requirement cannot be fulfilled in heterogeneous platoons, since with different vehicles come along different shapes of back views. Furthermore, a lot of specific features are required to locate the object of interest [13]. This condition cannot be fulfilled, because the number of the reference points on the back view is very small at a far distance.

Therefore, other modifications were tested to detect and track the required objects in a dynamic environment. Leibe et al. [18] have improved method by combining the object detection and tracking in a non-Markovian hypothesis selection framework. Although their system used the optimization procedure and is kept efficient through incremental computation and conservative hypothesis pruning, the system as a whole is not yet capable of working under real-time constraints [18].The system had not been tested for automated highway system. Thus, their method has not been tested when the object is affected by external effects (i.e. sunshine reflection).

As mentioned above an agent is required to complete the detection process using AdaBoost [13]. The main function of that agent is to track BVPV in the next video stream frames without the need to check the whole frame again. Therefore, the tracking process has to be also applicable under real time constraints and in dynamic environment. Figure 1 shows the flowchart of the proposed approach and the details of the agent.

3.1   Detection Module

At the beginning, the BVPV is detected based on the agent introduced in [12]. Therein the two dimension coordinates of the detected back view are generated and sent to a manual confirmation process. Considering AHS, it represents the moment when the platoon is coupled. Meaning that when the driver of the following vehicle accepts the invitation to join the platoon he confirms simultaneously the detection process. In case the drivers response is negative the detection process is run again within 0.03–0.089 s [12].

3.2   Locating the Search Area

The main goal of this step is to locate the search area where the BVPV might be located in the next frame. The criterion used to locate this area is to extract the largest possible area, where its upper edge is the level of the vanishing point [13] and the center point of the new area is the same center point as of the area detected before. In addition, the size of the search area should be close to the size of the detected area. Thus the tracking method automatically cuts out the region in which the BVPV is located instead of checking the whole frame. The advantage is that the size of the search area is then close to the actual size of the back view. Thus, the running time of the developed method is decreased whenever the distance between the preceding and following vehicle is increased.

3.3   Generation of Factors F_K

The Factor F_K represents coefficients of the expected sizes of BVPV in the frame and is calculated based on Equation 1 with S being a fixed increment coefficient of $S = 1.2$ (calculated manually based on frame resolution).

$$F_{K+1} = F_K \cdot S \qquad K = 1, 2, 3, \ldots \qquad (1)$$

With these coefficients the different size of BVPV in the tested frame are determined. Since the K is a finite index, the maximum value of K is achieved when the following condition in Equation 2 is satisfied

$$F_K \cdot W > W_{Frame} \quad \text{and} \quad F_K \cdot H > H_{Frame} \qquad (2)$$

with W being the minimum assumed width of the object of interest, H the minimum assumed height of the object of interest, W_{FRAME} the width of current frame I and H_{FRAME}, the height of current frame I.

3.4 Calculation of $S_{Expected}$ for F_K

The expected size $S_{Expected}$ of the BVPV is calculated based on Equations 3, 4 and 5. Where $S_{Expected\,Width}$ is the expected width and $S_{Expected\,Height}$ is the expected height. This step is the beginning of the *Loop NextF$_K$* which determines the expected sizes of the BVPV. The initial value of the F_K is the values associated with the maximum index. The *Loop NextF$_K$* starts from the maximum value of the generated F_K which leads to the maximum size of $S_{Expected}$. The reason behind this selection is as follows; the criterion used to locate the search area indicates that the detected size of the back view in the previous frame is close to the size of the search area. Further, the short periods of time between captured frames lead to insignificant differences in the size of the back view. Thus, the size of the back view to be tracked in the next frame is close to the size of the search area. The maximum size of $S_{Expected}$ is close to the size of the search area. This method searches for the largest possible BVPV size at the beginning. The expected size is determined as follows:

$$S_{Expected} = S_{Expected\,Width}, S_{Expected\,Height} \tag{3}$$

$$S_{Expected\,Width} = F_K \cdot W \tag{4}$$

$$S_{Expected\,Height} = F_K \cdot H \tag{5}$$

3.5 Extraction of WS_J by Cropping WS Out of I

After having the expected size, the sub-region of the image associated with has to be cropped to be checked. Thus, a sub-region of the image WS_J corresponding to the expected size $S_{Expected}$ is extracted by cropping the $S_{Expected}$ out of the current Frame I. Many different sub-region of the image ($J = 1, 2, 3, ... n$) are extracted from the current frame and a *Loop NextJ* is started at this point to check the WS_J.

3.6 The Loop NextJ

Inside the *Loop NextJ*, the rejection cascades are used to check whether the WSJ shows a BVPV or not. If a BVPV is detected, the group rectangle (Section 3.7) is enabled because to exit the loop as fast as possible since there is no need to keep processing other WS_J. This way the developed method saves the running time when it detects enough features for back view.

3.7   Group Rectangle

In the previous step sometimes more than one sub-region of the image is detected and the developed method clusters the detected them. As a result, the parameter "NClasses" is generated. In this context, NClasses (0, 1, 2 or greater than 2) represents how many clusters are generated from the step Group rectangle. This step distinguishes itself from the similar step in the detection process [12] through the Automatic Confirmation Process. It is enabled automatically when the detected size is less than 80 pixels. This value is chosen since at 80 pixels the distance between the preceding and following vehicle is larger than or equal to 10 m based on the resolution of the video stream. Therefore, the appearance of BVPV is not clear enough to enable the developed method to detect the back view's features.

4   Results of Experiment and Discussion

The proposed method was run on experimental data to check whether it works reliable, meaning that it works under real time constraints and that it is able to track the BVPV whenever it is hidden by environmental effects. The experiments were performed using a non-optimized implementation and run on a PC with a 2 GHz Intel Duo Core CPU. The modified method was tested with over 1430 frames taken from a video stream captured by the Artificial Vision and Intelligent Systems Laboratory (VisLab) of Parma University in Italy.

The adaptable search area reduces the running time and since the developed method estimates successfully the expected location of the back view the accuracy of the results is improved. Instead of checking the whole next frame, the search area locates the expected location of the BVPV. Figure 2 shows many possible search area sizes. The maximum size associated with the maximum red rectangle is selected as a search area. Moreover, several expected sizes of the search area (other red rectangles) are indicated in order to show the center point of the search area.

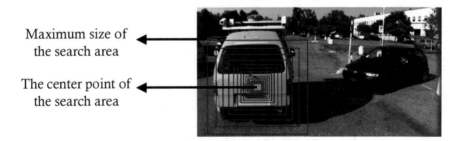

Figure 2   The expected sizes of the search area in the frame

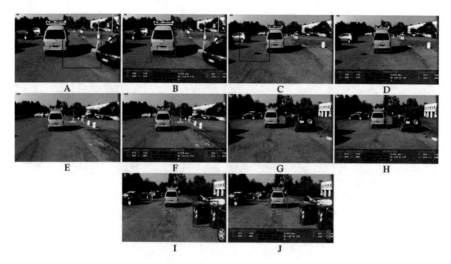

Figure 3 A comparison between the A, C, E, G, I: unsuccessful results of the detection process and the B, D, F, H, J: successful results of the development method

The developed tracking method has to be able to detect and track several shapes of the back view of a heterogeneous platoon. Moreover, with the reduced search area the algorithm only searches in the region of the frame where the back view was detected before. For the case in Figure 2 this means, that it searches only in the middle of the left side of the frame and not in the right bit. Therefore it cannot accidently detect the dark car on the right, since it does not search in this part of the frame. The developed method is thus able to eliminate the unsuccessful results achieved by the detection process. Figure 3 shows a comparison between unsuccessful results from the detection process [12] and successful tracking results. Figure 3a is an unsuccessful detection case for a semi-similar back view. Another case is shown in Figure 3c where the side of a vehicle is detected. The detection process sometimes does not detect anything at all as shown in Figure 3e. Although a back view is successfully detected in Figure 3g, it is not the desired one. In this case, the result is considered as unsuccessful. In Figure 3i an undesired object is detected by the detection process. The successful results for all those un-successful results, obtained with the tracking algorithm, are shown in Figure 3b, d, f, h, j respectively.

In order to show to which extend the search area can reduce the running time of the developed method, a comparison is extracted in Figure 4. The graph shows the relation between the running time of the detection process and the running time of the tracking process. As shown in Figure 4, the running time of the detection process increases whenever the distance between the BVPV and the following vehicle increases. The running time of the tracking process stays stable ranging from 0.015 to 0.0162 s ($\approx$ 66–16 frames/s) whereas the running time of the detection process varies strongly, ranging from 0.015 to 0.375 s ($\approx$66–3 frames/s).

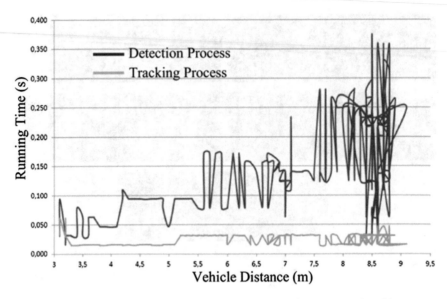

Figure 4 The relationship between the running time of the detection process and tracking process

To test the tracking method in different environments three scenarios were chosen: First (Scenario 1), the BVPV is 5–10 m away and surrounded by semi similar back views (e.g. the side views of vehicles). Scenario 2: The BVPV is 5–12 m away and environmental effects due to winding turns and reflection of the sunshine towards the camera inhibit a clear view onto the back view. Scenario 3: The BVPV is 10–12 m and above away without any surrounding cars.

For a better comparison the three scenarios were also tested with the detection process [12], Furthermore the results are split into those recorded by the left (Left) and right camera (Right) of the stereo vision system. The idea behind this is to figure out the impact of the viewing angle of the camera. However, the numbers show that the results from left and right camera are convergent. The number of successful results in percent and the number of frames tested for the three different scenarios for both processes (detection and tracking) are presented in Table 1.

As mentioned above, the tracking process is a supplement to the detection process and the results achieved from the detection process should be improved by the tracking process. This however was not achieved for the third scenario. Table 2 confirms this fact as the numbers achieved are very similar e.g. 100 % in the Scenario 1. The arithmetic average of success rate of the tracking process for Scenario 1 is 100 %. In contrast; the achieved result of the detection process is 76 %. The arithmetic average 90.5 and 77.4 % (of the tracking and detection process respectively) for Scenario 2 is discussed later in more details with Table 2.

As for Scenario 3, it is the opposite case. The arithmetic average of the detection process is 86.5 % and the arithmetic average of the tracking process is 76.5 %. The unsuccessful rate of the tracking process is due to the inability to function reliable

Table 1 The successful detection and tracking results shown separately for the left and the right camera of the stereo vision system for three different scenarios

Scenario ID	Detection process		Tracking process		Number of tested frames	
	Left (%)	Right (%)	Left (%)	Right (%)	Left	Right
Scenario 1	74	78	100	100	433	433
Scenario 2	77.4	77.4	87	94	115	115
Scenario 3	79.5	93.5	82	71	171	171

Table 2 Results for two different scenarios based on the distance parameter

Process	Detection process (%)		Tracking process (%)	
Distance (m) Scenario ID	>10 and <12	>12	>10 and <12	>12
Scenario 2	57	3	91	47
Scenario 3	100	85	100	74

at a distance larger than 12 m. However, the automated highway system needs about 10 m to keep coupled, otherwise the platoon is dissolved [7, 8].

To figure out up to which distance results are successfully achieved Table 2 shows a comparison of the detection and tracking process based on the distance between the BVPV and following vehicle for the Scenarios 2 and 3. Scenario 1 is neglected since here only measurements with distances smaller than 10 m were taken, for which the tracking method reached a success rate of 100 %. The distance is clustered into two groups. The first group (>10 and <12) represents the distance between 10 and 12 m. The distance of larger than 12 m is represented in the second group (>12).

In Scenario 2 and 3 for a distance ranging from 10–12 m the tracking method obtains 91 % and 100 % successful results respectively whereas the detection process obtained 57 % and 100 %. Hence the tracking process greatly improves the success rate for Scenario 2. A rate of 91 % is enough to warn the core system of the platoon in order to decrease the distance between preceding and following vehicle. As for a distance larger than 12 m, the tracking method improves the result of the detection process from Scenario 2 but only up to 47 %. For Scenario 3 it does not improve the results, it actually performs less well with just 74 % successful results opposed to 85 % of the detection method. This is due the fact that the tracking results are considered as false whenever the size of the detected back view is less than a threshold value (e.g. 80 pixels), which might occur if the BVPV moved slightly out of the search are. This is much more likely to arise for larger distances (>12 m). The detection process however searches over the whole frame and thus has a much higher possibility to find the whole BVPV and hence does not cross the 70 pixel limit. Although the result of tracking process is useable (i.e. about 60 % of the back view is detected), it is considered as unsuccessful as it just partially covers the BVPV.

Table 3 A comparison between the developed method and other methods

Method name	CEE	PS (GHz)	R (pixel)	RT (frame/s)
Tracking method	Yes	2.00	752 × 480	16–66
Detection process	Yes	2.00	752 × 480	12–30
On-line boosting	No	2.80	384 × 288	15–20
Coupled detection and tracking method	No	–	320 × 240	15–25

The results are compared with performances from On-line Boosting [15], detection process [12], tracking process itself and the coupled detection and tracking method by Leibe [18]. Table 3 shows that the comparison depends on four variables (The data of the second, third and fourth row is extracted based on the already published corresponding publications). The first parameter (Consideration of External Effects CEE) can take two different values (yes or no). If the method considers the external effects (i.e. sunshine reflections) the parameter takes the value yes. The second parameter (Processor Speed PS) is the processor speed of the computer used to run the method. The third parameter (Resolution R) represents the frame dimensions used. Finally, the last parameter (Running Time RT) represents the running time needed to detect the object of interest. Our described method distinguishes itself from others through detection and tracking of the back view using a normal processor, adapting with the highest resolution and considering external effects as well. Moreover, the developed method processes a frame faster than the other methods (16–66 fps).

5 Conclusions

The proposed method has been tested with over 1430 frames. Based on the distance of the preceding vehicle the results have been clustered. In context of reliable property (running under real time constraints and environmental effects), the successful results achieved for a distance less than 10 m were 100 %. For distance between 10 and 12 m, the successful results are 91 %. Regarding the safety distance, 91 % is enough to warn the core system of the platoon in order to decrease the distance to less than or equal 10 m. Problems occurred for a distance larger than 12 m because the back view is not clear enough to be detected. However, the successful results of the last cluster (arithmetic average 61 %) could be improved using the detection process whenever the tracking process did not track the BVPV. The proposed method is distinguished by its ability to rapidly track the BVPV. Moreover, it allows working under real time constraints, because the running time lies around 16–66 frames/s.

References

1. E. Shladover, S. Ahs research at the california path program and future ahs research needs. In: *IEEE International Conference on Vehicular Electronics and Safety, ICVES 2008*. 2008, pp. 4–5
2. S. Tsugawa, S. Kato, Energy its: another application of vehicular communications. IEEE Commun. Mag. **48** (11), 2010, pp. 120–126
3. A. Khodayari, A. Ghaffari, S. Ameli, J. Flahatgar, A historical review on lateral and longitudinal control of autonomous vehicle motions. In: *2010 2nd International Conference on Mechanical and Electrical Technology (ICMET)*. 2010, pp. 421–429
4. H. Fritz, Longitudinal and lateral control of heavy duty trucks for automated vehicle following in mixed traffic: experimental results from the chauffeur project. In: *Proceedings of the 1999 IEEE International Conference on Control Applications, 1999*, vol. 2. 1999, vol. 2, pp. 1348–1352
5. S. Tsugawa, A history of automated highway systems in japan and future issues. In: *IEEE International Conference on Vehicular Electronics and Safety, ICVES 2008*. pp. 2–3
6. H. Fritz, A. Gern, Chauffeur assistant: a driver assistance system for commercial vehicles based on fusion of advanced acc and lane keeping. In: *2004 IEEE Intelligent Vehicles Symposium*. 2004, pp. 495–500
7. R. Ramakers, K. Henning, S. Gies, D. Abel, M. Haberstroh, Electronically coupled truck platoons on german highways. In: *Automation, Communication and Cybernetics in Science and Engineering 2009/2010*, ed. by S. Jeschke, I. Isenhardt, K. Henning, Springer, Berlin, Heidelberg, 2011, pp. 441–451
8. R. Kunze, M. Haberstroh, R. Ramakers, K. Henning, S. Jeschke, Automated truck platoons on motorways – a contribution to the safety on roads. In: *Automation, Communication and Cybernetics in Science and Engineering 2009/2010*, ed. by S. Jeschke, I. Isenhardt, K. Henning, Springer, Berlin, Heidelberg, 2011, pp. 415–426
9. E. Chan, P. Gilhead, P. Jelínek, P. Krejci. Sartre cooperative control of fully automated platoon vehicles. Speech at 18th ITS World Congress, 2011
10. S. Solyom, E. Coelingh, Performance limitations in vehicle platoon control. In: *2012 15th International IEEE Conference on Intelligent Transportation Systems (ITSC)*. 2012, pp. 1–6
11. S. Jin, J. Cho, D. Pham, X. K. Lee, S. Park, M. Kim, W. Jeon, J. Fpga design and implementation of a real-time stereo vision system. IEEE Trans. Circuits Syst. Video Technol. **20** (1), 2010, pp. 15–26
12. M. Alfraheed, A. Dröge, R. Kunze, M. Klingender, D. Schilberg, S. Jeschke, Real time detection of the back view of a preceding vehicle for automated heterogenous platoons in unstructured environment using video. In: *2011 IEEE International Conference on Systems, Man, and Cybernetics (SMC)*. 2011, pp. 549–555
13. G. Bradski, A. Kaehler. Learning opencv, 2008
14. K. Okuma, A. Taleghani, N. de Freitas, J. Little, J. G. Lowe, D. A boosted particle filter: Multitarget detection and tracking. In: *Computer Vision – ECCV 2004*, ed. by T. Padjla, J. Matas, Springer, Berlin, Heidelberg, 2004, pp. 28–39
15. H. Grabner, H. Bischof, On-line boosting and vision. In: *2006 IEEE Computer Society Conference on Computer Vision and Pattern Recognition*, vol. 1, 2006, pp. 260–267
16. E. Schapire, R. *The Boosting Approach to Machine Learning: An Overview*. 2002
17. H. Bay, A. Ess, T. Tuytelaars, L. Van Gool, Speeded-up robust features (surf). Comput. Vis. Image Underst. **110** (3), 2008, pp. 346–359
18. B. Leibe, K. Schindler, N. Cornelis, L. Van Gool, Coupled object detection and tracking from static cameras and moving vehicles. IEEE Trans. Pattern Anal. mach. Initell. **30** (10), 2008, pp. 1683–1698

Präventiv Denken und Handeln für nachhaltige Beschäftigungsfähigkeit

Guido Becke, Peter Bleses, Claudia Jooß and Julia Eich

Zusammenfassung Nachhaltige Beschäftigungsfähigkeit ist notwendig, um die Beschäftigten mittel- und langfristig gesund und damit leistungsfähig zu erhalten. Soziale Innovationen für nachhaltige Beschäftigungsfähigkeit drücken sich in einem präventiven Denken und Handeln bei Beschäftigten und Unternehmen aus, das nicht mehr allein kurzfristige ökonomische Erfolge zum Ziel haben kann. Gefragt sind Gestaltungskonzepte einer auf Ressourcenerhaltung und -regeneration abzielenden Gesundheitsförderung, die insbesondere die Stärkung psychosozialer Ressourcen anstrebt und auch praktisch in kleinen und mittleren Unternehmen umsetzbar sein muss. Von zentraler Bedeutung ist dabei die Stärkung der Veränderungsfähigkeit von Menschen und Organisationen, denn Wandel wird die zukünftige Arbeitswelt auch weiterhin bestimmen. Welche Herausforderungen, Gestaltungsaufgaben und Forschungsbedarfe aus Sicht der Akteure des Förderschwerpunktes mit dieser Thematik einhergehen, werden in diesem Beitrag dargestellt.

Schlüsselwörter Nachhaltige Beschäftigungsfähigkeit · Prävention · Gesundheit · Ressourcen

1 Einleitung: Tagungssession „Präventiv Denken und Handeln für nachhaltige Beschäftigungsfähigkeit"

Im Rahmen der Aachener Tagung des BMBF-Förderschwerpunktes „Innovations-fähigkeit im demografischen Wandel" wurden die einzelnen Handlungsfelder für das geplante Memorandum [1] im Rahmen fünf paralleler Sessions mit den teil-nehmenden Akteuren diskutiert. Diskussionsgrundlage der Tagungssession

G. Becke (✉) · P. Bleses
artec, Forschungszentrum Nachhaltigkeit, Universität Bremen, Postfach 330440,
28334 Bremen, Germany
e-mail: becke@artec.uni-bremen.de

C. Jooß · J. Eich
IMA/ZLW & IfU, RWTH Aachen University, Dennewartstr. 27,
52068 Aachen, Germany

Originally published in "Exploring Demographics",
© Springer 2015. Reprint by Springer International Publishing Switzerland 2016,
DOI 10.1007/978-3-319-42620-4_2

„Präventiv Denken und Handeln für nachhaltige Beschäftigungsfähigkeit" bildet das in der folgenden Tabelle visualisierte Handlungsfeld (vgl. Tabelle 1).

Vor diesem Hintergrund wurden durch die Teilnehmenden Herausforderungen, Gestaltungsaufgaben und Forschungsbedarfe erarbeitet. Das Ziel dieses Beitrags liegt darin, die Inhalte sowie die zentralen Diskussionsergebnisse der Session zusammenzufassen.

Kapitel 2 stellt mittels eines kurzen theoretischen Inputs eine Einführung in das Handlungsfeld „Präventiv Denken und Handeln für nachhaltige Beschäftigungsfähigkeit" dar. Neben den Thesen des Memorandums werden Gründe für die Gefährdung nachhaltiger Beschäftigungsfähigkeit aufgeführt. Zudem wird der Bedarf nach innovativen Konzepten für die erwerbsbezogene Gesundheitsförderung erläutert. Kapitel 3 befasst sich mit dem Diskussionsergebnis der Gestaltungsaufgaben mit Blick auf Präventionskonzepte nachhaltiger Beschäftigungsfähigkeit. Dabei wurden drei zentrale Herausforderungen präventiven Denken und Handelns (Systemische Gestaltungskonzepte, Einsicht und Motivation sowie Digitalisierung) formuliert (vgl. Kapitel 4). Abschließend werden in Kapitel 5 aus der Diskussion abgeleitete Forschungsbedarfe formuliert.

Tabelle 1 Handlungsfeld „Präventiv Denken und Handeln für nachhaltige Beschäftigungsfähigkeit"

#1	Weitere Flexibilisierung und Virtualisierung der Arbeitswelt erfordert präventives Gestalten der Arbeitsbedingungen, um die Gesundheit und die Produktivität des Menschen in der neuen Arbeitswelt und im demografischen Wandel zu sichern. Dies erfordert neue und erweiterte Ansätze einer präventiven Arbeitsgestaltung
#2	Es sind weitergehende präventive Gestaltungskonzepte für bestimmte Branchen (wie der Pflegebereich), für kleine Unternehmen und für besonders vulnerable Zielgruppen wie Geringqualifizierte zu entwickeln, um Perspektiven für diese Gruppen und Menschen zu öffnen
#3	Innovative Konzepte der Gesundheitsförderung werden benötigt, die auf den permanenten Wandel von Unternehmen sowie die Flexibilisierung und Dynamisierung von Arbeit und Organisationen ausgerichtet sind
#4	Um soziale Innovationen zu ermöglichen, sind neue Konzepte der gesundheitlichen und sozialen Ressourcenstärkung erforderlich (wie Ethik, Werte, inspirierende Führungskultur, Achtsamkeit, Resilienz)
#5	Einflussmöglichkeiten und Grenzen von Führungskräften bei der präventiven Arbeitsgestaltung und Ressourcenstärkung sind weiter zu konkretisieren
#6	Spezielle Präventionskonzepte zur Förderung inkrementeller Innovationen vor allem in kleinen Unternehmen sind erforderlich, um diese oft vollkommen vernachlässigten und kaum beachteten Innovationen zu unterstützen
#7	Es sind Hilfsmittel für Förderung und Entwicklung sozialer Innovationen durch neue Formen des Zusammenhalts von Belegschaften angesichts zunehmender Vielfalt (Kulturen, Erwartungen, Interessenlagen) zu entwickeln

2 Theoretischer Input: Nachhaltige Beschäftigungsfähigkeit in der flexiblen Arbeitswelt

Die moderne Arbeitswelt ist durch hohe Leistungs- und Arbeitsanforderungen an Beschäftigte geprägt, die den Erhalt und die Regeneration ihrer Gesundheit gefährden können. Hierzu trägt maßgeblich die Flexibilisierung der Arbeit in räumlicher, zeitlicher sowie aufgaben- und organisationsbezogener Hinsicht bei. Die Digitalisierung und Virtualisierung von Erwerbsarbeit fördert nicht nur mobile Arbeit, sondern ist auch mit zunehmend erschwerten Grenzziehungen zwischen beruflichen Lebensbereichen und privater Lebensführung verbunden. Überdies sind flexible Arbeits- und Organisationsstrukturen oft durch eine steigende Ökonomisierung der Erwerbsarbeit geprägt, die auf einer ökonomischen Steigerungslogik zunehmend anspruchsvollerer Ziele und einer Intensivierung des Wettbewerbs basiert. Der daraus resultierende Leistungsdruck sowie die Arbeitsverdichtung begünstigen die Überschreitung individueller Belastungsgrenzen.

Die mehr oder weniger beständige Reorganisation von Unternehmen erhöht erwerbsbiografische Unsicherheiten. So ist die Reorganisation oft mit Personal- und Stellenabbau verbunden, der Karrierewege beeinträchtigen und längerfristige Beschäftigungsperspektiven in Frage stellen kann. Psychosoziale Gesundheitsgefährdungen entstehen hierbei etwa aufgrund von erlebter Arbeitsplatzunsicherheit und beruflichen Gratifikationskrisen, d. h. einem strukturell unausgeglichenem Verhältnis zwischen dem eigenen Arbeitsengagement in Veränderungsprozessen und den dafür erhaltenen materiellen wie immateriellen Belohnungen seitens der Unternehmen.

In Anbetracht dieser vorwiegend psychischen bzw. psychosozialen Gesundheitsgefährdungen mangelt es an praktikablen Präventionskonzepten und darauf bezogener valider Wirkungsforschung. Solche Präventionskonzepte gewinnen in Unternehmen oft erst an Bedeutung, wenn die nachhaltige Beschäftigungsfähigkeit von Mitarbeitenden und Führungskräften gefährdet ist, so dass Einbußen hinsichtlich der Leistungs- und Wettbewerbsfähigkeit drohen. Es werden daher zukünftig in wachsendem Maße präventive Gestaltungskonzepte benötigt, um Arbeitsbedingungen gesundheitsförderlich zu gestalten und darüber vermittelt auch die Leistungsfähigkeit und Produktivität von Erwerbspersonen und Unternehmen zu sichern.

Ein Beispiel für eine solche gesundheitsförderliche Gestaltung von Erwerbsarbeit bildet das *Konzept der nachhaltigen Beschäftigungsfähigkeit* ([2]: 8f.). Es geht davon aus, dass die Beschäftigungsfähigkeit von Erwerbspersonen gerade angesichts sich verändernder Arbeitsanforderungen und bedingungen zu erhalten ist. Eine innerhalb der Session erarbeitete Definition von Beschäftigungsfähigkeit zeigt Abbildung 1.

Nachhaltige Beschäftigungsfähigkeit ist in diesem Sinne eine dauerhafte Gestaltungsaufgabe für Erwerbspersonen, Unternehmen und außerbetriebliche Präventionsakteure. Im Konzept der nachhaltigen Beschäftigungsfähigkeit sind drei Komponenten miteinander verwoben: In gesundheitlicher Hinsicht zielt nachhaltige Beschäftigungsfähigkeit darauf ab, die physischen wie psychischen Ressourcen von Menschen zu erhalten und zu regenerieren. Mit Blick auf die Qualifikationen und

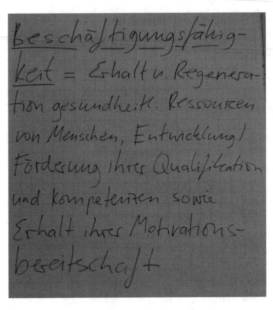

Abbildung 1 Begriffsbestimmung „Beschäftigungsfähigkeit" innerhalb der Session [3]

Kompetenzen von Erwerbspersonen intendiert nachhaltige Beschäftigungsfähigkeit die Entwicklung und Förderung der fachlichen Qualifikationen und Kompetenzen in Anbetracht sich verändernder Arbeitsanforderungen. Schließlich beinhaltet nachhaltige Beschäftigungsfähigkeit eine motivationale Komponente, d. h. sie verweist auf den Erhalt der Arbeitszufriedenheit und der Arbeitsmotivation von Erwerbspersonen.

Das *Konzept der nachhaltigen Beschäftigungsfähigkeit* verdeutlicht, dass hinsichtlich der Anforderungen der flexiblen Arbeitswelt innovative Konzepte einer erwerbsbezogenen Gesundheitsförderung zu entwickeln und umzusetzen sind:

- Präventive Gestaltungskonzepte müssen vor allem für bestimmte Branchen mit hohen psychosozialen bzw. physischen Gesundheitsrisiken (wie die Pflege und soziale Dienstleistungen) und für besonders verwundbare Zielgruppen (wie Geringqualifizierte und Alleinselbstständige) weiterentwickelt werden. Sie sind zudem stärker auf die Belange und Voraussetzungen kleiner Unternehmen auszurichten, da diese im Vergleich zu größeren Unternehmen in geringerem Maße über Ressourcen für eine betriebliche Gesundheitsförderung verfügen.
- Innovative Konzepte nachhaltiger Beschäftigungsfähigkeit müssen den permanenten Wandel von Unternehmen sowie die Flexibilisierung und Dynamisierung von Arbeit und Organisationen aufnehmen.
- Schließlich gilt es, unternehmensintern einen konstruktiven Umgang mit Erwartungskonflikten zwischen betrieblichen Zielen und Anforderungen auf der einen Seite und lebensweltlichen Ansprüchen und Erwartungen von Beschäftigten auf der anderen Seite zu fördern [4].

Es werden daher soziale Innovationen für nachhaltige Beschäftigungsfähigkeit in der flexiblen Arbeitswelt benötigt [5]. Damit sind neue oder veränderte Handlungspraktiken zur Förderung nachhaltiger Beschäftigungsfähigkeit gemeint, die auf eine Stärkung gesundheitlicher und sozialer Ressourcen abzielen. Solche Innovationen können sich auf folgende Aspekte beziehen:

• Gesundheitsförderliche Arbeitsorganisation und Unternehmenskultur.
• Führungskultur: Führungskräfte spielen eine zentrale Rolle für eine präventive Arbeits- und Organisationsgestaltung. Sie können den Erhalt und die Regeneration der gesundheitlichen Ressourcen von Beschäftigten unterstützen und müssen selbst auch in gesundheitlicher Hinsicht gestärkt werden.
• Individuelle und organisationale Achtsamkeit auf Gesundheitsrisiken und -ressourcen.
• Die Förderung organisationaler, sozialer und individueller Resilienz.
• Die Förderung inkrementeller Innovationen, die den Möglichkeiten und der Praxis kleinerer Unternehmen angepasst sind.
• Neue Formen des Zusammenhalts von Belegschaften angesichts zunehmender Vielfalt (z. B. Teil-Kulturen, Erwartungen, Interessenlagen).

3 Herausforderungen präventiven Denkens und Handelns

Bei der Entwicklung sozialer Innovationen zur Förderung nachhaltiger Beschäftigungsfähigkeit sind insbesondere drei zentrale Herausforderungen zu bewältigen, die im Rahmen der Session erarbeitet wurden und im Folgenden erläutert werden.

Erstens sind systemische Gestaltungskonzepte zu entwickeln, die Prävention und Gesundheit in Managementsystemen und in Unternehmensstrategien integrieren sowie auf den Handlungsebenen Individuen, Teams und Organisation ansetzen, um organisationale Lernprozesse zu ermöglichen. Systemische Präventionskonzepte im Sinne nachhaltiger Beschäftigungsfähigkeit sollten darüber hinaus Schnittstellen zu Politik und Gesellschaft definieren.

Soziale Innovationen im Sinne nachhaltiger Beschäftigungsfähigkeit erfordern zweitens, Akzeptanz und Verständnis für neue Präventionskonzepte bei Fachkräften des Arbeitsschutz- und Gesundheitsmanagements sowie bei Führungskräften und Mitarbeitenden zu schaffen. Diese unterschiedlichen Zielgruppen sollten vom Praxisnutzen der neuen Konzepte überzeugt werden. Hierfür ist eine zielgruppenspezifische Kommunikation und Übersetzung der Konzepte in Anwendungskontexte und in die Alltagssprache von Führungskräften und Beschäftigten erforderlich.

Schließlich sind drittens bei Fach- und Führungskräften wie Beschäftigten Gesundheitskompetenzen mit Blick auf die flexible Arbeitswelt zu entwickeln. Kernbestandteile dieser Gesundheitskompetenzen liegen darin, Veränderung als integralen Bestandteil der Erwerbsarbeit zu begreifen und den Umgang mit (erwerbsbiografischen) Unsicherheiten zu lernen. Gesundheitskompetenzen schließen zudem kooperative Kompetenz ein. Diese ist besonders hinsichtlich der Zusammenarbeit zwischen unterschiedlichen Akteuren im Arbeitsschutz- und Gesundheitsmanagement gefragt. In Anbetracht der hohen Bedeutung von Führung für den Gesundheitserhalt von Mitarbeitenden gilt es, ein wertschätzendes Führungsverhalten zu fördern, von dem eine Signalwirkung auf die Zusammenarbeit zwischen Beschäftigten ausgeht. Die Förderung von Gesundheitskompetenzen setzt voraus, dass diese systematisch in die betriebliche Aus- und Weiterbildung integriert werden.

4 Gestaltungsaufgaben

In einer zweiten Arbeitsphase wurden vor dem Hintergrund der formulierten Herausforderungen drei Gestaltungsaufgaben mit dem Blick auf Präventionskonzepte für nachhaltige Beschäftigungsfähigkeit aufgestellt.

Die erste Gestaltungsaufgabe richtet sich an die Digitalisierung von Arbeit. Hier wird für ganzheitliche Präventionsansätze plädiert, die Arbeits-, Organisations- und Technikgestaltung integrieren. Präventionskonzepte sollten zudem ,mit dem Takt der Technik Schritt halten', d. h. gesundheitsförderliche Optionen zur Beschleunigung von Arbeitsprozessen durch Digitalisierung entwickeln.

Ein zweites Bündel von Gestaltungsaufgaben bezieht sich auf die Frage, wie sich die Motivation von Führungskräften und Beschäftigten für innovative Präventionskonzepte und die Einsicht in gesundheitsförderliches Arbeitshandeln stärken lässt. Als eine Grundvoraussetzung für Präventionskonzepte nachhaltiger Beschäftigungsfähigkeit wird eine systematische Beteiligung von Beschäftigten an der Arbeits-, Organisations- und Technikgestaltung gesehen. Allerdings benötigen Beschäftigte hierfür Zeit und weitere Ressourcen, wie Autonomiespielräume, um die Beteiligungsmöglichkeiten wahrnehmen zu können. Flankierend hierzu ist eine Qualifizierung und Kompetenzentwicklung von Beschäftigten und Führungskräften anzuraten, die der Stärkung ihrer Partizipationskompetenz sowie ihrer Gesundheitskompetenz dienen sollen. Letztere bedeutet auch, Beschäftigte und Führungskräfte für gesundheitliche Risiken und den Erhalt ihrer Gesundheitsressourcen zu sensibilisieren, insbesondere mit Blick auf eine berufliche Mittel- und Langfristperspektive. Überdies sind finanzielle Ressourcen vorzuhalten, um Maßnahmen der Prävention auch realisieren und Präventionskonzepte institutionalisieren zu können. Auszuloten ist, inwieweit Präventionsansätze von Beschäftigten als Zwang bzw. Einschränkung ihrer Freiheitsgrade bei der Arbeit wahrgenommen werden.

Das erörterte dritte Feld von Gestaltungsaufgaben bezieht sich auf psychische Arbeitsbelastungen. Die jüngste Änderung des Arbeitsschutzgesetzes in 2013 sieht vor, dass sich die für alle Unternehmen durchzuführenden Gefährdungsbeurteilungen auch auf psychosoziale Belastungen erstrecken sollen. Präventionskonzepte, die psychosoziale Ressourcen stärken und psychosoziale Ressourcen reduzieren sollen, haben eine Schlüsselanforderung zu bewältigen: Sie müssen die Angst von Beschäftigten und Führungskräften vor sozialer Stigmatisierung abbauen helfen, um diese Zielgruppen dabei zu unterstützen, psychosoziale Gesundheitsrisiken auch betriebsöffentlich thematisieren zu können. Gerade in flexiblen Arbeitsstrukturen sind Selbstsorgekompetenzen auf Seiten von Beschäftigten wie Führungskräften zu entwickeln. Die Förderung von Selbstsorgefähigkeiten entbindet Unternehmen jedoch nicht von ihrer Verantwortung für eine gesundheitsförderliche Arbeitsgestaltung. Dies schließt auch ein, die betriebliche Leistungspolitik nach menschlichen Maßen zu gestalten, anstatt sich allein an der ökonomischen Steigerungslogik zu orientieren.

5 Forschungsbedarfe und Ausblick

Die zweite Arbeitsphase wurde mit einer Formulierung von drei zentralen Forschungsfragen abgeschlossen.

Mit Blick auf die *Motivation für gesundheitsförderliches Arbeitshandeln* ist zu untersuchen, wie sich Widerstände auf Seiten von Beschäftigten wie Führungskräften gegenüber Präventionskonzepten reduzieren lassen. Für eine gestaltungsorientierte Forschung stellt sich die methodische Frage, wie sich durch die Anlage von Projekten bzw. die Gestaltung des Projektsettings Arbeitsprozesse und -strukturen gesundheitsförderlich entwickeln lassen.

Es besteht Bedarf hinsichtlich einer *kontextsensibleren Forschung zu psychosozialen Belastungen und Ressourcen*. So ist weiter zu erforschen, unter welchen Voraussetzungen Arbeitsanforderungen von Menschen als psychische Belastungen oder als gesundheitliche Ressource erlebt werden. Zu entwickeln sind Frühindikatoren oder Marker für psychische Erkrankungen oder gesundheitliche Beeinträchtigungen in der Arbeitswelt (z. B. für Burnout), um auf der Verhaltens- und der Verhältnisebene möglichst integrierte und präventive Interventionen vornehmen zu können. Derzeit besteht ein deutliches Wissensdefizit hinsichtlich umsetzbarer Gestaltungskonzepte der Prävention psychosozialer Gesundheitsrisiken in flexiblen Arbeits- und Organisationsstrukturen. Zugleich fehlen niedrigschwellige Präventionsinstrumente, die in unterschiedlichen Arbeitskontexten auch ohne einschlägige Experten kompetent angewandt werden können.

Mit Blick auf die *Digitalisierung von Arbeit* sollte sich die Forschung nicht nur auf die gesundheitlichen Gefährdungspotenziale beziehen, sondern auch kontextsensibel analysieren, für welche Gruppen von Erwerbspersonen eine Entgrenzung von Beruf und privater Lebensführung gesundheitlich problematisch ist. Näher zu untersuchen wäre, wie sich ein gesundheitsförderlicher Umgang mit digitalen Medien unterstützen lässt. Mit Blick auf die Prävention ist zudem nicht hinreichend beantwortet, welches Potenzial neue digitale Medien bieten.

Abschließend lässt sich konstatieren, dass die Ergebnisse des Workshops gewinnbringend in den förderschwerpunktinternen Diskurs eingebracht werden konnten. Darüber hinaus eröffnet dieser Beitrag jedoch ebenfalls die Möglichkeit, die erarbeiteten Ergebnisse über den Förderschwerpunkt hinaus transparent zu machen. Zudem bietet der Beitrag eine inhaltliche Ergänzung der einzelnen Themen des Handlungsfeldes sowie eine inhaltliche Unterfütterung des Handlungsfeldes. Innerhalb der

Session konnten ebenfalls Anknüpfungspunkte für weiterführende Forschung und Konzepte erarbeitet werden und so ein Beitrag zur Überarbeitung bzw. der Erstellung des Thesenpapiers [1] sowie der Erstellung des Memorandums [6] geleistet werden.

Literaturverzeichnis

1. G. Becke, P. Bleses, O. Cernavin, A. Ducki, C. Jooß, R. Klatt, T. Langhoff, F.W. Nerdinger, Einführungsbeitrag: Von den Handlungsfeldern der Förderschwerpunkt-Tagung zum Memorandum. In: *Exploring Demographics*, ed. by S. Jeschke, A. Richert, F. Hees, C. Jooß, Springer Fachmedien Wiesbaden, 2015, pp. 55–66
2. P. Bleses, W. Ritter, Das Verbundprojekt ZUKUNFT:PFLEGE – Nachhaltige Beschäftigungsfähigkeit im Fokus. Nachhaltige Beschäftigungsfähigkeit in der ambulanten Pflege. Zwischenbericht des Verbundprojekts ZUKUNFT:PFLEGE, artec-paper Nr. 189, Forschungszentrum Nachhaltigkeit (artec), Universität Bremen, Bremen, 2013, pp. 7–32
3. Fotoprotokoll der Session „Präventiv Denken und Handeln für nachhaltige Beschäftigungsfähigkeit". URL http://demoscreen.de/demoscreen/beitraege/ Letzter Zugriff am: 2014-08-12
4. G. Becke, E. Senghaas-Knobloch, Erwartungskonflikte in betrieblichen Veränderungsprozessen – Psychosoziale Gesundheitsgefährdungen und Gestaltungsansätze. Mit 4 Expertisen von Rainer Müller, Antje Ducki, Christel Kumbruck und Walter Punke. artec-paper Nr. 198. Tech. rep., Forschungszentrum Nachhaltigkeit (artec), Universität Bremen, Bremen, 2014. http://www.artec.uni-bremen.de/papers/paper_198.pdf. Letzter Zugriff am: 12.08.2014
5. J. Howaldt, M. Schwarz, *'Soziale Innovation' im Fokus: Skizze eines gesellschaftstheoretisch inspirierten Forschungskonzepts*, 1st edn. transcript, 2010
6. D.G. Becke, P. Bleses, O. Cernavin, A. Ducki, R. Klatt, T. Langhoff, P.D.F.W. Nerdinger, Memorandum: Förderschwerpunkt Innovationsfähigkeit im demografischen Wandel. In: *Exploring Demographics*, ed. by S. Jeschke, A. Richert, F. Hees, C. Jooß, Springer Fachmedien Wiesbaden, 2015, pp. 119–121

Digitalisierung der Arbeit und demografischer Wandel

Oleg Cernavin, Thomas Thiele, Markus Kowalski and Stephanie Winter

Zusammenfassung Die Arbeitsforschung hat sich in den letzten fünfzehn Jahren intensiv mit dem Thema des demografischen Wandels befasst. Nun deutet sich ein neuer Megatrend an, der in das Zentrum der Arbeitsforschung geraten wird: Die zunehmende Digitalisierung der Arbeit, die mit dem Schlagwort Industrie 4.0 beschrieben wird. Personen und Unternehmen werden sich mit beiden Entwicklungen gleichermaßen auseinandersetzen und Gestaltungsfähigkeit für die sich weiter dynamisch verändernden Arbeits- und Lebenswelten entwickeln müssen. Die Aufgabe der Arbeitsforschung liegt dabei in der Erarbeitung von Strategien, Konzepten und Modellen, damit Menschen und Unternehmen die Entwicklungspotenziale der Digitalisierung der Arbeit nutzen und gleichzeitig lernen, mit den zunehmenden Ambivalenzen der Entwicklung umzugehen.

Schlüsselwörter Digitalisierung · Demografischer Wandel · Internet der Dinge · Internet der Dienste

1 Was ist unter Digitalisierung der Arbeitswelt (Industrie 4.0) zu verstehen?

Wir stehen am Beginn einer neuen Stufe einer Arbeitsentwicklung, die die reine Computerarbeit um neue Qualitäten ergänzt (siehe Abbildung 1). Neue Entwicklungen in der Informations- und Kommunikationstechnologie (IKT) öffnen den Weg zu einer neuen Qualität von Arbeits- und Lebenswelten. Diese nächste Stufe der Veränderung wird auch mit dem Begriff der Industrie 4.0 deklariert [1, 2]. Es ist zu vermuten, dass diese Veränderung weitreichende Auswirkungen auf die

O. Cernavin (✉)
BC Forschung Gmbh, Kaiser-Friedrich-Ring 53, 65185 Wiesbaden, Germany
e-mail: oleg.cernavin@bc-forschung.de

T. Thiele · M. Kowalski · S. Winter
IMA/ZLW & IfU, RWTH Aachen University, Dennewartstr. 27,
52068 Aachen, Germany

Originally published in "Exploring Demographics",
© Springer 2015. Reprint by Springer International Publishing Switzerland 2016,
DOI 10.1007/978-3-319-42620-4_3

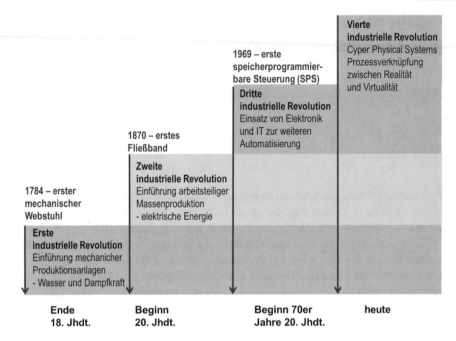

Abbildung 1 Stufen der technologischen Entwicklung ([2]: 10)

präventive Arbeitsgestaltung beinhaltet [3–5]. Aber wodurch zeichnen sich diese neuen Entwicklungen aus?

Die IKT bewirkte einen extrem dynamischen technologischen Fortschritt, der in eine qualitativ neue Entwicklungsstufe eintritt. Dieser Fortschritt basiert auf dem anhaltend rapiden Wachstum der Wirtschaft und der Verknüpfung von folgenden Entwicklungen:

• Zunehmende Rechenleistungen immer kleinerer Einheiten
• Ständig wachsende Bandbreite in Netzwerken
• Mobile Geräte in jeglicher Form
• Miniaturisierte, integrierte Schaltungen
• Intelligente Sensoren und Aktoren, die Prozesse eigenständig erfassen und steuern
• Semantische Technologien (z. B. Auswertung von Bedeutungszusammenhängen; Verbinden und Verknüpfen von Daten zu neuen Anwendungen; eigenständiges, flexibles und zielgerichtetes Erbringen von Leistungen; „intelligentes" Erschließen von Information; Integration einzelner Komponenten in komplexe Ablaufprozesse und Anwendungslösungen)

Durch diese Entwicklungen entstehen Systeme, in denen Arbeitsmittel, Prozesse, Objekte bis hin zu Alltagsgegenständen durch Programmierbarkeit, Speichervermögen, Sensoren, Aktoren und Kommunikationsfähigkeiten „intelligent" werden. Sie können über das Internet durch die miniaturisierte Technologie in hoher Geschwindigkeit eigenständig Informationen austauschen, Aktionen auslösen und sich wechselseitig steuern. Solche Systeme werden unter dem Terminus Cyber-Physical Systems (CPS) gefasst: Reale Arbeitsmittel, Menschen (!), soziale Prozesse (!) und Umgebungen (Arbeitsstätte, Raumumgebung generell), die über die miniaturisierten Technologien mit der virtuellen Welt verbunden sind. CPS (als Embedded Systems) kombinieren die „reale" mit der „virtuellen" Welt. Die „Dinge" und Leistungen werden Bestandteil des Internets (daher die Schlagworte „Internet der Dinge" und „Internet der Leistungen").

Die Industrie 4.0 beschreitet den Weg von der computerzentrierten Welt hin zu untereinander und über das Internet vernetzten, softwaretechnisch gesteuerten Prozessen, Geräten, Objekten und Umgebungen. Im Zuge dieses Trends entstehen offene, vernetzte, flexibel agierende und interaktive Systeme, die die physikalische Welt mit der virtuellen Welt der Informationstechnik verknüpfen (Abbildung 2).

Einige wenige Beispiele, die zeigen, in welche Richtung diese zu beobachtende Veränderungen gehen, sind:

- Produktions-Systeme und Komponenten überprüfen sich selbst und stellen gegebenenfalls Korrektheit und Qualität von Informations-, Kommunikations- und Steuerungsprozessen sicher
- Intelligente, vernetzte und mit Sensoren ausgestattete Komponenten übernehmen Funktionen wie Zustands- und Umgebungsbeobachtung, vernetzte Kontrolle, Koordination und Optimierung von Prozessen beispielsweise beim Warenfluss, in Wartungsprozessen oder für das Flotten- oder Gebäudemanagement
- Notfallsituation und Erstversorgung werden durch CPS erkannt und eingeleitet
- CPS steuern ortsunabhängiges Qualitäts- und Prozessmanagement
- CPS steuern Maschinen- und Fahrzeugführung

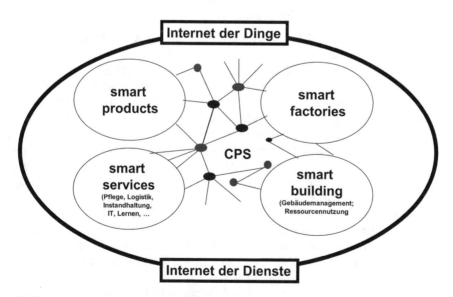

Abbildung 2 Cyber-Physical Systems (eigene Darstellung)

- Assistenzdienste („Assistenz-App") planen über ein mobiles Endgerät oder in der Cloud automatisch nach Vorgaben und Vorschlägen den Tagesablauf und übernehmen auch die Planung von individuellen Arbeitsabläufen
- Der Gesundheitsstatus einer Person wird bewertet durch den Abgleich von Daten unterschiedlicher Sensoren, Auswertungs- und Bestandsdaten mit Schnittstellen zu Mobilitätsdiensten, Apotheken, Therapeuten, Ärzten, Fallmanagern, Haus- und Gebäudemeistern und Servicekräften.

2 Handlungsfelder für die Arbeitsforschung

Die Arbeitsforschung hat die Aufgabe, die voranschreitenden Entwicklungen des demografischen Wandels mit denen der Digitalisierung der Arbeitswelt zu verknüpfen. Die Fragestellungen zu dem Forschungsfeld der Digitalisierung sind momentan insofern schwierig, da der Prozess der Digitalisierung erst am Anfang steht und noch keine eindeutigen Entwicklungen erkennbar sind. Gleichzeitig liegt hier auch eine Chance für die Arbeitsforschung, weil diese somit gleich zu Beginn der Entwicklung ihre Ansätze einer menschengerechten Arbeitsgestaltung einbringen kann und auf den Fortgang der Prozesse aktiv einwirken kann.

Die Arbeitsforschung steht vor der grundlegenden Aufgabe, in der so genannten „smart factory" u. a. die Gestaltung von humanen Arbeitsbedingungen zu unterstützen und zuverlässige, sichere sowie gesundheitsgerechte Prozesse zu entwickeln. Gleichzeitig hat die Arbeitsforschung die Potenziale des Internets der Dinge und Leistungen sowie die Innovationsmöglichkeiten in diesen Prozessen für den Arbeitsprozess und die Arbeitsgestaltung aufzuzeigen. Folgende Handlungsfelder zeichnen sich aktuell für die Arbeitsforschung ab:

- **Neue Produkte und Dienstleistungen im Bereich der Arbeitsgestaltung**: CPS ermöglichen auf allen Ebenen der Arbeit eine auf die individuellen Bedarfe zugeschnittene Arbeitsgestaltung. Um nur einige Beispiele zu nennen: Individuelle Arbeitsumgebungsgestaltung (Temperatur, Klima, Beleuchtung usw.), individuelle Prozesssteuerung von Arbeitsmitteln, Berücksichtigung des Gesundheitszustandes während der Arbeit, Überprüfung der Nutzung von persönlicher Schutzausrüstung und anderer Sicherheitseinrichtungen.
- **Neue Formen der Arbeitsorganisation und Führung**: CPS ermöglichen eine Individualisierung der Arbeitsprozessgestaltung und neue Formen der Arbeitsorganisation, neue Formen des Umgangs im Team, neue Führungskriterien und Controlling-Möglichkeiten.
- **Neue Denk-, Lern- und Verhaltens-Kulturen**: CPS führen zu einem Ineinanderfließen von realen und virtuellen Welten, für deren Bewältigung wir keine mentalen Muster besitzen. Hier sind neue Arbeitskulturen zu entwickeln, in denen auch die Frage nach der Werteorientierungen und den ethischen Grundlagen für die Wirtschaft eine Rolle spielen werden.
- **Der demografische Faktor in der Digitalisierung**: Es ist zu untersuchen, wie die Gestaltungsprozesse der Digitalisierung der Arbeitswelt und die des demografischen Wandels zusammenhängen. Zu untersuchen ist auch, wie sich sozialeInnovationen

in den Unternehmen auf die Bewältigung und Nutzung beider Entwicklungen gleichermaßen auswirken.

Die Digitalisierung der Arbeit wird die Ambivalenzen der Entwicklungen noch einmal deutlich erhöhen. Diese bieten jeweils erhebliche Chancen, sie sind aber gleichzeitig auch mit Risiken verbunden. Es kann beispielsweise außerordentlich hilfreich sein, wenn die Führungskraft feststellen kann, dass der Beschäftigte gerade psychisch belastet ist sie kann ihm dann eine Pause ermöglichen. Die Führungskraft würde in einer digitalisierten Arbeitswelt aber auch über ein Instrument verfügen, den Beschäftigten ganz aus dem Arbeitsprozess zu nehmen, falls er den Anforderungen nicht gewachsen sein sollte. Es hängt zunehmend vom Menschenbild, der Unternehmenskultur und den Entscheidungskriterien ab, wie die neuen Technologien genutzt werden. Die Ambivalenzen der möglichen Entwicklungen und der Nutzung der neuen Technologien liegen zwischen

- Einer Arbeitskultur, in der die Fähigkeiten der Beschäftigten intelligent für eine menschengerechte Arbeitsqualität genutzt werden (neue Beteiligungsprozesse an der Gestaltung der Arbeit und der Entwicklung von Produkten und Leistungen)

und

- Einer Arbeitskultur, in der CPS als restriktive, kontrollierende Mikrosteuerungen ausgelegt werden (Neo-Neo-Taylorismus), die zu einer weiteren Steigerung von Belastungen und Beanspruchungen sowie zu einer weiteren sozialen Polarisierung in der Arbeitswelt führen können.

Beispiele für mögliche ambivalente Gestaltungsfelder sind in der nachfolgenden Tabelle skizziert (vgl. Tabelle 1).

Mit den ambivalenten Entwicklungen der Digitalisierung der Arbeit und dem gleichzeitigen demografischen Wandel sind vor allem soziale Innovationen in den Prozessen der Arbeit gefordert, um die Anforderungen als Chance zu begreifen und um Menschen human in diese Prozesse integrieren zu können. Die Digitalisierung der Arbeit verstärkt noch einmal die Notwendigkeit, eine „Forschungsoffensive soziale Innovation in der Arbeitswelt" der Arbeitsforschung einzuleiten, wie sie in dem Memorandum des Förderschwerpunktes Innovationsfähigkeit im demografischen Wandel beschrieben wird [6].

3 Neue Forschungsfragen

Im Workshop im Rahmen der Förderschwerpunkt-Tagung 2014 wurde der Abschnitt „Digitalisierung der Arbeit und demografischer Wandel" des Förderschwerpunkt-Memorandums vertiefend mit den anwesenden Wissenschaftlern und Praktikern diskutiert und ergänzt. Die Ergebnisse sind in dem folgenden Abschnitt zusammengefasst.

Tabelle 1 Ambivalente Gestaltungsfelder (eigene Darstellung)

Beispiele für Gestaltungsfelder	Ambivalenzen + -
Eine kontinuierliche Beachturig der physischen, psychischen und mentalen Gesundheitdes Menschenim Arbeitsprozess	⬅➡
Ergonomische Gestaltung der umgebenden Situation und des Arbeitsumfeldes (wie Klima, Beleuchtung, Greifräume)	⬅➡
An den Menschen/an die Maschinen angepasste Gestaltung optimierter Prozesse	⬅➡
Kontinuierliche, in Prozesse integrierte Störungs- und Gefähr- dungsanalyse von Arbeitsmitteln	⬅➡
Steuerung und Wirksamkeitskontrollesicherheitsreievanten Verhaltens	⬅➡
AngepassteArbeits- und Fiihrüngskultur (bei fortschreitender Dematerialisierung und Virtuaiisierung)	⬅➡
an die neue Flexibilisierung, Komplexität und Intensivierung angepasste Arbeitsorganisation	⬅➡
Ergonomie-Konzepte, die den besonderen Anforderungen der fortschreitenden Verschmelzungrealer und virtuellerArbeitswel- ten für alle Beschäftigungsgruppen Rechnungtragen	⬅➡
Befähigungderarbeitenden Menschen (Wissen, Kompetenz, Qualifikation) mitden neuen Bedingungen psychisch, physisch (und kulturell) umzugehen, geradeunterden Bedingungendes demografischen Wandels	⬅➡

3.1 Gestaltung von Digitalisierung

CPS-Technologie ermöglicht neue Gestaltungsmöglichkeiten, die entweder reaktiv von den Akteuren angenommen oder aktiv selbst von diesen beeinflusst werden können. Dabei ergibt sich folgende Fragestellung:

- Welche Rolle spielt das Individuum in den CPS-Gestaltungsprozessen? Eine zugeteilte, eng begrenzte Rolle? Eine aktive gestaltende Rolle? Eine Rolle mit eigenen Handlungsfreiräumen (mit welchen Grenzen)?
- Welche Bedeutung hat die teilweise notwendige zentrale Steuerung in diesen Prozessen und wie ist die Balance zwischen zentraler Steuerung und den Beteiligungsprozessen?
- Wie wird die Kreativität der Individuen in die Gestaltungsprozesse über CPS-Technologie integriert?
- Welche Rolle spielen die unterschiedlichen Generationen in der Digitalisierung?
- Wie wird mit der zunehmenden sozialen Polarisierung im Arbeitsleben umgegangen?

Diese Forschungsfragen werden im Folgenden konkretisiert.

3.1.1 Das Menschenbild der Digitalisierung

Eine Voraussetzung für die Gestaltung der Digitalisierung ist ein Menschenbild, das allen eine Orientierung für ihr Verhalten und ihre Entscheidungen bietet. Ein solches Menschenbild ist Voraussetzung für die Dynamik sozialer Innovationen im Arbeitsprozess. Dazu sind unter anderem die folgenden Forschungsfragen zu klären:

- Wann macht Digitalisierung Sinn? Wann nicht? Wie kann es zu einer bewussten Entscheidung für oder gegen Digitalisierung kommen?
- Welches Menschenbild für wirtschaftliches Handeln liegt den CPS zugrunde und welches ist gewünscht/erforderlich?
- Welche Rolle spielt der Mensch in der digitalisierten Arbeitswelt und welche Rolle soll er in einer digitalisierten Gesellschaft einnehmen?
- Welches Menschenbild für die Digitalisierung der Arbeitswelt wollen wir in unserer Gesellschaft?
- Wie kann die Notwendigkeit eines solchen Menschenbildes allen Beteiligten vermittelt werden?

3.1.2 Ergonomische Kriterien für die Arbeitsgestaltung von CPS

Die Arbeitsgestaltung von CPS hat erhebliche Auswirkungen auf die Akzeptanz der CPS. Von der Gestaltung hängt auch ein Teil der Bereitschaft zur aktiven Mitgestaltung der Arbeitsprozesse ab. Dazu sind unter anderem die nachfolgenden Forschungsfragen zu lösen:

- Welche Kriterien für die Gebrauchstauglichkeit (Usability) und die Nutzungsqualität der Systeme gibt es?
- Welche ergonomischen Aspekte (Software-Ergonomie aber auch Gestaltungsergonomie generell) sind zu berücksichtigen?
- Welche Rolle spielen Intuition und Selbsterklärbarkeit bei der Gestaltung von CPS?
- Wie kann die Kompatibilität von Systemen sichergestellt werden? (Entwicklung und Bereitstellung eines Kriterienkatalogs zur Standardisierung bzw. Normierung digitaler Systeme)
- Welche ergonomischen Kriterien müssen generell erfüllt werden, damit ein Wandel zur Digitalisierung gestaltet und eine entsprechende Akzeptanz hervorgerufen werden kann?

3.1.3 Erweiterte technologisch-psychologische Systembetrachtung von CPS-Wirkprozessen

Bisher steht zumeist entweder die technologische oder psychologische Seite der Digitalisierung im Vordergrund. Insbesondere die Kombination beider Seiten wird jedoch bei der Bewältigung der CPS-Prozesse zunehmend wichtig. Dies betrifft sowohl Handlungskonzepte für Beschäftigte als auch für Führungskräfte. Schon durch die „Computerisierung der Arbeit" nahmen psychische Belastungen zu. Die systemischen Wirkungen von technologischen oder psychologischen Prozessen auf die Individuen werden im Umgang mit CPS noch einmal substantiell zunehmen. Forschungsbedarfe liegen hier unter anderem in den folgenden Bereichen vor:

- Entwicklung eines erweiterten technologisch-psychologischen Systemansatzes zur Analyse der Fragestellungen des Umgangs mit CPS-Systemen und mit ihrer sprunghaft gewachsenen Komplexität
- Entwicklung von Indikatoren zur Betrachtung und Analyse der komplexen CPS
- Entwicklung von Kriterien für die Frage: Wann ist ein digitales System und wann der Mensch die aktive steuernde Instanz?
- Entwicklung von Kriterien zur Schnittstellengestaltung (Mensch-Maschine)
- Entwicklung von Indikatoren zur Betrachtung und Analyse der integrierten und erheblich dynamisierten Lernprozesse
- Entwicklung von systemischen Kriterien, Strategien und Handlungskonzepten zur Reduktion der Komplexität der CPS-Wirkungs- und Lernprozesse (auch für unterschiedliche Altersgruppen)
- Entwicklung von systemischen Kriterien, Strategien und Handlungskonzepten zum Aufbau einer Infrastruktur der CPS-Prozesse (die alle Auswirkungen in der Wertschöpfungskette und im gesamten Arbeitssystem berücksichtigt)
- Entwicklung von systemischen Kriterien, Strategien und Handlungskonzepten (auch Datensicherheit und Arbeitsrecht) für den Umgang mit den zunehmend großen Datenmengen (Chancen und Grenzen)

3.1.4 Neue Anforderungen an die Führungsqualität

Die Digitalisierung der Arbeit stellt weitergehende und neue Anforderungen an die Führungskräfte auf unterschiedlichen Handlungsebenen. Dies umfasst beispielsweise die Reduktion zunehmender Handlungskomplexität, die neuen dynamisierten Innovations- und Lernprozesse, neue Möglichkeiten der Beteiligung und Einbindung der Beschäftigten, die neuen mentalen Anforderungen an die Beschäftigten, den Umgang mit differenzierten Kontrollsystemen oder eine angepasste Unternehmenskultur. Führung unter einem reifen CPS wird eine andere Führung sein, als sie es zum heutigen Zeitpunkt ist. Forschungsbedarfe können insbesondere in den nachfolgenden Bereichen identifiziert werden:

- Die neuen Anforderungen an Führungskräfte müssen identifiziert, definiert und mögliche Führungsstrategien für einen Umgang mit diesen entwickelt werden.
- Die Entwicklung eines neuen Selbstverständnisses von Führung in CPS-Prozessen: Die Anforderungen an Führungskräfte haben in den letzten zwei Jahrzehnten exorbitant zugenommen. Die neuen Anforderungen können nicht noch zusätzlich additiv hinzugefügt werden. Um den eigentlichen Führungsaufgaben gerecht zu werden, ist hier ein neues Anforderungsprofil von „CPS-Führung" zu entwickeln.

- Es sind Kriterien für das neue Führungsverhalten zu konstruieren (z. B. Werte, Ziele, Lernstrategien, Umgang mit unterschiedlichen Generationen im CPS).

3.1.5 Neue Anforderungen an die Interessenvertretungen und Arbeitnehmerrechte

Durch die Digitalisierung nimmt die örtliche und zeitliche Komplexität von Arbeit (die größtenteils noch im Unternehmen stattfindet) zu. Entsprechend verändern sich die Anforderungen an die Arbeitnehmer in allen Themenfeldern, so auch in denen, die in Kapitel 3.1.4 skizziert wurden. Dies schafft gleichzeitig neue Anforderungen an die entsprechende Interessenvertretung. Forschungsbedarfe liegen hier unter anderem in folgenden Bereichen vor:

- Identifikation, Definition und Entwicklung von Handlungsstrategien zum Umgang mit neuen Anforderungen an die fallspezifischen Interessenvertretungen
- Möglichkeiten der Beteiligung, der Mitwirkung und Mitbestimmung über CPS
- Schaffung von neuen Formen der Einbindung der Interessenvertretung über CPS (Kontrollproblematik in beide Richtungen)
- Weiterentwicklung des Selbstverständnisses der Interessenvertretung unter Berücksichtigung der zunehmenden Ambivalenz der CPS-Wirkprozesse (Werte, Ziele, Beteiligungsstrategien)
- Identifikation neuer Strategien der Absicherung (z. B. Zeit, Lohn, Verträge) für Arbeitnehmer bei einer fortschreitenden Digitalisierung im Unternehmen
- Entwicklung von Strategien für die zunehmende prekäre Beschäftigung im Rahmen der zunehmenden sozialen Polarisierung durch CPS-Prozesse
- Anpassung der vereinbarten Handlungsgrundlagen (rechtlich-normative Regelungen, Betriebsvereinbarungen)

3.1.6 Kriterien für die Entwickler zur Gestaltung digitaler Prozesse

Entscheidende Vorgaben für die Richtung der Arbeitsgestaltung werden bereits bei der Entwicklung von CPS gesetzt: Richten sich die Prozesse nach den menschlichen und wirtschaftlichen Anforderungen oder nach den Anforderungen der CPS-Software? Insofern sollten bereits für Entwickler von CPS entsprechende Kriterien und Vorgaben für die Arbeitsgestaltung konzipiert werden. Dabei sind gesundheitsgerechte und wirtschaftliche Aspekte gleichermaßen zu berücksichtigen. Forschungsbedarfe liegen hier unter anderem in folgenden Segmenten vor:

- Kriterien der Arbeitsgestaltung für Entwickler von CPS (Gesundheit, Sicherheit, Datenschutz, Produktivität, Leistungsbereitschaft/-fähigkeit, Unternehmenskultur, Kontrolle)
- Kriterien für Entwickler von CPS, wie die Kreativität der Beschäftigten für Innovations- und Lernprozesse in den jeweiligen Systemen gefördert wird

- Kriterien für Entwickler von CPS zur systematischen Berücksichtigung der erforderlichen Infrastruktur für die Digitalisierung (wenn dieser Faktor einer ausreichenden Infrastruktur nicht gegeben ist, scheitert die Digitalisierung; dieser Vorgang zeigte sich beispielsweise in Verwaltungen, in denen die technischen Mittel zwar eingesetzt, aber häufig durch eine fehlende Infrastruktur (was ist digital rechtlich gültig?) nicht nutzbar sind)
- Untersuchung, inwieweit neue rechtlich-normative Regelungen zur Absicherung der Entwicklung der gesundheitsgerechten und produktiven CPS erforderlich sind: Dabei müssen Kriterien für die zu akzeptierende Balance zwischen Innovationsförderung/-hemmung sowie humanen, sozialen und wirtschaftlichen Aspekten berücksichtigt werden

3.2 Forschungsbedarfe zugrundlegenden neuen Auswirkungen der Digitalisierung der Arbeit

Neben den Forschungsfragen zur Gestaltung von CPS-Arbeitssystemen wurden in dem Workshop von den Wissenschaftlern und Praktikern auch einige grundlegende neue Auswirkungen der Digitalisierung der Arbeit diskutiert. Zentrale Fragestellungen hierbei waren die beiden folgenden:

- Wie wirkt sich der Prozess der Digitalisierung auf die menschlichen Beziehungen und die sozialen Innovationen aus?
- Gibt es eine Adaption menschlicher Entwicklung/Sozialisation an digitale Prozesse und wie sieht diese aus?

Zu diesen Themenfeldern wurden folgende Forschungsbedarfe identifiziert und spezifiziert.

3.2.1 Veränderung von Vertrauensprozessen (in digitalisierten Prozessen)

Mentale Auswirkungen von Digitalisierung stehen in einer starken Abhängigkeit vom Level der Virtualität. Diese Aspekte wurden bereits in der Vergangenheit in Zusammenhang mit der „Computerisierung der Arbeit" (Industrie 3.0) untersucht. Digitalisierung ist prinzipiell kein neues Phänomen, sondern bereits seit den 1970er Jahren thematisiert worden. Maßgeblicher Treiber ist in der Industrie 4.0 eine neue technische Produkt- und Prozessqualität sowie eine hiermit verbundene erhöhte Komplexität und Geschwindigkeit der Prozesse durch die zunehmende Verschmelzung von realen und virtuellen Erfahrungen (siehe Kapitel 1 und Kapitel 2).

Zu klären ist die Frage, ob und wie die Digitalisierung analoge Strukturen in den Verhaltensweisen der Menschen verschiebt respektive verändert und wie der Mensch mit dieser zunehmenden neuen Komplexität umgeht. Forschungsbedarfe liegen hier unter anderem in den folgenden Bereichen:

- Wie verändert sich das Verantwortungsgefühl durch die Unsichtbarkeit und den mangelnden direkten Kontakt bei digitalen Prozessen?
- Wie können die Folgen von Digitalisierung bzw. die Digitalisierung selber sichtbar gemacht werden, um Vertrauen bei den Anwendern zu fördern?
- Wie kann das Gefühl der Kontrollierbarkeit von Prozessen vermittelt werden, damit Vertrauen aufgebaut werden kann? Wie kann eine tatsächliche Kontrollierbarkeit gesichert werden (nach welchen Kriterien)?
- Wie verändern sich Vertrauensprozesse durch Digitalisierung und wie kann dies bei der Gestaltung von Digitalisierung berücksichtigt werden?

3.2.2 Kontinuum zwischen privaten und beruflichen Lebenswelten

Schon die Computerisierung hat die Balance zwischen Arbeit und Freizeit verschoben (Work-Life-Balance). Die Digitalisierung der Arbeit in der nächsten Stufe (Industrie 4.0) wird diesen Prozess in der Dynamik noch weiter vorantreiben:

- CPS verschieben die Grenzen nicht nur zwischen Realität und Virtualität, sondern auch noch einmal deutlich die zwischen Arbeit und Freizeit
- Die Produkte der Industrie 4.0 werden gleichzeitig unser gesamtes Leben und unsere gesamte Lebensplanung fundamental beeinflussen (smart home, smart products, smart services)

Digitalisierungsprozesse in der Arbeit und in der Freizeit verlaufen reziprok und sind daher nicht voneinander trennbar. Forschungsbedarfe liegen hier unter anderem in folgenden Bereichen vor:

Abbildung 3 Forschungsbedarfe im Bereich „Digitalisierung der Arbeit und demografischer Wandel" aus Sicht der Förderschwerpunkt-Akteure

- Kriterien für die Abgrenzung zwischen Arbeit und Freizeit – auch mit Unterstützung von CPS
- Kriterien für die Schaffung von Freiräumen für private Lebenswelten
- Kriterien für Handlungskonzepte für unterschiedliche Lebensphasen und Altersgruppen

4 Fazit

Der vorliegende Beitrag fasst die Ergebnisse der Session „Digitalisierung der Arbeit und demografischer Wandel" auf der Förderschwerpunkt-Tagung 2014 zusammen. Aus der Perspektive der teilnehmenden Förderschwerpunktakteure haben sich die in Abbildung 3 dargestellten Forschungsbedarfe aus der Diskussion der Sessions als wesentlich erwiesen.

Sie basieren auf keiner fundierten systematischen Analyse. Dennoch öffnen sie einen Fächer hochinteressanter neuer Felder der Arbeitsforschung. Die Autoren bedanken sich herzlich bei allen Beteiligten am Workshop, die diese wertvollen Impulse zusammengetragen haben.

Literaturverzeichnis

1. E. Geisberger, M. Broy, eds., *Agenda CPS Integrierte Forschungsagenda Cyber-Physical Systems*. München, 2012
2. H. Kagermann, W. Wahlster, J. Helbig, Umsetzungsempfehlungen für das Zukunftsprojekt Industrie 4.0. Tech. rep., Berlin, 2012
3. O. Cernavin, Industrie 4.0 und Prävention. Sicherheitsingenieur **45** (6), 2014, pp. 18–21
4. S. Jeschke, R. Vossen, T. Thiele, I. Leisten, S. Fleischer, F. Welter, Industrie 4.0 als Treiber der demografischen Chancen. Tech. rep., Zentrum für Lern- und Wissensmanagement, Lehrstuhl für Informationsmanagement im Maschinenbau, IfU - Institut für Unternehmenskybernetik e.V., 2013
5. H. Kagermann, W. Wahlster, J. Helbig, Bericht der Promotorengruppe KOMMUNIKATION Im Fokus: Das Zukunftsprojekt Industrie 4.0. Handlungsempfehlungen zur Umsetzung. Tech. rep., Berlin, 2012
6. P.D.G. Becke, D.P. Bleses, O. Cernavin, P.D.A. Ducki, D.R. Klatt, P.D.T. Langhoff, P.D.F.W. Nerdinger, Memorandum: Förderschwerpunkt Innovationsfähigkeit im demografischen Wandel. In: *Exploring Demographics*, ed. by S. Jeschke, A. Richert, F. Hees, C. Jooß, Springer Fachmedien Wiesbaden, 2015, pp. 119–121
7. D. Spath, O. Ganschar, S. Gerlach, M. Hämmerle, T. Krause, S. Schlund, eds., *Produktionsarbeit der Zukunft Industrie 4.0*. Stuttgart, 2014

Ergebnistransfer nachhaltig gestalten – Eine strukturelle Übersicht

Oleg Cernavin, Stefan Schröder, Thomas Thiele and Claudia Jooß

Zusammenfassung Der nachhaltige Transfer von Ergebnissen aus der Forschung in die Praxis gewinnt in Zeiten von steigendem Wettbewerb und erhöhtem Bedarf an ganzheitlichen Innovationsprozessen zunehmend an Bedeutung (Leisten, Ingo (2012): Transfer Engineering in transdisziplinären Forschungsprojekten. Norderstedt: 5). Gleichzeitig werden aus den beiden letztgenannten Gründen Forschungsprozesse zunehmend unter transdisziplinären Rahmenbedingungen gemeinsam von Wissenschaft und Praxis gestaltet, was einerseits besondere Anforderungen an die Zusammenarbeit stellt, andererseits u. a. zahlreiche Potentiale hinsichtlich des Transfers der (Projekt-)Ergebnisse bietet. Der vorliegende Beitrag verfolgt daher das Ziel, eine strukturelle Übersicht über die Transferlandschaft, am Beispiel der Förderschwerpunkte im Kontext der Arbeitsforschung (z. B. Innovationsfähigkeit im demografischen Wandel), zu geben. Die komplexe Struktur der Förderschwerpunkte mit einer Vielzahl von kooperierenden wissenschaftlichen Disziplinen und Praxispartnern (insgesamt 27 Verbundprojekte und mehr als 80 beteiligte Institutionen forschen deutschlandweit gemeinsam im Rahmen des Förderschwerpunktes „Innovationsfähigkeit im demografischen Wandel" an der systematischen Erschließung von Innovationspotenzialen, welche aus dem demografischen Wandel entstehen, um diesen zur Steigerung der Wettbewerbsfähigkeit nutzbar zu machen.) erweist sich unter dem Fokus einer externen Adressierung der Forschungsergebnisse als besondere Herausforderung, die im Folgenden zunächst erörtert wird. Ferner werden Potentiale in diesem Kontext aufgezeigt. Im weiteren Verlauf erfolgt eine systematische Aufbereitung und Darstellung bereits bestehender Transferstrukturen. Abschließend werden Handlungsempfehlungen zur Gestaltung des Transfers unter Berücksichtigung der förderstrukturellen Rahmenbedingungen gegeben.

Schlüsselwörter Transfer · Nachhaltigkeit · Verwertung · Handlungsempfehlungen

O. Cernavin (✉)
BC Forschung GmbH, Kaiser-Friedrich-Ring 53, 65185 Wiesbaden, Germany
e-mail: oleg.cernavin@bc-forschung.de

S. Schröder · T. Thiele · C. Jooß
IMA/ZLW & IfU, RWTH Aachen University, Dennewartstr. 27, 52068 Aachen, Germany

Originally published in "Exploring Demographics",
© Springer 2015. Reprint by Springer International Publishing Switzerland 2016,
DOI 10.1007/978-3-319-42620-4_4

1 Einführung

Forschungsprojekte im Allgemeinen und transdisziplinäre Forschungsprojekte im Speziellen stehen oftmals vor der Herausforderung der Adressierung und Einbindung Externer, insbesondere im Sinne des nachhaltigen Transfers ihrer Ergebnisse ([1]: 99). Als Transfer wird dabei der Austauschprozess an den Schnittstellen zwischen Wissenschaft, Wirtschaft, Politik und Gesellschaft verstanden. Hierbei stehen im Besonderen Personen als Gestalter des Kommunikationsprozesses aus unterschiedlichen Kontexten im Vordergrund. „Die Generierung von Wissen und der Explizierung individuellen Wissens stellt die Akteure vor Herausforderungen, die nicht nur in Bezug auf die personale Wissenskommunikation zu sehen sind [...], sondern auch organisationale Prozesse und Strukturen beeinflussen" ([1]: 45). Dabei erstreckt sich die Kommunikation und damit auch der Transfer beispielsweise über Kontaktanbahnung, Dissemination zwischen potentiellen Partnern aus Wissenschaft und Wirtschaft, die Entwicklung gemeinsamer, zukünftiger Projektideen bis hin zum Einsatz und der Überführung neu entwickelter Produkte in wirtschaftliche Strukturen.

Viele Projekte bzw. Akteure im Kontext der Forschungsförderung stehen somit vor der Herausforderung des breitenwirksamen Transfers ihrer Forschungs- & Projektergebnisse. Die grundlegende Problemstellung liegt u. a. darin, praxisrelevante Ergebnisse zu generieren und gleichzeitig erfolgreich zu disseminieren. Speziell Projektträger und Ministerien verfolgen das Ziel, dass die von Ihnen begleiteten und geförderten Erkenntnisse, Strategien, Produkte und Leistungen nachhaltig in den Unternehmen Anwendung finden. Die grundlegende Forschungslücke kann darin gesehen werden, dass die Überführung fundierter Konzepte in die betriebliche Personal und Organisationsentwicklung oder in geeignete Geschäftsmodelle noch nicht vorliegt ([2]: 19).

So bestehen beispielsweise im Kontext der Arbeitsforschung viele Möglichkeiten die Projektergebnisse nach Projektabschluss weiter zu nutzen. Beispielsweise konnte das Projekt zur Bilanzierung der Arbeitsschutzforschung des Bundesministeriums für Bildung und Forschung (BMBF) bereits 2001 herausarbeiten, dass sich gerade der Transfer der gewonnen Erkenntnisse in die betriebliche Praxis auf Dauer als anspruchsvoll gestaltet [3].

Dennoch zeigen verschiedene Arbeiten im Kontext der Arbeitsforschung, dass umfassende und langfristige Wirkungen in der Entwicklung der Arbeit in Deutschland, zu verzeichnen sind [4–6]. Gleichzeitig bieten sich jedoch diverse Möglichkeiten, den Transfer innerhalb dieser Strukturen zu systematisieren und damit zu optimieren, um damit einen zielgerichteten Transfer zu unterstützen.

Gelingt es nicht, vorhandene Strukturen und Akteure einzubeziehen und zu nutzen, besteht das Risiko, dass innovative Konzepte, Ansätze und Instrumente in der Praxis nicht berücksichtigt werden. In diesem Beitrag soll daher auf die Möglichkeit des Transfers von Projektergebnissen über bestehende Institutionen und Netzwerke sowie die hieraus resultierenden Potentiale hingewiesen werden. Ferner werden, im Sinne eines lernenden Programms, Handlungsempfehlungen für die Akteure und neue Impulse für zukünftige Kooperationen im Förderschwerpunkt „Innovationsfähigkeit im demografischen Wandel" und darüber hinaus gegeben.

2 Potentiale und Herausforderungen

Ein möglicher Ansatz, die Nachhaltigkeit der Ergebnisse von Forschungsprojekten zu verbessern liegt in der Nutzung von Transferstrukturen bestehender Institutionen und Netzwerke. Diese bieten einen wesentlichen Vorteil: In etablierten Strukturen können die gewünschten Zielgruppen direkt adressiert werden, sodass ein passgenauer Transfer entstehen kann. Im Kontext der Arbeitsforschung sind die Zielgruppen meist kongruent mit den Transferzielgruppen der Forschungsprojekte – den Akteuren in Unternehmen, Intermediären und Verbänden. Diese umfassen beispielsweise die Fachkräfte für Arbeitssicherheit, die Präventionsberater der Krankenkassen, die Berater der Verbände der Sozialpartner und viele mehr. Als Unterstützer für Führungskräfte und Beschäftigte ist es u. a. die Aufgabe dieser Akteure, die Arbeit menschengerecht und innovationsfördernd zu gestalten und sich auf die zukünftigen Entwicklungen, im Kontext des demografischen Wandels, einzustellen.

Bei Institutionen, Netzwerken und Unternehmen besteht zudem fortlaufend Bedarf an Praxislösungen. Die Quellen für diese Lösungen sind vielfältig: So sind Konstellationen denkbar, in denen Lösungen sowohl aus der Wissenschaft (z. B. forschende Universitäten), der Praxis als auch von Intermediären entwickelt werden. Daher kommt dem Transfer zwischen diesen Entitäten eine besondere Bedeutung bei. Im Themenfeld der Arbeitsforschung ist zudem eine ausgeprägte Landschaft von Transferstrukturen vorhanden, durch deren langjährige Erfahrung in diesem Bereich ein großes Potenzial für die nachhaltige Nutzung und Umsetzung von Projektergebnissen liegt.

Die zentrale Herausforderung in diesem Zusammenhang liegt in der Wahrnehmung dieser Transferstrukturen: Nur wenn die Akteure aus Wissenschaft, Praxis und intermediären Organisationen Kenntnisse über die Zusammensetzung und die Verbindungen innerhalb der vorhandenen Transferstrukturen besitzen, lassen sich diese gewinnbringend nutzen. Eine weitere maßgebliche Herausforderung ist aber auch oftmals die beidseitige Wahrnehmung (Wissenschaft, Praxis und intermediäre Organisationen und Transferstrukturen), als basales Element des Transfers müssen die Akteure von ihren jeweiligen Bestrebungen in Kenntnis gesetzt werden.

Dabei sind eine Vielzahl von Potentialen für beide Seiten denkbar: Handlungsziele der bestehenden Institutionen und Netzwerke erweisen sich in vielen Fällen als identisch mit den Handlungszielen der Forschungsprojekte. Beide Seiten fokussieren die Verbesserung der Qualität der Arbeit. Im Detail konstatieren sich u. a. die in Tabelle 1 dargestellten Potentiale und Herausforderungen.

Da es im originären Interesse des Forschungsförderers liegt, die Nachhaltigkeit der Ergebnisse nach Projektende sicherzustellen, sollten die Forschungsprojekte intrinsisch motiviert sein, aktiv die beidseitige Wahrnehmung zu erhöhen. Dies hat vor allem zwei wesentliche Voraussetzungen:

Tabelle 1 Potentiale und Herausforderungen bei der Nutzung bestehender Transferstrukturen

Potentiale & Herausforderungen	
• Heterogene und große Anzahl an Institutionen, Netzwerken und Akteuren	• Heterogene und große Anzahl an Institutionen, Netzwerken und Akteuren
• Wirkungsgrad der bestehenden Institutionen und Netzwerke	• Weitere nachhaltige Nutzung der Projektergebnisse
• Transfer der Ergebnisse über die Strukturen der ergebnisgenerierenden Institutionen hinaus	• Auswahl geeigneter Intermediäre für den Transfer
• Integration der Transferpartner in Projektideen und -strukturen	• Zielgruppenadaptiver Transfer der (Projekt-)Ergebnisse

- *Institutionen und Netzwerke kennen*: Die Verbundprojekte sollten die Transferstrukturen der bestehenden Institutionen und Netzwerke zur Arbeitsqualität in ihrer Spezifik kennen.
- *Institutionen und Netzwerke frühzeitig einbinden*: Die Verbundprojekte sollten die Transferstrukturen nutzen und die entsprechenden Aktivitäten bereits im Projektantrag und im Projektverlauf über alle Arbeitsschritte mit berücksichtigen (Zeit, Personen, Kosten).

Im Folgenden wird eine erste Systematik intermediärer Organisationen im Bereich des demografischen Wandels und der Arbeitsforschung eingeführt. Diese soll eine thematische Zuordnung einzelner Institutionen und Netzwerke erleichtern und somit einen Transfer unterstützen.

3   Transferstrukturen Transparent

Die folgende Systematik orientiert sich an der jeweiligen Funktion der verschiedenen Institutionen und Einrichtungen. Dabei sei darauf hingewiesen, dass die einzelnen Felder nicht trennscharf sind und sich in vielen Bereichen überlappen. Dabei lassen sich fünf Felder von Institutionen und Netzwerken identifizieren, in denen (Beratungs-)Leistungen zur Arbeitsforschung und zum demografischen Wandel angeboten werden (vgl. Abbildung 1):

- **Transferstrukturen der Politik**
  Institutionen, die die staatlichen Aufgaben der Setzung von Arbeitsstandards, der Aus- und Weiterbildung, der Beschäftigungssicherung, der Wirtschaftsförderung und der Überwachung wahrnehmen. Zudem werden auch Aufgaben der Forschungsförderung und damit verbunden die strategische Ausrichtung von Forschungsprogrammen wahrgenommen (z. B. das F&E Programm „Arbeiten – Lernen – Kompetenzen entwickeln – Innovationsfähigkeit in einer modernen Arbeitswelt").

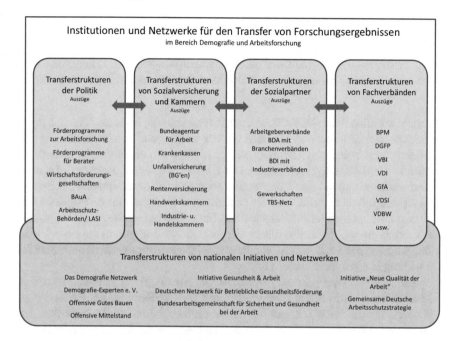

Abbildung 1   Institutionen und Netzwerke für den Transfer von Forschungsergebnissen

- **Transferstrukturen von Sozialversicherung und Kammern**
  Institutionen, die grundlegende soziale Aufgaben absichern und die gesetzliche vorgeschrieben und verbindlich sind. Bei diesen Institutionen müssen Personen oder Institutionen Mitglied sein („Zwangsmitgliedschaft").
- **Transferstrukturen der Sozialpartner**
  Institutionen, die die Interessen der Arbeitgeber und Arbeitnehmer wahrnehmen.
- **Transferstrukturen von Fachverbänden**
  Institutionen, die berufsständische Interessen vertreten und die Standards und Hilfen für beratende Berufe im Bereich der Demografie und einer präventiven Arbeitsgestaltung anbieten.
- **Transferstrukturen von nationalen Initiativen und Netzwerken**
  Zusammenschluss von Institutionen und Akteuren der ersten vier Transferbereiche mit dem Ziel spezifische bzw. übergeordnete Aspekte der Arbeitsgestaltung gemeinsam wirkungsvoll voranzubringen.

Diese einzelnen Felder sollen im Folgenden hinsichtlich ihrer Aktivitäten im Kontext des demografischen Wandels konkretisiert werden. Es werden auch Institutionen und Netzwerke dargestellt, die nicht direkt für den Transfer von Forschungsprojekten geeignet sind, die aber (politisch) wesentlich für die Zu-sammenhänge des Transfers sind. Die aufgeführten Institutionen haben ein begründetes Interesse an Ergebnissen aus Forschungsprojekten und können u. a. als Multiplikatoren eingebunden werden.

3.1 Die Transferstrukturen der Politik

Diese Strukturen werden vor allem durch staatliche Angebote und Dienstleistungen charakterisiert. Eine exemplarische Auswahl umfasst die nachfolgenden Institutionen:

- Zu den staatlich finanzierten Transferstrukturen gehören auch die *Forschungsprogramme zur Arbeitsforschung*, wie sie vor allem vom Bundesministerium für Bildung und Forschung (BMBF) und vom Bundesministerium für Arbeit und Soziales (BMAS) aufgelegt werden. Das BMBF-Programm „Arbeiten – Lernen – Kompetenzen entwickeln – Innovationsfähigkeit in einer modernen Arbeitswelt" bietet beispielsweise vielfältige Möglichkeiten der Kooperation von Projekten zum Thema Demografie und Arbeitsgestaltung (auch über beispielsweise Metaprojekte und Fokusgruppen). Ebenso untersucht das aktuelle Programm die Innovationsfähigkeit aus der Verknüpfung von Personal-, Organisations- und Kompetenzentwicklung in einer modernen Arbeitswelt.
- Spezifische *Förderprogramme für betriebliche Beratung* (z.B. unternehmensWert: Mensch des BMAS, Bundesförderprogramm Vor-Ort-Beratung der BAFA, Potenzialberatung in NRW). Diese Förderprogramme betreiben Datenbanken mit autorisierten Beratern. In die Anforderungen bzw. in die Informationen der Förderprogramme können ggf. Instrumente der Projekte mit integriert werden.
- *Wirtschaftsförderungsgesellschaften*, haben die Aufgabe den Wirtschaftsstandort durch Unterstützung der Unternehmen zu stärken, dessen Attraktivität zu stärken und Existenzgründungen zu fördern. Viele Wirtschaftsförderungsgesellschaften unterstützen Unternehmen auch darin, den demografischen Wandel erfolgreich zu bewältigen. Hier besteht oft auch Interesse für die Nutzung von Projektergebnissen. Wirtschaftsförderungsgesellschaften sind kommunal oder GmbHs (meist von staatlichen Trägern).
- Die *Bundesanstalt für Arbeitsschutz und Arbeitsmedizin (BAuA)* forscht und entwickelt im Themenfeld Sicherheit und Gesundheit bei der Arbeit, fördert den Wissenstransfer in die Praxis, berät die Politik und erfüllt hoheitliche Aufgaben – im Gefahrstoffrecht, bei der Produktsicherheit und mit dem Gesundheitsdatenarchiv.[1] Die Projekte sollten der BAUA die Ergebnisse bekannt geben, damit diese in Publikationen mit berücksichtigt werden können.
- Wichtige staatliche Einrichtungen im Bereich der Arbeitsgestaltung sind die *Arbeitsschutzbehörden* der Länder. Technische Aufsichtspersonen dieser Behörden überwachen die Einhaltung der Arbeitsschutzvorschriften in den Betrieben und helfen den Unternehmern sowie Arbeitnehmern bei der Umsetzung. Aufsichtspersonen des Staates weisen Unternehmen auf Instrumente hin, die helfen die Arbeitsaufgaben sicherer zu gestalten und die auch helfen, die Anforderungen des demografischen Wandels aus dieser Perspektive zu bewältigen.

---

[1] www.baua.de.

- Im LASI, dem *Länderausschuss für Arbeitsschutz und Sicherheitstechnik*[2] sind die Vertreter der Arbeitsschutzbehörden der Länder vertreten. Hier wird das gemeinsame Vorgehen abgestimmt und es werden gemeinsame Veröffentlichungen herausgegeben.

3.2 Die Transferstrukturen von Sozialversicherungen und Kammern

Die Sozialversicherungen und Kammern sind Institutionen, bei denen Unternehmen oder Arbeitnehmer per gesellschaftlichen Konsens Mitglied sein müssen („Zwangsmitgliedschaften"). Diesen Institutionen bzw. ihren regionalen Vertretungen sollten die Projektergebnisse zur Verfügung gestellt werden, weil Berater dieser Institutionen auf unterstützende Maßnahmen, die sich als praxisrelevantes Gestaltungselement erwiesen haben, verweisen und diese auch einsetzen. Zu diesen gehören zunächst die vier Zweige der Sozialversicherungsträger:

- Die *Bundesagentur für Arbeit*, die neben der Arbeitsvermittlung über ihren Arbeitgeber-Service (AG-S) mit 5000 Beratern zunehmend auch direkt Unternehmen zum Thema Personalentwicklung und Personalführung berät.[3]
- Die *gesetzlichen Krankenversicherungen*, die vielfältige Angebote zur gesundheitsgerechten Arbeitsgestaltung entwickelt haben und die über Ihre Präventionsberater die Betriebe und ihre Versicherten direkt berät. Der GKV-Spitzenverband vertritt die gemeinsamen Interessen der Krankenversicherungen.[4]
- Die *gesetzlichen Unfallversicherungen* (Berufsgenossenschaften), bei denen alle Unternehmen gegen Arbeitsunfälle und Berufskrankheiten versichert sind. Die Technischen Aufsichtspersonen der Unfallversicherung nehmen jährlich rund 500.000 Besichtigungen in Unternehmen vor und beraten die Unternehmer und Beschäftigten in Fragen des Arbeitsschutzes. Der Spitzenverband der Träger der gesetzlichen Unfallversicherung (UVT) ist die DGUV – die Deutsche gesetzliche Unfallversicherung.[5]
- Die *gesetzliche Rentenversicherung*, die mittlerweile über eigene Berater ebenfalls die Betriebe direkt berät – zum Beispiel über die „GeniAL"-Berater zur Bewältigung des demografischen Wandels.[6]

Neben den vier Sozialversicherungsträgern bestehen diese Transferstrukturen auch aus Kammern, bei denen ebenfalls eine Zwangsmitgliedschaft für Unternehmen besteht. Hierzu zählen

---

[2] www.lasi.osha.de.

[3] www.arbeitsagentur.de.

[4] www.gkv-spitzenverband.de.

[5] www.dguv.de.

[6] www.deutsche-rentenversicherung.de.

- Die *Handwerkskammern*, die eigene Berater haben, die sich vor allem mit betriebs-wirtschaftlichen Fragen befassen, die aber auch zunehmend zu Fragen der Arbeits-gestaltung und des demografischen Wandels beraten. Auf Bundesebene bündelt der Zentralverband des Deutschen Handwerks e. V. (ZDH) die Arbeit von 53 Handwerkskammern und 48 Fachverbänden des Handwerks.[7]
- Die *Industrie- und Handelskammern (IHK)*, die ebenfalls eigene Berater haben, die die Betriebe vor allem in Fragen der Aus- und Weiterbildung und der Existenz-gründung beraten. Der Deutsche Industrie- und Handelskammertag (DIHK) ist die Spitzenorganisation der insgesamt 82 Industrie- und Handelskammern.[8]

3.3 Die Transferstrukturen der Sozialpartner

Auch die Verbände und Einrichtungen der Sozialpartner besitzen Beraterstrukturen, die die Betriebe in Fragen der Arbeitsgestaltung und des demografischen Wandels unterstützen. Diesen Einrichtungen bzw. ihren regionalen Vertretungen erweisen sich als vielfältige Disseminations- und Transferpartner, da bei der Beratung direkt auf innovative Maßnahmen aus der Forschung verwiesen werden kann und diese auf diesem Weg zur Anwendung in der Praxis gelangen können.

- Viele *Arbeitgeberverbände* haben Berater bzw. Beratergesellschaften, die die Unternehmen in allen Fragen des Managements und der Arbeitsgestaltung beraten. Da es eine große Anzahl von einzelnen Verbänden gibt, die in der Regel nach Branchen und Region gegliedert sind, empfiehlt es sich, sich direkt vor Ort oder in der Branche nach den vorhandenen Beratungsangeboten zu erkundigen. Die Dachorganisation der kleinen und mittleren Unternehmen ist die BDA – die Bun-desvereinigung der Deutschen Arbeitgeberverbände, die über 50 Einzelverbände unterschiedlicher Branchen vertritt.[9] Die nationale Vertretung der Industrieun-ternehmen ist der BDI – Bundesverband der Deutschen Industrie e. V., der 37 Branchenverbände vertritt.[10]
- Die *Gewerkschaften* besitzen ebenfalls Berater, die vor allem die Betriebs- und Personalräte beraten. Diese Beratungen werden in der Regel von den Technolo-gieberatungsstellen (TBS) der Gewerkschaften übernommen.[11] Die 13 Technolo-gieberatungsstellen beim DGB bieten Beratung, Weiterbildung und Fachinforma-tion zu allen Fragen rund um Beschäftigung, Arbeitsorganisation, Technik und Gesundheit an.

---

[7] www.zdh.de.
[8] www.dihk.de.
[9] www.arbeitgeber.de.
[10] www.bdi.eu.
[11] www.tbs-netz.de.

3.4   Die Transferstrukturen von Fachverbänden

Es gibt eine große Anzahl von Fachverbänden, die Berater in unterschiedlichen Bereichen der Prävention und der Arbeitsgestaltung vertreten. Im Folgenden werden einige dieser Verbände vorgestellt: Diese Verbände sind für Forschungsprojekte insofern interessant, da sie entweder thematisch oder von der Anzahl der Berater die Projekte beim Transfer unterstützen können. Die folgende Auflistung kann aufgrund der vielfältigen Transferstruktur nur eine exemplarische Auswahl aus der Perspektive der Autoren sein und erhebt daher keinen Anspruch auf Vollständigkeit. Zu den Fachverbänden gehören beispielsweise:

- Der *Bundesverband der Personalmanager (BPM)* ist die berufsständische Vereinigung für Personalmanager und Personalverantwortliche aus Unternehmen, Verbänden und anderen Organisationen.[12] Der BPM besitzt Fachgruppen zu klassischen Themen der Arbeitsgestaltung. Die Personalmanager sind wesentliche Adressaten für die Ergebnisse von Projekten zum demografischen Wandel.
- Die *Deutsche Gesellschaft für Personalführung (DGFP) e. V.* unterstützt die Mitglieder in vielen Personalfragen zum demografischen Wandel (2.500 Mitgliedsunternehmen und Mitgliedern, mit insgesamt über 40.000 Personalverantwortlichen).[13]
- Der *Verband Beratender Ingenieure (VBI)* ist die Berufsorganisation unabhängig beratender und planender Ingenieure und Ingenieurunternehmen in Deutschland (3.500 Mitglieder).[14]
- Der *VDI Verein Deutscher Ingenieure e.V.* ist ein technisch-wissenschaftlicher Verein, der die Ingenieure in Deutschland vertritt (rund 150 000 Mitglieder).[15] Die Ingenieure beraten oft auch zur Prozessgestaltung, zum Qualitätsmanagement und zur Arbeitsorganisation. Zu diesen Themenfeldern gibt es Verknüpfungsmöglichkeiten zu den Projektergebnissen zum demografischen Wandel.
- Die *GfA – Gesellschaft für Arbeitswissenschaft* – ist eine Vereinigung von Wissenschaftlern und anderen Interessierten mit dem Ziel, die Arbeitswissenschaft zu fördern.[16] Sie versteht sich dabei auch als Plattform für den Austausch zwischen Wissenschaft und Praxis sowie allen interessierten gesellschaftlichen Gruppen. Hier besteht immer Interesse neue Forschungsergebnisse kennenzulernen und weiterzutragen.
- Der *VDSI – Verband für Sicherheit, Gesundheit und Umweltschutz bei der Arbeit e.V.* vertritt die *Fachkräfte für Arbeitssicherheit* in Deutschland.[17] Jedes Unternehmen ist per Gesetz verpflichtet, sich von einer Fachkraft für Arbeitssicherheit in Fragen des Arbeitsschutzes beraten zu lassen. Dazu gehört auch die

---

[12] www.bpm.de.

[13] www.dgfp.de.

[14] www.vbi.de.

[15] www.vdi.de.

[16] www.gesellschaft-fuer-arbeitswissenschaft.de.

[17] www.vdsi.de.

Beratung zu Maßnahmen zum demografischen Wandel. Ein anderer Fachverband in diesem Bereich ist der Bundesverband freiberuflicher Sicherheitsingenieure und überbetrieblicher Dienste e. V.[18] Viele Fachkräfte für Arbeitssicherheit und auch die genannten Verbände sind offen für Forschungsergebnisse vor allem zu den Themen Ergonomie, Arbeitsorganisation und psychische Gesundheit.

- Der VDBW ist der Berufsverband Deutscher Arbeitsmediziner, der die *Betriebsärzte* vertritt.[19] Jedes Unternehmen ist per Gesetz auch verpflichtet, sich von einem Betriebsarzt in Fragen des Gesundheitsschutzes beraten zu lassen. Auch viele Betriebsärzte greifen auf Forschungsergebnisse zurück und nutzen diese für Ihre Beratungen.

3.5 Die Transferstrukturen von nationalen Initiativen und Netzwerken

Neben einzelnen, oben beschriebenen Organisationen und Einrichtungen sind auch nationale und regionale Netzwerke und Initiativen für den Transfer von Projektergebnissen relevant. Im Folgenden werden einige wesentliche Initiativen und Netzwerke in Deutschland zum Themenbereich Arbeitsgestaltung und demografischer Wandel vorgestellt. Zu den Netzwerken und Initiativen gehören beispielsweise:

- *Das Demografie Netzwerk (ddn)* hat rund 400 Unternehmen und Institutionen mit einer Personalverantwortung zusammengeschlossen.[20] In Facharbeitskreisen und rund 20 regionalen Netzwerken werden Aktivitäten zur Bewältigung des demografischen Wandels koordiniert und vorangetrieben.
- *Der Demografie-Experten e. V. (DEx)* hat rund 200 Demografieberater qualifiziert und adressiert im Besonderen Fragen der Qualitätssicherung.[21] Als Verein unterstützt DEx vor allem Berater in ihrer Arbeit mit weiteren Hilfsmitteln und Instrumenten.
- In der *Offensive Gutes Bauen* engagieren sich über 100 Institutionen für eine gute Bauqualität.[22] Die Offensive Gutes Bauen hat ebenfalls Referenzinstrumente entwickelt, bei denen weiterführende Instrumente hinterlegt sind – auch hier können Projektergebnisse breitenwirksam transferiert werden. Außerdem werden auch im Baubereich Berater autorisiert und regionale Netzwerke zur Mitarbeit initiiert.
- Die *Offensive Mittelstand* ist eine eigenständige Initiative, in der sich über 250 Partnerorganisationen für Arbeitsqualität im Mittelstand einsetzen.[23] Die

---

[18] www.bfsi.de.

[19] www.vdbw.de.

[20] www.demographie-netzwerk.de.

[21] www.demografie-experten.de.

[22] www.offensive-gutes-bauen.de.

[23] www.offensive-mittelstand.de.

Offensive Mittelstand hat eigene Praxisinstrumente als gemeinsame Qualitäts-
standards erarbeitet wie den INQA-Unternehmens-check oder den INQA-Check-
Personalführung. In den Online-Tools der Checks sind Praxisinstrumente der Partnerorganisationen hinterlegt. Hier könnten auch Instrumente aus Forschungsprojekten hinterlegt werden. Die Offensive Mittelstand qualifiziert außerdem Berater,
die Checks und die hinterlegten Instrumente kompetent umzusetzen. Auch die rund
20 regionalen Netzwerke ermöglichen eine nachhaltige Umsetzung von Instrumenten aus Projekten.

- *Die Initiative „Neue Qualität der Arbeit"* ist eine vom Bundesministerium für
  Arbeit und Soziales initiierte Initiative, die die Qualität der Arbeit fördert.[24]
  Es gibt einen Steuerkreis auf Bundesebene (mit Sozialpartnern und anderen
  großen Intermediären) und Themenbotschaftern zu vier zentralen Themen (Personalführung, Diversity, Gesundheit, Wissen und Kompetenz). Für die Umsetzung
  von Projektergebnissen sei auf die oben beschriebenen vier Initiativen verwiesen,
  die als INQA-Netzwerke arbeiten.
- Die *Gemeinsame Deutsche Arbeitsschutzstrategie – GDA* ist eine auf Dauer
  angelegte nationale Strategie von Bund, Ländern, Unfallversicherungsträgern und
  Sozialpartnern zur Stärkung von Sicherheit und Gesundheit bei der Arbeit in
  Deutschland. Eines der Ziele ist es, die Betriebe bei der Umsetzung von Arbeits-
  schutzmaßnahmen zu unterstützen und damit auch die Wettbewerbsfähigkeit der
  deutschen Wirtschaft zu fördern.[25] Einzelne Projektergebnisse können über die
  Träger der GDA in die Diskussionen eingebracht werden, für die Dissemination
  vollständiger Projekte sei auf die vier erstgenannten Organisationen verwiesen.
- In der *Bundesarbeitsgemeinschaft für Sicherheit und Gesundheit bei der Arbeit
  (Basi) e.V.* arbeiten 80 Organisationen und Einrichtungen mit der Zielsetzung
  zusammen, die Sicherheit und Gesundheit bei der Arbeit in der Bundesrepub-
  lik Deutschland zu verbessern.[26] Die BASI veranstaltet die Messe Arbeitsschutz
  und Arbeitsmedizin in Düsseldorf. Die Messe mit dem angeschlossenen Kongress
  bietet eine konkrete Plattform für den Transfer von Projektergebnissen.
- In der *iga – Initiative Gesundheit & Arbeit* arbeiten der BKK Dachverband, der
  AOK-Bundesverband, der Verband der Ersatzkassen e. V. (vdek) und die Deutsche
  Gesetzliche Unfallversicherung in Fragen der Prävention und der betrieblichen
  Gesundheitsförderung zusammen und stimmen Ihre Maßnahmen ab.[27]
- Ein Netzwerk im Rahmen der iga ist das *Deutsche Netzwerk für Betriebliche
  Gesundheitsförderung DNBGF*.[28] Ziel des Netzwerks ist es, die Kooperation
  zwischen allen nationalen Akteuren zum Thema der betrieblichen Gesundheits-
  förderung zu verbessern. Hier finden sich Anknüpfungspunkte für Forschungspro-
  jekte und -kooperationen.

---

[24] www.inqa.de.

[25] www.gda-portal.de.

[26] www.basi.de.

[27] www.iga-info.de.

[28] www.dnbgf.de.

4 Fazit

Das Wissen um die Transferinstitutionen und Netzwerke ist als zentrale Voraussetzung für einen nachhaltigen Forschungstransfer genannt worden. Darüber hinaus ist
eine zweite Voraussetzung das frühzeitige Einbinden dieser Institutionen und Netzwerke in das Projekt. Dies ist eine Erfahrung aus Projekten, in denen Projektergebnisse in die Transferstrukturen bestehender Institutionen und Netzwerke integriert
werden konnten und somit auch lange nach Projektende weiter genutzt wurden.
Dabei hat sich aus der Perspektive der Autoren folgendes Vorgehen als hilfreich
herausgestellt:

- Schon in der Skizze und Antragsphase sollte bei der Auswahl von Valuepartnern
  oder sogar bei Projektpartnern berücksichtigt werden, welche Institutionen für den
  nachhaltigen Transfer der Projektergebnisse hilfreich sein könnten.
- Die Transferinstitutionen und Netzwerke sollen in allen Arbeitsphasen des Projektes mit einbezogen werden. Sie sollen ihre Erfahrungen bei der Umsetzung
  des Projektthemas mit einbringen können. Dies führt in der Regel dazu, dass
  die Projektergebnisse einen zusätzlichen Praxisbezug erhalten. Hierbei gilt es zu
  beachten, potentiell innovative Lösungen nicht durch einen „pessimistischen Rea-
  lismus" zu verwerfen.
- Die frühzeitige Einbindung von Transferinstitutionen und Netzwerken ist auch
  sinnvoll, damit diese das Projektthema zu ihrem Thema machen. Eine frühzeitige
  Einbeziehung führt zur Bindung und Identifizierung der Transferinstitutionen an
  die Projektergebnisse und Instrumente. Auf diese Weise gelingt die Verankerung
  der Informationen im individuellen Anwendungskontext. Das ist eine wichtige
  Voraussetzung, damit die Ergebnisse und Instrumente der Projekte nicht nur an
  die Laufzeit des Projektes gebunden sind, sondern sie darüber hinaus auch zu
  Inhalten der jeweiligen Institution werden.
- Bei der Evaluation der Projektergebnisse und Instrumente sollte die Transferin-
  stitution mit eingebunden werden, um auch hier noch ihre Erfahrungen in der
  konkreten Anwendung mit einbringen zu können.
- Projektbestandteil sollte auch die Implementierung der Forschungsergebnisse und
  Instrumente in die Transferstrukturen der beteiligten Institutionen und Netzwerke
  sein (z. B. Kommunikationsstrategien, Qualifizierungsunterlagen, Motivation und
  Qualifizierung von Beratern).

Die zuvor beschriebenen Empfehlungen lassen sich ebenfalls aus empirisch
begründeten Ansätzen ableiten. Hierbei werden außerdem weitere komplexe und differenzierte Ansätze des Forschungstransfers zur wissenschaftlichen Begleitforschung
der Förderprogramme des BMBF aufgegriffen [7–9].

Literaturverzeichnis

1. I. Leisten, U. Bach, F. Hees, Transferbar – Transfermethoden im Präventiven Arbeits- und Gesundheitsschutz. Tech. rep., Aachen, 2010
2. BMBF-Broschüre „Arbeiten – Lernen – Kompetenzen entwickeln. Innovationsfähigkeit in einer modernen Arbeitswelt", 2007. http://pt-ad.pt-dlr.de/de/94.php. Zuletzt aufgerufen: 25.09.2014
3. O. Cernavin, H. Luczak, K. Scheuch, K. Sonntag, Arbeitsschutzforschung als Innovation : eine Bilanzierung von 20 Jahren Arbeitsschutzforschung. In: *Arbeitsschutz - Bilanz und Zukunftsperspektiven des Forschungsfeldes*, ed. by H. Luczak, M. Rötting, K. Scheuch, K. Sonntag, O. Cernavin, Bremerhaven, 2001, pp. 9–74
4. G. Ernst. Von der Humanisierung zu Arbeitsgestaltung und Dienstleistungen – 40 Jahre Arbeitsforschung, 2008
5. P. Brödner, M. Knuth, *Nachhaltige Arbeitsgestaltung: Trendreports zur Entwicklung und Nutzung von Humanressourcen*. Hampp, München, 2002
6. G. Fuchs, K. Schönberger, Springer, eds., *Wissenstransfer in der Arbeitsforschung: Perspektiven und Probleme*. Stuttgart, 2003
7. U. Bach, *Entwicklung eines deliberativen Governance-Ansatzes für die Arbeitsforschung*. Budrich UniPress, Leverkusen, 2013
8. I. Leisten, *Transfer Engineering in transdisziplinären Forschungsprojekten*. Norderstedt, 2012
9. O. Cernavin, Von der Forschung zum Erfolg - „Prävention online". In: *Innovationsfähigkeit stärken – Wettbewerbsfähigkeit erhalten. Präventiver Arbeits- und Gesundheitsschutz als Treiber. Tagungsband zur 2. Jahrestagung des BMBF-Förderschwerpunkts*, ed. by K. Henning, I. Leisten, F. Hees, Aachen, 2009, pp. 31–48

Neue Kooperationsformen und Regionale Identitäten

Antje Ducki, Florian Welter and Julia Günther

Zusammenfassung Eine Perspektive, aus der sich der demografische Wandel analysieren lässt, stellt die regionale Perspektive bzw. die Region als multidimensionales Konstrukt dar. So geben unter anderem neue Kooperationsformen und regionale Identitäten gegenwärtig Antworten auf die Frage, wie Regionen innovativ bleiben und dem demografischen Wandel proaktiv begegnen können. Der Beitrag diskutiert den zunehmenden Zusammenschluss von Unternehmen und anderen Organisationen in regionalen Netzwerken am Beispiel von Ergebnissen eines BMBF Förderschwerpunkts. Inwiefern räumliche Nähe und eine gemeinsame Identität als vielversprechende Faktoren zum erfolgreichen Lernen in Netzwerken dienen, wird diesbezüglich erläutert. Im Sinne weiterer Gestaltungsaufgaben und Forschungsfragen, u. a. zur Koordination von Netzwerken, wird der Bedarf nach einer zunehmenden Systematisierung und Bündelung von regionalen Ergebnissen seitens der Wissenschaft und Praxis deutlich, durch die in Zukunft ein intensiveres Lernen zwischen Netzwerken ermöglicht werden kann.

Schlüsselwörter Kooperation · Region · Identität · Demografischer Wandel

1 Einführung

Dieser Beitrag ist das Resultat aus unterschiedlichen Diskussionen der Fokusgruppe „Regionale Aspekte des demografischen Wandels" im BMBF-Förderschwerpunkt „Innovationsfähigkeit im demografischen Wandel" sowie des Workshops „Neue Kooperationsformen und regionale Identitäten", der auf der zweiten Förderschwerpunkttagung im Mai 2014 in Aachen stattgefunden hat. Der Beitrag versteht sich vor diesem Hintergrund als programmatische Zusammenfassung des

A. Ducki (✉)
Fachbereich I: Wirtschafts- Und Gesellschaftswissenschaften, Beuth Hochschule
für Technik Berlin, Luxemburger Str. 10, 13353 Berlin, Germany
e-mail: ducki@beuth-hochschule.de

F. Welter · J. Günther
IMA/ZLW & IfU, RWTH Aachen University, Dennewartstr. 27,
52068 Aachen, Germany

Originally published in "Exploring Demographics",
© Springer 2015. Reprint by Springer International Publishing Switzerland 2016,
DOI 10.1007/978-3-319-42620-4_5

51

aktuellen Diskussionsstandes zu neuen Kooperationsformen und regionalen Identitäten. Für einen tiefergehenden Überblick über die konkreten Möglichkeiten regionaler Zugänge zur Gestaltung des demografischen Wandels vgl. Miosga et al. [1].

2 Grundproblematik

Der demografische Wandel in Deutschland verändert unsere Arbeitswelt tiefgreifend. Private Unternehmen wie auch Organisationen der öffentlichen Hand sind mit einem knapperem Fachkräfteangebot und alternden Belegschaften konfrontiert. Gleichzeitig bietet der demografische Wandel zahlreiche Innovationsimpulse: Produkte und Dienstleistungen müssen den veränderten Bedürfnissen einer alternden und gleichzeitig bunter werdenden Bevölkerung angepasst werden, intern verändern sich mit den Altersstrukturen der Belegschaften die Anforderungen an gute Arbeitsplatzgestaltung und Gesundheitsschutz, an Qualifizierung und Weiterbildung [1]. Soziale Innovationen, verstanden als gezielte Verbesserungen der Arbeitsgestaltung, der Führungsqualität und der Unternehmenskultur gewinnen an Bedeutung, denn sie sind wichtige Voraussetzungen für die Gesundheit, Leistungsfähigkeit und Motivation der Beschäftigten und bilden damit die Grundlage für technologische Innovationen.

Die Auswirkungen des demografischen Wandels wie auch die wirtschaftliche Innovationskraft sind regional stark unterschiedlich ausgeprägt. Metropolregionen wachsen seit Jahren und ihre Bevölkerungsstrukturen werden infolge internationaler Zuwanderung sozial und kulturell immer heterogener (vgl. [2]: 3). Gleichzeitig sind einige ländliche Regionen mit Problemen extremer Schrumpfung konfrontiert, die u. a. durch den Wegzug qualifizierter junger Menschen zustande kommen (vgl. [3]: 79). Aufgrund dieser Unterschiede empfiehlt sich eine regionale Herangehensweise bei der Gestaltung des demografischen Wandels und der Förderung sozialer Innovationen. Sie ist auch deshalb sinnvoll, weil über die Region gute Zugänge zu den Unternehmen und weiteren Organisationen gefunden werden können. Kleine und mittlere Unternehmen sowie Handwerksbetriebe identifizieren sich in der Regel mit ihrer Region. Sie sind vor Ort in vielfältige soziale Beziehungen eingebunden und fühlen sich mitverantwortlich für die Entwicklung ihrer Region. Räumliche Nähe erleichtert Kontaktmöglichkeiten und die Bildung von Vertrauen auch über Unternehmensgrenzen hinweg (vgl. [4]: 28). Nicht selten entstehen daher insbesondere in regionalen Verbünden vielfältige Kontakte, die den Erfahrungsaustausch erleichtern und zur raschen Entwicklung und Verbreitung von Innovationen beitragen. Diese Eigenschaft von Regionen als Interaktions- und Identifikationsraum kann hilfreich sein, um Innovationsprozesse zur Bearbeitung der demografischen Herausforderungen anzustoßen und ihre dauerhafte Umsetzung sicherzustellen. Die Identifikation mit der eigenen Region dient demnach oftmals dazu, gemeinschaftliches Handeln zu stärken und die Motivation der Akteure zu erleichtern [1].

Aber nicht jede regionale Kooperation funktioniert effektiv und wirkt nachhaltig. Konkurrierende Interessen der Akteure, gewachsene Machtstrukturen, unterschiedliche Deutungsmuster und Zuschreibungsprozesse in der Analyse und Ableitung geeigneter Interventionen, fehlende Fokussierung, Abstimmung und Koordination der regionalen Akteure können die Wirksamkeit und den Nutzen regionaler Kooperationen beschränken (vgl. u. a. [5]: 1344f.; [4]: 88f.; [6]: 53).

3 Herausforderungen und Anknüpfpunkte

Wenn Interventionen zur Gestaltung des demografischen Wandels und zur Stärkung der Innovationsfähigkeit in einer Region nachhaltig wirksam werden sollen, ist es erforderlich, die Besonderheiten einer Region, die sie in demografischer, sozioökonomischer, regionalpolitischer und wirtschaftlicher Hinsicht auszeichnet, zu ermitteln und in der Maßnahmengestaltung zu berücksichtigen. Ansatzpunkte in der Region sind somit gewachsene Strukturen, bestehende Institutionen und Betriebe. Regionale Politik, Intermediäre und betriebliche Akteure müssen zusammengeführt werden, ihre Interessenlagen austauschen und Kristallisationspunkte für gemeinsames Handelns identifizieren [7]. Bedingungen, Zielsetzungen und Nutzen eines regionalen Verbundes, organisatorische Fragen, soziale Strukturen und Machtkonstellationen innerhalb der Region müssen für eine erfolgreiche regionale Kooperation von den Akteuren zu Beginn einer Kooperation geklärt und fortlaufend angepasst werden. Häufig kennen Akteure in einer Region nicht die Maßnahmen und Aktivitäten der anderen Akteure, was dazu führen kann, dass z.B. Betriebe mehrfach zu einem Thema adressiert werden oder ihnen Beurteilungskriterien fehlen, um zu entscheiden, welche Netzwerk- oder Maßnahmenteilnahme für sie vorteilhaft sein könnte. Zusammengefasst benötigen regionale Zugänge eine strategische Ausrichtung und gute Koordination [8].

4 Gestaltungsaufgaben

Politik, Wirtschaft und Wissenschaft organisieren zunehmend auf regionaler Ebene neue Formen der Kooperation. Clusterorganisationen mit unterschiedlicher inhaltlicher Ausrichtung besitzen regionale Schwerpunkte, wirtschaftliche Netzwerke formieren sich und öffentliche Wirtschaftsförderungen arbeiten verstärkt auf regionaler Ebene mit wirtschaftlichen und wissenschaftlichen Akteuren zusammen und erarbeiten neue Koordinationsmechanismen (vgl. u. a. [9]: 215ff.; [10]: 18f.). Regionen können über diese Akteursgemeinschaften eine eigenständige Handlungsfähigkeit entfalten und zu Impulsgebern und Moderatoren der notwendigen Innovationen und Anpassungsprozesse werden. Auch staatliche Politik legt vermehrt Förderprogramme auf, die auf regionaler Ebene wirksam werden. Damit werden Regionen zu einem Programmraum für eine gezielte und systematische Förderung von Innovationen in Richtung Demografiefestigkeit.

Um den Programmraum nachhaltig zu entwickeln, besteht ein erster Schritt darin, die Vielfalt bereits bestehender Kooperationen zu systematisieren, Schnittstellen und Doppelangebote zu identifizieren, um sie dann beispielsweise in Interventionsarenen zu bündeln. Diese Aufgabe kann von der regionalen Wirtschaftsförderung oder auch von anderen Institutionen wie Kammern, Krankenkassen, bereits bestehenden (über-)regionalen Netzwerken oder auch extra für diesen Zweck neu gegründeten Netzwerken realisiert werden. Wem die Aufgabe der Koordination der regionalen Aktivitäten zukommen sollte, ist u. a. von der Durchdringungstiefe bzw. dem Vernetzungsgrad der Institution in der Region, ihrer verfügbaren finanziellen und personellen Ressourcen, aber auch von einfachen Fragen wie der zentralen örtlichen Lage (gute Erreichbarkeit für viele), und ihrem technologischem Know How abhängig. Darüber hinaus ist eine strategische Ausrichtung und größtmögliche Neutralität hilfreich, um die vielen unterschiedlichen Interessenslagen der Intermediäre und Betriebe zusammenzuführen [11]. Um den Erfolg regionaler Kooperationen weiter abzusichern, sind zielbezogene und maßnahmenspezifische Qualitätskriterien für das Netzwerk zu entwickeln, die eine Auswahl, bzw. Teilnahmeentscheidung erleichtern. Darüber hinaus wird in der Literatur immer wieder auf die besondere Bedeutung eines Netzwerkkoordinators und seiner fachlichen, methodischen und sozialen Kompetenzen hingewiesen. Hier sollten bereits bestehende Anforderungsprofile [12] auf die regionalen Besonderheiten angepasst werden. Zentrale Aspekte sind hier die Fähigkeiten des Koordinators eine Vertrauenskultur im Netzwerk aufzubauen, die Vereinbarkeit des Eigeninteresses mit der Netzwerkarbeit für Unternehmen aufzeigen, regionales Wissen ins Netzwerk einbringen und die Teilnahmemotivation durch innovative Angebote zu stabilisieren.

5   Forschungsfrage

Neben den vielen Gestaltungsaufgaben lassen sich ungeklärte Fragen benennen, die in Zukunft verstärkt erforscht werden sollen.

5.1   *Dissonante Netzwerke erforschen*

Regionale Netzwerke brauchen Dissonanz und Einklang gleichzeitig. Einklang, um die Grundlagen für gemeinsame Identität zu schaffen, Dissonanz um Innovationen hervorzubringen. Wieviel Einheit in Ansichten, Positionen, Interessenlagen der Akteure ist erforderlich, um eine regionale Identität, Wir-Gefühl und Vertrauen zu entwickeln? Wieviel Diversität und Dissonanz ist erforderlich, um Innovationen und Veränderungen zuzulassen? Hierbei ist zunächst zu klären, ob Einheit und Vielfalt Antipole sind oder ob sie sich auf unterschiedliche Handlungsdimensionen und -ebenen beziehen. So ist es aus einer handlungstheoretischen Perspektive sinnvoll, dass Einheit in Bezug auf übergeordnete Ziele regionaler Verbünde besteht, auf der Ebene der konkreten Maßnamengestaltung kann jedoch Vielfalt im Sinne des viele Wege führen zum Ziel durchaus wünschenswert sein. Allerdings müssen Grenzen der Vielfalt in Bezug auf die richtigen Vorgehensweisen oder die Art der Ressourcennutzung immer in Abhängigkeit von den übergeordneten Zielen spezifisch festgelegt werden. Dies kann nur kollektiv im Netzwerk geschehen und wird im Sinne eines iterativen Prozesses im Laufe der Netzwerkarbeit immer wieder neu auszuhandeln sein.

5.2 Netzwerksteuerung als Innovationsmotor und Identitätsmanagement

Es kann vermutet werden, dass in heterogen zusammengesetzten regionalen Netzwerken ähnliche Handlungslogiken wirksam werden, wie sie aus der Team- und Führungsforschung bekannt sind. Aus der Teamforschung ist bekannt, dass soziale Ähnlichkeiten, gemeinsame Sichtweisen und Haltungen nur unter bestimmten Voraussetzungen identitätsstiftend sind und dass Heterogenität bewusst gestaltet werden muss, damit sich z. B. positive Innovationseffekte entfalten können [13]. Führung spielt hier eine herausragende Rolle. Führung kann soziale Identität einer Gruppe positiv beeinflussen ([14]: 140ff.) und das Innovationsverhalten der Gruppenmitglieder fördern oder behindern ([15]: 3).

Regionale Netzwerke müssen identitätsstiftend und innovationsförderlich geführt und koordiniert werden. In Anlehnung an die Team- und Führungsforschung sollten zukünftig Erklärungsmodelle weiterentwickelt werden, die Netzwerk- und Kooperationserfolge auf regionaler Ebene prognostizieren können. Kriterien für langlebige und effizient arbeitende Netzwerke sollten weiter entwickelt werden, hierbei muss die Rolle des Netzwerkkoordinators hervorgehoben analysiert werden. Neuere Ansätze der Führungsforschung wie das Modell von Kerschreiter zum Identitätsmanagement sind hier auf ihre Übertragbarkeit zu prüfen und mit den bereits bestehenden Ansätzen zum Innovations- und Gesundheitsmanagement in Beziehung zu setzen. Erste Überlegungen hierzu bestehen ([16]: 173).

5.3 (Über-)regionales Netzwerklernen anstoßen

Es gibtseit vielen Jahrzehnten in vielen Regionen Deutschland unterschiedliche Netzwerke mit unterschiedlichsten Erfahrungen. Insbesondere die Anpassung der Netzwerke an die regionalen Besonderheiten variiert stark. Dieses Erfahrungswissen ist zu systematisieren und zusammenzuführen [17]. Bedingungen sind zu untersuchen, wie Kooperationen und Lernprozesse über verschiedene Netzwerke und Regionen hinweg weiter entwickelt und gefördert werden können. Zu diesem Zweck gilt es, u. a. auch in Zukunft Programmräume zu ermöglichen, die den netzwerkübergreifenden Austausch von Akteuren erlauben und good practices des Lernens zwischen Netz-

werken ermöglichen. Auf Seite der Wissenschaft sind hier insbesondere Fragen der
formativen und summativen Evaluation jener Programmräume wie auch Fragen des
Wissenstransfers zwischen den Akteuren verschiedener Netzwerke, Cluster und ver-
wandter (regionaler) Verbundformen zu erforschen.

6 Fazit und Ausblick

Die Untersuchung und Gestaltung neuer Kooperationsformen, darunter Netzwerke, Cluster und ähnliche Verbundformen, sind im Spannungsfeld gegenwärtiger
Trends, wie z. B. einer anhaltenden Globalisierung von ökonomischen Aktivitäten
bei einer gleichzeitig zunehmenden Bildung regionaler Netzwerke und dazuge-
höriger Identitäten, von hoher Bedeutung für Wissenschaft, Wirtschaft, Politik
und Gesellschaft. Der vorliegende Beitrag griff diesbezüglich die aktuelle Diskussion zu Grundsatzproblemen, Herausforderungen und Anknüpfungspunkten, Gestaltungsaufgaben wie auch Forschungsfragen mit Blick auf Ergebnisse des BMBF-
Förderschwerpunkts „Innovationsfähigkeit im demografischen Wandel" auf. Hierbei wurde deutlich, dass einerseits viele Forschungs- und Praxisergebnisse vor-
liegen, die in unterschiedlichen regionalen Kooperationsformen entwickelt wurden,
um z. B. dem demografischen Wandel erfolgreich zu begegnen. Andererseits können
diese Ergebnisse nicht unmittelbar in andere Regionen transferiert werden, sondern
sind stets einer regionsspezifischen Anpassung zu unterziehen. Diese Tatsache unter
streicht den weiteren Bedarf nach einer kontinuierlichen Forschung zur Systematisierung und Bündelung von regional verteilten Wissensbeständen sowie nach
geeigneten Formen den Wissenstransfers zwischen verschiedenen regionalen
Kooperationsformen.

Literaturverzeichnis

1. M. Miosga, A. Ducki, A. Ihm, A. Krauß, F. Welter, Regional innovativ – die Potenziale vor Ort
   nutzen, um Veränderung anzustoßen. præview – Zeitschrift für innovative Arbeitsgestaltung
   und Prävention **5** (2), 2014
2. H.J. Bullinger, Forschen für die Stadt von morgen. weiter.vorn Das Fraunhofer-Magazin (4),
   2012
3. H. Michel, Schrumpfung für Kommunen teurer als Wachstum. Die trügerische Dividende.
   Gastkommentar. In: *Unternehmerin Kommune + Forum Neue Länder*, ed. by M. Schäfer,
   FORUM NEUE LÄNDER-Verlag, Berlin, 2011, p. 79
4. F. Lerch, *Netzwerkdynamiken im Cluster: Optische Technologien in der Region Berlin-
   Brandenburg.* Dissertation-Online der Freien Universität Berlin, 2009
5. M. Fromhold-Eisebith, Die „Wissensregion" als Chance der Neukonzeption eines zukunfts-
   fähigen Leitbilds der Regionalentwicklung. Raumforschung und Raumordnung **67** (3), 2009,
   pp. 215–227
6. A. Malmberg, D. Power, True Clusters – A Severe Case of Conceptual Headache. In: *Clusters
   and Regional Development: Critical Reflections and Explorations*, ed. by B. Asheim, P. Cooke,
   R. Martin, Routledge, London, New York, 2006, pp. 50–68

7. A. Thiel, K. Joel, L. Dallner, Die strategische Allianz ADMIRe A^3 in der Umsetzung. præview – Zeitschrift für innovative Arbeitsgestaltung und Prävention **5** (2), 2014

8. S. Hafner, T. Engelmann, T. Merten, Mit einer Strategischen Allianz Transformationsaufgaben des demografischen Wandels, der Steigerung der Ressourceneffizienz und der Innovationsfähigkeit im Wirtschaftsraum Augsburg ganzheitlich anpacken. præview – Zeitschrift für innovative Arbeitsgestaltung und Prävention **5** (2), 2014

9. M. Fromhold-Eisebith, G. Eisebith, Looking Behind Facades: Evaluating Effects of (Automotive) Cluster Promotion. Regional Studies **42** (8), 2010, pp. 1343–1356

10. F. Welter, *Regelung wissenschaftlicher Exzellenzcluster mittels scorecardbasierter Performancemessung.* Books on Demand, Norderstedt, 2013

11. A. Ritter, R. Osranek, E. Jaschinski, Mit Strategie die Zukunft sichern – Erschließung von Innovationspotentialen im Handwerk durch regionale Allianzen. præview – Zeitschrift für innovative Arbeitsgestaltung und Prävention **5** (2), 2014

12. A. Ihm, A. Baumann, J. Schuler, Gute Netzwerkarbeit fordert gegenseitiges Vertrauen – vertrauensförderliche Maßnahmen in regionalen Unternehmensnetzwerken. præview – Zeitschrift für innovative Arbeitsgestaltung und Prävention **5** (2), 2014

13. F. Jungmann, P. Bilinska, J. Wegge, Alter(n)sgerechte Führung. In: *Trends der psychologischen Führungsforschung – Neue Konzepte, Methoden und Erkenntnisse*, ed. by J. Felfe, Hogrefe, Göttingen, 2014

14. R. Kerschreiter, Eine neue Sichtweise der Führungsaufgabe: Führung als Identitätsmanagement. In: *Führen lernen – Standortbestimmung, Kompetenzen, Entwicklung*, ed. by S. Litzcke, K. Häring, Schäffer-Poeschel Verlag, Stuttgart, 2013, pp. 137–159

15. D. Herrmann, J. Felfe, J. Hardt, Transformationale Führung und Veränderungsbereitschaft: Stressoren und Ressourcen als relevante Kontextbedingungen. Zeitschrift für Arbeits- und Organisationspsychologie **56**, 2012, pp. 70–86

16. N.K. Steffens, A.S. Haslam, R. Kerschreiter, S.C. Schuh, R. van Dick, Leaders Enhance Group Members' Work Engagement and Reduce Their Burnout by Crafting Social Identity. Zeitschrift für Personalforschung **28** (1–2), 2014, pp. 173–195

17. B. Luger, A. Krauß, Stadt – Land – ArbeitsFluss: Der Unternehmensstandort als strategiebestimmender Faktor bei der Fachkräftegewinnung. præview – Zeitschrift für innovative Arbeitsgestaltung und Prävention **5** (2), 2014

Menschen entwickeln Potenzial für neue Technologien – 30 Jahre Industriegeschichte

Klaus Henning and Ursula Bach

Zusammenfassung Was sind die einschlägigen und nachhaltigen Entwicklungen der Industrie, die heute noch unser Arbeiten in Industrie und Produktion prägen und verändert haben? Wenn wir zurück blicken zeigt die Erfahrung, dass es weder möglich ist, vollautomatische Fabriken zu entwickeln, noch alles Wissen der Menschheit mit Hilfe riesiger Datenbanken zu konservieren. Diese Ideen, die früher Trends setzten, sind nun überholt. Durch den Misserfolg dieser Ideen haben wir jedoch ein nützliches Konzept gelernt, den sogenannten „HOT Approach": zuerst der Mensch, als zweites die Organisation und drittens die Technologie. Auf diese Weise kann menschliches Wissen und Potential in Kombination mit adäquater Organisation zu technischen Innovationen führen. Im Moment besteht die Herausforderung darin, dieses Konzept auf jüngste und zukünftige Ideen zu transferieren; z. B. das vollautomatische Auto oder die Veränderung in der Nutzung erneuerbarer Energien. Dabei sollte bedacht werden, dass technische Innovation immer menschliche und organisationale Aspekte enthalten muss, um wirklich nachhaltig zu sein.

Schlüsselwörter HCIM · Wissensmanagement · HOT Approach · Mensch-Organisation-Technik · Industrieentwicklung · Technikentwicklung · Innovationsprozesse

1 Einführung

Was sind die einschlägigen und nachhaltigen Entwicklungen der Industrie, die heute noch unsere Arbeiten in Industrie und Produktion prägen und verändert haben? Die Idee der vollautomatischen Fabrik (für Büros die papierlose Firma), die Idee der Konservierung allen möglichen Wissens im Wissensmanagement und – ein Blick in die Zukunft – die Idee des autonomen Fahrens oder die Idee des flächendeckenden Einsatzes erneuerbarer Energie. Die Distanz zwischen der technisch-möglichen Umset-

K. Henning (✉) · U. Bach
IMA/ZLW & IfU, RWTH Aachen University, Dennewartstr. 27, 52068
Aachen, Germany
e-mail: klaus.henning@ima-zlw-ifu.rwth-aachen.de

Originally published in "Industrie-Management - Zeitschrift für industrielle Geschäftsprozesse 1/2014",
© GITO mbH Verlag für Industrielle Informationstechnik und Organisation 2014.
Reprint by Springer International Publishing Switzerland 2016,
DOI 10.1007/978-3-319-42620-4_6

zung der Idee und der nachhaltig-akzeptierten Realisierung in der Praxis weist eine hohe Diskrepanz auf, die einer Erklärung bedarf. Um im Bild zu bleiben: Das Uhrwerk und damit der Glaube an die Maschine hakt. Warum?

Vor 30 Jahren schlug sich die Vision des Maschinenmodells in der Idee der vollautomatischen Fabrik nieder: Computer Aided Manufactoring (CIM) hieß das Zauberwort, das schon in wenigen Jahren die Visionen der vollautomatischen Fabrik erfüllen sollte. Doch dies gelang nicht. Denn vorher wären organisatorische Maßnahmen erforderlich gewesen, um die Struktur der Fabrik an die neuen Technologien und Möglichkeiten anzupassen. Aber auch dies war noch nicht der finale Schlüssel zur erfolgreichen Umsetzung der vollautomatisierten Fabrik. Schließlich war der Mensch an entscheidenden Stellen unersetzlich, um zumindest eine teilweise Automatisierung zu realisieren. Das Konzept HCIM – Human and Computer Integrated Manufactoring – war geboren [1]. Die ganzheitlichen Ansätze – nicht zuletzt aus dem Programm Humanisierung des Arbeitslebens in der Tradition des BMB – führte dann zu einer erfolgreichen Kombination von Mensch, Organisation und Technik [2]:

- Beginne bei den Entwicklungspotenzialen des Menschen,
- stelle dann die organisatorischen Strukturfragen und
- suche anschließend nach der geeigneten Automatisierungstechnik [3].

Dieser HOT Approach „First Human, Second Organisation and Third Technology" ist bis heute der entscheidende Ansatz für erfolgreiche technische Innovationen. Die technischen Möglichkeiten setzen diesen Dreischritt in Gang (Abbildung 1).

Diese Erfahrungen und Erkenntnisse sind aus der „vollautomatischen Fabrik" der 1980er Jahre hinreichend belegt und durch umfangreiche wissenschaftliche Untersuchungen validiert [4]. Aber können wir eine Lernkurve in dieser Richtung erkennen? Agieren wir nun effizienter?

Abbildung 1 Dreischritt Mensch – Organisation – Technik zur Erhöhung der Durchsetzung technischer Innovationen

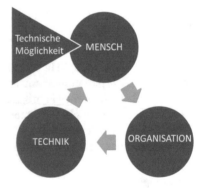

2 Wissensmanagement

Leider nein, denn es kam wieder anders. In einer neuen Welle der Technologisierung spielte das Schlagwort „Wissensmanagement" in den 1990er Jahren eine zentrale Rolle. Der damalige Ansatz besagte, mithilfe von Wissensingenieuren kann die spezielle Expertise der Facharbeiter und Techniker aufgeschrieben und dann in eine Wissensdatenbank gefüttert werden. Wenn dies gelingt, werden die Wissensträger als Person nicht mehr benötigt. Deren Know-how liege sicher in der Datenbank.

Es wiederholte sich der Irrglaube in das Maschinenmodell, in die Idee, dass die Welt und die Menschen wie Maschinen funktionieren. Elementare Zusammenhänge aus den Geisteswissenschaften waren in den Ingenieurwissenschaften noch nicht antizipiert worden oder in Vergessenheit geraten. Bald setzte sich die Ansicht durch, dass auf Dauer Spezialisten einzusetzen wären, die diese Wissensdatenbanken organisieren. Diese Menschen fanden ihren Platz in den jeweiligen Organisationen, indem spezielle Wissensmanagement-Stabsstellen oder Beauftragte für Fragen des Wissensmanagements ins Leben gerufen wurden. Mit solchen organisatorischen Maßnahmen versuchte man das Paradigma „Wissensdatenbank löst unsere Probleme" zu retten – erfolglos. Es bedurfte einer umfangreichen Diskussion, indiziert aus Japan, die uns in Erinnerung rief, was der Unterschied zwischen impliziten und expliziten Wissen meint [5] und welche Folgen dies für Unternehmen mit ihren neuen Wissensmanagementsystemen hat. Es kam (wieder) zur Renaissance der Rolle des Menschen im Umgang mit Wissen. Die Industrie schaffte die meisten der neuen Strukturen nach und nach ab und es wurde nach Konzepten gesucht, die die menschliche Komponente des Wissensmanagements integrieren und komplettieren. So hat sich die moderne Organisationsentwicklung mit der Sicherung der individuellen Wissensträger als Kern des industriellen Wissensmanagements etabliert. Wir lernten neu zu unterscheiden zwischen Informationen, die man auf einem Computer ablegen kann und Wissen, das auf die lebendige Kommunikation – im Sinne der Geisteswissenschaften – zwischen Menschen angewiesen ist [6]. Doch bis heute sprechen wir zu den Begriffen Kommunikation, Information, Wissen auch in den Wissenschaften keine gemeinsame Sprache und sind weit entfernt von einem gemeinsamen Verständnis aller Disziplinen.

In der Zwischenzeit hatte die Praxis zu einem gesunden Maß zurück gefunden und weiß, dass auf den Fachexperten, sei es ein Ingenieur, sei es ein Facharbeiter, nicht verzichtet werden kann. Diese Notwendigkeit zu handeln, macht der demografische Wandel noch dringlicher [7]. Es gibt keine Zeit mehr zu verlieren, das Wissen von Menschen mit Menschen auf Menschen zu übertragen. „Potenzialentwicklung" auf allen Ebenen heißt das neue vielversprechende Konzept [8]. Erst auf dieser Basis machen zur Beherrschung des Themas netzwerkartige, organisatorische Strukturen und Wissensdatenbanken Sinn.

Der Umstellungsprozess von einem Ansatz ‚Erst die Technik, dann die Organisation und dann der Mensch' zu einem Ansatz ‚Erst der Mensch, dann die Organisation und dann die Technik' ging im Falle des Themas Wissensmanagement schneller von statten als beim Thema vollautomatische Fabrik.

Das lässt hoffen: Hat die Gesellschaft, die Wissenschaft und die Industrie aus den Erfahrungen gelernt? Hat sich das Bewusstsein von der Welt als Maschinenmodell zu dem Bild der Welt als einem lebenden auf den Menschen zentrierten Organismus gewandelt?

3 Das vollautomatische Auto

Schauen wir in die Gegenwart: Google, Mercedes und VW fahren mit vollautomatischen Autos durch halbe Städte. Die ganze Welt ist fasziniert. Das „Vollautomat-Auto" ist da. Viele glauben plötzlich, dass fahrerloses Autofahren in Kürze möglich sein wird. Betrachten wir die zurückhaltende Euphorie, scheint es so zu sein, dass viele relevante Akteure begriffen haben, dass die vollautomatische Vision eine Vision ist, die an dem langen Ende eines Entwicklungsprozesses unter Berücksichtigung von Mensch und Organisation und Technik steht. Ein erster Schritt ist eine teilautomatische Lösung im Sinne eines dualen Entwurfs. Der Fahrer erfährt Unterstützung durch Fahrerassistenzsysteme und eine umfangreiche Vernetzung mit der Infrastruktur und den ihn umgebenden Fahrzeugen [8]. Solche halbautomatischen Unterstützungssysteme werden schon bald die Regel sein. Es ist schon paradox, dass ausgerechnet beim Auto eine große Chance besteht, dass der „HOT Approach" weiterhin das zielführende und erfolgsversprechende Zukunftsszenario sein wird: Erst über die Bedürfnisse des Menschen im Fahrzeug nachdenken, dann geeignete Netzwerkund Organisationsstrukturen schaffen, angefangen von der Infrastruktur bis zu den IT Netzen, um dann die Verknüpfungstechniken und Automaten zu entwickeln.

4 Energiewende

Aber wie ist es bei der Energiewende? Haben wir nicht ohne Blick für das Ganze Hals über Kopf in großem Stil auf Solar- und Windenergie gesetzt? Es wurden On- und Offshore Parks gegründet oder ganze Landstriche in den Dienst einer Technik gestellt, wie z. B. Solar Valley [9]. Gerechtfertigt und geschürt wurde dieser Aktionismus durch den guten Zweck, den nachhaltige und grüne Technologien versprechen – verbunden mit großen Mengen von Subventionsmitteln, die heute schon den Bürgern in der Strompreisrechnung präsentiert werden. Nicht erst heute entdecken wir, dass es noch wesentliche organisatorische Probleme gibt, wie zum Beispiel die Organisationen der Verteilung der Energie, die Frage der Balance zwischen dezentraler und zentraler Energieerzeugung und -verteilung? Oder die verdrängte Frage, wie das Zusammenwirken von dezentralen und zentralen Akteuren der Energieversorgung aussehen soll? Es ist also gelungen erneuerbare Energiequellen anzuzapfen, aber die Einspeisung in Netzstrukturen ist schwierig und ebenso die Akzeptanz bei den Bürgern für die damit verbundenen Veränderungsprozesse und höheren Kosten.

Legt dies nicht den Schluss nahe, dass es an der Zeit ist, eine Wende der Energiewende herbeizuführen: Erst über die Veränderungsfähigkeit und der Veränderungsbereitschaft der Menschen nachzudenken, dann über geeignete Strukturen und Organisationsformen der Energieerzeugung und Verwendung. Anschließend erst über die Kombination geeigneter Technologien, mit der wir die Energiewende tatsächlich ,auf die Straße' bekommen.

5   Ausblick

Diese angeführten Beispiele zeigen, dass es möglich ist, technische Neuerungen um menschliche und organisationale Aspekte zu erweitern und damit deren Durchsetzungskraft und Durchdringung des Markts zu ermöglichen. Sie zeigen aber auch, dass immer wieder der Glaube an die Maschine als Wundermittel die Erfahrungen ähnlicher Konzepte überfrachtet. Hier handeln wir wie Frischverliebte: Es wird nur das Rosarote gesehen und die ungeahnten Möglichkeiten, aber selten potenzielle Mängel oder Probleme angesprochen.

Gelingt es aber Mensch, Organisation und Technik gemeinsam zu denken, dann ist dies ein entscheidender Schlüsselfaktor für Deutschland und ein wertvolles Exportgut. Wir können somit die Fähigkeit verstärkt exportieren, in Systemen zu denken und komplexe Systeme zu entwickeln. Dann wird es z. B. auch gelingen, die Energiewende in Deutschland als ,Exportartikel' weltweit zu vermarkten.

Das alles wird besser funktionieren, wenn die Ingenieure und Naturwissenschaftler von Anfang an nicht nur das ,Technische System' sehen, die Betriebswirte und Controller sich von dem Glauben lösen, alles über Zahlen zu beurteilen und zu steuern und die Sozial- und Geisterwissenschaftler ihre Technikskepsis aufgeben.

Nur wenn wir das implizite Wissen unserer Facharbeiter, Ingenieure und Kaufleute als zentrale Ressource wahrnehmen, wertschätzen und nutzen, nur wenn wir die Beziehung der verschiedenen Akteure untereinander optimieren, haben wir eine Chance, unsere derzeitige exponiert gute Stellung als einer der reichsten Nationen der Welt im Weltmarkt zu halten [10].

Nur wenn wir in unserem Familienunternehmen wertschätzen, in denen ein großer Vorrang auf Beziehungen zwischen Menschen liegt, aus denen heraus dann Innovation wächst und neue Technik; nur wenn wir solchen Strukturen in allen Bereichen unserer Gesellschaft den Vorrang geben vor Organisationsstrukturen, die durch zentral gesteuerte Innovationsprozesse über umfangreiche Controlling und Kennzahlenprozeduren geprägt sind, werden wir unsere einzigartige Stellung im Weltmarkt behaupten.

Und obwohl wir immer wieder zu sehr in Technik verliebt sind, haben wir in der Vergangenheit immer noch rechtzeitig ,die Kurve' bekommen. Wir sollten es in Zukunft schneller beherrschen neue Technologien mit dem HOT Approach weiterzuentwickeln: Erst den Menschen mit seinen Potenzialen in den Blick nehmen, dann die Organisation und dann die Technik. Die dann entstehende Technik verdient dann auch die Bezeichnung „sustainable – nachhaltig" und wird auf den Weltmärkten unsere Zukunft sichern.

Literaturverzeichnis

1. K. Henning, D. Brandt, I. Tschiersch, The design of human-centered manufacturing systems. In: *Computer-Aided Design, Engineering and Manufacturing-Systems and Applications*, ed. by C. Leondes, Boca Raton, USA, 2001, pp. 1–56
2. U. Bach, *Deliberative Governance in der Arbeitsforschung. Ein Ansatz zur Demokratisierung von Forschungsprozessen in der anwendungsorientierten Forschung.* Opladen, 2013
3. I. Isenhardt, F. Hees, eds., *Der Mensch in der Kommunikation mit der Technik.* Aachen, 2005
4. L. Burckardt, Die rückkehr des facharbeiters. Gewerkschaftliche Monatshefte 7 , 1990, pp. 42–437
5. I. Nonaka, *Die Organisation des Wissens: wie japanische Unternehmen eine brachliegende Ressource nutzbar machen,* 2nd edn. Frankfurt am Main, 2012
6. K. North, *Wissensorientierte Unternehmensführung. Wertschöpfung durch Wissen.* Wiesbaden, 2002
7. S. Jeschke, *Der Demografie- Atlas. Deutschland Land der demografischen Chancen.* Aachen, 2013
8. A. Friedrichs, H. K., T. Tiltmann, L. Petry, Entwicklung und Untersuchung von Fahrerassistenzsystemen für elektronisch gekoppelte LKW-Konvois. VDI-Gesellschaft Fahrzeug- und Verkehrstechnik: Integrierte Sicherheit und Fahrerassistenzsysteme (VDI Bericht Nr. 1960), 2006, pp. 223–238
9. S. Mitteldeutschland. www.solarvalley.org. Abrufdatum 18.10.2013
10. P. und Informationsamt der Bundesregierung. Dialog über Deutschlands Zukunft, Ergebnisbericht des Expertendialogs der Bundeskanzlerin 2011/2012, 2012. www.dialog-ueberdeutschland.de/ergebnisbericht-kurz. Abrufdatum 18.10.2013
11. K. Henning, Von „Made in Germany" zu „Enabled by Germany". In: *Wirtschaftsrat Deutschland: Deutschland im Jahr 2035*, Berlin, 2013, pp. 67–78

Genderation BeSt – Investigation of Gender Neutral and Gender Sensitive Academic Recruiting Strategies

Yves Jeanrenaud, Larissa Müller, Esther Borowski, Anja Richert, Susanne Ihsen and Sabina Jeschke

Abstract Careers of women in science run along a gendered "leaky pipeline" (cf. European Commission 2001: 12). With each career level the percentage of women decreases. The German average of female professors currently reaches only about 20 % (cf. European Commission 2013: 90). On average, the percentage of female professors in EU-25 is 18 % (cf. European Commission 2013: 90). Recent studies about the marginalization of women in the German science system suggest that the under-representation of female scientists has to be analysed through deep insight of the scientific system and its structures in order to understand and explain the background. The main objective of "Genderation BeSt" was to analyse the habits and rules of appointment procedures at universities and to develop gender sensitive and gender neutral methods for assignment and personnel recruitment at German universities, especially in the fields of Science, Engineering and Technology (SET). The project "Genderation BeSt" was conducted in a cooperation between RWTH Aachen (IMA/ZLW & IfU) and the Technical University of Munich (Gender Studies in Science and Engineering) and funded by the ESF and the BMBF from December 2011 until February 2013. A first work package included qualitative interviews with members of appointment committees to evaluate which structural and cultural mechanisms exist and influence the proportion of female professors at German universities. Since the amount of external funding serves as an essential quality feature in appointment processes – particularly within STEM faculties –, the next step was to analyse whether and to what extent external funding calls of the leading German research institutions contribute to a gender asymmetry. This was achieved via text mining procedures and expert interviews. The investigations resulted in targeted recommendations to higher education, research institutions and science policy regarding

Y. Jeanrenaud (✉) · S. Ihsen
Gender Studies in Ingenieurwissenschaften, Technische Universität München,
Arcisstr. 21, 80290 München, Germany
e-mail: yves.jeanrenaud@tum.de

L. Müller · E. Borowski · A. Richert · S. Jeschke
IMA/ZLW & IfU, RWTH Aachen University, Dennewartstr. 27,
52068 Aachen, Germany

Originally published in "Proceedings of the 8th European Conference on Gender
Quality in Higher Education, Vienna, Austria, 2-5 September 2014",
© Vienna University of Technology 2014. Reprint by Springer International
Publishing Switzerland, DOI 10.1007/978-3-319-42620-4_7

65

gender-sensitive and gender-neutral appointment methods and recruitment strategies
for the promoted appointment of women in academia.

Keywords Genderation BeSt · Leaky Pipeline · Gender Sensitive · Gender Neutral ·
Academic Recruiting Strategies

1 Background and State of the Art

Women's careers in science are influenced by the phenomenon of a gender-specific
"leaky pipeline" [1]: the percentage of women decreases disproportionately the
higher the career level. This "dropout" of female academics [2] especially occurs
in the transitional postdoc phase between doctoral degree and habilitation. It continues with the appointment of professorships thus reproducing the gender asymmetry
at this level. This asymmetry varies strongly concerning the different subject areas.
At the beginning of their studies male students have a chance of 2 % to ever become a
professor. On average considering all faculties, the probability for their female counterparts is at only a 0.6 % [3]. Today the gender ratio among new students enrolling at
universities is balanced [4] – depending on the faculty female freshmen even prevail.
However, on average women's chances of becoming a professor are at only a third
of the chances of men.

Research on the subject of the underrepresentation of women in academia published since the beginning of the 1990s indicate the key reasons are to be found in
the scientific system and its structures. As a result the focus has shifted to a field
considerably influenced by the approach of the "Gendered Organization": asymmetrical gender relations exist and are reproduced in organizations and institutions and
are therefore to be found in universities and research facilities [5–8]. Additionally,
this asymmetry can be observed within appointment committees [9].

Many consider the low degrees of standardization and formalization of appointment procedures to constitute a structural problem concerning the dropout of female
academics [10]. Women are often assumed to drop out of academia due to family
responsibilities. Funding and promoting women thus seems riskier. In consequence,
it may unconsciously occur in a more defensive way than the funding and promoting
of men [10].

Moreover, there exists a variety of informal barriers whose influence is of crucial
importance. These are vague expectations in terms of job performance and availability[1] as well as a disadvantageous culture of confirmation and appreciation[2] towards
female academics [10], especially in the fields of SET [12]. Such behavioural patterns

---

[1]Due to restricted temporal resources such as childcare and family responsibilities the output funding
relations of women are not secured [11].

[2]During their studies female students feel less strengthened than their male counterparts when it
comes to their professional competence which negatively influences the tendency to strive for a
doctoral degree [11].

have been termed the withholding of the capital "acknowledgement" [2] and constitute a central obstacle on the way to considering an academic career.

A look at the gender asymmetry at professorial level provides a first impression of the problem of gender-specific career-development. This circumstance is even more crucial considering regional and subject-specific differences. Furthermore, there is a distinctive dispersion and variance concerning the representation of women at professorial level depending on the region as shown by the University Ranking based on gender equality aspects [13]. It is evident that the percentage of women is widely dispersed depending on the subject orientation of universities.

Gender-specific study choices make it difficult for technical universities in particular to raise the percentage of female students and staff in their characteristic disciplines, especially in the fields of SET [14]. Universities focusing on social sciences and economics are able to achieve this more easily. Although the percentage of female students and graduates is higher in these disciplines technical universities have a relatively higher percentage of women at higher career levels – measured by the scientific potential at hand [13]. A closer examination of these disciplines with regard to the percentage of female professors shows that the number of women decreases depending on the hierarchy level. However, differences exist: both, the percentage of female freshmen as well as the dropout at different levels of qualification are relatively low. By the time of graduation and the appointment of professorships the percentage falls by 13 %. When it comes to linguistics, cultural studies and art studies there is a high percentage of female freshmen. Nevertheless, the dropout of women in these disciplines is also relatively high. Compared to engineering the reduction amounts to 55 % [11].

In consequence, a simple causal connection between the percentage of female students and the percentage of female professors cannot be confirmed. Thus, it is presumed that one reason for the low percentage of female professors can be found in the processes of universities.

The amount of acquired external funds[3] serves as a benchmark during the selection process – as it is supposed to be an objective criterion. Although this measure substantiates one's scientific productivity and reputation [15] and thus influences the decisions made in appointment procedures, no research results were to be found regarding an unintended gender bias concerning the award of external funds and its scientific justifications.

As a matter of fact the number of female scientists that apply for external funds is much lower than that of their male colleagues [16]. Moreover, they have lower chances of success than male scientists [17]. One reason might be the fact that women are less likely to have permanent contracts thus not meeting the formal (or informal) application requirements for external funding of several research funding institutions [18]. Instead of examining the individual, structural and contextual differences

---

[3]External funds are defined as funds that are obtained for the funding of research and development in addition to the regular university budget and are acquired by public or private bodies. External funds can be provided for universities, one of their institutions (faculties, departments) or individual researchers (WR 2000).

between men and women [19], the gender specific choice of research topics is still being used as an explanation for their different chances of success in external funding. The Funding Ranking of the German Research Foundation [20] shows that between 2005 and 2007 18'159 persons from universities participated in research proposals addressed to the DFG. The percentage of proposals with female participants that were granted external funds amounts to only 16% (2'862 women). The Funding Atlas of 2012 by the German Research Foundation [21] reveals an identical quote of 16% in 2010.

The turnaround concerning university funding is a factor that structurally and informally affects academic recruitment strategies and is closely connected to the strategies at professorial level. Thiem [22] observes a different style of research between men and women and describes the reasons in the different socialization experiences as well as the long-practiced societal gendered division of labour. Furthermore, women tend to attribute greater importance to interdisciplinary and do research in less established fields [23] than their male colleagues. In summary, it can be presumed that this fact is to be discovered in regards to the low quota of applications by women on external funds and job postings. Furthermore, it can be supposed that not only the postings by research funding organizations but also those of professorships unconsciously contain a gender bias and put women at a disadvantage even before the application. As external funds grow more and more important, other benchmarks such as the number of publications and the volume of acquired external funds outweigh further criteria [24]. Often, female scientists do research in less popular research areas [25]. Thus, their chances are low in peer review systems, unaffected by their qualitative performance. This is closely related to the "Matthew Effect" in science described by Gross et al. [26]: According to that, performance is always implicitly considered in the light of known reputation and quantitative research achievements.

Structural barriers in appointment procedures[4] already been identified are the low degrees of standardization and formalization as evidenced on the examples of strongly varying lengths of processes and a slow flow of information in the course of proceedings [27]. Common countermeasures to these aspects are transparency in the course of procedure [28] or taking into account diverse scientific biographies. These measures, however, can only rarely be found in appointment regulations or guidance documents.

The chances for a successful scientific career are much higher for those scientists who develop consistent attitudes [29] than for those who live and work in a constant ambivalence between person and role [29]. In reference to male dominated behavioural patterns that are characteristic for the scientific habitus[5] and also for appointment procedures, male applicants may profit from a certain ease in their acts

---

[4]The rather less formalized access to academic qualification positions has created growing concerns during the last years. There is good evidence that women can profit from standardised, externally advertised recruitment procedures than from internal recruitment practice [10].

[5]Habitus being a system of incorporated internalised patterns of a specific culture which allow one to (re)produce system typical thoughts, perceptions and actions [30].

[31]. As womanhood is still perceived as in dissonance to the scientific working culture, women cannot profit from such attributions [10]. A seamless and disciplinary habitus development can only take place if role consistency is given. The socially and institutionally conditioned image of professors contributes considerably to the habitus-related ambivalence of female scientists on a cultural level. Moreover, seamless biographies are difficult to achieve for women (and men) who wish to reconcile job and family. These disruptions lead to the rigorous exclusion from the scientific community [2]. The formal equality of treatment may also lead to the reproduction of the asymmetrical potential for the compatible development of a scientific habitus. This applies not only to the scientific elite but also to other areas of education and profession, like schools, universities, organizations and associations [32].

2 Methods and Design of the Study

To investigate the question what inherent factors lead structures, cultures and organizations to the asymmetric distribution of scientific staff at professorial level in terms of gender, our study analyses various universities and disciplines to identify possible implications between these factors and the respective percentage of female professors. Nationwide qualitative and quantitative investigations were made concerning gender sensitive and gender neutral academic recruiting strategies (cf. Figure 1).

The second object of the study was the investigation of gender sensitive and gender neutral elements in postings of German public research funding organizations[6] and professorial job postings (cf. Figure 1) via problem-focused interviews and text mining.

The nationwide qualitative investigation via guided expert interviews in Germany was divided into a preliminary study, data collection and data analysis as well as the validation by experts. The collection and analysis of data was performed based on the preliminary study. The sampling included persons (e. g. internal and external commissioners, university management, gender and equal opportunities officers, professors, etc.) who were chosen due to their expertise and knowledge based on experience regarding academic recruitment strategies and appointment procedures in expectation of valuable contributions. Another criterion was the percentage of female professors of the investigated universities to examine possible implications between the arrangements of appointment procedures and the percentage of female professors. The interviews were accessed by means of content analysis (via MAX-QDA). The following expert workshop had the aim to validate the compiled data with the expertise and (implicit) knowledge of different actors and players (e. g. gender and equal opportunities officers, appointment commissioners, scientists and other multipliers from the scientific community).

---

[6]e. g. German Research Foundation (DFG), German Council of Science and Humanities (WR), Federal Ministry of Education and Research (BMBF), Federal Ministry of Economics and Technology (BMWi), Volkswagen Foundation, Academy of Sciences etc.

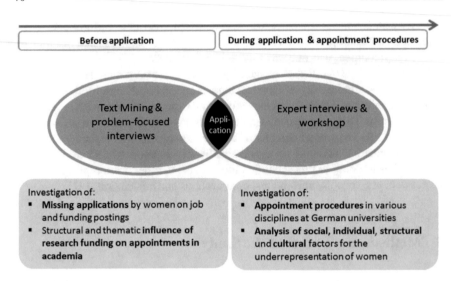

Figure 1 Subjects of investigation

The investigation of gender sensitive or gender neutral elements of the postings by German public research funding organizations and professorial job postings was carried out via text mining and problem-focused interviews. The text mining method was used to discover meaningful patterns and structures in weakly structured text data. Two large text corpora consisting of postings of funding organizations between 2007 and 2008 and professorial job postings between 2009 and 2010 were examined and compared. The data was analyzed to determine standard formulations, frequently used technical terms and frequently advertised research trends, thus indirectly looking for research trends in the postings. The problem-focused interviews aimed to portray the time course between research trend and funding as well as the qualification profiles of professorial applicants showing the connections in the decision-making process of qualified persons on applying for a professorship or not. The focus was on the identification of selection processes before the habilitation or – depending on the discipline – at other crucial links.

3   Investigation of Gender Sensitive and Gender Neutral Academic Recruiting Strategies

3.1   Preliminary Study

Eight guided interviews were used to deduce the five essential topics: "institutional and personal requirements", "basic conditions of appointment procedures", "advancement of women in academia", and "networks and process and completion of appointment procedures".

The interviewees of the preliminary study commented on the differences between disciplines and their importance for appointment procedures. The non-binding nature or even the entire lack of guidelines has to be emphasized in particular. Differences are to be found not only between different universities but also between universities located in the same federal state. Moreover, there were varieties between faculties of the same university. This heterogeneity was taken into consideration for the interview guide of the main study to scrutinize its impact on the appointment procedures themselves. Above that, the aspects of transparency and procedural documentation were factored in.

The problem of balancing academic work and family responsibilities was critically examined in consideration of the significant extension of childcare possibilities during the last years (e. g. [33]) and in consequence included in the main study. The issue was addressed subtly in the form of asking for part time work and part time possibilities as it can provoke answers that are uttered because of a certain social desirability.

The interviewees of the main study were primarily interviewed on their own experiences about the relevance of networks during the application process of female professors. Of particular interest were the criteria the interviewees used to choose the jobs they applied for. Furthermore, the inclusion of (personal) networks important. In the analysis the results could be contrasted to the point of view of the interviewed appointment commissioners.

3.2   Main Study

As there are inter-university differences in the percentage of female professors – as the data of the University Ranking shows [13] – the qualitative study examined the organizationally inherent, structural, cultural mechanisms taking place in appointment procedures that contribute to the gender-asymmetric distribution in the scientific system. The aim was to reconstruct and explain the causes of the low percentage of female professors.

All in all, 48 persons from 18 German universities, biased by their percentage of female full professors and regional distribution, were interviewed via brief questionnaires and guided expert interviews. The 48 expert interviews were recorded, transcribed and encoded in order to access them by means of content analysis followed by the expert validation within the frame of a workshop.

4 Investigation of Gender Sensitivity of the Postings by German Public Research Funding Organizations and Job Postings

4.1 Problem-focused Interviews

The guided, problem-focused interviews [34] were held with scientists of different disciplines who met the formal requirements for the appointment of a professorship in their respective disciplines. Above that, post-docs who were in a project management function were interviewed in order to cover the selection process before the habilitation.

The aims of the problem-focused interviews were

- to trace the **time course** from research trend to the qualification profiles of professorial applicants in order to
- point out the **connections** in the **decision-making process of qualified persons on applying for a professorship or not**.

This interview method is recommended because of its open approach as well as the centering on a certain problem. During the interviews the respondents had the possibility to comment on their subjective point of view. The respondents' remarks were led back to the leading questions of the investigation via inquiry and repeated questioning. Communication strategies [35] that generate narratives acted as important stimuli in the course of conversation.

In order to answer the main research questions fifteen interviews of post-docs from MINT fields selected by various aspects (gender, area of research, position, professional experience, career plans, family responsibility) were transcribed, encoded and analysed by making use of MAXQDA. A closer look into the coded passages shows the following motives for an application on a professorship:

- Passion for research
- Hierarchy
- Prestige
- Salary/ Money as a form of recognition
- Equipment
- Personnel responsibility

This leads to the conclusion that there is a divergence between intrinsic and extrinsic motivational factors when it comes to academic careers. The various motives are not easily compatible and mediating factors are to be found in the recognition of work: the equipment of the working environment, the salary grade and personnel responsibility are named by almost all of the respondents as a form of appreciation for their scientific achievements and leadership qualities.

5 Results

The following findings and recommendations are based on the empirical data collected in the project. We hereby present a summary of the main aspects that in view of the structural and cultural opportunities and barriers exist in the German scientific system. These are reflecting the statements of the experts interviewed. Based on the reported results of the investigation recommendations for action are listed in Section 6 following hereafter.

5.1 Structural and Cultural Barriers

As "Gendered Organization" universities proven to be rather resistant to change and reproduce asymmetrical gender relations system-internally. Universities as social spaces continue to be characterized by "patriarchal" structures only by high individual effort have access to the women. A self-selection, however, can only limitedly be observed. Certain external decision moments dominate the career paths in science. The appeal process itself is dominated by a few structural and formal specifications. Hence there is a lot of room for actions for the actors involved remains, but this does not contribute to appoint more women to professorships so far.

5.2 Cultural Chances and Barriers

Informal barriers such non concrete performance and availability expectations as well as an unfavourable acknowledgment and appreciation of culture against scientists comprising the same recognition before and provide a crucial hurdle on the way to a career in science. As cultural capital, however, networks and informal exchange relations are widely used by the female scientists who have decided to apply for a professorship despite the aforementioned obstacles.

Furthermore the "Matthew Effect" consistently leads to negative effects for women in science. First, due to the patriarchal heritage of "Gendered Organizational" the possibility for alternative scientific biographies will hardly be opened. Secondly, the ability to pass, though being viewed in the light of already known by the commission is neither objectively nor specifiable, ultimately depending on the personal composition of the commissions. It cannot be rejected on principle to pay attention to the "Pass Ability" of the candidates, because even group processes within the faculties and universities are to be taken into consideration. Nevertheless the question must be posed, to what extend these effects are reflected within the appointment procedures.

A systematic habitus ambivalence among the surveyed scientists has only positive impacts for men only positively by a write-up of habitus typical behaviour patterns.

For women, however, it would lead to exclusion from the system. This could not be observed. Scientists did not express divergence, which could be based on a non-matching attribution of science versus gender. Instead, they express their view on the appointment committees' work as an attempt to resort to apparent objective categories and decision criteria.

Another factor which has both structural and cultural impact is the growing importance of third-party funding of research projects at German universities. From one hand, this gives rise to a growing importance of publication numbers and volume of external funding that are gender asymmetric due to e.g. family times. On the other hand, the choice of research topics is increasingly relevant. To conduct research outside the mainstream thus entails increased opportunities to make themselves unfit for appointed to a professorship. Women are attributed to have a greater affinity for the less popular research topics [25]. However, it was found, that the importance of externally funded projects in respective job diverges is culturally specific. Not in every subject third-party funds appear to be applicable as a criterion of scientific productivity.

6 Recommendations for Gender Neutral and Gender Sensitive Academic Recruiting Strategies

Based on the collected empirical data and the previously presented results recommendations were generated which are discussed and reflected critically.

6.1 Under-representation of Women in the Appointment Committees

The under-representation of women in the appointment committees was emphasized by many interviewees. One possible solution to this problem is the introduction of subject-specific quotas by the university administration and national or federal administrations as a realistic approximation to the desired gender balance which is often defined by a guideline in appointment committees. By the development and implementation of a database of potential committee members, the subject-specific quota can be determined, monitored at regular intervals and updated if necessary. This recommendation would be targeted to the faculties and/or the dean's office.

To counteract the risk of a reduced number of women in appointment committees by a subject-specific quota, there should be implemented a minimum. Furthermore, there is the possibility of introducing this ratio as a "soft quota": e. g. the proportion of female students is at 20 % in a given subject, the subject-specific quota of women in the appointment committee could be 15 %. To gradually increase the proportion of women in appointment committees, it is helpful to check these regularly within the disciplines, depending on the University and, where appropriate, to increase the minimum. Moreover, it is recommended to introduce the subject-specific quota only for members with the rights to vote, in order to prevent the filling of the quota by non-voting members in appointment committees. Negative consequences of the introduction of a subject-specific rate, however, may also arise. It is conceivable that after the introduction of such a quota in the appointment committees of individual subjects fewer women are present than before. One way to counteract this is to expand the pool of potential members of the committee by not only appealing to members of each faculty, but also include external members of the scientific community obligatory. In this way, the risk of a possible overload of the women who are potential appointment committee members can be counteracted. In some cases it may be difficult, however, to find appropriate external members who are able to evaluate the applicants for the job as appropriate or inappropriate.

The question remains whether a higher representation of women can actually lead to an increase in the proportion of female professors in the ranks of the appointment committee members. A first step towards a higher proportion of women among professors may therefore be an awareness for the under-representation of women among all the members, or at least with the chairperson of the appointment committee. Problematic biases of reviewers and informal processes in conjunction with the review process were highlighted. These problems these are enforced by the practice of letting candidates choose or even suggest their reviewers by themselves. This is accompanied by the difficulty - depending on the subject – of having very scientific communities. In a correspondingly large community, one suggestion to approach this problem would be to set up the potential reviewers depending on denomination and community reputation, to regularly assess and select them randomly for each appointment process.

For this purpose, a pool can be created of possible reviewers before creating the list of candidates to be reviewed by the appointment committee. In addition, it is necessary to raise awareness among prospective referees for confidentiality within appeal. A disadvantage of the randomized selection of reviewers is the fact that there are certain areas with only a very small number of qualified reviewers. This can be countered by an increasing the amount of international reviewers. This has other positive side effects such as a higher attractiveness of German universities for foreign applicants.

6.2 Comparability of the Reviewers' Report

As part of the evaluation, the comparison of the reviewers' reports was emphasized as a problematic field, since often there are no guidelines or criteria stipulated for these reports available. Support could be offered by the creation of a standardized template for reviewers by the faculties or deaneries depending on the professorship to be appointed, which then can be fitted by the appointment committee with an evaluation matrix corresponding to the vacant post. Due to the high standardization of processes within appointments the risk has to be considered that a benchmarking system pass through that already have a gender bias. To avert this, that particular evaluation matrix can be checked in advance, for example, the Equal Opportunities Officer for any critical points. It should be noted however that complex selection process controlled by standards and rules is ought to be very difficult.

6.3   Mechanisms of Discrimination at the Receipt of Applications

The pre-sorting and/or sorting of applications on their reception was identified as a problem. The implementation of a cross-faculty professorial appointment management could support the avoidance of the sorting of applications without professional justification. This approach to appointment procedures was positively assessed by the universities that already have an appointment management at their command. Moreover, it is recommended for this step of the procedure to implement an evaluation matrix again as a guideline for the committee. This evaluation matrix should is partially standardized but also specialized on the specific professorship to be appointed and could be verified by an Equal Opportunities Officer in order to avoid discrimination mechanisms.

6.4   Structural Insecurity in the Position of Equality Officers

Particularly the position of the Equality Officer was mentioned during the interviews. The structural uncertainties and insecurities, the overload by committee work and the often missing subject specific knowledge and thus the lack of qualification for the proper assessment of candidates were viewed as problematic. Through the acquisition of participation in appointment committees through decentralized, specialized Equality Officer, this could be resolved. A structural strengthening of the Office of Equal Opportunity can be promoted by the federal- and inter-university institution of voting rights and of specialized decentralized deputies. Further structural strengthening could be a dedicated workplace and personnel support to provide support at the working level.

In order to promote cooperation within the appointment committee, depending on the professorship to be appointed, a professional coaching for Equality Officer (and for other non-specialist appointment committee members) should take place. In addition, there is the possibility for the appointment committee members, even before the actual process starts, to be coached by the Equal Opportunities Officers or gender issues brought to awareness.

6.5　Selection Process within the Professorial Appointment Committee

As part of the selection process within the appointment committee the inequalities within the structures as well as in dealing with applicants became visible. To solve this problem it is necessary to create standardized structures in appointment procedures that lead to a better traceability of the choice of the candidates and greater transparency of the procedures. This can be done by means of detailed appointment directives or manuals, in which individual process steps are clearly structured by the university administration. Furthermore, the individual components within the process could be limited by the means of time in order to avoid the unnecessary extension of the appointment process. The creation of multidisciplinary guidelines for appointment lectures and subsequent reliance conversations could also help to improve an equal treatment of male and female applicants.

6.6　Call for Applications

The call for application and thus the addressing of potential candidates for a professorship was shown by many interviewees as inadequate and improvement. To appeal to more potential candidates, it is possible that advise the actors who are responsible for creating the job advertisement by the Equal Opportunities Officer to attain adequate and gender sensitive language. Particularly increasingly relevant it also appears to be to sensitise active recruiting professors about the different offers for finding their active role in the (international) search for suitable candidates. Looking at the job postings, especially the width or narrowness of the tender was discussed divers.

While broader tenders lead to larger numbers applicants and thus possibly to a larger pool of female candidates, this increase, however, the number of unsuitable applicants leading to greater effort and may require a longer duration of the procedure. Closer but tender can be tailored to any particular person or address only a select group of applicants. In addition, applicants may lack the academic aspects of particularly specifically narrow tenders. Although the majority of respondents tend to be arguing for broader job postings, this way the call for applications is not clearly more advantageous for women than short-listed tender of professorships.

6.7　Structural Dependencies and Insecurities of Appointment Commissioners

During the occupation of the appointment committee structural dependencies between the different members within the commission were identified as problematic. In particular, the membership of doctoral students and the accompanying carers in appointment committees was mentioned, as these can have a negative impact in case of disagreement in the selection process. In addition, while the fixed-term employment contracts of Equality Officers and doctoral students was considered to have an important influence, as they may lead to a less well representation of their opinion during the commission's work.

6.8   Voting with the Appointment Committee

The vote on the appointment list which differs among universities is to be assessed as problematic. The interviews showed a dissent with regard to the anonymity of the vote. The advantage of an anonymous vote is, that structural dependencies are not effective or only to a small extent. As within appointment committees however generally open discussions were held about the candidates, an anonymous vote has no added value.

6.9   Sensitization on Gender Aspects of Structural Commissions

Structural Commissions of universities are usually consisting of deans, educational deans and other representatives of the faculties. They are responsible for the planning and the alignment of professorships in advance of the appointment committee who work on the basis of the work of the Structural Commissions. The interviews suggested that gender issues usually do not play a role at that stage of the procedure. Here, gender issues should be implemented as a cross-cutting issue, aware of the problem and each of the Commission's decision can be questioned on possible gendered effects. The implementation and controlling of gender sensitization of the structure (planning) commission should be carried out by the Equality Officer in cooperation with the university administration and the faculties. This can be anchored together with a subject-specific rate, as recommended for the appointment committees. It would also be advantageous to make the work of the Structural Commissions transparent and documented. Possible negative effects of these recommendations could be assumed with respect to the increasing bureaucracy.

These directives and guidelines will have to be developed, validated, implemented, monitored and reviewed at regular intervals. The proposed controlling also leads to more administrative overhead in the process. The denomination of chairs is an indispensable design tool for schools and universities, even to stand out from other research sites and to raise its profile. Here are only limited options to be identified in order not to restricting the possibilities of arranging professorships which might have negative effects on the German research and teaching landscape. Nevertheless, the varying alignment of professorships can lead to a higher diversity in research and the scientific community.

7 Summary

The improvement and transformation of German universities appointment procedures in terms of gender sensitivity and gender issues is now the subject of research for several years. When asked about improvement within the structural framework, it has been frequently pointed out how important the standardization of structures is. This need could be supported by the project "Genderation BeSt". Although often been addressed, the standardization of appointment processed and steps in this direction have been under way, still some problematic processes have been revealed in this regard.

This suggests that a radical restructuring and very strong standardization of such processes may not be the best way. Therefore, it is worth rethinking to have appointment processes tentatively run less standardized and then examine them for their earnings afterwards. As a deeper area of problems, there also emerges the orientation and alignment of professorships. At the same time the number of students in recent years has steadily increased, while the number of professorships has increased only slightly. This results in a rather small margin in the denominations of these professorships and also results in wider tenders, since teach must be guaranteed. The negative consequence of the wider opportunities as well as those promoted by wider professorships would be that the teaching which is required by program regulations, could not be executed.

The burden of female appointment committee members as well as the Equal Opportunities Officer by the additional work of an appointment process is an important issue. Because often several appointment processes run simultaneously and the Equal Opportunity Officer is a member of other committees, it is necessary to support Equality Officer. In each appointment committee, a person can be coached from the ranks of professors from the Equality Officers in such cases to act accordingly in the respective appointment process. An advantage of this would be the combination of expertise to the representation of gender aspects within the process.

In order to ensure equal treatment of all applicants in appointment processes, training for all actors involved in the process is one way of dealing not only with the structural barriers of the process, but also with the awareness on gender issues. The exchange of best practice examples from the German higher education landscape between university boards could propose a positive orientation for future appointment committees and may be involved in the training mentioned above. For possible objective or impartial control of appointment, the establishment of a general appointment or application management at all universities would be advantageous.

References

1. E. Commission, *Wissenschaftspolitik in der Europäischen Union. Förderung herausragender wissenschaftlicher Leistung durch Gender Mainstreaming. Bericht der ETAN- Expertinnengruppe "Frauen und Wissenschaft"* . Luxemburg, 2001
2. S. Metz-Göckel, P. Selent, R. Schürmann, Integration und Selektion. Dem Dropout von Wissenschaftlerinnen auf der Spur. Beiträge zur Hochschulforschung **32** (1), 2010, pp. 8–35
3. I. Lind, *Kurzexpertise zum Themenfeld Frauen in Wissenschaft und Forschung*. Center of Excellence Women and Science (CEWS), Bonn, 2006
4. F.O. of Statistics, *Bildung und Kultur. Schnellmeldungsergebnisse der Hochschulstatistik zu Studierenden und Studienanfänger/-innen*. Wiesbaden, 2012
5. E. Geenen, *Blockierte Karrieren. Frauen in der Hochschule*. Leske und Budrich, Opladen, 1994
6. C. Roloff, ed., *Personalentwicklung, Geschlechtergerechtigkeit und Qualitätsmanagement an der Hochschule, Wissenschaftliche Reihe*, vol. 142. Kleine, Bielefeld, 2002
7. J. Allmendinger, Hochschulen auf dem Weg ins 21. Jahrhundert - Abschied von der Männerdominanz? Beiträge zur Hochschulforschung **21** (4), 1999, pp. 247–256
8. H. Matthies, E. Kuhlmann, M. Oppen, D. Simon, *Karrieren und Barrieren im Wissenschaftsbetrieb: geschlechterdifferente Teilhabechancen in ausseruniversitären Forschungseinrichtungen*. Edition Sigma, Berlin, 2001
9. J. Allmendinger, E. Kienzle, K. Felker, S. Fuchs, *Und dann geht's stück für Stück weiter hoch oder auch nicht*. Abschlussbericht eines Forschungsprojekts über die Karrierewege von Männern und Frauen an der Tierärzt lichen Fakultät der Ludwig-Maximilians- Universität München. unbubl., München, 2004
10. I. Lind, Ursachen der Unterrepräsentanz von Wissenschaftlerinnen – Individuelle Entscheidungen oder Strukturelle Barrieren? In: *Exzellenz in Wissenschaft und Forschung. Neue Wege in der Gleichstellungspolitik. Dokumentation der Tagung am 28./29. November 2006 in Köln*, ed. by Wissenschaftsrat. Wissenschaftsrat, Köln, 2007, pp. 59–86
11. J. Allmendinger, ed., *Karriere ohne Vorlage. Junge Akademiker zwischen Hochschule und Beruf*. Edition Körber-Stiftung, Hamburg, 2005
12. S. Ihsen, E.A. Höhle, D. Baldin, *Spurensuche! Entscheidungskriterien für Natur- bzw. Ingenieurwissenschaften und mögliche Ursachen für frühe Studienabbrüche von Frauen und Männern an den TU9-Universitäten, TUM Gender- und Diversity-Studies*, vol. 1. LIT Verlag, Münster, 2013
13. A. Löther, *Hochschulranking nach Gleichstellungsaspekten. GESIS – Leibniz-Institut für Sozialwissenschaften. Fachinformation für Sozialwissen schaften*. Center of Excellence Women and Science (CEWS), Bonn, 2011
14. A. Schlüter, M. Harmeier, *Gender-Analyse der fachspezifischen Nachwuchsförderung an Hochschulen in NRW*. Essen, 2012
15. D. Jansen, A. Wald, K. Franke, U. Schmoch, T. Schubert, Drittmittel als Performanzindikator der wissenschaftlichen Forschung. Zum Einfluss von Rahmenbedingungen auf Forschungsleistung. Kölner Zeitschrift für Soziologie und Sozialpsychologie **59** (1), 2007, pp. 125–149
16. H. Etzkowitz, N. Gupta, M. Ranga, *Gender Effects in Research Funding. A review of the scientific discussion on the gender-specific aspects of the evaluation of funding proposals and the awarding of funding*. DFG, Bonn, 2012
17. K. Auspurg, T. Hinz, *Antragsaktivität und Förderchancen von Wissenschaftlerinnen bei Einzelanträgen auf DFG-Einzelförderung im Zeitraum 2005–2008*. DFG, Bonn, 2010
18. D. Forschungsgemeinschaft, *DFG-Stellungnahme zur Literaturstudie "Gendereffekte in der Forschungsförderung"*. DFG, Bonn, 2012
19. J. Allmendinger, T. Hinz, Perspektiven der Organisationssoziologie. Organisationssoziologie (Kölner Zeitschrift für Soziologie und Sozialpsychologie. Sonderheft, 42) , 2002, pp. 9–28
20. D. Forschungsgemeinschaft, *Förder-Ranking. Institutionen – Regionen – Netzwerke. Fachliche Profile von Hochschulen und außeruniversitären Forschungseinrichtungen im Licht öffentlich geförderter Forschung*. Wiley-VCH Verlag GmbH & Co. KGaA, Weinheim, 2009

21. D. Forschungsgemeinschaft, *Förderatlas. Kennzahlen zur öffentlich finanzierten Forschung in Deutschland*. Wiley-VCH Verlag GmbH & Co. KGaA, Weinheim, 2012
22. A. Thiem. Kompetenzprofile, Erfolgskriterien und Qualifizierungsanforderungen für Wissenschaftlerinnen im Forschungsfeld Umwelt- und Nachhaltigkeitswissenschaften, 2013. http://www.leuphana.de/professuren/umweltplanung/forschung/forschungsprojekte/wissenschaftlerinnen-im-forschungsfeld-umwelt-und-nachhaltigkeitswissenschaften.html
23. S. Bührer. Neue Kulturen der Wissenschaft: Forschen Frauen anders?, 2010. http://www.clubresearch.at/p-49814.html
24. S. Beaufaÿs, B. Krais, Doing science - doing gender : die Produktion von WissenschaftlerInnen und die Reproduktion von Machtverhältnissen im wissenschaftlichen Feld. Feministische Studien (Zeitschrift für interdisziplinäre Frauen- und Geschlechterforschung) **23** (1), 2005, pp. 82–99
25. S. Beaufaÿs, *Wie werden Wissenschaftler gemacht? Beobachtungen zur wechselseitigen Konstitution von Geschlecht und Wissenschaft*. transcript, Bielefeld, 2003
26. C. Gross, M. Jungbauer-Gans, P. Kriwy, Die Bedeutung meritokratischer und sozialer Kriterien für wissenschaftliche Karrieren. Ergebnisse von Expertengesprächen in ausgewählten Disziplinen. Beiträge zur Hochschulforschung **30** (4), 2008, pp. 8–32
27. T. Schmitt, N. Arnold, M. Rüde, *Berufungsverfahren im internationalen Vergleich*. Arbeitspapier Nr. 53. CHE Centrum für Hochschulentwicklung GmbH, Gütersloh, 2004
28. C. Färber, U. Spangenberg, *Wie werden Professuren besetzt? Chancengleichheit in Berufungsverfahren*. Campus, Frankfurt am Main, 2008
29. D. Janshen, H. Rudolph, *Ingenieurinnen. Frauen für die Zukunft*. de Gruyter, Berlin/New York, 1987
30. P. Bourdieu, *Die feinen Unterschiede. Kritik der gesellschaftlichen Urteilskraft*. Suhrkamp, Frankfurt am Main, 1982
31. S. Beaufaÿs, Alltag der Exzellenz : Konstruktionen von Leistung und Geschlecht in der Förderung wissenschaftlichen Nachwuchses. In: *Willkommen im Club? Frauen und Männer in Eliten, Forum Frauen- und Geschlechterforschung*, vol. 20, ed. by R.M. Dackweiler, Verl. Westfäl. Dampfboot, Münster, 2007, pp. 145–165
32. S. Ihsen, Special gender studies for engineering? European Journal of Engineering Education **30** (4), 2005, pp. 487–494
33. A. Althaber, J. Hess, L. Pfahl, Karriere mit Kind in der Wissenschaft : egalitärer Anspruch und tradierte Wirklichkeit der familiären Betreuungsarrangements von erfolgreichen Frauen und ihren Partnern. B. Budrich, Opladen, 2011, pp. 83–116
34. U. Flick, *Handbuch Qualitative Sozialforschung*. Beritz, Weinheim, 1995
35. A. Witzel, Das problemzentrierte Interview. In: *Qualitative Forschung in der Psychologie : Grundfragen, Verfahrensweisen, Anwendungsfelder*, ed. by G. Jüttemann, Beltz, Weinheim, 1985, pp. 227–255
36. S. Berghoff, G. Federkeil, P. Giebisch, C.D. Hachmeister, M. Hennings, D. Müller-Böling, *Das CHE-ForschungsRanking deutscher Universitäten 2007*. Arbeitspapier Nr. 102. CHE Centrum für Hochschulentwicklung gGmbH, Gütersloh, 2007
37. U. Bock, A. Braszeit, C. Schmerl, eds., *Frauen an den Universitäten. Zur Situation von Studentinnen und Hochschullehrerinnen in der männlichen Wissenschaftshierarchie*. Campus, Frankfurt am Main, 1983
38. P. Bourdieu, *Homo academicus*. Suhrkamp Verlag, Frankfurt am Main, 1988
39. P. Bourdieu, Physischer, sozialer und angeeigneter physischer Raum. In: *Stadt-Räume, Die Zukunft des Städtischen : Frankfurter Beiträge*, vol. 2, Campus, Frankfurt am Main, 1991, pp. 25–34
40. M. Brothun, Ursachen der Unterrepräsentanz von Frauen in universitären Spitzenpositionen. Kölner Zeitschrift für Soziologie und Sozialpsychologie **40** (2), 1988, pp. 316–336
41. E. Commission, *She Figures*. Brüssel, 2012
42. R. Gunter, A. Stambach, Differences in men and woman scientists' perception of workplace climate. Journal of Women and Minorities in Science and Engineering **11** (1), 2005, pp. 97–116

43. K. Hausen, H. Nowotny, eds., *Wie männlich ist die Wissenschaft?* Suhrkamp, Frankfurt am Main, 1990
44. S. Ihsen, *Zur Entwicklung einer neuen Qualitätskultur ingenieurwissenschaftlichen Studiengängen. Ein prozessbegleitendes Intervention skonzept, VDI-Fortschrittsberichte*, vol. 12. VDI, Düsseldorf, 1999
45. M. Löffler, Formalisierte Informalität? Wie das Leitbild Wissenschaftsgesellschaft Karrierebedingungen von Frauen an Universitäten verändert. SWS-Rundschau **48** (4), 2008, pp. 413–431
46. K.R. Malone, N.J. Nersessian, W. Newstetter, Gendert writ small: Gender enactments and gendered narratives about lab organizations and knowledge transmission in a biomedical engineering research setting. Journal of Women and Minorities in Science and Engineering **11** (1), 2005, pp. 61–82
47. S. Metz-Göckel, Exzellente Bildung von Frauen – Die Internationale Frauenuniversität als Forum einer Weltelite gebildeter Frauen. In: *Willkommen im Club? Frauen und Männer in Eliten, Forum Frauen- und Geschlechterforschung*, vol. 20, ed. by R.M. Dackweiler, Verl. Westfäl. Dampfboot, Münster, 2007, pp. 145–165
48. M. Meuser. Geschlecht und Arbeitswelt - Doing Gender in Organisationen, 2006. http://www.dji.de/index.php?id=40844
49. L. Müller, E. Borowski, Y. Jeanrenaud, A. Richert, S. Jeschke, S. Ihsen, Investigating gender-neutral and gender-sensitive academic recruiting strategies. In: *Proceedings of the 13th Annual IAS-STS Conference: Critical Issues in Science and Technology Studies. 5-6 May 2014.* STS, Graz, 2014
50. F.O. Ramirez, Frauen in der Wissenschaft – Frauen und Wissenschaft. Liberale und radikale Perspektiven in einem globalen Rahmen. In: *Zwischen Vorderbühne und Hinterbühne : Beiträge zum Wandel der Geschlechterbeziehungen in der Wissenschaft vom 17. Jahrhundert bis zur Gegenwart*, ed. by T. Wobbe, transcript Verl., Bielefeld, 2003
51. M.W. Rossiter, Der Matthäus Matthilda-Effekt in der Wissenschaft. In: *Zwischen Vorderbühne und Hinterbühne : Beiträge zum Wandel der Geschlechterbeziehungen in der Wissenschaft vom 17. Jahrhundert bis zur Gegenwart*, ed. by T. Wobbe, transcript Verl., Bielefeld, 2003
52. M. Schreier, Fallauswahl. In: *Handbuch Qualitative Forschung in der Psychologie*, ed. by G. Mey, K. Mruck, VS Verlag für Sozialwissenschaften, 2010, pp. 238–251
53. N.v. Stebut, *Eine Frage der Zeit? Zur Integration von Frauen in die Wissenschaft: Eine empirische Untersuchung der Max-Planck-Gesellschaft*. Opladen, 2003
54. Wissenschaftsrat, *Drittmittel und Grundmittel der Hochschulen 1993 bis 1998*. Geschäftsstelle d. Wissenschaftsrates, Köln, 2000
55. Wissenschaftsrat, ed., *Exzellenz in Wissenschaft und Forschung. Neue Wege in der Gleichstellungspolitik*. Neue Wege in der Gleichstellungspolitik. Dokumentation der Tagung am 28./29. November 2006 in Köln. Wissenschaftsrat, Köln, 2007

Integrative Knowledge Management in Interdisciplinary Research Clusters

Claudia Jooß, Thomas Thiele, René Vossen, Anja Richert
and Sabina Jeschke

Abstract Interdisciplinary Research (IDR) is described as a specific mode of collaboration: Besides the clash of different institutional cultures (e.g. different expectations/working processes), there is a clash of epistemic cultures (i.e. styles of thinking, different languages) (Jooß, C. (2014): Interdisciplinary Research collaboration: Critical Incidents from the researchers' perspective, in Gestaltung von Kooperationsprozessen interdisziplinärer Forschungsnetzwerke, Norderstedt: BOD-Verlag, pp. 146–179). Former research shows that the involved researchers demand an integrative knowledge management to support the expected integration of cultures (ibid.). In this paper two major aspects regarding integrative knowledge management for IDR are discussed. On the one hand the need for integrative knowledge management based on the researchers' perspective is depicted in the context of a use case. On the other hand the concept of a virtual mean is elaborated, which supports these needs. Both aspects underline the importance of a process-accompanying support in close coordination with the respective needs of the involved researchers.

Keywords Interdisciplinary Research (IDR) · Integration · Knowledge Management · Terminology-based Interfaces

1 Introduction

Advocates of interdisciplinary research (IDR) argue that this special form of research collaboration has a positive influence on problem solving and knowledge production [1–5]. IDR is considered to meet today's complex problems – whether the motives derive from inner-scientific needs or society. Despite the argument of problem solving and knowledge production, advocates of IDR also point out challenges of the expected integration. Although terms of interdisciplinarity vary and expectations of this form of collaboration are heterogeneous, a consensus shows within

C. Jooß (✉) · T. Thiele · R. Vossen · A. Richert · S. Jeschke
IMA/ZLW & IfU, RWTH Aachen University, Dennewartstr. 27,
52068 Aachen, Germany
e-mail: claudia.jooss@ima-zlw-ifu.rwth-aachen.de

Originally published in "Proceedings of the 8th International Conference
on E-Learning in the Workplace", © ICELW 2015. Reprint by Springer
International Publishing Switzerland 2016, DOI 10.1007/978-3-319-42620-4_8

83

the international discourse. This consensus displays integration as a central feature of interdisciplinarity [6]. The National Academies, for instance, state that interdisciplinarity is achieved only if an integration and a synthesis of ideas and methods take place [7]. A combination of concepts, theories, and methods is expected [8, 9]. This means that gained insights are not only accumulated ex post, but are also integrated during the cooperation process by means of an overall view. Thus, "synergies" are expected to be formed [10]. Strina describes synergies as interaction of different powers, elements, and parts that, hence, allow more than would be possible due to the features of the single parts [11]. Even though the researchers involved have a significant role due to these conditions [3, 4, 12], the characteristics and assessments of the researchers involved in interdisciplinary research networks are hardly investigated. We believe that this is an important gap to address, since it is the individual researcher who has to engage in research collaborations and develop scientific results [4]. Moreover, it is the individual researcher who is supposed to make an innovative contribution to solve today's complex problems. In contrast, the focus on the individual researchers' perspective provides wide-ranging implications for the design and effectiveness of interdisciplinary collaborations.

Against this background, different researchers of a complex interdisciplinary research cluster have been examined from various perspectives over an investigation period of five years [6]. A central result of this study shows that the researchers involved demand a knowledge management approach to support the expected integration. This so-called integrative knowledge management is a specific knowledge management in terms of the interdisciplinary vision. It comprises the continuous support of synchronization and adaption of the individual research activities with regard to the vision [6]. Integrative describes the assumption that a combination of various measures in a process-accompanying manner are of importance. These are developed, implemented and evaluated in a use case of an interdisciplinary research cluster (cf. Figure 1) and can be grouped e.g. in virtual and physical measures.

Based on this concept the question arises, which needs have to be fulfilled by the depicted measures in order to achieve an interdisciplinary integration.

As primary aim, this paper pursues the research question: Which needs of an integrative knowledge management are required from the researchers' perspective? To do so, Section 2 describes both the choice of the research field, the research design and method as well as central results regarding needs from the involved researchers. The second aim of this paper is to present an exemplary virtual mean to support integrative knowledge management. Within the interdisciplinary research cluster, our object of investigation, the involved researchers demand a mean to initiate the exchange on acquiring a common understanding of terminologies. Thus, we outline a concept of a technical solution in the third part of the paper.

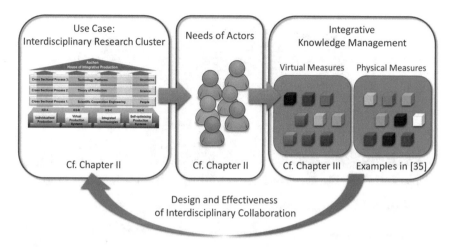

Figure 1 Structure of the paper

2 The Need of Integrative Knowledge Management in Interdisciplinary Research Clusters

2.1 Choice of Research Field and Organizational Structure

For our single case analysis, we selected an interdisciplinary research cluster as our object of investigation: The Cluster of Excellence (CoE) "Integrative Production Technology for High-Wage countries" at RWTH Aachen University. It was initiated by the German Research Foundation (DFG) and the German Council of Science and Humanities (WR) as part of the German excellence initiative. The consortium is located in Aachen and investigates the resolution of the polylemma of production [13] with various interdisciplinary partners from different faculties of RWTH Aachen University. The objective comprised ways to solve the tradeoff between scale and scope and between plan and value oriented production [14]. Within this CoE, many researchers from different university institutes, associated institutes, further non-university research institutions as well as different industrial and scientific advisors do research on a common vision. The aim is to develop a holistic theory of production by means of integrating economic, ecological, and social aspects. The CoE (cf. Figure 2 as an overview of its structure) consists of twelve subprojects with about 180 researchers and 200 student assistants in total. Various scientific disciplines (mechanical engineering, material science, mathematics, business studies, communication science, computer science and psychology) are brought together by these researchers who have also obtained various degrees as far as their academic education is concerned. A common understanding of e.g. terminology, language, methods, competences and perceptions of success is needed in order to enable these diverse personnel to cooperate. For overcoming these challenges of interdisciplinary

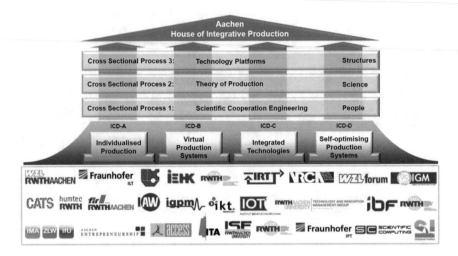

Figure 2 Structure of the CoE with its four Integrative Cluster Domains (ICDs) and three Cross Sectional Processes (CSPs)

cooperation and for supporting the performance of interdisciplinary research consortia in general new approaches are needed in order to cross link the amount of different researchers and institutes and to transfer solutions between them [15].

Next to several collaborative projects from the field of production technology, the CoE therefore additionally comprises cross-sectional projects entrusted with this task, the "Cross Sectional Processes" (CSPs). Their task is to design, implement and constantly evaluate concepts of supporting the integration of the different disciplines into the CoE on a physical and virtual level [15, 16].

2.2 Research Gap and Research Design

Within the framework of this continuous evaluation within the CSPs we draw on a research gap. It is perceived that researchers take on a special role within the integration but they have not been interrogated about their perception yet [6]. The majority of empirical studies on IDR are based on bibliometric data [2, 4, 7, 9, 17–21]. This database forms the main core for evaluating the design and effectiveness of interdisciplinary collaborations. This aspect, however, is increasingly challenged with criticism within the current discourse. Millar, for instance, claims it is about time to interrogate the researchers involved as well as to investigate how this form of collaboration affects the cooperation process [22]. Long-term studies, thus, aim at gathering insights on how to develop and design the dynamic and complex cooperation processes [12].

Our research contributes to this need by the use of a specific research design. It aims to obtain a view as extensive as possible on the heterogeneous researchers involved in interdisciplinary collaboration. Thus, the CoE has been examined from various perspectives over an investigation period of five years (2009–2013). In order to distinguish our study from quantitative procedures, neither antecedent nor theoretical concepts are examined, but rather concepts on the researchers' perception of interdisciplinary collaboration are reconstructed from the qualitative data [23]. We advocate the use of different datasets. In order to depict a sample that is as heterogeneous as possible following the principle of maximum structural variation, different researchers have been interrogated across all hierarchical levels in the course of the data collection process:

- Dataset 1: Structured guideline-based interviews
- Dataset 2: Partially standardized employee survey
- Dataset 3: Evaluation of the conducted measures
- Dataset 4: Participant observations

This characteristic provides the advantage that a holistic view on the object of investigation can be obtained, since it is observed from various perspectives (cf. Lamnek 1995 in [24]). In order to establish access to the data as well as to identify patterns several procedural steps of analysis have been conducted:

- As a first step, different phenomena of interdisciplinary collaboration are worked out that result from the researchers' perception.
- In another step of the analysis, critical factors (critical incidents – CIs) – in reference to the Critical Incidents Technique [25] – are identified from each dataset.
- After having empirically reasoned a theoretical saturation 30 CIs have been transferred to three patterns.

Through these three steps we are able to discuss the topic in a broader way. We have no indication, however, that our single case represents general CIs of IDR. But there are other empirical outcomes that corroborate our findings. These findings and all results are elaborately presented and discussed in "Interdisciplinary Research collaboration: Critical Incidents from the researchers' perspective" [6]. In this paper we focus on one selected pattern that deals with the elaboration of an integrated knowledge management with respect to reaching the common interdisciplinary vision. This is presented in the next section.

2.3 Results for Integrative Knowledge Management

Our research activities aimed at empirically exploring how researchers involved in the CoE perceive this mode of research. The research was also supposed to investigate what they need to challenge this mode of research. To reach an interdisciplinary integration during the research process the researchers involved demanded an integrative knowledge management. Its importance is shown in its support to accomplish the common interdisciplinary vision. The researchers involved demanded the continuous support of synchronization and adaptation of the individual research activities with regard to this vision. In this context, integrative describes the assumption that a combination of different supportive measures is of importance.

As far as the researchers' perception of interdisciplinary collaboration is concerned, it cannot be conducted by the mere addition of researchers from various disciplines. Hence, a continuous initiation of a project-, level-, and discipline-extensive networking is considered to be crucial. With regard to the common vision, an identification of interfaces is required, which, however, have to be continuously edited by means of an interactive exchange. It is also assumed that regular meetings are necessary to enable networking. In this context, it is important to provide a sufficient allocation of time for both the content-related negotiation processes, e.g. with regards to the common vision, and the social networking. The researchers involved also have to use these dates to meet and learn to appreciate each other in order to successfully work together in a team. Moreover, the researchers ascribe a significant role to so-called key persons who, however, are only capable of pushing ahead the active networking if a constant exchange between them and a reflection of research findings take place. The key persons are also supposed to have an overview on the contents and they need to be aware of their role. Furthermore, the researchers involved need a high motivation and willingness to participate in interdisciplinary collaborations. Both motivation and willingness can be supported by incentives, which is in line with the demands. In order to work on the common interdisciplinary vision, it is of importance to raise awareness for it. This raise of awareness can be supported by the researchers' demand for a visualization. With it, it is possible to localize researchers and projects within an overall image, which helps to identify and further process interfaces. Beyond that, it enables the researchers to consciously recognize and know their position and contribution to the overall goal.

A technical support is claimed to be necessary by the researchers involved, as the challenges to successful communication increase with working in an interdisciplinary collaboration. This technical support can, for instance, be realized by implementing a clearly structured, central communication platform. This technical, particularly virtual support is supposed to exceed a data management system. Furthermore, both a communication and cooperation platform and the visualization of the networking are considered to be important measures to handle staff turnover and to support the flow of communication and cooperation.

This demand of integrative knowledge management by the researchers of the CoE can be considered as being genera. But their elaboration is specific to each cluster and has to take place in close coordination with the respective needs of the involved researchers. To demonstrate how we deal with the virtual support within the CoE we present our Scientific Cooperation Portal in the following, which is jointly developed by the team of the CSPs.

3   Supporting Integrative Knowledge Management by Technical Means

The so-called Scientific Cooperation Portal is comprised of various apps, ranging from project management support to target-group-specific implementations. Based on the findings mentioned in Section 2 these apps have in common that they foster information dissemination on different levels within the CoE [15] and therefore support an integrative knowledge management by the connection of various information sources [26]. Thus, management support is given by basic features, e.g. a news section, a publication database and a calendar application.

More individualized aspects are addressed by a web app, which enables the user to identify interfaces between entities (e.g. projects or other persons) in the research cluster. This mainly aims at two identified CIs of IDR: The focus on communication and terminology as well as the identification and visualizing of connections [6]. These connections between entities are based on the usage of terminologies in this example: e.g. if two projects have a given set of terminologies in common, a connection between these two projects is displayed in the web application.

On the basis of this example a three step concept is outlined to address this idea: On the one hand the extraction of terminologies has to be discussed. On the other hand the mapping of entities has to be realized. The last step represents the visualization of the connection mentioned above.

3.1   Extraction of Terminologies

The first step in the identification process of connections can be seen in the extraction of terminologies from a given data source. As publications of all projects are stored in a common database on the Scientific Cooperation Portal, a process is necessary that allows the web application to access these publications and extract terminologies from the data source.

As Text Mining can be described as "a range of technologies for analyzing and processing semi-structured and unstructured text data" [27] this methodology can serve as one possible solution for the above mentioned challenge. After various pre-processing steps, including e.g. the tokenization of the publications as well as the filtering of stop words, POS-tagging is used to derive nouns from the publications as these word types are considered as main source for terminologies in the CoE. After that a standard tf-idf algorithm is applied to determine the frequencies of words within the publications. This aims at a vectorial description of the publications, which can be used in further processes. By combining several publications, which e.g. have been issued in the context of a project, a vectorial description of an entity within the research cluster becomes possible.

Within this vectorial description of frequencies various challenges concerning the linguistic properties of the terminologies have to be addressed. For example, so-called collocations are a typical challenge. These can be described as phrases which are considered to have an existence beyond the sum of the parts. This includes compounds (disk drive) or stock phrases (bacon and eggs) [28]. With this in mind the given publications are analyzed not only focusing the frequencies of words but also regarding words that are often used in combinations.

3.2 Mapping of Entities

After a vectorial description of terminologies has been created the second step represents the mapping of entities in order to derive statistically based connections. By using classification algorithms "the task is to classify a given data instance into a pre-specified set of categories" [29]. With the regard to the last chapter the pre-specified set of categories can been seen in the different entities (e.g. projects) in the CoE, which are described by the combined vectors of several publications. The web application then extracts terminologies from e.g. a new publication and uses a cluster algorithm to determine how likely this publication fits into the set of categories. The result is a set of probabilities, which can be used to describe the distance of a publication to the set of categories.

This leads to another thought in the context of distance between terminologies: the semantic similarity of words. The acquisition of meaning by automated systems is quite a challenge in Natural Language Processing. Approaches in this context focus mainly on a relative measure for semantic similarity, which can be used to determine how similar a word is to known words [28]. Thus, the relative measure enables the web application to classify the meaning of words by the use of vector space measures and binary similarity measures (e.g. Jaccard coefficient).

These similarity measures serve as mathematical method to describe the semantic meaning of a word in relation to other words. This is based on the analysis of frequency matrixes, which contain information about the context in which a word is used. Different matches e.g. between two contexts lead to higher ranking of semantic meaning between two words and therefore allow the web application to link these two words.

3.3 Visual Data Analysis

The third step in the concept of the web application is the visualization of the above mentioned results. The main goal is to depict new connections for the user on the basis of common terminologies and, therefore, common (research) topics between the user's project and other entities in the CoE. Hence, the visualization has to map two major aspects:

- The extracted terminologies from the user's project to the terminologies of other projects.
- The semantic similarity of words in relation to other words used in the terminologies of the CoE.

The first issue can be addressed by a metric that is based on the probabilities as described in Section 3.2 of this chapter. As a consequence the probabilities serve as a measure of distance between entities in the CoE: If the classification reveals a higher probability, the depicted distance is closer and vice versa.

As the semantic similarity is represented by relative measures of one word in relation to another word (cf. in Section 3.2 of this chapter), this aspect has also to be realized in the visualization. Therefore, this visualization has to depict word-to-word connections with regard to the relative measure of semantic similarity. One possible solution is a so-called tree graph (cf. Figure 3).

This tree graph allows the user to detect easily in how far a semantic similarity persists between its own terminologies and the terminologies used in other projects. Following this idea one goal of visual analytics, the synthesis of information and to communicate this assessment effectively for action is fulfilled [30]. Further aspects include separate pages for each terminology in a Wikipedia style: key-persons (e.g. experts) can contribute to the added-value of the application by integrating their knowledge in definitions and the recommendation of further sources (e.g. literature).

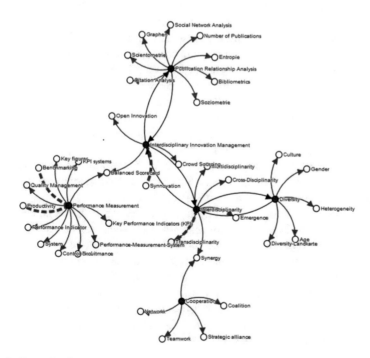

Figure 3 Example of a tree graph used in the web application

The next step can be seen in further exchange on acquiring a common understanding of terminologies (e.g. workshops).

In this context the web application addresses the demands outlined in Section 2. On a shallow level the visualization part of the app enables the researchers to be aware of other activities within the CoE. Key-persons for each terminology become identifiable and potential synergies visible. These can be used to foster further exchange processes regarding the semantic differentiation of terminologies. Going more into detail, the visualization can be seen as the individual part of the researchers' contribution to the vision of the CoE. As the depicted terminologies represent the researchers' field of action the visualization of connections show the embedding in related fields. Therefore, this allows the researcher to recognize his placing within the CoE and the placing of his field of action in an overall context.

4 Experiences and Outlook

The selected results give some indication of which aspects of interdisciplinary collaboration are important from the researchers' point of view to support the expected integration. Moreover, IDR is associated with high expectations, since synergies are supposed to be achieved. With regard to the common vision they demand a visualization since it is possible to localize researchers and projects within an overall image. Beyond that, the data have shown that it is essential to facilitate the expediting of the content-related networking by key persons. It is crucial for the key persons to be conscious about their role and that they are correspondingly promoted.

All these results underline the consensus on the insufficiency of merely adding researchers from various disciplines to make a research effort interdisciplinary. Therefore, integrative knowledge management has to be evaluated in a process-accompanying manner. In order to reinforce the findings it may be important to examine further research clusters to explore whether similar patterns can be identified. Additionally, to gain more insights into how CIs may vary in different scientific fields represents another important starting point for further research. The subsequent step is a quantification of the CIs. This aims at the development of quantified correlations between current qualitative statements.

As the web application is currently in conceptual stage the technical proof of concept has yet to be done. The major challenge behind this can be seen in the investigation of technical parameters, which enable the application to derive meaningful data from the Scientific Cooperation Platform database. If this has been finalized following steps will include support measures for the CoE researchers in order evaluate and stabilize the results shown in the application on a semantical level (like workshops etc.). On a technical level subsequent steps will include experiments with clustering algorithms in order to reveal thematic groups based on publication within the CoE, which are yet not revealed.

Acknowledgments The authors would like to thank the German Research Foundation DFG for the kind support within the Cluster of Excellence "Integrative Production Technology for High-Wage Countries".

References

1. J.A. Jacobs, S. Frickel, Interdisciplinarity: A Critical Assessment. Annual Review of Sociology **35**, 2009, pp. 43–65
2. J.T. Klein, *Interdisciplinarity : history, theory, and practice*. Wayne State University Press, Detroit, 1990
3. D. Rhoten, S. Pfirman, Women in interdisciplinary science: Exploring preferences and consequences. Research Policy **36** (1), 2007, pp. 56–75
4. F.J. van Rijnsoever, L.K. Hessels, Factors associated with disciplinary and interdisciplinary research collaboration. Research Policy **40** (3), 2011, pp. 463–472
5. M. Gibbons, C. Limoges, H. Nowotny, S. Schwartzman, P. Scott, M. Trow, *The new production of knowledge : the dynamics of science and research in contemporary societies*. Sage Publications, London, 1994
6. C. Jooß, Interdisciplinary Research collaboration: Critical Incidents from the researchers' perspective. In: *Gestaltung von Kooperationsprozessen interdisziplinärer Forschungsnetzwerke*, Books on Demand, Norderstedt, 2014, pp. 146–179. Pp. 10 ff, p. 177, p. 169, p. 149, p. 97
7. National Academies (U.S.), Committee on Facilitating Interdisciplinary Research, N.A.o.S.U.N.A.o.E.E. Committee on Science, and Public Policy (U.S.), Institute of Medicine (U.S.), *Facilitating interdisciplinary research*. National Academies Press, Washington, D.C., 2005
8. S.W. Aboelela, E. Larson, S. Bakken, O. Carrasquillo, A. Formicola, S.A. Glied, J. Haas, K.M. Gebbie, Defining Interdisciplinary Research: Conclusions from a Critical Review of the Literature. Health Services Research **42** (1p1), 2007, pp. 329–346
9. A.L. Porter, I. Rafols, Is science becoming more interdisciplinary? Measuring and mapping six research fields over time. Scientometrics **81** (3), 2009, pp. 719–745
10. R. Defila, A. Di Giulio, Interdisziplinarität und Disziplinarität. Zwischen den Fächern über den Dingen? Universalisierung versus Spezialisierung akademischer Bildung. In: *Zwischen den Fächern - über den Dingen?*, ed. by J.H. Olbertz, Leske und Budrich, Opladen, 1998, pp. 111–137
11. G. Strina, *Zur Messbarkeit nicht-quantitativer Größen im Rahmen unternehmenskybernetischer Prozesse. In der Vorbereitung zum Druck*. Aachen
12. G. Melin, Pragmatism and self-organization: Research collaboration on the individual level. Research Policy **29** (1), 2000, pp. 31–40
13. Integrative Production Technology for High-wage Countries. In: *Integrative Production Technology for High-Wage Countries*, ed. by C. Brecher, Springer Berlin Heidelberg, 2012
14. C. of Excellence 'Integrative Production Technology for High-Wage Countries', Renewal Proposal for a Cluster of Excellence "Integrative Production Technology for High-Wage Countries" – excellence initiative by the german federal and state governments to promote science and research at german universities, unpublished. Tech. rep.
15. T. Vaegs, A. Calero Valdez, A.K. Schaar, A. Braekling, S. Aghassi, U. Jansen, T. Thiele, F. Welter, C. Jooß, A. Richert, W. Schulz, G. Schuh, M. Ziefle, S. Jeschke, Enhancing Scientific Cooperation of an Interdisciplinary Cluster of Excellence via a Scientific Cooperation Portal. In: *Proceedings of the ICELW 2014, New York, June 11th–13th*
16. C. Jooß, F. Welter, I. Leisten, A. Richert, A.K. Schaar, A.C. Valdez, E.M. Nick, U. Prahl, U. Jansen, W. Schulz, M. Ziefle, S. Jeschke, Scientific Cooperation Engineering in the Cluster of Excellence Integrative Production Technology for High-Wage Countries at RWTH Aachen University. In: *accepted at 5th International Conference of Education, Research and Innovation (ICERI) 2012*

17. N. Carayol, T.U.N. Thi, Why do academic scientists engage in interdisciplinary research? Research Evaluation **14** (1), 2005, pp. 70–79
18. A.F.J. Van Raan, Scientometrics: State-of-the-art. Scientometrics **38** (1), 1997, pp. 205–218
19. D. Rhoten, Interdisciplinary Research: Trend or Transition. Items & Issues **5** (1–2), 2004, pp. 6–11
20. A.L. Porter, J.D. Roessner, A.S. Cohen, M. Perreault, Interdisciplinary research: meaning, metrics and nurture. Research Evaluation **15** (3), 2006, pp. 187–195
21. C. Raasch, V. Lee, S. Spaeth, C. Herstatt, The rise and fall of interdisciplinary research: The case of open source innovation. Research Policy **42** (5), 2013, pp. 1138–1151
22. M.M. Millar, Interdisciplinary research and the early career: The effect of interdisciplinary dissertation research on career placement and publication productivity of doctoral graduates in the sciences. Research Policy **42** (5), 2013, pp. 1152–1164
23. J. Kruse, Reader "Einführung in die Qualitative Interview-Forschung". Tech. rep., Freiburg, 2011
24. U. Kelle, *Die Integration qualitativer und quantitativer Methoden in der empirischen Sozialforschung. Theoretische Grundlagen und methodologische Konzepte*, 2nd edn. VS Verlag für Sozialwissenschaften, Wiesbaden, 2008
25. J.C. Flanagan, The critical incident technique. Psychological Bulletin **51** (4), 1954, pp. 327–358
26. H. Krcmar, *Einführung in das Informationsmanagement*. Springer Berlin Heidelberg, 2011
27. G. Miner, D. Delen, J. Elder, A. Fast, T. Hill, R.A. Nisbet, The Seven Practice Areas of Text Mining. In: *Practical Text Mining and Statistical Analysis for Non-structured Text Data Applications*, Academic Press, 2012, pp. 29–41
28. C.D. Manning, H. Schütze, *Foundations of Statistical Natural Language Processing*. MIT Press, Cambridge, Massachusetts, London, 2003
29. R. Feldman, J. Sanger, *The Text Mining Handbook: Advanced Approaches in Analyzing Unstructured Data*. Cambridge University Press, Cambridge, 2006
30. D.A. Keim, J. Kohlhammer, G. Ellis, F. Mansmann, *Mastering the Information Age - Solving Problems with Visual Analytics*. Thomas Müntzer, Goslar, 2010

Futures Studies Methods for Knowledge Management in Academic Research

Sabine Kadlubek, Stella Schulte-Cörne, Florian Welter,
Anja Richert and Sabina Jeschke

Abstract The management of academic knowledge is a relatively young area of attention. Universities and other higher education institutions accumulate a great deal of knowledge and the management of this asset is more than ever crucial for the strategic alignment. Hence, this paper aims at showing that knowledge management in academic research should work hand in hand with futures studies to develop and foster a strategic orientation. For this purpose the knowledge management model by Probst et al. (1998) with its eight building blocks serves as a framework. The focus of this paper lies on the processes of *knowledge goals* (ibid.) and *knowledge identification* (ibid.). Here it will be discussed that the futures studies methods monitoring, scenario technique and forecasting are suitable to complement knowledge management methods within academic research due to their ability to identify and concentrate information and knowledge relevant to the future.

Keywords Futures Studies · Knowledge Management · Academic Research · Monitoring · Forecasting · Scenario Technique

1 Introduction

The assumption that science and research are independent from material incentives [1] is especially for universities out-of-date. Universities and other higher education institutions are increasingly exposed to market pressures caused by the demand to become more productive and competitive [2, 3]. Hence, academic entities rely on external funding to realize innovative scientific and technological projects and ideas. Thereby, knowledge is the main capital of these entities. Hence, the implementation of knowledge management in academic research appears to become increasingly important [4]. Consequently, it is necessary that knowledge management includes a future orientation to align a strategy well in advance in order to grant public research

S. Kadlubek (✉) · S. Schulte-Cörne · F. Welter · A. Richert · S. Jeschke
IMA/ZLW & IfU, RWTH Aachen University,
Dennewartstr. 27, 52068 Aachen, Germany
e-mail: sabine.kadlubek@ima-zlw-ifu.rwth-aachen.de

Originally published in "Knowledge Engineering and Knowledge Management",
© Springer 2014. Reprint by Springer International Publishing Switzerland 2016,
DOI 10.1007/978-3-319-42620-4_9

funds as well as to avoid redundancy. In that respect knowledge management and futures studies are predestined to work hand in hand to develop and foster a strategic orientation in terms of targeted investments and the forward-looking identification of funding opportunities.

Since "there is as yet no agreement on a model which distills the essence of knowledge management" [2], the knowledge management model by Probst et al. [5] with its eight building blocks serves here as a framework. These building blocks "provide an outline of the areas where active knowledge management is possible" [5]. Considering that in research the pivotal interest foremost lays in future requirements and challenges, the creation of new and yet unknown knowledge is fundamental. This paper is devoted to the building blocks knowledge goals and knowledge identification due to their future orientation and their focus on the relevance of future knowledge. In the course of this paper it will be discussed in how far the futures studies methods monitoring, scenario technique and forecasting support the knowledge management of these two building blocks and complement the already established knowledge management methods and instruments [6] by continuously monitoring and evaluating the performance (monitoring), identifying diverse future challenges and consequences (scenario technique, forecasting) and thus initiating the changes required to move from the current to the desired state.

2   New Challenges for Knowledge Management in Academic Research

Since several decades modelling the generation, exchange as well as diffusion of knowledge depicts a major field of research in various scientific communities. With regard to the growing research community of knowledge management, a couple of authors developed principles and models that are dealing with normative, strategic, tactic and operative knowledge management elements. Hence, among others e.g. Nonaka and Takeuchi [7] focus on the entire knowledge creation in compa-nies, Bhatt [8] considers the organization of knowledge by describing a knowledge (management) cycle and Amin/Cohendet [9] explore knowledge architectures with regard to the management of respective knowledge communities. A further framework for knowledge management, which is worth mentioning here, is provided by Wiig et al. [10]. It describes four activities of knowledge management – Review, Conceptualize, Reflect and Act – which are performed sequentially [10]. As a central model of knowledge management in German literature the knowledge management model by Gilbert Probst et al. [11] has to be named due to its broad popularity in research and practice. The authors stress that the model was firstly developed to connect results of action research and organizational learning with real world problems of managers in companies [11]. The contemporary form of the model as described in 2003 comprises eight so called "building blocks" [11] – integrating strategic and operative management levels – that are connected by an iteration loop (cf. Figure 1).

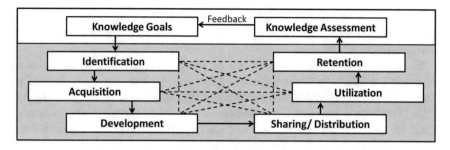

Figure 1 Eight building blocks of knowledge management by Probst et al. [5]

On a strategic level the iteration loop of the model begins with the building block knowledge goals [11]. This is followed by the building blocks: knowledge identification, knowledge acquisition, knowledge development, knowledge sharing/distribution, knowledge utilization and knowledge retention on an operative level. In this operative level systemic interconnections and cause-effect-relations of core processes are suggested by the authors [11]. Ending with the building block knowledge assessment on the strategic level, the iteration loop is completed (cf. Figure 1).

Probst et al. [5, 11] developed their knowledge management model to be used specifically in business environments. Considering the expansion of different forms in new public management, such as a more entrepreneurial alignment of universities in single federal states of Germany (Hochschulfreiheitsgesetz NRW 2006) [12], a reflection of the model in terms of its transferability on academic research environments can be a promising next step. However, there are obvious differences between an economic and an academic system with regard to knowledge [13]. A business organization usually faces the problem that knowledge assets are rarely visible. Hence, the detection, distribution and utilization of knowledge are the crucial tasks and building blocks of organizational knowledge management. In contrast, the focus of academic research environments lies on the constant generation of knowledge [13].

3   The Impact of the Building Blocks Knowledge Goals and Knowledge Identification on the Generation of Academic Knowledge

"Agreement on strategic goals is the core element in strategic planning, which in turn provides the basis for implementation and monitoring" [5]. In the case of a university, for example, it has become crucial to focus on future funding opportunities. Consequently, strategic research goals have to be set in order to identify future spheres of competence. In doing so it is important that the forecast and therefore the generated knowledge is reliable, i.e. it needs to be related to concrete research objectives. Already applied knowledge management methods, which support the process of goal setting, are "Strategic Knowledge Assets" [14] or knowledge-based "SWOT Analysis" [10, 15].

The more detailed it is known, which knowledge is actually available, the more precise is the definition of the unknown. Therefore, the building block knowledge identification is highly linked to the previously introduced building block knowledge goals. "Our knowledge goals will point us towards the areas and sources of knowledge which we need in order to strengthen our existing competencies or create new ones" [5]. Knowledge identification should consider internal as well as external knowledge [5]. The challenge within academic research is to identify the boundaries of already existing knowledge in order to excel. Knowledge management methods such as "Competence Mining", which is based on data mining techniques to identify the employees' competences based on their publications, support the identification of individual competences [16] and enhance the transparency required to attain the previously set goals.

Complementary to the exemplary knowledge management methods listed above, futures studies methods can support identifying which knowledge is going to be crucial in order to stay competitive respectively excellent by acquiring relevant projects and thus discovering and shaping topics of relevance to the future. In this context an integration of futures studies methods in knowledge management is discussed in the next sections, as it comprises a potential added value for new forms of knowledge management in academic research environments.

4   Considering a Stronger Integration of Futures Studies in Knowledge Management

4.1   Futures Studies and Its Methods: Why and Wherefore?

Futures studies is historically linked to humans' desire to find out something about themselves, their past, their present and their future. A noticeable lead in knowledge is often equated with added value or even market leadership for enterprises, scientific breakthroughs for universities and progress and prosperity for society – it is therefore always future-oriented. Furthermore, knowledge that is relevant to future is a necessary condition for thinking and talking about future and future development in a free and responsible manner [17]. It implies knowledge relevant to prognosis, action and orientation for various players and institutions [17] and this also applies to the academic field. To provide such knowledge, relevant to different facets of futures, no one can pass the use of futures studies or rather futures studies methods. Considering the remarks about the features of the building blocks knowledge goals and knowledge identification, it is clear that they are characterized by their forward-looking perspective. But regardless of whether we look upon businesses or the academic field, the detection of future developments, concrete challenges and aims deriving from them has to be done by appropriate methods.

Futures studies asks how possible, probable and desired futures could be and what ought to happen to make a necessary, possible or desired future scenario come true [18]. There is no need to say that futures studies cannot investigate the future – however, this is not the point at issue. The question is rather to identify future spheres of competence and create knowledge, assured through structured and methodical procedures, that is able to support decision-making processes. Thus futures studies pursue different cognitive objectives [15] which are materialized in corresponding methods.

Futures studies methods are as heterogeneous as their object and thus combine qualitative and quantitative approaches. They also have different disciplinary back-grounds, act normative, communicative or participative [19] and are context-sensitive. Some methods are more suitable than others in special situations concerning i.e. their time horizon and the meaningfulness or significance of the prediction. But what they provide independently of one another is to identify and concentrate information and knowledge relevant to the future, and thus serve for the administrative work at educational providers [17]. That is why futures studies are able to support knowledge management processes not only in businesses but also in science.

4.2   Futures Studies Methods and Their Value for Knowledge Management

In the following section the paper asks for the value which different futures studies methods have for knowledge management processes. The methods described further on are characterized by the fact that they are used variably for different time scales. That qualifies them for different requests in knowledge management processes and especially for the future-related questions "Which knowledge is strictly required in the future?" or "Which research and development projects have to be launched or initiated to cope with the challenges the future holds in store?". The selected methods are monitoring, scenario technique and forecasting.

Monitoring can be defined as a directed observation of key indicators relevant in a special field and is usually coupled with an undirected search for clues and evidence linked to the indicators. It is a kind of formative evaluation [20], where special information about processes and findings are collected to stimulate activities that are able to enhance their efficiency and effectiveness. The purpose of monitoring-activities and processes is to steadily observe a particular object based on relevant information, to reflect it and to initiate change. With regard to knowledge goals and knowledge identification, a continuous monitoring helps to detect concrete and possible developments in (academic) research and is able to make statements whether it is advisable for an organization to follow a proven path in research or to strike a new path; and if an organization opts for an adjustment or reorientation in research they must ask themselves whether they have all the know-how to successfully pursue the new path. Thus monitoring rather serves to assess developments over time descriptively and on the basis of reliable and regularly collected data [21] to stimulate and plan future oriented actions and forward-looking programs. In the monitoring project 'International Monitoring', funded by the German Federal Ministry of Education and Research from 2007 to 2013, the monitoring method was exemplarily used and adapted to establish a continuous observation of fields of future action on the topic of innovative ability and thus to foster the sustainable competitiveness of Germany and Europe in the global market.[1]

Forecasting is likely to be understood as the prediction respectively the projection of future developments on the basis of both earlier developmental stages and the actual situation [22]. When the assumptions are known, explainable statements about fundamental trends and sequences of events are possible [18]. As it is usually aimed to find strategies suitable to influence the future developments in a positive and ecologically-friendly, economically and socially responsible way, the statements carried out take the form of conditional statements [17, 23]. Forecasts can be short-term, intermediate and long-term orientated. They can support the knowledge management building blocks by generating relatively assured data about future developments and therefore provide a reliable basis for decision making about future research developments. As an example within the meta project 'DemoScreen', funded by the German Federal Ministry of Education and Research, expert workshops were taking place in order to elaborate future needs for research in terms of social innovation. (http://demoscreen.de/metaprojekt-demoscreen/demoscreen-news) If a forecast is complete and self-contained it is also called a scenario.

Through scenarios future developments or states of a forecasting object under certain and alternative conditions will be investigated and an aggregated picture for a certain prognostic time horizon will be designed. Thus, scenarios are based on information, opinions, views and valuations which determine their probability of occurrence. Questions like "What will happen?", "What can happen?" and last but not least "How can a specific target be reached?" define the range in which different types of scenarios can be classified [24]. Assuming that, goals, role models, options of action, recommendations and measures can be developed to shape the future as well as the journey towards it. In academic research scenarios allow to identify research desiderata and to anticipate possible developments. They help to detect interferences and support strategy development as well as the definition of research respectively knowledge goals. The scenario technique allows developing a

---

[1]http://www.internationalmonitoring.com/home.html.

longterm strategy which also includes unprecedented phenomena and developments. This aspect is interesting for research since it aims at finding out the yet unknown and increasingly depends on funding. Thus the main benefit deriving from the scenario technique method with regard to the building blocks knowledge goals and knowledge identification is the set of alternative futures which offers a plausible and consistent framework for the development of a strategy concerning required future knowledge.

5 Conclusion

In the present paper three selected methods of futures studies were discussed in terms of their added value to knowledge management of academic research environ-ments with a special focus on their value for the building blocks knowledge goals and knowledge identification by Probst et al. [5]. This knowledge management model provides several development options when applied for academic research, whereby this paper focused on two out of eight building blocks due to their importance for the definition of an academic strategy. The proposed integration of futures studies meth-ods into this model has not been validated in general yet, but it illustrates a useful and promising support for detecting future challenges and demands such as exem-plarily highlighted for monitoring and meta projects, funded by the German Federal Ministry of Education and Research. The selected methods complement knowledge management and enhance the research strategy by examining future developments and their completion – an important aspect for academic research e.g. in terms of external funding opportunities. Hence, a current strategy can be adapted if profound future changes are detected (e.g. changes of political strategies) on the one hand and decisions for future research projects can be shaped on the other hand. The strengths of the foregoing methods of futures studies in the present case lay in thinking more systematically, anticipating a long-term perspective, considering alternatives when thinking about the future and last but not least improving the quality of decision-making.

Further potential for adaptation of the knowledge management model by Probst et al. [5] to academic research needs is seen in the building block knowledge acquisi-tion in terms of a distinction between an acquisition of completely new knowledge on the one hand and a further development of already existing knowledge on the other hand. For both the acquisition of necessary knowledge, which does not exist yet, is required. However, the starting point and thus the quality and effort are different and have to reflect in knowledge management processes.

References

1. T. Ma, S. Liu, Y. Nakamori, Roadmapping as a way of knowledge management for supporting scientific research in academia. Systems Research and Behavioral Science **23** (6), 2006, pp. 743–755
2. J. Rowley, Is higher education ready for knowledge management? International Journal of Educational Management **14** (7), 2000, pp. 325–333
3. A. Metcalfe, The Political Economy of Knowledge Management in Higher Education. In: *Knowledge Management and Higher Education: A Critical Analysis*, ed. by A. Metcalfe, Information Science Publishing, Hershey, 2006, pp. 1–20
4. C.A.A. Sousa, P.H.J. Hendriks, Connecting Knowledge to Management: The Case of Academic Research. Organization **15** (6), 2008, pp. 811–830
5. G. Probst, S. Raub, K. Romhardt, *Wissen managen: Wie Unternehmen ihre wertvollste Ressource optimal nutzen*. Gabler, Wiesbaden, 1998
6. F. Lehner, *Wissensmanagement: Grundlagen, Methoden und technische Unterstützung*. Carl Hanser Verlag, München, 2009
7. I. Nonaka, H. Takeuchi, *The Knowledge-Creating Company*. Oxford University Press, New York, 1995
8. G.D. Bhatt, Organizing knowledge in the knowledge development cycle. Journal of Knowledge Management **4** (1), 2000, pp. 15–26
9. A. Amin, P. Cohendet, *Architectures of Knowledge: Firms, Capabilities, and Communities*. Oxford University Press, New York, 2004
10. K.M. Wiig, R. de Hoog, R. van der Spek, Supporting knowledge management: A selection of methods and techniques. Expert Systems with Applications **13** (1), 1997, pp. 15–27
11. G.J. Probst, S. Raub, K. Romhardt, *Managing Knowledge: Building Blocks for Success*. John Wiley & Sons, Chichester, 1998
12. Ministerium für Inneres und Kommunales des Landes Nordrhein-Westfalen. Hochschulfreiheitsgesetz. https://recht.nrw.de/lmi/owa/br_vbl_detail_text?anw_nr=6&vd_id=1460&menu=1&sg=0&keyword=Hochschulfreiheitsgesetz
13. M. Kölbel, Wissensmanagement in der Wissenschaft. In: *Wissensmanagement in der Wissenschaft: Wissenschaftsforschung Jahrbuch*, ed. by K. Fuchs-Kittowski, W. Umstätter, R. Wagner-Döbler, Gesellschaft für Wissenschaftsforschung, Berlin, 2004, pp. 89–101
14. U. Remus, Prozessorientiertes Wissensmanagement. Konzepte und Modellierung. Ph.D. thesis, Universität Regensburg, Wirtschaftswissenschaftliche Fakultät, 2002. http://epub.uni-regensburg.de/9925/
15. M.H. Zack, Developing a Knowledge Strategy. California Management Review **41** (3), 1999, pp. 125–145
16. S. Rodrigues, J. Oliveira, J.M. de Souza, Competence mining for team formation and virtual community recommendation. In: *Proceedings of the 9th Conference on Computer Supported Work in Design, Conventry, UK, May 24-26*, vol. 1. IEEE Computer Society, 2005, vol. 1, pp. 44–49
17. V. Tiberius, *Hochschuldidaktik Der Zukunftsforschung*, 2011th edn. VS Verlag, Wiesbaden, 2011
18. A.W. Müller, G. Müller-Stewens, *Strategic Foresight: Trend- und Zukunftsforschung in Unternehmen - Instrumente, Prozesse, Fallstudien*. Schäffer-Poeschel, Stuttgart, 2009
19. R. Kreibich, *Zukunftsforschung für die gesellschaftliche Praxis*. No. 29 In: ArbeitsBericht. 2008
20. K. Joo, Monitoring as a Program Evaluation Model: A Case Study of Developing Evaluation Criteria for Monitoring in a Korean University Lifelong Education Center. In: *Midwest Research-to-Practice Conference in Adult, Continuing, Community and Extension Education, Northeastern Illinois University, Chicago, IL, October 21-23*. 2009
21. R. Stockmann, *Qualitätsmanagement und Evaluation - konkurrierende oder sich ergänzende Konzepte?* No. 3 In: CEval-Arbeitspapier. CEval, Saarbrücken, 2002

22. S. Löchtefeld, Backcasting – Ein Instrument zur Zukunftsgestaltung. In: *Werkstattbuch Familienzentrum. Methoden für die erfolgreiche Praxis*, ed. by S. Rietmann, G. Hensen, VS Verlag, Wiesbaden, 2009, pp. 109–117
23. U. Schrader, M. Bode, C. Pfeiffer, Der Blick zurück aus der Zukunft. ökologisches Wirtschaften **1**, 2001, pp. 29–31
24. L. Börjeson, M. Höjer, K.H. Dreborg, T. Ekvall, G. Finnveden, Scenario types and techniques: Towards a user's guide. Futures **38** (7), 2006, pp. 723–739

Neue Formen der Arbeit und die neuen Erwerbsbiografien

Rüdiger Klatt, Kurt-Georg Ciesinger, Thomas Thiele, Meike Bücker
and Saskia Bakuhn

Zusammenfassung Der seit dem Ende des 20. Jahrhunderts anhaltende Trend zur Flexibilisierung und Entgrenzung von Erwerbsverläufen wird derzeit durch die Digitalisierung von Arbeitsprozessen und Unternehmensformen weiter verschärft. Auch subjektive Sinnansprüche an Arbeit – vor allem der jüngeren Generation – führen dazu, dass Arbeitsbiografien bunter werden. Vor diesem Hintergrund hat die Session „Neue Formen der Arbeit und die neuen Erwerbsbiografien" auf der Tagung des Metaprojektes DEMOSCREEN im BMBF-Förderschwerpunkt „Innovationsfähigkeit im demografischen Wandel" den Stand der Debatte skizziert und die Gestaltungsaufgaben und Forschungsfragen der Zukunft auf diesem Feld entwickelt. Der Beitrag dokumentiert die Ergebnisse der Session.

Schlüsselwörter Arbeitsbiografien · Digitalisierte Arbeit · Lebensphasenmanagement · Generationenbalance

1 Einführung

Erwerbsverläufe und Arbeitsbiografien werden durch die Digitalisierung von Arbeitsprozessen und Unternehmensformen weiter flexibilisiert und entgrenzt. Phasen abhängiger Beschäftigung lösen sich ab mit Familienphasen, mit Phasen von

R. Klatt (✉)
Forschungsinstitut für innovative Arbeitsgestaltung und Prävention e.V. (FIAP e.V.),
Wissenschaftspark Gelsenkirchen, Munscheidstr. 14, 45886 Gelsenkirchen, Germany
e-mail: rklatt@fiap-ev.de

K.-G. Ciesinger
Gaus GmbH, Stockholmer Allee 24,
44269 Dortmund, Germany

T. Thiele · M. Bücker · S. Bakuhn
IMA/ZLW & IfU, RWTH Aachen University, Dennewartstr. 27,
52068 Aachen, Germany

Originally published in "Exploring Demographics",
© Springer 2015. Reprint by Springer International Publishing Switzerland 2016,
DOI 10.1007/978-3-319-42620-4_10

Arbeitslosigkeit, Selbständigkeit und Qualifizierung. Zunehmend werden Unternehmensformen ,normal', in denen Arbeitsort, Arbeitszeit und Arbeitsprozess durch das ,Projekt'oder die Ergebniserwartung des Marktes bestimmt ist. Die gut qualifizierten Beschäftigten werden zu Arbeitsnomaden, die die Gestaltung ihrer Arbeitsbiografie und ihrer Kompetenzentwicklung selbst übernehmen (können), dies aber häufig ohne große Rücksicht auf mögliche langfristige Effekte auf Beschäftigungsfähigkeit und Gesundheit tun, während die weniger gut Qualifizierten ohnehin häufig in einer Prekarisierungsspirale stecken (vgl. z. B. [1]).

In beiden Fällen können die Übergänge zwischen den Erwerbsphasen problematisch verlaufen, zu Brüchen führen und Dequalifizierungsrisiken in sich bergen (vgl. zusammenfassend [2]). Und mögliche Potenziale ,bunter' Karriereverläufe werden nicht nur in konventionellen, sondern auch in den ,smarteren' Unternehmen kaum wahrgenommen, sodass die offene Gestaltung der Erwerbsbiografie auch ihre ganz eigenen Risiken mitführt und die Rückkehr in das gesicherte System eine Festanstellung unter Umständen behindert.

Es bedarf daher noch weiterer erheblicher Anstrengungen der Forschung, intermediärer Institutionen, der Unternehmen und der Beschäftigten selbst, um die Auswirkungen diskontinuierlicher erwerbsbiografischer Verläufe auf mobile Arbeitsformen, neue Kollaborationstechnologien und dezentrale Organisationsmodelle zu untersuchen und zu gestalten. Notwendig ist die Erforschung von Unternehmensstrategien für gelungene erwerbsbiografische Verläufe in einer digitalen Arbeitswelt, die betriebliche Innovationsfähigkeit wie individuelle Existenzmöglichkeiten sichert.

Erwerbsarbeit wird unter den Bedingungen digitalisierter Wissensarbeit immer stärker mit anderen Tätigkeiten (Familienarbeit, ehrenamtliche Tätigkeiten, Freizeitaktivitäten) zusammenfließen [3]. Dies bietet Potenziale für eine aktive Gestaltung der Erwerbsbiografie und erfordert gleichzeitig neue Modelle der sozialen Absicherung und Anerkennung. Die lebensphasensensible Arbeitsgestaltung ist daher aus der Sicht der Session ein Schlüsselfaktor für die Gesundheit und Innovationsfähigkeit der Beschäftigten.

Unternehmen und Beschäftigte benötigen auch umfassende gestaltungsorientierte Forschungen für neue Modelle der reflexiven Gestaltung von wissensintensiver Arbeit, (privatem) Engagement und Existenzsicherung. In der Langzeitperspektive sind die Modelle von hochqualifizierten Jobnomaden wissensintensiver Branchen ebenso riskant wie die der ohnehin prekär in Zeitverträgen und atypischen Beschäftigungsformen und Mehrfachbeschäftigungen tätigen Arbeitnehmer mit weniger nachgefragtem oder weniger hohem Kompetenzniveau.

Wir sehen daher die Notwendigkeit einer breit angelegten Erwerbsverlaufsforschung, die für digitalisierte Unternehmen und intermediäre Akteure über einen längeren Zeitraum die Entwicklung von Potenzialen und Kompetenzen beobachtet, aber auch die Entstehung von Risiken für Gesundheit, Qualifizierung, Work-Life-Balance und sozialer Absicherung analysiert und auf dieser Basis Modelle einer Integration von Fremd- und Selbstbestimmung im Lebensphasenmanagement entwickelt.

Benötigt werden schließlich auch Instrumente für das aktive Erwerbsbiografiemanagement und zur Integration von lebensphasenabhängigen Potenzialen in die unternehmerische Praxis, die es bislang noch nicht in hinreichender Verbreitung gibt. Dabei ist die bislang vorherrschende punktuelle Personalentwicklung eines Unternehmens abzulösen von Unterstützungsmodellen, für die (auch) unternehmensübergreifende Langzeitplanung der Arbeits- oder Erwerbsbiografie, an der sich auch die intermediären Akteure aktiv beteiligt sollten.

2 Ein kurzer Rückblick

Bis vor 20 Jahren war die Entwicklung der Arbeitsbiografie geprägt von der Vorstellung, dass es dem deutschen Arbeitsmarkt und den Arbeitsverhältnissen an Flexibilität fehlt. Zuviel Normalarbeit, zu wenig flexible Beschäftigung, um den zunehmend globalisierten Wettbewerbsstrukturen etwas entgegenzusetzen. Die daraufhin einsetzende Welle an Flexibilisierungsprogrammen, von denen die in 2003/2004 umgesetzte Agenda 2010 sicher das prominenteste war, hat eine umfassende Umgestaltung des Arbeitsmarktes vollendet, die allerdings schon Ende des letzten Jahrhunderts eingeleitet wurde. Diskontinuierliche und prekäre Beschäftigungsverhältnisse waren seitdem auch Gegenstand der Arbeitsforschung, die auf mögliche Gefährdungen der Arbeitsfähigkeit durch Selbstausbeutung, Dequalifizierung und gesundheitliche Überforderung (z. B. Burnout) aufmerksam gemacht hat [4]. Sie wurden aber in den letzten Jahren auch zunehmend zum Spielfeld einer neuen Generation Beschäftigter, die die Chancen neuer Formen der Arbeit insbesondere in der Wissensökonomie in Bezug auf die Erweiterung von Autonomiespielräumen, Sinnhaftigkeit, Kompetenzerwerb und einer besseren Vereinbarkeit von Arbeits- und Privatleben ausprobierten [5].

Bislang ist es der Arbeitsforschung noch nicht hinreichend gelungen, herauszuarbeiten, wie – das heißt unter welchen gesellschaftlichen, betrieblichen und individuellen Bedingungen – diskontinuierliche (und kontinuierliche) Erwerbsbiografien hinsichtlich des Aufbaus von Kompetenzen, einer gelungenen Balance von Arbeit und Leben, der Gesundheitsförderlichkeit, der Sinnhaftigkeit von Arbeit usw. erfolgreich verlaufen oder scheitern. Die Herausforderung für die Forschung liegt darin, erwerbsbiografische Verläufe über einen längeren Zeitraum nachzuzeichnen, um aus der vergleichenden Analyse für alle Branchen und Beschäftigtengruppen differenzierte Erkenntnisse zu gewinnen und diese mit und für die unternehmerische Praxis in wirksame Unterstützungsmodelle umzusetzen.

Eine besondere Herausforderung stellt dabei – in den letzten Jahren verstärkt – die demografische Entwicklung dar, die durch eine unterdurchschnittliche Geburtenrate einerseits und eine unaufhaltsam alternde Arbeitsgesellschaft andererseits gekennzeichnet ist. Die Gesamtlage deutet auf ein gestörtes Verhältnis von Arbeit/ Beruf/Karriere und Familie: eine Herausforderung auch für die gestaltungsorientierte Arbeitsforschung. Sie induziert auch das Problem der Erhaltung von Kompetenzen, Gesundheit, Motivation, Innnovations- und Beschäftigungsfähigkeit für die Generation 50+, wofür die Arbeitsforschung ebenfalls Lösungsmodelle und Unterstützungsinstrumente für Unternehmen kreieren muss. Und dabei sind die möglichen

Gefährdungen der Generationenbalance in den Betrieben sowie die Aktivierung des Beschäftigungspotenzials in der Nacherwerbsphase noch gar nicht richtig in der Forschung und der Praxis angekommen.

Angesichts der demografischen Entwicklung gewinnt die Europäisierung und Globalisierung des Arbeitsmarktes eine neue Valenz [6]. Sie wird zunehmend als Chance begriffen, durch Arbeitsmigration in einem zusammenwachsenden Europa demografiebedingte Probleme auf einer europäischen Ebene zu lösen. Auch die Untersuchung von Erwerbsverläufen von Menschen mit Migrationshintergrund und der Bedingungen ihres Erfolges oder ihres Scheiterns gehört zu den Aufgaben, die auf die Agenda einer zukünftigen Arbeitsforschung gehören. Die Chancen diversitätsorientierter Unternehmensformen liegen auf der Hand, werden aber nur in den modernsten und zugleich international agierenden Unternehmen wirklich umgesetzt.

Angesichts der Digitalisierung nahezu aller Arbeitsformen und -prozesse, die die Mobilisierung und Flexibilisierung der Erwerbsarbeit und der Erwerbsverläufe durch die Entgrenzung von Arbeit (zeitlich, örtlich und sozial) noch befeuern werden, stellt sich die Frage, wie angesichts dieser Entwicklungstendenzen hin zu hochindividualisierter Arbeit und Patchwork-Biografien noch produktiv gearbeitet werden kann. Oder bieten die mobilen Arbeitsformen, die neue, digitalisierte Kollaborationstechnologien in dezentralen und vernetzten Organisationen nutzen, um in kundenintegrierenden Produktions- und Dienstleistungsprozessen Geld zu verdienen, eventuell sogar Chancen für ein demografiefestes, individuumzentriertes Biografiemanagement [7]?

Erwerbsarbeit wird jedenfalls unter den Bedingungen digitalisierter Wissensarbeit immer stärker mit anderen Tätigkeiten zusammenfließen. Dies bietet auch Potenziale für eine aktive Gestaltung der Erwerbsbiografie, erfordert aber gleichzeitig neue Modelle der sozialen Absicherung und Anerkennung. Nach wie vor bevorzugen Unternehmen Bewerber mit gradlinigen Verläufen, eher selten werden Ausnahmen gemacht. Bunte Karriereverläufe werden nach wie vor kaum als Chance für mehr Innovationsfähigkeit gesehen. Es gilt herauszufinden, welche Produktivitätspotenziale bei Quereinsteigern und älteren Beschäftigten vorhanden sind und produktiv genutzt werden können. Dabei müssen die Unternehmen von der gestaltungsorientierten Arbeitsforschung unterstützt werden.

Die lebensphasensensible Arbeitsgestaltung ist aus unserer Sicht ein Schlüsselfaktor für die Gesundheit und Innovationsfähigkeit der Beschäftigten. Doch dazu benötigen Unternehmen und Beschäftigte Forschungen für neue Modelle der reflexiven Gestaltung von wissensintensiver Arbeit, (privatem) Engagement und Existenzsicherung. Notwendig ist eine neue Erwerbsverlaufsforschung, die für digitalisierte Unternehmen und intermediäre Akteure Modelle einer Integration von Fremd- und Selbstbestimmung im Lebensphasenmanagement angesichts der demografischen Herausforderungen entwickelt. Benötigt werden praxisfähige Instrumente für das aktive Erwerbsbiografiemanagement und für die Integration von lebensphasenabhängigen Potenzialen in die unternehmerische Praxis.

In der Session wurden auf der Basis dieses Diskurses zahlreiche neue Forschungsfragen und Gestaltungsfelder im Handlungsfeld „Erwerbsbiografien" entwickelt, die zukünftig Einzug in die Forschungsagenda des BMBF finden sollten. Beispielhaft sind hier folgende Forschungsfragen genannt:

- Wie können die Unternehmen in einer entgrenzten Arbeitswelt in die Lage versetzt werden, lebensphasensensible Anforderungsstrukturen, Aufgabengestaltungen, Arbeitszeiten und Tätigkeitsmuster zu entwickeln?
- Welche Instrumente zur Bewertung von Biografien (für Unternehmen und Arbeitnehmer) und zur Absicherung von diskontinuierlichen Verläufen sind zu entwickeln?
- Wie können Unternehmen Mitarbeiter befähigen, sich mit diskontinuierlichen Verläufen produktiv in Unternehmen zu integrieren?
- Wie können unterschiedliche Lebensphasen als Diversity-Kategorie begriffen werden? Welche neuen Möglichkeiten ergeben sich daraus für das Personalmanagement?
- Welche Generationsspezifika lassen sich in einer digitalisierten Arbeitswelt feststellen und wie können sie produktiv gestaltet werden?
- Sind die heutigen Arbeitszeitmodelle in einer digitalisierten Wirtschaft und von dem Hintergrund des demografischen Wandels noch zeitgemäß?

3 Fazit/Ausblick

Die Session machte deutlich, dass die Gestaltung der Erwerbsbiografie in Unternehmen nicht nur eine zentrale Herausforderung der Arbeitsforschung und der unternehmerischen Praxis ist, sondern auch eine wesentliche Stellschraube bei der innovations- und gesundheitsförderlichen Arbeitsgestaltung in der digitalen Ökonomie.

Die Ergebnisse der Session zeigen, dass es noch weiterer Anstrengungen der Forschung, intermediärer Institutionen, der Unternehmen und der Beschäftigten selbst bedarf, um die Chancen und Risiken diskontinuierlicher erwerbsbiografischer Verläufe für mobile Arbeitsformen, für neue Kollaborationstechnologien und für dezentrale Organisationsmodelle zu untersuchen und zu gestalten. Notwendig ist die Erforschung und Entwicklung von Unternehmensstrategien für gelungene erwerbsbiografische Verläufe in einer digitalen Arbeitswelt, die betriebliche Innovationsfähigkeit wie individuelle Existenzmöglichkeiten sichert.

Literaturverzeichnis

1. K.G. Ciesinger, R. Klatt, D. Siebecke, Janusköpfige Wissensarbeit. Die zwei Seiten neuer Arbeitsformen in der IT-Branche. Journal Arbeit 1(10), 2011, pp. 6–7
2. H. Keupp, H. Dill, eds., *Erschöpfende Arbeit - Gesundheit und Prävention in der flexiblen Arbeitswelt.* Transcript, Bielefeld, 2010
3. R. Klatt, Work-Life-Balance - Der belastende Spagat für Unternehmen und Beschäftigte. præview - Zeitschrift für innovative Arbeitsgestaltung und Prävention (2), 2010

4. R. Castel, K. Dörre, *Prekarität, Abstieg, Ausgrenzung: Die soziale Frage am Beginn des 21. Jahrhunderts*. Campus Verlag, 2009
5. H. Friebe, S. Lobo, *Wir nennen es Arbeit: die digitale Boheme oder: intelligentes Leben jenseits der Festanstellung*. Heyne, München, 2008
6. R. Klatt, S. Steinberg, „Mobile Worker" in Europa: Erschöpfte Wanderer zwischen den Welten. præview – Zeitschrift für innovative Arbeitsgestaltung und Prävention (2), 2010, pp. 22–23
7. R. Klatt, K.G. Ciesinger, Wege zu einem aktiven Erwerbsbiografie-Management in Unternehmen: Praktische Ansätze und internationale Erfahrungen. In: *Innovationsfähigkeit im demografischen Wandel: Beiträge der Demografietagung des BMBF im Wissenschaftsjahr 2013*, ed. by S. Jeschke, Campus Verlag, Frankfurt am Main, New York, 2013, pp. 79–86

Managing Interdisciplinary Research Clusters

Sarah L. Müller, Thomas Thiele, Claudia Jooß, Anja Richert,
René Vossen, Ingrid Isenhardt and Sabina Jeschke

Abstract The complexity and dynamicity of interdisciplinary research clusters requires an efficient management in order to ensure a good performance. Therefore, this paper presents an iterative regulatory process for managing interdisciplinary research clusters which has been implemented at the two clusters of excellence at RWTH Aachen University. Thereby, an annual evaluation of the cluster performance through a cluster-specific employee survey forms the basis for the derivation and implementation of different measures. By evaluating the performance again, the loop of organizational learning starts anew. As one example of these measures, the colloquia of employees are described in further details because the adaptation and optimization of concepts illustrate the effects of the continuous improvement. Furthermore, the general performance development of the cluster was analyzed to show the comprehensive effects of the regulatory process.

Keywords Interdisciplinarity · Clusters of Excellence · Cluster-specific Balanced Scorecard · Colloquia of Employees · Organizational Learning · Learning Loop

1 Introduction

Science is claimed to be relevant for society, but an asymmetry between the kind of scientific, economical and societal problems and the problem solutions of single disciplines can be observed [1]. Interdisciplinary research gives an answer to this disbalance through the integration of "information, data, techniques, tools, perspectives, concepts, and/or theories from two or more disciplines (...) to solve problems which solutions are beyond of the scope of a single discipline" [2]. Therefore, it has become very popular over the last years [3]. As part of the nationwide excellence initiative of the federal government of Germany the *Cluster of Excellence "Integrative*

S.L. Müller (✉) · T. Thiele · C. Jooß · A. Richert · R. Vossen · I. Isenhardt · S. Jeschke
IMA/ZLW & IfU, RWTH Aachen University, Dennewartstr. 27,
52068 Aachen, Germany
e-mail: sarah.mueller@ima-zlw-ifu.rwth-aachen.de

Originally published in "Proceedings of the 22nd International Conference
on Industrial Engineering and Engineering Management 2015",
© IEEE 2016. Reprint by Springer International Publishing Switzerland 2016,
DOI 10.1007/978-3-319-42620-4_11

Production Technology for High-Wage countries"[1] (CoE 1) and the *Cluster of Excellence "Tailor-Made Fuels from Biomass"*[2] (CoE 2) were initiated at RWTH Aachen University. Although there is no uniform definition of a cluster of excellence [4], it can be defined as (local) bundled activities of groups of actors with the aim at gaining a competitive advantage [5]. A cluster is in general characterized by an increased density of relational structures, and with that, network adequate processes [6]. Therefore, the German CoE can be compared to the Industry/University Cooperative Research Centers in the US [7]. CoE 1 comprises scientists from more than twenty five institutes of RWTH Aachen University from the areas of production, material sciences, economics as well as natural and social sciences. The research aims at developing sustainable solutions for production in the future in cooperation with companies[3] of different branches [9]. CoE 2 involves more than twenty RWTH Aachen University institutes working in the fields of chemistry, bio-technology, process engineering, and mechanical engineering. Their goal is to determine the optimal combination of fuel components based on renewable materials and their production processes [10]. Both clusters bundle their focus research activities in research fields in which various disciplines work together to answer research questions which require the methods and theories of more than a single discipline. Therefore, actors with different backgrounds have to integrate their theoretical and practical knowledge to achieve a good performance. The cooperation between different disciplines can be challenging because researchers prone to see problems, its causes, consequences and solutions only out of their disciplinary perspective [11]. This is why it is crucial to sensitize for the other perspectives and to create a joint understanding of terminologies, methods, and theories. In addition to content- and method-based challenges, the organizational structures of research clusters have to be taken into account: the heterogeneity of organizational cultures of the involved institutes, the different hierarchical levels in a cluster, the range of age between the researchers, the surplus of men and the employee fluctuation in scientific communities which is quite high [9]. Thus, the work processes have to be managed continuously to achieve a good cluster performance. That is why both CoE have a cross sectional project [12], which is called *Cross Sectional Processes or Supplementary Cluster Activities*, respectively. These projects are working in close consultation with the cluster management on a sustainable personal and organizational development supporting the process of interdisciplinary cooperation. The cluster-specific management approach as well as the performance measurement tool will be introduced in Section 2. Section 3 follows with an overview of permanently implemented measures which counteract the described challenges of the clusters. Hereby, the colloquia of employees are described exemplarily in further detail. In Section 4 the outcomes of the entire management approach are described and discussed in Section 5, and outlook is given in Section 6.

---

[1] http://www.produktionstechnik.rwth-aachen.de/.

[2] http://www.fuelcenter.rwth-aachen.de/.

[3] Interdisciplinary research which is conducted together with non-scientific organizations is often called transdisciplinary [8]. Because in this study, only the researchers are objective of study, the term interdisciplinary is used.

2 Cluster Performance Management

Resulting from the described challenges, the processes and structures of both CoE are characterized through a high dynamicity and complexity [13]. Therefore, a continuous performance management, which considers the cluster-specific features, is required to evoke organizational learning processes which represent the basis for interdisciplinary cooperation and knowledge production [14–17]. Consequently, an iterative regulatory process, modelled after [18–21], was implemented to improve the cluster performance and the management itself (see Figure 1). The circulatory system starts with evaluation of the cluster performance (step 1 and 2) on an annual base. After the condensation and analysis of the data (step 3), a content-related analysis and reflection takes place for back-coupling the information (step 4). Due to the analysis and the comparison of the different questions and their development over time, positive and negative changes can be identified and reflected. The analysis of the data enables the derivation of action recommendations for the cluster management in order to improve the negative or to enhance the positive developments (step 5). Through measuring the performance again on an annual repetition, the overall effect of the (maintained, newly implemented or improved) measures can be observed and further improvements can be derived.

For measuring the cluster performance a *cluster-specific Balanced Scorecard* (BSC) was developed [22]. The BSC is an established strategy performance management tool which was originally developed by [23]. Considering the special demands of an interdisciplinary research cluster, this approach was adapted and the prototype

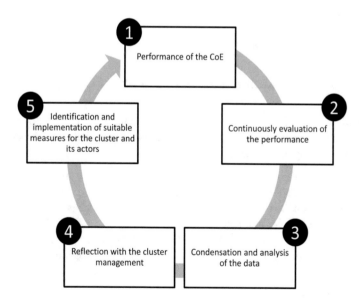

Figure 1 Iterative regulatory process of the clusters of excellence to manage the performance

was implemented in both CoE at RWTH Aachen University. The implementation of the BSC forms the basis for managing the described challenges and meets the need for developing further concepts to steer and regulate interdisciplinary cooperation which is required by the CoE sponsors, the *German Research Foundation* (DFG)[4] and the *German Council of Science and Humanity*[5] [24]. Out of the visions of the CoE at RWTH Aachen University, primary aims were derived by researchers of the cross sectional projects in cooperation with the cluster management. Equivalent to the generic basic model of the BSC, these aims were clustered within four perspectives which influence each other:

- internal perspective/research cooperation
- perspective on learning and development
- output/ client perspective and
- financial perspective.

All of these perspectives were operationalized through a scorecard-project team of the CoE in close consultation with the cluster management and representatives of the different cluster projects who derived the questions from the evaluation criteria of the DFG for interdisciplinary research clusters [25]. The actor-oriented performance of both CoE is evaluated by cluster actors on various hierarchical levels reaching from the research assistants to the cluster management. Out of this repetitive evaluation, measures can be derived to meet the current or continuous challenges. Examples for continuously implemented measures are presented in the following section. They meet in particular the need for transparency of research results and support the comprehensive communication.

3 Implemented Measures

3.1 Measures for Managing Interdisciplinary Clusters

As a result of the BSC evaluation and the back-coupling process different measures were implemented to foster interdisciplinary cooperation and communication between the heterogeneous researchers.

- General Meetings are internal meeting for the whole cluster in which the current status of the different projects is reported.
- The CoE-Conference is a cluster hosted conference and addresses the presentation and discussion of scientific results.
- The International Conferences address both international external researchers as well as non-university organizations to support research dissemination in a broader perspective.

[4]http://www.dfg.de/.

[5]http://www.wissenschaftsrat.de/home.html.

- The Scientific Advisory Board as a structural element for the steering of the CoE. This board consists of the leading professorial researchers.
- The Industrial Advisory Board can be seen as an industry interface. Scientists meet with the industry partners of the CoE to present and demonstrate the research results.
- Strategy Workshops are meetings for the higher hierarchical levels of the CoE to define and clarify the long-term strategy of the clusters (vision, mission, objectives) as well as to substantiate research questions.
- Meetings of the Core Team and Meetings of the Projects Leaders are implemented for discussing organizational and strategic questions on different hierarchical levels.
- Project Meetings are used to discuss the status of the projects in detail and to plan and manage the daily processes.
- The Virtual Cooperation Platform facilitates the sharing of information between researchers themselves and the management. Furthermore, terminologies are explained to create a common language.
- Informal Networking Events create a common identity and promote the private and professional exchange.
- Workshops and Trainings are conducted to sensitize for interdisciplinary cooperation and to develop further skills like project management and scientific presenting and writing.

Exemplarily for all of these measures, the colloquia of employees are described in further details in the next section. The colloquia were chosen because the effects of the iterative improvement circle can be seen by the change of the colloquia' concept.

3.2 Colloquia of Employees

The colloquium for employees is a meeting on researcher-level. The researchers are the largest group of actors in the clusters[6] which already makes them a group of special interest. Moreover, [13] pointed out that it is of particular importance for the quality of cooperation that all actors of an interdisciplinary research cluster participate bottom-up in the development of scientific contents and aims. Organized by the cross sectional projects upon consultation with the cluster management, this event was created to increase the information flow within the sub-projects of the clusters as well as to increase the atmosphere and the networking. The colloquia have taken place twice a year since 2008. Different methods and topics are used due to the current demands which are derived from the iterative regulatory process of the CoE: The performance of the CoE in general (step 1) is evaluated with the BSC (step 2). The results are condensed (step 3) and reflected by the cluster management and the cross sectional project, considering also the evaluation of the colloquium itself

---

[6]In CoE 1 has 66 % and in CoE 2 84.7 % of the actors are research assistances.

(step 4). Due to this reflection, the topics and methods for the following colloquium are identified and implemented (step 5). Analyzing the topics and contents of the colloquia, the development of the cluster can be illustrated and grouped into four main phases:

- Initiation: The first colloquia took place during the finding phase of the clusters. Therefore, they aimed at creating a common sense regarding a joint vision and mission of the line of research.
- Interfaces: In the following colloquia, the aim was to highlight the interfaces between the different disciplines and projects as well as to improve the interdisciplinary communication. As a result, synergies and ideas for new cooperation should be evoked.
- Learning: In the third period of colloquia, the interdisciplinary mutual learning was promoted through presentations or microteachings. The presentations give information about the different methods and contents of the diverse disciplines, so the employees get an impression of the other projects and related topics. Thus, the transparency within the CoE can be improved, researchers are sensitized for the other disciplines and new research ideas can be created through highlighting intersections between the sub-projects.
- Sustainability: Approaching the end of the second funding period, the on-going colloquia aim at developing visions for the future.

Most of the colloquia close with an informal part. The informal part, which is most often a get-together, should encourage private and professional networking. Furthermore, it supports the idea of a community. Both factors support the success of team work [26]. Besides the shifts of the contents, a transition from a top-down to a bottom-up management can be observed. In the beginning of the colloquia, the topics were predetermined by the cluster management and designed with regard to the needs of the management. This means the iterative regulatory process, in particular the reflection of the evaluation (step 4) and the derivation of changes (step 5), was first completed only by the management. By and by, the project leaders and actors of the higher level were integrated in the improvement circle. Starting from 2009, some colloquia were evaluated by the employees directly and since 2012 every colloquia has been evaluated. Consequently, the employees' influence on the topics and organization of the colloquia increased steadily.

4 Effects of Management

The cluster-specific BSC is conducted annually in winter and carried out with the web-based software "Unipark". In CoE 1, which is reported representatively for both CoE, every year between 60 and 121 participants filled in the survey.[7] To analyze the

---

[7]The variations result from the steadily decreasing participation rate from 60.5 % in 2009 to 30.0 % in 2014.

effect of the cluster management, time series of each question are plotted referring to the annual arithmetic averages. The items which are exemplarily taken into account in this paper and represent the perspective "internal perspective/ research cooperation" of the BSC are: "How do you rate the quality of existing scientific cooperation of the chairs and research institutions?" which is an indicator for the perceived quality of cooperation between the different disciplines and institutes which is a crucial challenge for both CoE as pointed out. "I am able to present the content and the scientific argumentation publicly and to embed our research into the context." was taken into analysis because it is an indicator for the understanding of the entire cluster project and thus for the interdisciplinary integration. "How do you rate the atmosphere?" was selected as an overall indicator for the personnel satisfaction with the entire situation in the CoE. All of the items are rated at three different levels: "within the whole CoE", "within your research field" and "within your project", but for further analysis these subdivisions are combined because the overall effect of the management measures should be analyzed. All of the questions were evaluated on a five point Likert-scale reaching from 1 (good rating) to 5 (bad rating).

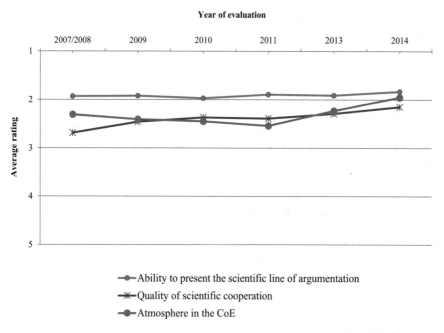

Figure 2 Time series of the results of the cluster-specific Balanced Scorecard. The scale reaches from 1 (good rating) to 5 (bad rating). Because of the transition from the first to the second funding period, now survey took place in 2012

As shown in Figure 2, all of the ratings indicate a constant or progressive development. The quality of cooperation as well as the perceived atmosphere increase during the entire period of time (lower numbers indicate better ratings), starting from $\overline{x} = 2.69$, $\sigma = 0.98$, to $\overline{x} = 2.15$, $\sigma = 0.74$, and $\overline{x} = 2.31$, $\sigma = 0.56$ to $\overline{x} = 1.95$, $\sigma = 0.55$, respectively. The ability to present the scientific line of argumentation remained nearly constantly ($\overline{x} = 1.93$, $\sigma = 0.93$ to $\overline{x} = 1.83$, $\sigma = 0.55$). The slightly increasing curves show the positive effects of the management, but it is also apparent that interdisciplinary cooperation needs time. Furthermore, cooperation processes do not take place by themselves [13], wherefore these effects result from the promotion of integration and learning processes, tolerance towards other disciplines, self-reflection and communication – especially through the colloquia.

5 Discussion

Despite of all of the challenges of both CoE, which are namely the high employee fluctuation, the integration of different disciplines and organizational structures and cultures as well as the cooperation between researchers of different age and hierarchical level, the employee survey does not show negative progressions as one can suspect. On the contrary, the quality of cooperation as well as the atmosphere shows positive progressions and the ability to present the scientific line of argumentations remains constant. Therefore, it seems likely that the derived measures of the cluster management were successful countermeasures and the continuous annual regulatory process leads to steady improvement of the interdisciplinary cooperation and identification. According to this, the introduced measures do not meet the challenges, but induce a positive process of organizational learning. The colloquia of employees as one example of these measures illustrate the ongoing change of needs of the cluster and the corresponding adaption through a change of contents and methods. Therefore, they are an adequate example for the need and the results of an iterative improvement process. Nevertheless, it has to be taken into account, that the cluster-specific BSC is mainly a management tool, wherefore data were not collected for further statistical analysis. This is why data were collected anonymously without a codification, and due to this fact it cannot be evaluated how individual persons answered over time. Therefore, an inference-statistical analysis of the data is not possible.

6 Conclusion and Outlook

For managing interdisciplinary research clusters, an iterative regulatory process is necessary to meet the dynamic and complex needs of the actors and the management. The management loop of both CoE at RWTH Aachen starts with measuring the cluster performance using a cluster-specific BSC. The results of this employee survey are analyzed and suitable measures are derived. By the subsequent periodic evaluation the circle is completed. This paper contributes to the implementation of different measures which were derived from the employee survey. The effects of these measures are analyzed by looking deeper at the time series of the quality of cooperation, the atmosphere and the ability to present the scientific line of argumentation. The time series show that these measures play a critical role for the described factors within research clusters. Due to the ongoing iterative process, further ways of supporting the actors of the clusters will be derived and implemented. One example can be seen in an online visualization of the employee survey's results which should be accessible for every researcher through the Virtual Cooperation Platform. Thus, the transparency of the employee survey gets improved to enhance the transparence of the entire management. In addition to that, the participation rate of the employee survey should be increased by illustrate the importance of the researchers' feedback. Furthermore, the virtualization of results in combination with various filtering techniques allows the individual assessment of projects. Following this idea, the visualization represents a virtual management tool, which is able to support the different hierarchical levels by fulfilling individual information needs. The iterative regulatory process as well as the cluster-specific BSC are well established in both CoE at RWTH Aachen University and consequently can be transferred to other interdisciplinary research clusters, too. However, the questionnaires used in this study were developed specifically to the needs of the particular CoE and are therefore highly individual. That is why the employee survey has to be adapted to the special needs of the concerned cluster before it can be transferred.

Acknowledgments The authors would like to thank the German Research Foundation DFG for the kind support within the Clusters of Excellence "Integrative Production Technology for High-Wage Countries" and "Tailor-Made Fuels from Biomass" at RWTH Aachen University.

References

1. J. Mittelstraß, Methodische Transdisziplinarität. Technologiefolgenabschätzung – Theorie und Praxis **2** (14), 2005, pp. 18–23. https://www.tatup-journal.de/tatup052_mitt05a.php
2. N. Academies, *Facilitating Interdisciplinary Research*. The National Academies Press, Washington, DC, 2005. http://www.nap.edu/catalog/11153/facilitating-interdisciplinary-research
3. M.M. Millar, Interdisciplinary research and the early career: The effect of interdisciplinary dissertation research on career placement and publication productivity of doctoral graduates in the sciences. Research Policy **42** (5), 2013, pp. 1152–1164. doi:10.1016/j.respol.2013.02.004. http://www.sciencedirect.com/science/article/pii/S0048733313000401
4. B. Alecke, G. Untiedt, *Zur Förderung von Clustern. „Heilsbringer" oder „Wolf im Schafspelz"?* GEFRA, Münster, 2005. http://doku.iab.de/veranstaltungen/2005/gfr_2005_alecke_untiedt.pdf

5. M. Sondermann, D. Simon, A.M. Scholz, S. Hornbostel, *Die Exzellenzinitiative: Beobachtungen aus der Implementierungsphase*. No. 5 In: iFQ-Working paper. 2008. http://www.bildungsserver.de/db/mlesen.html?Id=42076

6. R. Häußling, *Grenzen von Netzwerken*, 1st edn. VS Verlag für Sozialwissenschaften, Wiesbaden, 2009

7. C. Boardman, D. Gray, The new science and engineering management: cooperative research centers as government policies, industry strategies, and organizations. The Journal of Technology Transfer **35** (5), 2010, pp. 445–459. doi:10.1007/s10961-010-9162-y. http://link.springer.com/article/10.1007/s10961-010-9162-y

8. R. Defila, A. DiGiulio, Interdisziplinarität und Disziplinarität. In: *Zwischen den Fächern - über den Dingen?*, Leske u. Budrich, Opladen, 1998, pp. 111–137. http://sowiport.gesis.org/search/id/gesis-solis-00231872/Cite

9. C. Brecher, *Integrative Production Technology for High-Wage Countries*, 1st edn. Springer, Berlin, 2012

10. Cluster of Excellence Tailor-Made Fuels from Biomass/Maßgeschneiderte Kraftstoffe aus Biomasse: Informationsmaterial. http://www.fuelcenter.rwth-aachen.de/index.php?id=341

11. C. Pohl, G. Hadorn, Methodenentwicklung in der transdisziplinären Forschung. In: *Transdisziplinäre Forschung: Integrative Forschungsprozesse verstehen und bewerten*, ed. by M. Bergmann, E. Schramm, Campus Verlag, Frankfurt/New York, 2008, pp. 69–93

12. C. Jooß, F. Welter, A. Richert, S. Jeschke, C. Brecher, A Management approach for interdisciplinary Research networks in a knowledge-based Society – Case study of the cluster of Excellence "Integrative Production technology for high-wage countries". ICERI2010 Proceedings, 2010, pp. 138–143. http://library.iated.org/view/JOOSS2010AMA

13. C. Jooß, *Gestaltung von Kooperationsprozessen interdisziplinärer Forschungsnetzwerke*, 1st edn. Books on Demand, Norderstedt, 2014

14. J. Sydow, S. Duschek, *Management interorganisationaler Beziehungen; Netzwerke - Cluster - Allianzen*, 1st edn. Kohlhammer, Stuttgart, 2011

15. P. Pawlowsky, Betriebliche Qualifikationsstrategien und organisationales Lernen. In: *Managementforschung*, vol. 2, ed. by W. Staehle, P. Conrad, DeGruyter, Berlin, 1992, pp. 177–238

16. C. Argyris, D.A. Schon, *Organizational Learning: A Theory of Action Perspective*. Addison Wesley Longman Publishing Co, Reading, MA, 1978

17. B. Liebsch, *Phänomen Organisationales Lernen: Kompendium der Theorien individuellen, sozialen und organisationalen Lernens sowie interorganisationalen Lernens in Netzwerken*, 1st edn. Hampp, München, 2011

18. A. Neely, *Measuring Business Performance*. Bloomberg Press, London, 1998

19. C. Jansen, *Scorecard für die Wissensmanagement-Performance in heterogenen Unternehmensnetzwerken, Meß-, Steuerungs- und Regelungstechnik*, vol. 1024. VDI, Düsseldorf, 2003

20. P. Horváth, M. Seiter, Performance Measurement. In: *Die Betriebswirtschaft*, no. 3 In: 69, 2009, pp. 393–414

21. F. Welter, R. Vossen, A. Richert, I. Isenhardt, Network Management for Clusters of Excellence - A Balanced-Scorecard Approach as a Performance Measurement Tool. In: *Automation, Communication and Cybernetics in Science and Engineering 2009/2010*, ed. by S. Jeschke, I. Isenhardt, K. Henning, Springer Berlin Heidelberg, 2011, pp. 195–207. http://link.springer.com/chapter/10.1007/978-3-642-16208-4_17

22. F. Welter, *Regelung wissenschaftlicher Exzellenzcluster mittels scorecardbasierter Performancemessung*, 1st edn. Books on Demand, 2013

23. R.S. Kaplan, D. Norton, The Balanced Scorecard: Measures that Drive Performance. Harvard Business Review **70** (1), 1992, pp. 71–79. http://www.hbs.edu/faculty/Pages/item.aspx?num=9161

24. D.D.F.a. Wissenschaftsrat, *Bericht der Gemeinsamen Kommission zur Exzellenzinitiative an die Gemeinsame Wissenschaftskonferenz*. Bonn, 2008. http://www.wissenschaftsrat.de/download/archiv/exini_GWK-Bericht-%5B1%5D.pdf

25. D.D. Forschungsgemeinschaft, *1. Ausschreibung in der Exzellenzinitiative: Auswahl der Antragsteller. Bewertungskriterien für die 3. Förderlinie*. Köln, 2006. https://www.google.de/search?q=1.+Ausschreibung+in+der+Exzellenz-+initiative:+Auswahl+der+Antragsteller&ie=utf-8&oe=utf-8&gws_rd=cr&ei=mH5PVbHEEoa9swGhqICYBA
26. L. Gratton, T.J. Erickson, 8 ways to build collaborative teams. Harvard Business Review **85** (11), pp. 100–109, 153

Ein kybernetisches Modell beschaffungsinduzierter Störgrößen

Stephan Printz, Johann Philipp von Cube, René Vossen, Robert Schmitt and Sabina Jeschke

Zusammenfassung Mit der Globalisierung wächst der Kostendruck für Unternehmen. Die Spezialisierung von produzierenden Unternehmen auf einen Bereich der Wertschöpfungskette führt zu einer Reduktion der internen Wertschöpfung (BDI & Z_Punkt GmbH (Hrsg.) (2012). Deutschland 2030 - Zukunftperspektiven der Wertschöpfung.). Insbesondere die geographische Verteilung der Produktionsnetzwerke aufgrund der Globalisierung verstärkt die Abhängigkeit von Lieferanten und ausländischen Produktionsstandorten (Schatz, A., Hermann, M., & Mandel, J. (2010). Risikomanagement in der Beschaffung eingesetzte Strategien und Methoden, organisatorische Verankerung, Bedeutung und Reifegrad des Risikomanagements in der Beschaffung in der Industrie. Stuttgart: Fraunhofer Inst.). Dies führt zu länder- und branchenübergreifenden Risiken. Daher ist für produzierende Unternehmen die Identifikation und Bewertung dieser Risiken zur Erhaltung der Wettbewerbsfähigkeit erforderlich. In der Literatur existieren unterschiedliche Ansätze und Methoden für die Risikobewertung. Allerdings sind für die Anwendung dieser Methoden unterschiedliche quantitative und qualitative Informationen und Verfahren verfügbar, aus denen der Anwender eine Auswahl zu treffen hat. Im Rahmen der Risikobewertung wird eine stärkere Berücksichtigung von Unsicherheit in Risikomanagement-Modellen gefordert (Gleißner, W. (2011). Grundlagen des Risikomanagements im Unternehmen: Controlling, Unternehmensstrategie und wertorientiertes Management. Vahlen.). Dies bedingt eine Einbeziehung moderierender Effekte, was eine potentielle Reduktion von kognitiven Verzerrungen zur Folge hat. In diesem Artikel wird neben der Herleitung des Risikomanagements aus dem Informationsmanagement und ein Literaturreview bestehender Risikomanagement-

S. Printz (✉) · R. Vossen · S. Jeschke
IMA/ZLW & IfU, RWTH Aachen University, Dennewartstr. 27,
52068 Aachen, Germany
e-mail: stephan.printz@ima-zlw-ifu.rwth-aachen.de

J.P. von Cube · R. Schmitt
Fraunhofer-Institut für Produktionstechnologie (IPT),
Steinbachstraße 17, 52074 Aachen, Germany

Originally published in "Exploring Cybernetics - Kybernetik im Interdisz). Springer 2015. Reprint by Springer International Publishing Switzerland 2016, DOI 10.1007/978-3-319-42620-4_12

Modelle und Risikobewertungstechniken durchgeführt. Neben einer Definition der Begriffe Unsicherheit, Ungewissheit und Risiken, werden Risikobewertungstechniken im Hinblick auf ihre Eignung zum Einsatz in Managementmodellen nach der Systematik von Ziegenbein analysiert. Auf Basis dieser Ergebnisse wird ein kybernetisches Modell beschaffungsinduzierter Störgrößen erstellt. Dieses Modell ist durch eine Kombination bestehender Techniken und einen szenariobasierten Ansatz gekennzeichnet. Abschließend erfolgt ein Ausblick auf die notwendigen Schritte zur Entwicklung einer Software.

Schlüsselwörter Unsicherheit · Risiko · Risikomanagement · Kybernetik

1 Einleitung

Im Zuge der Globalisierung und der damit einhergehenden Dynamik der Märkte wächst die Komplexität und Unsicherheit in ökonomischen Systemen [1]. Ein Beispiel eines solchen ökonomischen Systems ist die Beschaffungsabteilung eines Unternehmens. Diese Abteilung ist aufgrund wechselnder geographischer Lieferantenstandorte und operationeller Risiken in Form interner und externer Störgrößen mit der Herausforderung der Bewertung von Unsicherheiten konfrontiert [1, 2]. Insbesondere Lieferketten unterliegen der Herausforderung der Bewertung und Erfassung komplexer, wechselseitiger Beschaffungsrisiken. Dies stellt für Entscheidungsträger sowohl Chancen (z. B. erhaltene Lieferfähigkeit trotz einer Störung) als auch Risiken dar [3]. Es besteht in diesem Zusammenhang der Bedarf nach einer Risikoanalyse des Beschaffungsprozesses zur Entscheidungsunterstützung des Managements. Diese Risikoanalyse ermöglicht eine Simulation der beschaffungsinduzierten Störgrößen.

Die Kybernetik (altgr. kybernétes: steuermännisch, Steuermannskunst) als Metawissenschaft bietet einen geeigneten Lösungsansatz. Aus dem historischen Kontext wird die Kybernetik mit biologischen Konzepten wie der Schwarmintelligenz, als Fähigkeit zur Selbstorganisation, der Autopoiesis und der Bionik in Verbindung gebracht [4]. Daher wird die Kybernetik auch als Wissenschaft der Modelle beschrieben [5]. In diesem Zusammenhang steht insbesondere der Forschungsschwerpunkt der Konzeptentwicklung zur Steuerung und Regelung komplexer, hybrider Systeme im Fokus. Die abstrakte Darstellung komplexer Zusammenhänge ermöglicht eine erkenntnistheoretische Auswertung der Modelle durch Experten. Auf Grundlage dieses kybernetischen Grundgedankens wird zur Bewältigung des komplexen Systems der Beschaffungsrisiken ein Modell zur simulativen Quantifizierung von beschaffungsinduzierten Störgrößen entwickelt.

Über das Informationsmanagement und dessen Implikationen für die Lieferkette werden der vorherrschende Informationsbedarf und resultierende Beschaffungsrisiken eines produzierenden Unternehmens verdeutlicht. Neben einer Begriffsdifferenzierung zwischen Unsicherheit, Ungewissheit und Risiko wird der Bedarf nach einem Risikomanagement-Modell aufgezeigt. Im anschließenden Teil werden bestehende Risikomanagement-Modelle hinsichtlich des Mehrwertes für Unternehmen und deren Ergebnisdimensionen nach Ziegenbein bewertet. Abschließend wird auf Basis dieser Bewertung ein kybernetisches Modells zur Quantifizierung von Risiken in komplexen Lieferketten erstellt.

2 Informationsmanagement

Informationen werden aus einer Verknüpfung von quantitativen oder qualitativen Daten in einem Kontextbezug gewonnen [6]. „Quantitative Daten besitzen einen Wertebereich und gestatten die Durchführung arithmetischer Operationen. Beispiele sind sensorisch erfasste physikalische Kenngrößen oder Unternehmenskennzahlen. " [7]. Qualitative Daten oder kategorische Daten hingegen werden durch einen nicht-metrischen Wertebereich beschrieben. Sie dienen der Gruppierung und Ordnung [8]. Ein Beispiel für qualitative Daten sind ordinale oder nominale Daten, die jeweils nur einen vergleichenden Test im Sinne von Gleichheit und Ungleichheit ermöglichen [7].

"An Daten fehlt es heute kaum in einem Unternehmen. Wir haben eher zu viel davon. Information hingegen ist noch immer Mangelware, und man kann sich nicht darauf verlassen, dass alle Manager wissen, wie man von Daten zu Informationen kommt." [9]. Informationen bezeichnen „eine gegenwarts- und praxisbezogene Mitteilung über Dinge, die uns im Augenblick zu wissen wichtig sind." [10]. Fehlendes methodisches Wissen von Managern zur Verarbeitung der vorhandenen Daten zu Informationen wird auch als interne Unsicherheit bezeichnet. Im Gegensatz dazu werden Veränderungen der Umweltzustände als externe Unsicherheiten definiert [11]. Die Einbeziehung der Unsicherheiten in den Management Prozess generiert Wettbewerbsvorteile [12, 13]. Zur Integration der Unsicherheiten werden Informationssysteme als Entscheidungsunterstützungssysteme für die Interaktion in unstrukturierten Entscheidungssituationen durch Methoden, Modelle oder Daten verwendet [14]. Ein solches Informationsmanagement-Modell ist in Abbildung 1 dargestellt (nach [6]).

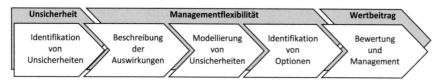

Abbildung 1 Wertbeitrag durch Managementflexibilität (nach [6])

Neben der Identifikation von Unsicherheiten ist die exakte Beschreibung und Modellierung der Auswirkungen erforderlich. Die Informationsmodellierung ist die Grundlage eines jeden Informationssystems zur Bereitstellung der gewünschten Aufgaben und Informationen. Dieser Informationsmodellierungsprozess ist in den Kontext des durchgängigen Informationsmanagement zu integrieren. Neben der Identifikation der Informations-Nutzer ist eine zusätzliche Identifikation und Sammlung von Informationen außerhalb der bestehenden Informationsstruktur erforderlich. Im anschließenden Prozessschritt sind potentielle Datenquellen zu identifizieren und in die Informationsstruktur zu integrieren. Der Zugang zu potentiellen Informationsquellen erfordert domänen-spezifische Analysen und Evaluierungsmethoden zur Anreicherung der vorhandenen Informationen. Der Fokus liegt auf der benutzerorientierten Bewertung und dem Management als Ergebnis einer Analyse aus den vorhandenen Informationen. Die Herausforderung in der Unternehmensanwendungsintegration eines solchen Managementprozesses besteht in der Datenintegration von unterschiedlichen Datenquellen in unterschiedlichen Datenformaten. Diese Interoperabilität zwischen Informationsquellen basiert auf der Integration von Informationsmodellen des gesamten Prozesses. Dies erfordert eine Formalisierung der Domäne und der gesammelten Daten [6].

Darüber hinaus besteht neben der Notwendigkeit eines durchgängigen Informationsmanagement die Herausforderung der Informationsgewinnung. Zur Generierung von Wettbewerbsvorteilen fokussieren sich Unternehmen auf die Auswertung vorhandener interner Daten [15]. Insbesondere sind in diesem Zusammenhang „Data Science", „Predictive Analysis" und „Big Data" zu nennen. Im Zuge des „Data Science" werden mathematische statistische Methoden sowie Methoden der Informatik und der Verhaltensforschung zur Gewinnung von Daten miteinander verknüpft [15]. Als Erweiterung werden durch „Predictive Analytics "Datensätze zur Prognose zukünftiger Ereignisse angewandt [16]. Während „Big Data" ursprünglich durch die Verwendung einer großen Anzahl von Datensätzen mit einem großen Volumen gekennzeichnet ist, wird heute auch die Geschwindigkeit und die Art der vorliegenden Daten in der Analyse berücksichtigt [17]. Aufgrund der Ausbreitung dieser neuen Ansätze im Umgang mit Unsicherheiten im Management von Risiken und der teilweise fehlerhaft zur Verfügung stehenden Daten, nimmt der Stellenwert der Informationsqualität bezüglich der Lieferkette stetig zu [15, 18]. Die einhergehenden Herausforderungen der Informationsbeschaffung und des durchgängigen Informationsmanagement werden anhand eines Fallbeispiels aus der Beschaffung verdeutlicht.

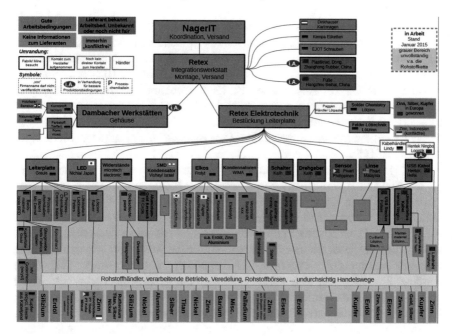

Abbildung 2 Lieferkette der *Nager IT* [59]

3 Implikationen für die Lieferanten

Als ein Beispiel für eine Lieferkette ist die Produktion einer Computermaus der Fair Trade Organisation *NAGER IT e.V., 2011, Bichl* in Abbildung 2 dargestellt. Dieses vergleichsweise kleine Produkt benötigt für die Produktion neben Zulieferern aus Deutschland Rohstoffe und Komponenten aus zehn weiteren Nationen.

Diese Nationen sind Österreich, die Benelux Staaten, Indonesien, Japan, Israel, China, die Philippinen und Malaysia. Sowohl die geographische Streuung als auch die Abhängigkeit von Rohstoffen und Systemlieferanten, wie beispielsweise Dambacher Werkstätten oder Retex Elektrotechnik, verdeutlichen die Herausforderungen für die Beschaffung. Diese Herausforderungen sind neben der Erfassung der Lieferkettenstruktur und der jeweiligen Unternehmensprozesse die Identifikation und das Management der unterschiedlichen Risiken [19, 20]. Insbesondere besteht der Bedarf nach einem Quantifizierung und Modellierung von Beschaffungsrisiken [1]. Bisherige Versuche der Implementierung eines Risikomanagements in produzierenden Unternehmen scheiterten entweder am Zugang zu Informationen, am Wissen über die Verarbeitung der Informationen oder an unpräzisen Informationssystemen [21]. Zur Ableitung von Gegenmaßnahmen sind vollständige Informationen über alle potentiellen Risiken erforderlich [22].

Zum aktuellen Zeitpunkt werden zum Informationsaustausch computer-gestützte Technologien und Enterprise Resource Planning (ERP) Systeme mit offenen Datenformaten und -strukturen als Datenquellen angewendet [23]. In der Praxis scheitert die Implementierung von Risikomanagement-Modellen an nicht definierten Standards zum Informationsaustausch, der Bereitschaft der Manager zur Preisgabe von unternehmensinternen Daten sowie der Komplexität der Risikomanagement-Modelle selbst [24–26]. Neben diesen Faktoren ist ebenfalls die Informationsgewinnung und deren Qualität eine zentrale Herausforderung für das Management von Risiken [27]. Insbesondere die Ergänzung bestehender Risikomanagement-Modelle durch die aufkommenden Ansätze des „Data Science" bietet an dieser Stelle einen Ansatzpunkt [15].

Im Zuge des Risikomanagement von komplexen Lieferketten wird Data Science als die Anwendung quantitativer und qualitativer Methoden aus unterschiedlichen wissenschaftlichen Disziplinen definiert. Insbesondere die Einbeziehung bekannter Lösungsansätze und die Generierung von Prognosen unter der Verwendung der Attribute Datenqualität und -verfügbarkeit liefern einen Mehrwert [15]. Als Erweiterung des Data Science liefert der „Predictive Analytics" Ansatz Prognosen unter Einbeziehung der Prozesse und Zielgrößen im Unternehmen [28]. Diese Ansätze führen zu einer Reduktion der Prognoseunsicherheiten und zur Erhöhung der Ergebnisvalidität. In diesem Kontext ist eine Definition der Begriffe Unsicherheit, Ungewissheit und Risiko erforderlich. Risiko beschreibt nach ISO 31000 [58] die Ungewissheit bezüglich der Zielerreichung und wird als eine Teilmenge aus messbaren quantitativen Größen und nicht-messbaren qualitativen Größen beschrieben [29, 30]. Im Rahmen der Entscheidungstheorie wird der Unsicherheitsbegriff in zwei Aspekte gegliedert. Es wird zwischen Entscheidungen unter Risiko und Entscheidungen unter Ungewissheit differenziert [29, 31, 32]. Dieser Zusammenhang ist in Abbildung 3 dargestellt. Risiken entstehen durch Unsicherheiten über zukünftige unerwartete Ereignisse [33, 34]. Der Zustand Unsicherheit ist dementsprechend durch nicht-mess- und nicht-quantifizierbare Größen gekennzeichnet [30]. Diese Unsicherheiten werden in zwei Kategorien unterteilt. Zum einen endogene Unsicherheiten, welche im Unternehmen existieren und durch gezieltes Management verringert werden können. Und zum anderen exogene Unsicherheiten, welche außerhalb des Einflussbereiches von Unternehmen liegen [2, 35].

Entscheidungen unter Risiko werden durch die Beschreibung potentieller Umweltzustände mittels Eintrittswahrscheinlichkeiten charakterisiert. Im Gegensatz dazu ist eine Zuordnung von Eintrittswahrscheinlichkeiten bei Entscheidungen unter Ungewissheit nicht möglich [32]. Unter Verwendung dieser Definitionen ist jede Entscheidungssituation im unternehmerischen Umfeld mit Unsicherheiten belegt [2]. Eine Entscheidung unter Unsicherheit liegt dann vor, wenn eine Entscheidung nicht zu einem einzigen vorhersehbaren Ergebnis führt, sondern wenn keine vollständige Information der Folgen der zur Auswahl stehenden Alternative vorhanden ist [36]. In Bezug auf den Entscheidungsprozess werden Unsicherheiten als verbleibendes Defizit nach der Informationsbeschaffung und -verarbeitung definiert [37]. Insbesondere die Informationsverarbeitung als rationaler Prozess wird hierbei als entscheidende Größe zur Reduzierung von Unsicherheiten angesehen [38, 39].

Aufgrund unterschiedlicher Interpretation und Definitionen des Risikobegriffes in der Literatur, wird in diesem Beitrag eine Definition der Beschaffungsrisiken aus

Abbildung 3 Risiko und
Ungewissheit als Konzept
der Unsicherheit

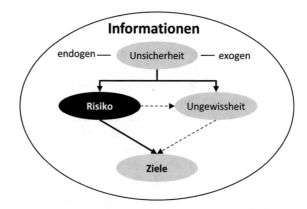

Tabelle 1 Klassifizierung von Beschaffungsrisiken [3, 41]

Risikoklasse	Beispielhaftes Risiko	Gegenmaßnahme	Informationsbedarf aller Risikoklassen
Umwelt	Politik, Wetter, Regularien	Rücksprache mit Vertretern, Versicherung, Pufferzeiten vergrößern zur Ableitung von weiteren spezifischen Gegenmaßnahmen.	• Entdeckungszeitpunkt • Ursache • Eintrittswahrscheinlichkeit
Lieferung	Bearbeitungszeit, Qualität, Transport	Pufferbestände, Lieferanten-Audits, Konventionalstrafen	• Schadensausmaß,
Prozessrisiken	Herstellungsrate, Kapazität, Informationsverzögerung	Nachplanung der Produktion, Nutzung von Vertragspartner, Pufferzeit vergrößern	• aggregiertes Gesamt-Risiko • alternative Strategien • gewählte Strategie
Nachfrage	Vorhersagefehler, Verzögerung, Kunden-Preis	Sicherheitszuschlag, Neuplanung, Sonderfreigabe	• und die Gesamt-Wirkung

unterschiedlichen Quellen hergeleitet [1]. Im Prozess der Beschaffung wird eine Differenzierung in Nachfrage- und Lieferunsicherheiten vorgenommen [40]. Die Reduzierung dieser Unsicherheiten führt zu einer ökonomischen Wertsteigerung [2]. Daher ist eine kontinuierliche Beobachtung und Bewältigung der Risiken unter Berücksichtigung ihrer Kosten- und Nutzenfaktoren erforderlich [11]. Im Kontext der Beschaffung werden die zu betrachtenden Risiken in Klassen untergliedert. Diese Klassen und entsprechende Beispiele mit potentiellen Gegenmaßnahmen und dem entsprechenden Informationsbedarf sind in Tabelle 1 dargestellt.

4 Risikomanagement

Die ISO 31000 bietet Anwendern einen systematischen Rahmen zum Management von Risiken. Der Prozess des systematischen Managements von Risiken nach ISO 31000 ist in Abbildung 4 dargestellt. Dieser Prozess beruht auf einer Kombination von faktischen und prognostischen Informationen.

Als elementare Prozesse des Risikomanagements werden sowohl die Kommunikation und Beratung als auch die Überwachung und Kontrolle von Risiken im Sinne eines kontinuierlichen Verbesserungsprozesses angesehen. In einem ersten Schritt wird das zu betrachtende System festgelegt. Im Zuge der Risikobeurteilung wird eine Differenzierung in die Schritte der Risikoidentifikation, der Risikoanalyse und der Risikobewertung vorgenommen. Der erste Schritt der Risikoidentifikation dient neben der Erfassung interner Risiken auch der Betrachtung externer Risiken, wie beispielsweise Störungen in der Lieferkette. Störungen entstehen aufgrund von sozialpolitischen, ökonomischen, technologischen oder geographischen Gegebenheiten [20] Die Risiken der einzelnen Risikoklassen sind im Folgeschritt der Risikoanalyse auf die Wirkung für das Unternehmen zu erfassen und zu bewerten.

Im Prozess der Risikobewertung wird einem Risiko, basierend auf dessen Analyse, ein Wert zugewiesen. Mittels welcher Technik Risiken zu bewerten sind, hängt von den Rahmenbedingungen des Bewertungsobjektes ab. Das Dilemma der Risikobewertung liegt in dem Zielkonflikt der qualitativen Informations- und quantitativen Bewertungsgrundlage [42]. Generell ist ein Einklang zwischen dem Unternehmensziel und der festgelegten Risikostrategie durch das Risikomanagement und die entsprechende Bewertung herzustellen [6]. Die Bewertung von Risiken erfolgt anhand der Informationen zur Eintrittswahrscheinlichkeit und potentiellem Schadensausmaß [41]. Die zu verwendende Technik ist abhängig vom Entscheider und der vorliegenden Daten bzw. Informationsqualität. Letztlich unterliegen Unternehmen auf externer Seite gesetzlichen und vertraglichen Regularien [43]. Die Rahmenbedingungen der Risikobewertung lassen sich auf vier

Abbildung 4 Risiko-
management Prozess (ISO
31000]

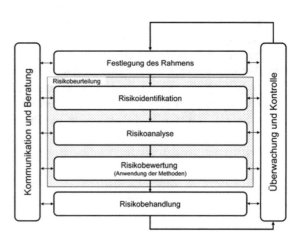

Kerneigenschaften herunterbrechen, die bei der Auswahl der geeigneten Technik zu berücksichtigen sind [43]:

- Komplexität des Bewertungsobjektes
- Unsicherheit der verfügbaren Informationen
- Verfügbarkeit von Ressourcen
- Quantitatives Ergebnis

In der IEC 31010 [43] sind Bewertungstechniken zusammengefasst und bewertend verglichen [43]. In der Praxis wird der Fundus der verschiedenen Bewertungstechniken allerdings nicht ausgeschöpft. Gut 90 % der Unternehmen setzt auf Risikoportfolios, Teamabsprache, die Fehlermöglichkeits- und -einflussanalyse (FMEA) oder Experteninterviews. In der Risikoanalyse herrschen qualitative Ansätze, wie beispielsweise die FMEA, Brainstormings, Ishikawa-Diagramme und SWOT-Analysen vor, wobei die Fehlerbaumanalyse als Technik der quantitativen Ursachenanalyse herangezogen wird [44]. Dieser Umstand ist maßgeblich der Unsicherheit von verfügbaren Informationen und dem vergleichsweise hohen Aufwand geschuldet, quantitative Techniken gewinnbringend einzusetzen. Letztere Ursache ist vorwiegend auf die unzureichende Verfügbarkeit von Ressourcen zurückzuführen.

Neben der Akzeptanz bestimmter Bewertungstechniken ist ein weiterer Punkt die Eignung der Techniken für die Bewertung von Risiken im Kontext der Supply Chain. In Tabelle 2 sind Techniken zur Bewertung von Risiken der Supply Chain zusammenfassend dargestellt [45].

Die Ergebnisse der Risikobewertung werden anhand zwei unterschiedlicher Datentypen klassifiziert. Qualitative Ergebnisse sind beispielsweise eine wertende Aussage auf einer Ordinalskala bezüglich der Wahrscheinlichkeits- und Schadenshöhe [42]. Quantitative Ergebnisse sind im Gegensatz durch messbare Werte zu Eintrittschancen und dem korrespondierenden (monetären) Ausmaß der Schäden charakterisiert [42]. Die Bewertung erfolgt anhand einer Literaturstudie durch Ziegenbein und berücksichtigt die Möglichkeiten zur Bestimmung von Eintrittswahrscheinlichkeit und das Schadensausmaß. Sowohl für das Risikomanagement als auch für die Unternehmensführung ist die Formulierung von spezifischen, messbaren, aktionsorientierten, realisierbaren und terminierten (SMART) Zielen erforderlich [46]. Im Sinne der Messbarkeit von Risiken sind quantitative Verfahren zur Bewertung von Risiken zu präferieren. Allerdings stellt der sowohl technische als auch personelle Aufwand eine Herausforderung bei der Implementierung eines quantitativen Verfahrens dar. Im Vergleich zu den quantitativen Verfahren sind qualitative Verfahren in der Regel schneller anzuwenden.

Sowohl die von Ziegenbein aufgeführte Expertenschätzung als auch die FMEA sind in der Beschaffung verbreitete Techniken. Bei der Fehlerbaumanalyse (FTA) sowie der Ereignisbaumanalyse (ETA) handelt es sich um Techniken, die sowohl zur quantitativen als auch zur qualitativen Bewertung herangezogen werden können. Hierbei ist die FTA zur Bestimmung der Eintrittswahrscheinlichkeiten geeignet, wohingegen die ETA zur Schadensbestimmung verwendet wird. Zur quantitativen Bewertung werden statistische Risiko-Datenbanken oder Simulationsmodelle

Tabelle 2 Auszug zu Techniken der Risikobewertung [45]

Ausgewählte Techniken zur Risikobewertung	Charakteristiken der Bewertung			weitere Bemerkungen
	Eintrittswahr-scheinlichkeit	Schadens-ausmaß	qualitativ, quantitativ	
Expertenschätzung	••	••	qualitativ	effektive und effiziente, aber stellenweise unpräzise Bewertung
Simulationsmodelle	••	••••	meist quantitativ	präzise Bewertung, aber großer zeitlicher Aufwand und Modellierungs-erfahrung notwendig
Risiko-Datenbanken, statistische Auswertungen	•••	•	quantitativ	Exakte Bewertung, aber nur für wenige Supply Chain Risiken geeignet
Fehlermöglichkeits- und Einflussanalyse (FMEA)	•••	•••	meist qualitativ	systematisches Vorgehen, Bewertung teilweise unpräzise
Fehlerbaumanalyse (FTA)	••••	•	quantitativ und qualitativ	aufwendige Technik, aber sehr gut geeignet zur Bestimmung der Eintrittswahrscheinlichkeit
Ereignisbaumanalyse (ETA)	•	•••	quantitativ und qualitativ	Technik erfordert Detailkenntnisse, aber sehr gut geeignet, um zeitliche Reaktionsketten und ihre Folgen zu analysieren

Legende: ° nicht geeignet, • kaum geeignet, •• teilweise geeignet, ••• größtenteils geeignet, •••• vollständig geeignet

verwendet. In den Datenbanken sind dabei die statistischen Eintrittswahrschein-lichkeiten zu bereits erfassten Risiken hinterlegt. Die resultierenden Schadensaus-maße sind individuell vom betroffenen Unternehmen und dem betrachteten Fall abhängig. Zur Bestimmung der Schadensausmaße werden Simulationsmodelle ver-wendet, welche anhand variabler Parameter fallspezifische Schäden innerhalb eines Toleranzbereiches monetär berechnen. Gängige Ansätze des Supply-Chain-Risiko-Managements sind in Tabelle 3 dargestellt und im Hinblick auf den Mehrwert für das Unternehmen und deren Ergebnisdimension (quantitativ, qualitativ) bewertet. Der Mehrwert wird anhand folgender Faktoren beurteilt: Einfluss auf das opera-tive Geschäftsergebnis, Einschätzung der Eintrittswahrscheinlichkeit, Objektivität der Ergebnisse, Anwendbarkeit und notwendige Erfahrung, Komplexität des Modells und Entscheidungsunterstützung im Hinblick auf zukünftige Maßnahmen zur Stabilisierung der Supply Chain [47].

Die beschriebenen Ansätze von Norrman und Jansson sowie von Zsidisin et al. aus dem Jahr 2004 stellen qualitative Vorgehensweisen dar. Zur Erreichung einer messbaren Größe ist eine Kombination mit quantitativen Verfahren anzuwenden. Die Probabilistic Risk Analysis (PRA) von Deleris et al. ebenfalls aus 2004 bietet quan-titative Ergebnisse, jedoch basieren diese auf Wahrscheinlichkeitsverteilungen und potentiellen Auswirkungen auf die Leistungsindikatoren als Eingangsparameter. Die Cranfield School of Management stellt ein Vorgehen zur Identifizierung und Bewer-

Tabelle 3 Literaturreview Modelle des Risikomanagement in der Beschaffung [47]

Referenz	Beschreibung	Vor- und Nachteile
Norman und Jansson, 2004	Die Autoren beschreiben qualitativ die Kernaspekte von *Ericsson's supply chain risk management* Ansatz. Risikoquellen sind auf Lieferantenbasis klassifiziert und anhand ihrer Business Recovery Time im Fall einer Störung bemessen. Zudem wird die Eintrittswahrscheinlichkeit mit Hilfe der Fehlerbaumanalyse berechnet. Die Zahlen werden auf Unternehmensebene in den Business Interruption Value zusammengeführt, der nach Eintritt eines unerwünschten Ereignisses als wirtschaftlicher Schaden für den Kernbetrieb erachtet wird.	Vorteil: Überblick über die Einzelheiten der wirtschaftlichen Schäden nach einer Versorgungsstörung. Nachteil: Die Ergebnisse sind qualitativ. Weitere Literatur (z.B. Produktionsplanung, Kostenrechnung) ist nötig, um ein quantitatives Modell abzuleiten.
Zsidisin et al. 2004	Zsidisin et al. legen empirische Ergebnisse einer Untersuchung bisheriger Versorgungsstörungen vor. Als Ergebnis zeigen sie Vorgehensweisen zur Reduzierung von Supply Chain Risiken auf. Das anwendungsbezogene Maß ist die Auswirkung auf das EBIT.	Vorteil: Überblick über die Einzelheiten der wirtschaftlichen Schäden nach einer Versorgungsstörung. Nachteil: Die Ergebnisse sind qualitativ und nicht unabhängig messbar.
Deleris et al. 2004	Das Modell zur Bestimmung der Supply Chain Risiken ist abgeleitet aus *Engineering probabilistic risk analysis (PRA)*. Die Auswirkung auf die Supply Chain Leistungsindikatoren kann für unterschiedliche Szenarien simuliert werden.	Vorteil: Das angewandte Modell ist quantitativ und durch jeden Anwender reproduzierbar. Nachteil: Der Ansatz benötigt Wahrscheinlichkeitsverteilungen und potentielle Auswirkungen auf die Leistungsindikatoren als Eingangsparameter.
Cranfield School of Management, 2003	Der Autor stellt ein Vorgehen zur Identifizierung und Bewertung von Risiken in Supply Chains in einem *self-assessment workbook* dar. Die Risikoaufdeckung wird konkretisiert durch die Total Costs of Risk (TRC), welche das Ergebnis der kombinierten Ordnungszahlen der einzelnen Risikoquellen darstellen.	Vorteil: Die Methode ist direkt in der Praxis anwendbar. Nachteil: Die Rangskala ermöglicht unpräzise Bewertung der Relevanz einer Risikoquelle.

(continued)

Tabelle 3 (continued)

Referenz	Beschreibung	Vor- und Nachteile
Jegliche TQM Literatur	*Failure modes and effects analysis (FMEA)* wird üblicherweise im TQM genutzt und bietet einen Ansatz zur Identifizierung potentieller Ursachen von Systemausfällen. Nach der Auswahl potentieller Ereignisse, die zu einem Systemausfall führen könnten, werden Präventionsstrategien vorgeschlagen und anhand eines Vorher-Nachher-Vergleiches der Risikoerkennung bewertet. Das eingeführte Risikomaß ist die Risk Priority Number (RPN), welche die Fehlereintrittswahrscheinlichkeit (O), die Fehlerentdeckungswahrscheinlichkeit (D) und die Gewichtung (S) enthält. Die RPN ergibt sich aus dem Produkt dieser drei Variablen, welche jeweils einen Wert auf einer Skala von 1 bis 10 besitzen. Demnach liegt der Ergebnisbereich für die RPN zwischen 1 und 1000.	Vorteil: FMEA ist eine strukturierte Methode zur Beurteilung des Risikos potentieller Systemausfälle. Die Wareneingangslogistik bzw. die gesamte Supply Chain eines Unternehmens stellt ein System dar, für das die zugehörige Risikoabschätzung berechnet werden kann. Nachteil: Die RPN bietet eine Einstufung der Risikoquellen.
Kmenta und Kosuke 2000	Aufgrund der Nachteile der FMEA wurde die *Scenario Based FMEA* entwickelt. In diesem Modell bestehen Szenarien aus abhängigen Ereignisketten aus Ursache und Wirkung. Für jedes Szenario muss die Wahrscheinlichkeit und die Auswirkung auf Erträge bewertet werden. Das empfohlene Risikomaß ist der Erwartungswert der Gesamtkosten aller Szenarien. Die Wahrscheinlichkeit eines Szenarios kann in die Wahrscheinlichkeit einer Ursache und ihrer möglichen Auswirkungen unterteilt werden. Die Kosten sind hierbei ein Maß für die finanziellen Auswirkungen.	Vorteil: Die szenariobasierte FMEA reduziert die dargestellten Nachteile der konventionellen FMEA durch den Einsatz von Wahrscheinlichkeiten und Kosten als Risikomaß. Nachteil: Die Wahrscheinlichkeiten müssen basierend auf historischen Daten bewertet werden. Jedoch liegen für schwerwiegende Versorgungsausfälle vermutlich keine Daten vor.
Modarres 1993	Die *Fault Tree Analysis (FTA)* findet üblicherweise in der Bewertung technischer Risiken Anwendung. Die Methode beginnt mit der Untersuchung potentieller Fehlerursachen im Bezugssystem. Im nächsten Schritt wird der Fehlerbaum konstruiert und die Eintrittswahrscheinlichkeiten der hochrangigen Ereignisse werden bewertet, basierend auf den Wahrscheinlichkeiten der Ursachen.	Vorteil: Das Ziel der Bewertung von Wahrscheinlichkeiten für unerwünschte Ereignisse wird bewerkstelligt durch die Zuschreibung des gesamten Ausfallrisikos zu Ursachen grundlegender Ereignisse. Das Verfahren ist objektiv und reduziert die Komplexität, da nur Ursachen betrachtet werden, die hochrangige Ereignisse auslösen könnten. FTA effektiv und intuitiv durchführbar und ist sind nur grundlegende Statistiken nötig. Nachteil: Keine.

tung von Risiken in Supply Chains in einem self-assessment workbook dar. Dieses Verfahren liefert aufgrund ihrer Ergebnisauslegung in Form einer Rangskala eine eingeschränkte Bewertung der Risikorelevanz. Eine Optimierung dieser Eigenschaft wird durch die szenariobasierte FMEA geboten. Die szenariobasierte FMEA verwendet Wahrscheinlichkeiten und Kosten als Risikomaß. Die Abschätzung dieser Wahrscheinlichkeiten und Kosten basiert auf einer Analyse historischer Datensätze, wobei die Verfügbarkeit umfangreicher historischer Datensätze diesen Ansatz limitiert. Bei der von Modarres beschriebenen Fehlerbaumanalyse (FTA) wird ein quantitatives Ergebnis bezüglich der Eintrittswahrscheinlichkeiten generiert. Allerdings verfügt dieses Verfahren nicht über die Möglichkeit, das monetäre Schadensausmaß zu quantifizieren.

Bestehende Simulationsmodelle, wie z. B. von Deleris beschrieben, vernachlässigen allerdings die Herausforderungen der Praxis: Verfügbarkeit valider Informationen, Verfügbarkeit von Ressourcen und einfache/anwendungsorientierte Abbildung komplexer Bewertungsobjekte. Die Simulationen stützen sich dabei auf eine Basis an Informationen bezüglich des Supply Chain Netzwerkes, des Produktes und der möglichen Risiken. Die Zusammenhänge dieser Aspekte werden anschließend in entsprechenden Modellen, wie zum Beispiel dem General Semi Markov Process (GSMP) (Deleris et al. 2004), aufbereitet und nachfolgend mittels Monte Carlo Simulation berechnet. Als Resultat entsteht eine Risikokurve, welche die monetären Schäden anhand ihrer Eintrittswahrscheinlichkeiten abbildet. Im Rahmen des Beschaffungsprozesses werden die Risiken jedoch ohne eine Einbeziehung eventueller Kontroll- oder Gegenmaßnahmen zur ihrer eventuellen Behandlung simuliert. Diese fehlende Modellierung verwehrt eine Abbildung und Bewertung der Auswirkungen solcher Maßnahmen des Risikomanagements. Folglich dienen die beschriebenen Modelle der Identifizierung potentieller Vorgehensweisen zur Risikominimierung, jedoch nicht ihrer Überprüfung. Neben der Vernachlässigung der Risikobehandlung werden auch Vereinfachungen im Bereich der Risikoauswirkungen getroffen. In den von Deleris beschriebenen Modellen wird davon ausgegangen, dass sich Fehler in der Produktion bzw. nicht produzierte Teile direkt auf die Verkaufserlöse auswirken. Mögliche Puffer aufgrund von Lagerbeständen oder verzögerten Fehlmengenpunkten bleiben unberücksichtigt (Deleris et al. 2004; Deleris und Erhun 2005). Diese Kritikpunkte an den bestehenden Modellen erfordern eine Anpassung bestehender Modelle.

5 Kritik an den bestehenden Techniken und Modellen zur Risikobewertung

Weder die von Ziegenbein untersuchten Modelle zum Risikomanagement von Supply Chains noch die in der ISO 31010 gelisteten Techniken gewährleisten, sowohl Eintrittswahrscheinlichkeiten als auch den Schaden quantitativ mit geringem Aufwand zu erfassen. Die Kombination von bestehenden Methoden zur Risikobewertung und die Einbindung in ein Simulationsmodell sind ein potentieller Ansatz, unterschiedliche Szenarien quantitativ im Hinblick auf Risikoeintrittswahrscheinlichkeit und Schadensausmaß zu bewerten [41]. Zum aktuellen Zeitpunkt ist den Autoren kein Ansatz zum Management von Beschaffungsrisiken bekannt, der sowohl eine Simulation der Störgrößen als auch eine Simulation des Schadensausmaßes in der Produktion ermöglicht. Tabelle 4 stellt einen direkten Vergleich zwischen den Risikobewertungstechniken und den Anforderungen an ein Risikomanagement-Modell dar. [45, 48] Die Bewertungsdimensionen sind in Anlehnung an [45, 48] gewählt worden.

Kriterien für die Bewertung sind unter anderem der administrative Aufwand, die Menge an benötigten Informationen, die Genauigkeit und Vollständigkeit der Ergebnisse sowie die Fähigkeit, direkte bzw. Folgeschäden und ihre Eintrittswahrscheinlichkeiten zu erfassen. Dabei verfügen die unterschiedlichen Techniken über unter-

Tabelle 4 Anforderungen und Techniken des Risikomanagements [45, 48]

Fortlaufende Nummerierung der Techniken	1 Expertenschätzung	2 Simulationsmodelle	3 Risiko-Datenbanken, statistische Auswertungen	4 Fehlermöglichkeits-und Einflussanalyse (FMEA)	5 Fehlerbaumanalyse (FTA)	6 Ereignisbaumanalyse (ETA)
administrativer Aufwand	•••	•	••••	••	••	•
Informationsbedarf	••	•••	•••	•••	••••	•••
Genauigkeit bzw. Vermeidung von Scheingenauigkeiten	•	•••	•••	••	••••	•••
Vollständigkeit und durchgängige Quantifizierung	••	••••	•	••	•••	•••
Bewertung hochfrequenter Kleinstschäden und „Black Swans"	•	••••	•	••	••	••
Nachvollziehbarkeit und Transparenz der Ergebnisse	•	•••	•••	•••	••••	••••
Kosten-Nutzen-Betrachtung von Gegenmaßnahmen/-strategien	•	••••	•	••	••	•••
Schäden durch Betriebsunterbrechungen	•••	••••	••	••	•••	•••
Abbildung von Folgekosten	••	•••	••	••	•••	•••
Investitionskosten ins Umlauf- oder Anlagevermögen	••	•••	••	••	•••	•••
Ergebniseignung als Risikomaß	••	••••	••	•••	••••	••••
Fazit:	effiziente Erstellung. Unpräzise Ergebnisse	präzise Bewertung. Gefahr von Scheingenauigkeit	Präzises histor. Abbild. Bei Prognosen unpräzise	Abhängig von hist. Werten.	Präzise Bewertung von Risikofolgen	Präzise Bewertung von Folgeschäden

Legende: ○ nicht geeignet/sehr gering, • kaum geeignet/gering, •• teilweise geeignet/mittel, ••• größtenteils geeignet/hoch, •••• vollständig geeignet/sehr hoch

schiedliche Stärken und Schwächen. Daher ist eine fallspezifische Betrachtung des Einsatzgebietes notwendig. Das Fazit besteht aus einem deskriptiven Teil und einer Bewertung der Gesamtpunktzahl der jeweiligen Technik. In Abhängigkeit des zu bewertenden Risikos und der vorhandenen Kompetenzen im Unternehmen zur Anwendung der Risikotechniken ist auf Grundlage dieser Darstellung eine geeignete Auswahl qualitativer oder quantitativer Techniken zu erfolgen. Die Auswahl ist insbesondere von der Unternehmensstruktur und größe abhängig.

Qualitative Ergebnisse sind aufgrund der fehlenden Spezifizierung nach den SMART Kriterien nur bedingt geeignet zur Steuerung von Risiken. Damit diese qualitativen Ergebnisse zur Steuerung von Risiken verwendet werden können, ist eine Interpretation durch Experten notwendig. Die Kombination aus qualitativen Daten und Expertenschätzung bietet einen Trade-Off zwischen Ergebnisvalidität und dem minimalen administrativen Aufwand zur Erstellung eines Risikomanagement-Modells für die Beschaffung. Im Vergleich bieten Simulationsmodelle eine präzise Bewertung mit dem Nachteil eines großen administrativen Aufwandes. Ein wesentlicher Vorteil von Simulationsmodellen ist der Einbezug von Kosten- und Nutzenbetrachtung potentieller Risikostrategien. Es besteht jedoch die Gefahr von Scheingenauigkeiten. Diese Scheingenauigkeiten werden konzeptionell durch eine Kombination von Simulationsmodellen und Risikodatenbanken reduziert bzw. eliminiert. Aufgrund umfassender historischer Daten werden Korrelationen identifiziert und im Hinblick auf die Risikobewertung kritisch überprüft. Insbesondere die Nachvollziehbarkeit und Transparenz ist bei der Verwendung von Datenbanken hervorzuheben. Die FMEA beruht ebenfalls auf der Auswertung und Analyse historischer Daten. Allerdings ist der administrative Aufwand größer als bei der Verwendung einer Datenbank. Im Gegenzug ist die Prognosegenauigkeit bezüglich der Erfassung von Schäden für den Umsatz und die Kundenbindung durch die Darstellung von Fehlerfolgen valider. Sowohl die FTA und ETA sind zur präzisen Bewertung von Risiken geeignet. Während die FTA insbesondere zur Bewertung von Risikofolgen verwendet wird, besteht das Hauptanwendungsfeld der ETA in der Bewertung von Folgeschäden.

6   Das kybernetische Simulationsmodell zur Risikobewertung

Die Bewertungstechniken werden dem vorherrschenden Informationsbedarf nicht gerecht (vgl. Tabelle 1). Im Hinblick auf die notwendigen Informationen sind quantitative Verfahren nicht ausreichend. Während Entdeckungszeitpunkte, Ursachen, Eintrittswahrscheinlichkeiten sowie aggregiertes Gesamtrisiko und die gesamte Wirkung quantitativ darzustellen sind, besteht der Bedarf zur Beschreibung von alternativen und gewählten Strategien nach qualitativen Informationen. Die Autoren haben auf dieser Basis eine Kombination aus qualitativen und quantitativen Verfahren entwickelt. Im Hinblick auf den Wertbeitrag durch Managementflexibilität und die

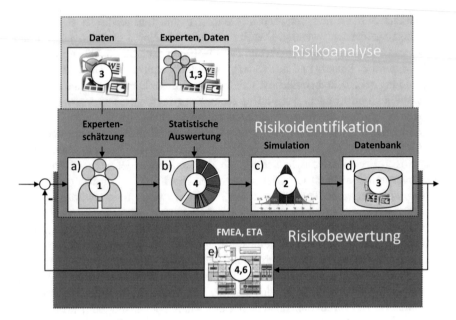

Abbildung 5 Ein kybernetisches Modell zur simulativen Quantifizierung von Risikofolgen in komplexen Prozessketten

ISO 31000 sind folgende Voraussetzungen für ein Risikomanagement zu erfüllen (vgl. Abbildung 1).

- Identifikation von Unsicherheiten
- Beschreibung der Auswirkungen
- Modellbildung
- Identifikation von Optionen.

Aufbauend auf diesen Grundvoraussetzungen und dem Informationsbedarf der einzelnen Risikoklassen wird ein kybernetisches Modell zur simulativen Quantifizierung von beschaffungsinduzierten Störgrößen erstellt (vgl. Abbildung 5). Als Grundlage dienen sowohl die Ergebnisse der Untersuchung etablierter Bewertungstechniken als auch die kybernetische Modellierung am Beispiel von System Dynamics. System Dynamics ist ein Modellierung- und Simulationsansatz auf Basis der Rückkopplungstheorie zur Ableitung von Systemverhalten und Handlungsempfehlungen [49].

Die Bewertungstechniken werden im Hinblick auf einen Trade-Off zwischen Nutzen und Aufwand eingesetzt. Die Nummerierung erfolgt analog anhand der in Tabelle 4 festgelegten Zuordnung. Ziel des Modells ist es im Sinne eines kontinuierlichen Verbesserungsprozesses, ein dauerhaftes Risikomanagement von Beschaffungsrisiken zu implementieren. Dieses Modell besteht aus fünf Schritten:

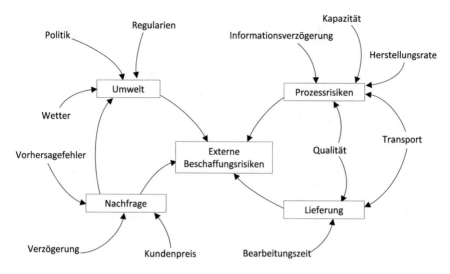

Abbildung 6 Beispiel eines aggregierten System Dynamics Modells für beschaffungsinduzierte Störgrößen

(a) Erstellung eines System Dynamics Modells
(b) Statistische Auswertung der Datenbasis zur Bestimmung der Korrelationen und Wechselwirkungen.
(c) Simulation der Risiken mittels System Dynamics Modell
(d) Überführung der Ergebnisse in eine Datenbank
(e) Bewertung der Simulationsergebnisse

Als Ausgangspunkt (a) wird ein qualitatives System Dynamics Modell durch die Verwendung von Expertenschätzungen (1) und mittels Unterstützung durch eine Risikodatenbank (3) erstellt. Die beispielhafte Modellierung der negativ wirkenden Risiken ist in Abbildung 6 dargestellt. Als zu modellierende Größen werden die Risikoklassen und deren Detailausprägungen der Beschaffung verwendet (vgl. Tabelle 1). Die dargestellten Risiken stammen aus einer Literaturreview der Schnittmenge folgender Autoren [50–55]. Jede Veränderung des Einzelrisikos verursacht eine potentielle Verzögerung des Liefertermins. Im Prozess der Risikoanalyse werden durch Experten unternehmensspezifische Daten untersucht. Durch die Ableitung von Informationen wird eine Risikoidentifikation ermöglicht und in ein Modell überführt. Dieses qualitative Modell erfüllt sowohl die Forderung nach einem geringen administrativen Aufwandes zur Erstellung als auch die Grundvoraussetzung für eine anschließende Simulation.

Für die Risikoklasse Umwelt sind beispielsweise die jeweilige Landespolitik und extreme Wetterbedingungen einzubeziehen. Analog ist für vorhandene Prozessrisiken der Zulieferer, eine verzögerte Ankunft der Informationen sowie die Kapazität und aktuelle Herstellungsraten zu berücksichtigen. Für die Risikoklasse „Lieferung" existieren Schnittmengen zur Risikoklasse Prozessrisiken. Diese sind neben dem Qualitätsrisiko auch Transportrisiken. Die Klasse „Nachfrage" wird durch interne Vorhersagefehler bzw. Verzögerungen der Produkteinführung und den Preis verursacht. Durch die Visualisierung der Wechselwirkung potentieller Risiken wird eine einheitliche Informationsbasis geschaffen.

In dieser Phase besteht die Möglichkeit der Verwendung statistisch erfasster Korrelationen aus den Unternehmensdaten (b). Diese statistisch erfassten Daten stammen aus einer Risikodatenbank (3) und werden ggf. durch Expertenschätzung (1) ergänzt und in eine FMEA (4) überführt. Durch die Verwendung der Korrelationen bzw. der simulierten und dokumentierten Wechselwirkung wird das qualitative System in ein quantitatives dynamisches Simulationsmodell der Beschaffungsrisiken überführt. System Dynamics bietet in diesem Zusammenhang den Vorteil, dass von den Experten identifizierte, aber nicht quantifizierte Korrelationen simuliert werden können. Als Beispiel sei an dieser Stelle die Monte Carlo Simulation zur Ermittlung einer Verteilungsfunktion aufgeführt.

Im folgenden Schritt (c) wird eine Simulation (3) auf Grundlage des quantitativen System Dynamics Modells durchgeführt. Über die aggregierte Variable „Externe Beschaffungsrisiken" werden alle Einzelrisiken (vgl. Tabelle 1) hinsichtlich der Zielgrößen ‚Wahrscheinlichkeit einer Verzögerung' und ‚Maximale Verzögerung' bewertet. Als zentrales Ergebnis wird ein Wert für die Liefertermintreue des gesamten externen Risikos berechnet. Die Operationalisierung der Einzelrisiken erfolgt anhand der durchgeführten statistischen Auswertung mittels Korrelationsanalyse. Potentielle Startwerte einzelner aggregierter Risikoklassen sind durch Expertenschätzung möglich. Durch die Möglichkeit einer individuellen Betrachtung der Beschaffungsrisiken und deren Verlauf über die Zeit findet automatisch eine Dokumentation der Einzelparameter statt. Die Einzelwerte werden im nächsten Schritt (d) an eine Datenbank (3) übergeben und gespeichert. Im Anschluss an die Speicherung der Daten erfolgt eine Auswertung und Informationsbearbeitung mittels FMEA (4) und ETA (6). Durch die Darstellung der Zusammenhänge im System Dynamics Modell und die Ergebnisse der Simulation sind die Eintrittswahrscheinlichkeiten potentieller Störfaktoren und deren Wechselwirkungen bekannt. Für die Abschätzung des Schadensausmaßes ist eine Expertenschätzung notwendig. Diese Abschätzung ist anhand eines vordefinierten Rankings im Sinne der Firmenstrategie vorzunehmen. Durch den generischen Charakter des Modells ist eine Simulation der Eintrittswahrscheinlichkeiten für einzelne Beschaffungsgüter unabhängig voneinander möglich. Die Rückkopplung der Bewertungsergebnisse in das Modell ermöglicht eine iterative Anpassung und Verbesserung der Informations- und Ergebnisqualität.

7 Fazit & Ausblick

Der vorliegende Beitrag stellt die Herausforderungen eines durchgängigen Informationsmanagement in Kontext des Managements komplexer Lieferketten dar. Neben der Darstellung eines Fallbeispiels, sind die Begriffe Unsicherheit, Risiko und Ungewissheit definiert worden. In diesem Kontext wurden Anforderungen an ein Risikomanagement-Modell für die Beschaffung erhoben und mit bestehenden Modellen aus der Literatur verglichen. Auf Basis der Ergebnisse wurde ein kybernetisches Modell zur simulativen Quantifizierung von beschaffungsinduzierten Störgrößen.

Im Hinblick auf die Reform zur Zertifizierung nach der DIN ISO 9001:2015 [56] ist ein Umgang mit Risiken verpflichtend. Der bestehende Ansatz bietet sowohl KMU als auch Großunternehmen einen Ansatz zum Risikomanagement. Das dargestellte kybernetische Modell zur simulativen Quantifizierung von Risikofolgen in komplexen Lieferketten liefert einen Ansatz zum szenariobasierten Risikomanagement. Dies bietet die Möglichkeit Strategien für ein flexibles Risikomanagement zu entwickeln. Für potentielle Risiken sind Schadensausmaße quantifiziert dargestellt und Handlungsalternativen im System hinterlegt. Im Falle eines Schadenseintritts sind entsprechende Prozeduren aus der Datenbank aufzurufen und die hinterlegte Handlungsempfehlung durchzuführen. Dynamische Veränderungen werden durch die simulative Quantifizierung von externen Risiken erfasst und die Parameter aufgezeichnet. Über eine Analyse der aufgezeichneten Daten und einen Vergleich wird die Evaluation der unmittelbaren Auswirkungen von potentiellen Risiken auf den Produktionsprozess ermöglicht. Dies bedeutet einen Mehrwert zur Identifizierung von kritischen Einflussfaktoren für einzelne Beschaffungsgüter und potentielle kritische Lieferanten. Im Vergleich zu bestehenden Ansätzen wird durch die Abbildung in einem vernetzten Modell eine Quantifizierung des Schadensausmaßes ermöglicht. Die verbundenen Wirkzusammenhänge dienen als Hilfestellung zur Berechnung des Schadensausmaßes.

Neben dem Vorteil eines flexiblen Risikomanagements besteht ein weiterer Vorteil des Modells in dem gewählten Methodenmix. Das Gesamtergebnis hat einen quantitativen Charakter und erfüllt somit die SMART-Kriterien zur Steuerung eines Unternehmens. Allerdings werden durch die Einbindung von Expertenschätzungen ebenfalls qualitative Informationen und Bewertungen hinsichtlich der Risikobewertung berücksichtigt. Der dynamische Wechsel zwischen qualitativen und quantitativen Verfahren bewirkt eine zeit- und kostenoptimierte Steuerung der für das Risikomanagement eingesetzten Ressourcen. Ebenso wird durch die Einbindung von Experten in den Prozess der statische Charakter existierender Modelle aufgebrochen und die Erkennung/Einbindung von Sonderereignissen (z.B. Naturkatastrophen) in den Management Prozess eingebunden. Diese Sonderereignisse sind durch ein hohes Schadensausmaß und eine nicht existente Prognostizierbarkeit gekennzeichnet.

Die Einbindung der Experten in die Entwicklung des Modells bewirkt die Abbildung individueller Unternehmensrisiken und der vorherrschenden Risikostrategie. Durch die anschließende statistische Auswertung des vorhandenen Datenmaterials zu den bestehenden Risiken in der Beschaffung wird eine unternehmensweite Grundlage zur Betrachtung von Beschaffungsrisiken gelegt. Jedoch entziehen sich z. B. Umweltrisiken einer exakten Prognosefähigkeit. Daher biete die anschließende Simulation die Möglichkeit Zufallsverteilung für diese Risiken zu verwenden. Durch Speicherung der Simulationsergebnisse wird eine Verbindung der Zufallsverteilungen von Risiken und des auftretenden Schadensausmaßes hergestellt. Dementsprechend können auch „Worst-Case" Szenarien simuliert und entsprechende Handlungsdirektiven abgeleitet werden. Die abschließende Aufbereitung in einer FMEA stellt die ermittelten Zusammenhänge kompakt dar und bietet eine Grundlage zur Unternehmensteuerung. Die Rückführung in das Modell ermöglicht somit unterschiedliche Szenarien zu entwickeln und entsprechende Gegenmaßnahmen im Sinne eines kontinuierlichen Verbesserungsprozess abzuleiten.

Der in diesem kybernetischen Risikomanagement (vgl. Abbildung 5) dargestellte Ansatz ist durch eine Schnittstelle zum internen Risikomanagement zu erweitern. Hierfür ist die Definition und Implementierung einer Schnittstelle erforderlich. Als Möglichkeit einer Schnittstelle ist die Modellierung des kritischen Pfades für die Produktion denkbar. Es erfolgt eine Priorisierung der zu modellierenden Beschaffungswege anhand der auf dem kritischen Pfad verwendeten Materialien. Hierbei bilden die simulierten Verzugszeiten aus dem System Dynamics Model die Grundlage zur Kopplung mit den kritischen Materialien aus der kritischen Pfadanalyse. In diesem Zusammenhang ist die Entwicklung eines Gesamtkonzeptes und die Überführung der Ergebnisse in einen Software-Demonstrator erforderlich. Dieser Software-Demonstrator erfordert zur Prognose des Schadensausmaßes eine Schnittstelle zu den vorgehaltenen Produktionsdaten. Insbesondere das vorgehaltene Datenformat und die entsprechende Datenmenge stellen hierbei die zentrale Herausforderung dar. Zur Steigerung der gesamten Simulationsergebnisse ist die Informationsqualität durch objektive Kriterien und Bewertungsmethoden zu ergänzen.

Danksagung Das IGF-Vorhaben 127 E der Forschungsvereinigung Institut für Unternehmenskybernetik e. V. wurde über die AiF im Rahmen des Programms zur Förderung der Industriellen Gemeinschaftsforschung (IGF) vom Bundesministerium für Wirtschaft und Energie aufgrund eines Beschlusses des Deutschen Bundestages gefördert.

Literaturverzeichnis

1. I. Heckmann, T. Comes, S. Nickel, A critical review on supply chain risk–Definition, measure and modeling. Omega **52**, 2015, pp. 119–132. doi:10.1016/j.omega.2014.10.004
2. F. Aqlan, S. Lam, A fuzzy-based integrated framework for supply chain risk assessment. International Journal of Production Economics **161**, 2015, pp. 54–63
3. G. Schlegel, R. Trent, *Supply Chain Risk Management: An Emerging Discipline*. CRC Press, 2015

4. C. Pias. Geschichte und Theorie der Kybernetik, 2006. http://www.medientheorie.com/veranstaltungen/06_ws_kybernetik.html. Zugegriffen 19. November 2014
5. *Epoche Atom und Automation; Enzyklopädie des technischen Jahrhunderts*. W. Limpert-Verlag, Frankfurt am Main, 1958
6. H. Krcmar, *Informationsmanagement*. Springer, Berlin, Heidelberg, 2010
7. B. Preim, R. Dachselt, *Interaktive Systeme*, vol. 1, 2nd edn. Springer, Berlin, 2010
8. H. Schumann, W. Müller, *Visualisierung Grundlagen und allgemeine Methoden*. Springer, Berlin, 2000
9. F. Malik, *Führen, Leisten, Leben: Wirksames Management für eine neue Zeit*. Campus Verlag GmbH, Frankfurt am Main, 2006
10. H. Seiffert, *Information über die Information*. Verlag C H Beck, Munich, 1971
11. I. Durbach, T. Stewart, A comparison of simplified value function approaches for treating uncertainty in multi-criteria decision analysis. Omega **40** (4), 2012, pp. 456–464
12. R. Gottwald, *Entscheidung unter Unsicherheit: Informationsdefizite und unklare Präferenzen*. Gabler, Wiesbaden, 1990
13. M. Janz, *Erfolgsfaktoren der Beschaffung im Einzelhandel*. Dt. Univ.-Verl., Wiesbaden, 2004
14. P. Gluchowski, R. Gabriel, C. Dittmar, *Management-Support-Systeme und Business intelligence: computergestützte Informationssysteme für Fach- und Führungskräfte*. Springer, Berlin, Heidelberg, 2008
15. B. Hazen, C. Boone, J. Ezell, L. Jones-Farmer, Data quality for data science, predictive analytics, and big data in supply chain management: An introduction to the problem and suggestions for research and applications. International Journal of Production Economics **154**, 2014, pp. 72–80
16. D. Barton, D. Court, Making advanced analytics work for you. Harvard business review **90** (10), 2012, pp. 78–83
17. F. Megahed, L. Jones-Farmer, *A statistical process monitoring perspective on big data*. Frontiers in Statistical Quality Control. Springer, New York, 2013
18. S. Li, B. Lin, Accessing information sharing and information quality in supply chain management. Decision Support Systems **42** (3), 2006, pp. 1641–1656
19. D. Lambert, M. Cooper, J. Pagh, Supply Chain Management: Implementation Issues and Research Opportunities. The International Journal of Logistics Management **9** (2), 1998, pp. 1–20
20. J. Thun, D. Hoenig, An empirical analysis of supply chain risk management in the German automotive industry. International Journal of Production Economics **131** (1), 2011, pp. 242–249
21. J. Montoya-Torres, D. Ortiz-Vargas, Collaboration and information sharing in dyadic supply chains: A literature review over the period 2000–2012. Estudios Gerenciales **30** (133), 2014, pp. 343–354
22. M. Sodhi, C. Tang, *Managing Supply Chain Risk*, vol. 172. Springer US, Boston, MA, 2012
23. S. Boyson, Cyber supply chain risk management: Revolutionizing the strategic control of critical IT systems. Technovation **34** (7), 2014, pp. 342–353
24. S. Fawcett, G. Magnan, M. McCarter, Benefits, barriers, and bridges to effective supply chain management. Supply Chain Management: An International Journal **13** (1), 2008, pp. 35–48
25. I. Manuj, J. Mentzer, Global supply chain risk management strategies. International Journal of Physical Distribution & Logistics Management **38** (3), 2008, pp. 192–223
26. S. Qrunfleh, M. Tarafdar, Supply chain information systems strategy: Impacts on supply chain performance and firm performance. International Journal of Production Economics **147**, 2014, pp. 340–350
27. O. Kwon, N. Lee, B. Shin, Data quality management, data usage experience and acquisition intention of big data analytics. International Journal of Information Management **34** (3), 2014, pp. 387–394
28. M. Waller, S. Fawcett, Data science, predictive analytics, and big data: a revolution that will transform supply chain design and management. Journal of Business Logistics **34** (2), 2013, pp. 77–84
29. F. Knight, *Risk, Uncertainty and Profit*. Beard Books, 2002

30. M. Wiebel, T. Eifler, J. Mathias, H. Kloberdanz, A. Bohn, H. Birkhofer, Modellierung von Unsicherheit in der Produktentwicklung. In: *Exploring Uncertainty*, ed. by S. Jeschke, E. Jakobs, A. Dröge, Springer Fachmedien, Wiesbaden, 2013, pp. 245–269
31. H. Bitz, *Risikomanagement nach KonTraG*. Schäffer-Poeschel Verlag, Stuttgart, 2000
32. H. Laux, R. Gillenkirch, H. Schenk-Mathes, *Entscheidungstheorie*. Springer, 2014
33. H. Hanselka, R. Platz, Berechnung Ansätze und Massnahmen zur Beherrschung von Unsicherheit in lasttragenden Systemen des Maschinenbaus. Konstruktion **62** (11), 2010, p. 55
34. D. Waters, *Supply Chain Risk Management: Vulnerability and Resilience in Logistics*. Kogan Page Publishers, 2011
35. G. Hult, C. Craighead, J. Ketchen, D. J., Risk Uncertainty and Supply Chain Decisions: A Real Options Perspective. Decision Sciences **41** (3), 2010, pp. 435–458
36. E. Hildt, Zur Rolle von Risiko, Ungewissheit und Nichtwissen in der Bioethik. In: *Exploring Uncertainty*, ed. by S. Jeschke, E. Jakobs, A. Dröge, Springer Fachmedien Wiesbaden, 2013, pp. 37–53
37. R. Ireland, M. Hitt, D. Vaidyanath, Alliance Management as a Source of Competitive Advantage. Journal of Management **28** (3), 2002, pp. 413–446
38. H. Simon, Invariants of Human Behavior. Annual Review of Psychology **41** (1), 1990, pp. 1–20
39. M. Wouters, J. Anderson, J. Narus, F. Wynstra, Improving sourcing decisions in NPD projects: Monetary quantification of points of difference. Journal of Operations Management **27** (1), 2009, pp. 64–77
40. T. Cheong, S. Song, The value of information on supply risk under random yields. Transportation Research Part E: Logistics and Transportation Review **60**, 2013, pp. 27–38
41. EVP, C. L. P., PMP, PMI-RMP, *Risk Management: Concepts and Guidance*, 5th edn. CRC Press, 2014
42. M. Schermann, *Risk Service Engineering: Informationsmodelle für das Risikomanagement*, 1st edn. Gabler, Wiesbaden, 2011
43. DIN EN 31010:2010-11 Risikomanagement - Verfahren zur Risikobeurteilung (IEC/ISO 31010:2009)
44. T. Zentis, A. Czech, T. Prefi, S. R., eds., *Technisches Risikomanagement in produzierenden Unternehmen*. Fraunhofer Inst. (IPT), Aachen, 2011
45. A. Ziegenbein, *Supply Chain Risiken: Identifikation, Bewertung und Steuerung*. vdf Hochschulverlag AG, Zürich, 2007
46. G. Doran, There's a S.M.A.R.T. way to write managements's goals and objectives. Management Review **70** (11), 1981, p. 35
47. A. Ziegenbein, *Supply chain risk assessment: a quantitative approach*. ETH-Zentrum für Unternehmenswissenschaften, Zürich, 2006
48. K. Hardy, *Enterprise risk management: a guide for government professionals*. 2014
49. A. Assad, S. Gass, *Profiles in operations research pioneers and innovators*. Springer, New York, London, 2011
50. A. Eberle, *Risikomanagement in der Beschaffungslogistik: Gestaltungsempfehlungen für ein System*. 2005
51. J. Hoffmann, *Risikomanagement für mittelständische Unternehmen: Risikopotenziale erkennen und erfolgreich bewältigen - mit zahlreichen Praxissituationen*. BoD - Books on Demand, 2012
52. R. Meierbeck, *Strategisches Risikomanagement der Beschaffung Entwicklung eines ganzheitlichen Modells am Beispiel der Automobilindustrie*. Eul, Lohmar, Köln, 2010
53. M. Moder, E. Hartmann, C. Jahns, *Supply Frühwarnsysteme: Ergebnisse einer Studie zur Identifikation und Analyse von Risiken im Supply Management*. Sternenfels: Wissenschaft & Praxis, 2008
54. O. Schneider, *Adding enterprise value: mitigating investment decision risks by assessing the economic value of supply chain initatives*. vdf Hochschulverl, Zürich, 2010
55. H. Wildemann, *Risikomanagement und Rating*. TCW, Transfer-Centrum, München, 2006
56. BDI, ed.,
57. DIN EN ISO 9001:2015 Qualitätsmanagementsysteme - Anforderungen (ISO/DIS 9001:2015)

58. W. Gleißner, *Grundlagen des Risikomanagements im Unternehmen: Controlling, Unternehmensstrategie und wertorientiertes Management*. Vahlen, 2011
59. ISO 31000:2009-11 Risikomanagement - Allgemeine Anleitung zu den Grundsätzen und zur Implementierung eines Risikomanagements
60. S. Jordan. Lieferkette der nager it, 2015. https://www.nager-it.de/static/pdf/lieferkette.pdf. Erschienen: Januar 31, letzter Zugriff: 11. März 2015

Measuring the Quality of Cooperation in Interdisciplinary Research Clusters

Stefan Schröder, Markus Kowalski, Claudia Jooß, René Vossen, Anja Richert and Sabina Jeschke

Abstract Research in the challenging field of industrial engineering and engineering management often needs the expertise from more than one discipline. Therefore interdisciplinary research has taken on continuously greater significance. Over the last years various research clusters – such as cluster of excellence – were initiated in the course of the German excellence initiative. Thereby researchers with different disciplinary backgrounds are brought together, to investigate research questions with societal and economical relevance. These new research clusters cause various challenges, especially on communication, because communication is one decisive factor of success in interdisciplinary cooperation. Approaches to investigate communication, performance and interdisciplinarity within these clusters are compared and a communication oriented measurement approach is explained. Especially the operationalization and validation of the indicator communication, giving conclusions onto the quality of cooperation, is depicted as well as measures to react on.

Keywords Interdisciplinary Communication · Cluster-specific Balanced Scorecard · Interdisciplinarity · Cluster of Excellence (CoE) · Interdisciplinary Research Clusters

1 Introduction

The establishment of interdisciplinary research clusters has been rising especially over the last decade. Reasons for this development are given by the insight that for a large number of scientific questions coming up, disciplinary approaches are insufficient [1, 2].

In Germany, so called clusters of excellence (CoE) are initiated to focus on current issues and research questions which cannot be processed by only one discipline and

S. Schröder (✉) · M. Kowalski · C. Jooß · R. Vossen · A. Richert · S. Jeschke
IMA/ZLW & IfU, RWTH Aachen University, Dennewartstr. 27,
52068 Aachen, Germany
e-mail: stefan.schroeder@ima-zlw-ifu.rwth-aachen.de

Originally published in "2014 IEEE International Conference
on Industrial Engineering and Engineering Management",
© IEEE 2014. Reprint by Springer International Publishing Switzerland 2016,
DOI 10.1007/978-3-319-42620-4_13

their researchers respectively [1]. Due to the initiation of CoE the complexity of cooperation increases. Cooperation of actors with different disciplinary backgrounds is built to deal with these societal and economic research questions. Section 2 is concerned with these challenges of interdisciplinary research and with the importance of communication (e. g. specific disciplinary terminologies).

Consequently, the shaping of cooperation (to foster the quality of cooperation) becomes particularly challenging for the researchers as well as for the management [3] and in particular, the difficulty to measure the quality of cooperation. For that reason, various measurement approaches have occurred on the agenda of interdisciplinary research over the last years (Section 3) [4–8]. Different existing methods and tools are used as groundwork to adapt the specific needs of interdisciplinary research. As these measurement approaches mainly focus on measurable, chiefly quantitative indicators, weak interpersonal parameters such as communication to describe the quality of cooperation are still almost neglected (c. f. Section 3). Hence, as in the context of the objectives of establishing CoE by the German Research Foundation (DFG), the aspect of measuring and fostering the *quality of cooperation* is in the course of this paper considered.

The paper elucidates reasons for shaping interdisciplinary cooperation in research clusters and how to (measureable) deal with interdisciplinary challenges (mainly beyond the classical quantitative key figures). While communication (e. g. common language and understanding of all terms, objects and problems) is one factor of success in cooperation [9–11] Section 4 addresses the operationalization of communication (in cooperation with the cluster management) and its validation. Furthermore measures to improve communication and cooperation within the CoE are depicted. Section 5 gives a brief outlook on further research activities.

2 Challenges of Interdisciplinary Research

The effort resulting from interdisciplinary research clusters causes challenges to be solved. This includes the fusion of different bodies of knowledge and heterogeneous objectives but also the embodiment of the operative work under the consideration of different cultures and specific disciplines [12]. Further challenges within these research clusters are for example the heterogeneity of actors/researchers, their specific disciplinary background and associated terminologies, their understanding and use of methods as well as on-professional competences in project work and steering of the CoE through the cluster management. All this becomes especially challenging in an interdisciplinary environment [13].

In the context of a heterogeneous mix of disciplines, new cooperation abilities need to be enabled. Communication and negotiation of difference are the linchpins of cooperative interdisciplinary research. In the essay *Interdisciplinary and Transdisciplinary Landscape studies*, published by Aenis and Nagel in 2003, there are two axiomatic considerations for inter- and transdisciplinary evaluation indicated:

participation (communication between researchers and regional actors) and the met-alevel of interdisciplinarity (communication among researchers) [14].

'At the heart of interdisciplinarity is communication – the conversations, connections and combinations that bring new insights to virtually every kind of scientist and engineer' [15]. This quote elucidates that communication is essential for the delivery of high performance and the quality of cooperation within interdisciplinary research cluster. Against this backdrop, how performance is actually measured and which focus these measurement approaches contain is outlined in the following chapter.

3   Current Measurement Approaches

3.1   State of the Art

While communication is an important factor for the quality of cooperation in inter-disciplinary research clusters [16] and the measurement of performance of research clusters occurs on the agenda over the last years, the following approaches are briefly introduced and their measuring-focus is outlined. Thereby an overview is given, in which ways communication and interdisciplinarity (cooperation) are analyzed by other researchers.

Kaufmann clarifies in his study (1987) that interdisciplinarity is not a given fact but rather has to be build up gradually [17]. Due to the difficulty in communication outside the disciplinary language there is a need for a specific form of scientific communication to support the interdisciplinary approach [17]. There are certain indicators of interdisciplinary research and the most commonly used is co-authorship of scholarly output [18]. Another approach analyses the nature of journals in which individuals or interdisciplinary groups publish (Thomson-Isi 244) or examines the citations in research papers [18].

The literature appropriates some aspects about weaknesses and strengths about *interdisciplinarity* and measures they imply. Klein (1999) highlights this *special form of interaction* that this may be the preferred indicator of interdisciplinary activities. However, Leydesdorff and Cozzens have accomplished studies of journal categorization in the 1990s. The key aspect of this approach is the fact, that journal sets are dynamic and cross citations help reveal changing research specialization [19]. In addition to this, van Raan (1999) pointed out a confluence of three approaches for *investigating the interdisciplinary nature of science*: (a) research activity profiling, (b) research influence profiling and (c) bibliometric mapping.

Studies on interdisciplinary cooperation, which are mainly based on bibliometric- or citations-based data [12, 20] and the evaluation of interdisciplinary research clusters [12, 20] describe experiences with interdisciplinarity from a superior perspective [20]. On the basis of literature it becomes clear that the indicator *communication* [10, 21–23] is a very important factor concerning the measurement of the quality of cooperation in interdisciplinary research [14].

Moreover, further approaches exist in literature to support communication in inter-disciplinary research, for example Klavans and Boyack (2006) are mapping science overall using journal citation interactions [4]. Rinia et al. (2001) illustrates the significance of considering exactly how bibliometric measures, which count the scientific activity, are crafted [5]. Various bibliometric approaches [6] exist concerning this research field of such measures. Klein (1999) suggests that interactions may be the preferred indicator of interdisciplinary activity [7] whereas Kostoff (2002) uses co-word analyses to understand topical emphases [8]. As a conclusion it is obvious that there does not exist one general measurement approach.

3.2 Questioning CoE in Germany

With respect to the importance of measuring communication and quality of cooperation Welter (2012) stated a study focusing on all existing (n = 37) CoE in Germany [24]. Thereby all speakers and CEO were asked by the aid of which indicators they evaluate the quality of cooperation and the CoE performance. About 37 percent (n = 13) of all CoE of the 1st funding period (2007–2012) – except the two CoE of RWTH Aachen University which the responsible authors of the study are involved in – took part in the survey.

By means of the term *evaluate*, the participants were asked to judge by using marks. Answers were supposed to be given by an interval scale ranging from 1=very appropriate to 5 = inappropriate. Especially the interval scaling of this closed questions allowed data analyses considering arithmetic averages ($\bar{x}$) and the corresponding standard deviation (δ) to analyze the dispersion of answers.

The *top three* performance indicators (out of eleven) of the CoE constitute: "Number of international visiting researchers" ($\bar{x} = 1.75$; $\delta = 0.71$), "Quota of publications" ($\bar{x} = 1.86$; $\delta = 0.69$), and "The number of awards" ($\bar{x} = 2.00$; $\delta = 0.58$).

The least appropriate performance indicators for the CoE depict "The number of patent applications" ($\bar{x} = 3.42$; $\delta = 1.29$), the "Share of female researcher in the whole research and development-staff" ($\bar{x} = 3.13$; $\delta = 1.13$), and the "citation-index" ($\bar{x} = 3.00$; $\delta = 1.29$). It is noticeable, however, that the level of disagreement among the participants is relatively high [22].

3.3 Conclusion

As a conclusion drawn from these insights, one can state that bibliometric indicators are assessed as the most common indicators for the scientific performance in average to evaluate the quality of cooperation. Considering the top performance indicators of the CoE, stated in the study of Welter (2012), it appears as if they emphasize more classical scientific performance indicators such as citations or the number of awards [25]. This way of proceeding is also prioritized by the other stated approaches.

With regard to the relevance of communication, it seems surprising that communication is not a main point in the investigations of interdisciplinary cooperation. Indicators concerning communication and its influence on the quality of cooperation have not been evaluated at all (either in free text comments). Particularly this thematic is increasingly demanded in the current discussions about interdisciplinarity [12, 14, 26]. If the interdisciplinarity discourse continues to play a crucial role in research it reveals an urgent need to examine how communication influences quality of cooperation. An approach, to deal and shape with this challenge of communication within the CoE, is described in the following.

4 New Communication Shaping Approach

The strong focus on bibliometric or citations-based data [12, 20] and the evaluation of interdisciplinary constructs elucidates that the indicator communication is still not investigated in CoE. In particular the fact that in science the cooperation in interdisciplinary research clusters increases supports this possible shift in measuring the quality of cooperation in interdisciplinary research clusters. Therefore a communication shaping approach and recommendations of action in two CoE at RWTH Aachen University were developed. Hence the adaption of a Balanced Scorecard approach is depicted (Section 4.1) and afterwards the items of operationalizing communication are described (Section 4.2) and validated (Section 4.3). At least measures (Section 4.4) to support the quality of cooperation are presented.

4.1 Adaption of a Prototypical Measurement Instrument in Order to Deal with Communications Within CoE

A necessary requirement concerning a successful management is a systematic control and reflection of all activities [27]. One approach, used in this context, is Kaplan and Norton's (1992) Balanced Scorecard (BSC) [28]. Referring to this classic scorecard, a *cluster-specific Balanced Scorecard* [25] that regards the characteristics of an interdisciplinary research cluster was adapted.

In order to deal with *communication* within these CoE, the BSC was implemented and tries to bridge the gap between hard bibliometric facts (e. g. publications, citations, guest researchers, awards etc.) and weak inter-personal parameters (e. g. frequency and quality of meetings, terminologies, forwarding of decisions etc.). This cluster-specific BSC has been implemented in two CoE- at RWTH Aachen University, namely *Integrative Production Technology for High-Wage-Countries* and *Tailor-Made Fuels from Biomass (TMFB)*, to qualitatively and quantitatively accompany the communication abilities of an interdisciplinary research cluster.

In cooperation with the management of the CoE, the visions and objectives of the interdisciplinary research cluster have been transferred into aims which have been operationalized and assigned to the individual perspectives of the classic BSC (*internal-, customer-, organizational- and financial-perspective*). The development of the cluster performance is retrieved with regard to the different aspects on basis of an annual evaluation over the entire period of the project-time. Along these iterations, data are gathered and analyzed to work out recommendations of action in cooperation with the cluster-management to preferably increase the performance and quality of cooperation of the CoE. Till now in both CoE at RWTH Aachen University, within the 1st and 2nd funding period, four evaluations took place.

In sum, the BSC is an approach which systematically enhances the measurement areas traditionally involved in accounting. Non-financial, weak interpersonal indicators were integrated in a strategic control framework so that they are no longer local, isolated systems but are merged together in a causal chain. Potentially, it contributes to sharpen communication in an interdisciplinary research cluster.

4.2 Cluster-specific Communication Indicators

Hence, to identify cluster-specific indicators, strategy workshops in cooperation with the cluster-management were conducted and in accordance with the cluster strategy and vision specific indicators were derived and operationalized. Objectives of these workshops were among others: defining and operationalizing suitable indicators to depict the communication within the CoE and their influences onto the quality of cooperation.

Exemplary indicators are depicted in Table 1 and a special focus in this context is on the weak interpersonal indicator *communication* and its items. Further indicators are also categorically listed.

To allow a more significant analysis of this data four evaluation series (years 2009 (n = 65), 2010 (n = 83), 2011 (n = 86) and 2013 (n = 87)) of the CoE TMFB are taken into account. The following validation of the communication items happens on this data basis.

4.3 Validating Communication Indicators

In order to validate the impact of the communication items of the quality of cooperation a regression analysis was conducted. To investigate the influence of different variables a multivariate linear regression is done. The target variable is explained by the aid of predictors X of a linear function:

$$Y = a + b_1 \cdot X_1 + b_2 \cdot X_2 + \cdots + b_n \cdot X_n$$

Table 1 Operationalizing the item "communication" in cooperation with the cluster management-dimensions of the cluster-specific balanced scorecard

Indicator (Y)	Appearance
	Predictor (X)
Communication	How do you feel informed about the progress and results of the research work within the Cluster of Excellence?
	How well are content wise and organizational decisions documented and forwarded to the staff of the Cluster of Excellence?
	How do you evaluate the quality of the meetings?
	How do you evaluate the frequency of the meetings?
	To what extent are the contents and results helpful concerning the transparency of decisions and results?
	To what extent are the contents and results helpful concerning the strengthening of cooperation within the Cluster?
Social	e. g. staff turnover, supervision of PhD-thesis, atmosphere of the CoE
Technical	e. g. use of knowledge map (infrastructure), central assignments
Cognitive	e. g. presentation of contents, scientific and organizational line
Publicity	e. g. exhibition appearance
Bibliometrics	e. g. publications, thesis

Each influencing variable X is estimated by a regression coefficient b of the regression model. As with univariate regression the coefficient of determination explains the overall context between predictor variables (X) and target variables (Y). The coefficient of determination corresponds to the square of the multiple correlation coefficients, which means the correlation between:

$$Y \text{ and } b_1 \cdot X_1 + \cdots + b_n \cdot X_n$$

This statistical process is used to estimate relationships among the selected variables [29]. The relationship between the quality of cooperation and the implemented communication predictors was investigated. In this case, the thesis is *that the quality of cooperation is essential influenced by communication abilities*. Consequently, the variable *quality of cooperation* was used as target variable (Y) and the communication predictors (c. f. Figure 1) as influencing variables (X). To allow a more sophisticated

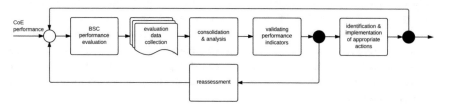

Figure 1 Balanced Scorecard specific control loop

Table 2 Regression quality of cooperation and communication indicators

Model	R	R-Square	Adjusted R-Square	Std. Error of the Estimate
1	0,710[a]	0,504	0,461	0,604

[a]target variables:progress_and_results; documentation_and_forwarding_of_decisions; quality_meetings; frequency_meetings; transparency_decisions; strengthening_cooperation

impression of this relationship the data from four evaluations (c. f. Section 4.2) were taken into account as a sample. Results are shown in Table 2.

It can be noticed that the quality of cooperation can be described by means of the listed indicators. The coefficient of determination amounts to $r^2 = 0,504$ with $\alpha <$ 0,05. This plainly shows a medium strengthened statistical coherence [30]. By means of using the communication operationalization, statements regarding the quality of cooperation can be successfully extrapolated. Nevertheless, these results show that there is further potential in validating of new or even superior apt indicators in future.

4.4   Measures and Immediate Back Coupling

Furthermore, a content-related analysis takes place right after each evaluation. Objective of this analysis is to directly react onto (chiefly negative) changes e. g. in dissemination of information regarding the whole CoE, discontent with research efforts or current needs. Thus, in accordance with the cluster-management, an immediate back coupling of information is given and suitable measures are implemented. This back coupling is visualized in Figure 1 (identification & implementation of appropriate actions). Furthermore the optimization of e. g. indicators is depicted (validating performance indicators).

This back coupling is also grounded on two ways of content analysis. The first one deals with results concerning one evaluation period and changes of analysis within specific questions and their sub-questions. The second analysis deals with changes across different or the whole time series. Over time series, variances are taken into account and measures are derived and discussed together with the cluster management.

One of the results of the evaluation is for instance a discrepancy regarding the level of information about one CoE at RWTH Aachen University as well as a cooperation divergence within the organizational levels of the CoE. While the level of information within the sub-projects is judged good ($\bar{x} = 2.05$; $\delta = 0.75$; ordinal-scaled from 1 (very good) to 5 (very bad)), there is a decrease towards the level of the overall CoE ($\bar{x} = 2.86$; $\delta = 0.96$). To increase the information flow between the different recursion layers, *colloquia for research assistants* are conducted and *general meetings* are planned. In this context, duty cycles with regard to project milestones and points of intersection (e. g. which project or research assistants need or deliver information to other projects) are defined.

Table 3 Portfolio of measures to increase interdisciplinary cooperation

	COLLOQUIA FOR RESEARCH ASSISTANTS	• Meetings on researcher level • Content demand-actuated
	GENERAL MEETINGS	• Internal conferences • Status reports of project results
	CONFERENCES	• Participants from international universities and non-university organizations • External addressing of CoE results
	WORKSHOPS AND ADVANCED TRAINING PROGRAM	• Sensitize for interdisciplinary cooperation, • Develop soft-skills as project management, time- and self-management, scientific presenting and writing
	VIRTUAL COOPERATION PLATFORM	• Share data, inform on current research topics • Visualize researcher expertise • Terminologies used in everyday work • Increase cluster-wide transparency
	INFORMAL NETWORK - MEETINGS	• Summer and barbecue revels • Fostering common identity

Table 3 gives an overview of exemplary measures to foster interdisciplinary cooperation and communication between heterogeneous researchers in an interdisciplinary research cluster. To what extent these measures influence the results of the evaluation will be investigated within the current 2nd funding period of the CoE.

5 Conclusion and Outlook

By continuously evaluating the interdisciplinary research clusters and implementing a cluster-specific Balanced Scorecard, performance of interdisciplinary research clusters, with special focus on communication and cooperation, can be shaped. The analysis of these data enables the derivation of recommendations of actions for the cluster-management. Thereby, the objective of enhancing the interdisciplinary cooperation can be fostered. In this context, communication measures have a high impact on the quality of cooperation. Consequently, recommendations of actions within the 1st funding period of the CoE at RWTH Aachen University take various communication-oriented actions into account: a tailor-made advanced training program for PhD-students, adaption and expansion of the cluster-internal knowledge map, initiation of informal summer fairs, thematic exchange by regular project meetings and the organization of Summer Schools can be listed.

Based on these results, a further differentiation of the operationalization according to the indicator *communication* will be taken into account. As a factor of success and quality of cooperation influencing predictor, cause-effect relationships between measures and measurement results will be investigated within the 2nd funding period in the CoE.

However, conclusions onto the quality of cooperation by the use of communication items can be given by means of the above-mentioned cluster-specific Balanced Scorecard approach [25]. An immediate correlation between derived and implemented measures and changes within the evaluation results are currently a challenge in these interdisciplinary research clusters and should be examined more in detail within further studies.

This paper mainly contributes to the development of a new approach – by analyzing challenges in interdisciplinary research, especially communication, in order to measure the quality of cooperation in interdisciplinary research clusters. Thereby this approach is transferable to other interdisciplinary networks. However the operationalization of indicators took place in accordance with the specific CoE aims and visions and is for that reason highly individual.

Acknowledgments This research has been funded by the German Research Foundation (DFG) as part of the Clusters of Excellences *Integrative Production Technology for High-Wage Countries* and *Tailor-Made Fuels from Biomass* at RWTH Aachen University.

Additionally, the work was supported by the German Federal Ministry of Education and Research (BMBF) under Grant 01HH11088 and was cofinanced by the European Social Funds (ESF).

References

1. G.R.F. DFG, *Exzellenzinitiative auf einen Blick – Der Wettbewerb des Bundes und der Länder zur Stärkung der universitären Spitzenforschung*, 5th edn. Bonn, 2013
2. G.R.F. DFG, *Eckpunkte zur Weiterentwicklung der Exzellenzinitiative*. Bonn, 2008
3. S. Hornbostel, M. von Ins, Leistungsmessung in der Forschung – Erfordernis oder Vermessenheit? Laborwelt **5**, 2009, pp. 34–35
4. R. Klavans, K. Boyack, Identifying a better measure of relatedness for mapping science. Journal of the American Society for Information Science and Technology **57**, 2006, pp. 251–263
5. E. Rinia, T. van Leuven, H. van Vuren, A. van Raan, Influence of interdisciplinarity on peer-review and bibliometric evaluations in physics. Research Policy **30**, 2001, pp. 357–361
6. A.F. van Raan, Scientometrics: State-of-the-art. Scientometrics **38** (1), 1997, pp. 205–218

7. J. Klein, A conceptual vocabulary of interdisciplinary science. In: *Practising Interdisciplinarity*, ed. by P. Weingart, N. Stehr, University of Toronto Press, Toronto, 1999, pp. 3–24
8. R. Kostoff, Citation analysis of research performer quality. Scientometrics **53** (1), 2002, pp. 49–71
9. J. Knape, Grenzen der Transdisziplinarität. Warum echte Grenzüberschreitungen selten passieren. attempto **30**, 2011, pp. 22–23
10. M. Bergmann, *Methoden transdisziplinärer Forschung ein Überblick mit Anwendungsbeispielen.* Frankfurt am Main, 2010
11. M. Jungert, Was zwischen wem und warum eigentlich? Grundsätzliche Fragen der Interdisziplinarität. In: *Interdisziplinarität: Theorie, Praxis, Probleme*, ed. by M. Jungert, Darmstadt, 2010, pp. 1–12
12. A. Porter, J. Roessner, A. Cohen, M. Perreault, Interdisciplinary research: meaning, metrics and nurture **15** (3), 2006, pp. 187–195
13. R. Frodeman, J. Klein, C. Mitcham, *The Oxford Handbook of Interdisciplinarity.* Oxford University Press, Oxford, 2010
14. J. Klein, Afterword: the emergent literature on interdisciplinary and transdisciplinary research evaluation. Research Evaluation **15** (1), 2006, pp. 75–80
15. N.N., *Facilitating Interdisciplinary Research. Committee of facilitating interdisciplinary research.* The national academies press, Washington, D.C., 2005
16. I. Leisten, *Transfer Engineering in transdisziplinären Forschungsprojekten.* Books on Demand, Norderstedt, 2012
17. F.X. Kaufmann, Interdisziplinäre Wissenschaftspraxis. Erfahrungen und Kriterien. In: *Interdisziplinarität. Praxis, Herausforderung, Ideologie*, ed. by J. Kocka, Frankfurt am Main, 1987, pp. 63–81
18. A. Porter, J. Roessner, A. Heberger, How interdisciplinary is a given body of research? **17** (4), 2008, pp. 273–282
19. L. Leydesdorff, S. Cozzens, P. van den Besselaar, Tracking areas of strategic importance using scientometric journal mappings. Research policy **23**, 1994, pp. 217–229
20. N. Metzger, R. Zare, Interdisciplinary research: From belief to reality. Science (5402), 1999, pp. 642–643
21. J. Mittelstraß, Transdisziplinarität oder? : von der schwachen zur starken Interdisziplinarität. Gegenworte: Hefte für den Disput über Wissen , 2012
22. M. Gibbons, C. Limoges, H. Nowotny, S. Schwartzman, P. Scott, M. Trow, *The new production of knowledge?: the dynamics of science and research in contemporary societies.* London, 1994
23. J. Klein, *Interdisciplinarity?: history, theory, and practice.* Detroit, 1990
24. F. Welter, S. Schröder, I. Leisten, A. Richert, S. Jeschke, Scientific performance indicators – empirical results from collaborative research centers and clusters of excellence in germany. In: *Proceedings of XVI IRSPM Conference. Rome, Italy, 11-13 April 2012.* 2012, pp. 1–22
25. F. Welter, *Regelung wissenschaftlicher Exzellenzcluster mittels Performancemessung.* Books on Demand GmbH, Norderstedt, 2013
26. C. Raasch, V. Lee, S. Spaeth, C. Herstatt, The rise and fall of interdisciplinary research: The case of open source innovation. Research Policy (5), 2013, pp. 1138–1151
27. R. Gleich, *Balanced Scorecard: Best-Practice-Lösungen für die strategische Unternehmenssteuerung.* Haufe-Lexware GmbH & Co. KG, Freiburg, 2012
28. R. Kaplan, D. Norton, The balanced scorecard: Measures that drive performance. Harvard-Business-Review, 1992, pp. 71–79
29. D. Kleinbaum, *Applied regression analysis and multi-variable methods.* CengageBrain.com, 2007
30. J. Bortz, C. Schuster, *Statistik für Human- und Sozialwissenschaftler.* Springer, Berlin, Heidelberg, 2010

Research Performance and Evaluation – Empirical Results from Collaborative Research Centers and Clusters of Excellence in Germany

Stefan Schröder, Florian Welter, Ingo Leisten, Anja Richert
and Sabina Jeschke

Abstract Collaborative research centers and clusters of excellence constitute public funded programs aspiring to advance research in interdisciplinary forms of collaboration throughout Germany. Due to emerging funding volumes and increasing expectations in results, concepts for performance measurement and management gain in importance. Results of an empirical study among all actively funded collaborative research centers and clusters of excellence make obvious that key performance indicators – such as the quota of publications or the number of international visiting researchers – are central. Nevertheless, holistic methods and concepts of performance measurement seem still not to be widespread among respective speakers and chief executive officers.

Keywords Scientific Performance · Performance Measurement · Key Performance Indicators · Collaborative Research Centers · Clusters of Excellence

1 Introduction

As stated by the German Ministry of Science and Technology (BMBF) in 2008, the aim of the government is to promote the country in order to become one of the most competitive global locations in terms of science, research and innovation ([1, 2]: 5). This is to be achieved by further exhausting its potential of creativity and innovative capability, especially with regard to research and development activities in interdisciplinary forms of collaboration. Against this background, public funding for collaborative research centers (CRC) was expanded in 1999 and public funding for clusters of excellence (CoE) was initiated by the German government and the federal states in 2006. Regarding the size of the consortia, the funding volume and the degree of interdisciplinarity, the two programs break with other research structures and constitute innovative instruments for the promotion of collaborative research.

S. Schröder (✉) · F. Welter · I. Leisten · A. Richert · S. Jeschke
IMA/ZLW & IfU, RWTH Aachen University, Dennewartstr. 27, 52068 Aachen, Germany
e-mail: stefan.schroeder@ima-zlw-ifu.rwth-aachen.de

Originally published in "Journal Research Evaluation",
© Oxford University Press 2014. Reprint by Springer International Publishing
Switzerland 2016, DOI 10.1007/978-3-319-42620-4_14

Hence framework conditions and aims of both programs will be described in the first sections of this paper.

Adequate concepts for performance measurement and performance management of interdisciplinary collaborative research depict new challenges for promoters and science managers in the research environment of the twenty-first century ([3]: 34). This can be explained by an augmenting degree of complexity concerning structures and behavior of interdisciplinary forms of collaborative research demanding new forms of organization and management [4, 5]. In order to measure and manage the performance, the definition and collection of scientific key performance indicators (KPIs) depicts an approach that can be used as a base for steering decisions in inter-disciplinary scientific collaborations. This includes indicators to measure e. g. the quality of scientific exchange, output and sharing of information in contrast to e. g. classical monetary indicators of profit-organizations. Whereas former and contem-porary scientific publications focus on university performance in terms of a single organization [6–8], research concerning performance on an inter-organizational level of scientific cooperation constitutes a new field of research what will be outlined in a following section of the paper [9–11].

The initial description of CRC and CoE as objects of research as well as the characterization of contemporary challenges in performance measurement and man-agement lead to the super-ordinated research question of the paper: Which KPIs are suitable to measure the performance of CRC and CoE? After presenting the design of an empirical study, the paper will provide results that base on a semi-standardized survey which was addressed to the speakers and chief executing officers of all active German CRC and CoE giving insights in the ranking and comparison of KPIs. The study provides answers about the frequency and the methodology used concerning the process of scientific performance measurement. Moreover, it reflects the impact of the programs CRC and CoE on the promotion of the research location Germany.

In the final sections of the paper the framework of the empirical study is discussed. This is followed by concluding remarks concerning the main results of the study. The latter also gives evidence about a further need for research in the topic of performance measurement and management in the context of CRC and CoE.

2 Research Performance and Evaluation in Europe

Due to the increasing demand for accountability, evaluation systems became a key science policy issue in Europe according to the OECD [12]. Public funding needs to be legitimated and therefore related to some measures of performance ([13]: 2). Consequently, many governments implemented research evaluation systems. These systems differ considerably due to the fact that each country has different university funding approaches and legislative systems ([14]: 81, [13]: 2). Geuna and Martin classify four evaluation system cultures in Europe: centralized frameworks (i. e. UK, and France), well-established but uncoordinated evaluation systems across ministries and agencies (i. e. Germany and the Netherlands), weak evaluation culture caused by rigid legislative frameworks for science policy (i. e. most of southern Europe) and finally well-established and distinctive evaluation systems which involve heavy use of overseas panelists (i. e. most of northern Europe) ([14]: 82).

As the evaluation culture differs considerably so does the use of methods. A very common form is the use of peer reviews to evaluate the quality of a paper for example as well as bibliometrics and a combination of these two ([13]: 3). The criteria vary from quality, quantity, impact and utility. A consensus of a set of indicators as well as a holistic evaluation approach for an entire university department or scientific collaboration has not been found yet ([13]: 3). Due to project-based and fundamental research the focus of this paper is on methods and concepts of performance measurement of German public funded CRC and CoE.

3 Forms of Public Funded Scientific Collaboration in Germany

Speaking of different forms of public funded scientific collaboration in Germany, especially funding instruments of the German Research Foundation (DFG) like groups of researchers, graduate schools, CRC and CoE are characteristic. Regarding e. g. structural aspects, the degree of interdisciplinarity and overall (strategic) targets of CRC and CoE, a comparison appears to be reasonable. Here studies and literature provided by the DFG as well as the German Council of Science and Humanities (WR) are used in particular.

For both forms the main funding volume is provided by the BMBF and respective ministries of the sixteen German federal states. The average number of employees of the actually 242 promoted CRC depict forty-one employees with an average public funding volume of 1.67 million Euros per year and CRC ([15, 16]: 6). Considering the four major groups of scientific disciplines (arts and social sciences; life sciences; natural sciences; engineering) which are classified by the DFG, it can be stated that life sciences represent the strongest group among CRC with a share of over 40 per cent of all funded CRC. Approximately 25 per cent of all funded CRC is located in natural sciences, about 20 per cent falls to engineering and about 10 per cent belongs to the group of arts and social sciences 00([16]: 6).

In comparison to that, the average annual funding volume in CoE amounts about 6.5 million Euros per year and CoE with an average number of forty-one employees ([15–17]: 21). Nearly 17 per cent of the funding volume is spent for CoE in arts and social sciences, about 31 per cent for CoE with the focus on life sciences, about 25 per cent for CoE with an emphasis on natural sciences and nearly 27 per cent to CoE focusing on engineering [15].

Considering structural and organizational aspects, e. g. management boards, speakers as well as chief executive officers and coordinators are prevailing in CRC and CoE [16, 18]. Both forms of collaborative research are characterized by a relatively heterogeneous composition of their actors which expresses itself e. g. in diversity of disciplines and the integration of several scientific institutes and affiliated partners – partly industrial partners and/or non-university research institutes – in a consortium. Due to strategic aims of the promoters, such as strengthening the existing structures of individual scientific institutions, promoting young scientists or enlarging the international reputation of Germany as a research location, interdisciplinary teams of scientists increasingly collaborate in common research topics [15, 18].

The overall aim of CRC is to support excellent research within the scientific network and to facilitate the universities' focus through a long-term funding lasting up to twelve years ([19]: 1). From 1968 till 1999 the composition of CRC was limited to one university which simultaneously occurred as applicant for public funding. From 1999 CRC program variants induced a diversified structure [20]. In this way so called CRC/Transregio, which have to demonstrate the participation of several leading universities (normally up to three universities) as a basis for their research work, can be promoted as well. Thereby the former local profile was changed [20] so that not only scientists have to be integrated in a CRC that work at the same location but also scientists from different research locations across Germany.

CoE constitute a funding program in the context of the German excellence initiative. The latter has been initiated by the BMBF as well as the German federal states in 2005 in order to strengthen university structures, promoting Germany as a location for excellent science. In the course of the first funding period of the excellence initiative, thirty-seven CoE were initiated in the years 2006 and 2007 for at least five years. The provision of an excellent research performance within the respective scientific fields represents a superior funding requirement ([21]: 2f). Besides the intention to create and exploit synergies at one university, the willingness for an interdisciplinary cooperation with other universities, institutes and (regional and international) partners from industry is desired. To establish structural sustainability of CoE, cooperative relations with the private sector are of great importance to reduce the dependence on public funding in the long-term ([17]: 51f.). The thematic coherence of the scientific program depicts another funding requirement that has to be internationally visible and relevant for social and economic interests. Moreover, the internal promotion of young researchers as well as gender equality are expected ([21]: 2f, cf. Table 1).

An ongoing public funding of CRC and CoE is in accordance with the overall strategy of BMBF and DFG to overcome the pillarization of single disciplines enabling cross-sectional innovations [22]. Due to the augmenting complexity of these new forms of cross-sectional collaboration, performance measurement aspires to contribute to steering and regulation of CRC and CoE what will be outlined in the following sections.

Table 1 Central aspects of comparing CRC with CoE

Similarities of CRC and CoE	Differences between CRC and CoE
• Aiming at improving the structural development of universities	• First initiation in Germany (CRC: 1968 vs. CoE: 2006)
• Aiming at promoting young scientists and women in science	• Average funding volume: 1.67 millions of Euros per year and CRC vs. 6.5 millions of Euros per year and CoE
• Cooperation with (partly non-university; international) research institutes	• Duration of respective funding periods (CRC: four years vs. CoE: five years)
• Objective to transfer scientific results (e. g. into practice)	• Stronger focus on basic research than on applied research in CRC in comparison to CoE
• Aiming at an increasing international visibility and reputation of the research location Germany	• Local profile of leading partners in classical CRC and CoE vs. leading partners from different regions in CRC/Transregio
• Central organizational structures (e. g. speakers and chief executive officers)	

4 About a Changing Understanding of Performance Measurement

The current term performance is ambiguously defined within literature. On the one hand performance is defined as the consequence of efficient and effective actions on all service and decision-making levels of an organization in the light of satisfying plural interests concerning multidimensional objectives such as finances, quality or processes ([23]: 33). On the other hand performance stands for the quality of benefits for the stakeholders of an organization ([24]: 49, [25]: 10). For Neely [26] performance measurement embodies a process of quantifying efficiency and effectiveness of an action. With help of performance measurement, organizational objectives shall be improved and the cross-hierarchical communication and cooperation shall be promoted ([27]: 239). Performance measurement has its roots in business administration – especially in the sub-areas: financial management, human resources and controlling. In this context it is directly linked to the phases of management: planning, steering and controlling and embodies the process of defining goals as well as setting up a feedback process for an organization in the sense of a cybernetic loop ([28]: 21).

It is notable that the understanding of performance measurement and respective KPIs has changed in theory and practice since the 1990s. In the conventional sense, KPIs represent quantitative data compressing a complex reality ([29]: 15, [30]: 5). As proponents of a changing understanding of performance measurement and associated indicators, Eccles [31] as well as Kaplan/Norton [32] point to the need to measure KPIs which go beyond financial aspects and which possess a multidimensional character [33]. To provide information, e. g. about output or processes of an organization, KPIs thus play a decisive role for persons in leading management positions. To develop and to collect a set of KPIs, different methods can be used ([34]: 167). Diverse dimensions have to be considered such as the past- or future-orientation of indicators, the organization-internal or -external use of indicators, and the nominal or interval scaling of indicators. Overall, KPIs are ideally allowing a certain indicator-flexibility and dynamic of change ([28]: 23).

With the ascent of the New Public Management in the 1980s which advanced the transfer of management methods and principles from the private sector to public administrations and institutions, further fields of application emerged for performance measurement such as research facilities, research associations, CRC or CoE ([35]: 37ff). Concerning the implementation of performance measurement in CRC or CoE, e. g. the challenge can be described by supporting the realization of defined objectives, providing information about ongoing processes and decisions as well as supplying a base for decision-making for the management and integration of interdisciplinary research partners [10].

Whereas performance measurement in a profit-oriented organization is targeted to profit maximization, performance measurement in the context of CRC and CoE primarily focus on the measurement of efficiency in interdisciplinary collaborations. Hence, the research output like (interdisciplinary) publications, awards, products, new projects and the quality of processes e. g. concerning cooperation, communication and research work have to be measured. Performance measurement within CRC and CoE has to be guided by the visions and objectives determined in the application of funding. In order to achieve the defined objectives as well as to initiate a continued improvement process among all stakeholders, the results of performance measurement are used to draw up recommendations for steering and regulating interdisciplinary forms of collaboration. At this point it has to be emphasized that the importance of financial KPIs is relativized. Finances represent the base of a continual improvement process and not the overall target in the context of a non-profit scientific organization. Innovation and learning processes – including associated KPIs – are focused instead in scientific collaborations [36].

The need for concepts to measure, steer and regulate the examined forms of scientific collaboration is expressed by organizations that are responsible for public funding such as the DFG and the WR ([18]: 44f, [17]: 7). Thus, the study aims at providing more empirical evidence about application and diffusion of KPIs as well as forms of performance measurement in CRC and CoE.

5   Design of the Empirical Study

To gain more evidence about the application of performance measurement approaches in CRC and CoE an empirical study was executed throughout Germany in 2010/2011. The study also aspired to receive an evaluation of scientific performance indicators by representatives of CRC and CoE and insights about the impact of CRC and CoE on the promotion of the research location Germany. For the design of the study an online-survey was carried out using a semi-standardized questionnaire which followed Dillman's [37] total design method. Not only questions with interval scaled variables but also questions with open answers were constructed to facilitate comments in form of free text. The range of the interval scaled variables covered five possible answers enabling neutral answers, too, plus the optional answer 'no indication'. The group of speakers and chief executive officers of all actively funded CRC and CoE (n = 278) was selected as the target group (except two CoE at RWTH Aachen University in which the authors of the study are involved as members) because – due to their high responsibility as science managers – it is the most capable group for statements concerning adequate KPIs and methods of performance measurement.

The selection of respective items of the questionnaire based on the following considerations: The success of a CRC or CoE is fundamentally determined by the attainment of objectives formulated in the respective funding agreement. Thus, the overall objectives of CRC and CoE (among other things: strengthening the science location, improving the international competitiveness or expanding excellent peaks in the science sector) as well as possible forms for measuring the attainment of objectives were requested in the questionnaire. The overall objectives of both forms of scientific collaboration were derived from the funding criteria of the DFG and the WR [16, 38]. These were:

- Research quality,
- Originality and coherence of the scientific program,
- Interdisciplinary collaboration,
- Influence on the field of research in future,
- Options for transferability into practice,
- Quality of scientists and their further options for development,
- Integration of local research capacities,
- Structural development of the university,
- International visibility.

6 Results

6.1 Selected Results Concerning Participants and Response Proportion

A total of about 40 per cent of all addressed speakers (main representative and coordinator of a CoE/CRC) and chief executive officers (n = 278) participated or at least started with the survey whereas the response quota was about 23 per cent (finished surveys: n = 65). The majority of those respondents that did not complete the survey cancelled the survey after the introduction and the general questions. The entire number of sixty-five exploitable surveys allowed analyses by means of descriptive statistics.

In detail about 21 per cent (n = 52) of all active CRC and about 37 per cent (n = 13) of all active CoE – except the two CoE of RWTH Aachen University in which the authors are involved – took part in the survey. With reference to the four major groups

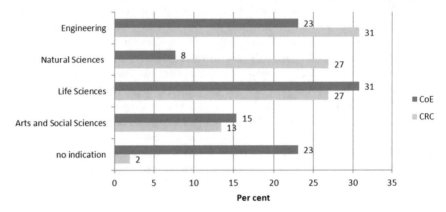

Figure 1 Categorization of responding CRC and CoE into major groups of scientific disciplines

of scientific disciplines (arts and social sciences; life sciences; natural sciences; engineering) it can be stated that representatives from all groups answered the survey (cf. Figure 1). Referring to the latter, 23 per cent of the responding CoE answered with 'no indication' what can be explained by a higher degree of interdisciplinarity in CoE in comparison to CRC.

Structural differences in terms of varying numbers of employees became obvious during the data analyses, too. Whereas all participating CoE indicated to have more than one-hundred employees, the numbers are varying among CRC. In detail about 39 per cent of the participating CRC have ten to fifty employees about 52 per cent have fifty to one-hundred employees and just about 6 per cent indicated to have more than one-hundred employees.

6.2 Selected Results Concerning the Importance of Performance Measurement and KPIs in the Context of CRC and CoE

The interval scaling of the closed questions (in style of the German school mark system) allowed data analyses e.g. considering arithmetic averages ($\bar{x}$) and corresponding standard deviations (σ) to analyze the dispersion of answers. Open questions enabled the addressed speakers and chief executing officers to add suggestions or criticism.

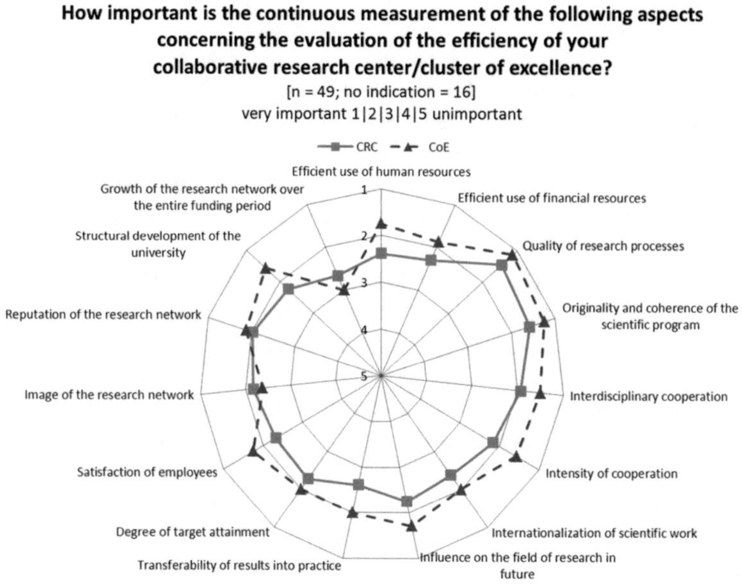

How important is the continuous measurement of the following aspects concerning the evaluation of the efficiency of your collaborative research center/cluster of excellence?
[n = 49; no indication = 16]
very important 1|2|3|4|5 unimportant

Figure 2 Importance of continuous performance measurement

In the following network diagrams results are represented as arithmetic averages to enable interpretations according to the German school mark system, including e. g. 'indifferent answers'. The results are symbolized by squares for CRC and by triangles for CoE (cf. Figure 2).

With reference to the fulfillment of external funding criteria – formulated by the DFG and the WR – the speakers and chief executive officers were asked to respond to the question: How important is the continuous measurement of the following aspects concerning the evaluation of the efficiency of your CRC/CoE? (cf. Figure 2). Answers could be given by an interval scale ranging from 1 = very important, 2 = important, 3 = neutral, 4 = less important to 5 = unimportant plus the optional answer 'no indication'.

Regarding the answers of the CRC in Figure 2, the top three items concerning the importance of a continuous performance measurement depict: quality of research processes ($\bar{x} = 1.44$; $\sigma = 0.73$), originality and coherence of the scientific program ($\bar{x} = 1.59$; $\sigma = 0.81$) and interdisciplinary cooperation ($\bar{x} = 1.93$; $\sigma = 0.85$). The growth of the research network over the entire funding period ($\bar{x} = 2.66$; $\sigma = 0.82$) constitutes the least important item concerning the answers of the CRC. Here one can assume that the expansion of a public funded CRC is rather associated with a phase of structural sustainability, e. g. including private promoters, than during the phase of public funding.

With reference to the three most important items concerning a continuous performance measurement in CoE, the same ranking as in the CRC can be described: quality of research processes ($\bar{x} = 1.13$; $\sigma = 0.35$), originality and coherence of the scientific program ($\bar{x} = 1.25$; $\sigma = 0.46$) and interdisciplinary cooperation ($\bar{x} = 1.50$; $\sigma = 0.53$). The answers among the speakers and chief executive officers just vary a few what can be underlined by low σ-values. Analyzing the least important item in the eyes of CoE representatives – and parallel to the results of the CRC – the growth of the research network over the entire funding period ($\bar{x} = 3.00$; $\sigma = 1.10$) is named. The dispersion of the corresponding answers is relatively high what can be due to a more sustainable strategic management of collaborative research in CoE than in CRC.

In summary thirteen from fifteen items concerning Figure 2 are evaluated as being important or very important among CoE representatives whereas the items were evaluated slightly worse among CRC representatives. Overall, one can assume that the importance of performance measurement seems to be more important among CoE representatives than among CRC representatives. The quality of research processes constitutes the most essential item for both forms of scientific collaboration. This emphasizes that e. g. an optimization of interdisciplinary research processes in terms of effectiveness and efficiency constitutes a central task within CRC and CoE.

In a following question the speakers and chief executive officers were asked: How do you evaluate the following indicators concerning an evaluation of the performance of your CRC/CoE? By the term "evaluate" the participants were asked to judge by using marks. In contrast to the previous questions, answers could be given by an interval scale ranging from 1 = very appropriate, 2 = appropriate, 3 = neutral, 4 = less appropriate to 5 = inappropriate and the optional answer 'no indication' (cf. Figure 3).

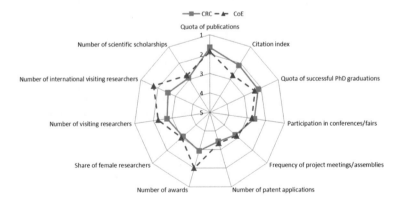

Figure 3 Evaluation of performance indicators

The top three performance indicators of the CRC are: quota of publications ($\bar{x} = 1.66$; $\sigma = 0.82$), citation index ($\bar{x} = 2.15$; $\sigma = 1.04$) and quota of successful PhD graduations ($\bar{x} = 2.18$; $\sigma = 0.87$). Searching for the least appropriate indicator among the CRC representatives, the number of patent applications ($\bar{x} = 3.44$; $\sigma = 1.33$) can be mentioned what is not really astonishing due to the fact that CRC classically focus more on basic research than on applied research. As a reason for the relatively high σ-value, a changing understanding of the CRC research alignment can be supposed, e. g. in terms of a growing share of practice-oriented research questions in CRC. Furthermore, the σ-value can be explained by different scientific alignments of CRC. Hence, patents are more probable in engineering than in arts and social sciences.

The top three performance indicators of CoE constitute: number of international visiting researchers ($\bar{x} = 1.75$; $\sigma = 0.71$), quota of publications ($\bar{x} = 1.86$; $\sigma = 0.69$) and number of awards ($\bar{x} = 2.00$; $\sigma = 0.58$). The least appropriate performance indicator for CoE depicts the number of patent applications ($\bar{x} = 3.42$; $\sigma = 1.29$), although the level of disagreement among the participants is relatively high. Like in CRC, this can be explained by different scientific alignments in CoE.

As a conclusion of Figure 3 one can state that the performance indicator: quota of publications is evaluated in average as the most important indicator for scientific performance in CRC and CoE. The indicator does not only receive the best evaluations in average, but also a high level of agreement by all respondents – elucidated by the low σ-value. Considering the other top performance indicators of CRC and CoE, it appears as if CRC emphasized more classical scientific performance indicators such as citations in comparison to CoE which e. g. highlighted the number of awards.

Table 2 Regression analysis of items concerning the importance of performance measurement and KPIs

15 items concerning the importance of performance measurement	Adjusted r^2
Efficient use of human resources	**0.533**
Efficient use of financial resources	**0.528**
Quality of research processes	0.284
Originality and coherence of the scientific program	0.175
Interdisciplinary cooperation	0.446
Intensity of cooperation	0.376
Internalization of scientific work	0.459
Influence of the field of research in future	**0.524**
Transferability of results into practice	**0.506**
Degree of target attainment	**0.664**
Satisfaction of employees	0.490
Image of the research network	**0.511**
Reputation of research network	0.390
Structural development of the university	0.310
Growth of the research network over the entire funding period	**0.562**

Additionally a regression analysis was done to estimate relationships among variables ([39]: 36). Therefore all 15 items concerning the importance of performance measurement were analyzed to get an impression of the typical value of a criteria variable when the independent variables (possible KPI) are fixed. The analysis shows that 7 out of 15 items can be predicted by the KPI (adjusted $r^2 > 0.5$): degree of target attainment ($r^2 = 0.664$), growth of the research network over the entire funding period ($r^2 = 0.562$), efficient use of financial resources ($r^2 = 0.528$), efficient use of human resources, influence on the field of research in future ($r^2 = 0.524$), image of the research network ($r^2 = 0.511$) and transferability of results into practice ($r^2 = 0.506$) (cf. Table 2).

Furthermore the regression analysis shows which KPIs have a high impact on the items concerning the importance of performance measurement. A correlation coefficient needs to be higher than 0.5 to have a significant impact ($\beta > 0.5$) ([28]: 284, [40]: 116). Therefore, it can be stated that the number of scientific scholarships

Table 3 Categorized additional performance indicators and comments

Category	Performance indicators/comments
Publications/thesis	• Common (interdisciplinary) publications of sub-projects • Quality of publications • Comparison of scientific results with the respective state-of-the-art • Share of finished PhD examinations graded with 'summa cum laude' • Number of supervised master theses
Scientists	• Personal reputation of the involved scientists • Number of scientists (and especially female scientists) appointed to professorships
University structures	• Structural development of the university
Internationalization, networking and public relations	• Internationalization of scientific concepts • Organization of (international) conferences • Global networking with other subject-specific research networks • Public relations (e. g. events and projects)
Benefits of results for practice and exploitation	• Benefit of scientific results for practice • Exploitation of scientific results in science/economy/society • Number of cooperations with industrial partners • Number of transfer projects (and application of results in economy)
Finances and funding	• Number of sub-projects receiving additional funding by programs of the European Union • Return on investment with regard to funding • Amount of additionally acquired project funding by involved scientists

has the highest impact as four items correlate significantly. Especially interdisciplinary cooperation is influenced significantly by number of visiting researchers ($\beta = -0.878$), number of international visiting researchers ($\beta = 0.579$) and number of scientific scholarships ($\beta = 0.514$).

All respondents were asked to add further performance indicators or comments from their point of view in a corresponding open question section of Figure 3. The performance indicators/comments are subjective, have not been evaluated by other participants and were divided into six categories (cf. Table 3).

The emphasized performance indicators/comments underlined a holistic, multi-dimensional understanding of performance measurement by the respondents. Hence, qualitative aspects like personal reputation of the involved scientists or benefit of scientific results for practice, as well as quantitative aspects such as return on investment with regard to funding or number of cooperations with industrial partners were named.

6.3 Selected Results Concerning the Impact of CRC and CoE on the Promotion of the Research Location Germany

The impact of CRC and CoE on the promotion of the research location Germany was evaluated in a subsequent question. The participants of the survey could answer by an interval scale ranging from 1 = very applicable, 2 = applicable, 3 = neutral, 4 = less applicable to 5 = not applicable plus the optional answer 'no indication' (cf. Figure 4).

Regarding the top three answers given by the speakers and chief executive officers of CRC, the respondents highlighted the relatively high added value of the funding program CRC for the German science system ($\bar{x} = 1.37$; $\sigma = 0.58$). This item is followed by the statement that the CRC promoted the funding of young scientists in Germany ($\bar{x} = 1.50$; $\sigma = 0.77$). In addition to that, innovative research concepts were realized ($\bar{x} = 1.63$; $\sigma = 0.72$) with help of the program CRC. Concerning the least applicable impact of the funding program CRC, the intensified cooperation with economy ($\bar{x} = 3.38$; $\sigma = 1.06$) was indicated what e.g. can be explained by a relatively strong basic research alignment of CRC and a weak involvement of industrial partners.

With reference to the CoE, the participants associated the highest impact with the image improvement of the research location Germany ($\bar{x} = 1.33$; $\sigma = 0.50$). This is followed by the added value for the German science system ($\bar{x} = 1.50$; $\sigma = 0.53$) as well as the statement that CoE promoted the funding of young scientists in Germany ($\bar{x} = 1.89$; $\sigma = 1.05$). Analyzing the least applicable aspect concerning the impact of the CoE, the added value for the regional economy ($\bar{x} = 3.00$; $\sigma = 0.71$) was indicated what could be traced back to the fact that CoE constitute a relatively new funding program, initiated in 2006.

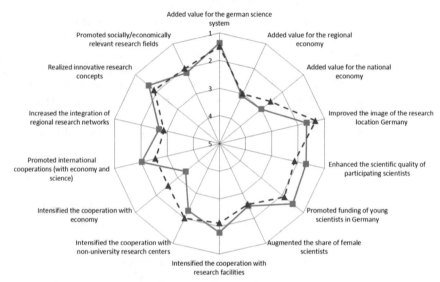

The funding programs collaborative research centers and clusters of excellence...
[n = 52; no indication: 13]
very applicable 1|2|3|4|5 not applicable

Figure 4 Impact of CRC and CoE on the promotion of the research location Germany

6.4 Selected Results Concerning Frequency and Methodology Used in the Context of Performance Measurement in CRC and CoE

Indicating the frequency and methodology used in the context of performance measurement in CRC and CoE was the focus of Figure 5. Analyzing the CRC answers, 32 per cent of the respondents stated to collect the performance indicators on a yearly basis, 21 per cent on a six-monthly basis, 20 per cent sporadically, 17 per cent on a quaterly basis and 11 per cent indicated to collect the indicators monthly. CoE-specific data revealed the following results: 56 per cent collect the performance indicators on a yearly basis, 22 per cent sporadically and 11 per cent monthly (cf. Figure 5). Referring to the yearly collection of indicators, one can assume that the majority of respondents from CoE and about one third of the respondents from CRC uses the results preliminary in terms of yearly status reports.

Asking for methodologies used in the context of performance measurement, the following items can be highlighted, because they are used/partly used in CRC. Personal internal conversations with employees and personal external conversations with the scientific community are broadly used as well as reporting and evaluations (cf. Figure 6). The two most common forms of obtaining performance indicators in CRC

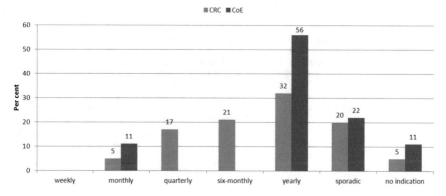

Figure 5 Frequency of collecting performance indicators in CRC and CoE

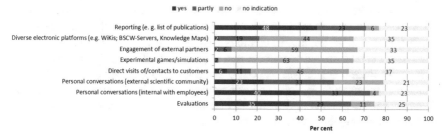

Figure 6 Methodology used in the context of performance measurement in CRC

are reporting (used by 48 per cent of all participating CRC) and personal internal conversations with employees (used by 40 per cent of all participating CRC).

With reference to the used/partly used methodologies among CoE participants, reporting, personal internal conversations with employees and personal external conversations with the scientific community as well as evaluations are indicated (cf. Figure 7). The two most common forms thus constitute reporting (used by 46 per cent of all participating CoE) and evaluations (used by 38 per cent of all participating CoE). In comparison to the answers of the CRC it is astonishing that 23 per cent of the CoE participants obtain performance indicators by the engagement of external partners and 15 per cent indicate to receive indicators via diverse electronic platforms such as WiKis or knowledge maps.

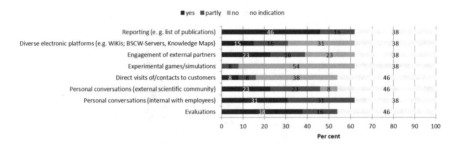

Figure 7 Methodology used in the context of performance measurement in CoE

In the context of an additional open question in this section of the survey, the participants could indicate a favorite method for the strategic target/performance comparison of their respective CRC/CoE. It has to be underlined that the majority of speakers and chief executive officers did not give a qualitative answer. Those participants who answered the question mentioned methods like: SWOT-analyses, project controlling, workshops to discuss results and targets formulated in the funding agreement as well as qualitative methods such as interviews with employees. Apart from that, keeping an eye on competitors (e. g. other CRC or CoE) and their super-ordinated research targets was named by one respondent, revealing the existence of benchmarking aspects. Overall, one can assume that holistic methods and concepts of target/performance comparison and performance measurement seem not to be widespread among the responding CRC and CoE.

7 Discussion of Methodological Approach

The study used a semi-standardized questionnaire to gain more empirical evidence about the application of performance measurement, the evaluation of scientific performance indicators by speakers and chief executive officers of CRC and CoE and the impact of both funding programs on the promotion of the research location Germany. The development of the entire survey – especially the interval scaled questions aiming at an evaluation of scientific performance indicators – based on contemporary findings in the topic of scientific performance measurement and on funding criteria for CRC and CoE. The study does not claim to be exhaustive in all aspects of scientific performance measurement. In order to counteract this fact, the survey provided the possibility of adding further aspects of performance measurement or scientific performance indicators in form of free text.

Considering the entire sample of all actively funded CRC and CoE (n = 278) the response quota of 23 per cent can be explained in particular by the following reasons: On the one hand, speakers and chief executive officers of CRC and CoE are confronted with enormous time restrictions regarding their daily business as science and management executives. On the other hand, questions about performance measurement and scientific performance indicators depict to some extent critical data which is not shared and disclosed by every participant voluntarily. Additionally, extend and design of the survey could be possible reasons for the response quota.

8 Concluding Remarks and Outlook

The increasing demand for legitimating public funding and therefore accountability in non-profit organizations in Europe can be met by establishing research evaluation systems ([13]: 2). As a holistic approach has not been found yet, this paper discusses performance measurement methods and concepts in the context of German research collaboration forms.

In the context of emerging New Public Management approaches as well as a changing understanding of performance measurement ([35]: 37ff), the latter is adapted and extended in order to transfer successful concepts – preliminary implemented in profit-oriented organizations – to non-profit-oriented organizations. Special forms of non-profit-oriented organizations thus depict CRC and CoE in Germany, which can be described as public funded scientific collaborations integrating in particular interdisciplinary scientific partners.

With the presentation of results of an empirical study – addressed to the speakers and chief executive officers of all actively funded German CRC and CoE in 2010/2011 – the paper gives insights in how far performance indicators and performance measurement concepts are used to promote scientific processes in CRC and CoE as well as the respective research output:

Concerning the importance of a continuous performance measurement (such as: quality of research processes, originality and coherence of the scientific program and interdisciplinary cooperation) thirteen from fifteen items are evaluated as being important or very important among the respondents (cf. Figure 2). The items are evaluated slightly worse among representatives of the CRC. One can assume that the importance of performance measurement seems to be greater among representatives of CoE than among representatives of CRC. A possible reason could be the average number of employees, because the responding CRC had fewer employees in average than the responding CoE.

Searching for KPIs that are suitable to measure the performance of CRC and CoE and fulfil the demand to compress a complex reality (cf. Lachnit 1979: 15; cf. Gladen 2002: 5) one can state that in average the performance indicator: quota of publications is evaluated as the most important indicator for scientific performance in CRC and CoE. This is not surprising, because publications constitute a superior scientific output in general. In addition to that the quota of successful PhD graduations and the number of international visiting researchers are further KPIs for both forms. As a differentiating example the number of awards is a KPI for CoE which is not as decisive for CRC. Vice versa the citation index is a KPI for CRC but not as crucial among CoE. At this point, one can assume that CRC are more performing in a classical scientific manner than CoE which perform slightly more in a project-oriented manner. Reflecting an overall interpretation of Figures 2 and 3 one can consider that continuous performance measurement is of great importance in CRC and CoE whereas the listed performance indicators (cf. Figure 3) seem not to completely meet the understanding of scientific performance indicators in the eyes of all respondents. However, the demand to measure KPIs which go beyond financial aspects and which possess a multidimensional character [33] can be met with the investigated items.

Considering the impact of the funding programs CRC and CoE on different aspects, the added value of both funding programs for the German science system is highlighted (cf. Figure 4). Especially, both programs promoted the funding of young scientists what can be evaluated as a central achievement of the overall funding strategy formulated by BMFB, DFG and WR.

With regard to the frequency of collecting performance indicators, the majority of respondents indicated to collect data on a yearly basis. In CoE this frequency seems to be prevailing instead of a more diversified frequency among CRC (cf. Figure 5). This could mean that performance indicators are particularly used in terms of a yearly status report.

Concerning methods implemented for collecting performance indicators, reporting, personal internal conversations with employees, personal external conversations with the scientific community as well as evaluations are noted by the majority of respondents. Hence, qualitative aspects of performance measurement such as conversations or interviews can be assumed as being important.

Overall, the results of the study revealed that a lot of speakers and chief executive officers are aware of the importance of performance measurement and respective KPIs of their CRC and CoE although holistic methods and concepts of target/performance comparison and performance measurement seem not to be widespread. Hence, more research is necessary, considering the development of performance measurement and performance management concepts in the context of CRC and CoE in order to set up frameworks that provide orientation for the management level also in terms of steering and regulation collaborative research [10].

Acknowledgments This work was performed as part of the cluster of excellences "Integrative Production Technology for High-Wage Countries" and "Tailor-Made Fuels from Biomass", which are funded by the excellence initiative by the German federal and state governments to promote science and research at German universities.

References

1. BMBF – Bundesministerium für Bildung und Forschung. Bundesbericht Forschung und Innovation, 2008. http://www.bmbf.de/pub/bufi_2008.pdf
2. BMBF – Bundesministerium für Bildung und Forschung. Exzellenzinitiative, 2008. http://www.bmbf.de/de/1321.php
3. S. Hornbostel, M.v. Ins, Leistungsmessung in der Forschung - Erfordernis oder Vermessenheit? Laborwelt **10** (5), 2009, pp. 34–35
4. R. Vossen, F. Welter, I. Leisten, A. Richert, I. Isenhardt, Making scientific performance measurable - experiences from a german cluster of excellence. In: *ICERI 2011*, ed. by L. Gómez Chova, A. López Martínez, I. Candel Torres, International Association of Technology, Education and Development, IATED, Valencia, 2011, pp. 3878–3886
5. Welter, F., Jooß, C., Richert, A., Jeschke, S., C. Brecher, Organisation and management of integrative research. In: *Integrative production technology for high-wage countries*, ed. by C. Brecher, Springer, Berlin and New York, 2012, pp. 64–73
6. J.C. Higgins, Performance measurement in universities. European Journal of Operational Research **38** (3), 1989, pp. 358–368
7. J. Guthrie, R. Neumann, Economic and non-financial performance indicators in universities. Public Management Review **9** (2), 2007, pp. 231–252
8. S. Rassenhövel, *Performancemessung im Hochschulbereich: Theoretische Grundlagen und empirische Befunde*. Gabler, 2010
9. C. Jooß, F. Welter, A. Richert, S. Jeschke, C. Brecher, A management approach for interdisciplinary research networks in a knowledge-based society – case study of the cluster of excellence integrative production technology for high-wage countries. In: *ICERI2010 Proceedings*. pp. 138–143
10. Welter, F., Vossen, R., Richert, A., I. Isenhardt, Network management for clusters of excellence: A balanced-scorecard approach as a performance measurement tool. The Business Review **15** (1), Summer 2010, pp. 171–178
11. J. Sydow, S. Duschek, *Management interorganisationaler Beziehungen*. Kohlhammer Edition Management. Kohlhammer, Stuttgart, 2010
12. OECD. The evaluation of scientific research: Selected experiences, 1997. http://www.oecd.org/science/sci-tech/2754549.pdf
13. T. Luukkonen, Research evaluation in europe: state of the art. Research Evaluation **11** (2), 2002, pp. 81–84. doi:10.3152/147154402781776871
14. A. Geuna, B.R. Martin, University research evaluation and funding: an international comparison. Minerva **41**, 2003, pp. 277–304
15. DFG – Deutsche Forschungsgemeinschaft, ed., *40 Jahre Sonderforschungsbereiche*. Bonn, 2008
16. DFG – Deutsche Forschungsgemeinschaft. Monitoring des Förderprogramms Sonderforschungsbereiche, 2010
17. M. Sondermann, D. Simon, A.M. Scholz, S. Hornbostel, eds., *Die Exzellenzinitiative: Beobachtungen aus der Implementierungsphase*. 5. Bonn, 2008
18. DFG/WR – Deutsche Forschungsgemeinschaft and Wissenschaftsrat. Bericht der gemeinsamen Kommission zur Exzellenzinitiative an die gemeinsame Wissenschaftskonferenz, 2008. http://www.gwk-bonn.de/fileadmin/Papers/GWK-Bericht-Exzellenzinitiative.pdf
19. A. Reinhardt. Mehr als die summe der teile - die Sonderforschungsbereiche/Transregio: Zentrale Befunde der Evaluation der Variante des Programms Sonderforschungsbereiche, 2009. http://www.dfg.de/download/pdf/dfg_im_profil/evaluation_statistik/programm_evaluation/ib01_2009.pdf
20. DFG – Deutsche Forschungsgemeinschaft. Sonderforschungsbereiche, 2012. http://www.dfg.de/foerderung/programme/koordinierte_programme/sfb/
21. DFG – Deutsche Forschungsgemeinschaft. Merkblatt Exzellenzcluster, 2006. http://www.dfg.de/forschungsfoerderung/formulare/download/exin1.pdf

22. Jooß, C. and Welter, F. and Richert, A. and Jeschke, S., The challenge of scientific cooperation in large complex research clusters - experiences from the cluster of excellence "integrative production technology for high-wage countries". In: *Proceedings of the 12th European Conference on Knowledge Management.* pp. 481–487

23. D. Hilgers, *Performance-Management: Leistungserfassung und Leistungssteuerung in Unternehmen und öffentlichen Verwaltungen*, 1st edn. Gabler-Edition Wissenschaft. Gabler, Wiesbaden, 2008

24. S. Kozielski, *Integratives Kennzahlensystem für den Werkzeugbau.* Apprimus Verlag, Aachen, 2010

25. T. Wettstein. Gesamtheitliches performance-measurement - Vorgehensmodell und Informationstechnische Ausgestaltung., 2002

26. A. Neely, *Measuring Business Performance.* Bloomberg Press, London, 1989

27. P. Horváth, R. Gleich, D. Voggenreiter, *Controlling umsetzen: Fallstudien, Lösungen und Basiswissen*, 3rd edn. Schäffer-Poeschel, Stuttgart, 2001

28. R. Gleich, *Das System des Performance measurement: Theoretisches Grundkonzept, Entwicklungs- und Anwendungsstand.* Vahlen, München, 2001

29. L. Lachnit, *Systemorientierte Jahresabschlussanalyse.* Wiesbaden, 1979

30. W. Gladen, Performance measurement als Methode der Unternehmenssteuerung. In: *Performance Measurement*, ed. by H.P. Fröschle, Praxis der Wirtschaftsinformatik, 2002, pp. 5–16

31. R.G. Eccles, The performance measurement manifesto. Harvard Business Review **43**, 1991, pp. 131–137

32. R.S. Kaplan, D.P. Norton, The balanced scorecard measures that drive performance. Harvard Business Review **Jan-Feb**, 1992, pp. 71–79

33. P. Horváth, M. Seiter, Performance measurement. Die Betriebswirtschaft **69** (3), 2009, pp. 393–414

34. J. Weber, U. Schäffer, *Einführung in das Controlling*, 11th edn. Schäffer-Poeschel, Stuttgart, 2006

35. K. Schedler, *New Public Management, UTB für Wissenschaft. Uni-Taschenbücher*, vol. 2132 . Kleine Reihe. Haupt and UTB, Bern and Stuttgart, 2000

36. N. Klingebiel, *Performance measurement & balanced scorecard.* Controlling. Vahlen, München, 2001

37. D.A. Dillman, *Mail and telephone surveys: The total design method.* Wiley, New York, 1978

38. DFG – Deutsche Forschungsgemeinschaft. Zweite Phase der Exzellenzinitiative - Förderlinie Exzellenzcluster, 2010

39. D. Kleinbaum, L. Kupper, A. Nizam, E. Rosenberg, *Applied regression analysis and other multivariable methods*, fifth edition edn.

40. J. Cohen, *Statistical power analysis for the behavioral sciences*, 2nd edn. L. Erlbaum Associates, Hillsdale and N.J, 1988

Shaping the Future Through Cybernetic Approaches of Social Media Monitoring

Sebastian Stiehm, Florian Welter, Anja Richert and Sabina Jeschke

Abstract Scientific research and development programs (R&D programs) are national instruments to sustainably secure innovative capability and competitiveness. Due to an increasing rate of change in all societal functional areas, these programs have to be continuously advanced, but also new R&D programs have to be tendered. Prospectively more societal impulses have to be taken into account for the advancement of R&D programs and the ex-ante determination of program contents. Here, the methodical basis is characterized by the analysis of social needs. In terms of substance, sources of Social Media (SM) work out perfectly as data or text corpora: Everyday life is becoming increasingly digitally networked and a large part of interpersonal communication is realized via SM. SM represent a pool of qualitative and quantitative data in order to reflect societal moods. It can be regarded as untouched, raw and unevaluated data. Existing methods of Social Media Monitoring (SMMO) use this information as a basis for trend analysis, issue monitoring and the detection of influencers. SMMO is no temporal specific action, but rather an open-ended task. The conventional application fields of SMMO primarily relate to commercial market research, corporate communications and public relations. In this context SMMO is used with the intent of an overall social and political use, interest or benefit. A new approach is currently being developed by considering methods of system theory and cybernetics. Using this theoretical, system-oriented framework, R&D programs can be constructed as socio-technical, complex living systems. Finally, cybernetic SMMO allows for a continuous and active involvement of the society into politics. It supports program management and research promoters of publicly funded R&D projects by taking into account social impulses for the advancement of R&D programs and the ex-ante determination of program contents. Cybernetic SMMO enables an active shaping of the future according to societal developments, trends and needs.

S. Stiehm (✉) · F. Welter · A. Richert · S. Jeschke
IMA/ZLW & IfU, RWTH Aachen University,
Dennewartstr. 27, 52068 Aachen, Germany
e-mail: sebastian.stiehm@ima-zlw-ifu.rwth-aachen.de

Originally published in "Proceedings of the European Conference on Social Media (ECSM 2014)", © Academic Conferences and Publishing International Limited 2014. Reprint by Springer International Publishing Switzerland 2016, DOI 10.1007/978-3-319-42620-4_15

179

Keywords Social Media Monitoring · Social Media · Cybernetic Approach · Trends and Pattern Identification

1   Introduction and Problem Statement

Large scientific R&D programs with a duration of five up to ten years are instruments of the German federal government to sustainably secure Germany's innovative capability and competitiveness [1, 2]. Shaping and steering these programs depicts a complex challenge due to their strong application reference on the one hand and due to an increasing rate of change within the respective thematic focus on the other hand [1]. Trends like digitalization, automation etc. especially enforce short cyclic changes, wherefore R&D programs have to be continuously adapted [3]. A complete ex-ante determination of program contents and the output of funded research projects hence cannot stand the dynamics of socio-economic developments. Flexibility and adaptability, as crucial characteristics of contemporary individuals and organizations [4], thus must be reflected in the requirements for R&D programs.

As Trantow [1] explains, permanent organizational learning processes are required to allow a utilization of generated research results and other program outputs in order to shape funding activities. In order to enable learning processes, a continuous monitoring and reflecting of relevant, R&D program-related transdisciplinary information is necessary. This allows going beyond own program results and particularly considering the international research context. Relevant information is processed by monitoring and reported back to different functional units of the R&D program. So far, monitoring activities are basic elements for the constitution of learning processes in R&D programs.

In addition to the previously mentioned monitoring activities, more social impulses will prospectively be taken into account for the advancement of R&D programs and the ex-ante determination of program contents. Approaches such as the citizen dialogues by the Federal Government, EU citizens dialogues, Google Hangouts with the German Chancellor or events such as "Dialog und Kongress – Fortschritt gestalten" by the state government of North Rhine-Westphalia [5, 6] illustrate how the dialogue between politics and society increasingly gains importance. Moreover, SM also play a fundamental role in this context, because everyday life is becoming increasingly digitally networked and a large part of interpersonal communication is realized via SM [3]. On typical online platforms such as wikis, forums, blogs, microblogs and social networks, people depict current snapshots as well as opinions. Hereby they digitally discuss issues, problems and developments of politics and certain other societal areas such as labour, education or economy etc. [7]. Thus, SM clearly reflect societal moods. Here, monitoring SM will allow for a continuous and active involvement of the society into politics. To approach SMMO, a delimiting theoretical framework is needed. Feed backed steering-, control- and regulation-processes in complex systems have to be considered, wherefore a cybernetic and system-oriented approach is being provided. This results in the following research question:

- How can a cybernetic Social Media Monitoring (SMMO) be designed in order to support the continuous advancement of existing as well as the development of future R&D programs?

The main objective of this paper is the development of a cybernetic approach of SMMO in order to identify and reflect societal moods from SM for an ongoing advancement of R&D programs. A cybernetic approach of SMMO shall therefore allow an active shaping of future according to societal developments, trends and needs.

2 Social Media, Monitoring and Cybernetics

2.1 Social Media

"Social Media are online tools and platforms, that allow internet users to collaborate on content, share insights and experiences, and connect for business and pleasure" [8].

Communication via internet has fundamentally changed during the last years. The classical one-way communication represented by an exclusive availability of static websites belongs to the past. With Web 2.0 applications, respectively the social web, users have many opportunities to communicate across SM platforms such as wikis, forums, blogs and social networks with each other and various stakeholders. Within the social web, a large amount of content is being created that is being replicated by a large number of authors and commented on by an even larger number of readers [9]. This results in user-generated content that is detectable and accessible to anyone on the web. Three factors of the social web are particularly noteworthy [9, 10]:

- *Trust*: User-generated information is often subjective and expresses a personal opinion, which is perceived as credible by other users.
- *Time*: Information is created and spread within a few moments.
- *Mass*: Anyone can create and distribute information. User generated information is constantly newly combined and networked [10].

The contemporary society is characterized by so called prosumers – simultaneous producers and consumers – who determine the content in the social web. The content is being shared and spread and nearly each action is (critically) assessed by the social web. Especially striking actions that are rated critically are virally spreading (i. e. even more quickly) on the social web [11].

Content on the social web is very heterogeneous: There are platforms that specialize in certain media formats such as images, video, and music, while others are defined rather by their functions, e. g. public relations, evaluation or information. In most cases, a proper distinction is not possible and the boundary between user generated content and editorial content is blurring. Nevertheless, Table 1 provides an

Table 1 German conversation channels on the social web (source representation based on Ethority [12])

Examples	Conversation Channels		Examples
Facebook.de GooglePlus.de	Social Networks	Interest and Curated Networks	Xing.de Linked.in
IOFF.de Gulli:Board.de	Forums	Collaboration	Doodle.de Mindmister.de
Wikipedia.de Wikileaks.org	Crowdsourced Content/Wikis	Reviews/Ratings	HolidayCheck.de TripAdvisor.de
Wordpress.de eBlogger.de Tumblr.de	Blog Platforms and Communities		Idealo.de
		Social Bookmarks	MisterWong.de Delicious.com
Twitalyzer.de Kred.com	Influence	Instant Messaging	Skype.de Icq.de
Twitter.de Bleeper.de	Micromedia	Live Casting/Lifestreams	lifestream.fm Justin.tv
Twitpic.de Tweepi.de	Twitter Ecosystems	Social Shopping/Social Commerce	Groupon.de DaWanda.de Brands4Friends.de
GooglePlaces.de Foursquare.de	Location Based Services		
Wer-Weiss-Was.de GuteFrage.net	Question & Answer Sites	Reputation	Yasni.de 123people.de
Youtube.de		Documents/Content	Dropbox.de Slideshare.net
Veoh.de Vimeo.de	Video	Gaming	Farmville.de Comunio.de
Last.fm Spotify.de	Music	Mobile Apps	Runtastic Evernote
Flickr.de Picasa.de	Picture	Social Media Tools	Addthis.com Netvibes.com

overview summarizing various types of conversation channels according to different categories.

2.2 Social Media Monitoring

The increasing importance of social networks and associated platforms in daily communication explains an increasing demand to follow and understand the conversations within the networks. SMMO allows just that [...] [11]. SMMO is characterized by the observation and analysis of conversations plus the resulting opinion-forming on the social web. Here, the focus is primarily set on blogs, social networks, microblogs and platforms that allow users to easily publish content online. Hence, trends can be captured and polarizing platforms as well as influencers can be identified with SMMO [7]. SMMO refers exclusively to user generated content. The core of SMMO is the identification of relevant content, whereas its potential to gather quantitative and qualitative data becomes obvious. User generated content is not only being listed, but also analyzed and interpreted. In comparison to that, web monitoring refers to the mentioning of topics, products or comparable aspects across the whole web [11].

SMMO can be operated as a specific static action, but, as well as most of all web mining applications, provides the best results by a continuous operation. Moreover, the observation of the social web and the derivation of findings can be understood as interaction of an individual, repetitive and coherent set of steps (cf. Section 3) [13].

The conventional fields of application for SMMO primarily relate to commercial market research, corporate communications and public relations [7]. In this case, non-commercial, scientific and political intentions are being focused on. In the following part, three major fields of application are being identified.

Influencer Detection: SM allow personal dialogues. Users post their opinions through various channels [13]. Through SMMO, especially in forums and among bloggers, highly active individuals who influence the opinions of other readers can be identified. These opinion leaders can be involved in policy-making or could even be gained as advocates in case of crises. They often enjoy greater confidence in their web community then conventional political representatives [10].

Trend Analysis: Trend analyses provide the potential to cluster content of SM and filter out relevant information. The special feature of SMMO is a continuous online monitoring. Thus, developments can be monitored and analyzed, which can evolve into trends or trends that can be derived from these developments [10]. Here, also deficit analyses are applied, which derive societal needs on the basis of available postings. In the original, conventional field of market research these types of analyses focus on the development and reputation of brands and certain products, whereas the search queries in this research project are unattached to products and brands.

Issue Monitoring: Many topics are being discussed firstly on the social web, before being published by mass media such as newspapers and television. Through a continuous SMMO, emerging crises and possible critical issues can be identified more quickly, which facilitates a better response or defense. SM can for example be used to forecast epidemics or pandemics in order to prevent mass outbreaks, as the EU-founded project MECO shows [14, 15]. Thus, the findings from user generated content can support analyses of emerging issues and optional reactions. Important multipliers or influencers, should especially be monitored constantly, because they can have a positive or a negative impact on other participants of the social web [7, 10].

To sum up, all of the three application fields show how SMMO can affect the management or advancement of R&D programs. The respective thematic focus of a certain R&D program determines which field to focus on.

2.3   Cybernetics and System Theory

To start developing an approach of SMMO in order to support the ongoing advancement of R&D programs (cf. Section 1), a scientific theoretical framework is needed. A system-oriented perspective enables a holistic and integrative consideration [16] of relevant factors for the development of a SMMO approach that focuses on R&D programs. In terms of a system-oriented and cybernetic understanding, R&D programs are considered as complex socio-technical or living systems. Such an approach makes it possible to treat individual entities of R&D programs as abstract units so that an adequate monitoring approach can be developed. As monitoring supports the management of such systems, aspects of cybernetics are being taken into account that concern feed backed steering-, control- and regulation-processes in complex systems [17–23]. In order to promote the development of a *cybernetic* SMMO approach, central concepts of system theory and cybernetics will be described in the following paragraph.

System and Environment: As an interdisciplinary and integrative constructivist insight model, the modern system theory primarily works out as a system-environment theory [24]. System theory offers the possibility to regard any entity as a system and to distinguish it from its environment. Thus, for example, the human body is to be constructed as a biochemical system, a central heating as a technical system and a football team as a social system [1]. From the perspective of system theory, these very different entities initially have in common that they consist of different elements (e. g. cells and hormones, radiators and pipes, players and coaches) that have certain relationships to each other [1]. A system is fundamentally characterized by a holistic context of parts and their relationship with each other, which are quantitatively more intense and qualitatively more productive than their relationships to other elements. This diversity of relations constitutes a system boundary that separates the system and the environment from each other [25]. Henning [26] also specifies characteristics of living systems: although these are operationally closed by a system boundary, there are certain relationships to its environment, which are crucial for the existence of the system. This existential dependence implies that a system may increase its own capability of adaptation and survival through the continuous monitoring of changes in the relevant environment [1].

Transformation and Feedback: While the system-oriented perspective primarily aims at structural aspects, a cybernetic perspective emphasizes in particular the functional unit of a system [20, 21]. These processes of exchange with the environment can be described by a functional input-output relation. The system accepts a variety of influences from the environment, it processes and transforms these influences and creates an output that clearly differs from the input. A hot water heater, for example, requires fuel as input and transforms it over the water warming into the output, namely space heating [1]. The focused R&D programs in this paper particularly have financial resources as an input and in the context of funded projects transform it into research results as an output. Even if the input from the environment is necessary for the system, not all of the effects are desirable. The same applies to the outputs of the system: a boiler also produces combustion residues and so does an R&D program, which e. g. to some extent produces obsolete information [1].

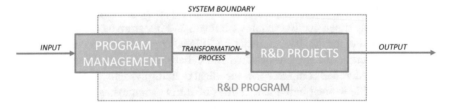

Figure 1 Linear transformation process (source: own representation based on [1])

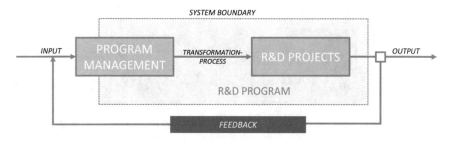

Figure 2 Transformation process with feedback loop (source: own representation based on [1])

Figure 1 explains the input-output relation as a linear transformation process. The represented idea shows the linearity of input, transformation and output that initially corresponds to a mono-causal effect relationship [26]. Thus, an impact or effect is always a consequence of a specific cause so that a change of the effect can only be achieved by changing the cause. This, however, does not include the influence of the effect of the cause, the so-called repercussions or feedback. With respect to R&D programs, this means that the system input not only consists of financial resources, which lead to research results through transformation, but that this output again is an input to the system. If these feedback loops are used by the system, it can be prevented that funds for research activities that have already been worked on are issued. By this means, systems are able to change their behavior through targeted monitoring and feedback of their outputs and thus to initiate learning processes [1]. According to cybernetics, feedback processes are an intrinsic characteristic of self-regulating, adaptive and thus learning capable systems [20, 26]. Figure 2 illustrates a system integrating a feedback loop.

3 Cybernetic Approaches of Social Media Monitoring

R&D programs are construed as socio-technical, complex living systems. Elements like single R&D projects or the program management itself are parts of this living system, which strongly depends on its environment (cf. Figures 1 and 2). This environment is characterized by socio-technical developments, political impulses, natural influences or force majeures (cf. Figure 3). Figure 3 shows an extended version of the previous figures including SMMO.

With regard to Figure 3, research promoters, such as the Federal Ministry of Education and Research in Germany, are part of several systems, i.e. different R&D programs. Here, the dark grey systems represent current R&D programs and the light grey systems show future R&D programs. The society represents a wide number of stakeholders, who can also be a part of certain systems or R&D programs. However, the majority of the society is not part of these systems and strongly characterizes the system environment. Of course, political impulses, the relevant envi-

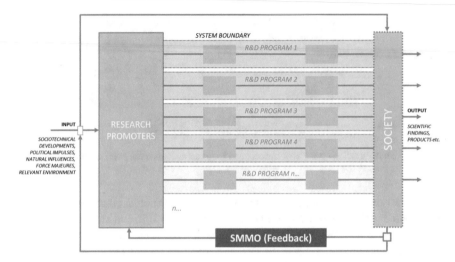

Figure 3 Cybernetic approach of SMMO (source: own representation)

ronment, socio-technical developments etc. (Figure 3) are interwoven with society, too. Here, SMMO takes additional supportive functions regarding of program management. SMMO continuously identifies societal snapshots, opinions and moods concerning content of current R&D programs or future R&D programs and reports these back to the research promoters. In this manner, current R&D programs can be continuously advanced and future programs can be shaped. After the placement of SMMO into a conceptual and thematic framework, the following chapter thoroughly describes SMMO with its several processing steps (Figure 4). Figure 4 explains the black SMMO-Box of Figure 3 in detail.

Information Demand: Taking into account the societal developments and trends, R&D programs should be continuously advanced or re-tendered. Citizens, for example, get more involved in this advancement process. The relevant stakeholders are policy makers, such as the German Federal Ministry of Education and Research. Thus target agreements will be made concerning the form of results of SMMO.

Identification & Integration: Here, the need for information of policy makers is initially identified and overarching strategic goals are defined, as these affect all subsequent steps [9]. Categories, such as keywords are set for a targeted monitoring in the thematic framework of the corresponding R&D program [10]. In addition, a specific corpus of data sources respectively SM (cf. Table 1) is defined. For the detection of trends, especially high-traffic posts (Treads) [27] are suitable. A continuous review and extension of the source is indispensable. In addition, the selection of appropriate social media monitoring tools according to specific requirements is essential (Table 2). After the selection of suitable tools, previously developed categories are integrated into these tools as well as the limited corpus of data. Here, crucial stakeholders come from from Futurology and other complementary studies.

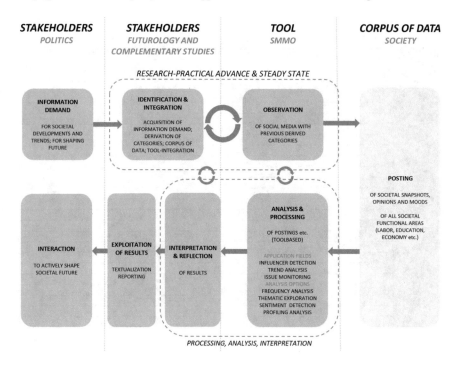

Figure 4 Processing steps of SMMO (source: own representation)

Observation: The basis of SMMO is generated by sources that are automatically crawled regarding the predefined search queries and categories [10]. The goal is to extract structured data: Depending on the specific subject conditions, for example, certain keywords are included or excluded. Search queries are described in terms of Boolean syntax and are gradually refined. The amount of used search words can vary greatly depending on the topic. By means of creating various queries, various topics can be compared simultaneously [13].

The processes of *Identification & Integration*, as well as *Observation* are summarized as research-practical advance and steady state, as the steps are mutually dependent. So, for example, during the continuous surveillance, categories or the data corpus have to be adapted continuously, as well.

Analysis & Processing: The main task of this step is the tool-based matching of the specified information demand (e. g. a query) with the amount of documents of the data corpus [9]. Filters can further refine the results of a search query. Further analysis can be applied to parts of the results, e. g. regarding time of posting, language, source or geography [13]. In order to serve the mentioned fields of application (cf. Section 2.2), there are various analysis options. The analysis options described in the following part are based on different mining methods.

Using *Extraction* methods, contents of documents are being summarized, e. g. in terms of automatic extraction. The Feature *Extraction* only refers to information

Table 2 Tool evaluation categories [9]

MISCELLANEOUS	OPERATING MODE (WEBBASED OR INSTALLED VERSION)
	KIND-OF-USE (SELF-SERVICE, SELF SERVICE WITH CONSULTING, FULL SERVICE)
	EMAIL ALERTS
	ACCESS FOR DIFFERENT USERS
	USER MANAGEMENT (PROVIDER ODER CUSTOMER)
	DATA TRANSFER & EXPORT
ANALYSIS TECHNIQUE	INFLUENCER DETECTION
	TREND ANALYSIS
	ISSUE-MONITORING
	FREQUENCY ANALYSIS
	THEMATIC EXPLORATION
	SENTIMENT DETECTION (GERMAN)
	SENTIMENT DETECTION (ENGLISH)
	PROFILING ANALYSIS
	MINING PROCEDURES
	CONSOLIDATED DATA
	CORPUS OF DATA ADAPTABLE
	SEARCH QUERY HISTORY
	FULL TEXT ACCESS
	GEOGRAPHY BASED QUERIES
	RELEVANCE EVALUATION
	PROFILES OF AUTHORS
INTERFACE	DASHBOARD WITH DRILLDOWN-FUNCTION
	USER INTERFACE (CLARITY, USABILITY)
	COMMENT- AND FORWARD-FUNCTION
	DASHBOARD ADAPTABLE
	COSTS

regarding particular keywords. As a Bag of Words model, (vector-based) grammar and sentence structure are being ignored, articles and conjunctions are being sorted out and weighted according to frequency of appearance of the rest. In contrast, the automatic compression provides the most important parts of an original document in form of a summary (e. g. Google search results) [28]. The compression in terms of an extraction provides the most important parts of the original document with phrases, sentences or paragraphs. Using *Classification* and *Clustering*, information is assigned to the classes of a given classification or taxonomy. If the classification should not be given, but is even a mining result itself, it is called Clustering or *Segmentation* [9]. *Topic Detection* and *Tracking* is being used for the detection of important issues of corpus of data by methods of text clustering. Thus, the identification of future trends is possible through the extrapolation of measured data over a period of time [9].

   In addition to the mining methods mentioned above, there are various analysis options.

- *Frequency Analysis*: The frequency analysis is a very common method of analysis in SMMO: Counting the number of posts on a certain topic which are being published and indexed within a certain time. Bar or line graphs can demonstrate how topics develop and whether there are significant abnormalities [10, 13].
- *Thematic Exploration*: Using the keyword "Extraction", relevant or frequently occurring words are extracted from a text corpus. In this way, a faster thematic overview is enabled and unfamiliar thematic links show up. Here, tag clouds rep-

resent the most common form for theme exploration [13]. The more often a word appears in the examined text corpus, the greater it gets in the tag cloud. Since the size of the word is only a snapshot of the number of mentions, coloration is also used to indicate the time trend of a word. Besides tag clouds, association graphs present another important tool for theme exploration: Relationships between objects can be visually illustrated. Important words are being highlighted like in a tag cloud, but additionally relationships are being represented by lines of different thickness. The semantic information of a compound is often limited to the common designation in the text [10].

- *Sentiment Detection*: The Sentiment Detection allows a display of the development of moods and opinions over a certain time on a given topic. For this purpose, texts are annotated with characteristic values of the individual authors' attitude to suggest whether it is a positive, neutral or negative contribution. This qualitative analysis is one of the most informative features in SMMO, although the automated assignment does not always work correctly [13]. It is still difficult to grasp irony, sarcasm or affable youth language.

- *Profiling Analysis*: To identify opinion leaders, the collection and analysis of personal data is essential. Information such as location, gender, age etc. additionally increases the quality of statements [10]. Using profiling analysis, it is possible to show where certain developments and trends take place.

- *Export and Integration*: Graphics, Excel sheets, structured text files etc. are created automatically and can be further processed into summary reports or other media [13]. Using a dashboard it summarizes the main results of analysis and relevant posts at a glance. This (graphical) visualization has a high degree of interactivity so that the data corpus is easily accessible [13].

Interpretation & Reflection: During the previously described step, information is being analyzed in a tool-based manner and subsequently processed. Now, the main focus of this step is the individual, human-led interpretation and reflection of the results. Depending on the results and the respective significance, there are also feedback loops back to previous steps like *Identification & Integration* – in some cases there is need for further adjustment. Overall, together with the tool-based step *Analysis & Processing*, these steps are summarized as *Processing, Analysis* and *Interpretation* (Figure 4).

Exploitation of Results: After the results have been interpreted and reflected, this step is aspiring to exploit those for the fulfilling of the identification demands given in the first step. Depending on the pre-established target agreements, media such as continuous reports, quarterly reports, booklets or other formats are being created.

Interaction: Due to the continuous processing of the results of SMMO, overall societal developments are mirrored into policy-making. Ongoing R&D programs can be advanced as learning systems or new R&D programs can be tendered according to the overall societal demands. In this way, future can be shaped more actively.

4   Summary and Outlook

SM are a promising pool of qualitative and quantitative data to reflect societal moods. To exploit these data, a cybernetic SMMO is currently being developed. Cybernetic SMMO allows for a continuous and active involvement of the society into politics. Using this theoretical, system-oriented framework R&D programs can be constructed as socio-technical, complex living systems. It supports the program management and research promoters of publicly funded R&D projects and programs by taking into account social impulses for the advancement of R&D programs and the ex-ante determination of program contents. Therefore, an active shaping of the future according to societal developments, trends and needs is aspired through cybernetic SMMO. Thus, the interaction between politics and society can improve continuously. Nevertheless, further evaluation studies need to reveal how cybernetic SMMO can be applied within different thematic settings.

References

1. S. Trantow, *Ein kybernetisches Modell für das Internationale Monitoring von F&E-Programmen im Bereich der Arbeitsforschung.* Books on Demand, Norderstedt, 2012
2. D. Hartmann, H. Brentel, H. Rohn, *Lern- und Innovationsfähigkeit von Unternehmen und Organisationen. Kriterien und Indikatoren.* Wuppertal Institut für Klima, Umwelt und Energie in Kooperation mit der Triforum-Beratungsgesellschaft mbH, Wuppertal, 2006
3. Z. GmbH. Megatrends - Update, 2013. http://www.z-punkt.de/fileadmin/be_user/D_ Publikationen/D_Giveaways/Megatrends_Update_DE.pdf
4. R. Sennett, *Der flexible Mensch - Die Kultur des neuen Kapitalismus.* Berliner Taschenbuch-Verlag, Berlin, 2006
5. Bundesregierung. EU-bürgerdialog, 2013. http://eu-buergerdialog.de/16-0-Impressum.html
6. MIWF (Ministerium für Innovation, Wissenschaft und Forschung des Landes Nordrhein-Westfalen). Dialog und Kongress - Fortschritt gestalten, online 2013
7. S. Osswald, Social media monitoring. In: *Leitfaden WOM Marketing*, ed. by A.M. Schüller, T. Schwarz, marketing-BÖRSE GmbH, Waghäusel, 2010, pp. 389–394
8. A. Beal, J. Strauss, *Radically Transparent - Monitoring and Managing Reputations Online.* Wiley, Indianapolis, 2008
9. J. Finzen, H. Kasper, M. Kintz, *Innovation Mining.* Fraunhofer IAO, Stuttgart, 2010
10. H. Kasper, M. Dausinger, H. Kett, T. Renner, *Social Media Monitoring Tools.* Fraunhofer IAO, Stuttgart, 2010
11. M. Lange, Social Media Monitoring. In: *Leitfaden Online Marketing*, ed. by T. Schwarz, marketing-BÖRSE GmbH, Waghäusel, 2011, pp. 655–659
12. Ethority. Conversations in social media - german edition v5.0, 2012. http://www.ethority.de/ weblog/social-media-prisma/
13. H. Kasper, H. Kett, *Social Media Monitoring-Tools*, marketing-BÖRSE GmbH, Waghäusel, 2011, pp. 662–669
14. P. Laage. Epidemien früher erkennen mit Twitter & Co, 2012. http://www.welt.de/gesundheit/ article106735414/Epidemien-frueher-erkennen-mit-Twitter-amp-Co.html
15. M.M. Ecosystem). About m-eco, 2010. http://www.meco-project.eu/about
16. K. Henning, *Kybernetische Verfahren der Ingenieurwissenschaften.* Hochschuldidaktisches Zentrum der RWTH Aachen, Aachen, 1985
17. W. Ashby, *An Introduction to Cybernetics.* Chapman and Hall, London, 1956

18. J. Lattwein, *Wertorientierte strategische Steuerung*. DUV, Wiesbaden, 2002
19. E.A. Hartmann, *Arbeitssysteme und Arbeitsprozesse*. vdf Hochschulverlag, Zürich, 2005
20. G. Strina, *Zur Messbarkeit nicht-quantitativer Größen im Rahmen unternehmens-kybernetischer Prozesse*. RWTH Aachen University, Aachen, 2006
21. T. Brosze, *Kybernetisches Management wandlungsfähiger Produktionssysteme*. Apprimus, Aachen, 2011
22. S. Trantow, F. Hees, A. Richert, Monitoring als konstitutive Funktion in lernenden F&E-Programmen. In: *: Kybernetik und Wissensgesellschaft. Vernetztes Denken, geteiltes Wissen—auf dem Weg in die Wissensökonomie*, Stuttgart, 2012
23. F. Welter, *Regelung wissenschaftlicher Exzellenzcluster mittels scorecardbasierter Performancemessung*. Books on Demand, Norderstedt, 2013
24. H. Willke, *Systemisches Wissensmanagement*. Lucius & Lucius, Stuttgart, 2001
25. H. Willke, *Systemtheorie I - Grundlagen. Eine Einführung in die Grundprobleme der Theorie sozialer Systeme*. UTB, Stuttgart, 2006
26. K. Henning, *Spuren im Chaos*. Olzog, München, 1993
27. B. Steimel, C. Halemba, T. Dimitrova, *Social Media Monitoring*. MIND Business Consultants, Meerbusch, 2010
28. A. Ahlemeyer-Stubbe, *Social Media Monitoring*, Springer, Wiesbaden, 2013

Unterstützung interdisziplinärer integration am Beispiel einer Exzellenzcluster-Community

Thomas Thiele, Stefan Schröder, André Calero-Valdez, Claudia Jooß,
Anja Richert, Martina Ziefle, Ingrid Isenhardt and Sabina Jeschke

Zusammenfassung Aktuelle wissenschaftliche Fragestellungen sind häufig nur durch eine Vielzahl unterschiedlicher wissenschaftlicher Fachdisziplinen zu beantworten. Dies gilt besonders für Forschung im Bereich des Maschinenbaus und speziell im Bereich der Produktionstechnik. Zielsetzung ist dabei die Kombination verschiedener Kompetenzen aus unterschiedlichen Fachdisziplinen zur Lösung von Forschungsfragen und Problemstellungen einzusetzen, die sich an Schnittstellen der Fachdisziplinen ergeben. Voraussetzung hierfür ist die Integration dieser verschiedenen Fachdisziplinen in einer interdisziplinären Community. Im Rahmen dieses Beitrags wird zunächst ein Analyseinstrument vorgestellt, dass zur Erfassung von Bedarfen zur Integration innerhalb der interdisziplinären Community dient. Drei Maßnahmen zur Unterstützung dieser Integration werden vorgestellt, deren Entwicklung auf Basis von Bedarfen der Community begonnen wurde und auf dieser kontinuierlich aktualisierten Grundlage iterativ fortgeführt wird. Das Ziel dieses Vorgehens ist die Unterstützung einer kontinuierlich optimierten Integration der Community im Exzellenzcluster „Integrative Produktionstechnik für Hochlohnländer".

Schlüsselwörter Interdisziplinäre Kooperation · Integration · Unterstützung wissenschaftlicher Communities

1 Herausforderungen und Motivation

Die ansteigende Komplexität von Problemstellungen und denen sich daraus entsprechend ergebenden Forschungsfragen im Bereich der Produktionstechnik und im Maschinenbau lässt sich seit Jahren feststellen ([1]: 6ff.). Der wachsende Komplexitätsgrad lässt sich dabei nicht nur auf den höheren Innovations- und

T. Thiele (✉) · S. Schröder · C. Jooß · A. Richert ·
I. Isenhardt · S. Jeschke
IMA/ZLW & IfU, RWTH Aachen University, Dennewartstr. 27, 52068 Aachen, Germany
e-mail: thomas.thiele@ima-zlw-ifu.rwth-aachen.de

A. Calero-Valdez · M. Ziefle
Human-Computer Interaction Center, RWTH Aachen University, Dennewartstr. 27, 52068 Aachen, Germany

Originally published in "Zukunft gestalten: Soziale Technologien in Organisationen in Zeiten des demografischen Wandels, FIR-Edition Forschung, Bd. 15",
© FIR e. V. an der RWTH Aachen 2015 2015. Reprint by Springer International Publishing Switzerland 2016, DOI 10.1007/978-3-319-42620-4_16

Wettbewerbsdruck am Markt, sondern auch auf immer vielfältigere Produktfunktionen[1] zurückführen. Diese gestalten sich vielfach fachdisziplin-übergreifend, woraus resultiert, dass „der Gewinn an Erkenntnissen heute nicht nur in immer mehr fachgebundenen Tiefbohrungen', sondern zunehmend in den Anregungen aus interdisziplinären Kontakten" liegt ([3]: 17). Somit ist eine isolierte Betrachtung aus disziplinärer Sicht für die komplexen Problemstellungen nicht mehr ausreichend [4, 5], was in der Bildung interdisziplinärer Communities resultiert, deren verschiedene Fachdisziplinen in einem Forschungsprozess integriert werden müssen. Als direkte Folge arbeiten WissenschaftlerInnen aus unterschiedlichen Fachdisziplinen gemeinsam an einer Problemstellung, zu deren Lösung sowohl das jeweilige Spezialwissen der Fachdisziplin als auch die Erfahrung der beteiligten Akteure benötigt werden. Werden zudem Partner aus der Wirtschaft in den Forschungsprozess – beispielsweise als direkter oder Transferpartner – eingebunden, führt dies zu einer weiteren Steigerung der Komplexität [6].

Die individuellen Bedarfe dieser verschiedenen Akteure müssen bei der Integration in eine gemeinsame wissenschaftliche Community berücksichtigt werden. Dabei stellt bereits die Frage nach einem tauglichen Verfahren zur Ermittlung dieser Bedarfe eine zentrale Herausforderung dar. Im weiteren Verlauf soll daher als eine mögliche Antwort eine Balanced Scorecard (BSC) basierte Evaluation zur Ermittlung von Bedarfen vorgestellt werden. Als Fallbeispiel dient das Exzellenzcluster „Integrative Produktionstechnik für Hochlohnländer", das seit 2006 von der Deutschen Forschungsgemeinschaft (DFG)[2] und dem Wissenschaftsrat (WR)[3] an der RWTH Aachen University gefördert wird.

2   Balanced scorecard basierte Evaluation als Analyseinstrument zur Bedarfsermittlung

In den vergangenen Jahren sprachen sich Forschungsförderer, wie die DFG und der WR, für die Entwicklung von Konzepten zur Messung, Steuerung und Regulierung interdisziplinärer Communities aus [7, 8].[4] Dies hatte zur Folge, dass u. a. diverse existierende Ansätze zur Performancemessung adaptiert und an die Bedarfe interdisziplinärer Communities angepasst wurden. Grundsätzliches Ziel dabei ist es, die interdisziplinären Communities in ihrer Kooperation zu befördern und die Zusammenarbeit zu intensivieren [9, 10]. Das (kontinuierliche) „Messen" der Performance dient dazu, beispielsweise den Fortschritt bzw. aktuellen Stand der

---

[1] Als ein typisches Beispiel weisen viele Produkte des Maschinenbaus informationstechnologische Komponenten auf, so erkennbar z. B. in sog. Cyber-Physical Systems [2]. Hieraus wird die enge Verzahnung von Informatik und klassischer Hardware-Entwicklung im Maschinenbau erkennbar.

[2] http://www.dfg.de/foerderung/programme/exzellenzinitiative/.

[3] http://www.wissenschaftsrat.de/arbeitsbereiche-arbeitsprogramm/exzellenzinitiative.html.

[4] Das im Folgenden vorgestellte Instrument kommt dabei in den beiden DFG-geförderten Exzellenzclustern „Integrative Produktionstechnik für Hochlohnländer" und „Tailor-made Fuels from Biomass" zum Einsatz.

Forschungsaktivitäten transparent zu machen, aber auch Bedarfe, Handlungsempfehlungen und – als Konsequenz hieraus – Maßnahmen zur Optimierung der Zusammenarbeit zu identifizieren und während der Förderlaufzeit umzusetzen. Aus diesem Grund ist eine prozessbegleitende Messung und Rückkopplung der Ergebnisse wichtig [11].

In diesen Zusammenhang wurde die im unternehmerischen Kontext bereits viele Jahre bewährte BSC an die Bedarfe wissenschaftlicher Forschungsverbünde angepasst [12, 13]. Dabei besteht die clusterspezifische BSC, welche in Kooperation mit dem Community-Management und den Akteuren entwickelt wurde, aus vier Perspektiven: Interne Perspektive/Forschungskooperation, Lern- und Entwicklungsperspektive, Output/Kundenperspektive und finanzielle Perspektive. Um die Performance von Forschungskooperationen zu messen, werden für jede Perspektive entsprechende Key Performance Indikatoren (KPIs) identifiziert, wie zum Beispiel Fragen zur Häufigkeit, Qualität und der Nutzen von Meetings, wissenschaftliche Kooperation und deren Schnittstellen sowie Anzahl an Publikationen etc. [12]. Anschließend werden einmal pro Jahr besagte Indikatoren erhoben und evaluiert. Durch die anschließende Datenanalyse können Handlungsempfehlungen für das Community-Management abgeleitet werden, mit dem Ziel die interdisziplinäre Integration zu befördern.

Als Reaktion auf die Evaluationsergebnisse sind konkret umgesetzte oder sich in der Entwicklung befindende Maßnahmen entstanden. Exemplarisch zu nennen sind in diesem Zusammenhang

- Die Einführung von Präsenz-Netzwerkveranstaltungen mit Möglichkeiten zur informellen Kommunikation, wie z. B. Mitarbeiterkolloquien,
- Eine virtuelle Kooperationsplattform,
- Die Analyse von Terminologien zur Beschreibung von Schnittstellen sowie zur Unterstützung der Entwicklung eines gemeinsamen fachsprachlichen Begriffsverständnisses und
- Die bibliometrische Analyse als Möglichkeit zur Beschreibung von Verbindungen innerhalb der Community.

Diese Maßnahmen werden im Folgenden Kapitel näher beschrieben.

3 Exemplarische Maßnahmen zur Community Unterstützung

In den nachfolgenden Kapiteln wird ein Auszug der Maßnahmen beschrieben, die bereits in der Exzellenzcluster-Community auf Basis der BSC-Evaluation umgesetzt wurden oder derzeit werden. Das Maßnahmenset stellt eine vergleichsweise heterogene Mischung dar: Einerseits adressieren physische Netzwerktreffen ein vergleichsweise klassisches Instrument der Community Integration. Andererseits prägen virtuelle bzw. digitale Maßnahmen zur Unterstützung in (Business-) Communities bereits maßgeblich die Forschungs- und Unternehmenslandschaft [14]. Durch

die zielgerichtete und kombinierte Anwendung dieser Maßnahmen werden daher
die individuellen Bedarfe der Akteure in der interdisziplinären Exzellenzcluster-
Community adressiert [15].

3.1 Physische Community-Veranstaltungen

Die Maßnahmen im Rahmen der Präsenz- -Veranstaltungen der Community bestehen im Wesentlichen aus verschiedenen formellen und informellen Treffen mit
dem Ziel Transparenz im Rahmen des inhaltlichen Fortschritts und von bspw.
organisatorischen Entscheidungen zu schaffen, sowie der Informationsverteilung
zwischen den Akteuren in der Community. Diese Maßnahmen dienen als Austausch-
plattformen, die den Teilnehmenden die Möglichkeit zur Face-to-Face Kommunikation und Interaktion bieten und damit vor allem durch die Weiterentwicklung persönlicher Kontakte einen Gewinn für die Community darstellen [16].

Die mittels dieser Veranstaltungen unterstützte informelle Ebene der Kommunikation unterstützt den Wissensaustausch, da die hierbei genutzten kognitiven Zugänge
(z. B. im Kontext persönlicher Interaktion) vielfach diverser sind als bei jedem
anderen Medium [17]. So können wissenschaftliche Ergebnisse direkt im Rahmen
einer Präsentation ausgetauscht werden, gleichzeitig besteht die Möglichkeit die
vorgestellten Ergebnisse und damit verbundene Methoden und Konzepte aktiv zu
diskutieren. Vor dem interdisziplinären Hintergrund der Community erweist sich im
Besonderen der aktive Austausch und die gemeinsame Erarbeitung von Inhalten als
wichtiges Element zur Integration der unterschiedlichen Fachdisziplinen [6].

3.2 Virtuelle Communityplattform

Im Rahmen der Entwicklung von Maßnahmen zur virtuellen Kommunikation und
Vernetzung wurde eine zielgruppenspezifische Communityplattform entwickelt,
welche die Face-to-Face Maßnahmen der Community komplementär ergänzt. Das
sogenannte CoE-Portal (siehe Abbildung 1) unterstützt den Austausch zwischen den
Akteuren der Exzellenzcluster-Community und den internen und externen Transfer von Ergebnissen durch die Bereitstellung eines zeit- und ortsungebundenen
Informations- und Interaktionsangebots.

Die virtuelle Community setzt hierbei auf ein technisches Grundgerüst, um die
verschiedenen Akteure in der Community zu verbinden. Maßgebliches Kriterium bei
der Gestaltung ist die Integration in tägliche Arbeitsprozesse. Hierbei wird das Ziel
verfolgt, einen Mehrwert für die Akteure dadurch zu generieren, dass Informationen
in verschiedenen Formen aufbereitet und angeboten werden. Gleichzeitig tragen aus
sozialen Netzwerken überführte Maßnahmen zum Austausch bei, um so wiederum
auch auf der informellen Ebene einen Mehrwert zu generieren [17].

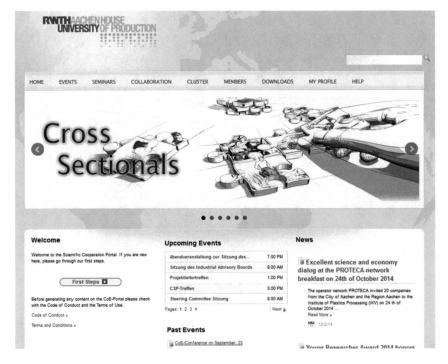

Abbildung 1 Screenshot des CoE-Portals

3.3 Terminologie-Analyse

Die Analyse von Terminologien zur Unterstützung der Exzellenzcluster-Community[5] beinhaltet die Ermittlung und die Visualisierung zentraler Begriffe in der Community. Das Ziel ist hierbei zum einen die Visualisierung von terminologie-basierten Schnittstellen zwischen den Akteuren in der Community. Zum anderen wird durch die Bereitstellung von Begriffen und deren Definitionen die Entwicklung eines disziplinübergreifenden Verständnisses von Terminologien unterstützt, was als ein zentraler kritischer Faktor interdisziplinärer Integration angesehen werden kann [6].

Hierbei baut die Terminologie-Analyse auf die technischen Rahmenbedingungen der virtuellen Communityplattform auf und stellt mittels eines Online Tools (siehe Abbildung 2) Funktionalitäten hinsichtlich der Erfassung von Definitionen im Stil eines Wikis zur Verfügung. Auf diese Weise wird es den Akteuren ermöglicht, an der Entwicklung eines gemeinsamen Verständnisses zu arbeiten und so beispielsweise Projektziele, Methodenwissen und Ergebnisse effizienter zu kommunizieren.

---

[5]Diese Community dient für den vorliegenden Beitrag als Fallbeispiel, jedoch ist eine breitere Anwendung, z.B. zur Unterstützung des Managements interdisziplinärer Kooperationen in der Industrie 4.0 denkbar.

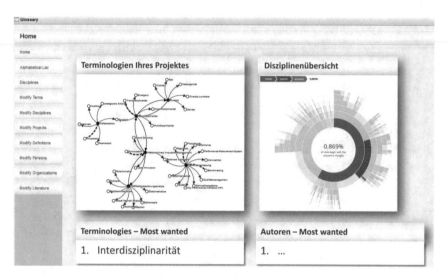

Abbildung 2 Mock-Up des Online Tools zur Terminologie-Analyse

Darüber hinaus werden zusätzlich Schnittstellen zwischen den zentralen Begriff-lichkeiten ermittelt und visualisiert, sodass für die Nutzer die Möglichkeit entsteht, Schnittstellen zu anderen Akteuren zu erkennen. Da diese Schnittstellen auf den in den Publikationen der Akteure verwendeten Begriffen basieren, wird auf diese Weise die Transparenz und die Wahrnehmung anderer Fachdisziplinen in der Community erhöht [18]. Auf diese Weise kann ein Austausch zum unterschiedlichen Verständnis von Begrifflichkeiten initiiert werden, was zu einem besseren Verständnis der Akteure führt. Dieses Verständnis trägt maßgeblich zu einer effizienteren Wissensgenerierung und der gemeinsamen Lösungsfindung bei [19–21].

3.4 Bibliometrische Analyse

Ein weiteres Mittel der Community-Unterstützung im wissenschaftlichen Kontext stellt die bibliometrische Analyse von Kooperationsprozessen anhand von Publika-tionen auf dem CoE-Portal dar. Hierfür werden neben den Interaktionen von Akteuren auf dem Portal auch Kooperationen, die sich aus der gemeinsamen Produktion von Dokumenten ermitteln lassen, visualisiert. Durch sog. Ko-Autorschafts-Analyse wird untersucht, welche Cluster Akteure zu welchen Themen gearbeitet haben. Auf-grund dieser Daten und der Profildaten des CoE-Portals werden Visualisierungen erstellt, welche den Kollaborationsverlauf über die Zeit darstellen [22].

Hierbei dient die Visualisierung mehreren Zielen. Für einzelne WissenschaftlerIn-nen bietet sie ein Werkzeug zur Selbstreflektion über Themen und Kollaborationen im Verlauf des Forschungsprojekts. Für die Verwaltung bietet die Visualisierung

Überblickswissen über bereits durchgeführte Zusammenarbeit und somit zentrale Ansatzpunkte für Maßnahmen bei mangelnder Zusammenarbeit. Zusätzlich kann die Visualisierung zur Bewertung von interdisziplinärer Zusammenarbeit herangezogen werden [23].

Zentraler Aspekt dieser Visualisierung ist die Darstellung von Kollaborationsartefakten, welche auch außerhalb der Online-Community existieren (d. h. wissenschaftliche Veröffentlichungen). Denn erst durch den Abgleich mit Daten aus der Community (d. h. Profildaten) können zusätzliche Erkenntnisse gewonnen werden. Somit werden den Mitarbeitern langfristig im Exzellenzcluster vielversprechende Kollaborationen und noch benötigte Informationen empfohlen. Zentraler Ausgangspunkt solcher Empfehlungssysteme sind immer nutzerzentrierte Anforderungsanalysen, da die Darstellung und Verwendung von Informationen im Kontext einer Online-Community andere Akzeptanzmuster zu Tage bringen als im Kontext einer Desktop-Anwendung. Denn gerade Aspekte wie Datenschutz und Privatsphäre spielen vor dem Hintergrund der Digitalisierung von Arbeit eine zentrale Rolle und müssen in Hinblick auf deren Nützlichkeit für den Arbeitsprozess mit Nutzerbedarfen harmonisiert werden.

4 Zusammenfassung und Ausblick

Während der Förderphase der Exzellenzcluster-Community wurden zur Unterstützung der interdisziplinären Integration eine Vielzahl von Maßnahmen entwickelt, von denen im Rahmen des vorliegenden Beitrags ein exemplarisches Maßnahmenset vorgestellt worden ist. Die Bedarfe für diese Maßnahmen wurden in Zusammenarbeit mit der Community durch die kontinuierliche Evaluation mittels der clusterspezifischen BSC erhoben. Mit Hilfe dieses Werkzeugs bzw. durch die prozessbegleitende Bedarfsermittlung wird gleichzeitig die kontinuierliche Weiterentwicklung der vorgestellten Maßnahmen realisiert.

Vor allem im Kontext der Terminologie-Analyse und den nachfolgenden Schritten sind weitere Entwicklungen notwendig: Wahrnehmung und Diskussion von Begrifflichkeiten ist nur ein erster Schritt. Im weiteren Verlauf ist die Weiterentwicklung des Online Tools insofern denkbar, als dass eine direkte Unterstützung in der täglichen Arbeit der Akteure erfolgen kann. So ist beispielsweise die Unterstützung von gemeinsamen interdisziplinären Publikationen zwischen Akteuren denkbar, die auf Basis von gemeinsam genutzten Begrifflichkeiten angestoßen werden kann. Die weitere Entwicklung dieser Form von Zusammenarbeit setzt aber den zuvor genannten Diskussionsprozess voraus, birgt jedoch ein großes Potential im Hinblick auf die Weiterentwicklung einer interdisziplinären wissenschaftlichen Community.

Im Kontext interdisziplinärer Integration zielt die Anwendung der genannten Maßnahmen maßgeblich auf die Entwicklung einer nachhaltigen Kooperation der Akteure ab, um so einen Beitrag zur Entwicklung von Strukturen zu leisten, die über derzeitige Förderungen hinaus gehen.

Danksagung Die diesem Beitrag zu Grunde liegende Forschung wurde mit Mitteln der DFG als Teil der Arbeiten im Exzellenzcluster „Integrative Produktionstechnik für Hochlohnländer" und „Tailor-Made Fuels from Biomass" an der RWTH Aachen University gefördert.

Literaturverzeichnis

1. J. Feldhusen, K.H. Grote, eds., *Konstruktionslehre. Methoden und Anwendung erfolgreicher Produktentwicklung*. Springer, 2013
2. S. Jeschke, Kybernetik und die intelligenz verteilter systeme, 2014
3. K. Ehrlenspiel, *Integrierte Produktentwicklung. Denkabläufe, Methodeneinsatz, Zusammenarbeit. 4., überarbeitete Auflage*. Hanser Verlag, 2009
4. J.R. Bryson, Hybrid manufacturing systems and hybrid products: Services, production and industrialisation. Studies for Innovation in a Modern Working Environment. International Monitoring , 2009
5. C. Jooß, F. Welter, A. Richert, S. Jeschke, Fostering innovative capability in germany – the role of interdisciplinary research networks. Enabling Innovation – Innovative Capability – German and International Views, 2011
6. C. Jooß. Gestaltung von Kooperationsprozessen interdisziplinärer Forschungsnetzwerke, 2014
7. DFG/WR – Deutsche Forschungsgemeinschaft and Wissenschaftsrat eds. Bericht der gemeinsamen Kommission zur Exzellenzinitiative an die gemeinsame Wissenschaftskonferenz, 2008. http://www.gwk-bonn.de/fileadmin/Papers/GWK-Bericht-Exzellenzinitiative.pdf
8. M. Sondermann, D. Simon, A.M. Scholz, S. Hornbostel, Die exzellenzinitiative: Beobachtungen aus der implementierungsphase. ifQ-Working Paper, 2008
9. S. Schröder, F. Welter, I. Leisten, A. Richert, S. Jeschke, Research performance and evaluation– empirical results from collaborative research centers and clusters of excellence in germany. Research Evaluation (2014), Oxford University Press, 2014
10. D. Greiling, Trust and performance management in non-profit organizations. The Innovation Journal: Public Sector Innovation Journal, 2007
11. C. Jooß, R. Vossen, I. Leisten, A. Richert, S. Jeschke. Knowledge engineering in interdisciplinary research clusters, 2012
12. F. Welter. Regelung wissenschaftlicher Exzellenzcluster mittels Scorecardbasierter Performancemessung, 2013
13. P. Horváth, R. Gleich, D. Voggenreiter, Controlling umsetzen – Fallstudien, Lösungen und Basiswissen. Stuttgart: Schäffer-Poeschel, 2001
14. U. Konradt, P. Köppel, *Erfolgsfaktoren virtueller Kooperationen*. 2008
15. T. Vaegs, Zimmer, Inna, Schröder, Stefan, I. Leisten, R. Vossen, S. Jeschke. Fostering interdisciplinary integration in engineering management., 2013
16. R. Defila, A. Di Giulio, M. Scheuermann, *Forschungsverbundmanagement. Handbuch für die Gestaltung inter- und transdisziplinärer Projekte*. Vdf Hochschulverlag, 2006
17. C. Mast, S. Huck, Hubbard, Unternehmenskommunikation: Ein Leitfaden. Lucius & Lucius, 2008
18. T. Thiele, C. Jooß, F. Welter, R. Vossen, A. Richert, S. Jeschke, Detecting central research results in research alliances through text mining on publications. Proceedings of the 8th International Technology, Education and Development Conference (INTED 2014), Valencia, Spain, 10–12 March 2014, Valencia. IATED, S. 7244–7252, 2014
19. H. Sonneveld, K. Loening, eds., *Terminology. Applications in interdisciplinary communication*. John Benjamins Publishing Co, 1993
20. T.W. Hellmuth, *Terminologiemanagement – Aspekte einer effizienten Kommunikation in der computerunterstützten Informationsverarbeitung*. Verlag für Wissenschaft und Forschung, 1997
21. G. Budin, *Kommunikation in Netzwerken – Terminologiemanagement*. Springer, 2006

22. A. Calero Valdez, A.K. Schaar, M. Ziefle, A. Holzinger, S. Jeschke, C. Brecher, Using mixed node publication network graphs for analyzing success in interdisciplinary teams. Lecture Notes in Computer Science (including subseries Lecture Notes in Artificial Intelligence and Lecture Notes in Bioinformatics), 2013

23. A. Holzinger, B. Ofner, C. Stocker, A. Calero Valdez, A.K. Schaar, M. Ziefle, M. Dehmer, On graph entropy measures for knowledge discovery from publication network data. Availability, Reliability, and Security in Information Systems and HCI, 2013

Enhancing Scientific Cooperation of an Interdisciplinary Cluster of Excellence via a Scientific Cooperation Portal

Tobias Vaegs, André Calero Valdez, Anne Kathrin Schaar,
André Breakling, Susanne Aghassi, Ulrich Jansen, Thomas Thiele,
Florian Welter, Claudia Jooß, Anja Richert, Wolfgang Schulz,
Günther Schuh, Martina Ziefle and Sabina Jeschke

Abstract In the Cluster of Excellence (CoE) "Integrative Production Technology for High-Wage countries" at RWTH Aachen University, scientists from different institutions investigate interdisciplinary ways to solve the polylemma's tradeoffs between scale and scope as well as between plan and value oriented production. Next to the CoE's four scientific subfields – the Integrative Cluster Domains (ICDs) – there are three additional subprojects performing cross sectional research and providing means for physical and virtual cross-linkage, the Cross Sectional Processes (CSP). Scientific cooperation in such a large and diverse consortium – as a meta-structure to the structures present in the member institutes – poses many challenges. To tackle these, an online learning and collaboration platform is developed, called the "Scientific Cooperation Portal", to optimize the cluster-wide cooperation process. Technically building on the Liferay framework, the portal provides basic features like a member list and an event calendar as well as functionalities to help cluster members to gain a deeper understanding of the CoE's current state regarding the diversity in

T. Vaegs (✉) · T. Thiele · F. Welter · C. Jooß · A. Richert · S. Jeschke
IMA/ZLW & IfU, RWTH Aachen University, Dennewartstr. 27, 52068 Aachen, Germany
e-mail: tobias.vaegs@ima-zlw-ifu.rwth-aachen.de

A.C. Valdez · A.K. Schaar · M. Ziefle
Human-Computer Interaction Center, RWTH Aachen University, Aachen, Germany

A. Breakling · S. Aghassi
Fraunhofer Institute for Production Technology, Aachen, Germany

U. Jansen · W. Schulz
Chair for Nonlinear Dynamics of Laser Processing,
RWTH Aachen University, Aachen, Germany

G. Schuh
Werkzeugmaschinenlabor, RWTH Aachen, 52056 Aachen, Germany

Originally published in "Proceedings of the Seventh International Conference on
E-Learning in the Workplace ICELW 2014", © ICELW 2014. Reprint by Springer
International Publishing Switzerland 2016, DOI 10.1007/978-3-319-42620-4_17

interdisciplinary terminology, patterns in publication relationships, knowledge management and developed technologies.

Keywords Cluster of Excellence · Interdisciplinary Integration · Scientific Cooperation · Social Media

1 Introduction

Modern research questions more and more require a collaborative and additionally an interdisciplinary approach, since they often originate from the interfaces between different disciplines [1]. In such a joint research process between different disciplines, however, the participants often face problems resulting from the clash of different cultures e.g. regarding publication behavior or terminology [2].

The Cluster of Excellence (CoE) "Integrative Production Technology for High-Wage countries" at RWTH Aachen University was initiated by the German Research Foundation (DFG) and the German Council of Science and Humanities (WR) as part of the German excellence initiative. The consortium is located in Aachen with various interdisciplinary partners from different faculties of RWTH Aachen University investigating the resolution of the polylemma of production [3], i.e. ways to solve the tradeoff between scale and scope and between plan and value oriented production [4]. The CoE (cf. Figure 1 for an overview over its structure) consists of twelve subprojects with a total of about 180 researchers and 200 student assistants. These

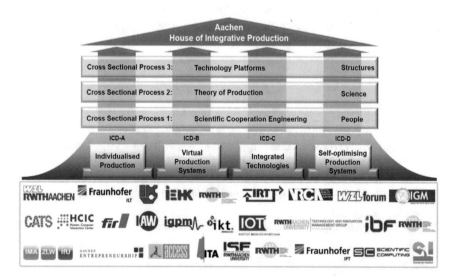

Figure 1 Structure of the CoE with its four Integrative Cluster Domains (ICDs) and three Cross Sectional Processes (CSPs)

researchers come from various scientific disciplines in varying degrees of completion regarding their education. Bringing all this personnel to the same table and enabling them to cooperate requires a common understanding of terminology, language, methods, competences, cognitive models, perceptions of success, and many more criteria.

Foremost it is necessary to ensure that all members are able to communicate effectively despite the different terminologies in their given discipline. Hence, one of the central challenges of interdisciplinary work is the disciplinary coloring of terminology, which hampers communication if handled unwittingly. Discussions about meanings of terms may occur afterwards when seemingly an agreement has been found. Even when a concept of linguistic/terminological diversity exists, differences in methodology and cognitive models may hinder effective scientific cooperation.

Second, even teams which seem to cooperate successfully may have diverging criteria for what counts as scientific success. Disciplinary differences emerge e. g. in different publication behavior (community size, citation frequency, citation half-life etc.) and a different understanding of what constitutes a "successful" publication. The same applies for other success criteria and the evaluation of their impact (e. g. patents, acquired third party-funding, supervision of theses and teaching). Increasing the awareness of the diversity of an interdisciplinary venture and establishing a culture of valuing this diversity are crucial for moving from a multidisciplinary perspective to an interdisciplinary one.

To overcome these and other challenges of interdisciplinary cooperation and to support the performance of such interdisciplinary research consortia, new approaches are needed for the cross linkage of the different researchers and institutes as well as for the transfer of solutions between them. Therefore, the CoE next to several collaborative projects from the field of production technology additionally comprises cross-sectional projects entrusted with this task, the "Cross Sectional Processes" (CSPs). Here, concepts of supporting the integration of the different disciplines into the CoE on a physical and virtual level are designed, implemented and constantly evaluated [5]. Among other approaches, an online platform is developed, which offers different applications to the cluster members.

This paper is structured as follows: In Section 2 this platform, the Scientific Cooperation Portal (SCP), is introduced. The section describes the technology behind the portal as well as its basic features to help researchers in their everyday work. Section 3 describes the applications offered by the portal that are used to advance the research efforts of the CSP, before Section 4 ends with first experiences with the portal and a short outlook onto further functionality.

2   The Scientific Cooperation Portal

In this section the Scientific Cooperation Portal (SCP) jointly developed by the team of the Cross Sectional Processes 1 and the Cross Sectional Processes 3 is presented with its basic functionality to assist cluster members in their daily work as well as its research functionality to advance the research of the CSPs.

2.1 Technology

The social software framework that is used to set up the SCP is a Liferay Server. Liferay is open-source enterprise portal software that is free of charge and runs on JavaEE Servers (e. g. Tomcat Server) connected to an SQL database (e. g. MySQL). It was chosen because expertise in Liferay Development was readily available in the team and the feature set of Liferay extends beyond the leading commercial competitor (i. e. Sharepoint). Liferay offers a social community platform, allowing the forming of communities of interest, which get access to community specific functions and content. Users of the software can become members of these communities and connect with other users, accessing their personal profile and their contact information. Typical social features such as blogs, messages, chats, message boards and wikis are also part of Liferay.

Liferay provides web-authoring mechanisms including workflow management (incl. roles such as authors, editors, etc.). This allows customizing the look and feel of Liferay to make users feel at home (cf. Figure 2). It furthermore supports document and media management with versioning. All content generated in Liferay can be tagged, categorized, commented, rated, and accessed from different applications (i. e. portlets) that interconnect the various forms of data available to the Liferay server. By writing own applications developers can leverage the framework of Liferay to make their own data accessible by other applications and access other information (e. g. profile data).

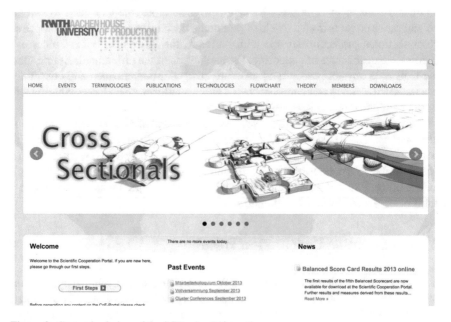

Figure 2 Customized view of the SCP using Liferay themes

2.2 Basic Features

The previously described Liferay technology already provides a set of basic features, which support the scientific cooperation of the cluster employees.

A full member list gives information about involved institutes and researchers as well as their contact details. In addition, each member has an own profile, which can contain a picture and information about further research interests outside the CoE. An extension to Liferay called "Social Office" also enables plenty of other social networking features, e. g., networking between different users, a display who is online at the moment combined with a simple chat and a messaging tool.

One feature to improve the scientists' collaboration is the calendar. It shows all CoE-related appointments and also sends reminder messages to participants upon request. The forthcoming Liferay update will come with a revised version of the calendar, which extends the current functionality by an improved user interface (similar to the well-known Google Calendar) and new technical interfaces to synchronize the CSPs' calendar with established tools and devices at the different participating institutions.

As in all knowledge intensive organizations transparency is a key need of employees within the CoE. The sum of all basic features therefore originates from the necessity for any large cooperation to disseminate information across all members of the cooperation and to avoid hampering flow of information across hierarchy. Furthermore, the portal offers different standard tools to organize contents, e. g. a simple content and file management system or a typical wiki application.

3 Research Applications of the Portal

The SCP serves as a dual purpose solution, beyond the aforementioned function; it generates valuable data for the research of the CSPs directly from the CoE itself, which in turn can be used as performance indicators for the cybernetic management approach of the CoE. These aspects are also addressed by specific features of the SCP (i. e. Terminologies, Publications, Technologies, and Project Management).

3.1 Terminologies

The Cluster Terminologies application is one of these applications and helps members to become aware of and cope with differing terminologies. Figure 3 depicts a view showing all definitions stored for a given term.

The application provides cluster members with the possibility of getting an overview over different terms, which are supposed to be central terms from research activities of the CoE as well as from the scientific fields involved. Together with these terms the application presents various definitions for each term reflecting the

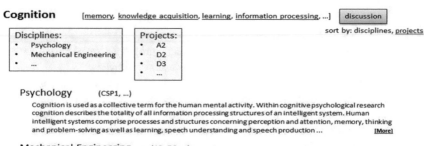

Figure 3 View on an example term in the SCP

fact that the same term (or very similar terms) can be defined differently depending on the scientific discipline. Every definition is assigned to one or more disciplines and to one or more subsidiary projects of the CoE, where the term is used as the definition indicates. Definitions can also be contrasted with common understandings of the term outside the CoE, if a research topic is examined from a different perspective or in a different way in the CoE.

Reading the data stored for a term, the user directly recognizes how many definitions exist for this term and how ambiguously it is used and understood in different disciplines. Thus, the user learns about definitions from other scientific fields leading to an integrated understanding of the diverse terminologies. Moreover, the application provides the possibility of starting online discussions about every definition. New opinions can be introduced, discussed and if necessary integrated into existing definitions. Thus, the application's database always contains the current working definition of terms inside the CoE.

The user is provided with different ways of accessing the definitions. Next to a simple alphabetical list or glossary function of the defined terms the user can browse through a structured list of the scientific disciplines, which define terms from their perspective. Moreover, the organizational structure of the CoE provides a graphical way of discovering which terms are used and defined in the different subsidiary projects. Additionally, the defined terms are tagged with keywords, which help to examine the terminology on a specific topic.

All this information is gathered, presented and discussed in a persistent way so that it can be established in the CoE instead of getting lost due to staff turnover. New members can always get an overview over the current state of the cluster terminologies.

3.2   Publications

Publications depict a form of scientific cluster output. They (ideally) contain information about research progress and cooperation. By looking at the author list of a publication one can understand who cooperates with whom. Additional information can be made available when using the SCP from the user profiles. By this, one can assess how much interdisciplinary cooperation is actually occurring [6, 7] in the regarding publication.

Nonetheless reducing scientific efforts to publications is an oversimplification when trying to assess researchers' performance. Performance measurement from bibliometric data (alone) is controversial since differences between disciplines impede comparing simple metrics such as citation counts.

In this approach the focus lies on assessing degrees of cooperation between researchers on a meta-level scale (i. e. the whole CoE). For this a graph-based approach for publication analysis is used (cf. Figure 4). By constructing co-authorship graphs the amount of cooperation (as portrayed in publications' author lists) can be measured [8]. Enriching the purely bibliometric data with sociometric data can help visualize the social precursors for successful scientific cooperation: e. g., are spatial nearness or the familiarity of authors a motor for successful cooperation? Gathering more information about social factors and their impact on cooperation might allow the derivation of supporting measures.

In this context the approach of publication visualization is focused on both providing general information about the whole CoE performance and an individualized

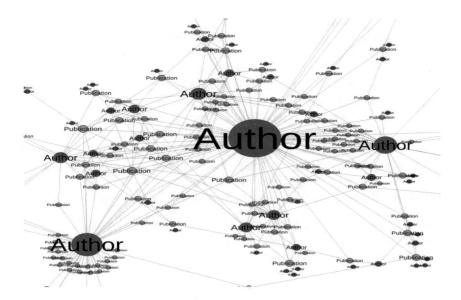

Figure 4 Exemplary publication graph – authors are denoted as

view on publications. Users will see with whom they have published, about what topics, and will get an egocentric view of their publication network. This allows users to understand their own publishing behavior and identify further authors in their thematic proximity who could enrich their personal work [9].

Extracting further information from publications allows additional analyses like grouping publications according to keywords used to find other publications from the CoE that deal with similar topics and might lead to further cooperation between the respective authors. On a meta-scale level, related topics can be identified if their respective publications appear in proximity without necessarily being directly inter-connected.

By providing the means of egocentric micro-level visualization of bibliometric data as well as meta-level visualizations to steering agents the publication relationship analysis application supports the management of the cooperation in the CoE.

3.3 Technologies

In order to bridge the gap between research and industry, promoting technology and knowledge transfer becomes increasingly important. Especially small- and medium-sized enterprises, having only little R&D resources themselves, depend on external technology development activities to remain innovative [10]. In contrast, academia research needs industrial partners, who are capable and willing to commercialize their technologies as in most public funding programs dissemination activities are required by the funding body.

Meanwhile, modern web technologies offer more and more "social" functionalities and open up new ways of user interaction. These social features offer a great potential for supporting technology transfer [11] especially in its early phases by bringing together technology demand and supply [12]. However, technology transfer portals have to be designed carefully in order to meet future users' needs and thus being successful in operation [13]. A technology transfer application is developed in the CSPs through an iterative implementation approach on the SCP.

Technology transfer beyond the CoE will be supported by the SCP by bundling of and simplifying access to the technologies developed within the CoE and the corresponding technology experts behind them. Furthermore, it should serve as a discussion forum and meeting point of expert communities for connecting people and technologies within the CoE. This will be achieved by the following core functions:

- Users are able to present themselves and their expertise in individual user profiles, connect their technologies to their individual profiles, define their interests and get updated about selected technologies.
- Users are able to find other experts within the CoE and be found by potential cooperation partners and discuss their technologies and possible applications in expert communities. Figure 5 shows one of the implemented functions of the SCP, where technologies can be described in an application oriented way, including a

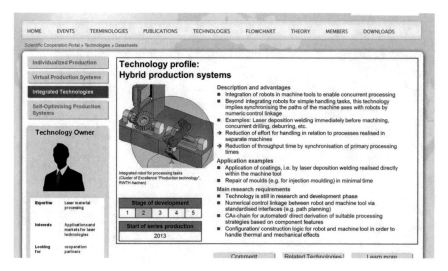

Figure 5   Technology profiles linked to user profiles

short description, future fields of applications, technology suppliers or experts as well as the current technology readiness level.

In a next step, these technology profiles will be linked to the already existing user profiles. In the future, users will also be able to define, which parts of the information about themselves and their technologies will by publicly accessible and thus be visible to possible external cooperation partners, such as interested companies or other research institutions.

3.4   Project Management

Subprojects in interdisciplinary research corporations usually show complex work package structures. To keep track of the different work packages and to ensure that all team members have the same knowledge base, the FlowChart App is developed and evaluated within the CoE and for the SCP. The main objective is to generate a web-based tool that is easy to understand and has an intuitively operable user interface, which creates a transparent view on the knowledge gaining process.

To achieve that, ideas from existing project planning tools are combined to a pictographic language approach which

(a) reflects the organizational structure inside interdisciplinary and interinstitutional projects,
(b) shows the initial situation and the main project goals, and
(c) shows the project work packages, its dependencies and interlinks on a time schedule.

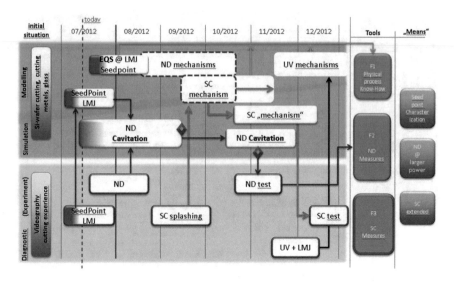

Figure 6 The FlowChart application

An example flow chart is shown in Figure 6. In the background different work package groups are indicated by different colors. Different work package groups may also indicate different institutions or different resources. Inside these work package groups the initial situation or needs are placed in the left column. On the right hand side two columns carry information about the project outcome. These columns are separated by practical project results or means, like new machines or new processes and results that reflect the knowledge or tools gained inside the project.

Inside the central area the work packages are situated on a time schedule. Interlinks between the work packages depict the dependencies between work packages and their contribution to the final results. The horizontal extent of a work package reflects the amount of time that is planned for this work package. The degree of fulfillment is drawn into the work package box using background gradient coloring.

The resulting image is a flow chart that shows the state of the project and the upcoming work packages on a single presentation slide, which is easy to explain to project members as well as customers. To make it easy to create, share and distribute flow charts, a web-based application that will be placed on the SCP is implemented.

To measure the acceptance and effect of the FlowChart approach inside inter-disciplinary and interinstitutional research groups, interviews and surveys will be performed inside the CoE. The resulting application is a project management tool that is reduced to the information that is essential to understand the knowledge gaining process inside the project.

3.5 Data Gathering

The SCP requires specific types of data to be useful to the researchers. Large-scale interdisciplinary research efforts bear the problem of lacking transparency regarding personal information of the many individual researchers. Thus they require an active management of information regarding human resources, individual research efforts, and interfaces between researchers (i. e. information on users' disciplines, scientific methods, publications, terminologies, and technologies). Only when these are made readily available, the strength of the weak-tie network can be leveraged and lead to an emergence of an innovative interdisciplinary research output [14].

The necessary data is collected using both online and offline approaches. Data for the various applications is collected automatically by accessing other data sources, such as the databases of the RWTH Aachen University Library. Additionally users have the ability to upload data or edit data live on the website, allowing them to customize the user experience to their desire.

In many other cases it is necessary to systematically collect data from all cluster members (e. g. sociometric data). For these cases either survey questionnaires or interview studies are conducted and the evaluated data is integrated into the SCP. One example of a reoccurring questionnaire study is the cluster-specific Balanced Score Card [15], which is performed annually. In this study critical measurements (e. g. evaluation of research quality, research cooperation, information policy, etc.) are assessed in regard to subproject and hierarchical level to understand where optimizations should be made.

Another way of generating data is using the colloquia of employees of the cluster staff. In these (usually) full day meetings certain topics that are relevant to all cluster domains are worked on, and the generated results are then systematically treated and improved for online integration on the SCP. The last colloquium dealt with the variety in technologies, terminologies and methods that are used in the CoE. For example, by identifying who uses what type of methods and creating online resources for the applied methods, users can find experts in the CoE in case they are looking for assistance on a certain topic.

Since the SCP offers the opportunity to setup special interest user groups with designated private data stores, these can also be used to generate data for groups. This data can then be used to enhance the user experience for these groups as well. For example, users can get recommended literature derived from their membership in special interest groups.

4 Experiences and Outlook

The SCP is available to the cluster members since November 2013. User profiles of the cluster employees were not prefilled; all user data had to be uploaded by the users themselves. The voluntariness of use and data disclosure leverages the possible negative experiences users have made with other forms of social media in professional contexts [16]. In the first step *transparency* was identified as a fundamental basis, which reflects in the cluster members using the portal predominantly as a kind of yellow pages, to get an overview about the forthcoming events and to download achieved results. In addition, information about former events was accessed and user profiles were filled.

Using Piwik[1] as an analytics tool which corresponds to German privacy laws, e. g., by storing data locally and disguising the users' IP addresses, one can get an overview of the portal usage and expect further findings about the users' requirements [17]. Piwik is used in order to identify intensities of usage of single portal functions so that the prototype of the SCP can be continuously improved and developed in a user-centered way. Frequently used functions can be kept, while rarely used functions can be discarded again.

Next, new features will be enabled one after another and existing functionalities improved. First, an update to the new Liferay version 6.2. is planned. This contains among other novelties the described calendar improvements but also overall interface innovations like responsive design, which increases the user experience on mobile devices. This has been determined to be a key factor in technology acceptance. Second, the different applications and features explained here will be interlinked. For example, the profiles of users who are experts in a specific field will link to a technology datasheet related to the same field and vice versa. This will allow users to identify required corresponding experts. Furthermore, the portal will be linked to external data sources like the university library so that for instance cluster-relevant publications can be pushed to the portal and users only have to monitor their publications instead of uploading them manually. This dynamic interconnection of information leverages all key benefits of social media, the effect of which remains to be studied by the CSPs.

To extend the SCP's benefits, sub-communities will become an important part of the portal. These special interest communities (SICs) relate to the specific structures of the CoE and assist researchers in their daily environment. These SICs can flexibly be assigned, customized in their feature set to match given needs, and can therefore react to changes in structures and requirements in a research setting. Accordingly, the SCP is envisioned to become the central information and collaboration platform of the CoE itself, the different cluster projects and groups and in the best of cases even of the single member institutes. In the future this will also constitute the basis for opening parts of the provided information to external interest groups, such as potential cooperation partners from industry.

Next to the support for the Cluster of Excellence the SCP is used to gather data for the research questions the CSP subprojects are concerned with. This data is then analyzed to approach the question about how interdisciplinary collaboration works and which factors influence its output.

One aspect of this is the influence of differing terminologies and the question if the acquisition of the terminology used inside such a consortium is a suitable means

---

[1]http://www.piwik.org.

of discovering possible synergies between researchers in different subprojects, who are not yet working together directly.

Furthermore, the SCP generates and collects data about the level of cooperation, which itself will be used as research data. By this we can measure the effectiveness of interventions (such as the colloquia of employees) for enhancing interdisciplinary cooperation, which will finally result in new structures emerging from the publication network of the cluster.

Acknowledgments This work was performed as part of the Cluster of Excellence Integrative Production Technology for High-Wage Countries, which is funded by the excellence initiative by the German federal and state governments to promote science and research at German universities.

References

1. T. Jahn. "Transdisziplinarität in der Forschungspraxis", Transdisziplinäre Forschung: integrative Forschungsprozesse verstehen und bewerten , 2008
2. J. R. Bryson. Hybrid Manufacturing Systems and Hybrid Products: Services, Production and Industrialisation: Studies for Innovation in a Modern Working Environment. International Monitoring. Ph.D. thesis, RWTH Aachen University, Aachen, 2009
3. C. Brecher. Integrative production technology for high-wage countries
4. Cluster of Excellence 'Integrative Production Technology for High-Wage Countries' Renewal Proposal for a Cluster of Excellence – Excellence Initiative by the German Federal and State Governments to Promote Science and Research at German Universities, Unpublished
5. C. Jooß, F. Welter, I. Leisten, A. Richert, A. K. Schaar, A. Calero, Valdez, E. -V. Nick, U. Prahl, U. Jansen, W. Schulz, M. Ziefle, S. Jeschke. Scientific cooperation engineering in the cluster of excellence integrative production technology for high-wage countries at rwth aachen university: accepted at 5th international conference of education, research and innovation (iceri) 2012
6. H. Snyder, S. Bonzi. Patterns of self-citation across disciplines, 1998
7. Alexander von Humboldt-Stiftung. Publikationsverhalten in unterschiedlichen wissenschaftlichen Disziplinen. Beiträge zurBeurteilung von Forschungsleistungen, Discussion paper
8. A. Calero Valdez, A. K. Schaar, M. Ziefle, A. Holzinger, S. Jeschke, C. Brecher. Using mixed node publication network graphs for analyzing success in interdisciplinary teams, 2012
9. A. K. Schaar, A. Calero Valdez, M. Ziefle. The impact of user diversity on the willingness to disclose personal information in social network services, 2013
10. T. Laube. Methodik des interorganisationalen Technologietransfers: (IPA-IAO Forschung und Praxis, bd. 483).heimsheim: Jost-jetter, 2009, p. 88, 2009
11. D. Czarnitzki, C. Rammer. Technology transfer via the internet: A way to link public science and enterprises? 2003
12. G. Schuh, S. Aghassi, A. Calero Valdez. Supporting technology transfer via web-based platforms: Proceedings of picmet, 2013
13. G. Schuh, S. Aghassi. Technology transfer portals: A design model for supporting technology transfer via social software solutions, 2013
14. M. Granovetter. The strength of weak ties, 1973
15. F. Welter, R. Vossen, A. Richert, I. Isenhardt. Network management for clusters of excellence-a balanced-scorecard approach as a performance measurement tool, 2009/2010
16. A. K. Schaar, A. Calero Valdez, M. Ziefle. Publication network visualization as an approach for interdisciplinary innovation management: Professional communication conference(ipcc), 2013, pp. 1-8
17. D. Waisberg, K. Avinash. Web analytics 2.0: empowering customer centricity, 2009

Scientific Cooperation Engineering Making Interdisciplinary Knowledge Available Within Research Facilities and to External Stakeholders

André Calero Valdez, Anne Kathrin Schaar, Tobias Vaegs, Thomas Thiele, Markus Kowalski, Susanne Aghassi, Ulrich Jansen, Wolfgang Schulz, Günther Schuh, Sabina Jeschke and Martina Ziefle

Abstract In this paper we introduce the Scientific Cooperation Portal (SCP), a social enterprise software, and how it is integrated into our process of Scientific Cooperation Engineering. This process is applied in a large-scale interdisciplinary research cluster to ensure and manage the success of the interdisciplinary cooperation of over 180 researchers in different qualification levels. We investigate the influence of shared method competencies as an exemplary driver for collaboration. From the results we address both offline and online measures to improve interdisciplinary collaboration. We show how the knowledge generated from offline measures such as colloquia are transferred to the SCP and connected with other data available on the portal. This includes the handling of interdisciplinary terminologies, the disposability of publications and technology data sheets. The portal fosters knowledge exchange, and interdisciplinary awareness within the research cluster as well as technology dissemination both within the cluster, across the university, and into industry. The effectiveness of the approach is continuously assessed using a traditional balanced scorecard approach as well as additional qualitative measures such as interviews and focus groups.

Keywords Virtual Collaboration · Publication Relationship Analysis · Network Analysis · Terminologies · Technology Transfer

A.C. Valdez (✉) · A.K. Schaar · M. Ziefle
Human-Computer Interaction Center, RWTH Aachen University,
Campus Boulevard 57, 52074 Aachen, Germany
e-mail: calero-valdez@comm.rwth-aachen.de

T. Vaegs · T. Thiele · M. Kowalski · S. Jeschke
IMA/ZLW & IfU, RWTH Aachen University, Dennewartstr. 27, 52068 Aachen, Germany

S. Aghassi · G. Schuh
Fraunhofer Institute for Production Technology, Steinbachstr. 17,
52074 Aachen, Germany

U. Jansen · W. Schulz
Department of Nonlinear Dynamics of Laser Processing, RWTH Aachen University,
Steinbachstr. 15, 52074 Aachen, Germany

Originally published in "Proceedings of the 10th International Conference
on Webometrics, Informetrics, and Scientometrics (WIS), 15th COLLNET Meeting",
© TU Ilmenau 2014. Reprint by Springer International Publishing Switzerland 2016,
DOI 10.1007/978-3-319-42620-4 _18

1   Introduction

Dealing with complex global challenges often requires interdisciplinary research approaches to find suitable solutions [1]. Staying within disciplinary boundaries may prevent researchers to get a holistic overview of the topic at hand. Although the term interdisciplinarity lacks a unified definition [2] it can be seen as the successful coop-eration of researchers trained in the methods and conceptual approaches of different disciplines. Interdisciplinary research integrates these various methods to create new insights and methods for complex problems. Yet, actually making interdisciplinary research happen can be cumbersome because of lacking a common language, method competencies and understanding of scientific success. This problem intensifies under conditions of high staff turnover, research group size [1], performance pressure, and increasing complexity of the research problem. How to measure interdisciplinary collaboration and finding reasons for this collaboration, and the deliberate steering of interdisciplinary groups are still largely unsolved questions. Thus active support for such collaboration requires various measures and a constant evaluation of these measures. We apply findings from bibliometrics and cybernetics to management principles of a research cluster in order support interdisciplinary collaboration and scientific success of the cluster.

1.1   Related Work

Collaboration trumps solo-efforts in generating knowledge [3]. Finding evidence of (interdisciplinary) collaboration can traditionally be done by analyzing coauthorship networks [4], although one must be careful not to mistake co-authorship for col-laboration and vice versa [5]. Investigating who publishes with whom can reveal collaboration patterns and thus be used to understand interdisciplinary cooperation. Glänzel & Schubert found that geopolitical location and language are determining factors for collaboration. Collaboration decreases exponentially with physical dis-tances [6, 7]. Kretschmer [8] found that similarity as well complementarity can be used to explain researchers' collaboration by analyzing co-authorship relationships. By applying this approach Kretschmer and Kretschmer [9] could explain up to 99 % of the variance for 77 % of the co-authorship relationships. De Solla Price and Gürsey [10] identified different types of authors according to their publishing behavior (i. e. continuants, transients, recruits, terminators) for which Braun et al. [11] identified differing author productivity and collaboration patterns. Newman [12] found pat-terns of small world phenomena (i. e. short paths between any two random authors). Co-author networks showed various levels of clustering and a fractal nature (e. g. self-similarity). Van Raan [13] developed a model to determine growth of scientific literature based on the fractal nature of science. Sub-systems grow individually and can be seen as self-organizing units. This reflects in the cybernetic nature of how universities are managed [14]. Cybernetics in this regard means that no centralized "premeditated" plan (for publications) is conceived by the management but, in the manner of a thermostat, a target output is defined and measures are taken to reach the target.

Using interviews Hara et al. [15] created a model for determining factors of collaboration in in a research center. From the interviews they found two different types of collaboration, "complementary" and "integrative" collaboration. Determining factors were compatibility (i. e. work style, priority, management style, approach to science, personality), work connections (i. e. work interests, expertise), incentives (i. e. external funding, publication, internal) and sociotechnical infrastructure (i. e. awareness, communication mechanism, organization culture and structure, access to collaborators). Overall they assume personal relationships beget professional relationships and thus collaboration. They suggest that technological support could enhance the process of collaboration and that it needs further investigation.

Various forms of these collaboration support systems exist. This new emerging field of EScience and E-Infrastructure draws on the tools and methods developed from Computer- Supported Cooperative Work [16]. Zheng et al. [17] present TSEP a social platform to assist collaboration between scientists. Li et al. [18] and Müller-Tomfelde et al. [19] strengthen the need for shared workspaces and audio-visual support of workgroups in a health laboratory, but also tailoring to the needs of the workgroup. Alves et al. [20] have suggested a system for finding possible collaborators in a scientific setting. Romano et al. [21] suggest the use of wikis and ontologies along with learning environments to support researchers in the field of bioinformatics. Above all tailoring a Social-Network-Solution (SNS) to the users needs is critical, as communicative preferences may depend on user characteristics [22].

2 Research Questions

In this paper we demonstrate the efforts undertaken in a research cluster to support interdisciplinary collaboration. For this purpose we look into both online and offline measures that support collaboration. We assume that shared method competencies may also be a driver of collaboration. Here we compare the shared method competencies of workgroups generated from both publication data and qualitative data collected at a member colloquium. Furthermore we show how the insights from the study are used as feedback to the researchers in the cluster. In the following sections we first describe the research cluster, the Scientific Cooperation Portal and then the analysis of methods used in the cluster.

3 The Scenario – The Aachen Cluster of Excellence

The challenge of keeping production industry sustainable in countries with high wages is also in interdisciplinary one. In the research cluster of excellence (CoE) "Aachen House of Integrated Production" researchers from various subfields of physics, material sciences, engineering, computer science, up to economics and social sciences are faced with the challenges of production on various levels of scale and their interfaces (i. e. from raw material properties to production processes to factory and logistics planning, with respect to human needs on all of these levels). Overcoming the stereotypic scale-scope dilemma (individualized products vs. mass production) of production [23] is one key goal of this research cluster. Additionally it faces the unification of the dilemma of plan- versus value-oriented production, in conjunction called the polylemma of production. In total about 180 researchers work on this holistic view on production technology, grouped in different working areas. These researchers work in four integrated cluster domains (ICDs), which are interconnected by so called cross-sectional processes (CSPs, see Figure 1). These CSPs ensure sustainability of the research cluster in regard to human resources, advancement of scientific theory and development of technology platforms [24]. Their research goal is to investigate, what methods work effectively to achieve said sustainability. Additionally they assist the steering committee of the cluster by providing insights on performance and recommending a course of action.

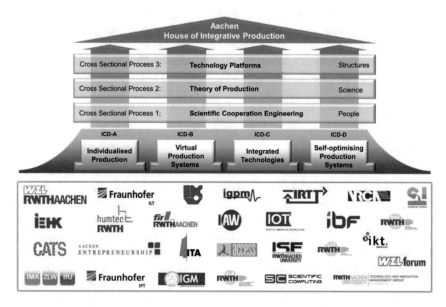

Figure 1   Research structure of the CoE, integrating institutes from five faculties of RWTH Aachen University and focusing on sustainability within the dimensions people, science and structure, incorporated within the Aachen House of Integrative Production [23]

3.1   Managing Collaboration

In order to ensure that the cluster works effectively key performance indicators (KPI) are established to measure performance for both internal (management) and external use (funding agency evaluation). This is done using a balanced-score-card approach [25] with typical performance measures as (peer-reviewed) publications, patents and third-party funding, but are also contrasted by criteria like knowledge dissemination, interdisciplinarity, quality of supervision, and many more. These are used to determine how well the cluster works and where it needs improvement.

Bringing researchers from so many scientific fields together requires management of many of these success criteria in an individualized fashion. Disciplines differ in regard to what is considered successful as a publication or as advancement in theory. In order to unify the dilemma of required disciplinary diversity and the need for a unified measure of success a cybernetic management approach is applied. For example, indicators are developed that measure the transfer of knowledge within the cluster, the development of interdisciplinary methods, the coherence of the research road map, or the transfer of technology within the cluster and into industry. Measuring performance in an interdisciplinary context is not a trivial task, but beyond that, steering performance is even harder. The cybernetic management approach incorporates various measures to both measure and steer performance.

A mix of offline and online measures is used to reach a maximum of potential cluster members. As offline steering measures the CSPs conduct member colloquia, cluster conferences, general assemblies, seminars, and workshops. In the member colloquia all partaking researchers spend a whole day dealing with topics that overarch the ICD-structure of the cluster, such as interdisciplinary communication skills (e. g. presenting research to non-experts), finding research partners (e. g. scientific speed dating) and developing a common research road map. On dedicated cluster conferences researchers present the results of their individual scientific research to the other members. In general assemblies principle investigators (PI) present the meta-level of research from their institutional point of view connecting the theory behind partaking institutes. These measures foster the interdisciplinary awareness, cooperation, communication and method skills. Some topics are addressed in seminars or workshop to address individual and sub-project based needs. For example a seminar on interdisciplinary publishing addresses the participants perception of the publishing process form their disciplinary perspective. Best-practices in cluster-typical cooperation are discussed and shared with the participants. An online method to enrich these offline approaches is the Scientific Cooperation Portal presented in this paper.

All measures are all evaluated in regard to the KPIs quantitatively (using a questionnaire method) but they are also addressed in interviews and focus groups with the researchers to ensure validity of the measurements.

4 The Scientific Cooperation Portal

As an online measure the CSPs introduced the Scientific Cooperation Portal (SCP) in 2013 [26]. The SCP is a social portal system used as a centralized knowledge storage system and was introduced to face the aspect of transparency of communication, which appeared in several evaluations. Voluntary access to the SCP is limited to cluster members and PIs exclusively (yet).

The SCP provides user profiles, yellow pages, a cluster based news feed, calendar and event system, and a centralized file storage system. Required forms for typical needs (e. g. travel expense forms) are available from this centralized storage system. All data on the SCP can be tagged and thus interconnected with each other. As specific features designed to match the cluster specific needs measured by the BSC, interviews, and focus groups, applications are built to address the challenges of interdisciplinary use of terminology, interdisciplinary publications, and technology transfer.

4.1 User Profiles

Members profiles can be found through the yellow page system and contain information about disciplinary background, method competencies, expertise in technology, publications, and participation on terminology definitions. Furthermore typical contact information is available.

4.2 Terminologies

One critical aspect mention in many evaluations is the lack of a unified language/terminology. Since different disciplines use terminology differently the approach of the CSPs is not to unify terminology, but to enhance awareness of disciplinary differences. For this purpose an application is developed that portrays the differing definitions of frequently used terms from the various perspectives, highlighting differences in understanding. Definitions are connected to their authors, publications in which they are used, and their technology data.

4.3 Publication Relationship Analysis

Publications are a peculiar aspect of scientific work, as they disseminate knowledge gain to the scientific community. They are often (wrongly) used as sole performance indicators overvaluing quantity above quality. The SCP uses publications to establish researcher profiles. This allows the CSPs to understand (and measure by proxy) the collaboration in the CoE. Furthermore we will use visualization and graph based approaches to understand and communicate publishing efforts of the CoE to its members [27]. User profile pages will be connected with their co-authors, but also with topics stemming for publications keywords. Furthermore used technology and terminology from publications are connected with their respective technology data sheets and terminology pages.

4.4 Technology Transfer

Technology developed in the CoE should be disseminated both within and to industry partners to be useful to a possible consumer of the technology. In order to simplify communication of advances, a technology transfer portal is integrated into the SCP [28]. Here technology data sheets present key advantages of developed technology and contact information of the provider of the technology (see Figure 2). They are also connected to their provider users as well as publications that relate to the technology. Technology data sheets can be customized to be viewable by external partners (e. g. industry) once they have achieved a sufficient level of stability.

Figure 2   Example technology data sheet on the SCP

5   Methodology – Assessing Method Competencies

In order to find out what methods are used in the cluster we approach that topic from two directions. First we pick full-text data from the cluster and manually scan the methodology sections of these papers for named-entities that refer to method-names. We then perform manual deletion of duplicates on synonyms on the data. We create a method graph connecting each workgroup with its methods. Since classical database coverage of engineering sciences is subpar [29], we collect publication data manually by requiring researchers to submit their work in order collect funding for travel expenses for instance.

In a second step, conducted during a member colloquium, we asked all workgroups to brainstorm on the methods that they used on a daily basis (see Figure 3). The time frame for this task was about 90 min, and instructions were given to collect methods that are both used in publications and methods that are available but have not been used yet. As a working definition what constitutes a method several definitions were given (US patent definition, a definition derived from philosophy of science, a definition from Computer Science) to heighten awareness of disciplinary differences in the meaning of the term "method". Methods are then again cleared for duplicates and synonyms. Another method graph is constructed. Both method graphs are then compared an evaluated in regard to graph statistics.

Figure 3   Exemplary results of a method workshop in a subproject

Not addressed in this paper are the workshops that address in a similar fashion the topics of interdisciplinary terminologies and technology data sheets.

6   Results and Interpretation

At this current timeframe full-text publications were available for 7 of 12 sub-projects. From over 500 publications 76 were selected (availability and containing a clear method section) and manually scanned for methods. Form these, 222 named-entities were recognized and reduced to 195 unique methods. The constructed method graph (see Figure 4) showed a graph density of *0.006*. Community detection [30] revealed 7 communities and a modularity of *0.773*.

The method collection from the member colloquia surprisingly also resulted in a sum of 195 methods (after deletion of duplicates and synonyms). The graph (see Figure 4) showed a graph density of *0.005* and also revealed 7 communities. Modularity of the graph was determined at *0.766*.

Interestingly the nodes connecting most sub-projects in both graphs are nodes that relate to "modelling", "FEM" and "Software Development". Method overlap in both cases is sparse, meaning that either shared methods are sparse, remain unmentioned (in both verbal an written communication) or that no unified terminology exists regarding applied methods. Both graphs show a structural symmetry between each other.

As a side note is worth mentioning that even the term "method" is far from having a shared understanding. During the member colloquium the need for clarification arose, in particular in regard to discerning it from the term "technology". In the various fields

(a)   (b)

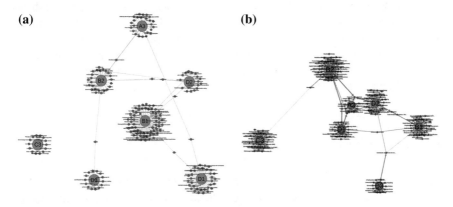

Figure 4   Method graphs constructed from member colloquium data (*left*) and from method sections of publications (*right*)

of engineering, clear differentiation is not always possible. One develops a technology that is used by others as a method. Discussions regarding this took substantial time off of brainstorming times.

7   Conclusion

The differences in terminology, in particular in regard to the term "method" itself, further underline the need for support in an interdisciplinary setting. As mentioned by Hara et al. [15] compatibility is essential for scientific collaboration.

Applying the approach from Alves et al. [20], we enrich researcher's profiles with method competencies to enable finding researchers within the cluster that share research interests. The terminology application must respect disciplinary differences in understanding of methods (that can also be technologies) and can be seen as a measure to broaden understanding of method competencies across disciplinary boarders. Furthermore technology transfer must be performed not only to external stakeholders but also within a research cluster. The findings from the member colloquium confirm the need for social software that integrates terminology, methodology, technology, and publications as an online support measure to our research cluster. This means when a user opens another user's profile, he will see a list of methods used by this researcher, which hyperlinks to an ontology-based wiki and also full-text publications (when available) that contain these methods. Furthermore technology used by a researcher is hyperlinked to technology data sheets, which in turn are linked to publications and terminology.

In the future, we are able to better understand interdisciplinary cooperation by following the individual as well as the work groups' usage behavior of information of the Portal. Both, the genesis of a novel cooperation can be retraced and related to the respective genesis conditions as well as the growing density of the collaboration's network in order to see growing novel topics or methodologies within and across work groups. Also, looking from the industry side and the analysis of industry's interest and search for information behavior can be also a promising approach for emerging topics and research fields.

7.1   Limitations

The procedures to generate graphs rely heavily on manual correction and synonym detection. We must assume that further unnoticed synonyms exist in the data as the author is no expert in all of the found methods. This limitation also applies to the manual named-entity search in the papers. Furthermore only a fraction of the actual publication output was used, due to availability of full texts.

The similarity of the graph could to a large extend be caused by the method of construction. For both graphs first workgroup nodes are created and then connected to their method nodes. This would in many cases lead to similar graphs, if methods were unrelated.

The presented approach was used as a starting point into the data. In the future users of the portal may choose to add their own synonyms to method definitions to enhance the analysis process in future iterations. The approach also only reflects collaboration of the similarity type. Complementary or integrative collaboration should in essence not contain the same set of methods. Nonetheless an overlap that enables communication should be found.

Furthermore we have not looked into interrelations between both graphs yet, as the methods are not in a single language. Finding adequate translations should also be a user driven task as well.

8 Summary and Outlook

In this paper we presented the scientific cooperation portal a social portal to support interdisciplinary collaboration in research clusters. The features of the portal were developed from systematic evaluation of researchers needs using both qualitative and quantitative methods [31]. Content for the portal is generated by both the users and the CSPS from at various events. Furthermore we looked into shared method competencies as a driver for collaboration by investigating the methods used in the sub-projects both from verbal and written evidence. We found low overlap between sub-projects in methods, but high similarity for both approaches. Interestingly when comparing the method overlap with actual collaboration from publication data [27], we find a similar graph density (0.005) but a higher level of clustering (27 communities, modularity 0.844). Further evaluation (e. g. graph isomorphism) will reveal whether this accurately reflects similarity between the different graphs. Furthermore looking into references and citation data could prove useful. Researchers sharing the same methodology should cite similar work. The hypothesis that ones technology is another's method could also be verified by looking into citations in method sections. From these findings we derive the need for collaboration support and underline the selection of features of the Scientific Cooperation Portal as well as conducting member colloquia which bring researchers together on a personal level and foster communication between sub-projects and across disciplinary boarders.

Connecting both offline with online measures has improved KPIs for scientific collaboration, which was established by a BSC-approach.

Acknowledgments We would like to thank the participants in the colloquium for the efforts. We would also like to thank Tatjana Hamann and Juliana Brell for their support. This research was funded by the Excellence Initiative of the German federal and state governments.

References

1. F. Repko. Interdisciplinary research: Process and theory, 2012
2. M. Jungert, E. Romfeld, T. Sukopp, U. Voigt. Interdisziplinarität: Theorie, Praxis, Probleme, 2010
3. S. Wuchty, B.F. Jones, B. Uzzi. The increasing dominance of teams in production of knowledge, 2007
4. W. Glänzel, A. Schubert. Analysing scientific networks through co-authorship, 2005
5. G. Melin, O. Persson. Studying research collaboration using co-authorships, 1996
6. J.S. Katz. Geographical proximity and scientific collaboration, 1994
7. J. Hoekman, K. Frenken, R.J. Tijssen. Research collaboration at a distance: Changing spatial patterns of scientific collaboration within europe, 2010
8. H. Kretschmer. A new model of scientific colloboration part 1. theoretical approach, 1999
9. H. Kretschmer, T. Kretschmer. Who is collaborating with whom in science? explanation of a fundamental principle, 2012
10. de Solla Price, D., S. Gürsey. Studies in scientometrics i transience and continuance in scientific authorship, 1975
11. T. Braun, W. Glänzel, A. Schubert. Publication and cooperation patterns of the authors of neuroscience journals, 2001
12. M.E. Newman. The structure of scientific collaboration networks. Proceedings of the national academy of sciences, 2001
13. A.F. van Raan. On growth, ageing, and fractal differentiation of science, 2000
14. R. Birnbaum, P.J. Edelson. How colleges work: The cybernetics of academic organization and leadership, 1989
15. N. Hara, P. Solomon, S.L. Kim, D.H. Sonnenwald. An emerging view of scientific collaboration: scientists' perspectives on collaboration and factors that impact collaboration, 2003
16. M. Jirotka, C.P. Lee, G.M. Olson. Supporting scientific collaboration: Methods, tools and concepts, 2013
17. X. Zheng, G. Ke, D.D. Zeng, S. Ram, H. Lu. Next-generation team-science platform for scientific collaboration, 2011
18. J. Li, C. Muller-Tomfelde, T. Robertson. Designing for distributed scientific collaboration: a case study in an animal health laboratory, 2012
19. C. Muller-Tomfelde, J. Li, A. Hyatt. An integrated communication and collaboration platform for distributed scientific workgroups, 2011
20. T.P. Alves, M.R. Borges, A.S. Vivacqua. An environment to support the discovery of potential partners in a research group., 2013
21. P. Romano, R. Giugno, A. Pulvirenti. Tools and collaborative environments for bioinformatics research, 2011
22. A. Calero Valdez, A.K. Schaar, M. Ziefle. State of the (net) work address developing criteria for applying social networking to the work environment, 2012
23. S. Brecher, S. Jeschke, G. Schuh, S. Aghassi, J. Arnoscht, F. Bauhoff, F. Welter. Integrative production technology for high-wage countries, 2012
24. C. Jooß, F. Welter, I. Leisten, I. Richert, A.K. Schaar, A. Calero Valdez, S. Jeschke. Scientific cooperation engineering in the cluster of excellence integrative production technology for high-wage countries at rwth aachen university, 2012
25. F. Welter, R. Vossen, A. Richert, I. Isenhardt. Network management for clusters of excellence-a balanced-scorecard approach as a performance measurement tool, 2011
26. Vaegs, T., Enhancing scientific cooperation of an interdisciplinary cluster of excellence via a scientific cooperation portal. Ph.D. thesis, 2014
27. A. Calero Valdez, A.K. Schaar, M. Ziefle, A. Holzinger, S. Jeschke, C. Brecher. Using mixed node publication network graphs for analyzing success in interdisciplinary teams, 2012
28. G. Schuh, S. Aghassi, A. Calero Valdez. Supporting technology transfer via web-based platforms, 2013

29. A.W. Harzing, R. Van der Wal. Google scholar: the democratization of citation analysis, 2007
30. V.D. Blondel, J.L. Guillaume, R. Lambiotte, E. Lefebvre. Fast unfolding of communities in large networks, 2008
31. A.K. Schaar, A. Calero Valdez, M. Ziefle. Publication network visualization as an approach for interdisciplinary innovation management, 2013

Part II
Next-Generation Teaching and Learning Concepts for Universities and the Economy

Sentiment Analysis of Social Media for Evaluating Universities

Anas Abdelrazeq, Daniela Janßen, Christian Tummel,
Sabina Jeschke and Anja Richert

Abstract In the age of digitalization, a huge amount of sentiments are expressed daily on university related topics using social media platforms. Particularly, posted statements from students and teachers can provide a potential source for evaluating universities. Twitter as one of the most popular microblogging platforms is a rich data resource for opinion mining. Stimulated by this fact, ways to analyze Twitter for information in the context of universities are sought. This paper looks at the analysis of social media sentiment as a complementary source for evaluating universities. The extracted results can support university rankings that experience criticism in terms of measuring vital indicators. This paper relays on sentiment analysis methods to analyze opinions published on Twitter. For this purpose, at first, tweets that are related to selected universities in Germany were collected. Second, the tweets were classified based on their sentiment into "Positive" and "Not Positive" tweets. At last, the results were analyzed providing information about the communicative topics at the universities. This paper gives an outlook to further research in context of an automated analysis of social media content in order to support the evaluation of universities.

Keywords Social Media · Universities · Higher Education · Twitter · Sentiment Analysis · Data Analysis · Natural Languages Processing · Learning Environments

1 Introduction

Nowadays, social media platforms form a substantial data source for opinions on any topic. This include statements that are related to universities' topics and events. Particularly, microblogs appeared as one of the most common social media tools, in which, countless users can participate and interact at any time. Using microblogging

A. Abdelrazeq (✉) · D. Janßen · C. Tummel · S. Jeschke · A. Richert
IMA/ZLW & IfU, RWTH Aachen University,
Dennewartstr. 27, 52068 Aachen, Germany
e-mail: anas.abdelrazeq@ima-zlw-ifu.rwth-aachen.de

Originally published in "Proceedings of Second International Conference on Digital Information Processing, Data Mining, and Wireless Communications (DIPDMWC2015), Dubai, UAE, 2015",
© Society of Digital Information and Wireless Communications (SDIWC) 2015.
Reprint by Springer International Publishing Switzerland 2016, DOI 10.1007/978-3-319-42620-4_19

platforms, content from every user can be published, read, commented, linked or forwarded as desired. As a result, an open communication medium is established. This open exchange with the participation of different users helps to bridge and interlock formal and informal learning contexts. Consequently, microblogging is a medium that offers real-time communication of its users thoughts and ideas.

Twitter is one of the most opinion-rich resources, where huge amounts of opinions on different topics are expressed. Such opinion-rich data can be used for extracting and analyzing the opinions in terms of specific questions.

In this context, data mining techniques have the potential to detect opinions from large amounts of data. With the help of existing tools from machine learning and natural languages processing fields, concepts of sentiment analysis can be applied on the collected data from social media. This can provide vital measurements and insights over the communicative topics in the university environment.

Analyzing students' and teachers' opinions as well as comparing and evaluating universities is important on different scales. On the one hand, universities would like to have a performance measurement mechanism for improvement and adjusting plans. On the other hand, students relay on these comparisons to support their decision for joining a specific university to proceed with their studies.

The standardized process of comparing and evaluating universities is based on university rankings on a national and international level. It operates on the basis of certain criteria (i. e. indicators). Recently, university rankings experienced a lot of criticism in terms of: measurement accuracy, measuring the university as a whole institution, and the way data is being collected for measuring specific indicators at universities.

Besides the official standardized university rankings, alternative and complementary approaches must be investigated. Standard rankings try to search for statements from teachers and students that are related to their experiences at the universities. A promising addition lies in extending the search to include new platforms such as social media.

The main contribution of this papers is to bring sentiment analysis tools in analyzing social media content – especially microblogs- to act as a supportive indicator for evaluating universities.

In Section 2, related work is presented. The paper concept of using social media as supplementary tool for evaluating university is discussed in Section 3. Afterwards, the paper method used for data processing is presented in Section 4. Later on, the results and discussion on using sentiment analysis for Twitter data in university context are shown in Section 5. Finally, a conclusion and outlook are presented in Section 6.

2 Related Work

This paper aims at investigating new indicators for universities evaluation and comparison. Such indicators can be provided by applying sentiment analysis concept and methods on the available data from social media. In the upcoming sections, the related work for using social media platform at universities, evaluating universities by university rankings and sentiment analysis concepts are presented.

2.1   Social Media at Universities

Recently, social media platforms have become a main medium for people to express their daily activities, reactions and emotions. Blogs and microblogs are the most common form of social media [1]. Blogs (the abbreviated form of Weblog) are informal sites on the worldwide web where users are used to post (i. e. publish) ideas, discussions, thoughts, etc. [2]. While, microblogs are smaller blogs with short posts up to a limited number of signs [3]. One thing they have in common is that they consist of entries that are listed in a chronologically descending order (i. e. the latest news is on top). They are tools that enable discussions and comments on information shared with other users. They are characterized by its dynamic and up-to-datedness.

Students use social media platforms and internet on a daily basis, often times more frequently than other mass media such as newspaper or television. Therefore, they are experienced in how to use social media [4]. Based on this fact and the increasingly usage of social media, it seems reasonable to use social media in university environments for the purpose of engaging the social media tools in the teaching process, as well as for the purpose of analyzing opinions regarding university topics like teaching and learning. On university level, microblogs are tools for simultaneous communication, knowledge management and publication service. In formal and informal teaching and learning contexts, microblogs can support individual and cooperative communication, knowledge management, reflection and feedback processes.

Teachers use microblogs to encourage motivation and participation of students. This is justified by two features [5]: First, it raises interactivity of students and creates the opportunity to implement social and team building aspects even in large classes. Second, it makes use of the media usage of smartphones, tablets and notebooks of today's students in learning contexts [1]. The fields of application of microblogs in teaching scenarios can be divided into three parts: collaboration, feedback-giving/discussion and public scientific communication. By adding the ability to share information and knowledge via social media, mentioning Twitter specifically, collaborative work and learning is enabled with an open feedback-giving space.

Furthermore, general and personal questions as well as lecture or seminar contents can be reflected. Twitter can be utilized to collect questions and feedback of students and discuss them [6]. One of the interesting use cases is the "Twitter wall", where the concept is realized by projecting Twitter's posts on a big screen during the class time [7]. In addition to discussing the most relevant questions at the end of the lecture, other questions and feedback can also be answered after the lecture via Twitter [6].

Besides the well-known applications of microblogs in different teaching and learning scenarios, semantic technologies are often put to use in order to analyze and evaluate automatically opinions made in social media, e. g. Twitter or Facebook. Brauer and Bernroider [8] conduct an international study analyzing the usage of Facebook within higher education institutes in Germany, Austria and Switzerland. In this case, the social media strategy that is related to Facebook of selected universities is analyzed. Another field of social media analytics in universities is the analysis of students and teachers opinions which are made in e. g. Twitter or Facebook [1].

2.2   Evaluating Universities via University Rankings

The comparable analysis of universities worldwide has been established over the past few decades. National and international university rankings try to compare and evaluate universities through various criteria and try to make differences in quality measurable.

According to a user study from the Center for Higher Education Development (CHE), the most common reason to make use of university rankings is to acquire information about universities current or potential status [9]. Therefore, the main goal of any university ranking is to process and present data in order to make comparing universities with each other clearer. It also helps in the decision making process towards the future of teaching and learning environment based on the data collected and information provided [10].

Via comparing the available study programs and conditions of universities, the transparency of performance is improved [11]. University rankings have to meet certain methodical standards to be helpful for the decision making process of students, as well as an orientation tool for university's improvements. At first, university rankings have to be disciplinary because universities are versatile and have different foci [11]. Another important requirement for university rankings is being multidimensional. Thus, other indicators, such as timetables, external funding or the state of the university's library as examples, have to be put in comparison.

On an international level, the most popular rankings in Germany are the "Ranking of World Universities" by the magazine "Times Higher Education (THE)", as well as the "QS World University Ranking" [9]. The THE reviews research-led universities across all their core missions in teaching, research, knowledge transfer and international outlook using 13 indicators. The "QS World University Ranking" compares universities worldwide on a basis of eight indicators. In addition to the state of research, publications and Nobel prizes are also emphasized in the QS. The QS ranking does not only compare universities on a global level, but also individually for each country or region.

Even though rankings have become an increasingly popular way to compare higher education performance and productivity, university rankings have been criticized over the recent years in three major aspects [12, 13].

First, most rankings measure universities as whole institution. Thereby, only the average quality of a university is measured. Individual subjects are not taken into account. Based on weighing single indicators a total value (i. e. "composed indicator") is calculated [12].

The second aspect relates to the way different aspects and dimensions have been measured in some indicators, and the way the related data is collected. In total, some indicators (e. g. university reputation) have a higher influence value than some others. [12].

Third, the measurement of educational quality such as the quality of teaching and learning or the quality of student experience is underrepresented in the existing ranking [13].

2.3   Sentiment Analysis

Sentiment analysis is a method that analyzes how opinions, reactions, impressions, emotions and perspectives are expressed in a language. Its algorithms can extract evaluative information from large text databases and summarize it [14].

In order to analyze the opinion of people and customers, sentiment analysis appears as the main tool in different contexts. As an example, sentiment analysis has been used to measure customers satisfaction via statements they comment on a specific product they bought or a service they were delivered [15]. It also appeared in the extend of detecting different opinions regarding political events such as elections [14].

Sentiment analysis methods are well developed in the domain of blogs and product reviews [16, 17]. Researchers have been working on detecting sentiment in text via presenting different algorithms for detecting semantic orientation [18].

In favor of producing meaningful information from tweets, sentiment analysis has been also used [14, 15, 19]. Different features selection techniques have been investigated, establishing a comparison between different ones such as n-grams, part of speech, lexicons, etc. [20]. Besides, different classifiers with their learning performance have been tested in different contexts [21].

This paper applies the existing developed approaches in sentiment analysis to microblogging platforms data such as Twitter in order to explore complimentary resources for university evaluation and comparison.

3 Twitter Sentiment Analysis for Evaluating Universities in Germany

This paper suggests that social media content is a vital source for collecting feedback and reactions on the daily events and activities that is related to universities. To prove this hypothesis, a case study is established which evaluates the reactions and feedback from the social media data that is related to nine universities in Germany that are part of the TU9[1] German Institutes of Technology. The TU9 universities are: RWTH Aachen University, TU[2] Berlin, TU Braunschweig. TU Darmstadt, TU Dresden, Leibniz Universität Hannover, Karlsruher Institut für Technologie, TU München and Universität Stuttgart.

3.1 Data Collection

Out of many social media platforms, Twitter is the most popular microblogging platform [22]. Therefore, it was chosen to be the source of the social media data for this paper.

Students in universities message about anything and everything in their day-to-day lives. Twitter users post about their reactions and feedback in a form of microblogs, each is a text up to 140 letters that is called a tweet. The tweets that are related to the TU9 form the data set.

The Twitter API[3] enables different ways to search and filter the tweets. In the literature, different ways for extracting the tweets are adapted. For example, Pak and Paroubek [23] and Bifet and Frank [24] have extracted the tweets based on specific mentioned emoticons. Others, such as Davidov et al. [25], have extracted the tweets based on a list of mentioned hashtags in the tweets.

To collect tweets which are related to the TU9, all tweets which have a matching word from a list of keywords were extracted via Twitter API. The keyword list contains combinations of the universities names and titles.

The data extraction script began running on October 1, 2014 till March 31, 2015. In the German academic calendar, this time span is the entire winter semester of 2014/2015. The script catches every tweet that matches any of the keywords and saves it in a database. As a result, the script collected 16488 tweets.

The collected tweets were posted in both English and German. Also, it included original tweets and the re-posted (i. e. retweeted) ones. Table 1 shows how tweets are divided over the TU9.[4]

---

[1] TU9 is an incorporated society of the nine most prestigious, oldest, and largest universities focusing on engineering and technology in Germany.

[2] TU: Technical University.

[3] API: Application Programming Interface.

[4] Few tweets belong to more than one university, that leads to have the tweets summation larger than the total count of collected tweets.

Table 1 Tweets count for each of the TU9[4]

University	Count
RWTH Aachen	8073
TU Dresden	4075
Universität Stuttgart	1014
TU Darmstadt	951
Karlsruher Institut für Technologie	767
Leibniz Universität Hannover	450
TU Braunschweig	445
TU Berlin	401
TU München	300

Table 2 Tweets count for each language

Language	Count
German	12906
English	3582

Table 3 Number of original and retweeted Tweets

Tweet type	Count
Original Tweets	10189
Retweets	6287

Considering the fact that German universities – especially the TU9 – rely on multiple languages for communication and enroll a large number of international students, it makes sense to keep the tweets in both English and German. Table 2 shows the number of collected tweets in each language. Retweets begin with "RT" and are mostly copies of other original tweets, with some possible text at the beginning [26]. Still, retweets are useful to be considered as they emphasize the statement and facts included in the original tweet. Table 3 shows the number of original and retweeted tweets in the data set.

3.2 Defining Tweets Sentiment

This paper's approach classifies the sentiment of each tweet to be either "Positive" or "Not Positive". This is known as a two way sentiment classification [15].

A "Positive" tweet refers to text that indicates a positive statement regarding an event such as a lecture, class or activity that is related to one of the TU9. A "Not Positive" tweet can be either a negative statement regarding an event, or a neutral one, such as an announcement or advertisement regarding an event in the university.

Adapting two way classification can be considered as a limitation. Nevertheless, it is easier to process in the classifier learning step. The same approach was adapted by Go et al. [20] who consider that the "Not Positive" tweets are actually "Negative" ones, ignoring the neutral nature of some tweets. Pak and Paroubek [23] proved that adapting a three way classification leads to bad performance which can be avoided by the two way classification [21].

3.3 Tweets Sentiment Analysis Challenges

Dealing with social media as a source of information – especially microblogging platforms such as Twitter – adds extra difficulties to the sentiment analysis process [19]. Tweets are plain text written in an informal manner and its processing face challenges such as:

- Length: Tweets have a limited text length, which is 140 characters. This forces users to start using some common and uncommon abbreviations and phrases. As an example, abbreviations such as OMG,[5] WTH,[6] DKDC,[7] TY,[8] etc. appears often in Twitter.
- Informality: Twitter is mostly used as a non-formal communication medium. This leads to many informal statements which probably contain errors such as mis-spellings, unstructured sentences and slang. Informality may also infer sarcasm, which adds an extra layer of difficulty in guessing the right sentiment of each tweet.
- Credibility: This paper's approach of gathering the tweets is based on a list of keywords. This does not guarantee the credibility of who and what tweets are generated on Twitter. This leaves the possibility that one anonymous user has generated all the content about a specific university with different usernames, rather than the students.
- Data availability: collecting the right data is always a challenge, but having enough data is another critical issue. The target data for this study is very specific, which can be problematic for collecting data over a six-month period of time. It is significant to note that more data leads to more trusted results.

4   Data Processing

Figure 1 illustrates the main steps of the data processing. The starting point is a set of tweets which was extracted via Twitter API. Based on sentiment analysis approach, a sentiment classifier will be build by learning from previously annotated subset of

---

[5]OMG: Oh My God!.

[6]WTH: What the hell!.

[7]DKDC: Don't know, don't care.

[8]TY: Thank you.

Figure 1 Data processing
flow

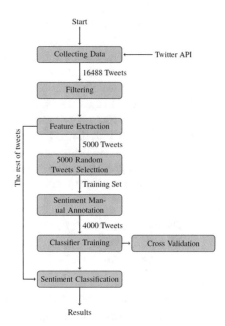

tweets in order to classify the rest of tweets. The classifier to be built will be able to learn the defined sentiment: "Positive" and "Not Positive". The processing steps are divided into three main steps: Tweets' text filtering, feature extraction, and sentiment classification.

4.1 Tweets Text Filtering

As previously mentioned, tweets are informal sentences that have to pass through a filtering stage before it can be processed for the up coming steps. Filtering is the process of cleaning the tweets text removing all irrelevant text for the sentiment classifier learning step. The following are tweets filtering steps mentioned in the order they were performed:

1. All text is switched to lowercase including those words which are completely capitalized. Despite the fact that some users tend to emphasize specific words with capitalization, this was not the general case with the collected tweets. Many names and sentences are found completely capitalized indicating no emphasizing on the meaning. Taking the capitalization into account in such cases might lead to false results. Therefore, all text is changed to lowercase.
2. All hyperlinks are removed. Tweets mostly contain hyperlinks to other sites and photos which does not contribute to the sentiment of the tweet.

3. All mentioned usernames (identified by words that start with @) are removed and all hashed words with the # symbol are replaced with the word itself. These specific symbols and markups mentioning usernames or include hashed words that tag a place, name, etc. are so general to contribute to a specific tweet sentiment.

4. The "RT" text which indicates a retweet is removed.

5. Repeated lettered are filtered. Often, users emphasize words by repeating letters such as: "I am Happyyyyy". Go et al. [20] suggest to remove out repeated letters leaving only two of them. This also guarantees that words such as "cool" with original double letters are left unaffected.

6. Common emoticons are replaced with their semantic. Emoticons are often used in social media language to indicate the users' emotions [23]. The found emoticons are classified as:

   • Happy emoticons: ":)", ":-)", ":D", ";)", ";']", etc. which are replaced by "HAPPY_ FACE".
   • Sad emoticons: ":-(", ":(", "=(", ";(", ":[", etc. which are replaced by "SAD_ FACE"

7. Negations are detected in the tweet. Depending on the language, negation appears in different forms. Accordingly, the sentiment of the words appear before and after the negation are changed. For example, "I don't like exams" is changed to "I NOT_ do NOT_ Like exams".

8. All words which do not start with a letter are removed. This eliminates all phone numbers and dates included in the tweet.

9. Extra spaces and punctuation marks are removed.

10. All stop words and keywords (including the universities' names) are removed based on the language of the tweet.

After these steps, every tweet text is left only with words that can play a role in indicating the sentiment of the tweet. Here is an example of a tweet after applying all filtering steps: "*I would tell you. That I loved you. If I thought ...Boys Don't Cry by Grant-Lee Phillips (at RWTH Bibliothek 2)* – https://t.co/qGoJm1bmV8" "tell loved thought boys NOT_ do NOT_ cry grantlee phillips bibliothek"

4.2 Features Selection and Extraction

An important part of the sentiment analysis process is features selection. Features are the sentence properties that are analyzed in an attempt to correlate it to the tweet sentiment (i. e. "Positive" or "Not Positive"). A feature can be the fact that the tweet contains a word, emoticon, a combination of words, etc. The selection of the features is very important as they act as the input for the classifier in the next step.

Different features selection approaches appear in the literature in the context of tweets sentiment analysis. Some algorithms use unigrams (i. e. single words), some others use bigrams or trigrams (i. e. two or three consecutive words, respectively) [20]. Also, many algorithms use part-of-speech tags and lexicons [19]. Hashtags and emoticons also appeared as features in some algorithms [25].

Unigram features give a wide coverage for the tweet's text. On the other hand, n-grams with $n \geq 2$ show a higher ability to capture the sentiment expression patterns. As larger as n gets, the sentiment is more specific. N-grams with $n \geq 2$ are used to find out the domain specific language. This is useful for this research, as the aim is to build a set of vocabulary that is related to the universities context.

Depending on the context, the chosen size of n-grams affects differently. For example, unigrams are a better choice than bigrams when performing the sentiment classification of movie reviews [16]. On the contrary, bigrams and trigrams worked better for the product review polarity classification [27]. In general, using only bigrams as features is not useful because the feature space becomes very sparse [20].

The best settings of the tweets features has to be determined. This paper adapts a combination of uni- and bigrams features to benefit from the unigrams' coverage of the data and the bigrams' ability to capture the sentiment expression patterns. The same approach was adapted by Pak and Paroubek [23].

Using part-of-speech features leads to a drop in the sentiment classification performance. These results were proven by Go et al. [20] and Kouloumpis et al. [19]. Therefore, the part-of-speech features were neglected.

The sentiment of the emoticons can be part of the n-grams features. On the one hand, stripping out the emoticons from the tweets leads the classifier to learn from other words that forms unigrams and bigrams [20]. On the other hand, including the emoticons data showed a performance improvement referring to the results of Kouloumpis et al. [19].

Based on the filtering step, emoticons are replaced by their sentiment (i. e. "HAPPY_ FACE" and "SAD_ FACE"). In the features extraction step, they take part in forming the unigrams and bigrams features of the tweets. This leads to a better performance for the classifier [20].

4.3 Sentiment Classification

Different classifiers have been presented in the literature for Twitter sentiment analysis. Supervised classifiers are the focus. They require a training set to be prepared forehead.

The training set has to be annotated. Pak and Paroubek [23] did this automatically based on the fact that all tweets in their dataset contains emoticons. They labeled each tweet based on the emoticon sentiment to be either "Positive" or "Negative".

For the training set, it got annotated manually by different people based on their own feeling whether the tweet indicate a "Positive" or "Not Positive" sentiment. Automatic labeling was not possible at this stage as the tweets lack a common feature for sentiment labeling.

Several classifiers can be used when it comes to the tweets sentiment analysis. Mainly, three common classifiers in the field of machine learning have been used in the literature: Naive Bayes classifier, Support Vector Machines (SVM), and Maximum Entropy.

Naive Bayes and SVM have been compared by Pak and Paroubek [23] and Go et al. [20, 21]; Naive Bayes has performed better.

Theoretically, Maximum Entropy performs better than Naive Bayes as it handles feature overlap better. However, in practice, Naive Bayes showed better performance on a variety of problems [20].

Naive Bayes classifier is adapted by this paper's approach. It is a common method for text categorization. It appeared often for solving the problem of determining the category or class of documents that belongs to using word frequencies as the features.

In machine learning, Naive Bayes classifier belongs to the family of probabilistic classifiers based on applying Bayes' theorem with the assumption that features are conditionally independence from each other given a specific class.

$$P(s|f) = \frac{P(s).P(f|s)}{P(f)} \tag{1}$$

Equation 1 shows the basic formula of Naive theorem where s is the sentiment class (i. e. "Positive" or "Not Positive") and f is a specific feature. This equation computes the probability of having a tweet with the sentiment s when it contains the feature f. It is calculated based on the probability of having a specific sentiment, probability of the feature existence in all tweets, and the probability of finding the feature in the tweets that belongs to that specific sentiment.

5 Results and Discussion

The data set holds 16488 tweets. Each tweet contains a statement regarding a university or more from the TU9 in Germany. For the training set, 5000 tweets were chosen randomly and got annotated manually by one sentiment either a "Positive" tweet or "Not Positive" tweets. From the 5000 tweets, 4000 tweets where chosen randomly divided equally between 2000 "Positive" and 2000 "Not Positive" tweets. They are the input for the training step of the Naive Bayes classifier (see Figure 1).

The results section evaluates three main aspects of the presented method:

1. Measuring the classifier efficiency based on the suggested filtering and features extraction steps.
2. Establishing a comparison between the TU9 based on each university's tweets trying to prove the hypothesis that social media content may act as an indicator for university comparisons.
3. Investigating the tweets sentiment on daily basis for each university to obtain feedback on different events and activities.

Each is presented in the following sections.

Table 4 The most informative features learned by the Naive Bayes classifier

Feature	Sentiment class	× Times
Pegida Demonstrant	Not Positive	27.0
Glückwunsch	Positive	16.3
Mittelschicht	Not Positive	14.6
Ojeu	Not Positive	13.0
Pegida Studie	Not Positive	13.0
Berufstaetig	Not Positive	12.3
Studie Pegida	Not Positive	12.1
Maennlich Jahre	Not Positive	11.7
Ausgebildet	Not Positive	11.0
Ausgebildet berufstätig	Not Positive	11.0
Schreiben	Positive	11.0
NOT_ gut ausgebildet	Not Positive	11.0
Kritik Pegida	Not Positive	11.0
Klausuren	Positive	10.3
Willkommen	Positive	10.3
Mittelschicht gut	Not Positive	9.7
Herzlichen	Positive	9.7
Demonstrant mittelschicht	Not Positive	9.7
Geheimdienst	Not Positive	9.0
Wünschen	Positive	8.3

5.1 Classifier Efficiency

Each tweet has been processed through the filtering step. Then, its unigram and bigram features were extracted. For the purpose of cross checking validation, 25 % (i. e. 1000) of the training set tweets were chosen randomly leaving out 3000 tweets for training. The 1000 tweets are used for testing the classifier performance, which were excluded from the training.

A script was written in Python using the Natural Languages Tool Kit (NLTK) python library. The NLTK classifier was used for performing this paper's approach. Part of the results of the trained classifier are shown in Table 4. The table shows the most informative features which were learned by the classifier. In other words, each listed feature was more found several times for the corresponding feature class.

The results show that some of these features have a dominant effect on the classification of the tweet. As an example, whenever the word "Pegida[9]" is mentioned, tweets are probably "Not Positive" 27 more times than being "Positive". On the

---

[9]Patriotic Europeans Against the Islamisation of the West.

Table 5 Classified tweets counts per university of the TU9

University	Positive	Not Positive
RWTH Aachen	4385	3688
TU Berlin	240	161
TU Braunschweig	241	204
TU Darmstadt	498	453
TU Dresden	1499	2576
Leibniz Universität Hannover	274	176
Karlsruher Institut für Technologie	601	166
TU München	122	178
Universität Stuttgart	603	411

other hand, whenever the word "Glückwunsch[10]" is mentioned, tweets are probably "Positive" 16.3 more times than being "Not Positive".

It can be also noticed that both unigrams and bigrams played a role as the most informative features.

The classier performance was verified with the rest of the annotated tweets (i. e. 1000 tweets). 73.6 % of the tweets sentiment were guessed correctly by the classifier. This result is considered to be good considering the data size and compared to the other developed classifier by Pak and Paroubek [23] and Go et al. [20], which achieved around 80 % success rate.

5.2   Tweets Sentiment for Universities Comparison

The rest of the tweets have been classified by the learned classifier. Each tweet belongs to one sentiment class. Tweets were divided over the TU9, showing how many "Positive" and "Not Positive" tweets each university had in the specified time span (see Table 5).

The number of the total tweets per university varied depending on how active Twitter users are at that university. This makes it difficult to establish a comparison between the universities. Nevertheless, to get an overview on how positive the given feedback was by Twitter users on each university, the percentage of the "Positive" tweets is considered for carrying out the comparison. Results are shown in Figure 2.

Karlsruher Institute für Technologie had the highest percentage of "Positive" tweets forming 78 % out of its total 767 tweets. RWTH Aachen University got the highest number of positive tweets with 4385 tweets. Meanwhile, TU Dresden had the lowest percentage of the positive tweets (37 %).

---

[10]Congratulations.

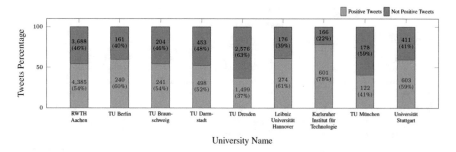

Figure 2 Percentage of "Positive" and "Not Positive" tweets per university in the time span from October 1, 2014 till March 31, 2015

The results might act as an indicator on how the higher education environment at each university is perceived by Twitter users. Such indicator which is supported by the social media content can play a role in enhancing the universities rankings.

5.3 Daily Scale Tweets Sentiment

The next goal is to build a more detailed evaluation by having a closer look into the tweet sentiments regarding daily events and activities at each university. The classified tweets are divided over time on daily basis, where the "Positive" and "Not Positive" tweets count and percentages are considered.

In order to present the case, the two universities with the highest number of tweets were chosen, namely: RWTH Aachen University and TU Dresden.

Figure 3 shows how RWTH Aachen University tweets are spread over the six month period (i. e. winter semester 2014/2015). The upper two figures show the number of "Positive" and "Not Positive" tweets respectively. The third figure shows the "Positive" tweets percentage on each day.

The "Positive" tweets in RWTH Aachen University data varied over the semester in different frequencies. The graph in Figure 3 can track when and what events or activities are interesting for the students.

Few local maxima points are investigated. The tweets at each point were extracted and word frequencies are analyzed. For RWTH Aachen University, the highest "Positive" tweets count occurred on October 6, 2014, when the word frequency analysis shows that many users that tweeted were excited about the beginning of the new semester. Table 6 shows the most frequent words appeared in that day's tweets. On October 6 the orientation sessions and welcoming event took place, in which many new and old students were tweeting "positively" about it.

Looking at other points similarly, on November 14, 2014 another peak was experienced where the event of the "Wissenschaftsnacht" (Science Night) took place. It is an annual event where students from different institutes present their work and

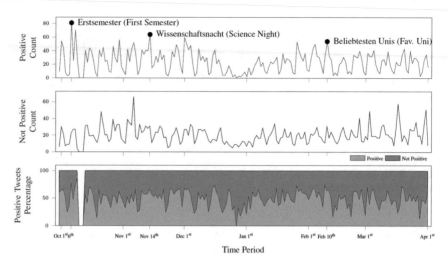

Figure 3 RWTH Aachen University tweets sentiment on daily bases over the period from October 1, 2014 till March 31, 2015

Table 6 Most frequent words which were mentioned in the "Positive" tweets for RWTH Aachen University on October 6, 2014

Word	Translation
Erstsemester	First Semester
Heute	Today
Ersti	Newbie
:)	:)
Alle	All
Viel	A lot
Filmstudio	Filmstudio
Wünschen	Wish
neuen	New
Erstis	Newbies
Semesterstart	Semester Start
Bietet	Offers
Herzlich [Willkomen]	Warmly Welcomed

achievements. Many users tweeted about the event, and the activities which took place at it. One more example is on February 10, 2015, when the RWTH Aachen University was chosen as one of the top three "most favorite" universities in Germany.

The same analysis steps were applied to TU Dresden tweets. In total, TU Dresden tweets experienced the most of the "Not Positive" sentiments. Figure 4 shows the tweet sentiments on a daily basis over the semester. The maximum number of the "Not Positive" tweets appeared on January 14, 2015. As shown in Table 7, the word frequency analysis indicates that many tweets were related to "Pegida demonstrations"

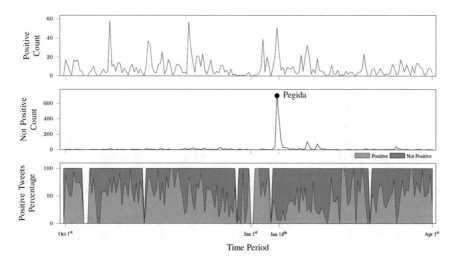

Figure 4 TU Dresden University tweets sentiment on daily bases over the period from October 1, 2014 till March 31, 2015

Table 7 Most frequent words which were mentioned in the "Positive" tweets for TU Dresden on January 14, 2014

Word	Translation	Count
Pegida	Pegida	1095
Studie	Study	895
Männlich	Male	243
Mittelschicht	Middle Class	196
Demonstrant	Demonstrator	187
Demonstranten	Demonstrators	181
Gut	Good	168
Typische	Typical	158
Typischen	Typical	155

that took place on January 2015 in the city of Dresden. This was a reason for many users to tweet "Not Positively" about the activities that happened regarding these demonstration.

Such analysis for the daily tweet sentiments in the context of higher education can give an insight over what activities and events which attract the attention of the Twitter users, and how they react based on it (i. e. positively or not). This analysis can be used by the university's administration to gather feedback. Based on it, actions can be delivered.

6   Conclusion and Outlook

Students nowadays use different social media platforms including Twitter to express their reactions, and tell about their daily activities. Such platforms can be a vital source for opinion mining related to the universities.

Standard university rankings are quite widespread in order to compare and evaluate universities. University rankings require complementary sources specially those which identify statements from students and teachers expressing their experiences.

This paper established a case study to prove the hypothesis that Twitter sentiment can potentially support the universities ranking system by analyzing posted statements and opinions of students and teachers in higher education institutions context.

The case study used sentiment analysis methods to analyze 16488 collected tweets about the TU9. 4000 annotated tweets between "Positive" and "Not Positive" were used to train a Naive Bayes classifier. The classifier performance achieved 73.6 % success rate. The classifier was used to guess the sentiment for the rest of the collected tweets.

The tweets sentiment was analyzed on two different scales. First, the percentage of "Positive" tweets was calculated to establish a comparison between the TU9. Second, the tweets were analyzed on daily basis for the RWTH Aachen and TU Dresden in order to discover the communicative topics and events at the universities.

The analysis established a new approach of using social media in comparing universities.

As a future improving steps, there is a lot of room in improving the classifier performance regarding the filtering, features selection and chosen classifier type. The sentiment classification can be extended to include a third neutral class. Also, the tweets can be classified based on topics to give a better insight on what attracts Twitter users in the context of higher education institutes.

References

1. I. Buchem, R. Appelt, S. Kaiser, S. Schön, M. Ebner, Blogging und microblogging-anwendungsmöglichkeiten im bildungskontext. booklet, 2011
2. D. Boyd, A blogger's blog: Exploring the definition of a medium. Reconstruction, online booklet **6** (4), 2006
3. A. Java, X. Song, T. Finin, B. Tseng, Why we twitter: understanding microblogging usage and communities. In: *Proceedings of the 9th WebKDD and 1st SNA-KDD 2007 workshop on Web mining and social network analysis*. ACM, 2007, pp. 56–65
4. H. Gapski, L. Gräßer, Medienkompetenz im Web 2.0 – Lebensqualität als Zielperspektive. Praxis Web **2**, 2007, pp. 11–34
5. J. Hisserich, J. Primsch, Wissensmanagement in 140 zeichen: Twitter in der Hochschullehre. webSquare, online booklet, 2010
6. M. Kerres, A. Preussler, Soziale Netzwerkbildung unterstützen mit Microblogs (twitter). Handbuch E-Learning, Köln: Deutscher Wirtschaftsdienst, online booklet, 2009

7. J. Herwig, A. Kittenberger, M. Nentwich, J. Schmirmund, Microblogging und die Wissenschaft. das Beispiel Twitter. Steckbrief IV im Rahmen de Projekts "'Interactive Science'" ITA-Reports **54-4**, 2009
8. C. Brauer, E.W. Bernroider, Social media analytics with facebook-the case of higher education institutions. In: *HCI in Business*, Springer, 2015, pp. 3–12
9. I. Roessler, Hochschulentwicklung, was war? Was bleibt? Was kommt? 15 Jahre Erfahrungen mit Rankings und Indikatoren im Hochschulbereich, Arbeitspapier Nr 167. booklet, 2013
10. C. Lebherz, C. Mohr, M. Henning, P. Sedlmeier, Wie brauchbar sind Hochschul-Rankings? Eine empirische Analyse. In: *Hochschullandschaft im Wandel*. 2005, pp. 188–208
11. S. Berghoff, G. Federkeil, P. Giebisch, C.D. Hachmeister, D. Müller-Böling, D.Rölle, Chehochschulranking. Vorgehensweise und Indikatoren, Arbeitspapier Nr 119. booklet, 2006
12. G. Federkeil, Internationale Hochschulrankings–eine kritische Bestandsaufnahme. Beiträge zur Hochschulforschung **35** (2), 2013, pp. 34–48
13. E. Hazelkorn, *World-Class Universities or World Class Systems?: Rankings and Higher Education Policy Choices*. Book of Dublin Institute of Technology, 2013
14. D. Maynard, A. Funk, Automatic detection of political opinions in tweets. In: *The semantic web: ESWC 2011 workshops*. Springer, 2012, pp. 88–99
15. A. Agarwal, B. Xie, I. Vovsha, O. Rambow, R. Passonneau, Sentiment analysis of twitter data. In: *Proceedings of the Workshop on Languages in Social Media*. Association for Computational Linguistics, 2011, pp. 30–38
16. B. Pang, L. Lee, Opinion mining and sentiment analysis. Foundations and trends in information retrieval **2** (1–2), 2008, pp. 1–135
17. B. Pang, L. Lee, A sentimental education: Sentiment analysis using subjectivity summarization based on minimum cuts. In: *Proceedings of the 42nd annual meeting on Association for Computational Linguistics*. Association for Computational Linguistics, 2004, pp. 271–279
18. P.D. Turney, Thumbs up or thumbs down?: semantic orientation applied to unsupervised classification of reviews. In: *Proceedings of the 40th annual meeting on association for computational linguistics*. Association for Computational Linguistics, 2002, pp. 417–424
19. E. Kouloumpis, T. Wilson, J. Moore, Twitter sentiment analysis: The good the bad and the omg! in Proceedings of the Fifth International AAAI Conference on Weblogs and Social Media (ICWSM-2011), 2011, pp. 538–541. http://www.aaai.org/ocs/index.php/ICWSM/ICWSM11/paper/view/2857/3251
20. A. Go, R. Bhayani, L. Huang, Twitter sentiment classification using distant supervision. CS224N Project Report, Stanford **1**, 2009, pp. 12–18
21. A. Go, L. Huang, R. Bhayani, Twitter sentiment analysis. Final Project Report, online booklet, 2009, pp. 3–9
22. F. Atefeh, W. Khreich, A survery of techniques for event detection in twitter. Computational Intelligence **31**, 2013, pp. 132–164
23. A. Pak, P. Paroubek, Twitter as a corpus for sentiment analysis and opinion mining. In: *In Proceedings of the Seventh Conference on International Language Resources and Evaluation*. 2010, pp. 1320–1326
24. A. Bifet, E. Frank, Sentiment knowledge discovery in twitter streaming data. In: *Discovery Science*. Springer, 2010, pp. 1–15
25. D. Davidov, O. Tsur, A. Rappoport, Enhanced sentiment learning using twitter hashtags and smileys. In: *Proceedings of the 23rd International Conference on Computational Linguistics: Posters*. Association for Computational Linguistics, 2010, pp. 241–249
26. A. Mollett, D. Moran, P. Dunleavy, Using twitter in university research, teaching and impact activities. in Impact of Social Sciences from LSE Public Policy Group, London School of Economics and Political Science, online booklet, 2011, pp. 3–4
27. K. Dave, S. Lawrence, D.M. Pennock, Mining the peanut gallery: Opinion extraction and semantic classification of product reviews. In: *Proceedings of the 12th international conference on World Wide Web*. ACM, 2003, pp. 519–528

Bridging the Gap Between Students and Laboratory Experiments

Max Hoffmann, Katharina Schuster, Daniel Schilberg
and Sabina Jeschke

Abstract After having finished studies, graduates need to apply their knowledge to a new environment. In order to professionally prepare students for new situations, virtual reality (VR) simulators can be utilized. During our research, such a simulator is applied in order to enable the visit of remote laboratories, which are designed through advanced computer graphics in order to create simulated representations of real world environments. That way, it is our aim to facilitate the access to practical engineering laboratories. Our goal is to enable a secure visit of elusive or dangerous places for students of technical studies. The first step towards the virtualization of engineering environments, e.g. a nuclear power plant, consists in the development of demonstrators. In the present paper, we describe the elaboration of an industry relevant demonstrator for the advanced teaching of engineering students. Within our approach, we use a virtual reality simulator that is called the "Virtual Theatre".

Keywords Virtual Reality · Virtual Theatre · Remote Laboratories · Immersion

1 Introduction

In terms of modern teaching methods within engineering classes, various different approaches can be utilized to impart knowledge to students. There are traditional teaching techniques, which are still suitable for most of the knowledge transfer. These methods are carried out by the use of written texts or the spoken word. However, due to the increasing number of study paths as well as the specialization of particularly technical oriented classes, there is a need for the integration of new media into the curriculum of most students [1]. Thus, the visualization of educational content in order to explain theory more concrete and tangible has gained in importance. Not least because of the progress in computer science and graphical visualization, the capabilities of visualizing objects of interest within an artificially designed context

M. Hoffmann (✉) · K. Schuster · D. Schilberg · S. Jeschke
IMA/ZLW & IfU, RWTH Aachen University, Dennewartstr. 27, 52068 Aachen, Germany
e-mail: max.hoffmann@ima-zlw-ifu.rwth-aachen.de

Originally published in "Virtual, Augmented and Mixed Reality.
Applications of Virtual and Augmented Reality", © Springer 2014.
Reprint by Springer International Publishing Switzerland 2016,
DOI 10.1007/978-3-319-42620-4_20

have grown to an exhaustive amount. However, not only the visualization techniques have emerged, the way of distributing knowledge through teaching media has also grown. One major improvement in reaching students independently to their location are E-Learning Platforms [2]. These technical possibilities of sharing and representing contents open up new opportunities in teaching and learning for students.

Thus, in nearly all courses of studies, new media have gained a high significance in the past decade. These new media are continuously replacing conventional media or in other words traditional, static teaching approaches using books and lecture notes. The new media are mostly based on methods of digital visualization [3], e. g. presentation applications like PowerPoint [4]. This switch from the traditional lecture speech to graphical representations have been performed, because this form of presentation enables focusing on the main points of educational content using illustrative representations and pictorial summaries [5]. Despite the positive [6], but also critical discussion about an overwhelming usage of PowerPoint [7–9] as primary teaching tool [10], the usage of presentation software in the classroom has grown constantly [11].

Applications like PowerPoint may be a far reaching advancement for most courses within university. However, even these IT-based teaching supports are limited to a certain kind of knowledge transfer. Especially practically oriented study paths like engineering courses have an urgent need for interaction possibilities. In these highly technical focused studies, the teaching personnel are facing more and more obstacles in imparting their knowledge tangible. Due to the advanced and complex technology level of the relevant applications [12], progressive methods have to be applied to fulfill the desired teaching goals. In order to make the problem based learning methodologies available [13], novel visualization techniques have to be carried out.

Studies of astronautics or nuclear research can serve as an incisive example for the need of innovative visualization capabilities. During astronomy studies, the teaching personnel will face insurmountable obstacles, if they want to impart practical knowledge about aerospace travelling to the students using theoretical approaches. In order to gain deep, experienced knowledge about real situations an astronaut has to face, realistic scenarios have to be carried out. This can for instance be performed by setting up expensive real-world demonstrators that facilitate practical experiences within aerospace travelling events, e. g. by making use of actual acceleration.

However, there is also a need for a visual representation of the situation. In order to fulfill the requirements of a holistic experience, these visualization techniques need to perform an immersive representation of the virtual world scenario. In this connection, the term immersion is defined according to Murray [14] as follow: "Immersion is a metaphorical term derived from the physical experience of being submerged in water. We seek the same feeling from a psychologically immersive experience that we do from a plunge in the ocean or swimming pool: the sensation of being surrounded by a completely other reality, as different as water is from air that takes over all of our attention, our whole perceptual apparatus."

It is obvious that experience can only be impressive enough to impart experienced knowledge, if the simulation of a virtual situation has an immersive effect on the perception of the user. Our latest research on creating virtual world scenarios has shown that immersion has got a high impact on the learning behavior of students [15]. Following the idea of facilitating the study circumstances for students of astronautics, our first demonstrator was carried out in terms of a Mars scenario [16]. Using novel visualization techniques in connection with realistic physics engines, we have carried out a realistic representation of a plateau located on the red planet.

In our next research phase, we want to go further to increase the interaction capabilities with the virtual environment the user is experiencing. In terms of the Mars representation, there were already few interaction possibilities like triggering of object movements or the navigation of vehicles [16]. However, this sort of interaction is based on rather artificial commands than on natural movements with realistic consequences in the representation of the virtual world scenario.

Hence, in the present paper, we want to introduce a more grounded scenario, which is based on the aforementioned idea of enabling the visit of elusive or dangerous places like an atomic plant. Accordingly, our first step in realizing an overall scenario of a detailed environment like a power plant consists in the development of single laboratory environments. In this context, our aim is to focus especially on the interaction capabilities within this demonstrator.

This target is pursued by carrying out a virtual prototype of an actual laboratory environment, which can be accessed virtually and in real-time by a user in a virtual reality simulator. The realization of these demonstrators is also known as the creation of "remote laboratories". In the present paper, we describe the development, optimization and testing of such a remote laboratory. After a brief introduction into the state-of-the-art of this comparatively new research field in Section 2, our special Virtual Reality simulator, which is used to simulate virtual environments in an immersive way, is described in Section 3. In Section 4, the technical design of the remote laboratory including its information and communication infrastructure is presented. In the Conclusion and Outlook, the next steps in realizing the overall goal of a virtual representation of an engineering environment like an atomic plant are pointed out.

2 State of the Art

In the introduction, we concluded that innovative teaching methodologies have to be adopted to be capable of imparting experienced knowledge to students. Thus, virtual reality teaching and learning approaches will be examined in the following.

Nowadays, an exhaustive number of applications can be found that make use of immersive elements within real-world scenarios. However, the immersive character of all these applications is based on two characteristics of the simulation: The first one is the quality of the three-dimensional representation; the second one is the user's identification with the avatar within the virtual world scenario.

The modeling quality of the three-dimensional representation of a virtual scenario is very important in order to be surrounded by a virtual reality that is realistic or even immersive. However, a high-quality graphical representation of the simulation is not sufficient for an intensive experience. Thus, according to Wolf and Perron [17], the following conditions have to be fulfilled in order to enable an immersive user experience within the scenario: "Three conditions create a sense of immersion in a virtual reality or 3-D computer game: The user's expectation of the game or environment must match the environment's conventions fairly closely. The user's actions must have a non-trivial impact on the environment. The conventions of the world must be consistent, even if they don't match those of the 'metaspace'."

The user's identification with virtual scenario is rather independent from the modeling of the environment. It is also depending on the user's empathy with the "avatar". Generally, an avatar is supposed to represent the user in a game or a virtual scenario. However, to fulfill its purposes according to the user's empathy, the avatar has to supply further characteristics. Accordingly, Bartle defines an avatar as follows: "An avatar is a player's representative in a world. [...] It does as it's told, it reports what happens to it, and it acts as a general conduit for the player and the world to interact. It may or may not have some graphical representation, it may or may not have a name. It refers to itself as a separate entity and communicates with the player."

There are already many technical solutions that are primarily focused on the creation of high-quality and complex three-dimensional environments, which are accurate to real-world scenarios in every detail. Flight Simulators, for example, provide vehicle tracking [18]. Thus, the flight virtual reality simulator is capable of tracking the locomotion of a flying vehicle within the virtual world, but does not take into account the head position of the user. Another VR simulator is the Omnimax Theater, which provides a large angle of view [19], but does not enable any tracking capabilities whatsoever. Head-tracked monitors were introduced by Codella et al. [20] and by Deering [21]. These special monitors provide an overall tracking system, but provide a rather limited angle of view [18]. The first attempt to create virtual reality in terms of a complete adjustment of the simulation to the user's position and head movements was introduced with the Boom Mounted Display by McDowall et al. [22]. However, these displays provided only poor resolutions and thus were not capable of a detailed graphical representation of the virtual environment [23].

In order to enable an extensive representation of the aimed remote laboratories, we are looking for representative scenarios that fit to immersive requirements using both a detailed graphical modeling as well as a realistic experience within the simulation. In this context, one highly advanced visualization technology was realized through the development of the Cave in 1991. In this context, the recursive acronym CAVE stands for Cave Automatic Virtual Environment [18] and was first mentioned in 1992 by Cruz-Neira [24]. Interestingly, the naming of the Cave is also inspired by Plato's Republic [25]. In this book, he "discusses inferring reality (ideal forms) form shadows (projections) on the cave wall" [18] within "The Smile of the Cave".

By making use of complex projection techniques combined with various projectors as well as six projection walls arranged in form of a cube, the developers of the Cave have redefined the standards in visualizing virtual reality scenarios. The Cave enables visualization techniques, which provide multi-screen stereo vision while reducing the effect of common tracking and system latency errors. Hence, in terms of resolution, color and flicker-free stereo vision the founders of the Cave have created a new level of immersion and virtual reality.

The Cave, which serves the ideal graphical representation of a virtual world, brings us further towards true Virtual Reality, which – according to Rheingold [26] – is described as an experience, in which a person is "surrounded by a three-dimensional computer-generated representation, and is able to move around in the virtual world and see it from different angles, to reach into it, grab it and reshape it." This enables various educational, but also industrial and technical applications. Hence, in the past the research already focused on the power of visualization in technical applications, e. g. for data visualizations purposes [27] or for the exploration and prototyping of complex systems like the visualization of air traffic simulation systems [28]. Furthermore, the Cave has also been used within medical or for other applications, which require annotations and labeling of objects, e. g. in teaching scenarios [29].

The founders of the Cave choose an even more specific definition of virtual reality: "A virtual reality system is one which provides real-time viewer-centered head-tracking perspective with a large angle of view, interactive control, and binocular display." [18] Cruz-Neira also mentions that – according to Bishop and Fuchs [30] – the competing term "virtual environment (VE)" has a "somewhat grander definition which also correctly encompasses touch, smell and sound." Hence, in order to gain a holistic VR experience, more interaction within the virtual environment is needed.

Though, it is our aim to turn Virtual Reality into a complete representation of a virtual environment by extending the needed interaction capabilities, which are, together with the according hardware, necessary to guarantee the immersion of the user into the virtual reality [31]. However, even the Cave has got restricted interaction capabilities as the user can only interact within the currently demonstrated perspectives. Furthermore, natural movement is very limited, as locomotion through the virtual environment is usually restricted to the currently shown spot of the scenario. Yet, natural movements including walking, running or even jumping through virtual reality are decisive for a highly immersive experience within the virtual environment.

This gap of limited interaction has to be filled by advanced technical devices without losing high-quality graphical representations of the virtual environment.

Hence, within this publication, we introduce the Virtual Theatre, which combines the visualization and interaction technique mentioned before. The technical setup and the application of the Virtual Theatre in virtual scenarios are described in the next chapter.

3   The Virtual Theatre – Enabling Virtual Reality in Action

The Virtual Theatre was developed by the MSEAB Weibull Company [32] and was originally carried out for military training purposes. However, as discovered by Ewert et al. [33], the usage of the Virtual Theatre can also be enhanced to meet educational requirements for teaching purposes of engineering students. It consists of four basic elements: The centerpiece, which is referred to as the omnidirectional treadmill, represents the Virtual Theatre's unique characteristics. Besides this moving floor, the Virtual Theatre also consists of a Head Mounted Display, a tracking system and a cyber glove. The interaction of these various technical devices composes a virtual reality simulator that combines the advantages of all conventional attempts to create virtual reality in one setup. This setup will be described in the following.

The Head Mounted Display (HMD) represents the visual perception part of the Virtual Theatre. This technical device consists of two screens that are located in a sort of helmet and enable stereo vision. These two screens – one for each eye of the user – enable a three-dimensional representation of the virtual environment in the perception of the user. HMDs were first mentioned in Fisher [34] and Teitel [35] as devices that use motion in order to create VR. Hence, the characteristic of the HMD consists in the fact that it has a perpendicular aligned to the user and thus adjusts the representation of the virtual environment to him. Each display of the HMD provides a 70° stereoscopic field with an SXGA resolution in order to create a gapless graphical representation of the virtualized scenario [33]. For our specific setup, we are using the Head Mounted Display from zSight [36]. An internal sound system in the HMD enables an acoustic accompaniment for the visualization to complete the immersive scenario.

As already mentioned, the ground part of the Virtual Theatre is the omnidirectional treadmill. This omnidirectional floor represents the navigation component of the Virtual Theatre. The moving floor consists of rigid rollers with increasing circumferences and a common origo [33]. The rotation direction of the rollers is oriented to the middle point of the floor, where a circular static area is located. The rollers are driven by a belt drive system, which is connected to all polygons of the treadmill through a system of coupled shafts and thus ensures the kinematic synchronization of all parts of the moving floor. The omnidirectional treadmill is depicted in Figure 1.

On the central area that is shown in the upper right corner of Figure 1, the user is able to stand without moving. As soon as he steps outside of this area, the rollers start moving and accelerate according to the distance of his position to the middle part. If the user returns to the middle area, the rotation of the rollers stops.

The tracking system of the Virtual Theatre is equipped with ten infrared cameras that are evenly distributed around the treadmill in 3 m above the floor. By recording the position of designated infrared markers attached to the HMD and the hand of

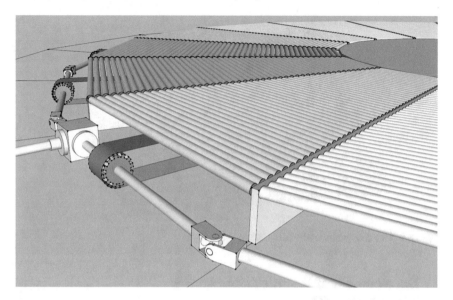

Figure 1 Technical design of the Virtual Theatre's omnidirectional treadmill

the user, the system is capable of tracking the user's movements [33]. Due to the unsymmetrical arrangement of the infrared markers the tracking system is not only capable of calculating the position of the user, but is also capable of determining looking directions. That way, the three-dimensional representation of the virtual scenario can be adjusted according to the user's current head position and orientation. Furthermore, the infrared tracking system is used in order to adjust the rotation speed of the rollers no only according to the user's distance from the middle point, but also according to the difference of these distances within a discrete time interval. Using these enhanced tracking techniques, the system can deal with situations, in which the user stands without moving while not being located in the middle of the omnidirectional floor.

The cyber glove ensures the tactile interaction capabilities. This special hand glove is equipped with 22 sensors, as indicated above, which are capable of determining the user's hand position and gestures [33]. This enables the triggering of gesture based events like the grasping of objects. Additionally, special programmable gestures can be utilized in order to implement specific interaction commands.

After setting up the required hardware of the Virtual Theatre, the user can plunge into different scenarios and can be immersed by virtual reality. After the development of learning and interaction scenarios as described in [16], our main interest here is focused on the development of remote laboratories, which represent the first step towards the realization of a virtual factory. The development, testing and evaluation of our first "Remote Lab" are described in the next chapter.

Figure 2 Two cooperating ABB IRB 120 six-axis robots

4   Development of Remote Laboratories in the Virtual Theatre

The described setup of the Virtual Theatre can be used to immerse the user into a virtual reality scenario not only for demonstration purposes, but especially for the application of scenarios, in which a distinctive interaction between the user and the simulation is required. One of these applications consists in the realization of remote laboratories, which represent the first step towards the creation of real-world demonstrators like a factory or an atomic plant into virtual reality.

The virtual remote laboratory described in this paper consists in a virtual representation of two cooperating robot arms that are setup within our laboratory environment (see Figure 2). These robots are located on a table in such a way that they can perform tasks by executing collaborative actions. For our information and communication infrastructure setup, it doesn't matter, if the robots are located in the same laboratory as our Virtual Theatre or in a distant respectively remote laboratory. In this context, our aim was to virtualize a virtual representation of the actual robot movements in the first step. In a second step, we want to control and to navigate the robots.

In order to visualize the movements of the robot arms in virtual reality, first, we had to design the three-dimensional models of the robots. The robot arms, which are installed within our laboratory setup are ABB IRB 120 six-axis robotic arms [37]. For the modeling purposes of the robots, we are using the 3-D optimization and rendering software Blender [38]. After modeling the single sections of the robot, which are connected by the joints of the six rotation axes, the full robot arm model had to be merged together using a bone structure. Using PhysX engine, the resulting mesh is capable of moving its joints in connection with the according bones in the same fashion as a real robot arm. This realistic modeling principally enables movements of the six-axis robot model in virtual reality according to the movements of the real robot. The virtual environment that contains the embedded robot arms is designed

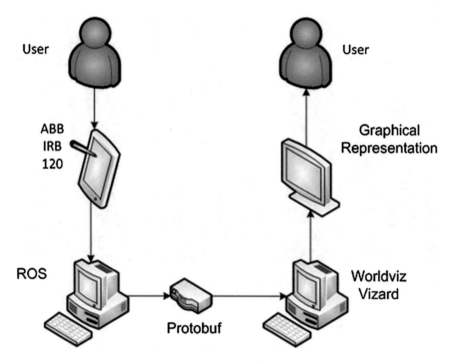

Figure 3 Information and communication infrastructure of the remote laboratory setup

using the WorldViz Vizard Framework [39], a toolkit for setting up virtual reality scenarios.

After the creation of the virtual representation of the robots, an information and communication infrastructure had to be set up in order to enable the exchange of information between the real laboratory and the simulation. The concept of the inter-communication as well as its practical realization is depicted in Figure 3.

As shown in the figure, the hardware of the remote laboratory setup is connected through an internal network. On the left side of the figure, a user is demonstrated, who operates the movements of the real robot arms manually through a control interface of the ABB IRB 120 robots. This data is processed by a computer using Linux with embedded Robot Operating System (ROS). The interconnection between the real laboratory and the virtual remote laboratory demonstrator is realized using the Protocol Buffers (Protobuf) serialization method for structured data. This interface description language, which was developed by Google [40], is capable of exchanging data between different applications in a structured form.

After the robots' position data is sent through the network interface, the information is interpreted by the WorldViz Vizard engine to visualize the movements of the actual robots in virtual reality. After first test phases and a technical optimization of the network configuration, the offset time between the robot arm motion in reality and in virtual reality could be reduced to 0.2 s. Due to the communication design of

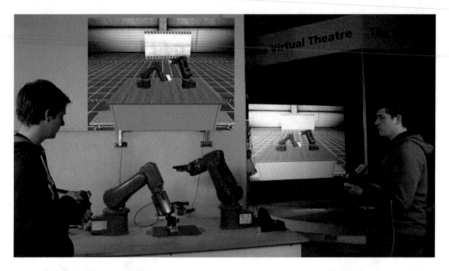

Figure 4 Manual control of the robots and visual representation in the Virtual Theatre

the network infrastructure in terms of internet-based communication methods, this value would not increase significantly, if the remote laboratory would be located in a distant place, for example in another city or on the other side of the globe.

The second user, which is depicted in the right upper part of Figure 3 and who is located in the Virtual Theatre, is immersed by the virtual reality scenario and can observe the positions and motions of the real robots in the virtual environment. In Figure 4, the full setup of the real and the remote laboratory is illustrated.

In the foreground of the figure, two users are controlling the movements of the actual robots in the real laboratory using manual control panels. In the background on the right side of the picture, the virtual representation of the two ABB IRB 120 robot arms is depicted. The picture on the right side of the wall is generated using two digital projectors, which are capable of creating a 3-D realistic picture by overlapping the pictures of both projections. The picture depicted on top of the robot arms table is a representation of the picture the user in the VR simulator is actually seeing during the simulation. It was artificially inserted into Figure 4 for demonstration purposes.

This virtual remote laboratory demonstrator shows impressively that it is already possible to create an interconnection between the real world and virtual reality.

5   Evaluation

The results of first evaluations within the test mode of our virtual remote laboratory demonstrator have shown that the immersive character of the virtual reality simulation has got a major impact on the learning behavior and especially on the motivation of the users. Within our test design, students were first encouraged to implemented specific movements of an ABB IRB 120 robot using the Python programming language. After this practical phase the students were divided into two groups.

The first group had the chance to watch a demonstration of the six axis robots carrying out a task using "LEGO" bricks. After seeing the actual movements of the robots within our laboratories, the students were fairly motivated to understand the way of automating the intelligent behavior of the two collaborating robots.

The second group of students had the possibility to take part in a remote laboratory experiment within the Virtual Theatre. After experiencing the robot movements in the simulated virtual environment performing the same task as the real world demonstrator, the students could observe the laboratory experiment they were just experiencing in the Virtual Theatre recorded on video. Their reaction on the video has shown that the immersion was more impressive than the observation of the actual robot's movements performed by the other group. Accordingly, the students of the second comparison group were even more motivated after their walk through the virtual laboratory. The students of the second group were actually aiming at staying in the laboratory until they finished automating the same robot tasks they just saw in virtual reality.

6   Conclusion and Outlook

In this paper, we have described the development of a virtual reality demonstrator for the visualization of remote laboratories. Through the demonstrated visualization techniques in the Virtual Theatre, we have shown that it is possible to impart experienced knowledge to any student independent of his current location. This enables new possibilities of experience-based and problem-based learning. As one major goal of our research project "ELLI – Exzellentes Lehren und Lernen in den Ingenieurwissenschaften (Excellent Teaching and Learning within engineering science)", which addresses this type of problem-based learning [13], the implemented demonstrator contributes to our aim of establishing advanced teaching methodologies. The visualization of real-world systems in virtual reality enables the training of problem-solving strategies within a virtual environment as well as on real objects at the same time.

The next steps of our research consist in advancing the existing demonstrator in terms of a bidirectional communication between the Virtual Theatre demonstrator and the remote laboratory. Through this bidirectional communication we want to enable a direct control of the real laboratory from the remote virtual reality demonstrator. First results in the testing phase of this bidirectional communication show that such a remote control will be realized in the near future. In order to enable a secure remote control of the remote laboratory, collision avoidance and other security systems for cooperating robots will be carried out and tested in the laboratory environment.

As the overall goal of our project consists in the development of virtual factories in order to enable the visit of an atomic plant or other elusive places, our research efforts will finally focus on the development of a detailed demonstrator for the realistic representation of an industrial environment.

Acknowledgments This work was supported by the project ELLI (Excellent Teaching and Learning within engineering science) as part of the excellence initiative at the RWTH Aachen.

References

1. M. Kerres, *Mediendidaktik: Konzeption und Entwicklung mediengestützer Lernangebote*. Oldenbourg, München, 2012
2. J. Handke, A.M. Schäfer, *E-Learning, E-Teaching and E-Assessment in der Hochschullehre: Eine Anleitung*. Oldenbourg, München, 2012
3. R.J. Craig, J.H. Amernic, Powerpoint presentation technology and the dynamics of teaching. Innovative Higher Education **31** (3), 2006, pp. 147–160
4. A. Szabo, N. Hastings, Using it in the undergraduate classroom: Should we replace the blackboard with powerpoint? Computer and Education **35**, 2000
5. T. Köhler, N. Kahnwald, M. Reitmaier, Lehren und Lernen mit Multimedia und Internet. In: *Medienpsychologie*, ed. by B. Batinic, M. Appel, Springer, Heidelberg, 2008, pp. 477–501
6. R.A. Bartsch, K.M. Cobern, Effectiveness of powerpoint presentation in lectures. Computer and Education **41**, 2003, pp. 77–86
7. T. Creed, Powerpoint, no! cyberspace, yes. The Nat. Teach. & Learn. F. **6** (4), 1997
8. D. Cyphert, The problems of powerpoint: Visual aid or visual rhetoric? Business Communication Quarterly **67**, 2004, pp. 80–83
9. P. Norvig, Powerpoint: Shot with its own bullets. The Lancet **362**, 2003, pp. 343–344
10. T. Simons, Does powerpoint make you stupid? Presentations **18** (3), 2005. http://global.factiva. com/
11. A.M. Jones, The use and abuse of powerpoint in teaching and learning in the life sciences: A personal view. BEE-j 2, 2003. http://www.bioscience.heacademy.ac.uk/journal/vol2/beej-2-3.pdf
12. E. André, Was ist eigentlich multimodale Mensch-Technik Interaktion? Anpassungen an den Faktor Mensch. Forschung und Lehre **21** (01/2014), 2014
13. M. Steffen, D. May, J. Deuse, The industrial engineering laboratory: Problem based learning in industrial eng. education at tu dortmund university. EDUCON, 2012
14. J.H. Murray, *Hamlet on the Holodeck: The Future of Narrative in Cyberspace*. MIT Press, Cambridge (Mass.), 1997
15. K. Schuster, D. Ewert, D. Johansson, U. Bach, R. Vossen, S. Jeschke, Verbesserung der Lernerfahrung durch die Integration des Virtual Theatres in die Ingenieurausbildung. In: *TeachING-LearnING.EU discussions*, ed. by A.E. Tekkaya, S. Jeschke, M. Petermann, D. May, N. Friese, C. Ernst, S. Lenz, K. Müller, K. Schuster, TeachING-LearnING.EU, Aachen, 2013
16. M. Hoffmann, K. Schuster, D. Schilberg, S. Jeschke, Next-generation teaching and learning using the virtual theatre. 4th Global Conference on Experiential Learning in Virtual Worlds, Prague, Czech Republic, 2014
17. Wolf, M. J. P, B. Perron, *The video game theory reader*. Routledge, NY, London, 2003
18. C. Cruz-Neira, D.J. Sandin, T.A. DeFanti, Surround-screen projection-based virtual reality. the design and implementation of the cave. SIGGRAPH'93 Proceedings of the 20th annual conference on Computer graphics and interactive techniques. ACM - New York, 1993, pp. 135–142. doi:10.1145/166117.166134
19. N. Max, Siggraph'84 call for omnimax films. Computer Graphics **16** (4), 1982, pp. 208–214
20. C. Codella, R. Jalili, L. Koved, B. Lewis, D.T. Ling, J.S. Lipscomb, D. Rabenhorst, C.P. Wang, A. Norton, P. Sweeny, G. Turk, Interactive simulation in a multi-person virtual world. ACM - Human Fact. in Comp. Syst. (CHI 1992 Conf.), 1992, pp. 329–334

21. M. Deering, High resolution virtual reality. Com. Graph. **26** (2), 1992, pp. 195–201
22. I.E. McDowall, M. Bolas, S. Pieper, S.S. Fisher, J. Humphries, Implementation and integration of a counterbalanced crt-based stereoscopic display for interactive viewpoint control in virtual environment applications. Proc. SPIE **1256** (16), 1990
23. S.R. Ellis, What are virtual environments? IEEE Computer Graphics and Applications **14** (1), 1994, pp. 17–22. doi:10.1109/38.250914
24. C. Cruz-Neira, D.J. Sandin, T.A. DeFanti, R.V. Kenyon, J.C. Hart, The cave: Audio visual experience automatic virtual environment. Communications of the ACM **35** (6), 1992, pp. 64–72. doi:10.1145/129888.129892
25. Plato, *The Republic*. The Academy Athens, Athens, 375 B.C.
26. H. Rheingold, *Virtual reality*. Summit Books, New York, 1991
27. C. Nowke, M. Schmidt, van Albada, S. J., J.M. Eppler, R. Bakker, M. Diesrnann, B. Hentschel, T. Kuhlen, Visnest – interactive analysis of neural activity data. 2013 IEEE Symposium on Biological Data Visualization (BioVis), 2013, pp. 65–72
28. S. Pick, F. Wefers, B. Hentschel, T. Kuhlen, Virtual air traffic system simulation – aiding the communication of air traffic effects. 2013 IEEE on Virtual Reality (VR), 2013, pp. 133–134
29. S. Pick, B. Hentschel, M. Wolter, I. Tedjo-Palczynski, T. Kuhlen, Automated positioning of annotations in immersive virtual environments. Proc. of the Joint Virtual Reality Conference of EuroVR - EGVE - VEC, 2010, pp. 1–8
30. G. Bishop, H. Fuchs, et al., Research directions in virtual environments. Computer Graphics **26** (3), 1992, pp. 153–177
31. D. Johansson, *Convergence in Mixed Reality-Virtuality Environments: Facilitating Natural User Behavior*. University of Örebro, Schweden, 2012
32. MSEAB Weibull, 2012. http://www.mseab.se/The-Virtual-Theatre.htm
33. D. Ewert, K. Schuster, D. Johansson, D. Schilberg, S. Jeschke, Intensifying learner's experience by incorporating the virtual theatre into engineering education. Proceedings of the 2013 IEEE Global Engineering Education Conference (EDUCON), 2013
34. S. Fisher, The ames virtual environment workstation (view). SIGGRAPH'89 (Course #29 Notes), 1989
35. M.A. Teitel, The eyephone: A head-mounted stereo display. Proc. SPIE **1256** (20), 1990, pp. 168–171
36. http://sensics.com/products/head-mounted-displays/zsight-integrated-sxga-hmd/specifications/
37. ABB. http://new.abb.com/products/robotics/industrial-robots/irb-120. Last checked: 27.01.2014
38. Blender. http://www.blender.org/. Last checked: 27.01.2014
39. WorldViz. http://www.worldviz.com/products/vizard. Last checked: 27.01.2014
40. Google. http://code.google.com/p/protobuf/wiki/thirdpartyaddons. Last checked: 27.01.2014

Enhancing the Learning Success of Engineering Students by Virtual Experiments

Max Hoffmann, Lana Plumanns, Laura Lenz, Katharina Schuster, Tobias Meisen and Sabina Jeschke

Abstract In a world that is characterized by highly specialized industry sectors, the demand for well-educated engineers increases significantly. Thus, the education of engineering students has become a major field of interest for universities. However, not every university is able to provide the required number of industry demonstrators to impart the needed practical knowledge to students. Our aim is to fill this gap by establishing Remote Labs. These laboratory experiments are performed in Virtual Reality environments which represent real laboratories accessible from different places. Following the implementation of such Remote Labs described within our past publications the aim of this contribution is to examine and evaluate possibilities of controlling Remote Labs from arbitrary locations. These control mechanisms are based on the virtualization of two concurrently working six-axis robots in combination with a game pad remote controller. The evaluation of the virtual demonstrator is carried out in terms of a study that is based on practical tests and questionnaires to the measure learning success.

Keywords Virtual Reality · Remote Laboratories · Game-based Learning · Experiential Learning · Virtual Theatre · Immersion

1 Introduction

The current developments within the industry and engineering sciences triggered by the Industry 4.0 pose major challenges for the education of engineering students in universities all over the world. Faster evolving technologies and rapidly changing requirements in industrial environments lead to rising demands in terms of practical education of engineering students. In the course of traditional training methods, the practical education of students is mostly performed by the attendance to laboratory

M. Hoffmann (✉) · L. Plumanns · L. Lenz · K. Schuster · T. Meisen · S. Jeschke
IMA/ZLW & IfU, RWTH Aachen University,
Dennewartstr. 27, 52068 Aachen, Germany
e-mail: max.hoffmann@ima-zlw-ifu.rwth-aachen.de

Originally published in "Learning and Collaboration Technologies. Second International Conference, LCT 2015, Held as Part of HCI International 2015, Los Angeles, CA, USA, August 2–7, 2015, Proceedings ", © Springer 2015. Reprint by Springer International Publishing Switzerland 2016, DOI 10.1007/978-3-319-42620-4_21

experiments or the visit of factories and production sites. However, in terms of changing circumstances and dynamically performed manufacturing execution the scope of laboratory experiments and practical education has to be adopted to these novel requirements as well. It is the aim of this paper to demonstrate novel methods of imparting practical knowledge to students considering the current developments within industrial reality.

One possibility of realizing these practical experiments without neglecting the demands of the Industry 4.0 is to virtualize the experience of visiting laboratory classes or manufacturing sites. In terms of these attempts, Virtual Reality simulations can be carried out in order to create virtual environments that can be adapted according to the current demands and demonstrator configurations. Another application of the described Virtual Reality solutions is to recreate existing laboratory environments from the real world and provide these environments as virtual demonstrators.

This application of Virtual Reality is referred to as Remote Laboratories and can be integrated into the curriculum of students in order to allow engineering students from arbitrary places to visit and experience laboratory environments that are not available at their university or place of study. Prototypical implementations of these Remote Labs have been carried out and examined in previous works of the author [1–3]. In terms of these developments the suitability of creating practical learning environments for engineering students were examined in order to deliver the basis for carrying out virtual experiments of real world demonstrators.

Based on our previous work, it is the aim of the current publications to describe, examine and evaluate ways of direct interaction with real world demonstrators through their virtual representation. Doing so, we extended an existing demonstrator with control mechanisms and implemented remote control solutions for active interaction of a user who is connected to the demonstrator by Virtual Reality tools. In order to evaluate these interaction capabilities the paper is divided into several parts.

In Section 2 we will discuss the state of the art in Game-based Learning in connection with laboratory experiments in the form of Remote Labs. Also, we will point out techniques to examine and create the didactical concepts needed to assess the learning success of students that perform experiments in game-like virtual environments. In Section 3, we will describe in detail the technical solutions that have been carried out and implemented to reach full remote control of distant laboratory environments from arbitrary places. In Section 4, the evaluation of the remote control capabilities takes place in form of a study that have been carried out with students from different universities in Germany. Section 5 summarizes the results and takes a look at further research opportunities in the field of Remote Labs.

2 State of the Art

Based on the existing Remote Lab demonstrator that has been carried out and described within our previous publications [1] the different mechanisms for the remote control of these labs are of primary interest in this publication.

Accordingly, the state of the art section of this work deals with evaluation methods that will be selected and implemented to evaluate the learning success of students that are surrounded by virtual environments, thus in terms of a situation comparable to game-based learning/serious gaming scenarios. The evaluation part is realized on the basis of questionnaires that, on the one hand analyzes general suitability of the learning methods for each test person, and on the other hand, assesses the learning success of each individual test person from the technical point of view while taking into account their experience with digital media.

Virtuality-based learning (VBL) is a recent trend not only in engineering education. It is closely related to game-based learning (GBL), which is defined as "[…] a type of game-play that has defined outcomes. Generally, GBL is designed to balance subject matter with game-play and the ability of the player to retain and apply said subject matter to the real world" [4]. What is equal here is the digitalization of a pre-given-subject matter, which has to be learned. The difference is that digitalized places do not necessarily need gamy elements in order to be useful. There is much more about using virtual environments in education: The main advantage is that mistakes can be made without any consequences, that contents are endlessly repeatable plus that it is extremely cost saving. Thus, in terms of Remote Labs, students learn how to use a robot in a virtual environment before actually using it.

Although the advantages of VBL seem to be obvious, the measurement of learning successes presents a major challenge for the parties in charge. It is not only that the learning effect per se needs to be measured, but whether the handling is so unproblematic that users experience a sense of flow [5, 6] (a spontaneous sense of joy while performing a not too easy, not too difficult task), (tele-) presence [7] (the feeling of being enabled to act in this case in a remote lab) and finally immersion [8], the sensation of fully diving into a virtual environment. Obviously, these possible experiences are highly dependent on the user's pre-knowledge (e. g., how to use the WASD plus mouse combination) and his intrinsic technical readiness. The reason is that only users who can forget about the handling of for example a controller can experience a sense of immersion. If they need to look at it and think about the usage again and again, they will constantly be reminded that they are solely performing a virtual task, which is non-existent in reality and might thus attach less importance/meaning to it.

Another big problem in the measurement of subjective virtuality experiences is the question whether to perform the tests quantitatively or qualitatively and which influence the corresponding decision will have on the validity, transparency, causal interrelations and reliability of the results. The usage of the questionnaires on subjective user sentiments and self-assessment is a necessary step since these facts are not objectively observable. The self-assessment questions help to relate the produced results to behavior-parameters, which then lead to tentative conclusions concerning whether there is an interrelation between user preferences/habits and VBL success.

For the pre-assessment of test-persons, the BIG Five questionnaire is named as the most useful way to assess a test person's personality traits. The entailed items cover *neuroticism*, meaning emotional instabilities like fears and sadness, *extraversion*, the willingness to be in the center of attention, *openness to experience*, meaning the willingness to learn, *agreeableness*, the general need to socialize and lastly *conscientiousness*, the willingness to be disciplined [9]. For psychologists, alternative methods to assess personality traits exist; however, in the end, they all come back to the big five although they may be named differently [9].

Another of the most contemporary assessment questionnaires is the MEC-SPQ on general media exposure [10]. The main advantage is that it is highly flexible and may entail eight, six or only four items per scale. It has been used in studies on mobile gaming [11], in the realm of computer gaming [12] and serious games, thus, in the area of game-based learning [13]. So far, it is the only validated and highly consistent measurement instrument on spatial thinking [10].

In addition to this, recent studies by Witte showed that the locus for control of technology (KUT) questionnaire is a validated instrument to measure the performance of test persons while being confronted with technical problems [14]. Burde and Blankertz proved, that there is a correlation between a high score in the KUT and the performance in technical handling [15].

However, besides assessing test persons, an overall system evaluation and technical assessment of all technological devices is of utmost importance. Are software and hardware stabile? Do all components run as desired? Are there any known errors or problems and can the program run 'fluently'? [16]. Secondly, special attention must be paid to the users: how is their first reaction to the virtual robot? Did they spontaneously know what do to? Was there a lot of explanation necessary?

In sum, it must be concluded that the evaluation of virtuality-based learning is partly problematic because of subjective user assessment, talent and perception, which cannot be measured objectively. There is always the risk of users being afraid to truthfully state their abilities or that they even overestimate their capabilities. Our approach addresses this issue by creating an interplay between the estimated technical readiness of individual test persons and their actual real-time learning progress. Accordingly, the risk of falsified results due to inaccurate self-assessment of the test persons can be minimized.

3 Active Interaction for Remote Labs in Virtual Reality Environments

The creation of fully interactive virtual environments is based on the VR techniques that have been utilized by carrying out the technical and virtual environments of the remote labs. To realize a fully capable Remote Lab several steps were performed, i.e.:

1. Virtualization of machines and plants in every detail for three-dimensional representation within virtual environments.
2. Embedding of three-dimensional objects into virtual environments to create a virtual scenario, in which users can move around to exploit objects and the environment.
3. Setup and implementation of an information and communication infrastructure for data exchange between real and virtual laboratory environments.
4. Enabling uni-directional communication between the real laboratory environment and its virtual representation in order to reproduce movements of the real world demonstrator within the virtual demonstrator in real-time.
5. Enabling bi-directional communication by embedding control mechanisms and devices for the real laboratory from VR experiments into the scope of the Remote Lab.

The user can interact with the Remote Lab through various interfaces, e. g. the Virtual Theatre described in [1] or other immersive technologies like the Oculus Rift. Figure 1 shows that a notebook together with a Head Mounted Display is a suitable environment.

The first step of this procedure has already been described by Hoffmann et al. [2]. The virtual demonstrator that is used within the current work consists of two cooperating six-axis robots that are placed on a table in order to perform concurrent tasks. The virtual representation of the robots has been designed using modeling tools for computer graphics and design. The modelling of the robots is performed by integrating a bone structure into the virtual representation whereas the bones of the robot are connected through joints. The meshing of this bone-joint-structure ensures the correct assignment of the single parts in terms of parent and child nodes in order to recreate physically realistic movements of the whole robot, i. e. if the root joint is moved, all subsequent child nodes of the robot (bones and joints) are moved accordingly as well.

The embedding of these robots into a virtual environment is performed by the use of a VR tool for virtual worlds, i. e. WorldViz Vizard as described in [1]. In terms of the modeling, the different components, e. g. the robot table, both robots, the objects to be treated by the robots as well as other elements like avatars or screens are included into this virtual environment to create an immersive scenario for the user experiment.

The information and communication infrastructure (ICT) for Remote Labs has been described in detail in [3] and is an integral part of the virtual laboratory experiment. The ICT consists of the two cooperating robots, which are controlled by two manual control panels and a computer that contains the Robot Operating System (ROS) environment. Over a network architecture this operating computer is connected to other computers that run the Virtual Reality simulation programs and are connected to VR simulators like the Virtual Theatre as described in [17] or the Oculus Rift in combination with a local client computer [18] as depicted in Figure 1. The connection between the robot operating computer and the VR simulation systems

Figure 1   Remote Lab environment – The user is immersed into the scenario via the Oculus Rift

is established by making use of the Protobuf Protocol interface for the exchange of robot information [19].

The ICT as described allows the uni-directional communication between a robot-focused laboratory experiment and a distant representation of this laboratory in terms of a Remote Lab. Using the Protobuf interface standard the angles and joint positions of the robots can be transferred over the network in real-time. Internal tests on the real-time capabilities of such Remote Lab, which allows the observation of distant experiments, determined the maximum lag between reality and virtuality to 0.1–0.2 s.

Besides the graphical interface for the visualization of Remote Labs at distant places, e. g. by making use of the Virtual Theatre, there are also interaction devices embedded into the ICT. For our scenario we have chosen a common game pad controller, the Nintendo Wii™, as remote control device for the interaction of the user with the real world demonstrator within the virtual environment. Using this game device, the robots of the simulation and accordingly the real robots can be successfully manipulated. The basic control functions are highlighted in Figure 2.

Figure 2   Nintendo Wii™ controller and basic functions for robot control

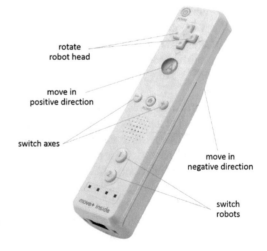

There are two control mechanisms that have been carried out for robot control, and which are both based on the usage of the Wii™ gaming controller:

1. Direct kinematics for direct control of the joint angles for each robots.
2. Inverse kinematics for user control of the movement axis (X, Y, Z) whereas the joint angles for the current robot position or moving trajectory are dynamically calculated during the experiments.

In terms of the direct kinematics robot control method, each of the six angles of the selected robot can be individually controlled using the "A" button for positive moving direction and the "B" button on the back of the remote control for negative moving direction. Using the "+" and "−" signs the axes of the robot joints can be subsequently selected. Using the buttons "1" and "2" the according robot can be selected. For direct kinematics the cross on the top of the Wii™ is not used, as the head rotation is represented by the sixth robot joint angle.

Concerning the inverse kinematics the "A" and "B" buttons are used to move the robot claw in positive respectively negative direction of the X, Y or Z axis. The axes are switched again using the "+" and "−" signs on the controller. The "1" and "2" also change the selected robot. The rotation of the robot head for inverse kinematics is implemented using the control cross at the top of the remote control. For the dynamic calculation of the single joint angle values suitable for the goal position or trajectory, an inverse calculation method is used for determining the joint parameters. For our use-case a MATLAB™ Toolbox has been adapted to the needs of the robot demonstrator.

The scenario, in which the described robot control methods are being applied, consists of a setup, where one of the robots has to be moved along a fixed path. This path is represented by a wire. For conducting the experiment an eyelet is attached to one of the robots. The task of the controlling person is to move the eyelet attached to the robot along the wire, which forms a certain curve (see Figure 3).

The described task is performed through the Wii™ remote control either by making use of the direct kinematic mechanism or by making use of inverse kinematics. In order to assess these methods against each other, user studies were performed, which are described in the following chapter.

4 User Studies for Examination and Evaluation of Control Mechanisms for Remote Labs

4.1 Design of Experiment and Expectations

The aim of this study is to examine which of the previously described control mechanisms for six-axis robots is the most beneficial for the implementation in Remote Labs especially with regard to an intuitive control and progress of learning as well as investigating the effects of the sequent comparison of both mechanism.

Figure 3 The eyelet attached to the right robot has to be driven along the steel wire in the middle

We expect that students – especially those who are used to gaming – will prefer the inverse control mechanism over the direct one, because it resembles their gaming experience. Furthermore we expect that successful practice experience through the trainings session will enhance the feelings of self-confidence and thus flow.

A representative number of engineering students from different advanced information science courses at multiple sophisticated, technical universities participated in the study in order to evaluate the learning progress of the concurrent methods for remote controlling the robots. The objective of conducting the study is to assess the different methodologies of control mechanisms suitable for engineering students.

The first part of the study is performed in cooperation with the Technical University of Dortmund, where test persons were recruited. The second part of the study is carried out in the course of a lecture with engineering students at the RWTH Aachen University. The study consists of three questionnaires and two practical tests, namely the remote control of the cooperating robots using direct and inverse kinematics. The sequence of the tests (direct and inverse mechanism) is randomized, the participants are accordingly assigned to either Group A or Group B. Participants who are assigned to group A start with the inverse kinematics test whereas participants of group B start with the direct kinematics test. Both user groups conduct both experiments, however in reverse order. The intention of this approach is on the one hand to examine the learning progress of the students during the experiments and on the other hand to equalize the effects of test order. The study is implemented in six steps:

1. Theoretical input in terms of the study design and methods of examination.
2. Pre-questionnaire for general assessment concerning the personal background of the test persons in terms of video game experiences and spatial thinking abilities.
3. First experiment using either inverse kinematics (Group A) or direct kinematics (Group B).

4. Questionnaire for the assessment of the previous test.
5. Second experiment using either direct kinematics (Group A) or inverse kinematics (Group B).
6. Questionnaire for the assessment of the previous test.

The first questionnaire is given to the participants before the experiment and is used for a general classification of the test person. Whereas questions such as the frequency of confrontation with digital games, the frequency of handling a console, whether the participants are active member of a digital sodality and the amount of hours spend on computer games a week are used to assess participants experience of gaming, individuals visual-spatial imagination (in virtual surroundings) are examined by items of the FRS [20] und questions of the subscale DSI of the MEC-Spatial Presence Questionnaire (MEC-SPQ) [10] adapted to computer games. This scale was already used successfully in previous studies and is characterized by fair quality criteria [21].

Besides this scale, items of the KUT [22] are used to assess participants locus of control when confronted with technical problems. Additionally, questions of the BIG Five Inventory [23] are used to assess subjects' personality and psychological biases, to get a broad picture of the participants.

The second and third questionnaire are used to assess the students' technical evaluation of the currently performed tests as well as their experience of learning progress while working. Participants are asked to rate the feasibility, advantages and disadvantages and the control of the just practiced remote mechanism as well as adapted questions concerning the experience of absorption due to the experience of flow [24]. A mental state of operation in which the individual, who is performing the task, is fully involved and immersed by feelings of energized focus [25]. All questions are presented on a seven-point scale, ranging from 1 = *total disagree* to 7 = *total agree*.

4.2 Correlational Approach

To gain further inside of the relationship between the individual gaming experience, such as hours of gaming per week and the evaluation of the control mechanism as well as learning progress during the experiment, the correlation between the pretest data was calculated. Data were analyzed with IBM SPSS statistics software. The results are visualized in Figure 4 in form of a graph that shows the strength of each correlation.

The correlational approach shows that subjects with more playing hours per week evaluate the inverse kinematics approach better than the direct one, but only if the inverse control mechanism is the initial one. There is no significant correlation between hours played a week and an appreciation of the direct mechanism, neither as first test nor as second test.

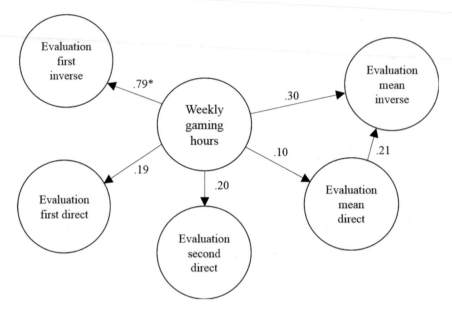

Figure 4 Correlation between amount of weekly gaming hours and evaluation of remote control

Further differences between the two participant groups were analyzed with a multivariate ANOVA, where each group served as a between-subject factor. The assumption of homogeneity of variances is investigated with Levene's test and shows no significant violations of the assumption for the dependent variable. Inspection of histograms show no significant deviations from normality for the rating of two groups. The analysis shows no main effect of rating due to group assignment, $F(1, 12) = 1.49$, $p = .266$, but additional analyses of the within-subject factor task-order show significant differences ($p = < .05$) between the two tests in both groups (see Table 1).

Statistical analyses reveal that the participants show no significant differences in preference due to both remote mechanisms. In both groups, the second remote

Table 1 Results of the significance analysis of Remote Lab control mechanisms

		Group A	Group B
		Inverse	Direct
Test 1	*M*	4.44	4.20
	SD	.98	1.38
		Direct	Inverse
Test 2	*M*	4.84*	4.84*
	SD	.99	1.63

* $= p < .05$, M = Mean, SD = Standard deviation

mechanism is rated significantly higher in preference than the first one, regardless of group membership. The subscale experienced learning progress is rated above the mean, in particular after the second testing session for the group that starts with the inverse mechanism ($M = 5.33$; $SD = 1.75$) respectively the group that starts with the direct mechanism ($M = 4.45$; $SD = 2.05$). These present findings do not confirm the hypothesis that students prefer the inverse mechanism in statistical terms, despite the fact that the mean values of the inverse mechanism are slightly higher than those of the direct mechanism. However, the results emphasize the importance of learning experience in both groups.

Thus, it can be concluded from these results, that the experience of flow and students' valuing of technical mechanism increase over time and are depending on practical experience and learning progress rather than a specific task mechanism per se.

5   Conclusion and Outlook

The aim of this work is to assess different mechanisms for the remote control of laboratory environments at arbitrary places. Based on an existing Remote Lab environment, direct and inverse kinematics control schemes have been carried out and implemented in order to enable the control of two cooperating six-axis robots.

The assessment in terms of the learning success lead to the result that there is not a significant preference for one of the two control mechanisms. However, the inverse kinematics – as expected – has been evaluated slightly better in comparison to the direct specification of joint angles. The study has also shown that the learning effect is equally good using both control methods, hence, both user groups evaluated the second test as preferable to the first one as they gained more self-confidence in controlling the robots during the progress of the study.

During the next steps in enhancing the usability and application of Remote Labs, it is our aim to enable a direct manipulation of the laboratory environment that can be located at arbitrary places. This real laboratory will be moved in real-time and accordingly to the exact digital representation, thus unexpected states of the experiment can be reached in the simulation similarly to the real-world demonstrator. In order to ensure the safety during these remote operations, a collision avoidance system based on the inverse kinematics implementation will be carried. Using this security layer, Remote Labs at arbitrary places can be independently controlled by users from Virtual Reality simulators from various locations. This will enable a holistic coverage of laboratory experiments for universities all over the world.

Acknowledgments This work was supported by the German Research Foundation (DFG) within the project ELLI (Excellent Teaching and Learning within engineering science) at RWTH Aachen University in terms of investigating laboratory experiments and Remote Labs.

References

1. M. Hoffmann, K. Schuster, D. Schilberg, T. Meisen, Next-generation teaching and learning using the virtual theatre. In: *At the Edge of the Rift*, ed. by S. Gregory, P. Jerry, N. Taveres Jones, 2014
2. M. Hoffmann, K. Schuster, D. Schilberg, S. Jeschke, Bridging the gap between students and laboratory experiments. In: *Virtual, Augmented and Mixed Reality, Lecture notes in computer science*, vol. Heraklion, Crete, Greece, ed. by R. Shumaker, Springer, Cham, 2014, pp. 39–50
3. M. Hoffmann, T. Meisen, S. Jeschke, Shifting virtual reality to the next level: Experiencing remote laboratories through mixed reality. The International Conference on Computer Science, Computer Engineering, and Education Technologies (CSCEET2014), 2014
4. C. Meier, S. Seufert, Game-based learning: Erfahrungen mit und Perspektiven für digitale Lernspiele in der betrieblichen Bildung. In: *Handbuch E-Learning*, ed. by A. Hohenstein, K. Wilbers, Fachverlag Deutscher Wirtschaftsdienst, Köln, 2005
5. M. Csikszentmihalyi, *Finding Flow: The Psychology of Engagement with Everyday Life*. Basic Books, New York, 1997
6. M. Csikszentmihalyi. Creativity: Flow and the psychology of discovery and invention, May 6th, 2013. http://books.google.de/books/about/Creativity.html?id=aci_Ea4c6woC&redir_esc=y
7. C. Bracken, P. Skalski. Telepresence and video games. the impact of image quality. http://www.psychology.org/File/PNJ7(1)/PSYCHNOLOGY_JOURNAL_7_1_BRACKEN.pdf
8. C. Jennett, A.L. Cox, P. Cairns, S. Dhoparee, A. Epps, T. Tijs, A. Walton, Measuring and defining the experience of immersion in games. International Journal of Human-Computer Studies **66** (9), 2008, pp. 641–661. 10.1016/j.ijhcs.2008.04.004
9. G. Matthews, I.J. Deary, M.C. Whiteman, *Personality Traits*, 2nd edn. Cambridge University Press, Cambridge, U.K. and New York, 2003
10. P. Vorderer, W. Wirth, F.R. Gouveia, F. Biocca, T. Saari, F. Jäncke, S. Böcking, H. Schramm, A. Gysbers, T. Hartmann, C. Klimmt, J. Laarni, N. Ravaja, A. Sacau, T. Baumgartner, P. Jäncke, Mec spatial presence questionnaire (mec-spq): Short documentation and instructions for application. report to the european community, project presence. MEC (IST-2001-37661)
11. J. Laarni, N. Ravaja, T. Saari, Presence experience in mobile gaming. Proceedings of DiGRA 2005 Conference: Changing Views – Worlds in Play, 2005
12. D. Weibel, B. Wissmath, Immersion in computer games: The role of spatial presence and flow. International Journal of Computer Games Technology **2011** (3), 2011, pp. 1–14. 10.1155/2011/282345
13. S. Göbel, *E-learning and games for training, education, health and sports: 7th international conference, Edutainment 2012, and 3rd international conference, GameDays 2012, Darmstadt, Germany, September 18 - 20, 2012; proceedings, Lecture notes in computer science*, vol. 7516. Springer, Berlin [u.a.], 2012
14. M. Witte, S.E. Kober, M. Ninaus, C. Neuper, G. Wood, Control beliefs can predict the ability to up-regulate sensorimotor rhythm during neurofeedback training. Frontiers in human neuroscience **7**, 2013, p. 478. 10.3389/fnhum.2013.00478
15. W. Burde, B. Blankertz, Is the locus of control of reinforcement a predictor of brain-computer interface performance. Proceedings of the 3rd International Braincomputer Inferface Workshop and Training Course, Graz, 2005, pp. 76–77
16. J. Wakolbinger, P. Kirchner. Netavatar – interaktion mit einem humanoiden roboter, 2004. www.hs-augsburg.de/~tr/prj/ss10-IP/07_WK/07_WK_NetAvatar.pdf
17. M. Hoffmann, K. Schuster, D. Schilberg, S. Jeschke, Next-generation teaching and learning using the virtual theatre. In: *4th Global Conference on Experiential Learning in Virtual Worlds Prague, Czech Republic*. 2014
18. OculusVR. https://www.oculus.com/dk2/
19. Google. http://code.google.com/p/protobuf/wiki/thirdpartyaddons
20. S. Münzer, C. Hölscher, Entwicklung und Validierung eines Fragebogens zu räumlichen Strategien. Diagnostica **57** (3), 2011, pp. 111–125

21. K. Schuster, M. Hoffmann, U. Bach, A. Richert, S. Jeschke, Diving in? how users experience virtual environments using the virtual theatre. In: *Virtual, Augmented and Mixed Reality, Lecture notes in computer science*, vol. Heraklion, Crete, Greece, ed. by R. Shumaker, Springer, Cham, 2014, pp. 636–646
22. G. Beier, Kontrollüberzeugungen im Umgang mit Technik: Ein Persönlichkeitsmerkmal mit Relevanz für die Gestaltung technischer Systeme. Disseration, Humboldt Universität, Berlin, 2004. www.dissertation.de
23. B. Rammstedt, O.P. John, Kurzversion des big five inventory (bfi-k). Diagnostica **51** (4), 2005, pp. 195–206
24. F. Rheinberg, S. Engeser, R. Vollmeyer, eds., *Measuring components of flow: the Flow- Short-Scale*. Proceedings of the 1st International Positive Psychology Summit. Washington DC, 2002
25. F. Rheinberg, Vollmeyer, R. Engeser, S., Die Erfassung des Flow-Erlebens. In: *Diagnostik von Motivation und Selbstkonzept*, ed. by J. Stiensmeier-Pelster, F. Rheinberg, pp. 261–279

Next-Generation Teaching and Learning Using the Virtual Theatre

Max Hoffmann, Katharina Schuster, Daniel Schilberg
and Sabina Jeschke

Abstract When graduates enter the working world, they have to apply their knowledge gained during their studies to new situations. Virtual Reality bears a great potential to simulate difficult situations, e. g. in dangerous environments. However, a major drawback of many Virtual Reality environments is the lack of natural navigation and free locomotion within the artificially designed world. Compared to a driving simulator, where users are sitting in a mock-up holding an actual wheel, users need to be able to move around freely if such situations are being simulated. Mixed Reality Simulators like the "Virtual Theatre" combine various technical devices. A head mounted display enables a three dimensional visualization of the simulation. The Virtual Theatre defines its unique characteristics through the omnidirectional treadmill. This omnidirectional floor consists of rollers, which are embedded centric to the middle point. Through this floor, the user is able to perform natural movements. By making use of a data glove, the user can actively take part in the events of his virtual experience based on hand movements. In the present work, an application of the Virtual Theatre is demonstrated based on a use-case, in which a plateau on "Mars" was implemented in terms of learning and exploring scenarios. The user is able to explore the Mars surface containing "Mars Rover" vehicles, orbiters and satellites. In the second major scenario, the user can maneuver a Mars rover through an obstacle course. In further development steps, the Virtual Theatre will be utilized for teaching purposes and to realize applications in terms of remote laboratories. Based on this, it is either possible to visit elusive points of interests like a nuclear power plant or to use experimental setups that are located at other universities remotely. These applications allow a holistic usage of innovative teaching approaches.

Keywords Virtual Theatre · Experiential Learning · Virtual Reality · Immersion

M. Hoffmann (✉) · K. Schuster · D. Schilberg · S. Jeschke
IMA/ZLW & IfU, RWTH Aachen University, Dennewartstr. 27,
52068 Aachen, Germany
e-mail: max.hoffmann@ima-zlw-ifu.rwth-aachen.de
Originally published in "At the Edge of the Rift",
© Inter-Disciplinary Press 2014. Reprint by Springer International Publishing
Switzerland 2016, DOI 10.1007/978-3-319-42620-4_22

1 Introduction

In terms of modern teaching, multiple approaches can be utilized to impart knowledge. In addition to classical teaching techniques, which are primarily based on the spoken word and on written texts, the visualization of educational content has gained in importance. Through progress in computer graphics, the possibilities of visualizing objects of interest have grown to an exhaustive amount. This opens up various new approaches in teaching and learning opportunities for students [1, 2].

In modern and innovative courses of studies, teaching media has evolved from static approaches using books and lecture notes to novel methods based on digital visualizations [3]. Hereby, the form of visualizing contents is mainly based on PowerPoint or similar presentation forms [4]. This can be explained as follows: The teaching person chooses a form of presentation that allows him to focus on his educational contents using illustrations and pictorial summaries [5]. Despite the critical discussion of PowerPoint as the primary teaching tool [6], the usage of presentation software, especially in teaching classes, has grown constantly [7].

However, even IT-based teaching methods with interactive and exploratory functions such as simulations or intelligent tutorial systems are limited to a certain kind of knowledge transfer. Especially in engineering or other technically focused courses the teaching personnel is facing more and more obstacles due to the advanced and complex technology level of the relevant applications [8], where problem based learning approaches are needed [9]. As incisive examples, studies of astronautics or nuclear research can be pointed out. During studies of astronautics, it is nearly impossible to bring the students in touch with a realistic experience of aerospace travelling using traditional teaching methods. Even high quality pictures or movies are incapable of transferring experienced knowledge. With regards to the second example, the problem is quite similar. As the visit of a nuclear power plant is generally too dangerous or is connected with a huge amount of organizational efforts, the transfer of experiential knowledge about the working atmosphere in an atomic plant is difficult [10] However, the usage of books, lecture notes or presentations cannot replace the experience of a power plant visit [11].

Thus, novel ways of teaching have to be developed in order to make even critical educational contents accessible for an exhaustive number of students. One promising solution to the mentioned obstacles consists in the virtualization of real-world scenarios using modern computer applications. By the use of advanced computer graphics and visualization methods in connection with interfaces that allow natural locomotion, the user can be immersed into the scenario in a realistic way [12] As far as students are concerned, the idea of the immersion of the student is to intensify the student's learning experience during exercises. According to Murray [13], in this connection, the term immersion is defined as follows:

> "Immersion is a metaphorical term derived from the physical experience of being submerged in water. We seek the same feeling from a psychologically immersive experience that we do from a plunge in the ocean or swimming pool: the sensation of being surrounded by a completely other reality, as different as water is from air that takes over all of our attention, our whole perceptual apparatus." [13]

Thus, by aiming for the experience of immersion within exploratory scenarios, the students can have a more intensive experience of an actual working scenario. This can lead to a higher success during the exercise (encoding new information) and later on in a working context (retrieval of information and applying the knowledge to a practical situation). However, even if an immersion is not fully realizable within a virtual scenario or the immersion does not have the desired effects such as improved learning, in some scenarios virtualization is just necessary to create a sufficient representation of complex educational contents.

In Section 2 summarizing the state of the art, some virtualization techniques, which are particularly relevant, are pointed out. In Section 3, the Virtual Theatre as an example for immersive user interfaces is described in detail. Section 4 depicts one scenario, which was developed during our research on more effective teaching methodologies. In this context, possible learning scenarios are referred to. Based on the development of the "Mars" it is our aim to enable real-world scenarios of elusive places like an atomic plant. In the Conclusion and Outlook we focus on further learning and teaching opportunities enabled by using the Virtual Theatre.

2 State of the Art

There are an exhaustive number of applications, which make use of immersive elements within virtual real-world scenarios. However, the immersive character of all these applications is based on two characteristics of the simulation. The first is the quality of the three-dimensional simulation and the second is the user's identification with the "avatar" within the virtual world scenario.

A realistic three-dimensional representation of a real-world scenario within the simulation is very important in order to be surrounded by the virtual world in a realistic way. According to Wolf and Perron [14], three conditions have to be fulfilled in order to reach a high immersion of the user's experience within the scenario:

> "Three conditions create a sense of immersion in a virtual reality or 3-D computer game: The user's expectations of the game or environment must match the environment's conventions fairly closely. The user's actions must have a non-trivial impact on the environment. The conventions of the world must be consistent, even if they don't match those of 'metaspace'." [14]

Furthermore, the identification of the user with the virtual environment has to be ensured. This is done by using "avatars", which create a representation of the actual user within the scenario. Bartle [15] defines an avatar as follows:

> "An avatar is a player's representative in a world. [...] It does as it's told, it reports what happens to it, and it acts as a general conduit for the player and the world to interact. It may or may not have some graphical representation; it may or may not have a name. It refers to itself as a separate entity and communicates with the player as such: 'I can't open the door'. It's a mere convenience, a tool. Contrast this with a character. A character is a player's representation in a world. It's a whole level of immersion deeper. Your character is an extension of yourself [...]. The game reports things that happen to the character as if they were happening to you. [...]" [15]

There are many technical solutions that are primarily focused on creating a high-quality and complex three-dimensional environment, which is accurate to the real-world scenario in every detail. One highly advanced visualization technology was realized through the development of the Cave [16]. By making use of complex projection techniques the developers of the Cave could define new quality standards in visualization in terms of resolution, colour and flicker-free stereo. It enables a technique, which provides multi-screen stereo vision while reducing the effect of common tracking and system latency errors.

The cave is known as the exhausting realization of virtual reality (VR). According-ing to Rheingold, VR can be described as an experience, in which a person is "surrounded by a three-dimensional computer-generated representation, and is able to move around in the virtual world and see it from different angles, to reach into it, grab it, and reshape it" [17].

The developers of the cave choose an even more specific definition of virtual reality: "a virtual reality system is one which provides real-time viewer-centered head-tracking perspective with a large angle of view, interactive control, and binocular display" [18]. Through this definition the authors try to integrate the user into the virtual reality scenario. If the user is an active part of the VR, who is able to act and to interact with the environment, he can finally be integrated into the virtual scenario in an immersive way.

Nevertheless, immersion is not only supposed to be based on an authentic and realistic graphical representation of the virtual environment, but also on the hardware and through natural interaction possibilities of the user [12]. However, even the Cave allows only little interaction capabilities as the user can just interact in the currently demonstrated perspectives. Furthermore, natural movements are also very limited, as a natural locomotion through the virtual environment is usually restricted to the currently shown spot of the scenario. These natural movements though including walking, running or even jumping through the VR are decisive for a highly immersive experience of the user within the virtual environment.

This gap of limited interaction is filled by the Virtual Theatre with the help of which the user is enabled to interact with the virtual environment more frequently. Through a multi-stage process we aim at examining the effect of immersion on students through the development of basic scenarios. These steps are performed to finally enable the participation of students in remote laboratory experiments, e. g. the visit of a plant. The idea and the related applications of the Virtual Theatre are described in the following chapter.

3 Next-Generation Teaching and Learning Using the Virtual Theatre

The Virtual Theatre was developed by the MSEAB Weibull company [19] for military training purposes. However, as discovered by Ewert et al. [20] the usage of the Virtual Theatre can also be extended in order to meet educational requirements for engineering students. It consists of three basic elements, the omnidirectional treadmill, a head mounted display and a cyber glove.

The omnidirectional treadmill or the omnidirectional floor represents the navigation component of the Virtual Theatre. It comprises polygons of rigid rollers with increasing circumferences and a common origo [20]. The rollers' rotation direction is oriented to the middle point of the omnidirectional floor, where a circular static area is located. The rollers are driven by a belt drive system, which is connected to all polygons of the treadmill and thus ensures the kinematic synchronization of all parts of the moving floor (Figure. 1).

On the central part that is shown in the upper right corner of Figure 1, the user is able to stand without moving. As soon as the user steps outside of this area, the rollers start moving and accelerate according to participant's position from the middle part. If the user returns to the middle, the rotation of the rollers will stop.

The head mounted display (HMD) represents the visual perception part of the user in the Virtual Theatre. The HMD consists of an interface that contains two separate displays, one for each eye. Through stereo vision technique the simulation is able to create a three-dimensional picture within the perception of the user. Each display provides a 70° stereoscopic field with an SXGA resolution in order to create a gapless graphical representation of the virtual environment. For our specific setup, we are using a HMD from zSight [21]. This HMD also contains an internal sound system to complete the holistic perception of Virtual Reality.

The third part of the Virtual Theatre that ensures tactile interaction capabilities within the virtual environment is the data glove. This special hand glove is equipped with 22 sensors, which are capable of determining the user's hand position and

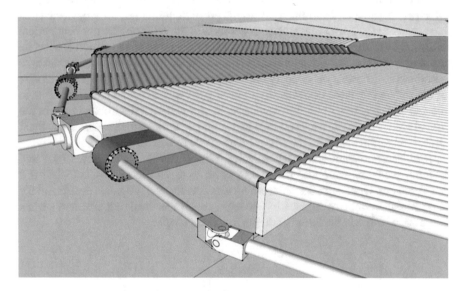

Figure 1   The drive engineering of the Virtual Theatre and its omnidirectional floor

Figure 2 A walk through virtual reality

gestures [20]. This enables gesture detection like the grasping of objects. Additionally, special gestures can be used for specific interaction commands.

In order to perform a walk through one of the scenarios, the different devices of the Virtual Theatre have to work perfectly together (see Figure 2).

On the left side of Figure 2, the user in action is visible while the scenario that he is just experiencing is visualized on the right side.

Attached to the omnidirectional treadmill, there is a frame carrying 10 cameras, which are capable of determining the user's position through an infrared device that is attached to the HMD of the user. In addition to the position, the tracking system is also able to determine walking and looking directions of the user. Thus, while the user is walking within the virtual environment, the picture he is seeing changes according to his actual movements and looking directions. Furthermore, the velocity of the rollers increases or decreases according to the user's walking speed. Due to the real-time reaction of the VR simulator the user gains a holistic experience of the virtual scenario. He can walk naturally and unlimited through the virtual environment while being immersed into the VR as being inside of a Cave.

Besides the technical realization of the Virtual Theatre, different teaching and learning scenarios were also realized in order to exploit the full potential of this VR simulator. One major scenario, which was developed during our research on the experiential learning of engineering students, consists in the development of a Mars scenario, which will be described in the following chapter.

4   The Walk on Mars – An Experiential Learning Scenario

The creation of scenarios that enable a virtual walk on the red planet came into mind due to the lack of adequate laboratory experiments for students of aerospace engineering. A realistic virtualization of an exemplary plateau on Mars should extend the student's learning experience to a higher level. The research results gained through these scenarios are needed for the development of remote labs.

The plateau on Mars was designed using highly detailed textures and three dimensional visualization techniques. In addition to the visualization features, a PhysX engine was implemented into the scenario. Thus, objects like stones or little rocks are moving in a physically realistic manner. Figure 3 shows one section from the plateau including rocks and light effects within the atmosphere above.

As depicted in Figure 3, the Mars scenario does not only contain environmental details, but is also enriched with exploration vehicles like Mars Rovers or satellites, which are flying on determined paths in the sky.

In a first stage of the development, the students were able to walk around on the Mars plateau in real time exploring environmental details of the surface. Within this simplified scenario vehicles could be approached in order to examine their technical construction. In a second step, these virtual exploration vehicles were provided with interactive elements in order to enable active information discovery by the students. Thus, the student can approach certain points of interest in order to gain detailed information about them One example is depicted in Figure 4.

As the user is approaching the Mars Rover vehicle on the left side, a sign with the inscription "MARS EXPLORATION ROVER" appears. If the user approaches

Figure 3   Screenshot of the walk on Mars plateau

Figure 4 Information of Mars Rovers (©NASA) and satellites while approaching

certain satellites, which are located in the sky above the plateau and steps into a certain radius around this satellite, it will hover down at the avatar's eye-level in order to facilitate a detailed observation.

The implementation of the overall exploration scenario containing various different objects to discover the plateau is called the "Mars Museum".

The second scenario, which was extensively developed during our research, is called "the Mars Parcours" or "the Mars Course". In contrast to the other scenarios, this scenario is not only targeted to learning. It focuses on a more intense interaction of the student with the virtual environment (Figure 5).

The Mars Parcours consists of an obstacle course that is located on top of the plateau. All technical objects are removed except for one exploration rover. The plateau surface is assembled with rocks, which serve as obstacles. During the experiment, the user has to navigate the rover through the obstacle course by using special gestures. These gestures can either be performed through head movements or through translations of the hand, but they can also be based on special gestures that are carried out using specific finger moves in the cyber glove.

The scenarios presented adopt the approach of serious gaming. According to Oblinger [22] and Prensky [23], who is – by the way – the originator of the term "digital native" [24], the connection of fun and learning can lead to higher attention and learning success of the student. Whitton [18] even talks about sound pedagogic reasons for employing computer games-based learning: "If games are experiential, active, problem-based and collaborative, then they have the potential to be effective environments for learning, not specifically because they are games, but because they exhibit the characteristics of constructivist learning environments."

Figure 5 The Mars Exploration Rover in the middle of the Mars Parcours

Especially this last scenario is one major step in the direction of enabling an interactive as well as experiential learning of the students. The evaluation of motoric capabilities to perform such scenario will have a major influence on the technical design of remote laboratories.

5 Conclusion and Outlook

Driven by the lack of suitable teaching resources for students of highly technical study courses, the discovery and development of novel teaching methods using the Virtual Theatre can be considered as a major improvement. The Virtual Theatre, which can be regarded as an "open Cave", serves novel learning opportunities even for students with a highly specialized academic background.

The development of the virtual Mars plateau including different scenarios has shown that learning and teaching methodologies for various purposes can be fulfilled through advanced visualization techniques in combination with intelligent user interaction. Furthermore, the walk on Mars enables and experienced discovery of elusive places in real time.

In further steps of our research activities, we will especially focus on extending interaction capabilities of our system within different scenarios. In order to be even more immersed by the virtual reality and to become a part of the virtual environment, the user needs the capability to influence the virtual environment in a non-trivial way. Thus, the development of further scenarios will concentrate on the elaboration of remote laboratories. Using the Virtual Theatre, it is the aim of our research to enable an interactive collaboration of different universities and to share real laboratories with students all over the world through a holistic virtualization.

References

1. M. Kerres, *Mediendidaktik: Konzeption und Entwicklung mediengestützer Lernangebote*. Oldenbourg, München, 2012
2. J. Handke, A.M. Schäfer, *E-Learning, E-Teaching and E-Assessment in der Hochschullehre: Eine Anleitung*. Oldenbourg, München, 2012
3. R.J. Craig, J.H. Amernic, Powerpoint presentation technology and the dynamics of teaching. Innovative Higher Education **31** (3), 2006, pp. 147–160
4. A. Szabo, N. Hastings, Using it in the undergraduate classroom: Should we replace the blackboard with powerpoint? Computer and Education **35**, 2000
5. T. Köhler, N. Kahnwald, M. Reitmaier, Lehren und lernen mit multimedia und internet. In: *Medienpsychologie*, ed. by B. Batinic, M. Appel, Springer, Heidelberg, 2008, pp. 477–501
6. T. Simons, Does powerpoint make you stupid? Presentations **18** (3), 2005. http://global.factiva.com/
7. A.M. Jones, The use and abuse of powerpoint in teaching and learning in the life sciences: A personal view. BEE-j 2, 2003. http://www.bioscience.heacademy.ac.uk/journal/vol2/beej-2-3.pdf
8. E. André, Was ist eigentlich Multimodale Mensch-Technik Interaktion? Anpassungen an den Faktor Mensch. Forschung und Lehre **21** (01/2014), 2014
9. M. Steffen, D. May, J. Deuse, The industrial engineering laboratory: Problem based learning in industrial eng. education at tu dortmund university. EDUCON, 2012
10. S. Malkawi, O. Al-Ariadah, Students' assessment of interactive distance experimentation in nuclear reactor physics laboratory education. European Journal of Engineering Education **Vol. 38** (No. 5), 2013, pp. 512–518
11. S. Hempen, Experiential learning in academic education: A teaching concept for efficient work system design. Proceedings of the 14th Workshop of the Special Interest Group in Experimental Interactive Learning in Industrial Management of the IFIP Working Group 5.7, 2010, pp. 71–78
12. D. Johansson, *Convergence in Mixed Reality-Virtuality Environments: Facilitating Natural User Behavior*. University of Örebro, Schweden, 2012
13. J.H. Murray, *Hamlet on the Holodeck: The Future of Narrative in Cyberspace*. MIT Press, Cambridge (Mass.), 1997
14. Wolf, M. J. P, B. Perron, *The video game theory reader*. Routledge, NY, London, 2003
15. R.A. Bartle. Avatar, character, persona. http://www.mud.co.uk/richard/acp.htm
16. M. Whitton, C. Cruz-Neira, D. Sandin, T. DeFanti, Surround-screen projection-based virtual reality. the design and implementation of the cave. In: *SIGGRAPH '93 Proceedings of the 20th annual conference on Computer graphics and interactive techniques (ACM - New York)*. 1993, pp. 135–142
17. H. Rheingold, *Virtual reality*. Summit Books, New York, 1991
18. N. Whitton, Motivation and computer game based learning. ICT: Providing choices for learners and learning. Proceedings of the Ascilite Singapore Conference, 2007, pp. 1063–1067

19. MSEAB Weibull. http://www.mseab.se/the-virtual-theatre.htm, 2012
20. D. Ewert, K. Schuster, D. Johansson, D. Schilberg, S. Jeschke, Intensifying learner's experience by incorporating the virtual theatre into engineering education. Proceedings of the 2013 IEEE Global Engineering Education Conference (EDUCON), 2013
21. http://sensics.com/products/head-mounted-displays/zsight-integrated-sxga-hmd/specifications/
22. D.G. Oblinger, The next generation of educational engagement. Journal of Interactive Media in Education **8**, 2004, pp. 1–18
23. M. Prensky, Digital natives, digital immigrants. On The Horizon. MCD University Press **Vol. 9** (No. 5), 2001
24. M. Prensky, *Digital Game-Based Learning*. McGraw Hill, New York, 2001
25. K. Schuster, D. Ewert, D. Johansson, U. Bach, R. Vossen, S. Jeschke, Verbesserung der Lerner-fahrung durch die Integration des Virtual Theatres in die Ingenieurausbildung. In: *TeachING-LearnING.EU discussions*, ed. by A.E. Tekkaya, S. Jeschke, M. Petermann, D. May, N. Friese, C. Ernst, S. Lenz, K. Müller, K. Schuster, TeachING-LearnING.EU, Aachen, 2013

Shifting Virtual Reality Education to the Next Level – Experiencing Remote Laboratories Through Mixed Reality

Max Hoffmann, Tobias Meisen and Sabina Jeschke

Abstract Technical universities are more and more focusing on engineering education as a primary discipline. All along with the integration of various innovative fields of application into the curriculum of prospective engineers the need for appropriate educational features into the studies also increases. Unlike exclusively theoretical studies as physics, mathematics or information sciences the education of engineers extensively relies on the integration of practical use-cases into the education process. However, not every university is able to provide technical demonstrators or laboratories for all of the various applications in the field of engineering. Thus, it is the aim of the current paper to propose a method that enables visiting a high variety of engineering laboratories based on Virtual Reality. A Virtual Reality simulator is used to create and emulate remote laboratories that can be located at arbitrary places far away from their Virtual Reality representation. This way, by melting real world demonstrators with virtual environments, we enable a physically and technically accurate simulation of various engineering applications. The proof of concept is performed by the implementation and testing of a laboratory experiment that consists of two six-axis robots performing collaborative tasks.

Keywords Virtual Reality · Mixed Reality · Augmented Virtuality · Virtual Learning Environments · Remote Laboratories · Engineering Education

1 Introduction

In a world that is increasingly based on scientific innovations and technological progress the education of engineering students constantly gains in importance. Furthermore, the field of studies and the variety of possible specializations in engineering classes also increase significantly. In order to qualify graduated engineers to enter

M. Hoffmann (✉) · T. Meisen · S. Jeschke
IMA/ZLW & IfU, RWTH Aachen University,
Dennewartstr. 27, 52068 Aachen, Germany
e-mail: max.hoffmann@ima-zlw-ifu.rwth-aachen.de

Originally published in "Proceedings of the International Conference on Computer Science, Computer Engineering, and Education Technologies (CSCEET2014), Kuala Lumpur, Malaysia, 17-19 November 2014", © The Society of Digital Information and Wireless Communications 2014. Reprint by Springer International Publishing Switzerland 2016, DOI 10.1007/978-3-319-42620-4_23

the working world successfully, the imparting of practical knowledge during studies plays a decisive role in education.

While theoretical knowledge is still transferred using written texts and the spoken word – normally through lecture notes and traditional readings – the experience of practical use-cases in engineering education commonly relies on the visit of laboratory experiments during the studies. However, since computer vision and digitalization techniques have grown to an extensive level, the integration of new media into the curriculum replacing the visit of real laboratories gains in importance [1]. Especially in terms of engineering applications, the use of high quality and realistic visualization techniques as a supplement to the attendance within practical laboratory experiments is of major importance for successfully impart basic concepts of engineering applications to students. Not at least due of the progress in computer science and graphical visualization techniques, the capabilities of visualizing objects of interest embedded into an artificially designed context have grown to an exhaustive amount. In this context, for example physical effects or technical subtleties of engineering applications can be presented in higher detail or in an amplified way in order to emphasize aspects that are not easily observable in reality. These novel potentials can be utilized to explain theoretical knowledge more concrete and tangible and help engineering students to understand the concepts of complex technical applications on different levels and from a practical point of view.

Another major trend that is emerging within the field of education and learning relies on the way of distributing information and knowledge through internet-based media among the students. The high significance of online platforms and social media during the everyday life of a student can be exploited in order to increase communication channels by the establishment of E-Learning-Platforms [2]. The technical possibilities of sharing and representing educational contents, spreading knowledge over the world-wide web and enabling the remote participation of students in engineering classes, open up new opportunities of teaching and learning within universities. In combination with modern visualization techniques these computer-based teaching concepts can for example be implemented through Virtual Reality or Mixed Reality applications [3].

Despite all these technological possibilities, universities are facing more and more obstacles to deal with the high variety of engineering courses and their different technical applications and needs. Thus, each university can only provide a limited amount of experimental or laboratory classes for the different fields of engineering during their curricula. However, the demand for enlarging the variety of engineering education contents is also growing constantly. This leads to a conflict of interests as universities are not supposed to advantage any particular engineering class in comparison to the others. Due to the limited laboratory and teaching capacities, the ability to satisfy the needs for experimental education of engineering students cannot be fulfilled adequately. Furthermore, most of the offered experimental simulations or virtual representations of laboratory environments are lacking the required quality standards in terms of graphical accuracy and interaction capabilities.

The goal of the current paper is to find ways for dealing with these obstacles and proposing possibilities to enable an extensive practical education of engineering classes by answering the following questions that are correlated to the described issues:

1. How can universities address the high variety of engineering related disciplines by ensuring the availability of suitable use-cases, experiments and laboratory exercises for students?
2. How is it possible to integrate concepts of Experiential Learning into engineering education against the background of limited laboratory resources?
3. What is the benefit of integrating Virtual Reality and Mixed Reality applications into engineering education from the technical/physical point-of-view?
4. Is it possible to address educational needs of engineering students in universities by enabling an active manipulation of the virtual laboratory environments?

The present publication intends to answer these questions by introducing a novel method for realizing use-cases for Virtual Reality exercises and laboratory experiments in terms of Remote Laboratories. In this context, the term Remote Laboratory introduces the idea of enabling the visit of distant places for conducting laboratory experiments based on its virtual representation. Hence, it is our aim to enable students to attend specific experimental environments suitable to their field of studies, even if these exercises are not provided at their particular university.

In the next section, the state of the art in teaching with the aid of new media and novel visualization applications is presented. Furthermore, educational advantages of Virtual Reality and Mixed Reality applications are carried out. In Section 3 the technical implementation of our next-generation Virtual Reality simulator is described together with former virtual scenarios that have been implemented using the simulator. Section 4 presents a scenario for enhancing students' learning behavior by the creation of a Remote Laboratory in connection with Mixed Reality approaches. In Section 5 first attempts in evaluating the advantages of such virtual representation of a laboratory experiment are carried out, assessing the learning capabilities of a group of students visiting such Remote Laboratory. Section 6 concludes the outcome of the current paper and specifies the next steps of enhancing the user's experience and creating larger Remote Laboratory environments.

2 State of the Art

New media have gained high significance in university studies and publications in the past decade. These new media – which are mostly based on computer visualization techniques – are continuously replacing traditional books and lecture notes. White boards and projectors are replaced by presentation software, which represents the new standard for visualizing text and pictures [4], with PowerPoint as market leader [5]. This switch from traditional lecture speech to graphical representation has been performed, because this form of presentation enables focusing on the main points of the educational content using illustrative representations and pictorial summaries [6]. Despite the positive, but also critical discussion about an overwhelming usage of PowerPoint [7–9] as primary teaching tool, the usage of presentation software

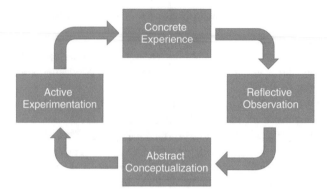

Figure 1   The Experiential Learning Cycle according to David A. Kolb (1984)

in the classroom has grown constantly [10, 11]. In connection with the entry of technological novelties into the classroom it is time to take the next steps from merely presenting pictures using presentation software to the usage of advanced graphical interfaces opening up interaction capabilities for the students involved in the engineering courses.

Presentation software like PowerPoint may be a far reaching advancement for most courses of studies. However, the usage of these meanwhile basic IT tools is also limited to a certain kind of knowledge transfer. Especially for practically oriented study paths like engineering classes active interaction capabilities within courses and exercises are inevitable. In these highly technical studies there is an urgent need for interactive laboratory experiments in order to impart practical and skill-based knowledge tangible to students. Against this background, David A. Kolb's traditional, well-established cycle on Experiential Learning is more up to date than ever [12]. The – almost classical – Learning Cycle is depicted in Figure 1.

In the picture, we see a never ending process of active experimentation, concrete experience, reflective observation and abstract conceptualization. Starting at active experimentation, there is a need for concrete experience in order to understand abstract concepts. The reflective observation following to the experience helps carrying out an abstract conceptualization that is based on a deeper understanding of the experienced content. Especially the practical part of the learning process in terms of the attendance to experimental courses cannot be replaced by any kind of theoretical knowledge transfer. Other learning related theories, which address the same matters as the experiential learning approach, are action learning, adventure learning, free choice learning, cooperative learning and service learning approaches [13]. In all of these theories, active interaction of the learning person plays an integral part in the learning process.

Due to the high complexity and advanced technological level of the relevant applications in engineering classes, sole static visualizations are not capable to serve as a medium for active experimentation or even concrete experience. In order to address these parts of the learning cycle, novel visualization concepts have to be applied. A key enabler that combines advanced visualization techniques with the experience of a certain scenario is Virtual Reality as shown in the relevant literature [14] and during prior studies in this field of application [15]. In the first step, Virtual Reality cannot serve direct interaction possibilities, but through the use of immersive effects, a Virtual Reality scenario is able to arouse an experienced reality within the perception of the user. This effect can be greatly characterized by the definition of immersion according to Murray [16]: "Immersion is a metaphorical term derived from the physical experience of being submerged in water. We seek the same feeling from a psychologically immersive experience that we do from a plunge in the ocean or swimming pool: the sensation of being surrounded by a completely other reality, as different as water is from air that takes over all of our attention, our whole perceptual apparatus."

In order to address the part of concrete experience during practical education accurately by virtual reality applications, the utilized tools have to fulfill a number of conditions to serve as a suitable complement for laboratory classes. In the following, we will therefore look at existing Virtual Reality applications and discuss if these are capable of bridging the gap of concrete experience to the students.

In terms of existing Virtual Reality technologies, there are already many technical solutions that are primarily focused on the creation of high-quality and complex three-dimensional environments, which are accurate to real-world scenarios in every detail. One example are flight simulators that are capable of tracking the locomotion of a flying vehicle in a virtual scenario [17]. However, these systems are usually not taking into account the position or the head movements of the user. Another Virtual Reality simulator is the well-known Omnimax Theatre, which provides a large angle of view [18], but does not allow any tracking capabilities whatsoever. First attempts to interact with Virtual Reality in a natural way were introduced by head-tracking monitors as conducted by Codella et al. [19] and Deering [20]. These specially designed monitors provide an overall tracking system, but are characterized by a rather limited angle of view [17]. The first mentionable approach to create a Virtual Reality environment with full tracking capabilities of movements and of the head position of the user was introduced by McDowall et al. [21] with the Boom Mounted Display. Despite advanced tracking capabilities these early attempts were characterized by poor resolutions and thus were not capable of a detailed graphical representation of a virtual environment [22].

Thus, in order to enable true user experience in simulated scenarios, Mixed Reality approaches have to be embedded into the Virtual Reality, were reality and virtuality are merged into each other [23]. One far reaching innovation in terms of enabling Virtual Reality and Mixed Reality applications was introduced with the CAVE in 1992 by Cruz-Neira et al. [24]. Hereby, the recursive acronym CAVE stands for Cave Automatic Virtual Environment. By making use of complex visualization techniques combined with various projectors and six projection walls arranged in form of a cube, the developers of the CAVE have redefined the standards in visualizing Virtual Reality scenarios by enabling a new level of immersion.

The CAVE reaches further towards true Virtual Reality which – according to Rheingold [25] – is described as an experience, in which a person is "surrounded by a three-dimensional computer-generated representation, and is able to move around in the virtual world and see it from different angles, to reach into it, grab it and reshape it." These active manipulation activities that can be performed by the user open up various new applications in education by rebuilding industrial use-cases. Thus, by enabling extended interaction capabilities with the scenarios in terms of providing relatively free manipulation of the virtual environment, the immersive effects of the scenarios are enhanced. This effects could lead to the desired impact on the learning behavior of students that consists of the ability to derive abstract conceptualizations on the basis of concrete or practical experience and active experimentation.

However, even the CAVE has got restricted interaction capabilities as the user can only interact in the currently demonstrated perspective of the scenario. Furthermore, natural movement is limited, as locomotion through different scenes of the scenario is usually performed by flying to the next spot. Yet, natural movement as walking, running or jumping through the Virtual Reality is decisive for a highly immersive experience in the virtual environment.

In order to fill this gap of limited interaction and accordingly deeper immersion into the scenario, additional devices and tracking systems for allowing such interaction have to be included into the scenario without losing the high quality of graphical representation of the virtual environment. One promising approach relies in the establishment of the Virtual Theatre that brings together a full-size stereo-vision view and various interaction devices and manipulation capabilities for the user.

3 The Virtual Theatre – Enabler for Extended Immersion

The Virtual Theatre represents a next-level Virtual Reality simulator that allows free locomotion in a virtual environment and active manipulation as well as the control of objects in the visualized scenario. The Virtual Theatre was carried out by a Swedish company [26] and was already described in detail in our previous publications [27] according to the scenarios that have been carried out during our latest research [28, 29]. The centerpiece of the Virtual Theatre is the omnidirectional treadmill, a moving floor that accelerates its centric arranged rollers according to the position of the user (Figure 2).

The user himself is tracked by an infrared tracking system; hence his head and hand movements are constantly observed and taken into account for a adaptation and manipulation of the virtual environment. The user is wears a Head Mounted Display (HMD) that is equipped with two screens – one for each eye – and enables highly immersive three-dimensional stereo vision for the exploration of the virtual space. The unique characteristic of a HMD is that this kind of devices are capable of measuring the user's head orientation through a perpendicular, which makes it possible to adjust the Virtual Reality according to the head's actual position and orientation. This enriches the immersive experience of the user as he is able to look around and explore the Virtual Reality in a similar way as he does in the real world. An embedded sound system into the HMD completes the plunge into virtuality. For

Figure 2 A user experiencing virtual environments in the Virtual Theatre

further information about the technical concept of the Virtual Theatre the reader is encouraged to refer to [28], where the hardware and technical setup is explained in detail.

Former scenarios that have been carried out using the Virtual Theatre were well received by the students of engineering classes [29]. One example for the creation of a huge sized virtual environment is our Mars project, where an extensive simulation of a plateau on the surface of the red planet was recreated to enable upcoming astronautics and aerospace students to perform a virtual visit and exploration of the Mars [15]. Another application of the Virtual Theatre was carried out in terms of a study in order to assess the learning behavior and learning efficiency of students while being surrounded in a virtual environment [29]. During the survey, the students were located in a virtual labyrinth, in which they needed to find objects and recognize their location and shape at a later point. Afterwards, the results of their learning efficiency were evaluated and compared to the efficiency using traditional computer screens for performing similar tasks.

However, former studies did not take into account deeper interaction capabilities as an active movement of objects or the remote control of devices for industrial use-cases. However, as mentioned earlier, exactly these interaction capabilities are strongly needed in order to create realistic virtual representation of experimentations, thus to enable Remote Laboratories. Hence, in the next step we present the inclusion extensive interaction capabilities into the use-case and we attempt to enable true immersion based on Mixed Reality concepts.

4 Enabling Remote Laboratories Through Mixed Reality

As part of a network of administrative computers the Virtual Theatre – including its integrated parts and the tracking system – can be expanded by additional hardware in order to enable a more natural user interaction within the virtual scenario.

Taking into account natural user behavior as well as the experience of students using well-established computer game devices, our team decided to carry out a remote control for virtual scenarios based on hardware of the ©Nintendo Wii™ Controller. The new conceptual design of the communication infrastructure for the Virtual Theatre and its surrounding hardware equipment is depicted in Figure 3.

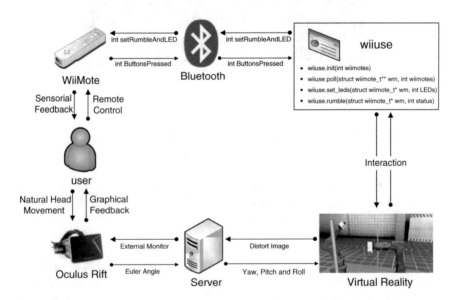

Figure 3 Communication Infrastructure of the Virtual Theatre with extended interaction capabilities

Figure 4 Visualization of two six-axis ABB™ robots performing collaborative tasks

In the middle part of the picture's bottom, the central server is visualized. The server deals with the signals of the Head Mounted Display, which is located on the left side, and processes its information for the user's movement, head position and orientation according to the virtual environment that is depicted on the right side at the bottom. The Wii™ remote controller is connected via Bluetooth and sends specific commands to the central server. The server processes these commands to manipulate the virtual environment and visualize the modified scenario in real-time.

A suitable application for including the described interaction device into a virtual scenario is based on an extended laboratory experiment that was virtualized by our team for education purposes. The setup consists of two six-axis robots that are placed on a table in order to perform collaborative tasks. The virtualized model is depicted in Figure 4.

In reality, the robots are located in the same distance as illustrated in the figure, enabling the ability to perform collaborative, interdepending tasks. As described in [28], the first attempts in carrying out a Remote Laboratory based on these robots consisted in the virtualization of the actual robot movements as well as a real-time alignment of the movements that were performed by the real world robots and the movements of our simulation. As shown in our previous publication, the full setup can be appropriately simulated in real time, i.e. the user inside of the virtual reality simulator can pursue the robot motion without any perceptible time lag. This real-time synchronization between the real world laboratory and its virtual representation enables active remote control of the robot setup as well as various use-cases:

- The actual position of the robots can be tracked and remotely manipulated from arbitrary locations.
- The control of the robot arms can be extended by additional security layers in order to assure save motion of the robots and to avoid collisions.
- The experimenter can easily work with dangerous materials or substances (e. g. chemicals) and is able to operate the robots if these are located at dangerous or non-accessible places.

In order to generate an added value to the remote control of a laboratory environment our research did not only concentrate on the development of remote control devices. We also focused on providing additional features that are enabled through Virtual Reality. One major progress of these efforts relies on the integration of Mixed Reality elements into the laboratory experimental context. In this connection, the term Mixed Reality is characterized as the merging of real and virtual worlds to produce new environments and visualizations, where physical and digital objects co-exist and interact in real-time [30]. Our use-cases including Mixed Reality approaches is able to address several aspects of the application:

- Systematic simulation of experiments with actual machines and components in real time. The exact simulation leads to a co-existence of real world objects in the laboratory and in Virtual Reality with interdepending system states.
- Feedback and manipulation capabilities of the user, which leads to an interaction with both, the objects in the virtual environment and the real components.
- Embedding of real world features into virtual environments by placing cameras into the laboratory environment. This enables the projection of detailed views of experimental insights that are captured by the camera onto a wall in the virtual environment. In terms of the Mixed Reality concept, this effect is also referred to as *Augmented Virtuality*.

Especially the last of the mentioned points of integrating Mixed Reality concept into our Virtual Reality scenarios bears a high potential to enhance the grade of immersion signifycantly. In terms of the user's perception the embedding of Augmented Virtuality – i. e. real pictures into the virtual scenario – is connected to effects of fuzziness between reality and virtuality, which leads to highly immersive impressions for the user. One possible application of integrating a camera in the real world scenario can for example consist in a placement of the camera in a bird's-eye perspective on top of the experimental setup to see the whole scene from a broader point-of-view.

Another possible scenario is to attach the camera onto a robot that is actually moving and thus to observe the scene from the robot point of view. An example for this extended perspective based on this Augmented Virtuality visualization technique is depicted in Figure 5.

On the bottom of the picture the simulation of the two six-axis robots is shown. The screen that is located above the table shows a video image that is taken from the perspective of the left robot in reality. This image is embedded into the Virtual Reality scenario as a sort video projection and hence represents an Augmented Virtuality element in the simulation.

This enrichment of the virtual scenario is connected to several improvements concerning the technical and the educational application of the scenario:

- The point of view perspective enables a detailed view of the simulated scene. Through the robot's perspective the tasks that have to be performed by the robot arms can be conducted with higher precision due to the overview based on more than one perspective.

Figure 5 The experimental setup in reality visualized in a Virtual Reality scenario

- The grade of immersion increases as the user of the simulation is able to see reality objects that are melting with the Virtual Reality scenario in real-time.
- For education purposes, multiple cameras can be attached at various locations of the demonstrator, which helps to explain the physical or technical effects of use-cases.

Besides the advantages for the single user that are connected to the embedding of Mixed Reality into the scenario there is also an added value for the remote control of a laboratory experiment from more than one user. Due to the placement of multiple cameras within the surroundings of the experiment, different users can perform collaborative tasks while observing a simulation of the actual system state in real-time, but from different perspectives. This enables a highly precise manipulation of the experimental conditions influenced by different users that can be located at arbitrary places. Especially due to this point, the far reaching benefits of Remote Laboratories as a new class of conducting experiments becomes known.

5 Evaluation of the Technical Implementation and Impact on the Students Learning Behavior

During the previous sections of this paper the added value of Remote Laboratories has been derived in terms of the overall usability and availability of laboratory experiments as well as the impact on the precise conduction of experiments using multiple information channels. In this section we would like to investigate the impact of the utilization of Remote Laboratories not only in terms of the availability of experimental setups in engineering classes, but also its effects on the learning behavior and motivation of students. In terms of the assessment according to this impact, we get back to initial research questions about enhancing the learning environment of students by remote setups or literally: "Is it possible to address the educational needs of engineering students in universities by enabling an active manipulation of the virtual laboratory environments?"

In order to answer this question, different facets of the educational needs of engineering students are taken into account:

- The impact of virtual experience and the grade of sensed immersion into the student's perception in virtual environments to build up a realistic scenario.
- The physical accuracy of simulations embedded into the virtual environment.
- The added value of enhancing the perspective of the user and the personalized view by emphasizing certain physical effects through amplified physical behavior or close-up views.

The first evaluation phase of these effects was performed in-house and based on the personnel and on the student employees of our institution. The visualization accuracy as well as the immersive effects of the simulation into the perception of the user could be verified during the testing phase. Especially the active motion of the six-axis robots through an easily manageable interface while being submerged in the virtual environment had clearly observable impacts on the understanding of robot motion and the need for automation.

In the next steps of the evaluation phase, the investigation of the scenario will be performed by a questionnaire that is carried out for laboratory classes of newcomer students. This evaluation, which will be further concentrating on didactical aspects of the experimentation environment, will take place in the following semester, in which the according students will assess their personal learning success after conducting several experiments with and without the help of the described Mixed Reality-related techniques. During this phase we will examine the effects of the virtual environment on the learning behavior of the engineering students by taken into account the following points:

- The impact of virtual experience and the grade of sensed immersion on the willingness and learning behavior of students in virtual environments.
- Correlations between the learning ability during laboratory experiments, gamification effects and fun in manipulating the laboratory environment in virtuality.
- The effect of hands-on experiments on the learning success of students in comparison to the mere observation of distant experiments that are not accessible in its real environment.

The study will show, if the different aspects like fun in learning, active involvement and free movement in virtual environments as well as the ability to manipulate a virtual representation of a real world demonstrator or – in other words – the reflective observation have a significant impact on the abstract conceptualization of complex engineering applications.

6 Conclusion and Outlook

The need for higher capacities in terms of the practical education of engineering students comprises a major challenge for today's universities. The constantly growing number of students with various educational backgrounds and different experiences as well as the wealth of study opportunities demands for innovative concepts in the organization of a profound engineering education.

In this paper, we have substantiated the idea of conducting real laboratory experiments through a Virtual Reality simulator by enabling Remote Laboratories. These laboratories can serve as an extensive supplement to real experimental setups, because they can be built up at arbitrary places and run simultaneously for multiple users. Various setups for virtual environments can be applied in order to emphasize immersive effects on the user with an expected impact on his learning behavior.

The next steps in connection with the presented scenarios will consist in a quantitative evaluation of the impact of Virtual Reality on the actual learning success of the students by assessing the conceptual knowledge of two different comparison groups, one that visits an actual laboratory experiment without any personal involvement or interaction with actual components, and the other group that visits a Virtual Reality based virtual environment of the laboratory experiment. Furthermore, we will discuss the effect of embedding Mixed Reality components into the Remote Laboratory on the students on the one hand in terms of their qualitative perception of being immersed into the virtual environment and on the other hand in terms of the advantages that are connected to the embedding of camera screens into the virtual scenario for additional perspectives.

On the technical side, the next steps concerning an extension of the Remote Laboratory environment consist in the development of a generic methodology to automate and to control robots of various kinds in virtual environments. In terms of this procedure, aspects of robot security, collision avoidance and inverse kinematics for robot control will be of major importance for an expedient experimentation environment. Next projects will concentrate on the implementation of complex scenarios with multiple robots and interaction devices in order to emphasize the idea of collaborative and concurrent engineering in virtual environments.

Acknowledgments The present work was conducted in terms of the investigation of virtual and remote laboratories in research and teaching and was supported by the project "ELLI – Excellent Teaching and Learning in Engineering Sciences" as part of the Excellence Initiative at the RWTH Aachen University.

References

1. M. Ebner, A. Holzinger, Successful implementation of user-centered game based learning in higher education: An example from civil engineering. Computers & Education **49** (3), 2007, pp. 873–890. http://www.sciencedirect.com/science/article/pii/S0360131505001910

2. M.J. Rosenberg, *E-learning: Strategies for delivering knowledge in the digital age*. McGraw-Hill, New York, 2001
3. Z. Pan, A.D. Cheok, H. Yang, J. Zhu, J. Shi, Virtual reality and mixed reality for virtual learning environments. Computers & Graphics **30** (1), 2006, pp. 20–28. http://www.sciencedirect.com/science/article/pii/S0097849305002025
4. A. Szabo, N. Hastings, Using it in the undergraduate classroom: Should we replace the blackboard with powerpoint? Computer and Education **35**, 2000
5. R.J. Craig, J.H. Amernic, Powerpoint presentation technology and the dynamics of teaching. Innovative Higher Education **31** (3), 2006, pp. 147–160
6. R.A. Bartsch, K.M. Cobern, Effectiveness of powerpoint presentation in lectures. Computer and Education **41**, 2003, pp. 77–86
7. T. Creed, Powerpoint, no! cyberspace, yes. The Nat. Teach. & Learn. F. **6** (4), 1997
8. D. Cyphert, The problems of powerpoint: Visual aid or visual rhetoric? Business Communication Quarterly **67**, 2004, pp. 80–83
9. P. Norvig, Powerpoint: Shot with its own bullets. The Lancet **362**, 2003, pp. 343–344
10. T. Simons, Does powerpoint make you stupid? Presentations **18** (3), 2005. http://global.factiva.com/
11. A.M. Jones, The use and abuse of powerpoint in teaching and learning in the life sciences: A personal view. BEE-j 2, 2003. http://www.bioscience.heacademy.ac.uk/journal/vol2/beej-2-3.pdf
12. D.A. Kolb, *Experiential learning: Experience as the source of learning and development*. Englewood Cliffs, NJ: Prentic Hall, 1984
13. C.M. Itin, Reasserting the philosophy of experiential education as a vehicle for change in the 21st century. Journal of Experiential Education **22** (2), 1999, pp. 91–98
14. D. Johansson, de Vin, L. J., Towards convergence in a virtual environment: Omnidirectional movement, physical feedback, social interaction and vision. Mechatronic Systems Journal (November 2011)
15. M. Hoffmann, K. Schuster, D. Schilberg, S. Jeschke, Next-generation teaching and learning using the virtual theatre. 4th Global Conference on Experiential Learning in Virtual Worlds **Prague, Czech Republic**, 2014
16. J.H. Murray, *Hamlet on the Holodeck: The Future of Narrative in Cyberspace*. MIT Press, Cambridge (Mass.), 1997
17. C. Cruz-Neira, D.J. Sandin, T.A. DeFanti, Surround-screen projection-based virtual reality. the design and implementation of the cave. SIGGRAPH '93 Proceedings of the 20th annual conference on Computer graphics and interactive techniques. ACM - New York, 1993, pp. 135–142
18. N. Max, Siggraph '84 call for omnimax films. Computer Graphics **16** (4), 1982, pp. 208–214
19. C. Codella, R. Jalili, L. Koved, B. Lewis, D.T. Ling, J.S. Lipscomb, D. Rabenhorst, C.P. Wang, A. Norton, P. Sweeny, G. Turk, Interactive simulation in a multi-person virtual world. ACM - Human Fact. in Comp. Syst. **CHI 1992 Conf.**, 1992, pp. 329–334
20. M. Deering, High resolution virtual reality. Com. Graph. **26** (2), 1992, pp. 195–201
21. I.E. McDowall, M. Bolas, S. Pieper, S.S. Fisher, J. Humphries, Implementation and integration of a counterbalanced crt-based stereoscopic display for interactive viewpoint control in virtual environment applications. Proc. SPIE **1256** (16), 1990
22. S.R. Ellis, What are virtual environments? IEEE Computer Graphics and Applications **14** (1), 1994, pp. 17–22
23. P. Milgram, A.F. Kishino, Taxonomy of mixed reality visual displays. IEICE Transactions on Information and Systems, 2013, pp. 1321–1329
24. C. Cruz-Neira, D.J. Sandin, T.A. DeFanti, R.V. Kenyon, J.C. Hart, The cave: Audio visual experience automatic virtual environment. Communications of the ACM **35** (6), 1992, pp. 64–72
25. H. Rheingold, *Virtual reality*. Summit Books, New York, 1991
26. MSEAB Weibull, 2012. http://www.mseab.se/The-Virtual-Theatre.htm

27. D. Ewert, K. Schuster, D. Johansson, D. Schilberg, S. Jeschke, Intensifying learner's experience by incorporating the virtual theatre into engineering education. Proceedings of the 2013 IEEE Global Engineering Education Conference (EDUCON), 2013
28. M. Hoffmann, K. Schuster, D. Schilberg, S. Jeschke, Bridging the gap between students and laboratory experiments. In: *Virtual, Augmented and Mixed Reality, Lecture notes in computer science*, vol. Heraklion, Crete, Greece, ed. by R. Shumaker, Springer, Cham, 2014, pp. 39–50
29. K. Schuster, M. Hoffmann, U. Bach, A. Richert, S. Jeschke, Diving in? how users experience virtual environments using the virtual theatre. In: *Virtual, Augmented and Mixed Reality, Lecture notes in computer science*, vol. Heraklion, Crete, Greece, ed. by R. Shumaker, Springer, Cham, 2014, pp. 636–646
30. Silva, Adriana de Souza e, D.M. Sutko, *Digital cityscapes: Merging digital and urban playspaces, Digital formations*, vol. v. 57. Peter Lang, New York, 2009

Pump it up! – An Online Game in the Lecture "Computer Science in Mechanical Engineering"

Daniela Janßen, Daniel Schilberg, Anja Richert and Sabina Jeschke

Abstract Evaluation results in the lecture "Computer Science in Mechanical Engineering" at RWTH Aachen University show that most students are unaware of the relevance of computer science in mechanical engineering as well as their importance and application in future careers. Therefore, the students' motivation to deal with computer science is often low. Over the past decade computer games have become very popular for educational purposes and are often used to facilitate learning. By playfully applying computer science skills through game-based learning, students can experience the importance of computer science for their future careers. The approach of game-based learning has proven to be suitable to motivate students. Therefore, the relevance of computer science for students of engineering education at the RWTH Aachen University is brought closer to them by the online computer game "Pump it up!". The game is part of the project-based task of the lecture "Computer Science in Mechanical Engineering". Students are given the task of manufacturing pump adapter pipes as well as programming industrial robots which is embedded in a storyline. In the game, the students are employees of a start-up company, which receive the order to manufacture pump adapter pipes for a brewery. The paper gives an outlook on the next steps of the application.

Keywords Game-based Learning · Computer Games · Students' Motivation · Learning Process · Computer Sciences · Mechanical Engineering

1 Introduction

The student dropout rate in mechanical engineering at German Universities is currently relatively high, approximately 40 %. A lacking sense of achievement and the lack of practical relevance for future careers are often named as reasons to quit studies prematurely. The missing of practical relevance results inter alia from a missing

D. Janßen (✉) · D. Schilberg · A. Richert · S. Jeschke
IMA/ZLW & IfU, RWTH Aachen University, Dennewartstr. 27, 52068 Aachen, Germany
e-mail: daniela.janssen@ima-zlw-ifu.rwth-aachen.de

Originally published in "The 8th European Conference on Games Based Learning; Proceedings of the 8th European Conference on Games Based Learning", © Academic Conferences and Publishing International Limited 2014. Reprint by Springer International Publishing Switzerland 2016, DOI 10.1007/978-3-319-42620-4_24

309

reference to the engineering profession. Evaluation results from the lecture "Computer Science in Mechanical Engineering" show that most students are unaware of the relevance of computer science in mechanical engineering as well as their importance and application in future career. Plus, 60.5 % of engineering students prefer situations in which they have to find their own practical solutions ([1]: 24). A promising approach combining self-controlled acting with fun is game-based learning. One third of the German population regularly plays computer games, two thirds of them play every day [2]. Although the group of so-called gamers is very heterogeneous regarding genres, user behaviour etc., it can be expected that the use of educational games increases the motivation regarding certain academic contents. The possibilities of entertainment and experience provided by serious games plays an important role.

2   Didactical Approach: Game-Based Learning

Over the past decade computer games have not only become very popular for educational purposes, but also are often used to facilitate learning, because playing games is related to having fun ([3]: 17) which is in turn related to emotions. According to Izard [4] enjoyment is referred to as one of ten emotions ([5]: 151). Emotions play an important role in the learning process and in relation to the learners' motivations. Wilkinson ([6]: 2) states that, "[w]hen we are emotionally aroused we may be more motivated to focus our concentration on learning, either because we find it enjoyable or due to a fear of possible failure". Events or situations, which have an emotional impact, are easier to remember than neutral ones ([7]: 139). Brandstätter et al. ([7]: 134) state that emotions have an important function in terms of motivation processes: motivated behaviour is aimed to gain positive emotions and to avoid negative ones. A well-known, popular model of motivation is the concept of flow by the psychologist Csikszentmihalyi ([8]: 29): "[h]is work on the theory of optimal experience led to the concept of 'flow'" ([6]: 2). Flow describes the involvement in an activity. Weibel and Wissmath ([9]: 11) also describe the feelings while playing games as "the sensation of being involved in the gaming action". According to Sherry ([10]: 328) computer games are especially suitable to induce flow. Therefore it can be assumed that playing games has the potential to evoke the sense of being involved in the game [11]. This emotional state can stimulate the students' learning process. The application of games with the aim of learning with digital media is also called game-based learning [12] and is accompanied by a new term since 2010, namely gamification [13]. Gamification describes the process of including game-based elements in a non-game context [14]. The gamified learning environment should foster the students' learning process. Students have the possibility to participate actively and experientially in the learning process. Therefore, the didactic concept of the online game *"Pump it up!"* is based on the game-based learning model by Garris et al. ([15]: 446). The learners are seen as active participants of the game who generate knowledge through the experience of playing. The overall aim is to maximize the motivation of those learners who perceive their task field to be too monotonous, theoretical or complex.

Gamification helps to make techniques attractive for users in the long term while supporting independence and freedom of choice at the same time. Giving frequent and transparent feedback regarding the level of completion provides the basis for long-term motivation. It is crucial for the users to know and fully understand the final desirable outcome or scenario beforehand. This enables users to evaluate their actions and personal benefit. Empirical studies on gamified applications indicate a significant improvement in the fields of motivation, learning outcomes and fixation of subject matters [16].

3   Development of the Online Game *"Pump it up!"*

The relevance of computer science is brought closer to the students via *"Pump it up!"*. The game will take up the already existent project-based task as part of the lecture "Computer Science in Mechanical Engineering". In the practical course, the students apply the lecture contents in hands-on work in teams. In addition, students get the possibility to deepen their knowledge from the lecture and of the practical course by playing the online game. Using the example of pump adapter pipes manufacturing as well as programming robots to manufacture the pipes, students experience the importance and application of basic knowledge on computer science for their later career in the field of mechanical engineering.

One important learning outcome is to illustrate the broad field of application of computer science in engineering education. For example, students have to program machine tools and industrial robots. Furthermore, students learn how to compile single commands into a complex program. The robots of each assembly cell must be programmed to precisely exercise the demanded task and to cooperate with other robots. Another learning outcome is the realization of further procedural steps typical for the industry as they are often found in the human-robot cooperation. This includes the gripping and holding of objects for manual assembly or the collaborative lifting and moving of elements by two robots. Assembly cells equipped with ABB RB 120 (Figure 1) serve as technical reference and are recreated virtually for the online game.

Different game modes and various levels make it possible to customize the game to the respective level of students' knowledge and rate of performance. However, the focus is on the best 10 % of the class as well as on those students who scored below two thirds of the points in the interim exams. Those students will have the possibility to play the game in the Virtual Theatre at least once. The Virtual Theatre is a Mixed-Reality Simulator allowing students to enter and explore virtual worlds in a physically limited room. The event status of this experience will further motivate strong students. Weaker students will be made aware of the practical relevance of the subject matters through the virtual scenario.

Figure 1 Assembly cell

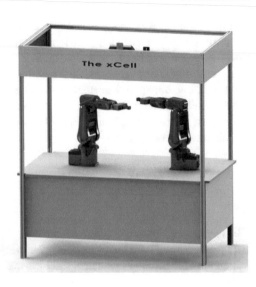

3.1 Storyline

The manufacturing of pump adapter pipes is embedded in a storyline which is based on a start-up company, called *"Pump it up!"*. The students can choose a character respectively an employee of the start-up company. The company gets the order to produce new pump adapter pipes for a brewery. In order to do so, different sub-steps are required.

The granularity of the implementation of the manufacturing process depends on the different game levels. The beginner level consists of the predetermination of the manufacturing process in which the students have to select the individual manufacturing steps in the correct order like milling, drilling, welding heat treatment and handling robots. On advanced levels the students define the process parameters like velocities, tools and materials. The students move from one level to another while they have successfully completed the lower level i.e. successfully manufactured the pump adopter pipes. Via a chat, students get the opportunity to interact with each other while playing the game.

Afterwards, they have to implement this information in the form of a pseudocode, the syntax of this code is given to the students. The student that delivers the highest quality result in the shortest period of time will win. Thus, the students will learn about the relation between the quality of the implemented program and the quality of the manufactured good.

Depending on the solution of the task of manufacturing pump adapter pipes, students can decide how the game end is designed. In case of a successful and high-quality manufacturing of the pipes, the start-up company receives follow-up orders by the brewery.

Figure 2 Virtual Theatre

3.2   *Technical Implementation*

The main technological objective is that students can play the game on different devices like their laptop, tablet pc or smart phone. To reach this goal, the implementation of the game must be independent from the OS e. g. Windows, IOS or Android. Therefore, the online game will be set up as a browser game. To gain the best performance and support for innovative concepts the game is constructed on HTML5, JQuery and CSS3 technologies. The remote control of the presented assembly cell will be realized by using ROS-Industrial. The assembly cell will be controlled via the Virtual Theatre [17] (Figure 2).

4   Outlook

The game "*Pump it up!*" is currently in the development stage. At the moment, the background story of the brewery and the start-up company are written and the didactical concept is specified. The programming of the game has already started in order to test the beta-test version in August and September.

In the beta-test version interested students will test the game, give feedback and take part in an online survey as a further form of evaluation of the beta-test version. Students can make suggestions or remarks on the application in designated comment sections which are taken into account for the didactical and technical adaption of the game. Highly dedicated or interested students receive the opportunity to take part in the advancement of the scenarios as part of the Virtual Theatre working group which is scheduled for November this year. The game will be applied in the project-based task as part of the lecture beginning in the winter semester in October 2014.

After the game has been developed and successfully applied in the project-based task for one semester, students receive an online-survey with standardised scales at the end of the semester which is divided in several question blogs concerning socio-demographic data, previous gaming experiences, students' learning motivation as well as the learning process. The analysis is conducted quantitatively by a frequency count in order to survey the students' attitude related to topics such as the application of the game in the project-based task, the didactical and technical implementation as well as the learning motivation. The results will be taken into account in the further development of the online game.

In order to assess the success of the teaching innovation, the examination results respectively the exam notes are to be compared to those from previous years.

Acknowledgments We would like to thank the Joachim Herz Foundation and the Federal Ministry of Education and Research for enabling us to realize this project.

References

1. W. Derboven, G. Winker, *Ingenieurwissenschaftliche Studiengänge attraktiver gestalten. Vorschläge für Hochschulen*. Springer, Heidelberg, 2009
2. K. Hampe. Jeder dritte deutsche ist ein gamer, 2011. http://www.bitkom.org/de/themen/54906_68946.aspx
3. M. Pivec, M. Moretti, *Game-based learning: Discover the pleasure of learning*. Pabst Science Publishers, 2008
4. C.E. Izard, Innate and universal facial expressions: Evidence from developmental and cross-cultural research. Psychol. Bull **115** (2), 1994, pp. 288–299
5. B. Batinic, M. Appel, *Medienpsychologie*. Springer, Heidelberg, 2008
6. P. Wilkinson, Affective educational games: Utilizing emotions in game-based learning. Games and Virtual Worlds for Serious Applications (VS-GAMES), 2013, pp. 1–8
7. V. Brandstätter, J. Schüler, R.M. Puca, L. Lozo, *Motivation und Emotion. Allgemeine Psychologie für Bachelor*. Springer, Heidelberg, 2013
8. M. Csikszentmihalyi, *Finding flow. The psychology of engagement with everyday life*. Basic Books, New York, 1997
9. D. Weibel, B. Wissmath, Immersion in computer games: The role of spatial presence and flow. International Journal of Computer Games Technology **2011**, 2011, pp. 1–14
10. J.L. Sherry, Flow and media enjoyment. CommunicationTheory **14** (4), 2004, pp. 328–347
11. K. Schuster, D. Ewert, D. Johansson, U. Bach, R. Vossen, S. Jeschke, Verbesserung der Lernerfahrung durch die Integration des Virtual Theatres in die Ingenieurausbildung. In: *TeachING-LearnING.EU discussions. Innovationen für die Zukunft der Lehre in den Ingenieur-wissenschaften*, ed. by A.E. Tekkaya, S. Jeschke, M. Petermann, D. May, N. Friese, C. Ernst, S. Lenz, M. K., K. Schuster, 2013
12. M. Prensky, *Digital Game-Based Learning*. Paragon House, St. Paul, Minnesota, 2007
13. K. Kapp, *The gamification of learning and instruction*. Pfeiffer, San Francisco, 2012

14. S. Deterding, R. Khaled, L.E. Nacke, D. Dixon, *Gamification: Towards a Definition*. ACM Press, Tampere, 2011
15. R. Garris, R. Ahlers, J.E. Diskrell, Games, motivation and learning. Simulating & Gaming. An Interdisciplinary Journal of Theory, Practice and Research **22** (4), 2002, pp. 43–56
16. M. Herger. Gamifizierung: Daten und Fakten, 2011. http://enterprise-gamification.com/index.php/de/fakten
17. D. Ewert, K. Schuster, D. Johansson, D. Schilberg, S. Jeschke, Intensifying learner's experience by incorporating the virtual theatre into engineering education. In: *Proceedings of the 2013 IEEE Global Engineering Education Conference*. Berlin, 2013

Pump it up! – Conception of a Serious Game Applying in Computer Science

Daniela Janßen, Christian Tummel, Anja Richert,
Daniel Schilberg and Sabina Jeschke

Abstract Student attrition in mechanical engineering at German universities currently lies at about 40 %. A lacking sense of practical relevance for a future career are often named as reasons to quit studies. Over the past decade online games have become very popular for educational purposes. The approach of game-based learning, however, has proven to be suitable to motivate students. At RWTH Aachen University engineering students are imparted the relevance of computer science for their field through an e-learning environment including the online game *Pump it up!* The paper describes the conception and game design of the game including didactical and technical requirements related to it.

Keywords Serious Games · Game-Based Learning · Virtual Worlds · Higher Education · Computer Sciences

1 Introduction

Digital media establish new possibilities of innovative ways of teaching and learning in higher education. With the arisen development of e-learning, the emerging tendency of Serious Games and the concept of game-based learning (the application of games with the aim of learning with digital media [1]) for educational purpose is an ongoing increasing trend. Mayer et al. talk about "a digital turn in the use of games and simulations for learning and training" [2]. At the same time, student attrition in mechanical engineering at German universities currently lies at a relatively high rate, approximately 40 %. A lacking sense of achievement and the lack of practical relevance for a future career are often named as reasons to quit studies. The missing of practical relevance is often embedded in the curriculum of engineering education because of a largely missing reference to the engineer profession. Evaluation results in the lecture "Computer Science in Mechanical Engineering" at RWTH Aachen University show that most students are unaware of the relevance of computer science in mechanical engineering as well as their importance and application in future

D. Janßen (✉) · C. Tummel · A. Richert · D. Schilberg · S. Jeschke
IMA/ZLW & IfU, RWTH Aachen University, Dennewartstr. 27, 52068 Aachen, Germany
e-mail: daniela.janssen@ima-zlw-ifu.rwth-aachen.de

Originally published in "HCI International 2015 - Posters' Extended Abstracts",   317
© Springer 2015. Reprint by Springer International Publishing Switzerland 2016,
DOI 10.1007/978-3-319-42620-4_25

career which has a negative effect on students' motivation. On the positive side, 60.5 % of engineering students prefer situations in which they have to find their own practical solutions [3]. A promising approach combining self-controlled acting with enjoyment is the game-based learning approach. By now one third of the German population regularly plays computer games, two thirds of them play every day [4]. Games have the power of engaging, motivating and emotionally involving students. Therefore, it can be expected that educational games increase students' learning motivation as well as demonstrate the practical relevance of academic content.

2 Related Work

An increasing number of studies have shown that Serious Games provide an effective environment for the purpose of learning, for example [5, 6]. This research field is also a controversially debated field when it comes to the effectiveness of Serious Games. On the one hand, a lot of researchers, e. g. [1, 7] point out that the application of Serious Games improves learning, motivation as well as performance. Several studies, e. g. [8] show the positive effect of educational games in math, science and military education. Van Eck [9] supports the positive effects of game-based learning by a review of experimental research. Mayer et al. [2] give an overview of several authors which examine the effects of game-based learning. Pivec [10] sees in the "application of games and simulations for learning [. . .] an opportunity for learners to apply acquired knowledge and to experiment, get feedback in form of consequences and thus gain experience in a "safe virtual world". Wilson et al. [7] examine the relationship between game attributes and learning outcomes. Furthermore, games encourage different learning approaches such as active learning, experimental learning and problem-based learning [5]. Consequently, it can be stated out that games have a positive impact as related to cognitive, skill-based and affective outcomes [7]. On the other hand there are several researchers [11, 12] who assess that the empirical effectiveness of game-based learning is quiet a mystery. Gunter et al. [13] cited in [14] report that "the effectiveness of an educational game is often based on enhancing learning motivation and social interactions rather than the effectiveness of knowledge acquisition" [14], an effect which is a quite unexplored [7]. Wilson et al. [7] claim a lack of research which combines game elements to learning outcomes. To summarize, it can be stated that despite the huge number of research of the effectiveness of game-based learning there are a lot of open research question. But in light of the positive learning outcomes of Serious Games in higher education, it is worth to continue further research of the effectiveness of Serious Games on learning outcomes as well as the relationship between games and their impact on learning as a direct or indirect process and what the mediating variables are [7].

3   Conception of the Online Game *Pump it up!*

According to De Gloria et al. [15] the following aspects are important in order to develop Serious Games: the underlying theory respectively the didactical concept, the academic content as well as the game design which are described in detail below.

3.1   Didactical Concept: Game-Based Learning Approach

Playing games is related to having fun [16]. Brandstätter et al. [17] state that emotions such as enjoyment have an important function in motivation processes: motivated behavior is aimed to gain positive emotions and to avoid negative ones. A well-known model of motivation is the concept of flow by Csikszentmihalyi [18]. Flow describes the involvement in an activity. Weibel and Wissmath [19] also describe the feelings while playing games as "the sensation of being involved in the gaming action". This emotional state fosters students' motivation in the learning process. Because of its activating and participating approach the game-based learning model by Garris et al. [20] is used as the underlying didactically concept of *Pump it up!* (Figure 1).

This model is based on constructivist learning theories after which students generate knowledge by making experience in a certain environment, in this context by playing the game [15]. Consequently students are seen as active participants of *Pump it up!*, who generate knowledge through the experience of playing. Giving frequent and transparent feedback regarding the level of completion provides the basis for long-term motivation. The key component of the model is the game cycle in which the students are in a "repeated judgment-behavior-feedback loop" [20]. This enables students to evaluate their actions and personal benefit. By playfully applying computer science skills through *Pump it up!*, students get feedback, gain practical experiences in terms of computer science and learn the importance of it for their future career.

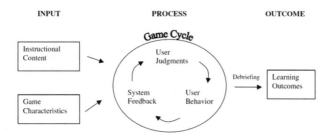

Figure 1   Game-based model (*Source* [20])

3.2   Game Design and Content of Pump it up!

The game design includes the theoretical conception of the game world, rules and characters with regard to a certain target group. According to Pivec and Moretti [16] *Pump it up!* is developed along the following game design steps: learning outcomes of the game, define target group, define the game and shape the game idea, elaborate the details (storyline), evaluate the game idea, (technical) implementation of the game. It is crucial for students to know and fully understand the final desirable outcome beforehand. This enables students to evaluate their actions and personal benefit. *Pump it up!* has the following **learning outcomes**: The overall aim is to maximize the motivation of those learners who perceive their task field to be too monotonous, theoretical or complex. Furthermore, students experience the importance and application of basic knowledge on computer science for their later career in the field of mechanical engineering. Students get a stronger practical relevance of computer science as well as deepen their knowledge from the lecture and of the practical course. A self-controlled, practical learning is fostered. Students learn how to integrate single commands into a complex program and realize further procedural steps typical for the industry as they are often found in the human-robot cooperation. The **target group** of *Pump it up!* are engineering students which attend the lecture "Computer Science in Mechanical Engineering". *Pump it up!* will take up the already existing project-based task as part of this lecture. In the practical course, students apply lecture contents in teams by hands-on work. Engineering students focus especially on content of mechanical engineering and machine construction. Therefore, the relevance of computer science is an underrepresented field for them. After defining the learning outcomes the **game idea** is defined as well as the game title and genre. The game is entitled *Pump it up!* because students have the task to program robots which in turn manufacture pump adapter pipes. Using this scenario, students experience the importance and application of basic knowledge on computer science in the field of mechanical engineering. *Pump it up!* is assigned to the computer game genre of simulation [21]. Simulations constitute a representation of a real in a virtual world and are a practical context for the reality without their consequences like manufacturing errors and are therefore especially valuable in engineer's hands-on education [22]. In context of *Pump it up!* students have to program the two robots of each assembly cell to precisely execute the demanded task and to cooperate with other robots. Thereby students learn the realization of further procedural steps. The tasks are embedded in a **storyline** along a brewery scenario. In this scenario the start-up company *Pump it up!* gets the order to produce new pump adapter pipes for a brewery. In order to do so, different sub-steps are required. The students move from one level to another once they have successfully completed the lower level i.e. successfully manufactured the pump adopter pipes meeting the geometric and mechanical requirements (Figure 2).

Figure 2 Screenshot of
programming in *Pump it up!*

Afterwards, they have to implement this information in the form of a pseudo code, the syntax of this code is given to the students. The student who delivers the highest quality result in the shortest period of time will win. Thus, students will learn about the relation between the quality of the implemented program and the quality of the manufactured good. Several possible game ends depending on the leaner's performance and various levels make it possible to customize the game to the respective level of students' knowledge, learning pace and rate of performance. Before the game goes into action, the beta-version will be tested with students in order to receive didactical and technical feedback. Up to **technical implementation** *Pump it up!* is set up as a browser game. The main technological objective is to play the game on different devices like notebook, tablet, pc or smart phones. For achieving platform independence the implementation of the game must be independent from the operating system e. g. Windows, IOS or Android as well as the game is constructed on HTML5, JQuery and CSS3 technologies.

4   Conclusion and Outlook

Driven by the lack of high student attrition in mechanical engineering at German Universities and the lack of practical relevance in computer science, the promising approach of game-based learning is applied in the engineering science at the RWTH Aachen University. Combining learning with engagement students get the possibility to experience practical relevance of computer science in a virtual learning environment in *Pump it up!* During the development of the game the students' requirements are collected and entered in the developing process in a didactical and technical way. The next steps include the programming and implementation of the game. Besides the online game, further research steps are planned in context of applying Serious Games into higher education. In order to provide students more complex tasks and a higher immersion, the game-based learning approach is extended to 3D virtual learning environments, e. g. using a so called virtual theatre [22] or different versions of mixed learning environments as suggested by Hoffmann et al. Therefore, in further research activities we will focus on virtual environment using 3D glasses in order to enter more virtual reality and to increase the immersion in the game to facilitate students learning as the next education level.

References

1. M. Prensky, *Digital Game-Based Learning*. Paragon House, Minneapolis, 2007
2. I. Mayer, H. Warmelink, G. Bekebrede, Learning in a game-based virtual environment: a comparative evaluation in higher education. European Journal of Engineering Education **38** (1), 2013, pp. 85–106
3. W. Derboven, G. Winker, *Ingenieurwissenschaftliche Studiengänge attraktiver gestalten. Vorschläge für Hochschulen*. Springer Verlag, Heidelberg, 2009
4. Bundesverband Informationswirtschaft. Telekommunikation und neue Medien e.V. http://www.bitkom.org/de/themen/54906_68946.aspx
5. D. Oblinger, The Next generation of Educational Engagement. Journal of Interactive Media in Education **8**, 2004, pp. 1–18
6. K. Kapp, *The gamification of learning and instruction*. Pfeiffer, San Francisco, 2012
7. K. Wilson, W. Bedwell, E. Lazzara, E. Salas, C. Burke, J. Estock, K. Orvis, C. Conkey, Relationships Between Game Attributes and Learning Outcomes: Review and Research Proposals. Simulation & Gaming Publications **40** (2), 2009, pp. 217–266
8. A. McFarlane, A. Sparrowhawk, Y. Heald. Report on the educational use of games: An exploration by TEEM of the contribution which games can make to the education process, 2002. https://pantherfile.uwm.edu/tjoosten/LTC/Gaming/teem_gamesined_full.pdf
9. R. Van Eck, Digital Game-Based Learning. Itis not just the digital natives who are restless. Educause Review **41** (2), 2006, pp. 17–30
10. M. Pivec, Editorial: Play and learn: potentials of game-based learning. British Journal of Educational Technology **38** (3), 2007, pp. 387–393
11. M. Papastergious, Digital game-based learning in high school computer science education: Impact on educational effectiveness and student motivation. Computers and Education **52** (1), 2009, pp. 1–12
12. F. Ke, A case study of computer gaming for math: Engaged learning for gameplay? Computers and Education **51** (4), 2008, pp. 1609–1620
13. G. Gunter, R. Kenny, E. Vick, Taking educational games seriously: Using the RETAIN model to design endogenous fantasy into standalone educational games. Educational Technology Research and Development **56** (5-6), 2008, pp. 511–537
14. F. Tsai, K. Yu, H. Hsiao, Exploring the Factors Influencing Learning Effectiveness in Digital-Game-based Learning. Educational Technology & Society **15** (3), 2012, pp. 240–250
15. A. De Gloria, F. Belloti, R. Berta, E. Lavagnino. Serious Games for education and training, 2014. http://dx.doi.org/10.17083/ijsg.v1i1.11
16. M. Pivec, M. Moretti, eds., *Game-based learning: Discover the pleasure of learning*. Pabst Science Publishers, Lengerich, 2008
17. V. Brandstätter, J. Schüler, R. Puca, L. Lozo, *Motivation und Emotion*. Springer Verlag, Heidelberg, 2013
18. M. Csikszentmihalyi, *Finding flow. The psychology of engagement with everyday life*. Basic Books, New York, 1997
19. D. Weibel, B. Wissmath, Immersion in Computer Games: The Role of Spatial Presence and Flow. International Journal of Computer Games Technology **2011**, 2011, pp. 1–14
20. R. Garris, R. Ahlers, J. Diskrell, Games, Motivation, and Learning: A Research and Pratice Model. Simulation Gaming **33** (4), 2009, pp. 441–467

21. D. Ewert, K. Schuster, D. Johansson, D. Schilberg, S. Jeschke, Intensifying learner's experience by incorporating the virtual theatre into engineering education. In: *Proceedings of the 2013 IEEE Global Engineering Education Conference, EDUCON.* 2013
22. M. Hoffmann, K. Schuster, D. Schilberg, S. Jeschke, Bridging the Gap between Students and Laboratory Experiments. In: *VAMR 2014, LNCS*, vol. 8526, ed. by R. Shumaker, S. Lackey, Springer International Publishing, Switzerland, 2014, pp. 39–50

Flipped Classroom on Top – Excellent Teaching Through a Method-Mix

Larissa Köttgen, Stefan Schröder, Esther Borowski, Anja Richert
and Ingrid Isenhardt

Abstract The continuing technological progress and changing conditions of studying have induced a demand for a realignment of academic teaching including the
imparting of knowledge and competence transfer. The question "How do students
learn?" gains a stronger focus of attention again. As a result, teachers take the ter-
tiary didactical interventions and innovations into account. These trends are geared
towards a "shift from teaching to learning" and call for a student-centred teaching
(The Bologna Declaration 1999, p. 4). To meet the previously described developments, the Center for Learning and Knowledge Management at RWTH Aachen
University presents a concept based on a didactical method mix. Within the framework of the course "Kommunikation und Organisationsentwicklung" (communication and organizational development, KOE) a combination of numerous methods
established, already scientifically validated and positive evaluated methods plus new,
innovative methods is used. The lecture is conceptualized mainly for first semester of
mechanical engineering at RWTH Aachen University. The sessions discuss various
content-related aspects of organizational development such as work and cooperation
processes in a company or repercussions and challenges of worldwide division of
labor. The didactical conception of the lecture combines the following elements:
Based on the "Flipped Classroom" approach a participatory learning curve emerges
for the students. The course consists of two central elements. First, there is the "KOE-
Online-Lecture", a video recording of the traditional lecture. In order to prepare for
the "KOE-Discussion-Forum" they are able to watch the "KOE-Online-Lecture"
in advance. The "KOE-Discussion-Forum" constitutes the second element; it takes
place every week and provides the opportunity to discuss open questions or to hold
lively debates. In practical-oriented tutorials for large groups the learning contents
can be reflected and deepened. The plenary "KOE-Discussion-Forum" addresses
several learning-types with different didactical concepts and demonstrates various
ways of studying (Flechsig, K.-H. 1996, Kleines Handbuch didaktischer Modelle,
Neuland). In the context of this paper the didactical concept of "KOE", which hosts
over 1,400 students, is described in detail in its complex methodological structure.

L. Köttgen (✉) · S. Schröder · E. Borowski · A. Richert · I. Isenhardt
IMA/ZLW & IfU, RWTH Aachen University,
Dennewartstr. 27, 52068 Aachen, Germany
e-mail: larissa.koettgen@ima-zlw-ifu.rwth-aachen.de

Originally published in "INTED2014 Proceedings",
© IATED 2014. Reprint by Springer International
Publishing Switzerland 2016, DOI 10.1007/978-3-319-42620-4_26

Future efforts for a continuous optimization of the concept and an adaption of the target audience are delineated.

Keywords Flipped Classroom · Excellent Teaching · Didactical Method-Mix

1 Excellent Teaching Through a Method-Mix – Introduction

"Students have been conditioned to take information and regurgitate facts on tests. We all know that learning is much deeper than only filling in facts" [1].

In order to provide a differentiated learning experience for the students, the Center for Learning and Knowledge Management (ZLW) at RWTH Aachen University employs a didactical method mix consisting of a supporting medial documentation, practical examples, using the 'Audience Response System' and the exchange- and data-platform L^2P, the laboratory tutorial plus the Flipped Classroom approach in context of the course KOE [2]. The aspiration is to form responsible and independent students who are interested in investing in their own learning [3–5]. The main objectives are to support them in achieving individual learning success and to ensure that they are able to make informed choices regarding their own education. The "shift from teaching to learning" [6] demonstrates the current research focus in higher education. The statements in this paper will illustrate that not only the teachers' role in lectures is changing. This shift also affects the students' role essentially [7]. In this context a new culture of learning develops. One key aspect of that new culture is that students start identifying self-reliant learning as their own goal [8]. The two cofounders of the Flipped Classroom approach, Jonathan Bergmann and Aaron Sams outline this new situation in the following terms.

"Class time is a learning experience for the student, not a download and upload of knowledge" [9]. They still use a technical language for their description, "downloading" and "uploading" of knowledge. This technical perspective already indicates another important fact, which cannot be ignored: the continuing technological progress. Bergmann and Sams know about these changing conditions and about complications as a consequence. "Today's students grew up with Internet access, Youtube, Facebook, MySpace and a host of other digital resources" [10]. Lecturers have the assignment to become supporters, who recognize these new conditions and observe them in their course conception [7]. An excellent higher education has to be determined by these requirements. Bergmann and Sams advocate an implementation of the Flipped Classroom approach in their following statement: "The students understand digital learning. To them, all we are doing is speaking their language" [9].

This paper presents the current design of the KOE lecture with its didactical concept based on a mix of different methods and the Flipped Classroom concept on top. This special method-mix takes account of the previously documented rationales.

Following the Flipped Classroom mentality facilitates the goal that lecturers and students start to create a place for efficient learning together.

2 The Object of Application: The Course "Kommunikation und Organisationsentwicklung" (KOE)

The lecture 'KOE' addresses basic knowledge in the context of communication and organizational development. It also includes learning- and knowledge management concepts and intercultural aspects of global work division management as well as system-theoretical approaches and practical inputs by industry experts. Altogether, the lecture consists of twelve modules. Each course ties in with practical lectures. Thus, an ideal transfer from theoretical knowledge impartation to practical relevance is established. Thereby, communication and organization are regarded as the most important requirements for the development of interaction between humans, technology and organization on the different organizational levels (e. g. management-, division- and employee-level). Until the last winter-term in 2012 the lecture 'KOE' has been working on the following methods (Figure 1).

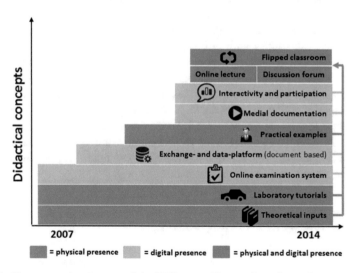

Figure 1 The concept development of the KOE course (Source: Own diagram)

2.1 Medial Documentation

The today's students are digital natives [10]. This fact should be useful promoted by using a medial documentation in KOE course [11]. The entire lecture was professionally recorded, edited and afterwards provided to the students via the online teaching and e-learning platform called "learning to people" (L^2P of RWTH Aachen University) (cf. Section 2.4) to support the learning process in auto-didactical-phases. By the aid of a hyperlinked directory modules and even chapters can be retrieved individually. Knowledge is thereby no longer appropriated as a lecturer-reserve, but accessible for students on demand [12, 13]. This online material was always presented to the students one week after the traditional lecture. They used it to rework the performed lessons and to prepare for the exam (cf. Section 3.3.1: the concept of learning on demand).

2.2 Practical Examples

The lecture forces the connection of practice applications with theoretical teaching. Additionally, it also aims at increasing the ability of understanding and coping with the complexity of working life. The inclusion of on-professional competences in studies is not only an issue since the Bologna reform [14]. Furthermore, it is important to connect the studies with professional practice, not only from the student's perspective, but also from the perspective of company representatives [15]. For this reason, it is more important to connect teaching with practical insights [14]. Based on this, guest speakers with professional industrial background give an insight into their companies and working experiences (e. g. Vodafone Group, Capgemini, p3 group, inside group). In this way, theories are linked to practical working life and relevance is outlined. In order to solve the area of conflict between the teaching of basic sciences and practical teaching at the expense of theoretical essentials [16] the distribution of proper contents and basics is kept reasonable.

2.3 Interactivity and Participation

As a reaction to the changed student behavior [11, 17–19], (new) technologies e. g. in the form of 'Audience Response Systems' (ARS) are implemented to improve the learning outcomes of students [17]. Due to the application of an ARS (questions are set by the lecturer and students are able to vote directly online using i. e. a smartphone or laptop), students are further involved in the education process. Through ARS students are able to interact with the lecture [18]. The use of ARS offers students quite diverse possibilities to participate in a lecture, which means that they can choose particular contents, recap their knowledge, interpret, reflect and prepare for examinations. Taxonomies of learning goals, like knowledge, understanding, application

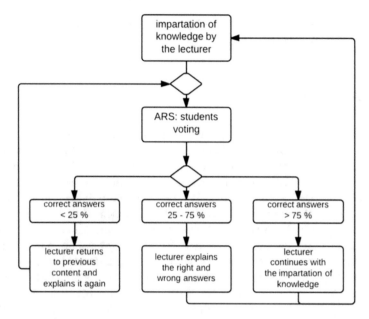

Figure 2 Didactical concept of using the ARS [19]

and analyzing [19, 20] are addressed holistically by the use of knowledge and comprehension checks. This variety of questions ranges from multiple-choice questions to the inquiry of calculation results etc. [21]. As a result problem-solving skills can be improved. The barriers for participation are marginal. Merely an end device with internet access is required [22].

Figure 2 illustrates the didactical concept of the Audience Response System, which is treated in the 'KOE'. Benefits of this way of proceeding are i. e. iteration-loops, to repeat misunderstood content and increase the learning-success. Repetitions are taken into account if there are less than 25 % correct answers. If the range of correct answers is between 25 and 75 % an explanation on the right answer is provided. Ensured that the content is understood (>75 % correct answers), the lecturer switches to the next topic.

2.4   Exchange- and Data-Platform

As a reaction to the various influencing factors like the new technical possibilities, the teaching and learning platform L^2P is used since the end of 2007. The learning platform is password secured and is only accessible by registered students. This platform is implemented in order to provide important contents like teaching materials (e. g. lecture notes, videos and presentations of experts) and (further) literature for the students on a web based server to support time-consuming and complex teaching and learning processes [23, 24].

2.5 Laboratory Tutorial

The theoretical conveyance of on-professional competences to students is challenging within extensive lectures. It is also impossible to teach all contents suitable within XL-Classes [25]. Therefore, parallel to the lecture 'KOE' a laboratory tutorial of one and a half days takes place. The laboratory tutorial is based on the concept of 'simulation-based learning' which is a research and training method that tries to create a realistic experience in a controlled environment [26]. According to this, simulation replaces or boosts real experiences [27] and thus offers enormous advantages in mediating knowledge long-acting [27, 28]. In groups of up to 40 students a foundation of a fictitious automotive company with different branches is simulated. Target systems and various strategies are developed, communication ways are defined and coordinated as well as key skills such as team building, time management and project management are trained. An innovative vehicle is constructed under the guidance of 40 professional coaches. Previously learned theoretical basics are applied practically during this company simulation in order to obtain first practical experiences in organizational communication and working processes.

2.6 Online Examination System

Since 2007 a digital online examination system is used to deal efficiently and content orientated with 1,400 students. This system is called OPS (Online-Prüfungssystem) and is externally developed especially for the lecture 'KOE' [29, 30]. Examination questions are derived from the lectures contents and distinguished in different taxonomy levels (degrees of knowledge reprocessing) [31]. Thereby, the whole taxonomy-spectrum by Bloom [32] on a cognitive base is addressed to the students. The taxonomy levels are: knowledge, comprehension, usage, analysis, synthesis and evaluation. For the preparation of the exam questions it is important to take into account that the taxonomy levels are hierarchically arranged [33]. Hence, the mediated competences for each taxonomy level must gain a specific manifestation before they can be applied to the next taxonomy level [34]. For example, without the knowledge the use of taught content (transfer capacity) is impossible [34]. Supported by a holistic handling of the mentioned difficulty degrees, the OPS enables to retrieve the student levels of awareness.

3 Flipped Classroom on Top

One of the overarching goals is to provide a long-lasting way of excellent academic teaching. In order to reach this objective it is necessary to offer a didactical concept that considers actual challenges and current trends in higher education. The success of combining different didactical concepts is already proven [35, 36]. Choosing the mentioned didactical concepts in chapter two is based on additional conveying various didactical methods to the students. Thus, they obtain, besides lecture content, working life experiences as e. g. presentation techniques and abilities to work in teams. In September 2013 the Flipped Classroom approach was included into the previous didactical concept for the first time. Before the reasons for using the Flipped Classroom concept are described in detail, an explanation of how the approach works will be given in a short excursus.

3.1 What Is the Flipped Classroom Approach

Basically, there is no clear definition of the concept in general. In his paper "Flipping 2.0" Jason Bretzmann explains "The beauty of flipping is it allows you more options, not just your students" [1]. The Flipped Classroom approach follows an ideological idea of excellent academic teaching and does not require the strict observance of a specific methodology. As related to the existing method-mix it is a perfect addition without any dictated set of rules or prototype to fit in [37]. Giving this quote should clarify that there is a possibility to use the concept in various ways. Comparing different implementations of the concept indicate that the majority employs video sequences to impart learning content. The attendance of the lecture enables the possibility to work together on different tutorials. The students have different options to benefit from this lecture time. They can solve and discuss problems they have with the learning content, they have the possibility to learn about various case studies and finally they have time to exchange and discuss ideas with other students [1]. A functional realization of this concept should consider the special focus orientation on the students.

3.2 Benefits of the Flipped Classroom Approach

The utilization of Flipped Classroom should guarantee a student-centered course [38]. "Students become stakeholders and active participants rather than passive buckets waiting to be filled" [1]. The approach concentrates on moving responsibility and workload from teacher to the student. In this context the traditional roles of teachers are changed. Students are supposed to learn on their own responsibility. This leads to the establishment of a new, mainly supporting, role of the teacher.

A decision to make use of this approach was made because of its open and individual practicability. "The flipped classroom is a simple concept that needs no title. Good teaching, regardless of discipline, should always limit passive transfer of knowledge in class, and promote learning environments built on the tenants of inquiry, collaboration and critical thinking" [39]. Students need an environment for different ways of learning. On the one hand they should work together with fellow students and on the other hand they should be able to take the responsibility for their own learning. The vision of perfection is determined by a course "where all students are engaged in their learning" [40].

Due to this new and innovative way of teaching the previous face-to-face interaction is optimized and offers new opportunities of an active exchange between lecturers and students but also students themselves. "Class becomes the place to work through problems, advance concepts, and engage in collaborative learning. Most importantly, all aspects of instruction can be rethought to best maximize the scarcest learning resource-time" [40].

As already mentioned, the new concept of the lecture itself generates various benefits. In addition to these advantages students benefit from a new way of consuming the learning content via online tools. They can integrate "doing homework" into their individual and changing timetable. According to the fact that all relevant documents are available via an online platform, students are able to access them at any time anywhere, as long as they dispose of an internet connection. Further benefits will be mentioned in the following Section 3.3.1.

3.3 The Implementation of Flipped Classroom in KOE

The implementation of Flipped Classroom in KOE includes different steps and requires various substructures and preparations, which are described in the following segments.

3.3.1 Preparing the Interaction: The KOE-Online-Lecture

According to the new didactical concept of "Flipped Classroom", the theoretical inputs are no longer presented within the scope of the attendance-session. The existing medial documentation (cf. Section 2.1) is shifted and converted into a Flipped Classroom compatible video documentation. The theoretical content is necessary and grounding for the KOE-Discussion-Forum. Content can be recapitulated without any limits and at any time. This form of making content available for students resembles the concept of learning on demand [32] and results in time saving potential for student learning [41]. This way offers the possibility to review important contents and especially the preparation of exam papers could be done in a more efficient way. The result is an approach to the teaching for "individual learning", because students can decide when, where and how fast they organize their learning process [42].

3.3.2   How to Use the KOE-Online-Lecture?

Since the KOE-Online-Lecture can be found on the RWTH e-learning-platform L^2P only signed in students can access the lecture related video documents. All files are exclusively available via a hyperlink. The user interface which presents the learning contents consists of three different elements. There is a Power Point slide show, which has been presented in last years' traditional lecture accompanied by last years' video-record of the lectures' presentation. These materials are structured by a table of contents, which helps to navigate between the different short video sequences. Figure 3 shows one of the various Layout possibilities consuming the KOE-Online-Lecture. The user can decide, whether he would like to see the Power Point slides on the bigger window and the video-record on the smaller window of the split screen or vice versa.

The KOE-Online-Lecture is available to the students one week prior the date of the related KOE-Discussion Forum till the end of the term. Since the students cannot save the KOE-Online-Lecture to their computers, they can download a pdf file with the Power Point slides from the e-learning-platform. This combination of supporting documents offers the opportunity to watch the KOE-Online-Lecture and take notes on the printed Power Point slides in parallel.

Consuming via video lecture "is not a big deal for today's students" [9]. They are usually digital natives and own a computer with internet connection in the majority of cases [11]. But it is possible that there is a little number of participants without this access to a computer with internet at home. To engage every student it is necessary

Figure 3   Screenshot KOE-Online-Lecture

to ensure, that every attendant has the possibility to watch the KOE-Online-Lecture. All participating students have the chance to use available computers in the Zuselab (Cip-Pool operated by the Center for Learning and Knowledge Management).

3.3.3 Active Learning Time with the Students: The KOE-Discussion-Forum

"Discussion is one of the oldest methods of instruction, and seems to be one that does not need the improvement of technology" [1]. During the conception of the new Flipped Classroom-influenced "KOE" the designation of the new "lecture time with the students" was a central topic. The new label should emphasize that it is not anymore a traditional lecture. The interaction between lecturer and student, between the students themselves and especially the planned discussions were focused. The combination with the term "forum" was conscious selected. Its meaning stands for a real or virtual place where opinions can be exchanged, questions can be asked and answered and for an open space for human interaction. Further comprehension questions can be submitted to a special email-contact as well as on the e-learning-platform. All requests are collected weekly and explained at the KOE-Discussion-Forum.

The KOE-Discussion-Forum should create such an open space. Handling with this high degree of creative development, some rules become necessary. "Students are responsible for viewing the videos and asking appropriate questions" [9]. Attending the KOE-Discussion-Forum is voluntary, those who visit the forum are supposed to watch the KOE-Online-Lecture and be prepared. The forum follows a roughly defined schedule. Every course starts with remembering "today's topics". The plenum starts with brainstorming aspects of the KOE-Online-Lecture which can be remembered by students. This procedure works like an activating method (breaking the ice). They gather answers: What have been taught in the video sequences? After creating a mind map of memories, every main topic gets successive addressed. Sometimes students' questions encourage a discussion, sometimes the lecturer presents a case study. Based on the method mix (e. g. the guest speakers and the application of the Audience Response System) the KOE-Discussion-Forum is presented as an exciting and diversified event.

Inputs for the discussion can be very divergent. Especially, the practical exercises are fascinating because of the group size. The KOE course hosts over 1,400 students, in the KOE-Discussion-Forum are ordinarily a few hundreds. Although handling such a big group poses a challenge to moderators the discussions were conducted successfully. Figure 4 shows how the students solve one of the exercises in the plenum. They receive selected terms chosen from a diagram and bring them in the right image arrangement. Those students without an own term paper should help their fellow students. After finishing the composition every student with a sheet containing one of the basic expressions must justify his choice of position in a few words. Conclusive the diagram is discussed by the whole group. The lecturer accompanies

Figure 4   Impression from the KOE-Discussion-Forum

this process. "Teachers become guides to understanding rather than dispensers of facts and students become active learners rather than receptacles of information" [37]. Students become the central actors in such a KOE-Discussion-Plenum.

4   Results of the KOE-Discussion-Forum Evaluation

All courses at RWTH Aachen University get evaluated through a partial standardized quantitative elicitation. This survey focuses on evaluating the course itself and the performance of the lecturer. The questionnaire presents statements and students can agree or disagree on a scale from one (agree) to five (disagree). The results of this evaluation yield that the KOE-Discussion-Forum has a clear recognisable structure and well defined learning targets (in detail 2,6 and 2,2 in arithmetic average). Nearly 100 % of the interviewed students approve that the KOE-Online-Lecture and the KOE-Discussion-Forum are geared to each other. The students perceive the difficulty level of the forum with nearby 80 % as appropriate.

Referring to the lecturers position the evaluation presents that a well done preparation is necessary and will be noticed by the students. More than 95 % of the students agree with the statement that the lecturer was good prepared (value 1,7 in detail). Another important requirement deals with handling comprehension questions. Once again almost 95 % of the samples confirm that the lecturer refers to the students' questions (likewise value 1,7 in detail). Beside these mentioned facts the assignment of the suitable mediums for improving the students' comprehension is a relevant factor for a successful KOE-Discussion-Forum. Representing a value of 1,8 interviewed students share the opinion that the lecture deploys the right medial resources for a great comprehension support.

Compared to the last years' traditional KOE lecture, the results of the actual evaluation are far better. For figuring out the reasons of this progress new data will be collected after the current KOE course period.

5 Conclusion and Prospects: What Will Come Next?

For developing an excellent academic teaching concept it must be clear that movement is an important fact according to the motto "The journey is its own reward". Involved teachers are always searching for a blend of tools and methods to fit the needs of well-organized course lessons [1]. The statements given in this paper illustrate that for the existing KOE method-mix the decision to extend the concept by the Flipped Classroom approach matches.

The description of the KOE-Discussion-Forum illustrates a case that already Confucius noticed: "I hear and I forget. I see and I remember. I do and I understand" [43]. Producing a comprehensive learning rests on action. The practical tutorials in the Forum make this required acting possible. The students have a space where they can "do" and "understand".

This observation is also supported by the demonstrated evaluation results. The whole concept including the mentioned method-mix with Flipped Classroom on top received a positive feedback (some selected individual results have been exemplified in Section 4). Considering the results of an additional evaluation, the Flipped Classroom approach gets adapted and will be continued afterwards. The benefits of the actual method-mix will become transparent by the additional evaluation. Furthermore, the comparison of the last years' and actual exam results should give informative indication of the method-efficiency.

Under the self-imposed paradigm of a continuous further development of the lecture 'KOE' in order to adopt the students demands and increase the learning process to an application-oriented usage for later working life other concepts are supposed to be implemented. By way of example the exchange and data platform L^2P can be used to discuss alongside of the presence-time (misunderstood, relevant for exams or/and hardly understandable) content. Questions can be asked by the students, the lecturer functions as a moderator, who gives right explanations and routes contributions within a discussion. Furthermore, the running time of video documentation is discussed [44]. The maximum length is supposed to be 4 min, rather less. Therefore, additional video clips will be prepared to impart particularly appreciable knowledge in a student demand-actuated and comprising way.

References

1. J. Bergmann, J. Overmeyer, B. Willie, The Flipped Class: Myths vs. Reality. The Daily Riff, 2013
2. R. Perkins, C. Camel, An Evaluation of the Flipped Classroom. Tech. Rep. 8/1/2011, EDTECH 505, 2011
3. C. Heine, Studienqualität nach Bologna aus Studierendensicht. Wahrnehmung und Bewertung von Studienbedingungen und Praxisbezug. Nach Bologna: Praktika im Studium – Pflicht oder Kür. In: *Empirische Analysen und Empfehlungen für die Hochschulpraxis*, Potsdam, 2011, pp. 45–78
4. T. Bargel, M. Ramm, F. Multrus, Studiensituation und studentische Orientierungen. 10. Studierendensurvey an Universitäten und Fachhochschulen. Tech. rep., Bundesministerium für Bildung und Forschung (BMBF), Bonn, Berlin, 2008
5. T. Bargel, F. Multrus, M. Ramm, H. Bargel, Bachelor-studierende: Erfahrungen in Studium und Lehre – eine Zwischenbilanz. Tech. rep., Bundesministerium für Bildung und Forschung (BMBF), Bonn, Berlin, 2009
6. B. Szczyrba, M. Wiemer, Lehrinnovation durch doppelten Perspektivenwechsel–Fachkulturell tradierte Lehrpraktiken und Hochschuldidaktik im Kontakt. In: *Fachbezogene und fachübergreifende Hochschuldidaktik*, W. Bertelsmann Verlag, Bielefeld, 2011, pp. 101–110
7. P. Wolters, J. Stieger, E. Hauck, S. Jeschke, Planspiel gestütztes Lernen—Ein Konzept zur Unterstützung vernetzten Denkens in universitären Lehrveranstaltungen. In: *Kybernetik und Wissensgemeinschaft Vernetztes Denken, geteiltes Wissen—auf dem Weg in die Wissensökonomie. Konferenz für Wirtschafts- und Sozialkybernetik (KyWi 2011), Stuttgart, Germany, 7-8 July 2011, Wirtschaftskybernetik und Systemanalyse*, vol. 28, ed. by GWS. Duncker & Humblot, 2011, *Wirtschaftskybernetik und Systemanalyse*, vol. 28
8. J. Bergmann, A. Sams, How the Flipped Classroom is Radically Transforming. The Daily Riff, 2012
9. J. Bergmann, A. Sams, *Flip your Classroom. Reach Every Student in Every Class Every Day*. International Society for Technology in Education, 2012
10. J. Palfrey, U. Gasser, *Born Digital: Understanding the First Generation of Digital Natives*. Basic Books, 2008
11. M. Prensky, Digital Natives, Digital Immigrants Part 1. On the Horizon **9** (5), 2001
12. H.M. Niegemann, S. Hessel, D. Hochscheid-Mauel, K. Aslanski, M. Deimann, G. Kreuzberger, *Kompendium E-Learning*. Springer, Berlin, 2004
13. R. Tozman, *Learning On Demand: How the Evolution of the Web Is Shaping the Future of Learning*. ASTD, 2012
14. M. Oechsle, G. Hessler, Praxis einbeziehen - Berufsorientierung und Studium. HDS Journal—Perspektiven guter Lehre **2** (2010), 2010, pp. 11–22
15. M. Winter, Praxis des Studierens und Praxisbezug im Studium. Ausgewählte Befunde der Hochschulforschung zum „neuen" und „alten" Studieren. In: *Nach Bologna: Praktika im Studium - Pflicht oder Kür? : empirische Analysen und Empfehlungen für die Hochschulpraxis*, ed. by W. Schubarth, K. Speck, A. Seidel, Postdam, 2011, pp. 7–44
16. E.J.E. Wegner, M. Nückles, Die Wirkung hochschuldidaktischer Weiterbildung auf den Umgang mit widersprüchlichen Handlungsanforderungen. Zeitschrift für Hochschulentwicklung **6** (3), 2011, pp. 171–188
17. A. Böss-Ostendorf, H. Senft, *Einführung in die Hochschul-Lehre: Ein Didaktik-Coach*. Regensburg, 2010
18. M. Kerres, *Mediendidaktik: Konzeption und Entwicklung mediengestützter Lernangebote*. München, 2012
19. I. Leisten, U. Bach, F. Hees, Everyone wants them - we enable them: communicative engineers. In: *Proceedings of the International Conference on Learning & Teaching (TIC). Enhancing Learning & Teaching in Higher Education*. 2008

20. M. Möhrle, Qualitätsverbesserung interaktiver Lehre durch das Lead-Learner-Konzept. In: *Hochschuldidaktik und Hochschulökonomie. Neue Konzepte und Erfahrungen,* ed. by H. Albach, P. Mertens, Wiesbaden, 1994, pp. 41–52

21. L.W. Anderson, ed., *A Taxonomy for Learning, Teaching, and Assessing: A Revision of Bloom's Taxonomy of Educational Objectives.* New York, 2000

22. V. Stehling, U. Bach, A. Richert, S. Jeschke, Teaching Professional Knowledge to XL-Classes with the Help of Digital Technologies. In: *Proceedings of the ProPEL International Conference, Stirling, UK, 9-11 May 2012.* 2012, p. 55

23. J. Handke, A.M. Schäfer, *E-Learning, E-Teaching und E-Assessment in der Hochschullehre: Eine Anleitung.* München, 2012

24. A.L. Dyckhoff, P. Rohde, P. Stalljohann, An Integrated Web-based Exercise Module. In: *Proceedings of the 11th IASTED International Conference on Computers and Advanced Technology in Education (CATE 2008).* ACTA Press, 2008

25. I. Leisten, S. Brall, F. Hees, Fostering Entrepreneurship in Engineering Education at RWTH Aachen University. In: *Proceedings of the International Conference on Global Cooperation in Engineering Education: Innovative Technologies, Studies and Professional Development, 4-6 October, 2007, Kaunas, Lithuania.* 2007

26. P. Mack, *Understanding Simulation-Based Learning.* SGH-Life Support Training Centre, Singapore, 2009

27. D.M. Gaba, The future vision of simulation in health care. Quality & Safety in Health Care **13 Suppl 1,** 2004, pp. i2–10

28. J. Castronova, Discovery learning for the 21st century: what is it and how does it compare to traditional learning in effectiveness in the 21st century. Literature Reviews, Action Research Exchange **1** (2), 2002

29. F. Hees, A. Hermanns, A. Huson, Prüfungserstellung mit Total Quality Management (TQM). In: *Prüfungen auf die Agenda! Hochschuldidaktische Perspektiven auf Reformen im Prüfungswesen, Blickpunkt Hochschuldidaktik,* vol. 118, W.Bertelsmann Verlag, 2008, pp. 129–141

30. P. Blum, *Ein inter-/intranet basiertes System zur Erstellung, Durchführung und automatisierten Bewertung von dynamischen Leistungstest in der medizinischen Lehre.* 2005

31. B.S. Bloom, *Taxonomie von Lernzielen im kognitiven Bereich.* Weinheim, 1976

32. T. Reglin, C. Speck, Zur Kosten-Nutzen-Analyse von eLearning. In: *VBM e.V., Leitfaden E-Learning,* München, 2003, pp. 221–235

33. W. Sitte, H. Wohlschlägl, *Beiträge zur Didaktik des "Geographie und Wirtschaftskunde" - Unterrichts,* 4th edn. Wien, 2006

34. F. Bruckmann, O. Reis, M. Scheidler, eds., *Kompetenzorientierte Lehre in der Theologie: Konkretion - Reflexion - Perspektiven.* Berlin, 2011

35. J. Bretzmann, *Flipping 2.0: Practical Strategies for Flipping Your Class.* Bretzmann Group LLC, 2013

36. S. Schroder, D. Janssen, I. Leisten, R. Vossen, I. Isenhardt, On-professional competences in engineering education for XL-Classes. In: *2013 IEEE Frontiers in Education Conference.* 2013, pp. 29–34

37. B. Bennett, D. Spencer, J. Bergmann, T. Cockrum, R. Musallam, A. Sams, K. Fisch, J. Overmyer, The flip class manifest. The Daily Riff, 2011

38. G. Gibbs, M. Coffey, The Impact of Training on University Teachers' Approaches to Teaching and on the Way their Students Learn. Das Hochschulwesen **50** (2), 2002, pp. 50–54

39. R. Musallam, Should You Flip Your Classroom? Edutopia, 2011

40. B. Tucker, The Flipped Classroom. Education Next **12** (1), 2012

41. S. Seufert, D. Euler, *Nachhaltigkeit von eLearning-Innovationen: Ergebnisse einer Delphi-Studie.* St. Gallen, 2004

42. P.D.O.K. Ferstl, D.K. Schmitz, Integrierte Lernumgebungen für virtuelle Hochschulen. Wirtschaftsinformatik **43** (1), 2001, pp. 13–22

43. Confucius, *Chinese philosopher & reformer (551 BC - 479 BC)*

44. A. Amresh, A.R. Carberry, J. Femiani, Evaluating the effectiveness of flipped classrooms for teaching CS1. In: *Proceedings of the IEEE Frontiers in Education Conference (FIE 2013).* 2013, pp. 733–735

Integrating Blended Learning – On the Way to an Excellent Didactical Method-Mix for Engineering Education

Larissa Köttgen, Stephanie Winter, Stefan Schröder, Anja Richert and Ingrid Isenhardt

Abstract University education has changed vastly in the last years. Constantly growing student numbers, the necessity of on-professional competences especially in engineering education and new technical possibilities have influenced the development of new didactical concepts. To cope with these circumstances a didactical set, which combines as many benefits as possible from different didactical methods would be preferred. Therefore, different field tested as well as tailored didactical elements are combined and integrated with the objective to enhance the quality of engineering education. This new approach is applied within the lecture Communication and Organisation Development (KOE), which is frequented by up to 1,400 first semester students of the faculty of mechanical engineering. The lecture is an inherent part of the bachelor-curriculum of mechanical engineering to fulfill the RWTH Aachen University's objective to foster on-professional competences from the very beginning of a student's career. After defining goals for restructuring the lecture from a blended learning perspective, the original and content-orientated lecture was redesigned in the winter term 2012/2013 and enhanced by implementing the flipped classroom concept with a discussion forum in the winter term 2013/2014. Students can access relevant learning documents via an online platform and can make use of the discussion forum in terms of a participatory interaction event. This method-mix is based on the cube model developed by Baumgartner and Payr (1996), which combines different perspectives and levels of a diverse teaching approach. The *KOE* lecture was evaluated twice (beside the common university-wide evaluation for lectures, the blended learning approach was assessed explicitly) and practical oriented elements were rated very positively. Making use of the cube model as well as the results of the evaluation led to an identification of further demands for optimisation. Following these demands, the related lab tutorial starting in the upcoming winter term will be redesigned. Students are supposed to work in a fictional mechanical engineering enterprise and different duty cycles shall be simulated. The overall plan is to construct a production robot using Lego Mindstorms. This paper – based on the applied cube model – describes the current didactical method-mix, presents

L. Köttgen (✉) · S. Winter · S. Schröder · A. Richert · I. Isenhardt
IMA/ZLW & IfU, RWTH Aachen University, Dennewartstr. 27,
52068 Aachen, Germany
e-mail: larissa.koettgen@ima-zlw-ifu.rwth-aachen.de

Originally published in "ICERI2014 Proceedings",
© IATED 2014. Reprint by Springer International Publishing Switzerland 2016,
DOI 10.1007/978-3-319-42620-4_27

339

demands identified by recent evaluation and introduces developments and optimizations following the blended learning approach for engineering education.

Keywords Engineering Education · Methodical and Didactical Concepts · Combination Approach · Active Learning · On-Professional Competences · Blended Learning · Flipped Classroom

1 Introduction

In response to new academic circumstances such as rising student numbers, the demand of on-professional competences, and the development of new technical possibilities, the lecture *KOE*, (Communication and Organisation Development) organized by the Center for Learning and Knowledge Management (ZLW) of the RWTH Aachen University and lectured by Professor Ingrid Isenhardt was continual redesigned in recent terms. The objective of this enduring redesign is to increase the quality of engineering education constantly.

As part of strategy planning and justification, changing circumstances have to be observed. Many universities, and RWTH Aachen University in particular because of its engineering focus, face a high number of engineering students [1]. Correlating to that fact the demand for engineers in Germany is growing [2]. While most lectures are traditionally designed and often content-oriented, the Bologna declaration leads to varied teaching concepts and overarching contents [3]. In this context learning outcomes in terms of the way of achieving results are focused. This new student-centered and participation-oriented design of teaching is characterized by innovative methods. The aim is to promote learning by communication with teachers and other learners and to take students as active participants in their own learning seriously. Moreover, transferable skills such as problem-solving, critical and reflective thinking need to be fostered to meet the demands of the contemporary employment market [4].

Another requirement is the transmission of on-professional competences [5]. While the conveyance of on-professional competences to students is often neglected because of a content oriented focus in vast audiences, soft skills like method-, self-, organizational- and social competences are expected from today's students [6]. Consequently, a reasonably adjustment of theoretical and competence-based teaching is demanded [7].

Against this background and additionally supported by the Bologna Process, which is "to create a European space for higher education in order to enhance the employability [...]" [3] the redesign of *KOE* was conducted. Referring to Bologna and the learning pattern blended learning, the vision () is to educate students in a way suitable for the employment market while taking the right combination of teaching and learning types into account. Therefore, a fast, efficient and adequately educated manner [8] is focused. This implies the students' ability to convert learned knowledge in higher education and on-professional competences in their later working life [9].

Concepts and methods for an excellent higher education approach are combinable and applicable in various ways. Böss-Ostendorf and Senft [10] confirm that useful methodical diversity is a factor of success for university education because of the multiplex access to teaching content. More reasons for integrating and combining different didactical concepts are mentioned by Flechsig [11] as follows:

- Various learning styles and types of students with different learning success,
- Diversity of study motivation and interest,
- Variety of competences and fields of knowledge,
- Variety of context in which learning is placed.

Thus, the students' learning process is supported through didactical multi-angulation [12]. The combination needs to allow a holistic teaching approach which meets different levels of knowledge and presents a diverse learning environment to the students. Most of the models considered like the one by Kerres and de Witt [13] or another one by Schulmeister [14] focus on learning and teaching strategies for e-Learning arrangements [15]. They are not extensive and flexible enough for analyzing the lecture *KOE*. Consequently, another didactical model has to be chosen in this case.

One model that combines different perspectives and levels of a diverse teaching approach is developed by Baumgartner and Payr [16]. This cube model offers guidance for reflections about the combination of diverse didactical methods and learning materials. Consequently, the diversity of learning types and levels of knowledge can be considered in a structured way and the demand for a student-centered teaching is observed. Therefore, the choice of didactical concepts was guided by the compliance of the three dimensions presented in the cube model: learning content, learning goals and teaching strategies [16]. With the cube model and the interpretation of the evaluation results the educational situation of *KOE* could be analyzed and the conclusion of the alignment allows further optimizations. For this purpose, in the next chapters the elements of *KOE* will be presented in detail, starting with the presentation and the application of the cube model.

2 The Lecture *KOE*

The compulsory lecture *KOE* is held every winter term. Almost 1,400 students, mostly engineering students but also students of communication studies or sociologists, participate in this weekly lecture. It addresses basic knowledge of communication and organizational development as well as knowledge of management concepts, intercultural aspects of global work division management from the perspective of system theory. The content is structured in twelve modules and since a few winter terms presented by an extending combination of didactical concepts.

The choice of the various elements was guided by the demand to meet different levels of knowledge and learning types and to create a diverse learning environment (for further information: Köttgen et al. [17]). Moreover, the teaching should include students as active learners and teaching contents need to be presented in orientation towards the target group.

Thus the following didactical concepts were chosen (in order of implementation): theoretical inputs, laboratory tutorials, online examination system, exchange- and data-platform, practical examples, medial documentation, interactivity and participation. All of them are applicable in a discussion forum and an online lecture (following the flipped classroom approach). The concepts require the students' physical and digital presence as well as a combination of both (see Figure 1, the concept development of the lecture *KOE* visualized in a timeline).

3 Cube Model by Baumgartner and Payr and Its Application

3.1 The Cube Model by Baumgartner and Payr

The model by Baumgartner and Payr [16] combines the three dimensions learning goals, learning contents and teaching strategies to show a heuristic for defining educational situations. The dimension learning goals (from bottom to top, Figure 1) represents various types of interaction between the learner and his or her environment

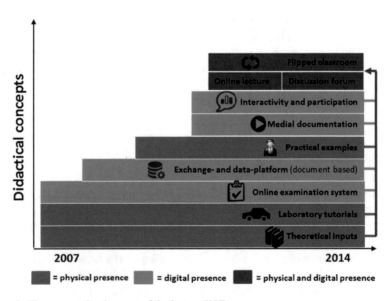

Figure 1 The concept development of the lecture *KOE*

that differs in complexity. The complexity of the learning goal ranges from perceiving, remembering to developing and acting and cover all levels of competence. Consequently, the whole taxonomy-spectrum by Bloom [18] is embedded (knowledge, comprehension, usage, analysis, synthesis and evaluation) [19]. The second dimension learning content is illustrated on a meta-level (from left to right, Figure 2) [16] and offers a detailed subject description. The learning contents are connected to the stages of a learning process from novice to expert. A novice for example is able to understand facts and context-free rules while an expert can manage complex situations. While the theoretical basis should be easy to understand for all students, their learning experience can go beyond by taking part in the additional didactical concepts. The third dimension of teaching strategies (from the front to the rear, Figure 2) refers to the role of the teacher [16]. He or she can firstly take the typical role of a teacher by explaining and demonstrating, secondly the role of a tutor by observing and correcting and thirdly the role of a coach by accompanying and participating.

These cube dimensions have been adapted to a part of the chosen didactical concept of the *KOE* lecture: the discussion forum, which is based on an online lecture (medially documented and following the flipped classroom approach). The theoretical inputs in the discussion forum are combined with practical examples, interactive and participative elements as well as an exchange- and data-platform. These concepts are broadened by a laboratory tutorial and an online examination system. The didactical concepts represent teaching techniques which have been implemented one by one and combined in the last years in education [7, 20].

Figure 2 The cube model – a heuristics for defining educational situations [16]

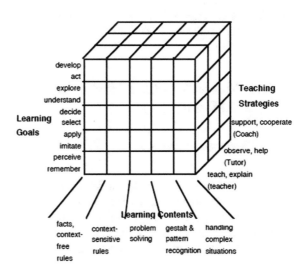

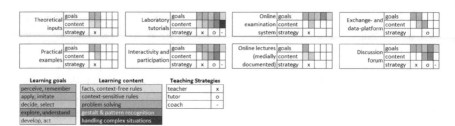

Figure 3 Didactical concept in context of the cube model

3.2 Application of the Cube Model

Each element of the didactical concept of the lecture *KOE* was chosen, because it can be allocated to at least one characteristic of each dimension, which is visualized in Figure 3. Moreover, the multi-angulation of concepts represents a compliance of all possible combinations of dimensions.

The colors represent the addressed learning goals. The range of greys visualize which learning content is taught within the teaching concept and the symbols, demonstrating the teaching strategy, show that each didactical concept covers different students' demands. Hence, the student's individuality regarding the level of competences is covered by a combination of didactical concepts. The results in detail of using the cube model for analyzing the lecture *KOE* (corresponding to the different elements of the lecture *KOE*) are described in the following chapter by presenting the elements of the lecture *KOE* in detail [17].

4 The Elements of the Didactical Concept of the Lecture *KOE* in Detail

4.1 Theoretical Inputs

As mentioned above the content of *KOE* comprises basics of communication and organization development as well as learning- and knowledge management, intercultural aspects of international division of labor and system-theoretical contents. Within this framework the importance of communication and organization as requirements for the development of interaction between humans, technique and organization on different organizational levels is highlighted. Thus, facts and context-free rules as well as context-sensitive rules are presented as the main learning content according to the cube model's dimension. By addressing on-professional competences theoretically and practically the students' employability can be advanced and one major goal of the Bologna Process is aspired to be accomplished [3]. The cube model analysis shows that this element serves as a basic learning content and learning goals level with the dominant teaching strategy "teacher". These inputs are part of the lectures content foundation.

4.2 Laboratory Tutorial

Due to the number of students, not all contents of *KOE* can be taught in a student-centred manner during the lecture [21]. Consequently, a laboratory tutorial was implemented parallel to the lecture and provides a realistic experience in a controlled environment. It follows a concept similar of 'simulation-based learning' [22] so that the simulation replaces real experiences on the one hand [23] and mediates long-acting knowledge on the other hand [23, 24].

Here students work in groups of max. 40 students during one and a half days and simulate a foundation of a fictitious automotive company with different departments. Each group is guided by professional coaches, who take the role of teachers, tutors and coaches so that all teaching strategies according to Baumgartner and Payr's cube model are combined. By means of a mixture of microteaching units for repeating and teaching basic knowledge and applying theoretically learned knowledge in practice, students can individually experience the importance of communication and development organization. The aim of the engineering students is to construct and present an innovative vehicle. Thereby, they need to work as a team and the communication between the different branches gives an insight to working life. Key skills like the ability to work in a team, time management and project management can be developed by this simulation based approach.

This element addresses all levels in all dimensions regarding the cube model analysis. And this result demonstrates that the laboratory tutorial constitutes an essential influence to a successful teaching experience in many different ways (as a teacher, tutor, coach, with the highest level of learning goals and all learning contents). Because of this, a contemporary redesign for the purpose of blended learning is planned for winter term 2014/2015 and will react to the various learning changes the laboratory tutorial enable. Hence, the students work in a mechanical engineering enterprise simulation game called ROBOFLEX. They will establish a fictional enterprise, calculate a marketing concept and finally construct a robot using Lego Mindstorms.

4.3 Online Examination System

The examinations of around 1,400 students face enormous challenges. The demand to retrieve the students' state of knowledge efficiently and content orientated, is met by the implementation of a digital online examination system (called OPS) since 2007 [25, 26]. It is externally developed and adjusted to the needs of *KOE*. By using

OPS, exams can be marked quickly and more objective. Moreover, the high amount of proofreading time and personnel effort can be reduced.

To receive meaningful information of the state of knowledge, the development of examination questions is guided by the presented theoretical and practical knowledge as well as the whole taxonomy-spectrum by Bloom [19]. This cognitive-based approach describes different taxonomy levels (degrees of knowledge reprocessing) which are hierarchically arranged [27]: knowledge, comprehension, usage, analysis, synthesis and evaluation. Therefore, the competences on a lower taxonomy-level must be achieved before the next taxonomy level can be reached [28] and all addressed learning goals according to the cube model are examined.

4.4 Exchange and Data-Platform

RWTH Aachen University provides the teaching and learning platform L^2P (learning to people) for students and teachers, which offers various functions and can be used inter alia for:

- Providing teaching materials (e.g. lecture notes, videos and presentations of experts) and literature,
- Conducting surveys,
- Simulating electronic tests,
- Making announcements,
- Discussing with professors, mates and/or coaches.

The students are tutored by the lecturer and can learn context-sensitive rules as well as problem-solving skills and contents (cf. Figure 3). The learning platform supports time-efficient and relieves complex teaching and learning processes [29, 30]. The students can learn, discuss and inform among themselves without limitations of time and place. The use of L^2P in *KOE* – for the first time in 2007 – fulfills the requests of a "web-enhanced course" according to Palloff and Pratt [31].

4.5 Practical Examples and Guest Lecturers

The combination of theory and professional practice allows a comprehensive understanding of the complexity of working life for the students. This demand is caused by the lack of practice-related content in higher education and the requested for on-professional competences, especially in engineering education [32]. By providing a company representative's perspective in the lecture, content is presented in different contexts and the relevance of the theoretical content in praxis is demonstrated.

Moreover, guest speakers with professional industrial background were invited to the lecture *KOE*. The guest speakers give an insight into their companies and working experience. Thereby, a connection between their theoretical and practical knowledge is adjusted [33]. These lectures were recorded and are available in digital form on the discussion forum. The practical application addresses the first two levels of learning goals perceive/remember and apply/imitate. Beyond that the evaluation of the practice orientated speeches was very positive (cf. Section 5).

4.6 Interactivity and Participation

For a further involvement of students in the learning and education process, various didactical tools can be used. As today's students "have changed radically", (new) technologies, like 'Audience Response Systems' (ARS) can support this objective [34].

Due to the fact that interactivity and participation can hardly be established in large student groups, ARS offer the possibility to meet this challenge. When closed questions with i. e. multiple-choice answers are presented, students only require an end device with access to the internet to participate and the cumulative answer of the group is presented immediately. Beside the direct interaction, students can check their learning success on all taxonomies according to Baumgartner and Payr's learning goals dimension (cf. Figure 3).

By this active learning approach the students are holistically addressed. Moreover, the teacher has the chance to get an impression of the level of knowledge and comprehension and can act appropriately.

ARS are used regularly since the winter term 2012/2013 in *KOE*. The questions aim for initiating discussions as well as comprehension checks. Depending on the outcome the teacher can react in different ways (cf. Figure 4): If more than 25 % of

Figure 4 Didactical concept of using the **Audience Response Systems** (ARS) [28]

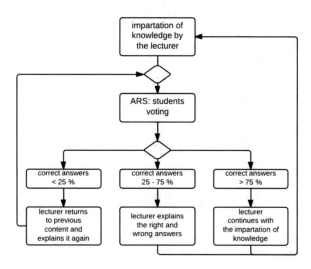

the given answers are wrong, the content has to be repeated. An explanation of the right and wrong answers has to be given between 25 and 75 % of correct answers. In case of more than 75 % correct answers, the content appears to be understood and the lecture continues. Beyond that the students are invited to look up the learning contents again and aid one another, if necessary.

4.7 Online Lecture and Medial Documentation

The online lecture is used as an exam preparation source for the students. The whole lecture was recorded and edited by a professional camera crew in the winter term 2012/2013. This medial documentation was provided to the students via the online teaching and learning platform L^2P of RWTH Aachen University (cf. Section 4.4) as a matter of a video documentation with various features. The chapters of each module can be individually retrieved. Moreover, presentation papers or manual sketches which were added by the lecturer during the explanation and animations are shown.

The use of different media channels offers the possibility to review important contents. Especially the preparation of both the discussion forum and the exam could be done in a more efficient way. By giving the students the opportunity to support the learning process with audiovisual aids, the theoretical content can be recapitulated without any limits and at any time. This form covers the requirements of the concept learning on demand [19].

4.8 Discussion Forum

In the winter term 2013/2014, the didactical approach was broadened by implementing a discussion forum on a weekly and voluntary basis. Supported by the lecturer, who assumes the role of a coach, students can enhance their recently learned knowledge with the help of the lecturer who assumes the role of a coach (cf. Figure 3). Students can work together on different tutorials, solve content related problems and exchange and discuss ideas with other students. The discussion forum is supported by various elements such as Audience Response Systems (cf. Section 4.6), case studies (cf. Section 4.5), practical exercises, and submitted comprehension questions, which were sent to a special email-contact as well as written on the e-learning-platform (cf. Section 4.4). Hence, the discussion forum represents a multifarious learning experience and allows a student-centered course [17].

5 Results of the *KOE* Evaluation

At RWTH Aachen University all lectures get evaluated through a partial standardized quantitative elicitation. This survey focuses on evaluating the performance of the

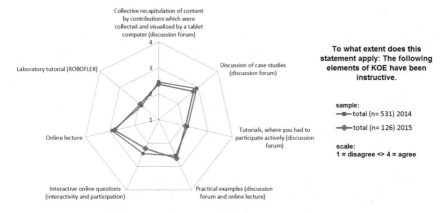

Figure 5 *KOE* evaluation result

lecturer and the conception of the lecture itself. The questionnaire presents statements and students can agree or disagree on a scale from one (agree) to five (disagree). In case of the *KOE* the in-class lecture, thus the discussion forum and the online lecture, have been evaluated (cf. Figure 5). The results of this evaluation yield that the *KOE* forum has a clear recognisable structure and well defined learning targets (in detail 2,6 and 2,2 in arithmetic average). Furthermore it is approved by nearly 100 % of the interviewed students that the *KOE* online lecture and the discussion forum are geared to each other. The students perceive the difficulty level of the discussion forum with nearby 80 % as appropriate.

An important finding refers to the lecturers' position and how he or she deals with comprehension questions. Almost 95 % of the samples confirm that the lecturer refers to the students' questions (likewise value 1,7 in detail). That the lecture deploys the right medial resources for a great comprehension support is represented by a value of 1,8 given by the interviewed students.

Besides the annual RWTH controlled evaluation, a second evaluation was accomplished by the *KOE* organization team for assessing the blended learning approach explicitly (Coding: 1 = disagree and 4 = agree). In this framework one of the questions inquires after the didacticism of the different *KOE* elements (cf. Figure 1). The response of the students covers with the recognition of the cube model analysis. All parts of the *KOE* in which the students get the opportunity to become an active part of the lecture receive a higher agreement (e. g. active tutorials, laboratory tutorial) than the passive elements (e. g. online lecture, practical examples).

6 Conclusion and Prospects: What Will Come Next?

The students' learning is supported by a student-centered and diverse educational process. The combination of didactical concepts follows the three dimensions of the cube model to address all levels of learning goals, learning contents and teaching strategies. An improvement of the lecture quality regarding to the combination of didactical concepts was investigated. To keep this quality, what new didactical elements and what kind of method-mix will come next? These questions need to be answered again and again over the following terms. Additionally, the method-mix itself needs to be evaluated consistently. In this context, interdependencies between the didactical concepts can be analyzed in detail.

Both characteristics need to be taken into account in the course of choosing and combining learning and teaching concepts. The importance of interaction and participation can also be underlined. Therefore existing concepts such as the data- and exchange platform L^2P will be used in a more interactive way by strengthening the discussion features of the platform and the implementation of the laboratory tutorial ROBOFLEX. Additionally and based on the positive evaluation results, the amount of practice oriented contents like case studies will be extended to enhance the employability of the students accessory.

In a next step, the theoretical attribution of the three cube model's dimensions to each didactical concept needs to be proven by an evaluation. Thus, learning and teaching gaps can be identified and the implementation of other didactical elements in the concept can be considered. Although the evaluation results show a positive development, potentials for optimization are given. Established as well as new concepts will be constantly developed and explored. This will be supported by a continuous monitoring of innovative teaching and learning concepts in higher education. Changing circumstances like e. g. student's preconditions and technical progress will underline this necessity.

Furthermore, the transferability of the multi-didactical concept of *KOE* needs to be further elaborated. This approach will be established in 2015 in *KOE* again, so that improvements can be implemented and results can be compared. The objective to "shift from teaching to learning" [35] and to motivate students in order to present contents only, will be further pursued.

References

1. R.A. University, Zahlenspiegel 2011. Tech. rep., 2012. Published brochure
2. VDI, Ingenieurmotor - der Arbeitsmarkt für Ingenieure im Februar 2013. Tech. rep., 2013. Published brochure
3. The Bologna Declaration. Tech. rep., 1999
4. T.E.S. Union, Student-centered learning – toolkit for students, staff and higher education institutions. Tech. rep., Brussels, 2010, pp. 5
5. VDI, VDI-Ingenieurstudie - Studie der VDI Wissensforum GmbH. Tech. rep., 2007. Published brochure
6. F. Pankow, *Die Studienreform zum Erfolg machen! Erwartungen der Wirtschaft an Hochschulabsolventen*. Deutscher Industrie- und Handelskammertag -DIHK-, Berlin, 2008
7. S. Schroder, D. Janssen, I. Leisten, R. Vossen, I. Isenhardt, On-professional competences in engineering education for XL-Classes. In: *2013 IEEE Frontiers in Education Conference*. 2013, pp. 29–34. 10.1109/FIE.2013.6684783

8. T. Jungmann, K. Müller, K. Schuster, Shift from TeachING to LearnING - Anforderungen an die Ingenieurausbildung in Deutschland. journal hochschuldidaktik **21** (2), 2010, pp. 6–8
9. N.V. Paetz, F. Ceylan, J. Fiehn, S. Schworm, C. Harteis, *Kompetenz in der Hochschuldidaktik. Ergebnisse einer Delphi-Studie über die Zukunft der Hochschullehre.* VS Verlag für Sozialwissenschaften, Wiesbaden, 2011
10. A. Böss-Ostendorf, H. Senft, *Einführung in die Hochschul-Lehre: Ein Didaktik-Coach.* UTB, Stuttgart, Opladen u.a., 2010
11. K.H. Flechsig, *Kleines Handbuch didaktischer Modelle.* Neuland, 1996
12. G. Macke, U. Hanke, P. Viehmann, *Hochschuldidaktik: Lehren - vortragen - prüfen - beraten. Mit Methodensammlung ≫Besser lehren≪, auch als Download.* Beltz, Weinheim/Basel, 2008
13. M. Kerres, C. de Witt, Zur (Neu-) Positionierung der Mediendidaktik: Handlungs- und Gestaltungsorientierung in der Medienpädagogik. In: *Medienbildung und Medienkompetenz. Beiträge zu Schlüsselbegriffen der Medienpädagogik*, ed. by H. Moser, P. Grell, H. Niesyto, KoPaed, München, 2011, pp. 239–249
14. R. Schulmeister, *Grundlagen hypermedialer Lernsysteme: Theorie – Didaktik – Design*, 4th edn. Oldenbourg, München/Wien, 2007
15. S. Seufert, D. Euler, *Learning Design: Gestaltung eLearning-gestützter Lernumgebungen in Hochschule und Unternehmen.* SCIL, Swiss Centre for Innovations in Learning, St. Gallen, 2005
16. P. Baumgärtner, S. Payr, Learning as action: A social science approach to the evaluation of interactive media. In: *In Proceedings of ED-MEDIA 96 - World Conference on Educational Multimedia and Hypermedia.* AACE, Charlottesville, 1996, pp. 31–37
17. L. Koettgen, S. Schröder, E. Borowski, A. Richert, I. Isenhardt, Flipped Classroom on Top - Excellent Teaching Through a Method-Mix. In: *Proceedings of the 8th International Technology, Education and Development Conference (INTED 2014), Valencia, Spain, 10-12 March 2014.* IATED, 2014, pp. 40–49
18. S. Bloom, B. D. Engelhart, M. J. Furst, E. H. Hill, W. R. Krathwohl, D. *Taxonomy of educational objectives: The classification of educational goals.* David McKay Company, New York, 1956
19. T. Reglin, C. Speck, Zur Kosten-Nutzen-Analyse von eLearning. In: *VBM e.V., Leitfaden E-Learning*, München, 2003, pp. 221–235
20. R.H. Kay, A. LeSage, Examining the benefits and challenges of using audience response systems: A review of the literature. Computers & Education **53**
21. I. Leisten, S. Brall, F. Hees, Fostering Entrepreneurship in Engineering Education at RWTH Aachen University. In: *Proceedings of the International Conference on Global Cooperation in Engineering Education: Innovative Technologies, Studies and Professional Development, 4-6 October, 2007, Kaunas, Lithuania.* 2007
22. P. Mack, *Understanding Simulation-Based Learning.* SGH-Life Support Training Centre, Singapore, 2009
23. D.M. Gaba, The future vision of simulation in health care. Quality & Safety in Health Care **13** Suppl 1 2004, pp. i2-10. 10.1136/qhc.13.suppl_1.i2
24. J. Castronova, Discovery learning for the 21st century: what is it and how does it compare to traditional learning in effectiveness in the 21st century. Literature Reviews, Action Research Exchange **1** (2), 2002
25. F. Hees, A. Hermanns, A. Huson, Prüfungserstellung mit Total Quality Management (TQM). In: *Prüfungen auf die Agenda! Hochschuldidaktische Perspektiven auf Reformen im Prüfungswesen, Blickpunkt Hochschuldidaktik*, vol. 118, W.Bertelsmann Verlag, 2008, pp. 129–141
26. P. Blum, Ein inter-/intranetbasiertes System zur Erstellung, Durchführung und automatisierten Bewertung von dynamischen Leistungstest in der medizinischen Lehre. Ph.D. thesis, Aachen
27. W. Sitte, H. Wohlschlägl, *Beiträge zur Didaktik des "Geographie und Wirtschaftskunde" - Unterrichts*, 4th edn. Wien, 2006
28. F. Bruckmann, O. Reis, M. Scheidler, eds., *Kompetenzorientierte Lehre in der Theologie: Konkretion - Reflexion - Perspektiven.* Berlin, 2011
29. J. Handke, A.M. Schäfer, *E-Learning, E-Teaching und E-Assessment in der Hochschullehre: Eine Anleitung.* München, 2012

30. A.L. Dyckhoff, P. Rohde, P. Stalljohann, An Integrated Web-based Exercise Module. In: *Proceedings of the 11th IASTED International Conference on Computers and Advanced Technology in Education (CATE 2008)*. ACTA Press, 2008

31. R.M. Palloff, K. Pratt, *Lessons from the Cyberspace Classroom: The Realities of Online Teaching*. Jossey-Bass, San Francisco, 2001

32. M. Oechsle, G. Hessler, Praxis einbeziehen - Berufsorientierung und Studium. HDS Journal - Perspektiven guter Lehre **2**

33. E.J.E. Wegner, M. Nückles, Die Wirkung hochschuldidaktischer Weiterbildung auf den Umgang mit widersprüchlichen Handlungsanforderungen. Zeitschrift für Hochschulentwicklung **6**

34. V. Stehling, U. Bach, A. Richert, S. Jeschke, Teaching Professional Knowledge to XL-Classes with the Help of Digital Technologies. In: *Proceedings of the ProPEL International Conference, Stirling, UK, 9-11 May 2012*. 2012, p. 55

35. J. Wildt, The Shift from Teaching to Learning' - Thesen zum Wandel der Lernkultur in modularisierten Studienstrukturen. In: *Unterwegs zu einem europäischen Bildungssystem*, Düsseldorf, 2003, pp. 14–18

Next Level Blended Learning for an Excellent Engineering Education

Larissa Köttgen, Sebastian Stiehm, Christian Tummel, Anja Richert and Ingrid Isenhardt

Abstract The digitalization of higher education in general and Blended Learning in particular have been focused for a long time. Next to applying new technology, focusing on interdisciplinary competences during studies has always been part of engineering studies. The lecture Communication and Organization Development (KOE) is a constituent part of the Bachelor's studies of the faculty of mechanical engineering, addressing 1500 freshmen of the RWTH Aachen University. Since decades the lecture KOE is part of frequently used revision along with an agile progression of applied mixed methods and new demands using the previously mentioned teaching development. The number of first semester students is still increasing and freshmen entering the university are partially minors (17 years old) due to the shortened school career. To master these and further challenges successfully, the didactical method mix is based on a cube model developed by Baumgartner and Payr (1996), which combines different levels and perspectives of a student-centred teaching approach. Elements such as the flipped classroom concept (Bretzmann, J. (2013): Flipping 2.0. Practical strategies for flipping your class, Bretzmann Group, LLC) (Bergmann. J., Sams, A. (2012): Flip your classroom. Reach every student in every class every day. International Society for Technology in Education, ASCD) along with the discussion forum, which were introduced during winter term 2013/2014, have become a permanent feature of KOE. Moreover, blended learning elements were increased by attaching a two-stage business simulation called ROBOFLEX to bring the theoretical elements to an immersive, problem-based learning approach. This paper presents the cube model's current status as part of the method mix of KOE. Additionally, based on last year's evaluation results, the paper examines student's reactions towards the tutorial class ROBOFLEX as well as its potential for optimization. Furthermore a general analysis of the evaluation results from winter term 2013/2014 compared with 2014/2015 is performed to identify scope for the further development of KOE.

Keywords Higher Education · Engineering Education · Blended Learning · Problem-based Learning

L. Köttgen (✉) · S. Stiehm · C. Tummel · A. Richert · I. Isenhardt
IMA/ZLW & IfU, RWTH Aachen University, Dennewartstr. 27,
52068 Aachen, Germany
e-mail: larissa.koettgen@ima-zlw-ifu.rwth-aachen.de

Originally published in " ICERI2015 Proceedings",
© IATED 2015. Reprint by Springer International Publishing Switzerland 2016,
DOI 10.1007/978-3-319-42620-4_28

1 Introduction: The Lecture KOE

The lecture KOE (Communication and Organization Development) - organized by the Center for Learning and Knowledge Management (ZLW) of the RWTH Aachen University – is a major class in engineering education, especially in Bachelor's studies. Through inherit interdisciplinary of ZLW, the institute provides a differentiated learning experience for their students. New opportunities for innovative and student-centered teaching are connected to a pedagogically and substantially well-founded combination of e-learning and classroom learning. The aim of further development of KOE is to increase the quality of engineering education constantly. By this time more than 1,600 students registered for KOE. Throughout winter term KOE is offered weekly being mandatory for mechanical engineering students. Besides, students from different subjects (e. g. sociology, political science) attend KOE as part of their interdisciplinary elective course. In twelve modules KOE teaches basic knowledge of communication and organizational development as well as knowledge of management concepts, intercultural aspects of global work division management from the perspective of system theory. For years all results of KOE's evaluation as well as the preliminary findings of the research area of "Higher Education", which is monitored steadily, have influenced the progress of further development of KOE (detailed descriptions of past progressions can be found in numerous publications [1, 2]). The actual output is a lecture, which addresses different levels of knowledge and learning types and therefore creates a diverse learning environment. Thus the following didactical methods and concepts were combined (in order of implementation): Sections 1.1–1.6. By following the green arrow on the right hand side in Figure 1, it becomes clear that all those methods are applicable in Sections 1.7 and 1.8 (all together are based on the flipped classroom approach). This individual concept requires the students' physical and digital presence as well as a combination of both (cf. Figure 1, the "timeline" of the lecture KOE). The individual concept elements are briefly introduced.

1.1 Theoretical Inputs

"Theoretical Inputs" intend teaching contents of classical lectures at university, which uses transparencies, readers and handouts. These contents exist from the beginning of KOE on and developed from "the" basis to "a" piece in the puzzle of the lecture. As seen from the perspective of the cube model analysis (cf. Figure 2), these theoretical inputs are the learning content on a basic learning goal level and are mainly conveyed by a teacher.

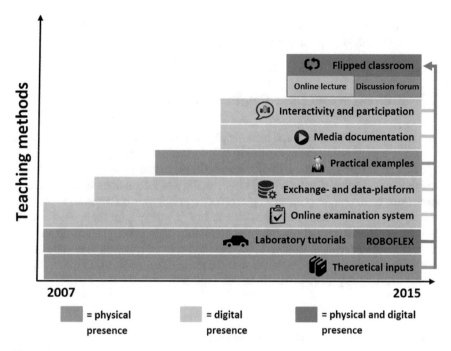

Figure 1 The concept development of the lecture KOE

1.2 Laboratory Tutorials (Now ROBOFLEX)

ROBOFLEX is a two-stage business simulation that enables students to experience realistic virtual communication within computer science and engineering disciplines. Students work in groups of about thirty people and become founders of start-ups that specialize in the production of innovative robots for the automotive industry. The start-ups organize themselves virtually using a web-based platform and physically, they create robots using Lego Mindstorms' NXT.

1.3 Online Examination System

Due to an increasing number of students the implementation of KOE's final written exam is a huge challenge. Therefore an online exam is the most effective method to test the knowledge of more than 1,600 students taking part in KOE. As early as 2007 the ZLW makes use of a digital online examination system (called OPS) which was generated specifically for the needs of KOE [3, 4]. However, the usage of OPS demands an accurate categorization of examination questions. For this purpose a particular taxonomy by Bloom [5] is used, by which exam-questions follow defined taxonomy levels (degrees of knowledge reprocessing) [6, 7].

1.4   Exchange- and Data-Platform (Document Based)

Due to the RWTH's teaching and learning platform L^2P (learning to people) students of KOE are provided with the capability of information exchange and intercommunication. In winter term 2014/15 L^2P was entirely revised with the result and self-learning units further developed and the possibility of interactions (especially with professors, mates and/or coaches in groups or separate) were elaborated [8, 9]. Most notably the offer of media documentation in L^2P is very helpful for student's exam preparation [10]. The students have the possibility of informing, learning and discussing in groups without limitations of time and place. Concerning the cube model L^2P's new feature increased in value, with the result that further learning content and greater learning targets can be pursued (cf. Figure 3).

1.5   Practical Examples (By Guest Lecturers)

Interdisciplinary competences, especially soft skills, figure prominently in engineering education. Next to communication and organization KOE's content also focusses on exactly this area of competence [11]. The complexity of professional life and the importance of communicational and organizational know-how are best demonstrated to the students by forcing them to talk to experts from practice. For this reason different guest speakers are invited to KOE every year [12]. This year's experts were asked to design their attendance interactively in order to use the input to create a common discussion forum rather than a classical presentation. The success of this further development is shown in Section 4.

1.6   Interactivity and Participation

In era of digitalizing higher education interactive and participative teaching has more to it than large-group-exercising and asking for opinion via hand signal. New technologies provide far greater offers in teaching methods. Hence KOE also applies, besides classical methods, an Audience Response System (ARS) [13] by "polleverywhere.com". Due to this software more than a hundred of student's opinions can be sampled simultaneously in a matter of seconds by using their smartphones. The use of ARS provides an enormous advantage for the KOE with approximately 1,600 participants to clarify multiple-choice questions or assignment-questions and so to gain an overall impression of the level of knowledge and comprehension of the students.

1.7   Online Lecture

The learning content of KOE's 12 modules is presented in professionally edited online lectures. These digitalized, educational videos are permanently available to students on L^2P (cf. Section 1.4 Exchange- and Data-platform). They are used as an exam preparation source for the students. In terms of the cube model the learning contents are broadened at this point by connecting online lectures with L^2P's new features. By giving the students the opportunity to support the learning process with audiovisual aids, the theoretical content can be recapitulated without any limits and at any time. This form covers the requirements of the concept learning on demand [5].

1.8   Discussion Forum

The KOE discussion forum was realized for the second time in winter term 2014/15. Following the "flipped-classroom-concept" the discussion forum provides the active attendance part of "Flips". The students receive the necessary input with the help of KOE's online lecture and L^2P that again is reflected in the weekly optional discussion forum. According to the cube model, the teacher considers himself as a coach in the discussion forum, who, in an ideal situation, structurally supports students in their interaction with each other. Many elements of the specific method mix (e.g. interactivity and participation or practical examples) as well as practical exercises and submitted comprehension questions, which were communicated on L^2P are applied. The discussion forum represents a complex learning experience and allows a student-centered course [1].

2   Implementing the Cube Model by Baumgartner and Payr

2.1   Short Summary: The Cube Model by Baumgartner and Payr

Since 2014 ZLW knowingly applies the cube model by Baumgartner and Payr to systematically illustrate particular teaching methods of KOE and to portray its further development (cf. Figure 2). The increment value of this model is that it combines different perspectives and levels of a diverse teaching approach. This cube model offers guidance for reflections about the combination of diverse didactical methods and learning materials. It considers three dimensions: learning content, learning goals and teaching strategies [14] to show a heuristic for defining educational situations. These cube dimensions have been adapted to a part of the chosen didactical concept of the KOE lecture. The mix of the different methods represent teaching techniques which have been implemented one by one and combined in the last years in education

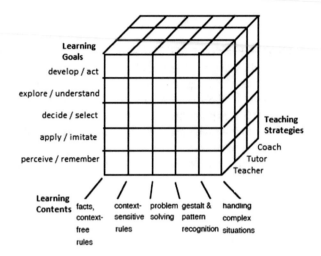

Figure 2   The cube model (cf. Baumgartner and Payr [14])

[15, 16]. In the next subchapter the current status of the cube's application in the elements of KOE will be presented in detail.

2.2   Current Status of the Cube's Application

In the course of last winter term's events ZLW attempted to develop single element of the didactical concept of the lecture KOE based on the dimensions of the cube model (cf. Figures 2 and 3). In comparison to last year, which was already mentioned in previous chapters, the elements Sections 1.4, 1.5 and 1.7 were improved by broadening learning goals as well as permitting a selection of possible roles among teaching strategies. The colors represent which learning goals are addressed, the range of greys visualizes which learning content is taught within the teaching concept and the symbols, demonstrating the teaching strategy.

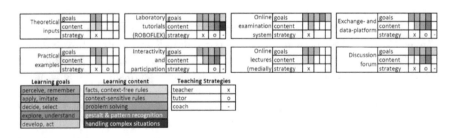

Figure 3   Actual didactical concept in context of the cube model

By analyzing the cube model and by interpreting the evaluation results (cf. the following Section 4) the ZLW is able to solve challenges solution-oriented and to integrate new trends from teaching and learning research into future lectures of KOE.

3 The KOE Tutorial Class ROBOFLEX

The KOE business simulation ROBOFLEX takes place parallel to the lectures of KOE. The simulation is based on the concept of 'simulation-based learning' which represents a research and training method that tries to create a realistic experience in a controlled environment [17]. According to this, simulation replaces or boosts real experiences [18] and thus offers enormous advantages in mediating knowledge long-acting [18, 19]. In the winter term 2014/15 the concept of ROBOFLEX was introduced. In ROBOFLEX, the simulation takes place over 9 weeks during the semester. Up until the 2013/2014 winter term the students went through an organizational development process during a 2 day tutorial. This former concept of the business simulation has faced criticism by students due to the expiration of the big arrangement on two consecutive weekends. On each Saturday higher default rates of students were identified. For this reasons, ROBOFLEX was introduced. It aims at the foundation of an engineering company for autonomously driving vehicles and it is divided in two phases: In the first phase of the ROBOFLEX scenario, the students create a 3 min business video to convince the fictitious local business development agency of their concept. They develop and define the corporate structure and corporate identity. The students also face other issues like self-organization and finding locations for their meetings. Divisions have to be founded, and decision making processes and communication methods are designed and presented. A first idea of their concept for an autonomously driving vehicle has to be presented. This concept has to consider sustainability, common usage, and energy saving. During the second phase, the actual prototype is built with Lego Mindstorms NXT robots (cf. Figure 4). A second 3 min promotional product video is then created and evaluated, which presents the mounted prototype and the overall concept to the local business development agency.

To support students that are not familiar with Lego Mindstorms NXT robots, tutorial courses with intensive support from scientific professionals are offered for each group. Then, each group has the possibility to book a timeslot for a 4 h workshop to freely build, program, test, and film their prototypes. For questions or concerns the two phases are accompanied by an intense online support via email, online-chat and physical consultation. To enhance the endurance of the different groups, every single group is encouraged to send three "tops and flops" weekly, reflecting the current group work. Both videos are rated and the best are honored at a final award ceremony by a jury that consists of scientific and industrial representatives.

Figure 4 Autonomously driving Lego Mindstorms NXT robot in ROBOFLEX

4 Selected Results of the KOE Evaluation

Similar to winter term 2013/14, the ZLW implemented a self-developed evaluation
of KOE lecture in the form of a survey. The evaluation's questions especially show
an interest in student's opinion and perception of teaching methods and how they rate
them (therewith a scale of one (disagree) to four (agree) is used). Additionally to basic
information such as age or number of semester, participants were asked to point out
their attitude towards innovative learning strategies. Due to organizational reasons the
time of evaluation in winter term 2014/15 was set to the last meeting before Christmas
holidays, which resulted in a lower number of participants compared to the preceding
year (2013/2014 = 531, 2014/2015 = 126). Nevertheless interesting findings can be
observed from this year's evaluation results. Regarding the profile of the group of
participating students, the data is almost the same compared to the previous winter
term. The vast majority of KOE students are 19 years old and freshman of mechanical
engineering. The proportion of male to female is 5 to 1. Students were asked to
name instructive elements of KOE (cf. Figure 5). Yet current results of 2015 confirm
assertions which were made 2014. The discussion of case studies and the online
lecture obtain most approval. The frequency of usage of ARS was reduced in the last
term, which directly resulted in a minimal worse-rated evaluation. At this point the
ZLW will increase the usage of ARS in order to find out whether the student's rating
will enhance. Basically it can be said concerning "instructive" the evaluation results
of the last 2 years are similar to each other.

Additionally the questionnaire includes an item asking students which different
KOE components they enjoyed and which not (cf. Figure 6). At this point a definite
progress is visible. Generally the ZLW ascertains that students enjoyed all KOE
components just as much as in the previous year. Though especially those elements,
which were refined by ZLW during the last years demonstrate a strong growth. As
shown in the previous chapter of implementing the cube model the practical examples

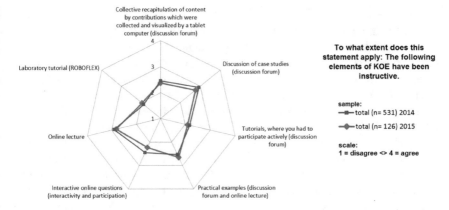

Figure 5 KOE evaluation result "instructive"

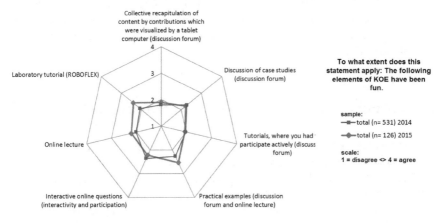

Figure 6 KOE evaluation result "fun"

and the online lecture increased in popularity. A reason might have been that the online lecture was revised and extended giving students the chance to work more interactively and therefore rich in variety. The improvement of practical examples can be seen in the request of guest speakers to design rather interactive discussion forums integrating students than classical lectures. The change in rating of the former laboratory tutorial and the current Roboflex organization simulation, demonstrated in Section 3, is exceptionally noticeable.

In addition to the results presented the survey focuses on evaluating the business simulation (Laboratory tutorial, ROBOFLEX) itself and the professional support. The questionnaire presents statements and students can agree or disagree on a scale from one (very appropriate) to four (not appropriate). Following there is a choice of significant statements. Thesis: The more the groups use the contents of the KOE lecture to communicate and solve problems, the more helpful the contents of the

KOE lecture are perceived by the students The two assertions have a correlation of $r = 0.674$ with a statistical significance of $p < 0.001$. Groups applied the contents of the lecture to communicate and solve problems among group members. It is an important outcome of the business simulation that the students see the practical use of the theoretical models and methods they learned during the lecture. Thesis: The better the information exchange is within the group, the better the group is in handling turbulences and solving problems. The two assertions have a correlation of $r = 0.616$ with a statistical significance of $p < .001$. Again this shows the importance of a good communication and information exchange to solve or avoid problems. The students were encouraged to reflect the behavior of their group to recognize such correlations for themselves. In summary for the evaluation results can be stated: The use of robotics as a "blended learning" tool in the context of engineering education has several advantages to combine theoretical and practical aspects. Students are able to transfer their theoretical knowledge to real engineering problems [20]. Furthermore working with robotics leads to an active participation and represents an effective student-centered teaching method which motivates students. They work in teams and incidentally improve soft skills such as teamwork, management, communication and presentation [21].

5 Conclusion and Scope for the Further Development

As the title of this paper suggests, the next level of realizing an excellent engineering education was successfully coped by an ongoing adjustment of different methods of KOE as well as the extensive revision of the former laboratory tutorials converted into the organizational simulation ROBOFLEX. Evaluation results emphasize that student-centered teaching in combination with highly interactive methods and elaborated participative components of target groups are successfully adopted. The strategy providing digital offers in selected fields supports the basis of Blended Learning and is of advantage to students. Ideally Blended Learning should be a successful combination of classical components (cf. Section 1.1) and modern influences (Section 1.4). To be in talk with students continuously show that they do not favor pure "online-teaching" but rather a meaningful exchange of knowledge with their teachers. The cube model established as a solid method, which lucid and structurally illustrates the current status of KOE. Moreover the cube model can be complemented with new introduced teaching methods. For the future the developed method mix has to be reviewed by evaluation continuously and new teaching-trends have to be implemented in order to keep KOE successful. Current findings demonstrate that approved methods such as interactive online questions will be intensively employed the following year to show whether popularity increases or not. The ZLW plans to elaborate the list of guest speakers for the upcoming winter term 2015/16 in order to enable anymore interaction with experts from professional practice. ROBOFLEX shows that teaching with robotics enhances student motivation because of the hands-on aspect of building and programming your own robot. To successfully build a robot, students have to improve their programming skills, be creative, and solve problems. Thus, students learn how to act and communicate in virtual as well as physical environments-considering a group size of about 30 persons. This represents a key competence, which increasingly gains in importance in modern working environments, especially in engineering fields [22, 23]. Courses in the past have shown that robotic projects are more attractive to students, because they are more appealing than learning programming without context. Thus, the timeline of ROBOFLEX will be shortened to 7 weeks in order to reduce student default rates.

References

1. L. Köttgen, S. Schröder, E. Borowski, A. Richert, I. Isenhardt, Flipped classroom on top - excellent teaching through a method-mix. In: *8th International Technology, Education and Development Conference (INTED). 10-12 March 2014*. IATED, Valencia, 2014, pp. 40–49
2. L. Köttgen, S. Winter, S. Schröder, A. Richert, I. Isenhardt, Integrating blended learning – on the way to an excellent didactical method-mix for engineering education. In: *Proceedings of the 7th International Conference of Education, Research and Innovation (ICERI). 17-19 November 2014*. IATED, Seville, 2014, pp. 4485–4494
3. F. Hees, H. A., A. Huson, "Prüfungserstellung mit Total Quality Management (TQM)" Prüfungen auf die Agenda! Hochschuldidaktische Perspektiven auf Reformen im Prüfungswesen. Blickpunkt Hochschuldidakti **118**, 2008, pp. 129–141
4. P. Blum, Ein inter-/intranetbasiertes System zur Erstellung, Durchführung und automatisierten Bewertung von dynamischen Leistungstest in der medizinischen Lehre. Ph.D. thesis, Aachen, 2005
5. T. Reglin, C. Speck, *Zur Kosten-Nutzen-Analyse von eLearning*. Leitfaden E-Learning. VBM e.V., 2003. pp. 221–235
6. W. Sitte, H. Wohlschlägl, *Beiträge zur Didaktik des "Geographie und Wirtschaftskunde"-Unterrichts*, vol. 4. Universität Wien Institut für Geographie und Regionalforschung, Wien, 2006
7. F. Bruckmann, O. Reis, M. Scheidler, *Kompetenzorientierte Lehre in der Theologie, Konkretion – Reflexion – Perspektiven*, vol. 3. LIT Verlag Münster, Münster, 2011
8. J. Handke, A. Schäfer, *E-Learning, E-Teaching und E-Assessment in der Hochschule*. Oldenburg Verlag, München, 2012. p. 10
9. A. Dyckhoff, P. Rohde, P. Stalljohann, An Integrated Web-based Exercise Module. In: *Computers and Advanced Technology in Education. 29 September – 1 October 2008. Proceedings of the 11th IASTED International Conference on Computers and Advanced Technology in Education*. Acta Press Calgary, Crete, 2008
10. R. Palloff, K. Pratt, *Lessons from the Cyberspace Classroom: The Realities of Online Teaching*. Jossey-Bass, San Francisco, 2001
11. M. Oechsle, G. Hessler, Praxis einbeziehen – Berufsorientierung und Studium. HDS Journal - Perspektiven guter Lehre **2**, 2010, pp. 11–22
12. E. Wegner, M. Nückles, Die Wirkung hochschuldidaktischer Weiterbildung auf den Umgang mit widersprüchlichen Handlungsanforderungen. Zeitschrift für Hochschulentwicklung **6** (3), 2011, pp. 171–188
13. V. Stehling, U. Bach, A. Richert, S. Jeschke, Teaching Professional Knowledge to XL-Classes with the Help of Digital Technologies. In: *The first international ProPEL Conference 2012 - Professions and Professional Learning in Troubling Times: Emerging Practices and Transgressive Knowledges, Stirling, UK, 9-11 May 2012*. 2012, p. 55

14. P. Baumgartner, S. Payr, Learning as action: A social science approach to the evaluation of interactive media. In: *Proceedings of ED-MEDIA 96 - World Conference on Educational Multimedia and Hypermedia.* AACE, Charlottesville, 1996, p. 33

15. S. Schröder, D. Janßen, I. Leisten, R. Vossen, I. Isenhardt, On-professional competences in engineering education for XL-Classes. In: *Frontiers in Education Conference. 23-26 October 2013. Frontiers in Education Conference, 2013 IEEE.* IEEE, 2013, p. 30

16. R. Kay, A. LeSage, Examining the benefits and challenges of using audience response systems: A review of the literature. Computers & Education 53 (3), 2009, p. 819

17. P. Mack, *Understanding Simulation-Based Learning.* SGH-Life Support Training Centre, Singapore, 2009

18. D. Gaba, The future vision of simulation in health care. Qual Saf Health Care 13 (**Suppl 1**), 2004, pp. i2–i10

19. J. Castronova, *Discovery Learning for the 21st Century: What is it and How Does it Compare to Traditional Learning in Effectiveness in the 21st Century.* Action Research Exchange, 2002

20. A. Cruz-Martín, J. Fernández-Madrigal, C. Galindo, J. González-Jiménez, C. Stockmans-Daou, J. Blanco-Claraco, A lego mindstorms nxt approach for teaching at data acquisition, control systems engineering and real-time systems undergraduate courses. Computers & Education 59 (3), 2012, pp. 974–988

21. A. Behrens, L. Atorf, R. Schwann, B. Neumann, R. Schnitzler, J. Ballé, T. Herold, A. Telle, T. Noll, K. Hameyer, T. Aach, MATLAB meets LEGO Mindstorms – A freshman introduction course into practical engineering. Education, IEEE Transactions on 53 (2), 2010, pp. 306–317

22. T. Jungmann, K. Müller, K. Schuster, Shift from TeachING to LearnING. Anforderungen an die Ingenieurausbildung in Deutschland. Journal Hochschuldidaktik 2, 2010, p. 7

23. N. Paetz, F. Ceylan, J. Fiehn, S. S., C. Harteis, *Kompetenz in der Hochschuldidaktik – Ergebnisse einer Delphi-Studie über die Zukunft der Hochschullehre.* VS Verlag für Sozialwissenschaften, Wiesbaden, 2011. pp. 7–12

Are Virtual Learning Environments Appropriate for Dyscalculic Students?

Laura Lenz, Katharina Schuster, Anja Richert and Sabina Jeschke

Abstract In Germany, there are more than four million people (almost 6% of the entire population) living with dyscalculia, a disorder which alludes numbers as well as general arithmetic and is closely related to dyslexia (Die Zeit Online, "Ziffern ohne Sinn". Retrieved online, May 21st, 2015, from http://www.zeit.de/2013/29/dyskalkulie-zahlenblind-teilleistungsstoerung). The estimated number of unreported cases is probably even higher. Medical researchers talk about a "forestalled elite" since these people are commonly not less intelligent than non-handicapped individuals. Still, they rarely make it to a university-entrance diploma; they get lost on the way because of missing standby facilities offered in primary and continuative schools (Die Zeit Online, "Verhinderte Elite". Retrieved online, May 21st, 2015, from http://www.zeit.de/2003/42/C-Legasthenie-Schule). They require special needs and attention in order to learn and show their de facto potential. This paper deals with the dyscalculic-friendliness of learning environments provided by Mixed-Reality Simulators. After a presentation of the scientific state of the art on the specific needs of affected students, it will be elaborated in how far virtual environments used in the education of mechanical engineering students can sufficiently not only meet those needs but support them in their study.

Keywords Dyscalculia · Dyslexics and Dyscalculics in Academia · Virtual Learning Environments · Learning Content Adaptation · Mixed Reality Simulator · DybusterCalcularis

1 Introduction

In autumn 2013/2014, more than 2.5 million students were enrolled at German universities – with a growing tendency [1]. The German Student Union claims that around 8.000 students of these are afflicted with dyslexia or dyscalculia. The affected

L. Lenz (✉) · K. Schuster · A. Richert · S. Jeschke
IMA/ZLW & IfU, RWTH Aachen University, Dennewartstr. 27,
52068 Aachen, Germany
e-mail: laura.lenz@ima-zlw-ifu.rwth-aachen.de

Originally published in "Games Entertainment Media Conference (GEM)",
© IEEE 2014. Reprint by Springer International Publishing Switzerland 2016,
DOI 10.1007/978-3-319-42620-4_29

students often suffer silently since many do not know about the opportunities their university offers for them [2]. Instead of informing their professor in advance, they fail exams because of numerous arithmetic errors although they are as well prepared as their fellow examinees. Institutions like the AStA Representatives for Disabled and Chronically Ill Students at RWTH Aachen University, which deal with all issues regarding studying with a handicap, provide manifold administrative prospects to assist their students throughout their daily study routine [3]. However, administrative acts are not where integrational processes and the aspiration for diversity and equal opportunities for students should come to a stop. Instead, the significance of particular (but often expensive) needs, such as a very personal educational mentoring, must be part of an inclusive, barrier-free curriculum design – independent of study field. At the IMA/ZLW & IfU institute cluster, which is part of RWTH University's Faculty of Mechanical Engineering, major seminal, didactic and curriculum related decisions on how a contemporary and attractive syllabus could be created are being made. Thus, it is inter alia their task to also include handicapped students into the designated curricula as far as possible. For all students, in order to make mechanical engineering 'touchable' instead of 'dry' as well as often extremely complex theory, virtual learning environments (VLEs) have been part of engineering education for several years.

For this sake, the RWTH acquired a worldwide unique mixed reality simulator. The so called 'Virtual Theatre' combines a head-mounted display (HMD) with an omnidirectional treadmill, so that students can physically walk through virtual worlds. But do these digital settings and their set-up sufficiently serve the special needs of dyscalculics? What happens if the possibility to physically move through VLEs is contingent? Can VLEs serve as a less-costly alternative for these target groups to learn, as they are available at any time, customizable and randomly often repeatable? And if they are not appropriate, which parts of their design and setting should and could be adapted and how? These are the core questions this paper seeks to answer.

To do so, the authors will proceed as follows: First, it will be summarized what exactly dyscalculia is. Second, the state of the art of academic provisions in connection with dyslexic and dyscalculic students will be elaborated. Third, in order to stress the specific needs of dyscalculics in education, the scientifically proven, dyscalculic-friendly Swiss concept of 'Dybuster Calcularis', a special learning software for this audience, will be examined. It shall serve as one example out of many, equally functioning software tools for the handicapped. Fifth, the relevance, assumptions and limits of this work will be outlined. Sixth, the researched hardware, the 'Virtual theatre' will be explained in technical terms. Seventh, it will be investigated in how far the VLEs used at IMA/ZLW & IfU suit the requirements for studying with dyscalculia as proposed by 'Dybuster'. In here, possible design adjustment suggestions concerning how to make the analyzed learning tools more optimal will be made. The conclusion will include a tentative outlook concerning how the academic education of dyscalculics in Germany could and must evolve.

2 What Is Dyscalculia

Dyscalculia is a difficulty in learning arithmetic correlations as well as fundamental mathematical cohesions in general. It is similar to dyslexia, but adverts to numbers rather than letters. Dyscalculics are 'blind' for numbers; they cannot understand or manipulate them. Like dyslexia, dyscalculia is not related to a high or low IQ. Besides arithmetic issues, patients do often have difficulties with time, measurement and spatial thinking. Their frustration level is relatively low, wherefore they become easily frustrated if they cannot complete tasks correctly and feel embarrassed. Dyscalculia often comes in combination with ADHD and concerns 3–6 % of the population [4]. If it is caused by a brain injury, the correct designation is acalculia. In contrast, dyscalculia has a developmental origin. In daily life, developmental dyscalculics furthermore face difficulties with recognizing the largest out of many numbers, budgeting, basic calculations, differentiating between left and right, mental visualization, the estimation of distances, the ability to complete mentally exhaustive tasks and the recollection of names as well as designations.

For a dyscalculic, the ideal learning environment contains as many as possible sensual stimuli (for example audible and visual) and is user- and context-adaptive [4]. The reason is that the three task-specific modules (verbal, symbolic and analogue magnitude, summarized in the so called 'triple-code model') are located in various parts of the brain and must thus be stimulated differently. In the context of number processing, a high overlap between these modules leads to an increased arithmetic understanding, which is less present in the minds of dyscalculics. However, in the brain of any person, the three modules develop hierarchically over time, wherefore each learner reacts variably to diverging mental attraction [4]. Still, the too little overlaps in a dyscalculic mind make individualization of learning crucial.

3 Provisions for Dyscalculics in German Academic Education

In terms of dyscalculic-friendly curriculum adaptation, there are currently no non-administrative provisions taken at German universities. However, there are some common guidelines to support handicapped persons; the measure is nationwide known as disadvantage-compensation and does inter alia concern dyslexics and dyscalculics. The disadvantage- compensation includes arrangements such as the non-consideration of spelling, punctuation and grammar mistakes in (take-home) exams, an extended exam-duration, the usage of a notebook with auto-correct instead of having to write with pen and paper, an oral instead of a written exam or several evaluation/correction loops before having to hand in an assignment. These provisions are taken by numerous well-known German universities [5]. Thus, the status quo demonstrates that the prevailing measures are exclusively of administrative, not syllabus-adaptive nature. Universities expect their dyslexic students to master teaching and learning contents, but do only make compromises when it comes to examinations. However, the actual problem starts much earlier – when dyslexic students must apprehend and internalize study-related information.

4 Learning Requirements of Dyscalculic Students: An Overview

After it has been clarified how a dyscalculic brain functions and that there a no curriculum-related measures taken by Germany universities yet, it must be examined which special learning requirements evolve hereof, which were the results of multitudinous studies [6, 7]. First of all, it must be ascertained to which of the three representational mental modules the respective dyscalculic predominantly reacts. This can be captured by a controlling algorithm embedded into a Bayes-Net, which will store, categorize and analyze individual learning performance as well as it recognizes the interconnectedness of skills and adapt the provided tasks accordingly. A dynamic, independent of age Bayesian model is part of most learning software tools for dyscalculics and targets the modelling of the intelligent mind [4]. Thus the selection of learning paths should be non-linear and flexible. Common learning software fulfills these requirements by offering the opportunities 'to stay', meaning to continue with the actual task, to 'go back', to return to the former task and to 'go forward', to proceed with the next (assembled) task. These opportunities benefit the adaptability of content in terms of individual needs, the locality, the configuration based on "[…] nodes and neighbors […]" [4] and generality, meaning the common applicability on any structure, model and content.

5 Dyscalculic, Adaptive Learning Through Gamification: An Example

In order to make the general theoretical requirements explicit, one example for dyscalculic learning software shall be given. Besides applications like 'Cool Math Games', 'ETA Cuinenaire' and 'NUMBER SENSE', 'Dybuster Calcularis' is one of the most commonly known tools [8]. In general, the mentioned software tools function equally and are relevant for persons of any age group since patients react to similar stimuli based on their module group as outlined in Section 2. 'Dybuster Calcularis' was nonetheless chosen here because it was created due to public funding by the ETH Zurich and the University of Zurich, Switzerland, scientifically proven and verified in two scientific studies and is nowadays used by more than 35.000 dyscalculics [9]. In addition, the parties in charge display their methods in a transparent way on their webpage and in numerous publications. However, what must be internalized is that Dyscalculia can solely be treated, not cured.

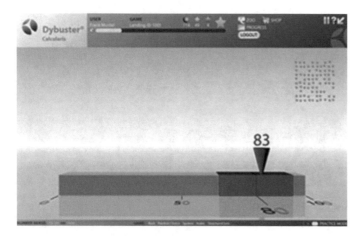

Figure 1 Dybuster Calcularis – The Landing Game

'Dybuster Calcularis' stimulates the interplay between the representational verbal, symbolic and analogue magnitude brain modules. It does so by offering alternative representations to existing arithmetic problems, which suit the individual mental peculiarity of the learning dyscalculic. This process, which is called transcoding, provides three principles of number understanding: cardinality, the (in-) finite numbering of elements, ordinality, the indexing of elements and relativity, the interrelatedness of elements [4].

In the 'Landing Game' (Figure 1), a basic level of 'Dybuster Calcularis', the user must position the displayed number (83). The completion of this task does not only give him a feeling for the correct ordering of numbers fosters the understanding of ordinality (which number is how much bigger/smaller?) and relativity (where between 80 and 90 is 83?). The numbers themselves are shown in an analogue magnitude (a block). Important is also that the numbers do have different colors in order for the patient not to exchange them.

On an advanced level, in the Plus-Minus game, analogue magnitude (eight blue and five green blocks of diverging form)) are combined with verbal (numbers) and symbolic (colored points) elements (Figure 2). The addition of 85 + 8 is hereby displayed on several modes in order to not only apply to one, but several mental schemes. Thus, 'Dybuster Calcularis' advances the connection building between the three mental modes of arithmetic thinking and does thereby accomplish improvement in the realm of mathematical thinking and understanding. If tasks are completed correctly, the user may adapt the background of the game as a kind of reward.

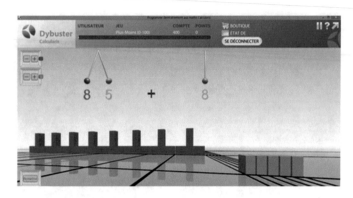

Figure 2 The Plus-Minus game

6 In-Between Summary: What Must Be Kept in Mind

For now, as an interim-summary, it must be kept in mind which the special requirements of a dyscalculic for successful learning are:

- The visualization of what must be learned. Dyscalculics are attracted by visual stimuli.
- Ideally, there should be a supportive auditory element to complement the visualization.
- If applicable, the stimulation of additional senses would be helpful, for example in the form of haptic feedback.
- The content must stimulate the verbal, symbolic and analogue magnitude modes of the brain and ideally combine those
- A stimulus to communicate an error, which should by no means be a punishment, but a short and simple information, for example a specific sound.
- Clear separation (for example due to complementary coloring) of numbers, which sound or look equal to a dyscalculic
- A Bayes-Net to not only analyze and store errors, but to categorize and interrelate them for the sake of self-controlled feedback.
- Forms in combination with numbers. If content is presented by the help of a visual depiction, it can easier be kept in mind [6, 9].
- The possibility to repeat new and old contents whenever desired (although this is crucial for any student). Ideally, there should be a mix of both to not only internalize what has already been learned, but to interconnect it with the new.
- There must be an opportunity to obtain a reward in order to increase (long term) motivation.

All in all, this makes 10 requirements, which must be checked in the 'Virtual Theatre' scenarios created by universities who want to design their virtual learning environments towards dyscalculic-friendly support.

7   Relevance, Assumption and Limits

Before starting with the analysis of the 'Virtual Theatre' scenarios, some relevant assumptions and limits of this work must be outlined. One important assumption of this paper is that the dyscalculic-friendly learning software 'Dybuster Calcularis' is indeed helpful for the learning process of dyscalculics. The scientific results they present to verify its impact must be taken for genuine because it is beyond the limits of a theoretical approach on a dyscalculic-friendly syllabus design to reappraise the medical results. Moreover 'Dybuster' is solely one example for many similarly functioning software tools and not the only solution for the problems of dyscalculics.

Another important assumption is that for the investigation of the learning environments' importance for the learning process of dyscalculics, one must suppose that the user has already had some training with 'Dybuster' or comparable dyslexia software. There must be tie points to this basic software in order to guarantee a continuous learning process based on equal principles. This is what this work seeks to do, identifying these connection points. It is beyond the limits of a university to (re-)teach arithmetic basics. Instead, one main ambition should be to make existing contents dyscalculic-friendly instead of creating entirely new learning scenarios.

As for additional limits, this paper shall solely serve as a theoretical fundament for further research on dyscalculia in academia. In the future, practical quantitative and qualitative research on and with affected students must be conducted in order to further prove and qualify the results. Thus, this work provides a subjective, tentative, self-critical assessment concerning in how far the special needs of dyscalculics are currently being dealt with throughout their learning process.

Lastly, it must be stated that RWTH Aachen University also uses digital games like 'Minecraft' for the education of engineers. However, this paper shall exclusively deal with virtual environments which were either created by RWTH itself or which they are enabled to adapt. It appears less meaningful do make fundamental design suggestions for learning games the university cannot adjust anyways.

8   Hardware Description: The Virtual
Theatre – a Mixed-Reality Simulator

To proceed with the evaluated hardware device, mixed reality simulators like the 'Virtual Theatre' combine the natural cut surface of a Head Mounted Display (HMD) with an omnidirectional floor. It can be connected to a hand tracker, which does however only support the position tracking. This is otherwise being done by ten infrared cameras. The omnidirectional floor consists of 16 trapezoidal elements. These are equipped with rolls, which have a common provenance in the middle of the 16 trapezes. In the middle, the user can stand still. As she leaves it, the motor and the rolls start moving. They allow for natural walking and will rotate faster as the user comes closer to the edge in order to not let her run out of the surrounding [10].

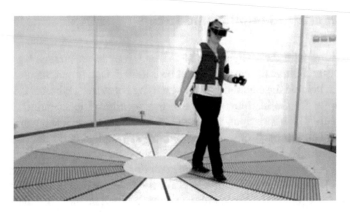

Figure 3 A test-person walking through a virtual world on the 'Virtual Theater'

The 'Virtual Theatre' makes a full 3D visualization of a virtual environment possible. The movements are captured in real time so that natural ways of acting and moving become possible. An infrared marker, which is attached to the HMD tracks the head movements. Although this is (not yet) possible, hand tracking will in the long run enable the user to actively interact with the virtual environment presented in the 'Virtual Theatre'. For now, the users can hold a cross-like hand tracker in their hand, which is also vested with infrared markers. If this hand tracker undercuts the height of 0.5 m, the simulation and motors stop, for example, if the user stumbles. Any interaction between user and 'Virtual Theatre' happens wirelessly (Figure 3).

One of the purposes of the 'Virtual Theatre' is its usage for the education of mechanical engineering students. It is supposed to depict repeatable, danger free scenarios in which young engineers may test their knowledge in a realistic way. Obviously, this also saves material costs [11]. Currently, there are four scenarios available for the 'Virtual Theatre', the test scenario 'Piazza', which depicts a typical Italian marketplace. Second there is the 'Maze', in which users need to find items such as a ball or a rubber duck. The task is to be able to locate the found items on a map afterwards in order to investigate the user's spatial abilities. Third there is another test scenario, the 'Gallery'. Users walk through a museum-like room with famous drawings in it. Fourth, there is the 'Mars' scenario, which was designed on the corresponding basis map material of the NASA. The 'Mars' is used as an explorative landscape, a space-museum or an obstacle parcours by the National Aeronautics and Space Research Center of the Federal Republic of Germany (DLR) School Lab for primary and secondary school pupils inter alia located at RWTH [12]. There are several space shuttles and crafts located on it, which are either drivable by the user or move freely. They are closely located to signs displaying their actual real life name/designation.

All scenarios are available for both computers and the 'Virtual Theatre'. Potentials and opportunities for dyscalculics shall already tentatively be traced by evaluating the status quo. It will be outlined whether and how the 'Virtual Theatre' gives rise to the inclusion of students with special needs from the genesis onwards and to which extent it could therefore be groundbreaking for dyscalculic academic education if executed carefully and correctly.

Before starting with the analysis of the 'Mars', 'Maze' 'Gallery' scenarios, it must be clarified, why only these two were chosen. The reason is that actually, only three of the four scenarios contain textual and or numerical elements. In order to analyze whether the virtual learning environments used at RWTH Aachen University are appropriate for dyscalculics, it does not make sense to investigate scenarios, which do not contain text or any other potential handicap-related content and do therefore not require any spatial thinking or arithmetic comprehension skills. The 'Piazza' shall be excluded because it depicts 'solely' a quadrangular place. It is trivial to talk about spatial thinking in a quadrangle. Nevertheless, it must be underlined that generally, any mathematical content could be included into any scenario. Additionally, the interaction with objects inside scenarios is generally an option due to a physics-engine, but not fully activated and declared yet.

9   Analysis: The 'Virtual Theatre' Scenarios 'Mars', 'Maze' & 'Gallery'

Ultimately, it is of utmost importance to present the reader of this paper what the analyzed scenarios look like. For this purpose, the authors will first give some impressions on the virtual environments (Figures 4–8). The three scenarios will be analyzed separately because they apply to varying modules and learning channels. At the end of the analysis, a score is calculated in terms of to which extent each scenario fulfills the ten previously mention dyscalculic requirements and whether it offers any extra potential.

On 'Mars' (Figures 4 and 5), it is only partly possible to interact with the presented shuttles. They either stand next to their name tag or move independently. Only one specific rover can be maneuvered by the user by Q, O, P, R control, but exclusively on laptops. Inside the 'Virtual Theatre', a Wii-controller can be attached to manipulate the VLE [13]. Apart from this, users may explore the virtual space and can learn about the shuttle's name. The aim is to enthuse pupils about engineering and inspire them. The crafts are empathized by their original paradigms and users can investigate their outer appearance and construction. In the 'Maze' (Figures 6 and 7), users walk around to find hidden subjects in it, but without interacting with these. Similarly, users can walk around the virtual 'Gallery' (Figure 8) and learn about the artworks and respective artists. Whenever the user virtually stands and looks right at the drawing in front of him, title and painter appear above it. The idea is to create an association between text and drawing in order to support the learning efficiency by not only saving information on a cognitive, visual level, but due to the physical movement in and around it.

Figure 4 A part of 'Virtual Theatre' scenario 'Mars' (1)

Figure 5 A part of the 'Virtual Theatre' scenario 'Mars' (2)

To begin with the requirement of visual elements/stimuli, this need is clearly being served on 'Mars', the 'Maze' and in the 'Gallery'. Not only that the numerous presented crafts and pictures can be investigated from all sides. Especially on 'Mars', although both scenarios are in 3D, the shuttles also move (either by themselves or

Figure 6 Gaze inside the maze. The user just found a car

by the user's manipulation) and hereby become credibly imaginable and alive. Thus, instead of seeing machines and drawings on book pages, they can be experienced and are therefore more attractive for dyscalculics; they can easier remember them. In addition to the depiction of the machines and artworks, a continuative positive effect is their name tags located nearby, either on a sign or (in case of the 'Gallery') as lettering on top of it. Independent from the character's color scheme, one can expect a positive connection between display (which is already targeting the 'translation into forms' criterion) and appellation. The 'Maze' is of special interest here because the finding of hidden items fosters spatial thinking as recent RWTH studies show [11]. As spatial thinking is a weakness of dyscalculics, the 'Maze' may serve as an especially stimulating environment here since it caters on numerous senses as it will be examined below.

As for the audible stimulus, there is none available (yet) in all scenarios. Generally, it is possible to attach headphones to the HMD, but this is currently not being done. Moreover, the roles of the 'Virtual Theatre' are relatively noisy, so the headphones would have to block external noises and foster the internal at the same time in order to create an immersive experience. Since 'Dybuster Calcularis' solely suggests the repetition of one certain tone in one specific context, it does not necessarily have to be an actual close to reality machine sound (in the case of 'Mars' or the 'Maze'), but a random one, affiliating one single item.

To continue with the stimulation of additional senses of dyscalculic users besides the auditory and the visual, 'Dybuster Calcularis' proposes the subjoining of additional tenors in order to create a more intense, immersive feeling. This is not crucial, but still to an extent desirable. It is not really one of the five senses (at best groping), but

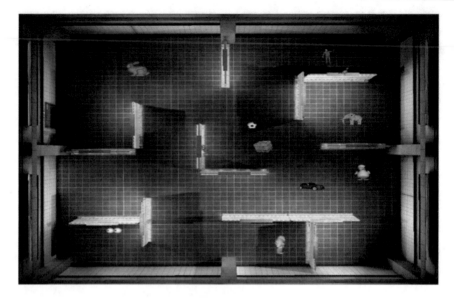

Figure 7 The 'Maze' from above with the all items to be found in it

Figure 8 The 'Gallery' scenario with Da Vinci's 'Mona Lisa'

presumably no disadvantage to offer test persons the opportunity to physically access the virtual learning environments to herein move naturally and freely. Besides this, one cannot talk about supplementary stimulation like for example (virtual/digital) haptic feedback. There is prevailingly no way the user can actively manipulate his surroundings (besides being able to move one shuttle by a portable keyboard), hear what happens inside of (in this case) the space craft or how the duck quacks (an item in the 'Maze').

Proceeding with a fixed color scheme, the letter color (brown-grey on 'Mars' and white in the 'Gallery') does not vary per machine or drawing, which is according to 'Dybuster Calcularis' highly problematic for the absorption process of a dyscalculic student. For her, it would make a major difference if each item's name tag had its own color and one respective sound. Although not implemented, the individual and adjustable coloring of numbers, letters and words is a change that can be undertaken with a marginal programming effort. The maze shall be left out here because it does not entail any signs, numbers or letters.

Currently, all scenarios are non-competitive, meaning that there are no errors to be made. However, if students would for example have to learn the space crafts' or artworks' technical terms or names and either allocate them from a selection of the same or type the names in directly, an analogical feedback tune could be created. It must be claimed that this ability is heavily dependent on the auditory feedback criterion and asks for its successful completion and implementation to be realized itself.

In two scenarios (the 'Mars' and the 'Gallery', but not in the 'Maze'), there are some technical appellations of crafts and drawings, which might be mixed up by dyscalculic students and cause confusion and disorientation depending how severe their handicap is and if it strikes dyslexic traits, too. One example could be the 'Surveyor' and the 'Sojourner' as depicted in Figures 4 and 5. 'Surveyor' and 'Sojourner' sound very akin. According to the color scheme of (oppositional) colors as proposed by 'Dybuster Calcularis', these two are not similar enough to demand fully complementary dyeing, but must by all means differ explicitly. As a side remark, there is a lot of potential for dyslexic students in the 'Gallery'. In accordance with 'Dybuster Ortograph', a dyslexia learning software, it is especially appealing to this target group if they first see the content and then its description. However, the analysis of dyslexia-friendliness goes beyond the limits of this paper [9, 14, 15].

Currently, as there is no audible stimulus integrated into the 'Mars', 'Maze' and 'Gallery' scenarios, there is in consequence no auditory feedback (whether positive or negative) hearable. Furthermore, the engineering contents are still under construction as it was elaborated above. Therefore, there is no Bayes-Net (yet) to store, analyze, interrelate and categorize possible errors of the dyscalculic exploring the virtual environment and for example locating items. This is why one can on the one hand not talk about a rudimental fulfillment of this demand and on the other hand not about a growing new potential.

In contrast, there is a translation of mechanical/artistic learning contents into 3D forms, meaning that the theoretical construction of a space shuttle or a picture frame are transferred into an experience the 'Virtual Theatre' user can actually undergo. In 'Dybuster Calcularis', this is being done in a more simplistic way, namely by depicting arithmetic elements as geometrical figures. Still, this fact underlines a dyscalculic's affection for imagination through accurate, 'touchable' forming, about which one can due to the credible digital realizations of shuttles and artworks talk here.

As for the fore-last criterion, the ability to arbitrarily repeat contents, the 'Virtual Theatre' appears to be a suitable technology. The repetition does not only reduce costs and dangers. The simulation always looks exactly the same and subject matters can theoretically be adjusted as the learning progress increases. At the same time, this leaves enough space for the adaptation in line with specific dyscalculic demands like individual focus points. If the VLEs are being adjusted slightly, dyscalculic students will have the opportunity to test their knowledge freely without being hesitant about breaking material and thus being ashamed in front of others, which will supposedly be a very positive experience for (not only) a dyscalculic.

Lastly, there are no tasks to pass or fail in all scenarios, which is why there are no rewards available (yet), which would be motivating for especially dyscalculics. However, in the long run, there appears to be a major potential here. An imaginable reward could be the random modification of the environment by the user (like in 'Dybuster Calcularis') or additional points for a final exam if the user completes certain assessment-relevant tasks in the virtual world (Tables 1, 2 and 3).

From the analysis of the RWTH's 'Virtual Theatre' scenarios, the following tables can be conducted (additional potential shall be covered under the 'Other' section, but not be part of the dyscalculic-friendliness calculation).

Table 1 Evaluation of the 'Mars' scenario

Mars	
Requirement	Application
Visual stimulus	✓
Audible stimulus	x
Additional senses	✓
Fixed color scheme	x
Error-feedback	x
Separation of similar words	x
Bayes-Net	x
Translation into forms	✓
Repetition	✓
Rewards	x
Other	✓ *Active manipulation*
Total	**4/10 (+1)**

Table 2 Evaluation of the 'Maze' scenario

Maze	
Requirement	Application
Visual stimulus	✓
Audible stimulus	x
Additional senses	✓
Fixed color scheme	x
Error-feedback	x
Separation of similar words	x
Bayes-Net	x
Translation into forms	✓
Repetition	✓
Rewards	x
Other	✓ *Spatial thinking*
Total	**5/10 (+1)**

Table 3 Evaluation of the 'Gallery' scenario

Gallery	
Requirement	Application
Visual stimulus	✓
Audible stimulus	x
Additional senses	✓
Fixed color scheme	x
Error-feedback	x
Separation of similar words	x
Bayes-Net	x
Translation into forms	x
Repetition	✓
Rewards	x
Other	✓ *Dyslexic-friendly*
Total	**3/10 (+1)**

10   Potentials for Dyscalculus

It has been shown that for dyscalculics, the 'Virtual Theatre' offers a lot of not fully fathomed possibilities. From the technical point of view, it is capable to provide free and unbound navigation, which means that the user is enabled to move the way she usually does and can thus immerse without further effort. It also benefits orientation and spatial capabilities as well as content-related visualization. The 'Virtual Theatre' might have some weaknesses in comparison to other mixed-reality simulators like the 'Cave', for example in terms of the visual stimulus quality, but offers in the long run more interactive ways to actively manipulate virtual environments. Due to this interactivity, any representational model, verbal, symbolic and analogue magnitude can be satisfied by attracting up to three senses (audible, visual, haptic). As for the status quo, the verbal is already served by the names of painters and drawings displayed in the 'Gallery'. Although it is currently 'only' letters, any scenario could be adapted by inserting arithmetic elements. Scenarios like the 'Mars' go far beyond the depiction of names and symbols. The 'Virtual Theatre' provides actual 3D machines (space shuttles) to manipulate, which is likely to increase the understanding of technical properties. Analogue magnitudes can be found in the 'Maze' where a specific number of items must be found and mapped.

For the didactical advantages, the 'Virtual Theatre' offers not only possibilities to arbitrarily repeat any content, but also self-determined learning (SDL). In SDL, students can freely decide when, how, where and what to learn, which leads to an increased learning success [7]. This does not only benefit non-handicapped students, which have a greater tendency to adapt to learning requirements and stimulated learning channels, but dyscalculics, who need to mainly focus on one specific channel, at least at the beginning. In the long run, module-overlaps must be fostered. Thus, if the sensual stimuli through the 'Virtual Theatre' are further refined, it will serve the representational mode of any dyscalculic and can thus serve as a means to transfer any learning content.

11 Conclusion

All in all, it has been shown that the 'Virtual Theatre' at RWTH Aachen University has a lot of potential to attract the learning 'channels' of dyscalculics. However, it is not at all fathomed. Currently, the creation of a surrounding, which enables the user to almost behave naturally in a virtual world which is accompanied with didactically valuable elements, for dyscalculics but also for non-handicapped students, is still in its initial stage. Nonetheless, the implementation of dyscalculic-equitable, academic learning environments is only one of the many measures which have to be taken in order to make the German education landscape more suitable for them. Ideally, they should be initiated in primary school and continued throughout continuative training of any kind. The question is whether other universities will in the long run offer equally interactive and multi-sensory learning conditions for marginal groups, since acquisitions like the 'Virtual Theatre' are not only costly, but require a lot of programming and didactic expertise. Until comparable learning conditions can be found in more academic institutions, large-scale cooperation is and will be key. What is crucial is that the facilitation of a dyscalculic's every-day study life is not only of administrative, but content-related nature.

Back to the micro-level, the non-handicapped studentship as well as the general public should be better informed about dyscalculics and their mental potential. They should internalize that this marginal group will contribute to technological and societal developments once they are enabled to be a fully-fledged part of academic education. Thus, there is a threefold interplay between public information and the thereof resulting acquaintance with dyscalculia, the early advancement and inclusion of dyscalculic pupils in schools and finally, ties into higher education needed. Only then, the "forestalled elite" will make it to where they belong, just like any other fringe group – into the middle of a creative, tolerant and forward thinking society 4.0.

Acknowledgments This work was supported by the project ELLI (Excellent Teaching and Learning within engineering science) at the RWTH Aachen University in terms of investigating laboratory experiment simulations and Remote Labs.

References

1. Spiegel Online – Unispiegel. Studenten in Deutschland: So viele gab's noch nie, 2014. http://www.spiegel.de/unispiegel/studium/studentenzahl-2-7-millionen-studieren-an-deutschen-hochschulen-a-1005107. Retrieved online, May 21st, 2015
2. Deutsches Studentenwerk. Nachteilsausgleich: Antragsverfahren und Nachweise. http://www.studentenwerke.de/de/content/nachteilsausgleich-antragsverfahren-und-nachweise. Retrieved online, May 21st, 2015
3. Allgemeiner Studierendenausschuss der RWTH Aachen. Gleichstellung. https://www.asta.rwth-aachen.de/de/startseite. Retrieved online, May 21st, 2015
4. T. Käser, A.G. Busetto, G.M. Baschera, J. Kohn, K. Kucian, M. von Aster, M. Gross. Modelling and optimizing the process of learning mathematics, 2012. http://link.springer.com/chapter/10.1007 Retrieved online, May 21st, 2015
5. Universität Würzburg. Informationen zum Nachteilsausgleich für Studierende mit Behinderung und chronischer Erkrankung. http://www.behindertenbeauftragter.uni-wuerzburg.de/fileadmin/32500250/_temp_/Broschuere_Nachteilsausgleich.pdf. Retrieved online, May 21st, 2015
6. L. Breiman. Bagging predictors, 1996. http://statistics.berkeley.edu/sites/default/files/tech-reports/421.pdf. Retrieved online, 21st, 2015
7. J. Haffner, K. Baro, P. Parzer, F. Resch. Heidelberger rechentest, 2005. http://www.testzentrale.de/programm/heidelberger-rechentest.html. Retrieved online, May 21st, 2015
8. Dyscalculia.org (n.d.). Best tools. http://www.dyscalculia.org/math-tools. Retrieved online, May 21st, 2015
9. Dybuster. Dybuster ortograph & Dybuster calcularis. http://www.dybuster.com/orthograph. Retrieved online, May 21st, 2015
10. K. Schuster, M. Hoffmann, U. Bach, A. Richert, S. Jeschke, Diving in? How users experience virtual environments using the virtual theatre. In: *Proceedings of the 3rd International Conference on Design, User Experience, and Usability (DUXU 2014), Heraklion, Crete, 22–27 June 2014, Lecture Notes in Computer Science Springer*, vol. 8518. Springer, 2014, *Lecture Notes in Computer Science Springer*, vol. 8518, pp. 636–646
11. K. Schuster, D. Ewert, D. Johansson, U. Bach, R. Vossen, S. Jeschke, Verbesserung der Lernerfahrung durch die Integration des Virtual Theatres in die Ingenieurausbildung. In: *Innovationen für die Zukunft der Lehre in den Ingenieurwissenschaften*, ed. by A.E. Tekkaya, S. Jeschke, M. Petermann, TeachING-LearnING.EU discussions, TeachING-LearnING.EU, 2013, pp. 246–260
12. M. Hoffmann, K. Schuster, D. Schilberg, S. Jeschke, Bridging the gap between students and laboratory experiments. In: *Virtual, Augmented and Mixed Reality. 6th International Conference, VAMR 2014, Held as Part of HCI International 2014, Heraklion, Crete, Greece, June*

22–27, 2014: proceedings, Heraklion, Crete, Greece, ed. by R. Shumaker. Springer, Cham, 2014, Lecture notes in computer science, 8525–8526, pp. 39–50

13. M. Hoffmann, K. Schuster, D. Schilberg, S. Jeschke, Next-generation teaching and learning using the virtual theatre. In: *4th Global Conference on Experiential Learning in Virtual Worlds Prague, Czech Republic*. 2014

14. A. Schabman, C. Klicpera, *Lehasthenie – LRS: Modelle, Diagnose, Therapie und Förderung*. Stuttgart, 2013

15. Bundesverband Legasthenie und Dyskalkulie. Legasthenie. http://www.bvl-legasthenie.de/legasthenie. Retrieved online, May 21st, 2015

Blended Learning and Beyond – Schlüsselfaktoren für Blended Learning am Beispiel der RWTH Aachen

Laura Lenz, Larissa Köttgen and Ingrid Isenhardt

Zusammenfassung Blended Learning – ein jeder versteht es anders. Dementsprechend gibt es unzählige Umsetzungsmöglichkeiten im Rahmen digitaler, interaktiver, interdisziplinärer und individualisierbarer Lehre. An der RWTH Aachen genießt die kontinuierliche Verbesserung der Lehre einen hohen Stellenwert. In diesem Sinne steht der Einsatz von didaktisch fundierten Methoden des Blended Learning ganz besonders im Fokus der Bemühungen. Eine Umsetzung moderner, hybrider Lehre, die Präsenz- und Selbstlernphasen Dank der Möglichkeiten der Neuen Medien miteinander vereint, wird engagiert durch das Rektorat und die 2014 ins Leben gerufene *Blended Learning Initiative* realisiert. Durch dieses Projekt erhalten motivierte Lehrende ein breites Angebot an Hilfestellungen. Neben zentralen Stützen in der Hochschulverwaltung und dem IT Center, fungieren Einrichtungen wie das CiL (Center for Innovative Learning Technologies), ExAcT (Center of Excellence in Academic Teaching) und der Service „Medien für die Lehre" an der RWTH für die Lehrenden als zentrale Ansprechpartner [1]. Dieser umfangreiche Support erleichtert den Lehrenden an der RWTH die Beantwortung zum Beispiel folgender Fragen: Wie gestalte ich Blended Learning passend zur eigenen Person und den Lehrinhalten? Welche technischen und organisatorischen Herausforderungen ergeben sich hieraus? Der vorliegende Beitrag gibt am Beispiel der Lehrveranstaltung Kommunikation und Organisationsentwicklung im Maschinenbaustudium einen Einblick in die Erfahrungen mit Blended Learning an der RWTH Aachen.

Schlüsselwörter Blended Learning · Universitäre Bildung · Flipped Classroom · Problembezogenes Lehren und Lernen · Ingenieurwissenschaften · Hochschullehre · RWTH Aachen

L. Lenz (✉) · L. Köttgen · I. Isenhardt
IMA/ZLW & IfU, RWTH Aachen University,
Dennewartstr. 27, 52068 Aachen, Germany
e-mail: laura.lenz@ima-zlw-ifu.rwth-aachen.de

Originally published in "Das Inverted Classroom Modell Begleitband zur 5. Konferenz "Inverted Classroom and Beyond"", ©ikon VerlagsGmbH 2016. Reprint by Springer International Publishing Switzerland 2016, DOI 10.1007/978-3-319-42620-4_30

1 Einführung

Der Begriff des Blended Learning beschreibt eine Lernform, bei der die Vorteile einer Präsenzveranstaltung mit denen des E-Learnings vereint werden [2]. Dieses Konzept ist seit rund zehn Jahren ein zentrales Thema in der universitären Bildung. Es beinhaltet insbesondere Interaktivität, Interdisziplinarität, Individualität und ortsunabhängige Abrufbarkeit der Inhalte. Dabei steht vor allem eins im Vordergrund: der Studierende als individuelle Persönlichkeit mit einzigartigen Präferenzen, Denk- und Lernmustern. An der RWTH Aachen in Deutschland, genauer in der verpflichtenden Erstsemester-Veranstaltung für Maschinenbau Studierende „Kommunikation und Organisationsentwicklung" (KOE) gehalten von Prof. Dr. Ingrid Isenhardt, wird agiles Blended Learning seit Jahren praktiziert und weiterentwickelt. Seit dem Wintersemester 2013/2014 wird die KOE konkret als *Flipped Classroom* [3] angeboten. Circa 1.500 Erstsemester-Studentinnen und Studenten finden in jedem Wintersemester ihren Weg in den (teils virtuellen) Hörsaal. Im Zentrum der Veranstaltung steht der Methoden-Mix. Dieser besteht in erster Linie aus einer Online-Vorlesung, die die Studierenden ortsungebunden an ihren Laptops, Tablets und Smartphones anhören, aber vor allem sehen können (Abbildung 1), sowie das hiermit kombinierte Diskussionsforum, bei dem Inhalte in einem analogen Hörsaal vertieft, Fragen gestellt und Diskussionen geführt werden (Abbildung 2). Hinzu kommen Praxis-Beiträge von Industriepartnern, die den Studierenden die Relevanz und

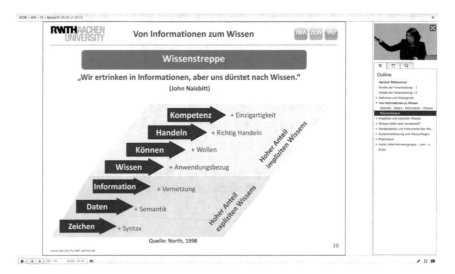

Abbildung 1 So wird eine Online-Vorlesung der KOE visualisiert. Die Studierenden sehen die Inhalte zentral, die Lehrenden sind parallel im Video zu sehen und zu hören (hier oben rechts). Es kann hierbei zwischen Inhalten/Kapiteln digital "geblättert" werden

Abbildung 2 Impressionen aus dem Diskussionsforum (Interaktion und Partizipation im Hörsaal)

Realität der gelernten Theorie nahebringen. Um einen möglichen Unternehmensalltag, aber auch hiermit verbundene Probleme zu erleben, werden diese Inhalte mit der Unternehmenssimulation *ROBOFLEX* gekoppelt, bei der es um die Generierung eines Businessmodells rund um ein autonomes Fahrzeug der Zukunft geht (Abbildung 3).

Abbildung 3 Impressionen aus der Unternehmenssimulation ROBOFLEX (Prototypenbau)

2 Fundament und Entwicklung der KOE

Die Bologna Reform brachte im Jahre 1999 zahlreiche Änderungen mit sich. Eine davon war eine stärkere Fokussierung auf kompetenzorientiertes Lehren [4]. Die Identifikation praxisrelevanter Inhalte ist hierbei nicht die größte Herausforderung. Viel mehr liegt sie in der ständigen Aktualisierung, besonders aber in der attraktiven Vermittlung in XL-Kurse, also zum Beispiel Vorlesungen mit besonders großen Hörerzahlen von mehr als 1500 Studierenden [5]. Auch in die kontinuierliche Überarbeitung der Vorlesung KOE sind diese Faktoren maßgeblich. Ein weiterer wichtiger Faktor ist die Attraktivität der Inhalte. Ingenieur-Studierende aus den unteren Semestern begeistern sich eher für „harte Fakten" rund um Maschinen und Mechanik, als für sogenannte „Soft Skills", wie zum Beispiel Wissensmanagement, Teamentwicklung und Kommunikationskultur [6]. Ein anderer Faktor ist das sinkende Alter der Studierenden im ersten Semester. Aufgrund der Schulreformen sind viele Zuhörende nicht volljährig. Auf diese sehr junge Zuhörerschaft muss in speziellem Maße eingegangen werden, zum Beispiel bei der Motivation der Studierenden sich in der Großgruppe an der Diskussion zu beteiligen.

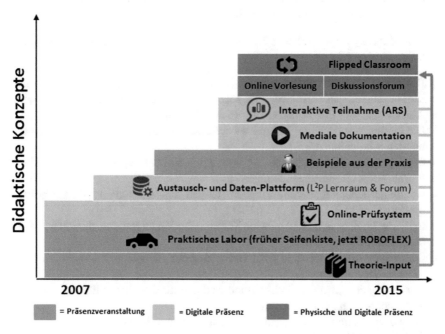

Abbildung 4 KOE an der RWTH Aachen - Didaktische Konzepte und ihre Weiterentwicklung

Die Implementierung eines passenden Blended Learning Modells ist komplex und ein langjähriger Prozess (Abbildung 4). In den letzten acht Jahren wurde die KOE ausgehend von der Theorievermittlung in einer klassischen Hochschulvorlesung und einer dazugehörigen Praxis-Übung (dem Bau einer Seifenkiste im Team) unter anderem durch folgende (interaktive) Elemente erweitert:

- Erstellung einer Online-Klausur mit Hilfe eines Online-Prüfsystems (OPS)
- Die Nutzung der RWTH-weiten, digitalen Plattform „L^2P", von der alle Lerninhalte jederzeit online abrufbar sind und in Foren kommentiert und diskutiert werden können („L^2P Lernraum")
- Praxis-Inputs von ausgewählten Experten aus der Wirtschaft inklusive moderierter Diskussion zwischen Gastreferent und Studierenden (Input: analog im Hörsaal und virtuell in der Online-Vorlesung)
- Die Nutzung eines Audience Response Systems (ARS), im Falle der KOE „Polleverywhere", zur Unterstützung der Partizipation in der Großgruppe
- Die Ersetzung des Seifenkisten-Baus durch die Unternehmenssimulation ROBOFLEX, in deren Rahmen sich autonom bewegende Fahrzeuge mithilfe von Lego Mindstorms NXT Robotern konstruiert, programmiert und gebaut werden. Ein dreiminütiges Businessvideo, produziert durch die Studierenden, vermittelt Konzept und Realisierung der Geschäftsidee.

Eine dauerhafte Optimierung der KOE beinhaltet, bezugnehmend auf die einzelnen Elemente, weiterhin:

- Die jährliche Weiterentwicklung des Lehrkonzepts anhand aktuellster Beispiele
- Monitoring relevanter Branchen zur Identifikation interessanter Praxis-Referenten
- Abstimmung der Praxis-Inputs mit Blick auf die Inhalte der Lehrveranstaltung
- Die Optimierung des L^2P Lernraums zur verbesserten Kommunikation unter und mit den Studierenden
- Die Erweiterung des Angebots zur Interaktion und Partizipation im Diskussionsforum (ARS, Großgruppenübungen etc.)
- Kontinuierliches Trendmonitoring zur stetigen Ausgestaltung des Diskussionsforums an sich
- Individuelle Betreuung und hybride Kommunikation im Rahmen der ROBOFLEX Unternehmenssimulation
- Einbindung der Expertenjury zur Bewertung der virtuellen Unternehmen durch Vertreter aus der Wirtschaft
- Anpassung der gewählten Lehrmethoden: die Methode selbst sollte Teil des Lernprozesses sein, nicht nur ihr Inhalt
- Relevanzprüfung und Anpassung der Klausurfragen (jährlich)
- Adaption der Evaluationsbögen an das aktuellste Lehrkonzept

3 Kritische Reflektion des Status Quo und was folgt

Die ständige Weiterentwicklung des Formats mit Blick auf Exzellenz, Realitätsnähe und Studierendenwünsche stellt jedes Semester die zentrale Herausforderung für die Dozentin der KOE und ihr Team dar. Die Motivation und der Lernwille der Studierenden sind hierbei immer wieder maßgeblich. Im aktuellen Wintersemester 2015/2016 zeichnete sich besonders deutlich ab, dass die Teilnehmenden intensiv mit den online herausgegebenen Materialien arbeiteten. Dies zeigte sich durch das dezidierte Nachfragen bezüglich der Lerninhalte, die allgemeine Qualität der Beiträge zum Diskussionsforum, sowie das stetige Mitbringen von zur Verfügung gestellten Materialien. Hierauf wurde unmittelbar agil reagiert. Die akzentuierte Wiederholung der (Praxis-) Inputs wurde stark gekürzt, die Praxisnähe anhand von Beispielen aus der Wirtschaft noch mehr in den Fokus gerückt. Zudem wurde das Diskussionsforum mit zahlreichen interaktiven Großgruppenübungen ergänzt, zum Beispiel mit der gemeinsamen Darstellung der Wissenstreppe nach North [7]. Der dazugehörige, direkte Austausch mit den Studierenden ermöglicht es außerdem, neben den Diskursen über Fachthemen, auch ein direktes Feedback zum jeweils gewählten Konzept zu erhalten. Die Resonanz, ob eine Übung lehrreich, ansprechend und verständlich ist oder nochmals überarbeitet werden muss erfolgt unmittelbar in jedem Diskussionsforum, beispielsweise durch ARS Fragen.

Auf Feedback Ebene muss das Format der Evaluation jährlich kritisch betrachtet werden. Da die Standard-Erhebung von Hochschulen, die häufig nur sehr allgemeine, strukturelle Fragen beinhalten, bei einem solch individuellen Lehrformat oft nicht ausreichen um den Gesamteindruck der Studierenden in Gänze zu erfassen. Eine Zusatzevaluation, ausgerichtet auf das Veranstaltungsformat an sich, ist gegebenenfalls von Vorteil, wie sie auch in der KOE durchgeführt wird (zusätzlich zum RWTH Evaluationssystem „EvaSys"). Im Gegensatz zu „EvaSys" beinhaltet die KOE-Zusatzevaluation unter anderem Fragen zum persönlichen Eindruck bezüglich des Diskussionsforums (zum Beispiel mögliche Gründe für eine verhaltene Beteiligung: „Ich war unsicher in der Gruppe zu diskutieren"), die die Studierenden auf einer Likert-Skala von „sehr zutreffend" bis „nicht zutreffend" bzw. „keine Angabe" beantworten.

Die Ergebnisse der Zusatzevaluation machen deutlich, dass gesteigerte Relevanz durch Praxisbezug keine idealisierte Erwartungshaltung ist, sondern die Wichtigkeit der Inhalte auf diesem Wege tatsächlich verinnerlicht wird. Die detaillierte Studierendenbefragung aus den Wintersemestern 2014/2015 und 2015/2016 zeigt diesbezüglich, dass mehr als 75% der Studierenden die Programmierungsschulung, sowie die ROBOFLEX-Roboterwerkstatt als nützlich und problemorientiert empfanden (WS 2014/2015). Zusätzlich waren signifikante Zusammenhänge zwischen der Qualität der Kommunikation innerhalb der Teams und der erfolgreichen Lösung von internen Problemen messbar. Je mehr die Studierenden die Inhalte der KOE reflektierten und nach Möglichkeit anwendeten, desto relevanter erschienen ihnen die umgesetzten Theorie-Inhalte. Insgesamt wurden besonders die diskutierten Fallbeispiele aus der Online Vorlesung, sowie die dazugehörigen Praxis-Inputs als

Ich habe die besprochenen
Fallbeispiele als lehrreich empfunden

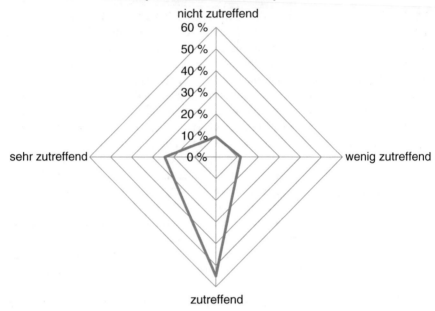

Abbildung 5 Auszug aus der KOE Zusatzevaluation im WS 2015/2016

wertvoll empfunden (Abbildung 5, WS 2015/2016). Weiterhin empfanden 51% der Befragten den Online Support über den L^2P Lernraum als hilfreich, sowie die Zeitfenster, die für das Sichten der zur Verfügung gestellten Materialen angesetzt waren (WS 2014/2015). 44% äußerten Wünsche nach Optimierung des L^2P Lernraums, der mit Blick auf die Benutzerfreundlichkeit noch Potential aufweist [8].

Die Frage, die sich aus dem Status Quo ergibt ist, welche Schritte in Hinblick auf eine zeitgemäße Weiterentwicklung die logische Konsequenz ist. Für Studierende ist Transparenz mit Blick auf die Klausur immer ein zentrales Thema [9]. Es ist daher geplant, den Studierenden in Form von Zusammenfassungen klar zu machen, welche Themen aus den 12 Modulen der KOE besonders prüfungsrelevant sind. In zwei- bis dreiminütigen Scribbles (Abbildung 6) werden ihnen die wichtigsten Inhalte nochmals anschaulich in Erinnerung gerufen. Die Verbildlichung von Lerninhalten hat sich als nicht nur beliebtes, sondern auch wirksames Instrument für die ansprechende und witzige Verinnerlichung von Inhalten bewährt [10]. Gemeinsam mit dem Service „Medien für die Lehre" werden hierzu aktuell Umsetzungsmöglichkeiten für das Wintersemester 2016/2017 geprüft.

Abbildung 6   ScribbleRohlinge zum Thema Teamarbeit

4   Zusammenfassung

Der vorliegende Beitrag hat dargelegt, dass Blended Learning in seiner individuellen Umsetzung nicht nur ein langfristiger Optimierungsprozess ist, sondern auch von individuellen Möglichkeiten, institutionellen Rahmenbedingungen, Ressourcen und professionellen Begleitevaluationen abhängt.

Realisierungs- und/oder Optimierungsschritte könnten beinhalten:

- Die Unterstützung neuer Lehr- und Lernformate durch die Hochschulausrichtung (Relevanz)
- Die in der Regel sehr kostenaufwändige Produktion von hochwertigen und aktuellen Lehrvideos
- Die Implementierung einer bedarfsorientierten, nutzerfreundlichen Online-Plattform und deren Betreuung
- Die ständige Analyse und Definition von Lehrplänen und Themen mit Blick auf die Anforderungen der Studierendenschaft (zum Beispiel Heterogenität), didaktische und fachbezogen
- Die Gestaltung und Umsetzung von passenden Zusatzevaluationen, die häufig sehr zeitintensiv sind

Die Strukturen und Möglichkeiten jeder Hochschule und der verschiedenen Disziplinen sind also so spezifisch, dass sich ein zentrales, allgemeingültiges Konzept kaum definieren lässt. Allenfalls lassen sich hier nützliche Beispiele finden, wie an der Umsetzung in der KOE an der RWTH Aachen illustriert (zu weiteren Beispielen, auch aus anderen Hochschulen, vgl. „Vorlesbar" [11]). Es gilt in jedem Fall, die eigenen Ressourcen, sowie die der jeweiligen Hochschule gut zu kennen, um ideale Ansprechpartner zu finden und das zielgruppenorientierte Angebot optimal in bereits bestehende Angebote einzubetten—und konstant zu optimieren.

„Erzähle mir und ich vergesse. Zeige mir und ich erinnere mich. Lasse es mich tun und ich verstehe". Dieses Zitat von Konfuzius ist das Credo der KOE. Es fließt ein in den Anspruch des Blended Learning, eine hybride Lehr- und Lernform mit aufeinander abgestimmten digitalen und analogen Elementen zu gestalten. Nur so kann das Interesse der Zuhörerschaft über ein ganzes Semester geweckt werden. Lehrende sollten also im Blick behalten, dass beim Blended Learning neben der guten Vorbereitung vor allem eines im Vordergrund stehen sollte: Der Spaß an der Arbeit mit den Studierenden.

Literaturverzeichnis

1. RWTH. Das Projekt Blended Learning und ETS 2014 bis 2017, 2016. http://www.rwth-aachen. de/go/id/hjgq
2. G. Reinmann-Rothmeier, *Didaktische Innovation durch Blended Learning: Leitlinien anhand eines Beispiels aus der Hochschule*. Huber, Bern, 2003
3. M. Stansbury. Teachers turn learning upside down, 2010. http://www.eschoolnews.com/2010/ 12/22/teachers-turn-learning-upside-down/
4. M. Schmidt, Wer ist Mister Bologna? Vor 15 Jahren wurde das Bachelor-Master-System beschlossen. Die Zeit (26), 18. Juni 2014
5. S. Schröder, D. Janßen, S. Leisten, R. Vossen, I. Isenhardt. On-professional competences in engineering education for xl-classes, 2013. http://ieeexplore.ieee.org/xpls/abs_all.jsp? arnumber=6684783&tag=1
6. L. Köttgen, S. Stiehm, C. Tummel, A. Richert, I. Isenhardt. Next level blended learning for an excellent engineering education, 2015. http://publications.rwth-aachen.de/record/227292?ln= de
7. K. North. Die Wissenstreppe, 2002. http://qib.f-bb.de/wissensmanagement/thema/wissen/ wissenstreppe.rsys
8. S. Stiehm, L. Koettgen, S. Thelen, M. Weisskopf, F. Welter, A. Richert, I. Isenhardt, S. Jeschke, Blended learning through integrating lego mindstorms nxt robots in engineering education. In: *Proceedings of the ASME International Mechanical Engineering Congress and Exposition (IMECE 2015)*. 2015
9. B.W. J. Wildt. Lernprozessorientiertes Prüfen im "Constructive Alignment", 2007. http://www. hrk-nexus.de/fileadmin/redaktion/hrk-nexus/07-Downloads/07-03-Material/pruefen.pdf
10. inside. Motion Scribble. Komplexes einfach dargestellt, 2012. http://www.checkpoint-elearning.de/article/11668.html
11. V. Stehling, K. Schuster, U. Bach, A. Richert, I. Isenhardt, ed., *Vorlesbar: Methodenhandbuch für Vorlesungen mit großen Hörerzahlen*. 2013
12. L. Köttgen, S. Schröder, E. Borowski, A. Richert, I. Isenhardt. Flipped classroom on top – excellent teaching through a method-mix, 2014. http://publications.rwth-aachen.de/record/ 227292?ln=de

Investigating Mixed-Reality Teaching and Learning Environments for Future Demands: The Trainers' Perspective

Lana Plumanns, Thorsten Sommer, Katharina Schuster, Anja Richert and Sabina Jeschke

Abstract The first three industrial revolutions were characterized by the invention of water and steam engine, centralized electric power infrastructure and mass production as well as digital computing and communications technology. The current developments caused by the fourth revolution, also known as "Industry 4.0", pose major challenges to almost every kind of work, workplace, and the employees. Due to the concepts of cyber-physical systems, Internet of Things and the increasing globalization, remote work is a fast-growing trend in the workplace, and educational strategies within virtual worlds become more important. Especially methods as teaching and learning within virtual worlds are expected to have an enormous impact on advanced education in the future. However, it is not trivial to transfer a reliable educational method from real to the virtual worlds. Therefore, it is important to adapt, check and change even small didactic elements to guarantee a sustainable learning success. As there is a lot of ongoing research about using virtual worlds for the training of hazardous situations, it has to be figured out which potential those environments bear for the everyday education of academic staff and which competencies and educational support trainers need to have respectively can give in those worlds. The used approach for this study was to investigate the trainers' didactic perspective on mixed-reality teaching and learning. A total of ten trainers from different areas in Germany took part in this study. Every participant pursued both roles: the teaching and the learning part in a virtual learning environment. In order to assess the learning success and important key factors the experiment yields data from the participants' behavior, their answers to a semi-structured interview and video analysis, recorded from the virtual world. Resulting data were analyzed by using different qualitative as well as quantitative methods. The findings of this explorative research suggest the potential for learning in virtual worlds and give inside into influencing variables. The online gaming experience and the age of participants can be shown to be related to participants' performance in the virtual world. It looks like the barriers for the

L. Plumanns (✉) · T. Sommer · K. Schuster · A. Richert · S. Jeschke
IMA/ZLW & IfU, RWTH Aachen University,
Dennewartstr. 27, 52068 Aachen, Germany
e-mail: lana.plumanns@ima-zlw-ifu.rwth-aachen.de

Originally published in "18th International Academic Conference, London, UK, 25-28 August 2015", © International Institute of Social and Economic Sciences (IISES) 2015.
Reprint by Springer International Publishing Switzerland 2016,
DOI 10.1007/978-3-319-42620-4_31

affected trainers are low regarding utilization of virtual worlds. Together with the mentioned advantages and possible usages, the potential of these setups is shown.

Keywords Education · Mixed-Reality · Teaching · Virtual World

1 Introduction

Recent examinations of 702 today's occupations show how many million tasks and areas are affected by the ongoing digitalization [1]. While some occupations will be ceased, others will change, and new ones will occur. Responsible for these change are today's concepts like e.g. "Industry 4.0" [2–5] or Internet of Things (IoT). The ongoing globalization will not end and therefore, employees have to follow the trend. Occupations like e.g. teachers and trainers change through the trend to massive and remote teaching as e.g. massive open online courses (MOOCs) [6]. Powered by serious games [7] and gamification concepts, virtual worlds push into the teaching and training activities. Further, the produced data by the usage of these technologies is an enabler for learning analytics [8] and general analysis-driven methods.

Regardless of these possibilities, a teacher and trainer must be able to reflect those options. The potential usage of media and technology depends on the learning subject [9]. Thus, for some learning subjects virtual worlds are suitable. Today, virtual worlds are used to train uncommon scenarios e.g. major incidents [10] or can be used to teach invisible processes e.g. the basics of a calculator. Besides incidents and inaccessible or non-existing places, also the training of dangerous activities is a possibility for such virtual worlds [11]. Researchers identified effects and advantages of virtual worlds as method for trainings, e.g. in some cases an increase of team performance of about 50 % or the fact that in case of 62 %, the usage of a virtual worlds had the same effectiveness as traditional methods [10].

Further, remote work is constantly pushing forward. Working remotely whether from home, a coffee bar or another place is booming. Research suggests that more than half of today's office-based employees will regularly be working remotely within the next decade, thanks to technological advances in the workplace [12]. Advantages are among others more efficient agreements due to avoided travel time and a reduction of costs and the enhanced comfortability for the user [13].

Due to current technological capacities it is possible to control machines and even whole factories remotely, so that no instructor has to be in place [14]. This technical development obviously shows the future requirements for such employees: While in the past an engineer was responsible for handling a specific local machine, tomorrow's engineer can control multiple factories remotely across the world. Hence, the remote collaboration is an important part of future companies and engineers might be confronted with e.g. intercultural issues. Prospectively, the future engineers must be aware of all processes, which are running at a factory instead of controlling one single process step locally. Further, the engineer has to understand and know all kind of machines at a factory and must be empowered to know their limits in order to control the whole factory.

Therefore, with the advent of Industry 4.0, a large market arises in the field of virtual training and settings for collaboration and schooling. But not every approach that is technically feasible improves users' learning outcomes; hence the danger of designing expensive virtual learning environments without having a positive effect on the users' learning is obvious. Thus, in order to ensure sustainable learning outcomes in virtual learning environments, people who provide professional skills in the physical world have to be involved. Based on previous studies about students' perspective on virtual education [5], this study analyses the trainers' perspective of teaching and training using a virtual world for educational purposes with immersive hardware.

Before such technologies can be used in everyday training, teaching and learning with groups, further research regarding the transfer of common methods is necessary. Is it suitable for a trainer to moderate a group of students in a virtual world just like in the real world, even if the simulated area is huge? Is it possible to transfer well-known methods such as e.g. think-pair-share into a virtual reality setting? Currently, answers to questions like this are unknown and object of further research activities.

Following the high expectations regarding learning and working within virtual worlds, this research assessed trainers' behavior in and opinion about virtual learning environments. Hence, this experiment yields data from the participants' spatial behaviour and movements, their answers to questions regarding education and their experience within mixed-reality virtual learning environments to answer the question: Is today's society ready for remote training by using these technical possibilities? This study tries to give a first answer to this question by investigating the trainers' perspective inside the mixed-reality with virtual learning environments by a threefold purpose. First, an overview of the experimental setup is given; then the experimental study shows the challenges that tomorrow's trainers have to face and variables that might affect their performance. Finally, some technical and conceptual limitations are shown to guide further research in this field.

2 Setup and Virtual Environment

In order to get inside of technological details of this investigation, this section gives an elaborate description of the setup, the used environment, and the technical conditions. Hence, today's minimal technical requirements to provide an immersive virtual learning environment are shown. As the essential structure of this study, only one location was used in which two persons were participating for each pass. Thus, no headset was required for the verbal communication. A head mounted display (Oculus Rift DK2) served as immersive hardware to enter the virtual learning environment. Head-mounted displays had been used successfully in previous studies and the usage of these displays within virtual worlds is connected to strengthen sensations of immersion, flow, and spatial presence [15]. Due to users' attention allocation

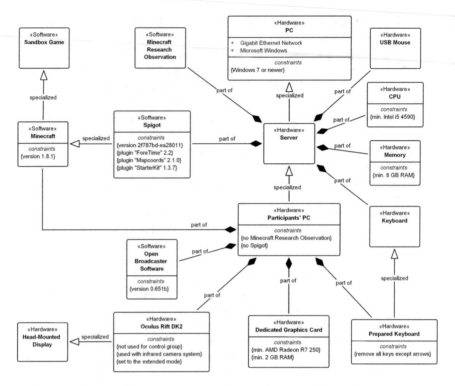

Figure 1 The UML (Unified Modeling Language) model about the hard- and software requirements for this study (*Source* Self-created model)

processes while wearing these displays it is possible to blend the physical with the virtual world and hence design virtual interaction and knowledge transfer as naturally as possible.

The hardware requirements for this investigation are almost standard, as the model in Figure 1 shows. A few components, however, need further consideration: Since this study was investigating trainers perspective on virtual education with the aid of immersive hardware, specifically the Oculus Rift, the participant's PC needed a dedicated graphics card in order to work with the head-mounted display. Further, for the convenient control of the participant's avatar in the virtual learning environments with equipped head-mounted display, a prepared keyboard was used. Except the arrow keys, all other keys were dismounted to prevent participants from pressing wrong or undesired keys. This modification is necessary because the participants cannot see the physical world while the head-mounted display is equipped. Additionally, a standard-sized USB mouse was used to enable a 360-degree turn, while sitting on in front of the PC.

The head-mounted display was used with the infrared camera system and was set to the extended mode. Thus, it behaves like an additional display. However, this setup is not the optimal setting regarding lowest system response time; it is very stable and durable. In comparison, the direct mode offers a lower response time, which should prevent the simulator sickness better, but it appears to be not very stable within forgone pretests. For the optimal immersion into the virtual world, the system was calibrated regarding participants' individual interpupillary distance (IPD) and their body height.

As setting for the virtual learning environment, the open-world and sandbox game Minecraft was chosen [16]. This program has been used successfully for learning and teaching purposes beforehand and allows participants among other things to freely explore a virtual environment [17]. With Minecraft, the creation and manipulation of virtual worlds are efficiently possible. For this study, two virtual worlds respectively virtual learning environments were necessary: The first world comprised a tutorial to teach the basics of Minecraft, which was the participants' trainee part. The second world served as the task that the participants should solve as a trainer. To prevent simulator sickness [18] as good as possible, both worlds are designed to be usable without jumping and climbing.

Figure 1 also shows, that "Spigot" was used on the server-side to provide the worlds. Its configuration was also important to make the worlds useful for scientific studies and virtual learning environments. The difficulty was changed to be friendly, which means that eventually present non-player characters do not attack the participants during the experiment. Further, the appearance of any non-player characters and the participants' possibility to attack each other were disabled.

3 Method

3.1 Participants

Ten professional trainers aged between 24 and 60 years ($M = 40.7$; $SD = 13.2$; $n = 2$ female) participated. Participants were recruited from different areas in Germany and active in various domains of personal development to cover a broad range of professions.

3.2 Assessment of Participants' Objective Behaviour

A screen capture tool recorded participants' behavior within the virtual learning environment. The assessment of these objective data is important to clarify whether the basic competencies for learning within virtual learning environments are given and to track participants' performance during the experiment [19]. A calibration with subsequent validation procedure was conducted prior to the experiment. To assess participants attention allocation processes, their gaze fixations, collaboration behavior, communication as well as fluency and speed of movement during the experiment was assessed with the OBS (Figure 2). These assessments were invisible to participants. Participants' movement fluency and speed allowed assumptions about their habituation progress, whereas the kind of communication and the progress of solving the subtasks reflected the efficiency as a trainer.

Participants' way of movements within the virtual learning environment was also assessed and extracted using Minecraft Research Observation tool to investigate whether they were e.g. rather following or autonomous and leading in the trainer part. These psychomotor skills can be classified into seven different categories, ranging from simple to complex: (1) perception, (2) readiness to act, (3) imitation, (4) habitual movement patterns, (5) complex overt response, (6) adapting the movement pattern to reach an aim and (7) creation of new movement patterns [20].

3.3 Semi-structured Interview

In order to assess trainers opinion and experience within the virtual learning environment, a semi-structured interview was conducted after the experiment. The quantitative questions were assessed anonymously on a laptop and covered items as questions regarding simulator sickness, sensations within the virtual world, potential use of these worlds as a training method and benefit of these (rated on a 5-point Likert scale with 1 = not helpful at all to 5 = very helpful). The qualitative questions of the expert interview covered among other things questions like the trainers experience within the virtual learning world, advantages, and disadvantages of this kind of mixed-reality education, further fields of application. Data of the interview were recorded for further transcription (see analysis section).

3.4 Procedure

Upon arrival, all participants signed the informed consent and filled in their demographic data (among other age, gender, previous gaming experience and used gaming modus, specific field of expertise). Each experiment started with a short introduction into the virtual learning environment to get the trainer habituated to the environment and the hardware. During the experiment, the participants' spatial behavior within the virtual learning environment was assessed with a screen capture tool and documented by a scientific researcher. In order to focus on modern engineering education, an engineering task was given to the participants, who entered the virtual environment by wearing a head-mounted display, displayed in Figure 2 (right side).

In order to get a deeper insight into the trainers' perspective of this new way of education as well as the discrepancy between learning and teaching within the virtual environment, every trainer pursued two roles: first the trainee and then the trainer part. The participants entered the virtual environment in groups of two. During this virtual meeting, both participants (the trainer and the confederate) were represented by

Figure 2 Screenshot of the teaching scenario (*left side*); participant entering the virtual learning environment (*right side*) (*Source* Own figure)

avatars. In the first part of the experiment (trainee-part), the participant was instructed by the research director to restore a broken electrical circuits based on the "Redstone" system [21]. This part of the experiment also enabled the trainer to get comfortable with the virtual learning environment and the handling of the unknown hardware. In the second part of the experiment, a similar problem was stated, but this time the participant had to instruct the confederate about how to solve the stated problem, without anticipating the problem-solving process (the trainer part).

This second world provides a small area with a house and a lighting system. The issue was similar to the first world, but, in this case, the electrical circuit was bigger, and the light was partially working. The participants must repair the circuit to activate all the lightings. This task can be divided into several sub-tasks: (1) Get an overview of the electrical setting of the building. (2) Find out, at which spots the electrical circuit is broken. (3) Remember the necessary steps to repair the electrical circuit. (4) Find the spot within the building from where the success of the problem solving can be controlled [5].

Starting with the briefing of the handling of the simplified keyboard and the mouse, the participants had to make sure that the other person (the trainee) feels comfortable within the virtual learning environment and solve the stated problem without further instructions. During the experiment, both screens (trainers and trainees) were recorded and gathered by the mentioned video capture tool for assessing the trainer's and the trainee's behavior in the virtual learning environment (Figure 3).

Figure 3 Screenshot of the OBS recordings with the trainees view (*right side*) and the trainers view (*left side*) (*Source* Own figure)

4 Analysis

The quantitative data were analyzed by using IBM SPSS, version 22 (www.spss. de). Three independent scientific raters coded all qualitative data and scored from 1 = low to 6 = high. Since the interrater variability was high the scores of the three rates were collapsed into one score for each variable and analysed by using SPSS, results of Spearman correlation (two-sided) are displayed in the parentheses below.

5 Results

The screening of the gathered video data (cf. Table 1) indicates that age and online-gaming experience were shown to be related to participants' spatial coordination within the virtual learning environment ($r_{age} = -0.60$, p < 0.05; $r_{gaming} = 0.78$, p < 0.01). Older participants, and those who had no gaming experience showed initial difficulties with the hand-cursor coordination in the habituation phase, which was indicated by more questions, slower and less fluid movements and spatial problems when it came to entering the buildings or reading instructions that were hanged on a wall. Furthermore, these participants were shown to stay longer in the more simple psychomotor categories as proposed by Simpson [20]. However, most difficulties diminished after around the first minute ($M = 43.4$ s; $SD = 23.8$ s) by all participants.

This study showed that participants who reported higher sensations of immersion, got used to the virtual world faster, as seen by their objective behavior within the virtual learning environment (e.g. number of gaze fixations ($r_{gaze} = 0.92$, p < 0.01), spatial coordination ($r_{spatial} = 0.94$, p < 0.01), general task performance (efficiency) as trainee, respectively trainer ($r_{trainee} = 0.91$, p < 0.01); ($r_{trainer} = 0.92$, p < 0.01)). However, whether this is a consequence of participants precondition and mixed-reality devices could not be analyzed in this study and deserves further investigation. All participants were able to solve the stated problem as a trainee and were capable of instructing another person verbally within the second virtual learning environment. Thereby, it was shown that participants who documented their course of action out loud as trainee performed more efficiently; hence they finished the sub-goals faster

Table 1 Inter-correlation matrix with Cronbach's alpha at the diagonals (*Source* Own analysis)

	1	2	3	4	5	6	7	8
1. Age	1							
2. Gaming	−0.65**	1						
3. Gaze	−0.63*	0.86**	0.91					
4. Spatial	−0.60	0.78**	0.98**	0.92				
5. Work	0.62	−0.25	0.02	0.06	1			
6. Immersion	−0.82*	0.74	0.92**	0.94**	−0.35	0.87		
7. Efficiency trainee	−0.72*	0.84**	0.98**	0.95**	0.04	0.91**	0.94	
8. Efficiency trainer	−0.50	0.77*	0.87*	0.85**	0.31	0.92**	0.93**	0.90

* Correlation is significant by level 0.05 (two-sided);
** Correlation is significant by level 0.01 (two-sided)

and transferred their knowledge into their role as a trainer. This behavior was particularly seen by trainers with extended working experience. However, the amount of work experience as trainer does not predict efficiency within the virtual world solely.

Other factors that seemed to influence the success of the trainee phase were among others the self-confidence appearance to the trainer within the virtual learning environment and the time they spend on reading the instructions. No participant got sick due to the simulation. The experimental session took around 35 min per participant. After the experiment, the participants were asked to evaluate their previous learning experience in the virtual learning environment anonymously. The participants' overall conclusion was very positive.

It was shown that even the barriers for the affected trainers are low regarding utilization of virtual learning environments for teaching. The benefit of virtual worlds in teaching got a mean score of 4.5 ($SD = 1$), rated on a 5-point Likert scale (with 1 = not helpful at all to 5 = very helpful). In reply to the question concerning the potential use of these worlds as a training method, all trainers answered with the highest rating (5 = very helpful). The additional expert interview yield more insight into the gathered data and the trainer's perspective. The training was rated as very adaptive, and the participants pronounced the feeling of immersion into the virtual world. The trainers' particularly mentioned the possibility to represent and adapt specific learning content and their feeling of deep and conscious learning as well as the fast familiarization with the virtual learning environment. The speed of movement and the visualization of the environment within the scenarios were stated as pleasant just like the navigation after the habituation phase. The level of difficulty was rated as appropriate for the purpose, and the setup of the virtual learning environment was rated as immediately intuitive for those with gaming experience ($n = 8$).

As possible fields of further application the trainer called among other things fields like emergency services, schooling of security staff and training of techniques, which are too hazardous for the training in field, as well as everyday schooling for higher education and development. As a particular advantage of learning in virtual worlds, the resource efficiency and flexibility, as well as the targeting of many senses at once and the consequential deep learning, were emphasized. Also, the possibility to change or adapt single parameters for training or learning success were mentioned as benefits of virtual learning environment as well as the exploration of environments or settings that are hard to visualize on plane surfaces.

The trainers emphasized the chance to visualize learning success immediately. Also, they told that it is forward-looking, to develop academic and personal needs with the aid of gamification of learning content. To the question of potential difficulties with virtual worlds for trainees and trainers, initial problems with the usage and the acceptance of technology were mentioned as well as a partial negative delay when looking into depth. However, the benefits exceeded the possible adverse effects. When asked about further suggestions, the trainers emphasized the importance of virtual learning environment for learning and future work forms and mentioned their interest in the progressive interlocking of economy and research due to the ongoing digitalization.

6   Conclusion and Outlook

This research gives an insight into trainers view on virtual education. Often, teenagers' and young adults' opinions are positive regarding modern technologies. However, the average age of the participants was relatively high ($M = 40.7$ years; $SD = 13.2$ years). Nevertheless, their opinion regarding training in virtual worlds with immersive hardware was positive. Though, a representative study would be interesting to get an average result from the trainers' and therefore, the teachers' cohorts. A shortcoming of this study was the cross-sectional design, a longitudinal investigation could be useful to give additional insight. Furthermore, a larger sample size would be needed in order to verify the current findings.

After the initial minute, the participants had no major problems with the hand-cursor coordination and everyone was able to show the necessary spatial movement patterns to complete the task. The possibility to verbal communicate within in the virtual learning environment was perceived as beneficial in the trainers self-report for both parts, the trainer, and the trainee part. Thus, the process of pointing toward e.g. an issue was possible by using the verbal communication as well as the avatar's gaze direction and arms. Furthermore, due to real-time communication any issues could be resolved directly. The trainers mentioned as an advantage of this setup that multiple senses are targeted at once. It would be interesting to investigate effects regarding learning, caused by targeting multiple senses in longitudinal studies.

The experience with this study shows possible improvements for further studies: The so-called "spawn point" is the point, where participants enter the world. Although the spawn point was predefined, the virtual worlds do not guarantee a particular point. Instead, a probable spawn area is used. The consequence was that the most participants enter at the right position, but some enter the world e.g. on the roof of the building. For further studies, the usage of an appropriate plugin is planned which allows the definition of a single static spawn point where also the point of view (the angel of the head) is pre-defined.

It is going to be a new challenge for trainers and users to teach, learn and work in the virtual world. Next to nowadays-required competencies, tomorrow's trainers need technical expertise to support users in case of technical problems, malfunctions, and digital literacy to tutor and collaborate appropriately within virtual environments. These requirements represent a major challenge for trainers especially when it comes to groups of users instead of a single one. Also, trainers must be aware if this technology is the right method for a particular subject and if they can transfer their didactic methods into these worlds.

For the successful usage of mixed-reality virtual learning environments for everyday teaching, training and learning for suitable topics, the corresponding industry must provide efficient programs for the creation of such settings. The creation and preparation of suitable worlds must be able and time-efficient as the creation of today's presentations. Some concepts are already promising: For example, some police restricts in Germany got already a suitable solution [22, 23]: They can utilize a pre-defined world with an editor that is customized for the police training to easily setup suitable scenarios. Sandbox games like Minecraft are another approach, which affords the changeability and openness of a suitable tool. However, often, these games are limited regarding related learning topics like e.g. physics, mathematics and training of major incidents, which is often caused by missing mechanical systems at the virtual worlds.

Regardless of the tooling, the virtual reality hardware needs more research regarding usage for training and teaching purposes: Currently, it is not suitable for the students to take or read notes while they are wearing the virtual-reality headsets. One promising approach is the usage of today's speech recognition to take notes. Reading of notes is potentially possible with the right tooling. Another approach is the upcoming augmented reality hardware. They combine the physical world with the virtual reality, respectively can blend the physical world with virtual elements. Because the students can also perceive the reality, notes can be written and read. Nevertheless, for both technologies are topics and scenarios suitable.

Due to new technological developments, the integration of activating learning elements in virtual environments is possible, even as working remotely and controlling machines from afar, offering training with the aid of avatars and many more. This study showed the potential of nowadays-recent progress regarding the ongoing digitalization, but it is necessary to take care that everyone can be involved in this developments. Therefore, continuing this kind of research is an important contribution towards tomorrow's proliferous and digitalized world.

Acknowledgments This work is part of the project "ELLI – Excellent Teaching and Learning in Engineering Sciences" and was funded by the federal ministry of education and research, Germany.

References

1. C.B. Frey, M.A. Osborne. The future of employment: how susceptible are jobs to computerisation?, 2013
2. Federal Ministry for Economic Affairs and Energy. Industrie 4.0 und Digitale Wirtschaft: Impulse für Wachstum, Beschäftigung und Innovation, 04/2015
3. R. Geissbauer, S. Schrauf, V. Koch, S. Kuge. Industrie 4.0: Chancen und Herausforderung der vierten industriellen Revolution, 10/2014
4. S. Jeschke. Everything 4.0 – drivers and challenges of cyber physical systems, 04.12.2013. http://www.ima-zlw-ifu.rwth-aachen.de/keynotes/Forschungsdialog4Dez2013.pdf
5. K. Schuster, K. Groß, R. Vossen, A. Richert, Preparing for industry 4.0 – collaborative virtual learning environments in engineering education. In: *The International Conference on E-Learning in the Workplace Conference Proceedings*, ed. by D. Guralnick. 2015
6. A. Sursock, *Trends 2015: Learning and Teaching in European Universities*. Brussels, Belgium, 2015
7. P. Moreno-Ger, I. Martinez-Ortiz, M. Freire, B. Manero, B. Fernandez-Manjon, Serious games: A journey from research to application. In: *Frontiers in Education Conference Proceedings*. 2014, pp. 1–4. 10.1109/FIE.2014.7044052
8. R.S. Baker, Educational data mining: An advance for intelligent systems in education. IEEE Intelligent Systems **29** (3), 2014, pp. 78–82. 10.1109/MIS.2014.42
9. M. Tesar, K. Stöckelmayr, R. Pucher, M. Ebner, J. Metscher, F. Vohle, Multimediale und interaktive Materialien: Gestaltung von Materialien zum Lernen und Lehren. In: *Lehrbuch für Lernen und Lehren mit Technologien*, ed. by M. Ebner, S. Schön, 2013
10. W. LeRoy Heinrichs, P. Youngblood, P.M. Harter, P. Dev, Simulation for team training and assessment: case studies of online training with virtual worlds. World journal of surgery **32** (2), 2008, pp. 161–170. 10.1007/s00268-007-9354-2
11. J.L. Encarnação. Serious games, ss 2008, 2008
12. P. Sawers. 60% of uk employees working remotely within a decade, 22.02.2012. http://thenextweb.com/uk/2012/02/22/home-sweet-home-60-of-uk-employees-could-be-working-remotely-within-a-decade/
13. R. Ubell, *Virtual Teamwork: Mastering the Art and Practice of Online Learning and Corporate Collaboration*. Wiley, New York, 2010
14. A. Höpner. Steuerungstechnik: Die ferngesteuerte Fabrik, 30.07.2012. http://www.handelsblatt.com/technik/forschung-innovation/steuerungstechnik-die-ferngesteuerte-fabrik/6913260-all.html
15. K. Schuster, Einfluss natürlicher Benutzerschnittstellen zur Steuerung des Sichtfeldes und der Fortbewegung auf Rezeptionsprozesse in virtuellen Lernumgebungen. Dissertation, RWTH Aachen University, Aachen, [Im Druck]
16. D. Short, Teaching scientific concepts using a virtual world - minecraft. Teaching Science (3), 2012, pp. 55–58
17. C. Schifter, M. Cipollone, Minecraft as a teaching tool: One case study. In: *Proceedings of Society for Information Technology & Teacher Education International Conference*, ed. by R. McBride, M. Searson. Association for the Advancement of Computing in Education (AACE), 2013, pp. 2951–2955
18. S. Höntzsch, U. Katzky, K. Bredl, F. Kappe, D. Krause, Simulationen und simulierte Welten: Lernen in immersiven Lernumgebungen. In: *Lehrbuch für Lernen und Lehren mit Technologien*, ed. by M. Ebner, S. Schön, 2013
19. K.A. Wilson, W.L. Bedwell, E.H. Lazzara, E. Salas, C.S. Burke, J.L. Estock, K.L. Orvis, C. Conkey, Relationships between game attributes and learning outcomes: Review and research proposals. Simulation & Gaming **40** (2), 2008, pp. 217–266. doi:10.1177/1046878108321866
20. E.J. Simpson, *The classification of educational objectives in the psychomotor domain*. Gryphon House, Washington, DC, 1972

21. M. Dezuanni, J. O'Mara, C. Beavis, 'redstone is like electricity': Children's performative representations in and around minecraft. E-Learning and Digital Media (12(2)), 2015, pp. 147–163. doi:10.1177/2042753014568176
22. M. Herkersdorf. Virtuell-interaktives Training (vipol) - eine bundesweit einmalige Lösung der Polizei BW, 15.10.2013. http://www.pfa.nrw.de/PTI_Internet/pti-intern.dhpol.local/TagSem/Seminar/Nr48_13/07_Herkersdorf_Internet/TriCAT_ViPol_15102013.pdf
23. C. Lecon, M. Herkersdorf, Virtual blended learning virtual 3d worlds and their integration in teaching scenarios. In: *Computer Science Education (ICCSE), 2014 9th International Conference on*. 2014, pp. 153–158. doi:10.1109/ICCSE.2014.6926446

New Perspectives for Engineering Education – About the Potential of Mixed Reality for Learning and Teaching Processes

Katharina Schuster, Anja Richert and Sabina Jeschke

Abstract The majority of mixed reality scenarios have been mainly the subject of game engines. 'Mixed Reality' describes the combination of virtual environments and natural user interfaces. Here, the user's field of view is controlled by his natural head movements via a head mounted display. Data gloves e.g. allow direct interaction with virtual objects and omnidirectional treadmills enable unrestricted navigation through a virtual environment by natural walking movements. To evaluate perspectives and potential for the use of mixed reality settings within engineering education an experimental study has been carried out, focusing on the impact of spatial presence and flow on cognitive processes. To assess the effects of natural user interfaces on cognitive processes, a two-group-plan (treatment and control group) was established. The mixed reality simulator was used as main stimulus of the treatment group whereas the control group used a laptop as interaction device. The learning environment was kept constant over both groups. The data were collected and interpreted with quantitative methods. Constraints of data collection exist since the influence of the hardware can only be evaluated within a set of independent variables, which consists of a combination of different user interfaces to a mixed reality simulator. Thereby not all of the disruptive factors could be eliminated. In this paper the study and the detailed results are described, which showed advantages especially regarding affective and motivational factors of virtual environments for cognitive processes. In particular, the depth of the resulting spatial presence and the phenomenon of flow are discussed. The paper closes with a discussion of the question, to what extend such innovative technologies establish new possibilities for educational sciences and pedagogics, especially focusing on engineering education and the field of virtual experiments.

Keywords Immersion · Spatial Presence · Flow · Learning · Natural User Interfaces · Engineering Education

---

K. Schuster (✉) · A. Richert · S. Jeschke
IMA/ZLW & IfU, RWTH Aachen University,
Dennewartstr. 27, 52068 Aachen, Germany
e-mail: katharina.schuster@ima-zlw-ifu.rwth-aachen.de

Originally published in "Proceedings of the ASSE's 122nd Annual
Conference & Exposition, Seattle, USA, 14-17 June 2015", © ASEE 2015.
Reprint by Springer International Publishing Switzerland 2016,
DOI 10.1007/978-3-319-42620-4_32

407

1   Introduction – New Perspectives for Engineering Education Through Mixed Reality

A main goal of engineering education is the development of professional skills, so that graduates can apply their knowledge in their later working environment. A proper knowledge transfer is an important precondition for engineers to act competently and to solve different kinds of problems. However, due to the increasing number of study paths as well as the specialization of particularly technical oriented classes, there is a need for the integration of new media into the curriculum of most students [1]. Thus, the visualization of educational content in order to explain theory more concrete and tangible has gained importance. To prepare students adequately for new situations in their work life, virtual reality (VR) can be an effective instrument for learning and teaching processes. By imitating real-world processes, professional skills can be developed, increased or maintained. Especially if the learning process requires expensive equipment or usually would take place in a hazardous environment, the use of simulations is not only advantageous but necessary [2, 3]. Apart from the virtual learning environment (VLE), the hardware of the given user interfaces applied in the simulation environment can affect the learning process [4]. One approach of improving learning with simulations is the development of natural user interfaces.

Classical memory theory claims that if the context in which knowledge is applied resembles the context in which the information has been learned initially, the memory works better. Moreover, how well we can retrieve knowledge from our long term memory depends on the quality of how well we encoded the information in the first place [5]. Sweller's Cognitive Load Theory postulates that learning is as a task, which is partitioned in at least two parallel sub-tasks: Dealing with the content and controlling the learning environment with the respected user interfaces [6]. Therefore a lot of research and development activities follow the assumption that if the user can interface with the system in a natural way, more cognitive resources are available for dealing with the actual content related exercise, which would increase the efficiency of VLEs [7].

However, to assume that hardware or software characteristics automatically lead to better learning outcomes is risky. Not every new approach which is technically feasible improves learning in the sense of task performance. The danger of designing complex and expensive VLEs without having a positive impact on learning outcomes is obvious. However, judging the value of a VLE simply by its effect on task performance misses out on other factors which support learning. Boosting the students' motivation to deal longer, more steady or more effectively with the given content is also an important goal of VLEs in engineering education [8, 9]. Apart from learning outcome and motivation, a peak to a different domain reveals a third intended effect of virtual environments. According to the entertainment sector, the extent to which a game or in general a virtual environment can "draw you in" functions as a quality seal [2]. This phenomenon is often referred to as immersion [10].

Enabling natural movement as the most basic form of interaction is considered an important hardware quality to create immersion [11]. Manufacturers of hardware

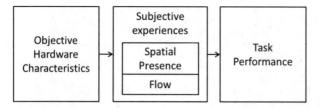

Figure 1 Expected relationship between hardware characteristics, subjective experiences and task performance

that are supposed to enhance immersion claim that "Moving naturally in virtual reality creates an unprecedented sense of immersion that cannot be experienced sitting down" [12]. Almost 20 years ago, this could already be confirmed by Slater [11]. Another basic assumption in the context of VLEs and natural user interfaces is that greater immersion means better learning and potentially higher training transfer [4, 7]. This suggests that immersion would be the precondition for better learning, caused by the qualities of the user interfaces. However, if virtual environments are used in educational contexts, those assumptions need to be confirmed by empirical evidence. The presented study therefore focuses on the following questions:

- Do natural user interfaces create a higher sense of immersion?
- Do natural user interfaces lead to better task performance?
- In what way do immersion and task performance interact in mixed reality learning environments?

If assessed in an experimental setting, the construct of immersion needs to be specified. Spatial presence and flow are considered key constructs to explain immersive experiences. In general, flow describes the involvement in an activity [13, 14], whilst spatial presence refers to the spatial sense in a mediated environment [11, 15]. Spatial presence, as indicated in the name of the construct, refers to the spatial component of being immersed, i.e. the spatial relation of oneself to the surrounding environment. If we experience spatial presence in a mediated environment, we shift our primary reference frame from physical to virtual reality [15].

2 Experimental Analysis of the Potential of Mixed Reality for Learning and Teaching Processes

2.1 Study Design – Focusing on Spatial Abilities

The study presented in this paper assesses the relationship between objective hardware characteristics, subjective experiences and task performance. Their expected relationship is visualized in Figure 1.

Figure 2 View of the virtual environment used in the study in the first and in the third task

All participants had to solve the same task in the same virtual environment, which was a large-scaled maze in a factory building. Within the maze, 11 different objects were located. The first task for the participants was to navigate through the maze and to imprint the positions of the objects to their memory. For that, they were given eight minutes of time. The second task was to recognize the objects seen before in the maze. The third task was to locate the positions of the objects on a map of the maze. This was done with a self-programmed application on a tablet (Nexus 10) with a drag-and-drop control mode. The view of the maze in the first and second task is pictured in Figure 2.

For both groups, the participants were given the chance to explore a test scenario (an italian piazza) freely for about three minutes before the actual task started. This was in order to get used to the respected control mode. All experimenters who conducted the experiments were trained in advance by experienced researchers. First they were being trained the functions of the hardware. In a second step, they took an observing position in a test run, and thirdly they conducted a test run on their own with the experienced researcher being the observer and giving feedback afterwards. Two groups of test persons were compared. The mixed reality simulator was used as main stimulus of the treatment group whereas the control group used a laptop as interaction device. The different hardware being used differed regarding the following characteristics:

- Control mode of the field of view,
- Control mode of locomotion,
- Display and
- Body posture of the user.

In the presented study, learning in a Mixed-Reality-Simulator was compared to a somehow conventional learning with a laptop. The technical equipment is described in more detail in the following:

Laptop. The type being used was a Fujitsu Lifebook S761 with a 13,3 inch display and a 1366 × 768 display resolution. The field of view was controlled with a mouse. Locomotion was controlled by WASD-keys, where W/S keys controlled forward and backward while A/D keys controlled left and right. The hardware usually results in a sitting body posture while using the device.

Figure 3 Head mounted
display and omnidirectional
floor of the virtual theatre

Mixed Reality Simulator. The Virtual Theatre is a mixed reality simulator which enables unrestricted movement through a virtual environment and therefore is used in an upright body posture. The user can move around within the environment by just walking in the desired direction. Therefore the control mode of locomotion is walking naturally. To track the movements of a user, the virtual theatre is equipped with 10 infrared cameras. They record the position of designated infrared markers attached to the HMD and an additional hand tracer. The components of the Virtual Theatre which came to use in the study are pictured in Figure 3. For a more detailed and complete description of the technical system see Ewert et al. [2] and Johansson [7, 16].

Due to the composition of the simulator which was applied in the study, the hardware characteristics could only be tested in a certain combination and could not be isolated any further. The whole experiment took one hour. The complete procedure is visualized in Figure 4.

2.2 Variables and Measurements

In this study, spatial presence was measured with elements of the MEC Spatial Presence Questionnaire of Vorderer et al. Several studies conducted by the authors strengthened the postulate of spatial presence being best explained as a two-level Model. This includes process factors (attention allocation, spatial situation model, self location, possible actions), variables referring to states and actions (higher cognitive involvement, suspension of disbelief), and variables addressing enduring personality factors (i.e. the trait-like constructs domain specific interest, visual spatial

Greeting, instruction of test person by trained experimenter, instructional video for use of virtual theatre				
Measurement of learner characteristics via online survey.	First task: Walking through maze and keeping in mind positions of objects.	Measurement of subjective experiences via online survey.	Second task: Recognizing objects via online survey.	Third task: Locating objects on a map of the maze.
End				

Figure 4 Procedure of the experiments

imagery, and absorption) [15]. Suspension of disbelief refers to the extent of how much a person pays attention to technical and content-related inconsistencies. The more a person can fade out the action of "looking for errors", the higher the feeling of spatial presence will be according to the theory. In the presented study, instead of the subscales attention allocation and absorption, the Flow-Shortscale of Rheinberg was used. In this scale, flow is operationalized as the mental state of operation in which a person performing an activity is fully immersed in a feeling of energized focus, full involvement, and enjoyment in the process of the activity. In essence, flow is characterized by complete absorption in what one does, as well as the feeling of smooth and automatic running of all task-relevant thoughts [13, 14].

The perception of a learning situation is highly likely not to be influenced just by objective criteria such as the technical configuration of the learning environment. The strength of spatial presence experienced in a VE is supposed to vary both as a function of individual differences and the characteristics of the VE [4]. A general interest in the topic appeals to a person's curiosity and the motivation to learn something new. If chances to learn or experience something new are low, the motivation to learn decreases [8, 17]. However, not only interest in a topic but also in the way of presenting it can influence subjective experiences during the learning situation as well as learning outcome. The subscale domain specific interest of the MEC-SPQ [18] refers to the topic of the medium, in this case the virtual environment. Because of the given considerations mentioned above and since interest in mazes didn't seem like a helpful operationalization for domain specific interest, it was adapted to interest in digital games. Based on all theoretical considerations, the general hypothesis of the study was that natural user interfaces should have a positive effect on subjective experiences during the learning situation as well as on learning outcome, in this case operationalized in task performance.

The set of hardware characteristics functioned as the first independent variable in the presented study. Furthermore, interest in digital games (second independent variable) was measured before the first task. As dependent variables, spatial presence and flow were measured after the first task which had to be fulfilled either in the Virtual Theatre or on the laptop. As dependent measures of task performance, three different variables were analyzed: The number of objects that were correctly recognized in the second task, the third task reaction time and the accuracy of locating the objects on the map in the third task.

2.3   Sample and Results

A total of 38 students between 20 and 33 years ($M = 24.71$; $SD = 3.06$; $n = 13$ female) volunteered to take part in the study. The sample therefore represents a potential user group of virtual environments in engineering education. They responded to a call for participation which was hung out at bulletin boards throughout the university but also posted on the front page of the virtual learning platform of the university and on several research and learning related blogs, social media platforms and news feeds. As an incentive and as a sign of appreciation, all participants took part in a drawing for a cordless screwdriver. All participants were healthy and highly interested in participating in the study. They did not report suffering from any physical or mental disorders. To rule out effects due to ametropia, participants were asked in advance to bring their corrective lenses just in case. If participants had been assigned to the mixed reality group, they were asked to wear sturdy shoes.

Hypotheses regarding influences of hardware conditions on subjective experiences and task performance measures were tested with analyses of variance (ANOVAs). With regard to the effects of the Virtual Theatre and the laptop on flow, significant differences were found (F (1, 36) = 4.18; $p < 0.05$). Thus more flow has been experienced in the Virtual Theatre (see Figure 5). Taking a closer look on subscales there is a highly significant difference between conditions in self-reported absorption (F (1, 36) = 10.63; $p < 0.01$), but not in smooth and automatic running. There are also effects of hardware conditions on spatial presence. Self Location in the Virtual Theatre was rated significantly higher (F (1, 36) = 15.79; $p < 0.001$), which refers to the feeling of actually being in the virtual environment. Similarly, participants in the Virtual Theatre showed higher scores on the possible actions subscale of spatial presence (F (1, 36) = 4.90; $p < 0.05$). There were no further significant effects regarding spatial presence (see Figure 5).

In addition to that the effects of hardware on task performance measures were calculated. In the recognition task of the objects from the virtual environment, participants in the Virtual Theatre condition made significantly more errors (F (1, 36) = 10.93; $p < 0.01$), which opposes the hypothesis that the use of natural user interfaces leads to better learning outcomes (see Figure 6). There were no differences regarding time on task and deviation between the two treatment conditions.

Regarding the question, in what way immersion and task performance interact in mixed reality learning environments, a significant interaction between flow and hardware characteristics was found in the case of the task performance indicators "deviation" (F (1, 36) = 9.53; $p < 0.01$, see Figure 7) and "total duration" (F (11, 36) = 4.65; $p < 0.05$, see Figure 7). In other words, high values of experienced flow subside with better task performance, if a laptop has been used for learning, but with worse task performance if a mixed reality simulator has been used.

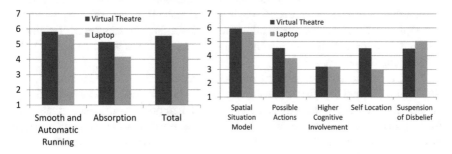

Figure 5 Effects of objective hardware characteristics on flow and spatial presence

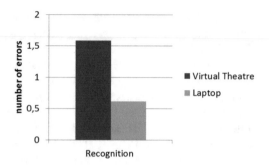

Figure 6 Effects of objective hardware characteristics on task performance, here: recognition of previously presented objects

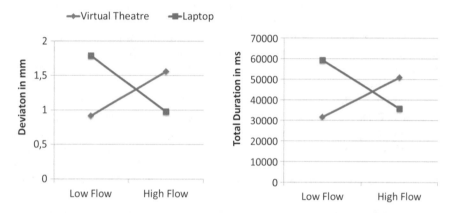

Figure 7 Interaction between flow and user interface according to deviation and total duration (shown in percentile split)

3 Discussion

The results show that mixed reality learning environments indeed lead to more flow. This is due to a higher self-reported level of absorption in the experimental group. Although flow is an activity related construct, this result is in line with the theoretical assumptions that hardware which allows natural walking can support the feeling of "diving" into the virtual environment, which in general terms is often referred to as immersion.

Next, the effects of the hardware on spatial presence are analyzed in more detail. Students who used the Virtual Theatre reported a higher self-location in the virtual environment which indicates that they had shifted their primary reference frame from the physical to the virtual world. Although the given task didn't require any further non-mental actions but navigating through the virtual environment, students in the

Virtual Theatre reported higher on the subscale of possible actions. However, for the other subscales, no differences were measured.

The interaction of flow and hardware characteristics in the case of the task performance indicators "deviation" and "total duration" leads to the assumption that in a mixed reality simulator, the subtasks of learning "switch roles": Controlling the mixed reality environment the phenomenon of "diving in" becomes the main task, while dealing with the actual content-related task moves gets less attention [6]. Although the given data set gives no insight to the question, if this is a conscious or a subconscious decision, this is a result of high importance in engineering education. Especially if the trend of developing immersive learning environments continues, the possibility for students to get used to the control mode is absolutely necessary in order not to inhibit the learning process.

According to the results of this study, immersion is not the precondition for better learning in virtual environments with natural user interfaces. Thus, the underlying model of the study (see Figure 1) needs to be adjusted for further research. The only effect of the Virtual Theatre on task performance was a negative influence on recognition. This result is contradictory to the assumption that immersion leads to better learning. It seems that controlling the hardware was less intuitive than expected. This probably led to the typical situation for learning with virtual environments: Dividing the available cognitive resources on the two parallel sub-tasks of dealing with the content and controlling the learning environment with the respected user interfaces [6]. Moreover, the combination of an HMD and real physical locomotion could lead to cognitive dissonance. When wearing the HMD, the user can see where he or she walks in virtuality, but not in physical reality. Therefore the user takes a risk and has to trust in the technology in order to continue his or her actions. Last but not least, walking on the omnidirectional floor is a new experience for users and therefore could result in the fear of falling. All interpretations for the given results are going to be addressed in a follow-up study, where previous participants of the study will be interviewed on their experiences.

Finally, some limitations of the present study are considered that should be pursued for future research. One limitation refers to the type of hardware examined. Since the different technical characteristics of the Virtual Theatre can only be tested in a set, it is not possible to isolate single effects. The other aspect concerns the task chosen for this experiment. Low levels of cognitive involvement in both groups indicate that the whole sample might not have been challenged enough. Since challenge is an important precondition for the motivation for learning, more challenging tasks are going to be tested in the future.

This exploratory study on the effects of mixed reality learning environments on subjective experiences and task performance confirmed a few theoretical assumptions but also contradicted others. In a next step, interviews with participants from both groups are going to be conducted. A deeper insight on the participants' experiences will allow a more differentiated view on the subject of our research. Since the digitalization and virtualization of engineering education will play an increasingly important part in the future, it is of high value to know the exact advantages and disadvantages of immerging teaching and learning technologies, before they come into practice.

Acknowledgments The present work is supported by the Federal Ministry of Education and Research within the project "Excellent teaching and learning in engineering sciences (ELLI)". The authors would like to thank Prof. Martina Ziefle for constant advice on the presented work.

References

1. M. Kerres, *Mediendidaktik: Konzeption und Entwicklung mediengestützter Lernangebote*, 3rd edn. Oldenbourg Verlag, München, 2012
2. D. Ewert, K. Schuster, D. Johansson, D. Schilberg, S. Jeschke, Intensifying learner's experience by incorporating the virtual theatre into engineering education. In: *Proceedings of the 2013 IEEE Global Engineering Education Conference*. Berlin, 2013, pp. 207–212
3. S. Malkawi, O. Al-Araidah, Students' assessment of interactive distance experimentation in nuclear reactor physics laboratory education. European Journal of Engineering Education **38** (5), 2013, pp. 512–518
4. B. Witmer, M. Singer, Measuring presence in virtual environments: A presence questionnaire. Presence: Teleoperators and Virtual Environments **7** (3), 1998, pp. 225–240
5. P. Zimbardo, R. Gerrig, *Psychologie*, 7th edn. Springer, Heidelberg, 2003
6. J. Sweller, Element Interactivity and Intrinsic, Extraneous, and Germane Cognitive Load. Educational Psychology Review **22** (2), 2010, pp. 123–138
7. D. Johansson, *Convergence in Mixed Reality-Virtuality Environments. Facilitating Natural User Behaviour*. No. 53 In: Örebro Studies in Technology. Örebro University, Örebro, Sweden, 2012
8. A. Hebbel-Seeger, Motiv: Motivation?!– Warum Lernen in virtuellen Welten trotzdem)funktionieren kann. Zeitschrift für E-Learning – Lernkultur und Bildungstechnologie **7** (1), 2012, pp. 23–35
9. F. Müller, Intresse und Lernen. Report – Zeitschrift für Weiterbildungsforschung **29** (1), 2006, pp. 48–62
10. J. Murray, *Hamlet on the Holodeck: The Future of Narrative in Cyberspace*. MIT Press, Cambridge, 1998
11. M. Slater, M. Usoh, A. Steed, Taking Steps: The Influence of a Walking Technique on Presence in Virtual Reality. ACM Transactions on Computer-Human Interaction **2** (3), 1995, pp. 201–219
12. Virtuix Technologies, 2015. http://www.virtuix.com/
13. M. Csikszentmihalyi, J. LeFevre, Optimal experience in work and leisure. Journal of Personality and Social Psychology **56** (5), 1989, pp. 815–822
14. F. Rheinberg, S. Engeser, R. Vollmeyer, Measuring Components of Flow: the Flow-Shot-Scale. In: *Proceedings of the 1st International Positive Psychology Summit*. Washington DC, USA, 2002
15. M. Slater, Measuring Presence: A Response to the Witmer and Singer Presence Questionnaire. Presence: Teleoperators and Virtual Environments **8** (5), 1999, pp. 560–565
16. D. Johansson, L. de Vin, Towards Convergence in a Virtual Environment: Omnidirectional Movement, Physical Feedback, Social Interaction and Vision. Mechatronic Systems Journal **2** (1), 2012, pp. 11–22
17. W. Edelmann, *Lernpsychologie*, 5th edn. Beltz PVU, Weinheim, 1996
18. P. Vorderer, W. Wirth, F. Gouveia, F. Biocca, T. Saari, F. Jäncke, S. Böcking, H. Schramm, A. Gysbers, T. Hartmann, C. Klimmt, J. Laarni, N. Ravaja, A. Sacau, T. Baumgartner, P. Jäncke. Mec spatial presence questionnaire (mec-spq): Short documentation and instructions for application. report to the european community, project presence: Mec (ist-2001-37661), 2004. http://www.ijk.hmt-hannover.de/presence

Preparing for Industry 4.0 – Collaborative Virtual Learning Environments in Engineering Education

Katharina Schuster, Kerstin Groß, René Vossen, Anja Richert and Sabina Jeschke

Abstract In consideration of future employment domains, engineering students should be prepared to meet the demands of society 4.0 and industry 4.0 – resulting from a fourth industrial revolution. Based on the technological concept of cyber-physical systems and the internet of things, it facilitates – among others - the vision of the smart factory. The vision of "industry 4.0" is characterized by highly individualized and at the same time cross-linked production processes. Physical reality and virtuality increasingly melt together and international teams collaborate across the globe within immersive virtual environments. In the context of the development from purely document based management systems to complex virtual learning environments (VLEs), a shift towards more interactive and collaborative components within higher educational e-learning can be noticed, but is still far from being called the state of the art. As a result, engineering education is faced with a large potential field of research, which ranges from the technical development and didactical conception of new VLEs to the investigation of students' acceptance or the proof of concept of the VLEs in terms of learning efficiency. This paper presents two corresponding qualitative studies: In a series of focus groups, it was investigated which kinds of VLEs students prefer in a higher education context. Building upon the results of the focus groups, a collaborative VLE was created within the open world game Minecraft. First screenings of the video material of the study indicate a connection between communicational behavior and successful collaborative problem solving in virtual environments.

Keywords Engineering Education · Minecraft · Oculus Rift · Virtual Collaboration · Virtual Learning Environments

K. Schuster (✉) · K. Groß · R. Vossen · A. Richert · S. Jeschke
IMA/ZLW & IfU, RWTH Aachen University,
Dennewartstr. 27, 52068 Aachen, Germany
e-mail: katharina.schuster@ima-zlw-ifu.rwth-aachen.de

Originally published in "Proceedings of the 8th International Conference on E-Learning in the Workplace", © ICELW 2015. Reprint by Springer International Publishing Switzerland 2016, DOI 10.1007/978-3-319-42620-4_33

1   Introduction: Today's Learning and Working in Preparation for Industry 4.0

Today's portfolio of e-learning solutions is as diverse as never before. Different kinds of media services, software for teaching and learning as well as innovative hardware solutions not only become a bigger part in higher education and the workplace but increasingly adapt to the massive changes our working world is going through. A common and frequently cited example is the use of learning management software, based amongst others on the open-source management system Moodle [1]. Today, Moodle counts 53.738 registered installations with 68.7 million users of 226 countries in 7.7 million courses [2]. Platforms like Moodle have different functions, which can also be viewed as e-learning solutions themselves: With chats, forums or messenger systems, students or workers can communicate in synchronous or in asynchronous ways. Wikis enable cooperative text production and different kinds of assessment modes or quizzes give teachers the chance to test the students whenever they want and as many times they want during the semester. Here, one of the biggest advantages is that the tests are rated automatically, which makes the frequent testing also suitable for large groups. Being tested frequently, the students get instant feedback about their current state of knowledge. Digitally supported learning brings direct individual advantages in terms of self-awareness of the content of the lecture.

The digitalization of education also means that learning becomes more collaborative [3]. The key word "user generated content" describes the fact that in times of web 2.0, content rarely is produced by just one single provider of content, but is generated by several users instead. Transferred to the context of higher education, the students' role changes. Whereas back in the days, when the teacher was more or less the only source who provided information, today students can get basically any information they want from the internet, but can also contribute actively within forums, wikis or blogs. The potential is there to switch the students' role from rather passive users of information to creators of knowledge in networked structures – with all accompanying advantages and disadvantages. With the goal in mind not only to boost the students' knowledge and to support them to strengthen their personality over the years, but also to develop crucial competences for the working world they are about to step in, various types of collaboration have to be trained and tested in learning scenarios.

In a first step, one can differentiate between cooperative and collaborative learning. In cooperative learning scenarios, each group member is given a sub-task e.g. reading and interpreting different parts of scientific literature, technical reports etc. The individually produced results, e.g. a presentation, are simply being added up. Therefore, the main result mostly doesn't represent the state of knowledge of each individual group member. It is more a question of how to divide the work in an efficient, but not necessarily in an effective way.

Collaboration instead focuses on the creation of a new knowledge baseline, which is built through interlinked and co-referenced work during the learning process [4]. Especially in engineering, collaborative learning in virtual environments is highly important in the context of a dynamic and digitalized working world. This can be realized by analyzing a defective machine, coming up with a logistics concept for a virtual factory or designing a virtual car. The last example points out the importance for engineers to link their own specific technical expertise with expertise of other domains. Working in interdisciplinary teams situated all over the world is standard practice. The increasing digitalization of economy and society links knowledge over borders of time, space and systems. In times of industry 4.0, physical reality melts with virtuality [5]. For almost decades now, e.g. finite element models, data models, analytical models or CAD-models of machine elements have been produced with software. The data is used, provided and linked within socio-technical working systems via clouds, ubiquitous computing, product-lifecycle management and product data management. Thus, in engineering, human work processes are increasingly being transferred to virtual spaces of an internationally networked world.

2 Collaborative Learning in Virtual Environments

In higher engineering education computer-supported cooperative and collaborative learning (CSCL) have long been established as methods which support self-driven and work-related learning processes. By further technical development as well as new requirements of the changing working world such common methods can be lifted to a new level. Virtual learning platforms like moodle can systematically be linked to virtual or teleoperative laboratories. Every student gets the opportunity to experiment with physically real equipment without the necessity to be physically present at the location of the machine [6]. With special booking systems, expensive equipment for teaching and training processes can be used more efficient, since students from different time zones (e.g. USA and Germany) can log in at different schedules. Thus, it is possible to introduce students to learning settings, which would otherwise be too dangerous (e.g. an atomic power plant), too hard to access (e.g. the surface of mars) or too big a risk for ongoing production (e.g. in a factory) [7].

Moreover, in massive open online courses (MOOCs), each student can learn at his own speed. Serious games offer the possibilities to learn in a playful manner, in single-player or in multi-player mode. Innovative virtual knowledge spaces therefore offer all kinds of possibilities for learning and working in times of industry 4.0. In order to use the new technologies for engineering education in a proper way, deeper insights in reception, cognition and communication in virtual environments are necessary. Simply providing the technical infrastructure doesn't automatically guarantee successful collaboration. Therefore, the analysis of key factors for successful collaboration in virtual environments is an important field of research in the context of the working world of the future. Linking the different fields of this research is a core point for its success.

In the project "Excellent Teaching and Learning in Engineering Sciences" the three large german universities RWTH Aachen University, Ruhr-Universität Bochum and Technical University Dortmund focus on the development of virtual and remote laboratories as well as non-experimental collaborative learning spaces. In order to show students the "bigger picture" of the engineering profession, but also in the light of increasing numbers of students in engineering, the necessity of experimental equipment is obvious.

When working on the development of virtual or remote laboratories, the focus is clearly laid upon the final product and its future way of use. From a different point of view, looking at the current media use of students can help to predict the steps that still need to be done, if one day collaboration in virtual learning environments is supposed to prepare for industry 4.0 on a large scale. But are today's students ready for innovative teaching methods? Current studies of digital media usage show a rather passive usership. The majority of students hardly uses media services which require an increased work load by generating content (e.g. wikis or blogs), as a long-term study on media use of students shows [8]. A study with german engineering students (n = 1587) focused on the frequency of usage of different kinds of media services. The results show that interactive and collaborative media services are not used very often by the majority of the sample. This conclusion also counts for blogs or tools for collaborative text production such as wikis. In other words: Absorbing content is still more popular than generating it [9].

But the same study reveals another important aspect. Although not many students have been in contact with innovative teaching formats such as serious games or virtual courses in real-time; those who have experience with such formats are highly satisfied with it. These results are of high relevance for the development of virtual collaboration spaces, but also for companies who wish to use them. Developers need to know which the crucial features of a virtual environment are that really solve students' problems and are not just "nice to have". Moreover, universities or companies who invest in virtual learning environments aim for some kind of return on invest, which is not likely to come if the VLE isn't used. However, there is still little evidence on the motifs of students for using, or better for not using media services which require active participation. Although today's students all grew up in a digital society, user profiles are highly diverse. Providers of virtual collaboration spaces such as universities or companies need deeper insights in actual user preferences of specific target groups. Which level of graphical precision is required to understand complex processes, which level of gamification is preferred or which kinds of narrative scenarios would motivate this user group to deal with the content longer or more often still needs to be answered. Therefore, in order to investigate the described quantitative research in the field of collaboration in virtual environments, a qualitative research design was chosen, which will be explained in more detail in the following chapter.

3   Experimental Setting for the Analysis of Collaboration in Virtual Environments

3.1   Lead User Workshops with Future Engineers

Since today's students are going to be working within industry 4.0 contexts it is important to integrate them in the research process on VLEs which are supposed to prepare for the corresponding requirements. The approach of user-centered design is well-known in the field of software development, but also under the label of open innovation in the case of new product development [10]. Within idea competitions or lead user workshops, the approach has also proven to be helpful in the development of new teaching methods or new formats of virtual learning in context of the Bologna Process [11]. As representatives of future user groups of such innovative learning and working spaces, students are questioned within focus groups. Two workshops with 23 students from Germany and one workshop with 13 participants of a European study program were conducted (compare Figure 1).

The two major aims of the workshop series were to collect the students' requirements but also their retentions on VLEs. The students were asked the following questions:

• Which scenarios would you like to experience within VLEs?
• Which didactical method would you prefer, e.g. game based learning, free exploration etc.?

The students first had to work on the questions in small teams and then presented the results to the whole group. Each idea had to be written down on a prompt card. For a deeper insight into the workshop results, the cards were analyzed with qualitative analysis [12]. The contributions to the two questions were therefore clustered into topics. Afterwards, the quantity of contributions in each category was counted. The

Figure 1 Lead user workshops with students on the topic of virtual learning environments

topics with the most contributions were considered the ones of greatest interest or greatest concern of the students.

The results of the analysis show that students equally prefer realistic (e.g. factory simulations) and fictional scenarios (e.g. traveling through a factory from the product's perspective). In case of fictional scenarios, the main principle is to exceed the limits of time, space and physics. The students like to be immersed by the virtual environment and to interact with it intuitively and naturally. The possibility to get instant feedback is valued very positive by the students. To combine learning with playing in terms of game based learning is welcomed by the students, but not necessary to enjoy the learning process within the VLE or to consider the VLE useful. The students had no major retentions to VLEs in general, but a few contributions referred to the concern that too many unnecessary features of the VLE could distract from the actual task and the content that should be learned or practiced.

Although surely not being the main contribution to learning success of students, the insights in students' preferences on VLEs delivered important information for the didactical design of future learning environments.

3.2   Work in Progress: Study on Collaboration in Virtual Environments

In line with the preference of students to be immersed in a VLE, a previous experimental study showed that students who used natural user interfaces for interacting with the virtual environment in individual learning scenarios experienced more immersion than students who solved the task on a laptop. Immersion generally referred to as "diving into the virtual environment" had been operationalized with the constructs of spatial presence and flow. An interaction of experienced flow and errors in task performance revealed the complexity of working in virtual environments: Being immersed by the environment unfortunately can also mean that one is absorbed more by the exploration of the environment than by solving the given task [13]. This finding stands in conflict with the user preferences found in the lead student workshops.

However, the nature of collaboration might help to compensate this problem. Since more people are involved in solving a given problem, more attention can be spent on problem-related details. Moreover, as mentioned in the introduction, the prediction of the working world of the future under the label of industry 4.0 specifically emphasizes the importance of controlling complex, geographically distributed industrial processes [5]. Collaborating in teams with diverse professional and cultural backgrounds is an important precondition for the success of such processes.

To understand the complex interactions of different human factors in situations of virtual collaboration, a current experimental study focuses on preconditions for successful collaborative problem solving in virtual environments. This study assesses the relationship between personal characteristics, objective hardware characteristics,

Figure 2 Expected relationship between personal characteristics, hardware characteristics, subjective experiences, objective collaboration behaviour and task performance

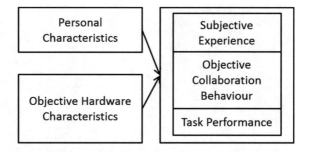

subjective experiences, objective collaboration behavior and task performance. Their expected relationship is visualized in Figure 2.

The virtual environment is based on the results of the lead user workshops. Therefore the VLE had to be immersive, interactive, give instant feedback on the task performance and have elements of gamification in it. Since in the context of higher education personal and financial resources are mostly small for the development of virtual learning environments, the open-world game Minecraft was chosen as the setting for the learning environment. Minecraft has already been used for teaching and learning settings in the USA and the UK [14] and provides many features which are crucial for virtual collaboration:

• Quick construction of simple learning settings without any programming skills,
• Possibility to build more complex technical environments with the use of blue prints, available mostly for free within the Minecraft community [15],
• Simple and easy to learn modes of interaction without sophisticated gaming skills,
• Possibility to move around freely and to explore the scenario as a user actively.

To link the study to an industry 4.0 scenario, pairs of students are given a task with an engineering background. All participants have to solve the same problem in the same virtual environment, which is to restore electricity in a virtual building. From no perspective within the VLE the whole electrical setting can be viewed completely, which leads to the necessity for the students to actively communicate with each other. The students only know the target state, but not the steps how to get there. The task of restoring electricity within the building can be divided into the following sub-tasks, which have to be encountered by the students without further instructions:

• Get overview of the complete electrical setting of the building,
• Find out, at which spots the electrical circuit is broken,
• Remember the necessary steps to repair the electrical circuit,
• Find the spot within the building from where the success of the problem solving can be controlled.

The process of the collaborative problem solving has to be organized by the students themselves. Before they start as a pair, each student has to run through a tutorial individually. A screenshot of the virtual environment is pictured in Figure 3.

Figure 3 Screenshot of the VLE, implemented in Minecraft

To analyze the possible interaction of immersion and task performance, the effect of natural user interfaces is integrated in the research design. The research plan consists of two groups. In both groups, the students work on laptops. Both groups use a simplified keyboard, where all keys except the arrows have been removed. With the arrow keys, the participants control horizontal movement. With a mouse, participants in both groups interact with the VLE. By clicking on the keys of the mouse, they can select different kinds of tools or resources they need to solve the problem. The experimental group fulfills the collaborative task wearing a head mounted display (Oculus Rift, DK 2). The field of view is therefore controlled by natural head movement. The control group controls the field of view by twisting the mouse.

Subjective experiences focus on immersion, operationalized with the constructs of spatial presence and flow. In this study, spatial presence is measured with elements of the MEC Spatial Presence Questionnaire of Vorderer et al. [18]. Flow as the mental state of operation in which a person performing an activity is fully immersed in a feeling of energized focus, full involvement, and enjoyment in the process of the activity, is measured with the shortscale of Rheinberg [17]. According to this instrument, in essence, flow is characterized by complete absorption in what one does, as well as the feeling of smooth and automatic running of all task-relevant thoughts [16].

Additionally to the questionnaire, the participants of the study are being interviewed about their experiences. Within the interviews, the experiences are linked with personal characteristics of the participants, e.g. their gaming experience. The subjects are also asked about the experienced quality of the collaboration itself, more precisely their strategies of problem solving, communication and task management. Since diffusion of responsibility within teams has proven to be an inhibiting factor for the success of group work [4] this aspect is also part of the interview. The behavior of the users is being captured by video camera, screen casts and spatial tracking systems. The task performance is measured in time needed for solving the problem.

A total of 8 students between 24 and 34 years ($M = 26$; $SD = 3.28$; $n = 5$ female) volunteered to take part in the pre-study. First screenings of the video material and the screen casts indicate a connection of problem-related speech-acts and task performance. Students who explicitly verbalize what they do and what they think the other one should do, get quickly to the point when they identify the necessary sub-tasks. For some students, especially those using the oculus rift, the tendency to "chit-chat" about the virtual environment from a meta-perspective and about its immersive effects was noted. For the analysis of the interviews it will be necessary to link this fact to the corresponding task performance. On the one hand, a strong interest in such "meta-information" can mean a positive effect regarding motivational aspects, but on the other hand it can mean some sort of distraction from the actual problem solving task, as it was indicated in previous studies on learning in virtual environments [13].

Comparing subjective experiences of the quality of collaboration with the actual task performance of the subjects will be another important aspect of the data analysis. However, corresponding conclusions will always have their limitations, since one of the greatest advantages of virtual learning environments is that people can use them at their own preferred speed.

4 Conclusion

The qualitative research on collaboration in virtual environments gives deeper insights into the relationship of personal preferences for VLEs, subjective experiences within them and actual task performance. To focus on virtual collaboration delivers important research results for universities and companies who wish to use virtual environments in the future with the vision of industry 4.0 in mind. The preliminary results of the video analysis of the pre-study indicate the importance of communication within virtual environments. If performance within VLEs was to be enhanced even from today's point of view, to train staff in virtual communication skills would be a promising point to start.

The approach of user-centered design helped to get deeper insights into specific design preferences of VLEs from a user group, who grew up in a digitalized society. However, the preferences can also be due to age instead of cohort. Especially in the light of demographic change and aging populations in Europe, it will be crucial to continue the kind of studies which have been presented in this paper. For the effective virtual collaboration within diverse teams, another research area covers collaboration of pairs with different skill levels: of novices and experts, of young and old co-workers or of IT-close vs. IT-distant people. Who can work best with whom, and in what kind of virtual environment will be an important aspect for effective human resources planning in companies.

Should research one day prove actual financial benefits of virtual collaboration, e.g. by reducing travel costs, this way of working will soon become established. Companies who know how to collaborate in virtual environments efficiently will have a strong competitive advantage compared to those who don't. Continuing this kind of research therefore is an important contribution towards a globalized, connected and digitalized working world in terms of industry 4.0.

References

1. B. Haerdle, Die Digitalisierung der Lehre. Wirtschaft und Wissenschaft **4**, 2013, pp. 10–16
2. Moodle Statistics, 2015. https://moodle.net/stats/
3. L. Johnson, S. Adams Becker, M. Estrada, A. Freeman, NMC Horizon Report: 2014 Higher Education Edition. Tech. rep., The New Media Consortium, Austin, Texas, 2014
4. H.M. Niegemann, S. Domagk, S. Hessel, A. Hein, M. Hupfer, A. Zobel, *Kompendium multimediales Lernen*. Springer-Verlag, Berlin; Heidelberg, 2008
5. Federal Ministry of Education and Research. Industrie 4.0, 2015. http://www.softwaresysteme.pt-dlr.de/de/industrie-4-0.php
6. C. Terkowsky, D. May, T. Haertel, C. Pleul, Experiential remote lab learning with e-portfolios: Integrating tele- operated experiments into environments for reflective learning. In: *15th International Conference on Interactive Collaborative Learning (ICL)*. 2012, pp. 1–7
7. D. Ewert, K. Schuster, D. Johansson, D. Schilberg, S. Jeschke, Intensifying learner's experience by incorporatig the virtual theatre into engineering education. In: *Proceedings of the IEEE EDUCON Conference*. Berlin, 2013, pp. 207–212
8. E. Dahlstrom, C. Dziuban, G. Morgan, ECAR Study of Undergraduate Students and Information Technology (Research Report). Tech. rep., EDUCAUSE Center for Analysis and Research, Louisville, CO, 2013. http://www.educause.edu/ecar
9. G. Gidion, M. Grosch, Welche Medien nutzen die Studierenden tatsächlich? Ergebnisse einer Umfrage zu den Mediennutzungsgewohnheiten von Studierenden. Forschung & Lehre **19** (6), 2012, pp. 450–451. http://www.forschung-und-lehre.de/wordpress/Archiv/2012/ful_06-2012.pdf
10. E. von Hippel, *Democratizing Innovation*. MIT Press, Cambridge, 2005
11. J. Koch, K. Schuster, F. Hees, S. Jeschke, Open Innovation - Kunden als Partner. Wie Hochschulen von Unternehmen lernen können. Wissensmanagement **17** (1), 2011, pp. 31–35
12. J. Gläser, G. Laudel, *Experteninterviews und qualitative Inhaltsanalyse*, 4th edn. Springer Verlag für Sozialwissenschaften, Wiesbaden, 2010
13. K. Schuster, M. Hoffmann, U. Bach, A. Richert, S. Jeschke, Diving in? How Users Experience Virtual Environments Using the Virtual Theatre. In: *Proceedings of the 3rd International Conference on Design, User Experience, and Usability*. Heraklion, Crete, 2014, pp. 636–646
14. The Minecraft Teacher. http://minecraftteacher.tumblr.com/
15. Elterpro: Realistisches Atomkraftwerk mit Kontrollzentrale. https://www.youtube.com/watch?v=l72VcTI4D88
16. F. Rheinberg, S. Engeser, R. Vollmeyer, Measuring Components of Flow: the Flow-Shot-Scale. In: *Proceedings of the 1st International Positive Psychology Summit*. Washington DC, USA, 2002

17. F. Rheinberg, R. Vollmeyer, S. Engeser, Die Erfassung des Flow-Erlebens. In: *Diagnostik von Motivation und Selbstkonzept. Test und Trends*, ed. by J. Stiensmeier-Pelster, F. Rheinberg, Hogrefe, Göttingen, 2003, pp. 261–279

18. P. Vorderer, W. Wirth, F. Gouveia, F. Biocca, T. Saari, F. Jäncke, S. Böcking, H. Schramm, A. Gysbers, T. Hartmann, C. Klimmt, J. Laarni, N. Ravaja, A. Sacau, T. Baumgartner, P. Jäncke. Mec spatial presence questionnaire (mec-spq): Short documentation and instructions for application. report to the european community, project presence: Mec (ist-2001-37661), 2004. http://www.ijk.hmt-hannover.de/presence

Status Quo of Media Usage and Mobile Learning in Engineering Education

Katharina Schuster, Kerstin Thöing, Dominik May, Karsten Lensing,
Michael Grosch, Anja Richert, A. Erman Tekkaya,
Marcus Petermann and Sabina Jeschke

Abstract The usage of different kinds of media is part and parcel of teaching and learning processes in higher education. According to today's possibilities of information and communication technologies, mobile devices and app-usage have become indispensable for a big share of the population in everyday life. However, there is little empirical evidence on how students use mobile devices for learning processes in higher education, especially in engineering education. Within the project "Excellent Teaching and Learning in Engineering Sciences (ELLI)", three large technical universities (RWTH Aachen University, Ruhr-University Bochum, Technical University Dortmund) follow different approaches in order to improve the current teaching and learning situation in engineering education. Many of the corresponding measures are media-related. In this context, a broad understanding of media is applied which includes hardware as well as software. Amongst others, research is conducted on the topics of mobile learning, virtual laboratories, virtual collaboration, social media services and e-learning recommendation systems for teaching staff. In order to match the literature and results of the project with the current habits of study related media usage of students, the three universities conducted a survey in cooperation with the Karlsruhe Institute of Technology (KIT). The KIT's questionnaire covers more than 50 education-related media and IT-Services and has been applied at over 20 universities

K. Schuster (✉) · K. Thöing · A. Richert · S. Jeschke
IMA/ZLW & IfU, RWTH Aachen University, Dennewartstr. 27, 52068 Aachen, Germany
e-mail: katharina.schuster@ima-zlw-ifu.rwth-aachen.de

D. May · K. Lensing
Center of Higher Education, TU Dortmund University, Hohe Str. 141, 44139 Dortmund, Germany

M. Grosch
Karlsruhe Institute of Technology, Kaiserstraße 12, 76131 Karlsruhe, Germany

A.E. Tekkaya
Institute of Forming Technology and Lightweight Construction, TU Dortmund University,
Baroper Str. 303, 44227 Dortmund, Germany

M. Petermann
Lehrstuhl für Feststoffverfahrenstechnik, Ruhr Universität Bochum, Universitätsstraße 150,
44801 Bochum, Germany

Originally published in "3th European Conference on e-Learning (ECEL 2014),
Copenhagen, Denmark, 30-31October 2014", © Academic Conferences
and Publishing International Limited 2014. Reprint by Springer International
Publishing Switzerland 2016, DOI 10.1007/978-3-319-42620-4_34

429

in 6 countries. For the survey conducted within the ELLI project, the topic of mobile learning was added to the questionnaire. Over 1.500 students were asked about their habits of study related media usage in terms of frequency of use and level of satisfaction. Regarding the topic of mobile learning, the students were asked for the kind of hardware and the kind of apps they use for higher education purposes. The 130 identified apps were clustered regarding subject and function. This paper presents the main results concerning the students' general habits of study related media usage and their mobile learning habits. It concludes with a special focus on the possibilities mobile devices offer to the improvement of engineering education.

Keywords Media Usage Habits · Social Media · Mobile Learning · Engineering Education

1 Introduction

The media usage at universities is highly diverse. Printed learning material, digital documents or complete online courses are just a few examples which illustrate the broad variety of media and IT-services offered to support the students' learning processes. Especially web-based services like search engines, facebook or special tools for online learning induced significant changes in society and in the landscape of higher education during the last years. In contrast to analogue media such as printed books, digital media and IT-services for learning purposes have been referred to as e-learning for the past 20 years. The term can be specified regarding the function of the corresponding service. Reinmann-Rothmeier [1] distinguishes between three lead functions: E-Learning (a) by distributing, (b) by interacting und (c) by collaborating. In the case of e-learning by distributing, the main function of the medium is to distribute information. E-learning by interacting applies to media and IT-services which allow the student to interact with the system without any additional personal help. E-learning by collaborating refers to processes of social problem solving and therefore also refers to the principle of "learning from different perspectives" [1]. Within the first focus of this paper, the research question of which function is needed most from the students' point of view was investigated.

In previous cycles of the KIT's study on habits of study related media usage, Gidion and Grosch [2] found that students prefer services which are linked to face-to-face learning settings, such as printed and electronical script or the online catalogue of the university's library. Media which require active participation are being used less frequently. Virtual teaching and learning environments are also being used not only seldom but also with a low rate of satisfaction [2].

Since one of the research foci of the ELLI project is learning with social media services, another question being investigated within this paper was whether students use services like facebook for studying. Social platforms such as facebook or information services like twitter provide an easy access as well as already implemented possibilities for communication by means of groups, private messages and forums for the students during their studies. Other web based media services for communication and collaboration such as wikis, weblogs and forums were implemented increasingly for the teamwork of students also in the teaching of engineering during the last years [3]. Surveys at other universities and faculties [4] as well as statements of teachers of the engineering sciences [5] show a rather hesitant use of this kind of media by students, which is in line with the results of Gidion and Grosch [2].

The second focus of this paper lies on the use of mobile devices and apps in teaching and learning contexts. The use of such devices and apps has become indispensible for many people's everyday life. According to the Federal Statistical Office of Germany, the number of mobile internet users in Germany increased in 2013 by 43 %. Amongst the 16 to 24 year olds, 81 % use mobile internet [6]. The trend towards mobile computers also applies to the type of technical devices being used: In 2013, 65 % of all German households were in possession of a mobile computer, e.g. a laptop, a notebook or a tablet-PC. The Federal Statistical Office asserts that the share has multiplied almost six-fold compared to surveys in 2003. Most households (33 %) use a mobile device additionally to the stationary PC, followed by almost as many households (32 %) who solely use a mobile device. The number of households, which solely use stationary PCs is decreasing: Whilst in 2003 the share was at 51 %, it dropped 20 % in 2013 [7]. Looking at these statistics, it is a logical consequence that the use of mobile devices is also increasing within the context of higher education. Current literature offers only little empirical data on the question how students are using mobile devices for their learning processes. In addition to that we do not know if, how often and in which contexts the students were asked by teachers to use their mobile devices in learning contexts in class or for the learning process in general.

In recent years the technology shift has changed the way how mobile learning is defined – from PDAs to smartphones and tablets. This is also visible by looking at different definitions over time. Quinn [8] offered a rather technology-centered view in 2001 by saying that mobile learning is "elearning through mobile computational devices: Palms, Windows CE machines, even your digital cellphone". Sharples et al. [9] have a more user-centered view by describing the learning environment and adding the learners' autonomy regarding the choice over time and place. In order to go even more into detail an extended literature review was conducted in the ELLI-Project. Thus over 100 definitions on mobile learning were found in about 240 different sources. The definitions could be divided into 3 different clusters, either highlighting the mobile devices itself, the flexibility for the learning process or new didactical approaches. The following quotations illustrate this variety and are taken from the most cited sources in the context of Mobile Learning:

> What is new in 'mobile learning' comes from the possibilities opened up by portable, lightweight devices that are sometimes small enough to fit in a pocket or in the palm of the one's hand [10].

> Mobile Learning devices are defined as handheld devices and "..." should be connected through wireless connections that ensure mobility and flexibility [8].

> [Mobile learning] provides the potential of personal mobile technologies that could improve lifelong learning programs and continuing adult educational opportunities [19].

Summing up all these perspectives, Crompton et al. (2013) published a description of their literature research regarding the mobile learning evolution over the years. According to them, mobile learning means "learning across multiple contexts, through social and content interactions, using personal electronic devices". In addition to that, four different but central characteristics of mobile learning can be identified based on literature [11]:

- Use of mobile devises
- Local independence for learning
- Contextualization of the learning process
- Informality of learning

In the following, the methodology of the study is described, including the description of the sample. The results are presented in two sections. The first section describes the ten most and least used media and IT services, as well as their corresponding values of satisfaction. In the second results section, the results on mobile learning are presented. The paper concludes with a broader view on all results presented and depicts further research questions.

2 Methodology and Description of the Sample

The survey used a questionnaire that was developed at Karlsruhe Institute of Technology in 2009 [18]. It was conducted several times before at 20 universities in 6 countries. One of the KIT's long-established survey's aims is to identify the potential of a university, which media and IT sectors still need to be supported further and which are already used by the students. The three universities RWTH Aachen University, Ruhr University Bochum and TU Dortmund University issued the questionnaire in co-operation with the Karlsruhe Institute of Technology in 2013. It focused on the students' habits of study related media usage, in order to expand the empirical database and to match it with the other topics focused in the ELLI project, such as mobile learning, virtual laboratories, virtual collaboration, social media services and e-learning-systems. More detailed, the research questions were:

- Which function of e-learning is needed most and least from a students' point of view?
- Which role do social media services play for study related media usage?
- Which kind of hardware and which kind of apps do students use for studying and what are the consequences for mobile learning activities of universities?

Therefore the survey asked engineering students from the three universities for frequency of use and level of satisfaction for 54 media and IT services as well as hardware devices. To deliver a complete picture of media usage habits, the items related to university internal as well as to external services. Students had to respond to each given item by answering the question "Which of the given devices or services do you use for your studies?" Each item, e.g. e-book reader, had to be responded using a five point scale, ranging from "very frequently" to "never". Due to the special focus on mobile learning, the questionnaire was modified and several items were added to the original KIT questionnaire. Going more into detail with regard to the use of software on mobile devices, the survey asked for apps, which are already used by the students in the context of learning.

The survey was synchronously carried out at the three universities. A total of 1587 engineering students answered the questionnaire between April and May 2013. Most of the students were first- (59.6%) or second-year (19.4%) students. 81.6% were male, 18.4% were female students. The results of media usage habits in general were conducted with the means of descriptive statistics, calculating the percentage of students who used a medium or an IT-service frequently or very frequently. The same calculation was performed for the least used media and IT-services. In addition, the corresponding satisfaction value was calculated for the most and least used media and IT services. To get deeper insights into the usage of mobile devices, the percentage of how many students used a specific device frequently or very frequently. Moreover, students were asked, which apps they use for study purposes. The given apps were analyzed qualitatively with an explorative approach, i.e. no categories had been conducted previously. After a first round of scanning the data, the apps were clustered by subject. In a second round of analysis, the apps were clustered by function. In a last step, the proportion of the categories was calculated for each cluster.

3 Results

3.1 General Media Usage of Students

The media and IT services most frequently used in the context of academic studies are presented in Figure 1. The service most frequently used is the Google search engine. This corresponds to the fact that Google is the most-used search-engine in the internet in general [20]. According to different web analytics services, also Wikipedia and Facebook are ranked among the most frequently used sites of the web [13–15].

As the results show, e-mail services of the university are being used frequently. It seems that communication via e-mail for learning purposes, e.g. to correspond with professors or with fellow students, is common and well established. The frequent use of course accompanying slides (78.7%) and course accompanying lecture notes (75.3%) online reflect media usage habits which fall in the category e-learning by distributing [1]. The use of analogue media in the form of course materials printed (68.4%) is less frequent, but still among the top 10 of all surveyed media and IT services. The fact that google books (77.1%) and other electronic books (39.5%) are also amongst the top 10 of study related media, underlines the function of e-learning by distributing as important once more from the students' point of view. 50.4% of all students stated that they use e-learning opportunities specifically provided by the

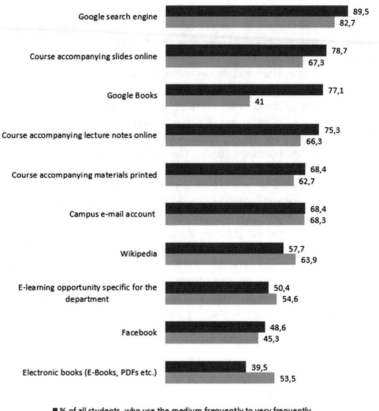

Figure 1 Most used media and IT-services

department frequently or very frequently. This reveals the fact that each e-learning solution provided has to fit the corresponding study field, an approach which is also followed within the ELLI project.

Apart from the top 10 of media and IT-services used most in a study related context, the data was also extracted for the least used media and IT-services. The percentage of frequency as well as the corresponding satisfaction rate is shown in Figure 2.

A very obvious difference is that the average discrepancy between frequency and satisfaction in this dataset is three times bigger than for the items of the most used media and IT services. Whilst the absolute average discrepancy in the most used media and IT services is 9.7 %, it is 23.1 % in the least used media and IT services. If we look at the services in detail, a first conclusion is that Facebook is the clear leader when it comes to social networks. Whilst Google+ is still used by 4.6 % of the students, other providers share only 1.8 % of the lot. This also corresponds to

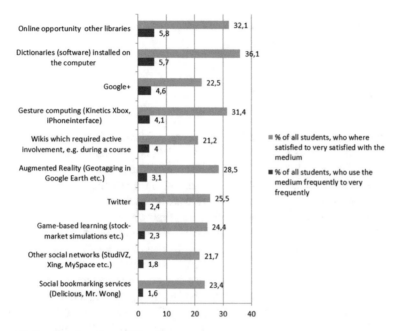

Figure 2 Least used media and IT-services

the results of web analytics services mentioned above. The microblogging system twitter is only used by 2.4 % of the students for study purposes.

Although twitter is also amongst the top ten web sites worldwide according to web analytic systems, it has not become common for study purposes yet. Dictionary software is also not used very frequently. This might be a result of frequently updated and well-established services which are also available online, such as leo.org or duden.de for the German speech community. Social bookmarks don't seem to be needed very often. Only 1.6 % of the students claim to use them frequently or very frequently.

The two second biggest values for the discrepancy between frequency of use and satisfaction occur for two innovative virtual reality technologies which already find their way into higher education: Gesture Computing and Augmented Reality. Both are not used very often (4.1 and 3.1 %), but the usage leads to high satisfaction. It can be assumed that the seldom usage of such technologies does not happen by choice. It seems more likely that students don't have enough possibilities yet to use the technology, or at least not as a regular part of their studying. The seldom use of the technologies might be the result of lack of supply, not of demand. This could also be the explanation for the seldom use of digital game-based learning. Wikis which require active involvement are not used very frequently, which is in line with previous results of the KIT's survey published by Gidion and Grosch [12]. However, it can not be assumed, that forms of learning with and by user generated content are being avoided in general. An aspect for further investigations is to filter all forms of media and IT services which involve user generated content and to analyse their acceptance on a deeper level.

3.2 Mobile Learning

In the following, all results regarding the topic of mobile learning are presented. Most of the questioned students (87.7 %) own a smartphone. The corresponding value for the tablet PC is much lower with 27.7 %.Therefore if teaching staff decides that the usage of tablet PCs is required for the course, students must be given the opportunity to borrow such a device. The results of hardware usage for study purposes regarding frequency of use and satisfaction are shown in Figure 3.

The results of the survey show that 63.9 % of the students use their smartphone often or very often in context of learning, whilst only 26 % use their tablet computer often or very often for studying. The latter is not surprising, as tablets are not very distributed among students yet. E-book readers are used even more rarely, as only 4 % use this device often or very often. The last question in this context was, how often the students use mobile apps for learning. The answers reveal that the use of apps explicitly for learning is not very common yet. Only 7.2 % of the students use mobile apps often or very often. This is a fairly low value. Competing conclusions of this result could be that there either are no adequate apps for studying, that existing apps are not usable or that students just do not know them.

Going more into detail the students were asked to name the apps they are already using for study. All in all 13.5 % of the students answered this question with "yes" and named an app which they used for study purposes. In total, the students named 357 apps (139 different ones). Based on the answers a ranking on how often the apps were named was worked out and the top five could be identified. In the following these are

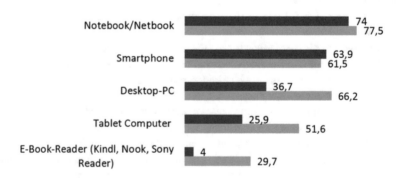

■ % of all students, who use the medium frequently to very frequently

■ % of all students, who where satisfied to very satisfied with the medium

Figure 3 Frequency and satisfaction of hardware usage for study purposes

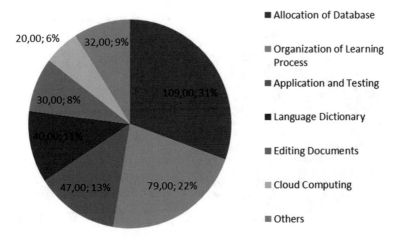

Figure 4 Apps clustered by function

explained briefly. Merck PSE is an interactive periodic table of chemical elements and was named 47 times (13 %). Wolfram Alpha, a computational knowledge search engine, was named 26 times (7.3 %). 22 times Schnittkraft-Meister, a game-based learning app to calculate cutting forces, was mentioned. For file hosting and as a cloud service mainly Dropbox is used (4.8 %). From those students who named an app, 4.2 % use Ankidroid, a program to design customized and personalized flashcards and use them for studying.

Furthermore the different apps were clustered, firstly by function and secondly by subject. In this context again all 357 answers were taken into account and duplications were ignored, in order to calculate the relative distribution among the app naming.

Function oriented cluster: Dividing the apps by the purpose they are used for, 7 different clusters could be identified. 109 (31 %) of the named apps serve as any kind of database. 79 (22 %) apps serve the organization of learning and 47 (13 %) times an app for any kind of application and testing was named. Furthermore the students named 40 (11 %) apps which fall in the category of language dictionaries, 30 (8 %) apps which assist the students to make notes or edit documents and 20 (6 %) apps for cloud computing. 32 (9 %) apps could not be allocated to any cluster. Hence these were summarized under "others". The results are visualized in Figure 4.

Subject oriented cluster: Looking at the related subjects the named apps could be allocated to six different clusters. The biggest cluster (160; 45 %) is formed by named apps, which had no connection to any special discipline. In this cluster apps like Dropbox or apps used to make notes are summarized. This cluster is followed by mentioned apps for mathematics and chemistry/physics (each 57; 16 %). 49 (14 % of the) times language learning apps, 30 (8 %) times mechanical engineering or logistics apps and finally 4 (1 %) times apps used in context with electrical engineering/informatics were named. These results are visualized in Figure 5.

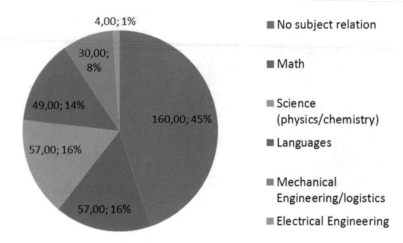

Figure 5 Apps clustered by subject

The results show that most of the used apps provide a general use without any subject relation. Looking into the list of named apps in this context shows that programs helping to organize the learning process, providing cloud computing and helping to edit documents dominate this kind of use. Hence, this is the area were students currently see the biggest advantages of app usage for study purposes. There are only some exceptions that provide the possibility for concrete knowledge application in a special subject context. Looking at those apps with a special subject connection it is visible that more apps are used in context with basic sciences in comparison to applied sciences. As in this study only engineering students were asked it is clear that even in engineering classes apps on basic sciences can be used. A second look into the app list shows that these apps in most cases are some kind of database (e.g. Merck PSE) – the biggest among the function oriented clusters. A question for further investigation can be, why only the minority of the mentioned apps could be used for subject related application. Is this a question of app quality or are there simply not that many apps? However the results indicate that are still high potentials lying in the development of specialized apps. Even if there are some apps, those are not used very often. The fact that in general only a little bit more than 13 % of the students could even name apps they are using for study purposes shows that very clearly. Considering furthermore that from all asked students (over 1500) only 47 named the most often mentioned app underlines this finding.

4   Conclusion

To summarize, students mostly use media and IT-services with the function of distribution of information. Facebook as probably the most common example of social media services is being used for study purposes, whilst other social media services like google+ or the microblogging service twitter have not found their way into universities yet. It can be stated that the study related media usage partly reflects the media usage in general. This is shown by the frequent use of popular sites like Facebook, Wikipedia or Google. Although with the given data, no statements can be made regarding cause and effect, it could be possible that especially in their first years at university, students use their already established mechanisms of media use as a starting point, before they develop other, more demanding and complex forms of e-learning. This can be interpreted as a form of reduction of complexity in the first years of studying. Especially to improve the introductory phase of the studies, which is affected by high dropout rates in engineering education, the relation between media usage and experience needs to be investigated on a deeper level.

E-learning services are being used more frequently than books or printed course materials. As a consequence for the ELLI project, this means that e-learning recommendation systems for teaching staff can be supportive for finding the tool that fits the teaching goals best. This could be of great help, especially for subject-specific e-learning opportunities, as in engineering education. Especially if the e-learning system is supposed to provide more functions than just distributing information, a personal recommendation is of great value. Since the conducted study focused on the students only, it cannot be determined whether the seldom usage of gesture computing or augmented reality is a consequence of the students' choice or of a lack of supply. Assuming that the latter is true, the ELLI project follows the right direction doing research on mixed-reality systems like the Virtual Theatre [16]. However, the application of such innovative hardware requires technical, psychological and subject-specific didactical skills in order to lead to the desired learning outcome [17]. Since "learning from different perspectives" is important for social problem solving [1], it seems to be an important precondition to teach students the competences they need in a globalized and digitalized working world. Why students don't use interactive or collaborative media and IT-services more often still needs to be investigated on a deeper level.

The results on mobile learning show that the use of mobile devices is not as common as it might be expected from general trends [7]. For example only a little bit more than a tenth of the students named apps they are using for study purposes. This could have at least two reasons. On the one hand it is possible that not enough appropriate apps exist, at least those which are usable in a learning context. Another explanation is that the students simply do not know which apps can be used effectively for which purpose. To conclude, the students must be supported in finding adequate apps. Even though the use of mobile apps at this point is not very common, mobile devices offer unique possibilities to support collaborative working processes. Especially the apps the students already named in the survey should stay in focus and it should be investigated why these apps are used for engineering education and what defines their unique selling point. In addition to that more qualitative studies are necessary on that point. In order to go more into detail student interviews should be conducted asking for reasons why they use or do not use their mobile devices for learning.

Acknowledgments The present work is supported by the Federal Ministry of Education and Research within the project "Excellent teaching and learning in engineering sciences (ELLI)".

References

1. G. Reinmann-Rothmeier, *Didaktische Innovation durch Blended Learning. Leitlinien anhand eines Beispiels aus der Hochschule.* Verlag Hans Huber, Bern, 2003
2. G. Gidion, M. Grosch, Welche Medien nutzen die Studierenden tatsächlich? Ergebnisse einer Untersuchung zu den Mediennutzungsgewohnheiten von Studierenden. Forschung und Lehre **6/12**, 2012, pp. 450–451
3. U. Dittler, E-Learning 2.0: Von den Hochschulen gehypt, aber von den Studierenden unerwünscht? In: *E-Learning: Eine Zwischenbilanz. Kritischer Rückblick als Basis eines Aufbruchs,* ed. by U. Dittler, J. Krameritsch, N. Nistor, C. Schwarz, A. Thillosen, Waxmann Verlag, Münster/New York/München/Berlin, 2009, pp. 205–218
4. M. Schiefner, M. Kerres, Web 2.0 in der Hochschullehre. In: *E-Learning: Einsatzkonzepte und Erfolgsfaktoren des Lernen mit interaktiven Medien,* ed. by U. Dittler, Oldenbourg Verlag, München, 2011, pp. 127–138
5. K. Thöing, U. Bach, R. Vossen, S. Jeschke, Herausforderungen kooperativen Lernens und Arbeitens im Web 2.0. In: *TeachING-LearnING.EU Tagungsband movING Forward – Engineering Education from vision to mission 18. und 19. Juni 2013,* ed. by A. Tekkaya, Dortmund, 2014, pp. 191–194
6. S. Bundesamt. Number of mobile internet users up 43% in 2013. Press Release Nr. 089, 2014. https://www.destatis.de/EN/PressServices/Press/pr/2014/03/PE14_089_63931.html
7. S. Bundesamt. Ob als Einzel- oder Zusatzgeräte: Mobile PCs setzen sich durch. Press Release Nr. 386, 2013. https://www.destatis.de/DE/PresseService/Presse/Pressemitteilungen/2013/11/PD13_386_632.html
8. C. Quinn, *mLearning: mobile, wireless, in-your-pocket learning.* LineZine, 2000. http://www.linezine.com
9. M. Sharples, J. Taylor, G. Vavoula, A Theory of Learning for the Mobile Age. In: *Medienbildung in neuen Kulturräumen,* ed. by B. Bachmann, VS Verlag für Sozialwissenschaften, Wiesbaden, 2010, pp. 87–99
10. A. Kukulska-Hulme, J. Traxler, *Mobile Learning: A handbook for educators and trainers.* Routledge, London, 2015
11. P. Maske, *Mobile Applikationen 1: Interdisziplinäre Entwicklung am Beispiel des Mobile Learning.* Gabler Verlag, Wiesbaden, 2012
12. M. Grosch, G. Gidion, *Mediennutzungsgewohnheiten im Wandel. Ergebnisse einer Befragung zur studiumsbezogenen Mediennutzung.* KIT Scientific Publishing, Karlsruhe, 2011
13. Alexa Internet Inc. "Top Sites in: All Categories > Computers > Internet > Searching > Search Engines", 2014. http://www.alexa.com/topsites/category/Top/Computers/Internet/Searching/Search_Engines
14. Quantcast Corporation. Rankings – Top sites, 2014. https://www.quantcast.com/top-sites
15. I. SEOmoz. The Moz Top 500, 2014. https://moz.com/top500
16. K. Schuster, D. Ewert, D. Johansson, U. Bach, A. Richert, S. Jeschke, Verbesserung der Lernerfahrung durch die Integration des Virtual Theatres in die Ingenieurausbildung. In: *TeachING-LearnING.EU discussions. Innovationen für die Zukunft der Lehre in den Ingenieurwissenschaften,* ed. by A. Tekkaya, S. Jeschke, M. Petermann, D. May, N. Friese, C. Ernst, S. Lenz, K. Müller, K. Schuster, 2013, pp. 246–260
17. K. Schuster, M. Hoffmann, U. Bach, A. Richert, S. Jeschke, Diving in? How Users Experience Virtual Environments Using the Virtual Theatre. In: *Proceedings of the 3rd International Conference on Design, User Experience, and Usability, Lecture Notes in Computer Science Springer,* vol. 8518. Springer, Heraklion, Crete, 2014, *Lecture Notes in Computer Science Springer,* vol. 8518, pp. 636–646

18. M. Grosch, *Mediennutzung im Studium. Eine empirische Untersuchung am Karlsruher Institut für Technologie*. Shaker, Aachen, 2012
19. M. Sharples, The design of personal mobile technologies for lifelong learning. Computers & Education **34** (3–4), 2000, pp. 177–193
20. Alexa Internet Inc. The top 500 sites on the web, 2014. http://www.alexa.com/topsites

A Web-Based Recommendation System for Engineering Education E-Learning Solutions

Thorsten Sommer, Ursula Bach, Anja Richert and Sabina Jeschke

Abstract The e-learning market consists of a wide variety of products, and it is still growing. To find an e-learning solution which fits the particular and situational demands is a very time consuming task, especially for teachers. Moreover, the technical and operational differences between the e-learning solutions are often not easy to understand from the product data and thereby consequences of choices are maybe not clear to the teacher. To solve these problems, a web-based recommendation system for teachers of engineering education is under development. This system is planned to support the decision making process of teachers about the use of an e-learning system. The precondition for setting up a recommendation system is that the desired entries (e.g. products, solutions, music or movies, etc.) are comparable to allow the algorithm to recommend: The current approach is to develop an e-learning scheme and compare the solutions based on this scheme. The determining of the necessary information for each e-learning solution has to be at least a half-automated process to keep the information up-to-date. Since expensive human time is needed to handle the necessary information, some approach with data and text mining as well as text analytics is promising. After the determining phase, each e-learning solution is represented by its data sheet. Apart from the e-learning solutions, also a teacher's requirements have to be comparable to e-learning solutions to allow the algorithm to recommend. That is why a web-based questionnaire is utilized to catch the teachers' requirements. A visual user-flow programming language is under development to provide an adequate environment for the development of the questionnaire and as an interface between the e-learning scheme, questionnaire and user-flow. The next step is to develop a functional prototype for the essential text analysis process to proof the concept. It is also required to analyze the current state of the e-learning scheme further to identify clusters of similar subjects, and to identify all critical properties to provide individual solutions for these.

T. Sommer (✉) · U. Bach · A. Richert · S. Jeschke
IMA/ZLW & IfU, RWTH Aachen University,
Dennewartstr. 27, 52068 Aachen, Germany
e-mail: thorsten.sommer@ima-zlw-ifu.rwth-aachen.de

Originally published in "The 9th International Conference on e-Learning ICEL-2014, 26-27 June 2014", © Academic Conferences and Publishing International 2014. Reprint by Springer International Publishing Switzerland 2016, DOI 10.1007/978-3-319-42620-4_35

443

Keywords E-Learning · Recommendation System · E-Learning Scheme · Text Mining · Visual Programming Language

1 Introduction

To assist teachers of engineering education in selecting an e-learning solution, a web-based recommendation system is under development: It enables teachers to reduce the amount of time finding an e-learning solution that matches their requirements.

The recommendation system requires a catalog with e-learning solutions. Unfortunately there is no common definition for "e-learning". Regarding the recommendation system, any technology aided solution a teacher for engineering education can use to teach or helps students to learn, is an e-learning solution. Therefore, the following definition was chosen: "E-learning is an approach to teaching and learning, representing all or part of the educational model applied, that is based on the use of electronic media and devices as tools for improving access to training, communication and interaction and that facilitates the adoption of new ways of understanding and developing learning." [1] This definition matches the expectations above and does not exclude any possible e-learning solution. Thus, this recommendation system will be recommend engineering education solutions like e.g. software and web-based solutions [2], hardware solutions [3] as well as game-based solutions [4], etc. Anyhow, this definition is broad but can be restricted if necessary.

A web-search for "e-learning" shows more than one billion results and a book-search for "e-learning" shows more than 60.000 books in English. Obviously, the teacher needs much time to handle this information overload. The consequences of a choice are perhaps not foreseeable: Is the chosen system able to handle the necessary amount of students, is the system prepared for the desired didactic method etc.?

The recommendation system proposed is parted into a public part (the website for teachers, called frontend) and a private administration part (the backend). The precondition for this recommendation system is that all e-learning solutions are comparable with each other, at least in respect to particular aspects: Otherwise it is not possible for algorithms to match the e-learning solutions to the teacher's requirements. The approach to provide a comparability is to find or define a general e-learning scheme. On the teachers' website, the teacher will receive questions out of a questionnaire to catch the teacher's requirements. The teacher's answers are the input for the recommendation algorithm: Together with the product data, the algorithm will be able to rate all solutions and present the teacher a weighted list of e-learning solutions.

To apply the e-learning scheme to the e-learning solutions, it is necessary to determine the required information for any solution. Both is not trivial and contains multiple challenges. This paper is focused to the approaches around the common scheme, the determining of the necessary information and how the teachers' questionnaire can be connected to the scheme.

2 E-Learning Scheme

Recommendation algorithms must be able to compare the desired entries (e.g. products, solutions, music or movies, etc.) among each other and with the input (e.g. answers, choices, requirements). In this case, all desired engineering educational e-learning solutions have to be comparable: It is desired to compare the technical aspects (e.g. hardware and software requirements, known scalability limits, supported data formats) operational behavior (e.g. availability of live interaction, availability of user-roles, possibility to comment, availability of access control) and didactic possibilities (e.g. possibility to declare learning objectives, availability of an interactive whiteboard, possibility of group work). Unfortunately, such a common e-learning scheme does not exist.

However, e-learning standards, specifications and reference models are available, e.g. IEEE P1484.1/D9 [5], SCORM [6], IMS CC [7]: But these are defining e.g. how learning units and content can be exchanged between different systems [6] or defining general e-learning system architectures [5]. Interesting is the "IMS Interoperability Conformance Certification Status" list [31]: These e-learning systems mentioned there are comparable because these systems are implementing the IMS CC standard e.g. at different versions. Nevertheless, this approach is not enough for this e-learning recommendation system: It would be a precondition, that interested teachers know and understand these standards and that all desired e-learning solutions and e-learning related systems are implementing these standard. Therefore, this approach is not sustainable for teachers nor for this recommendation system.

Further to these standards, e-learning classifications are available ([8]: Chapter 1.2): Schulmeister used the degree of interaction and Kolb used the categories "online" and "offline" to classify e-learning solutions. Richert allows also hybrids where an offline e-learning solutions can use additional an internet-connection to allow communication. Baumgartner and Payr define a model to classify the e-learning solutions by learning objective, learning content and teaching methods. None of these classifications compared by Richert [8] and none of the standards above covers the desired technical aspects, operational behavior and also the didactics: Nevertheless, the classifications suggested by Richert are covering parts of the technical aspects and parts of the didactics.

A modern alternative for a recommendation algorithm is to recommend based on e.g. social profiles [9, 10]: For each teacher who want a recommendation, the algorithm catches as much social information as possible and tries to recommend based on this data. Wherever this approach will work for the e-learning context (regarding the teachers' point of view, who is requesting an e-learning solution) is an open question. Especially, wherever the audience (teachers) is using the social networks enough for a good recommendation, is at least a demographic question [11]. Moreover, some aims are not related to the teacher, e.g. the technical aspects maybe related to the institute.

Therefore, the approach is to find or define an e-learning scheme that covers the necessary properties of e-learning solutions. This common e-learning scheme is subject of changes, because the e-learning market is changing [12]. It is preferred that also the teachers' questionnaire is connected to the scheme to ensure that the teachers' answers also comparable to the e-learning solutions: This mean, the e-learning scheme is also an interface between the teachers' answers and the e-learning solutions.

The current development state of the scheme contains more than 40 properties: Some parts of the classifications compared by Richert [8] are already included. However, the distinction by e.g. "online" and "offline" was no longer adequate and was replaced by "offline", "intranet" and "cloud": A "intranet" solution [13] can run on its own server at the institute, which occur also as an online solution to the students however the "cloud" solution runs on a third-party infrastructure and the institute has to pay continuously fees.

It is not possible to discuss more than 40 properties here: The following example properties are illustrating the general principle.

- Property #1: "Commercial solutions vs. free solutions"
- Property #2: "Number of students"
- Property #3: "Scheduled content vs. static content"
- Property #4: "Usability for teachers"
- Property #5: "Usability for students"

Some desired properties of the scheme are critical, e.g. #4 and #5. How to measure or rate the usability as human without to influence the result [14]? It is supposed, that the necessary degree of usability depends on the teacher's and student's experiences with such solutions. A naive approach (AP1) would be to count how often words or terms e.g. "problem", "complicated" and "does not understand" appears at the community for each e-learning solution to measure how difficult each solution is. However, this approach is unsuitable: Statistically, it is expected that for popular e-learning solutions also proportional more problems are reported.

It would be necessary to scale these measurements to a standardized scale e.g. from 0 to 10. Unfortunately is this not possible, because the population is unknown (e.g. the count of installations or the count of customers). Another approach is required to cover these critical desired properties. Perhaps it is enough to provide for each of these critical properties a suitable treatment: In case of this usability example, a few extra questions in the questionnaire are appropriate to proof or check the teacher's e-learning experience (AP2).

An alternative approach (AP3) uses the collected data from the vendors (see next section) and search for entries that are related to e.g. the usability subject. The questionnaire does not contain questions regarding the critical properties: Instead, the weighted list of results provide for each solution an abstract of related collected content. These abstracts are enabling the teacher to measure e.g. the usability to the own experiences: This is similar to classical customer reviews of online merchants and it is possible to analyze these data [15, 16].

Unfortunately, also this approach solves not all problems as the following thought experiment illustrate: Assume a teacher has completed his questionnaire and reads now the weighted list of results. Some solutions do not provide an abstract about the usability, because there were no records about it. A few solutions are providing abstracts with negative usability opinions, which is not interesting for the current teacher, because the reviews are related to e.g. Microsoft Windows, and the teacher wants to use e.g. a UNIX environment instead. Further, these solutions without abstracts are providing poor usability, which caused the absence of reviews. The current teacher chose an e-learning solution without an abstract, because of the assumption that none critical opinions are better as any critical opinions. At the conclusion, all three approaches have pros and cons but overall is the approach AP2 most convincing to satisfy the teachers.

3 Determining Solutions' Data

The current list of e-learning solutions was created by ourselves by researching public resources (e.g. websites, books, scientific journals). This list contains approximately 80 different solutions; each was selected according to the e-learning definition at Section 1. For each of these solutions, it is necessary to determine the required information to match the scheme: The result is a data sheet for each solution. It is not possible to produce these data sheets manually (purchase, install and evaluate each solution), because even with virtualization environments [17, 18] is this process to slow. It is obvious, that this manual task required many resources (e.g. time, money and staff). One alternative would be a crowdsourcing or community approach: The solutions' data to match the scheme would be mined by the crowd [19] or would be created by the community as user-generated content [32].

Moreover, the resulting data sheets are snapshots of one moment in time. It is required to keep these data sheets up-to-date to ensure that these cover the current state of each solution. Of course, the manual process could also utilize web and book resources instead of buying each solution, etc. Nevertheless, the major cost factor is the required staff costs: It is expected that the utilization of these web and book resources takes a long time. Therefore, an approach with data mining, text mining [20] and text analysis [21, 22] methods is promising: The needed runtime is usually short enough to execute the process continuously e.g. once a quarter. For each desired e-learning solution, the public vendor's information is captured. Additional, it is possible to capture also community information (e.g. discussion boards).

The determining process is split into data retrieval and the information retrieval. The data retrieval part fetches the raw data from the desired sources (e.g. vendors' public websites, communities) as HTML data [23–25]. A preprocessing of the raw data removes e.g. images and stores the data into a database. The information retrieval tries to extract specific information out of the preprocessed data [26]. In the case of this recommendation system, the main advantage is that the kind (vendors' information) and the context (e-learning) of the text data is known [20] and thereby the terminology that can occur is known. This is an important difference to the general text analytics purposes, where the context and kind of the text are unknown: In such a case, additional analytic methods are required to determine the kind and the context of the data.

Where the vendors information are at least not structured text data, the community text data is worse: Noisy text analysis is needed to process such data with issues like e.g. no case-sensitive wording, no punctuation, etc. because it is not possible to detect sentences or nouns [9]. However, a text mining method called "co-occurrences" is promising to get for a word related other words from its neighborhood [20]. It is even possible to get higher-order co-occurrences, if the co-occurrences for each vendor's data is consolidated together [20].

The current approach is to predefine one or more terms for any part of the common e-learning scheme and test these against the collected data. This approach based on the advantage, that the kind and context are known. A few examples to illustrate the principle: Assumed "Platform: Microsoft Windows?" is part of the scheme. The question is, if the solution operates on a Microsoft Windows server environment. If the predefined terms "Windows" and at least one of "not available" or "not ready" occur within one sentence, the process can assume with a certain probability that this e-learning solution is not available for Microsoft Windows environments. The property #1 from above, "Commercial solutions vs. free solutions", uses e.g. "dollar" and "euro" and supposed that both are not part of the vendor's information and at least one of "is free", "for free" or "free to use" occurs to assume together that a solution is not commercial. A dictionary with synonyms is not provided: the user who is defining these terms must provide alternatives as above (e.g. "not available", "not ready").

These are simple examples to show the principle, because with real text data, the analysis it not trivial. Use the example "Platform: Microsoft Windows?" once again: With the simple terms from above, the text "A version for Microsoft Windows is may available later." does not match the terms, because the semantic of the text is not known [20]. For this approach, further research and a technically proof of concept are required. With a functional prototype, it is possible to evaluate and compare the performance of this process with a sample (means in this case a test group of humans). In the case of the accuracy, this data mining and text analysis process must reach some degree of accuracy and must avoid serious errors but it is supposed that the sample performs better in such a complicated case: The humans are able to understand the semantic of a text easier. Nevertheless, it is also supposed that the sample is significant slower.

The expected result of this approach is, that at least half automated data sheets are producible with a good accuracy compared to the sample performance. In practice, it is perhaps possible to use natural language processing (NLP) [27] to solve some challenges e.g. understanding the semantic [28]. To work with NLP, programming libraries like e.g. NLTK available to speed up the development [29].

4 Visual User-Flow Language

To capture the teacher's requirements to an e-learning system, a questionnaire that is related to the e-learning scheme is necessary: Each question must be linked to the common scheme to ensure the comparability to the data sheets. To realize this requirement, an own visual programming language [30] was defined to model the user-flow through the questionnaire and to link the related parts of the scheme. The main advantage for a visual language is the visibility of the flow: The observer can directly view the flow through the graph because the flow is represented by e.g. lines or arrows to connect objects.

This language is represented by squares (called "function blocks") connected by arrows. The resulting visual program is linking the questionnaire, the common e-learning scheme and the user-flow together: For each inserted function block, the programmer has to choose the part of the common scheme and provide the related teachers' question.

Figure 1 shows an example to explain and illustrate the language: The program must read from left to right, from the unique start block to any of the end blocks. The starting point directs the teacher to question 1: If the teacher answers with "yes" or "neutral", the teacher is directed to the numeric query 1 - in case of the answer "no", the teacher reaches question 2. After the numeric query 1, the user-flow also reaches question 2. Further, if the teacher answer question 2 with "yes", question 3 follow up by end 1. If the answer for question 2 is "neutral" or "no", end 1 occurs directly. Instead of these example labels (e.g. "Question 1") the used part of the scheme is printed. Therefore, the observer can get a good overview about the program. With a double click, the details behind the function blocks are viewable: This is e.g. the question for the teachers, further explanation, numeric values, etc.

A first technical prototype, and thereby a technical proof of concept, exist: It is possible to program a user-flow with questions and subjects (to simulate the teachers)

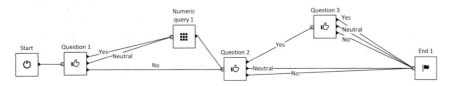

Figure 1 Example for a visual program

are able to going through these questions. Moreover, the visual program is changeable even if subjects are currently within the flow. Another thought experiment: Using the example at Figure 1 once again and assume there are four subjects (A, B, C and D): A is located at the starting block, B is located at "Question 1", C is located at "Question 3" and D is located in the end. Assume that "Question 2" is now changed: If subjects are located before the changed block (subject A and B), these subjects are receiving the changes directly without any interruption. Subject C and D are located behind the changed block: These subjects are not interrupted and effected. The recommendation system is, therefore, convenient to maintain.

If the common scheme is changed to cover changes on the e-learning market, the visual program must also change to provide related questions. Perhaps, also new function blocks for the language are required by some scheme changes. Therefore, this part of the recommendation system is permanently under development.

5   Results

The current results for this web-based recommendation system are promising: In case of the common e-learning scheme, the e-learning market currently knows multiple standards, specifications and classifications, but none covers the desired properties of technical requirements, operational behavior and didactics. Nevertheless, a common e-learning scheme is necessary to compare e-learning solutions. The approach is to define desired properties for the common e-learning scheme: Currently, approximate 40 plausible properties are part of the scheme. Some of these properties are difficult to measure or to rate because a standardized scale is missing: To define an own scale the underlying population is required - but often not available.

The manual determining of the necessary e-learning solution's data is expensive and time consuming. The scientific field of data mining, text mining, text analysis and NLP provide methods to retrieve information out of the text data. Depending on the text source, the handling of so called "noisy text" is required and e.g. it is not possible to detect nouns and sentences, which is the precondition of many algorithms and methods. Another problem is the semantic of the text data: Negating of the meaning or subjunctive sentences are difficult to analyze, but perhaps it is possible to cover this by using NLP.

Finally, the visual language is under development and allows it to program a user-flow with questions. Teachers are able to going through the questions according to the visual program. The visual program is even changeable if some teachers are currently within the flow: Teachers located before the changed blocks are receiving the changes directly without any interruption. In the case that some teachers are located behind the changed blocks, these teachers are not interrupted and effected. For the maintenance of the recommendation system, this behavior is convenient.

6 Conclusion & Outlook

From many discussions and conversations it was possible to infer that such a recommendation system for engineering education e-learning systems would be helpful to handle the information overload. The precondition of comparable solutions must be fulfilled to realize this recommendation system. The current approach is the common e-learning scheme: Because of the possibility that the mentioned challenges are not solvable, the research for an alternative to recommend is needed. The current over-all development state is promising to achieve this plan.

It is necessary for the determining of the data sheets to develop at least a half automated process to handle the amount of data. To ensure that the data set is up-to-date, this process must be able to run on a continuous basis. The current approach with data mining methods and textual analysis is promising but needs further investigation. After a technical proof of concept, the process must be evaluated: To be able to rate the evaluation results, a comparison with a sample (means here a test group of humans) is needed.

The current development state of the visual user-flow programming language is promising: the integration of the questionnaire, the user-flow and the scheme fulfills the expectations. The visual aspect of the language is also adequate to provide a good overview over the whole program and the flow through the questionnaire. The availability to change the program at runtime without interruption to the teachers is a good foundation for a convenient maintenance. The visual language is continuously under development to cover the current state of the scheme.

At the outlook, there are challenges but it is confident to solve these: Instead of asking the teacher about the own experiences with e-learning, a part of the questionnaire may need to proof the teacher's experiences. More research and different approaches are needed to cover this issue. Also needed is a further analysis to find dependencies between the properties of the scheme to build logical clusters.

The new aspect about the common e-learning scheme would be that different e-learning systems are comparable: It is then possible to compare a web-based solution with a game-based solution or an MOOC solution with a learning management system (LMS) and of course solutions of the same kind. Before this advantage is accessible, the determining of the necessary data for any solution is needed. Further investigation and research are required for the determining of the data, but the first approach is promising to get at least some basic information about any e-learning solution. Unlike a manual process of determining the data, this automated process can run continuously to keep the data up-to-date.

References

1. A. Sangrà, D. Vlachopoulos, N. Cabrera, Building an inclusive definition of e-learning: An approach to the conceptual framework. The International Review of Research in Open and Distance Learning **13** (2), 2012, pp. 145–159
2. S. Schön, M. Ebner, eds., *Lehrbuch für Lernen und Lehren mit Technologien*, 2nd edn. 2013. http://www.l3t.eu
3. K. Hamdan, N. Al-Qirim, M. Asmar, The effect of smart board on students behavior and motivation. In: *2012 International Conference on Innovations in Information Technology (IIT)*. 2012, pp. 162–166
4. D. Short, Teaching scientific concepts using a virtual world–Minecraft. Teaching Science-the Journal of the Australian Science Teachers Association **58** (3), 2012, p. 55
5. J. Tyler, B. Cheikes, *IEEE P1484.1/D9 - Draft Standard for Learning Technology - Learning Technology Systems Architecture (LTSA)*. IEEE Standards Activities Department, 2001
6. V. Devedzic, J. Jovanovic, D. Gasevic, The pragmatics of current e-learning standards. IEEE Internet Computing **11** (3), 2007, pp. 19–27
7. G. Durand, L. Belliveau, B. Craig, Simple learning design 2.0. In: *2010 IEEE 10th International Conference on Advanced Learning Technologies (ICALT)*. 2010, pp. 549–551
8. A. Richert, *Einfluss von Lernbiografien und subjektiven Theorien auf selbst gesteuertes Einzellernen mittels E-Learning am Beispiel Fremdsprachenlernen, Europäische Hochschulschriften / European University Studies / Publications Universitaires Européennes*, vol. 979. Peter Lang Publishing Group, Frankfurt am Main, Berlin, Bern, Bruxelles, New York, Oxford, Wien, 2009
9. M. Michelson, S.A. Macskassy, Discovering users' topics of interest on twitter: a first look. In: *Proceedings of the fourth workshop on Analytics for noisy unstructured text data*. 2010, pp. 73–80
10. F. Abel, E. Herder, G. Houben, N. Henze, D. Krause, Cross-system user modeling and personalization on the social web. User Modeling and User-Adapted Interaction **23** (2–3), 2013, pp. 169–209
11. M. Duggan, J. Brenner, The demographics of social media users, 2012. Tech. rep., Pew Research Center's Internet & American Life Project, 2013
12. L. Johnson, S. Adams Becker, V. Estrada, A. Freeman, *NMC Horizon Report: 2014 Higher Education Edition*. The New Media Consortium, Austin, Texas, 2014
13. P. Henry, E-learning technology, content and services. Education + Training **43** (4/5), 2001, pp. 249–255
14. J. Sauer, K. Seibel, B. Rüttinger, The influence of user expertise and prototype fidelity in usability tests. Applied ergonomics **41** (1), 2010, pp. 130–140
15. M. Hu, B. Liu, Mining and summarizing customer reviews. In: *Proceedings of the tenth ACM SIGKDD international conference on Knowledge discovery and data mining*. 2004, pp. 168–177
16. M. Hu, B. Liu, Mining opinion features in customer reviews. In: *AAAI*, vol. 4, 2004, pp. 755–760
17. M. Rosenblum, The reincarnation of virtual machines. Queue **2** (5), 2004, pp. 34–40
18. M. Fenn, M. Murphy, J. Martin, S. Goasguen, An evaluation of KVM for use in cloud computing. In: *Proc. 2nd International Conference on the Virtual Computing Initiative, RTP, NC, USA*. 2008
19. A. Doan, R. Ramakrishnan, A. Halevy, Crowdsourcing systems on the world-wide web. Commun. ACM **54** (4), 2011, pp. 86–96
20. G. Heyer, U. Quasthoff, T. Wittig, *Text Mining: Wissensrohstoff Text: Konzepte, Algorithmen, Ergebnisse*. W3L-Verl., Herdecke, 2006
21. I. Bierschenk, B. Bierschenk, Perspective text analysis: Tutorial to vertex. Kognitionsvetenskaplig forskning: Cognitive Science Research, 2011
22. T. Nasukawa, T. Nagano, Text analysis and knowledge mining system. IBM Systems Journal **40** (4), 2001, pp. 967–984

23. W. Jicheng, H. Yuan, W. Gangshan, Z. Fuyan, Web mining: knowledge discovery on the web. In: *1999 IEEE International Conference on Systems, Man, and Cybernetics, 1999. IEEE SMC '99 Conference Proceedings*, vol. 2. 1999, vol. 2, pp. 137–141

24. G. Pant, P. Srinivasan, F. Menczer, Crawling the web. In: *Web Dynamics*, Springer, 2004, pp. 153–177

25. C. Castillo, Effective web crawling. In: *ACM SIGIR Forum*, vol. 39. 2005, vol. 39, pp. 55–56

26. O. Egozi, S. Markovitch, E. Gabrilovich, Concept-based information retrieval using explicit semantic analysis. ACM Transactions on Information Systems (TOIS) **29** (2), 2011, p. 8

27. Y. Chen, Natural language processing in web data mining. In: *2010 IEEE 2nd Symposium on Web Society (SWS)*. 2010, pp. 388–391

28. S. M., S. Vranes, A natural language processing for semantic web services. In: *The International Conference on Computer as a Tool, 2005. EUROCON 2005*, vol. 1. 2005, vol. 1, pp. 229–232

29. M. Lobur, A. Romanyuk, M. Romanyshyn, Using NLTK for educational and scientific purposes. In: *2011 11th International Conference The Experience of Designing and Application of CAD Systems in Microelectronics (CADSM)*. 2011, pp. 426–428

30. D. Hils, Visual languages and computing survey: Data flow visual programming languages. Journal of Visual Languages & Computing **3** (1), 1992, pp. 69–101

31. IMS Global Learning Consortium, Inc. IMS interoperability conformance certification status, 2013. http://www.imsglobal.org/cc/statuschart.cfm

32. O. Nov, What motivates wikipedians? Communications of the ACM **50** (11), 2007, pp. 60–64

Access All Areas: Designing a Hands-On Robotics Course for Visually Impaired High School Students

Valerie Stehling, Katharina Schuster, Anja Richert and Sabina Jeschke

Abstract In recent years, student laboratories have been established as effective extracurricular learning areas for the promotion of educational processes in STEM fields. They provide various stimuli and potentials for enhancements and supplements in secondary school education (Reuter Sebastian et al.: Robotic Education in the DLR_SCHOOL_LAB RWTH AACHEN. Proceedings of the International Technology, Education and Development Conference INTED 2015, in Process (2015)). Most courses, however, do not offer full accessibility to all students. Those who e.g. suffer from visual impairment or even sightlessness find themselves not being able to participate in all tasks of the courses. On this account, the Center for Learning and Knowledge Management and Institute of Information Management in Mechanical Engineering at RWTH Aachen University have redesigned one of their robotics laboratory courses as a first step towards accessibility. This paper presents the work in progress of developing a barrier-free course design for visually impaired students. First feedback discussions with the training staff shows that even little changes can sometimes have a huge impact.

Keywords School Laboratories · Barrier Free · LEGO MindStorms · Visual Impairment · High-School Students

1 Introduction

Extracurricular school laboratories have proven to be an effective way to let students playfully experience the fundamentals of robotics, computer science or other technology-related topics. In combination with a hands-on approach, e.g. by working with LEGO MindStorms, they get a chance to learn on a cognitive, emotional and haptic level. While this is a widespread approach these days, not every pupil, however, is able to participate in courses like these due to a lack of accessibility. Ludi e.g. states that "awareness of potential career paths and access to adequate preparation

V. Stehling (✉) · K. Schuster · A. Richert · S. Jeschke
IMA/ZLW & IfU, RWTH Aachen University, Dennewartstr. 27, 52068 Aachen, Germany
e-mail: valerie.stehling@ima-zlw-ifu.rwth-aachen.de

Originally published in "HCI International 2015 - Posters? Extended Abstracts",
© Springer 2015. Reprint by Springer International Publishing Switzerland 2016,
DOI 10.1007/978-3-319-42620-4_36

remain barriers to students who are visually impaired" [1]. Due to their impairment or lack of sight it is rather impossible for them to fully participate in a programming process or when building a robot using e.g. LEGO MindStorms sets.

To overcome this sort of discrimination, the Center for Learning and Knowledge Management and Institute of Information Management in Mechanical Engineering of RWTH Aachen University have teamed up with a group of experts in order to develop a special barrier-free course design. This group of interdisciplinary researchers and practitioners – psychologists, school and university teachers, experts in the field of accessibility as well as robotics etc. – took the original course design from an existing robotics course for high school students and transformed it into an accessible course design by applying specific changes. Applying solely technical adjustments to the course, however, cannot be fully sufficient in the development of a new and adequate course design. Therefore, all changes applied to the course went hand in hand with an adjustment of teaching and learning strategies.

When designing a programming course for pupils with handicaps in a first step these strategies as well as required tools have to be thoroughly identified. The resulting new course design allows students with a handicap such as impairment of sight to access the same courses and benefit from the same experiences as their fellow pupils. This paper will present the original course design followed by results from the expert design workshops in terms of technical and didactical adjustments to the course. Finally it will present first indications through feedback discussions on the achievements made in first courses.

2 Original Course Design: "Roborescue" and "Rattlesnake"

In the original robotics course design high school students are given the chance to get an insight in building and programming robots using LEGO Mindstorms sets in a school laboratory. The main focus of the course lies on the construction and programming of various robot models with LEGO Mindstorms construction kits. By using the graphical programming interface NXT-G, which is also suitable for non-professionals, students find an easy access into the world of programming [2].

In order to prepare and motivate students for a future career in robotics, they can try their hands at building, programming and testing robots in a highly interactive and playful environment. The course allows them to experience the fascination of robotics by letting the students create either a "rescue robot" [3] that can search for virtual victims in a simulated rescue mission or a "rattlesnake" that snaps shut when someone crosses its field of vision. The choice of the scenario is subject to the age of the students – lower grades build a rattlesnake which is easier to build and to program while junior and senior classes go on a more complex rescue mission. Within this storyline, the four main tasks of the course are embedded: an introduction giving basic theoretical information, the construction phase, the programming process as well as the reflection or evaluation phase. The underlying didactical course concept focuses on own experiences made as a basis for all implicit learning processes. These processes primarily run playfully, practically and experimentally [2].

The school laboratory where the courses take place, however, is not located at school – it has been set up at RWTH Aachen University. This allows high school students to take a peek into the daily routine at University and is meant to facilitate the decision making process when it comes to choosing further steps after graduating from high school [4].

3 Enabling Higher Accessibility for Visually Impaired Students

3.1 Expert Design Workshops

In order to facilitate the process of redesigning the robotics course and reach a higher accessibility, researchers from RWTH Aachen University invited a team of interdisciplinary experts. In a series of expert design workshops, the roadmap of the redesign was created. The main goal of these workshops was to identify the key aspects of required adjustments in order to reach a distinctively higher level of accessibility.

In the course of the workshops, the participants gradually developed a grid of these requirements. In a first step they divided the course into its individual phases based on the established approach by Vieritz et al. [5]. They used the different phases of the course and analyzed the requirements and necessary adjustments for each individual part compared to those of the original course design. These phases are the introductory part, the construction phase, the programming phase as well as the phase for reflection. Combining their different experiences and testing single elements by simulating specifing eye disfunctions, the experts came to results in terms of requirements for each phase. These results are being presented and discussed in the chapters below divided into technical as well as didactical adjustments. At the end of chapter three, the developed grid gives a summarized overview of the results from the workshops.

3.2 Technical Requirements

Due to continuous research and rapid technical advancement, today, being visually impaired does not automatically exclude one from working on and with e.g. smaller objects or computers. It does, however, bring about specific technical requirements which have to be considered when designing a robotics course. According to the results of the design workshops, the identified requirements especially include auxiliary means which can be summed up as objects, software and computer settings. There are a lot of different eye dysfunctions which call for support by different objects. In order to increase accessibility these objects are e.g. magnifiers and common magnifying glasses. Other important objects for the different phases of the course are cameras and reading devices, printed handouts for every phase, additional lighting for the building process and sorting boxes for robot components.

In terms of software, screen readers such as JAWS or Dolphin, graphic programming using e.g. NXT-G [2] as well as textual programming using e.g. JBrick [6, 7] should be provided in the programming phase. Finally, the computers provided should allow for adjustments of graphic contrast on computer screens. These adjustments should also be possible on the provided work tables. Nevertheless, there is no "universal remedy" for increasing accessibility. In preparation of the course the teaching staff should therefore always acquaint themselves with the participants in order to be prepared for any special requirements the students might have.

3.3 Didactical Adjustments

Not every measure taken is helpful for every sort of handicap and not all changes can be made at once. In the presented case a fundamental distinction between different degrees of visual impairment up to sightlessness has been essential groundwork for further research and course development. Most advancements and adjustments have to be made gradually in order to reach full accessibility. This has proven to be a very helpful approach in the process of designing the new course. Some degrees of visual impairment, for example, are even contrary to one another [1], so there is a need for different technical as well as didactical approaches in one course to reduce or extinguish existing barriers for all participating students.

As a first result and requirement, printed manuals should be provided for the first three phases, the introductory part, the construction as well as the programming phase. This allows students to reread instructions at their individual pace.

Time has also proven to be one of the main but often underestimated factor [8]. Visually impaired students need to be given more time to work on their tasks in terms of reading instructions, following presentations as well as building and programming. The more severe the impairment is, the more time will be needed to finish a task. In addition to that, additional time needs to be invested in giving detailed information regarding the content of e.g. manuals or presentations, repeating this content, reflecting processes, practicing as well as post-processing. Practitioners from the workshop have come to the subjective conclusion that the time necessary for a traditional course design should be at least multiplied by four after monitoring their own ability to work through the tasks of the class by wearing glasses that simulate an eye dysfunction. On an average it took them four times as long to finish the assigned tasks. Further research and evaluations of the course will have to prove whether that factor needs to be adjusted.

Phase	Content	Original Course Design and equipment	Technical Requirements for a barrier free course	Didactical Requirements for a barrier free course
Introduction	Theoretical Input	Power Point Presentation	- Laptops with screen readers - Magnifying glasses	- Detailed explanations and descriptions of what the slide shows - Repetition of content - Simple slide design with high contrast - Printed Manuals
Construction	Building of the robot	Unsorted boxes	- Sorting boxes - Magnifying glasses - Reading Device - Graphic contrast on work tables	- Pre-sorting of components - Room for extra time and practice - Continuous supervision and support - Printed construction manuals
Programming	Programming of the robot	Laptops	- Contrast settings - Screenreader (JAWS/Dolphin) - Extra lighting - Printed Manual instead of beamer - On-screen magnifier - Graphic programming	- Continuous supervision and support - Room for extra time and practice - Printed programming manuals
Reflexion	Reflecting the Processes and Outcomes			Room for extra time

Figure 1 Results from the workshop: Requirements for the new course design

Another important adjustment relates to the teacher-student ratio. It has to be increased compared to traditional course designs which of course takes up additional time and resources on the teaching end. The required ratio can differ vastly as students have very diverse needs in terms of support. As we also know from Silva et al., even students without handicap perceive and process experiences in different preferred ways [9]. This has been confirmed also by the practitioners. Therefore, the supervisors need to provide a high level of flexibility regarding supervision and support throughout the course. Lastly the practitioners identified presorting the sorting boxes used in the construction phase as a helpful measure in the building process which does no longer exclude visually impaired students from the haptic and tangible experience of building a robot themselves.

Every course is highly influenced by diversity aspects and a thorough preparation and awareness of all possibilities and influences as well as a preanalysis of the expected target group of each course proves to be the key to a successful course design. Figure 1 sums up the results from the workshop in a grid.

4 Conclusion and Outlook

The paper has described the process of redesigning of a robotics course from an educational robotics laboratory. The redesign was performed in order to increase accessibility of the course for visually impaired students. The evaluation of an expert workshop has brought about a concept for the redesign which has been implemented and is currently being tested in a second run with various groups of visually impaired students. The developed grid of the workshop suggests that smaller as well as bigger adjustments to the designated phases of the lecture can lead to a higher level of accessibility. First anecdotal but enthusiastic feedback from the students leads to the gentle assumption that the applied changes suggested by the experts were successful.

Nevertheless, a huge part of the adjustments needs to be individually taken considering the needs and requirements that the specific dysfunctions of the target group bring about. At this point of research, there is no "one-fits-all"-solution to the challenge. As a consecutive step, evaluations of the designed courses will allow for a thorough analysis and serve the pursuit of continuous improvement. Additionally, it will be the key to future research. In order to broaden the range of accessibility, further research will have to focus on full accessibility also for blind students as well as other impairments such as hearing and e.g. physical disabilities.

References

1. S. Ludi, T. Reichlmayr, Developing inclusive outreach activities for students with visual impairments. In: *Proceedings of the 39th SIGCSE Technical Symposium on Computer Science Education 2008, USA*. 2008, pp. 439–443
2. A. Hansen, F. Hees, S. Jeschke, Hands on robotics. concept of a student laboratory on the basis of an experience-oriented learning model. In: *EDULEARN 2010, 5–7 July 2010*. IATED, Barcelona, Spain, 2010, Proceedings of the International Conference on Education and New Learning Technologies, pp. 6047–6057
3. S. Jeschke, F. Hees, N. Natho, O. Pfeiffer, A rescue robotics pbl course. In: *Proceedings of the ISCA 25th International Conference on Computers and their Applications (CATA), USA,*

24–26 March 2010. 2010, pp. 63–68
4. S. Jeschke, L. Knipping, M. Liebhardt, F. Muller, U. Vollmer, M. Wilke, X. Yan, What's it like to be an engineer? robotics in academic engineering education. In: *Proceedings of the Canadian Conference on Electrical and Computer Engineering (CCECE), Niagara Falls, Canada, 4–7 May 2008.* 2008, pp. 941–946
5. H. Vieritz, D. Schilberg, S. Jeschke, Early accessibility evaluation in web application development. universal access in human-computer interaction. user and context diversity. In: *Proceedings Part II of the 7th International Conference, UAHCI 2013, held as Part of HCI International, Las Vegas, NV, USA, 21–26 July 2013, Lecture notes in computer science*, vol. 8010. Springer, Berlin, 2013, *Lecture notes in computer science*, vol. 8010, pp. 726–733
6. S. Ludi, R. T., The use of robotics to promote computing to pre-college students with visual impairments. ACM Transactions on Computing Education **11** (3), 2011
7. S. Ludi, M. Abadi, Y. Fujiki, P. Sankaran, S. Herzberg, Jbrick: Accessible lego mindstorm programming tool for users who are visually impaired. In: *Proceedings of the 12th International ACM SIGACCESS Conference on Computers and Accessibility, ASSETS 2010, Orlando, FL, USA, October 25–27, 2010.* 2010
8. M. Kabátová, L. Jaskova, P. Lecky, V. Lassakova, Robotic activities for visually impaired secondary school children. In: *Proceedings of 3rd International Workshop Teaching Robotics, Teaching with Robotics Integrating Robotics in School Curriculum.* 2012, pp. 22–31
9. D.L. Silva, L.D. Sabino, E.M. Adina, D.M. Lanuza, O.C. Baluyot, Transforming diverse learners through a brain-based 4mat cycle of learning. In: *Proceedings of the World Congress on Engineering and Computer Science 2011, Vol 1, WCECS 2011, October 19–21, San Francisco, USA.* 2011

Please Vote Now! Evaluation of Audience Response Systems – First Results from a Flipped Classroom Setting

Valerie Stehling, Katharina Schuster, Anja Richert and Ingrid Isenhardt

Abstract Many University lecturers in Germany face the challenge of teaching very large classes, sometimes including 1000 or even more students. They often have to cope with a very high level of noise, bad room conditions, an extremely low level of participation as well as interaction and feedback. Some lecturers therefore try to overcome these challenges by using technology in their classroom. Previous research has already focused on evaluating the use of audience response systems (ARS) in a traditional but very large engineering lecture. This sort of technology has proven to be an effective tool in order to e. g. increase student motivation, give them additional support in the learning process and on the other hand give the lecturer feedback about the students' learning progress as well as possible crucial points of the lecture. This paper, however, goes one step further. It analyzes the use of ARS in a flipped classroom setting of a large engineering lecture for first-year-students. After having completed almost two thirds of the flipped classroom lecture, students were being questioned about their experiences and opinions about the use of ARS in this particular educational setting. The standardized questionnaire included questions issuing e. g. comprehension, motivation, frequency, enjoyment, interaction, involvement as well as usability aspects. First results show that e. g. the majority of the students feel that clicker questions foster their comprehension, motivate them to be attentive and increase the quality of the lecture. When comparing the results to findings from previous research in a traditional lecture, however, one thing becomes apparent: The evaluation of the use of ARS in the in a flipped classroom setting has turned out to be slightly less positive than that of the traditional lecture. This finding will be particularly discussed and may even call for further research in the designated field of interest. In a first step, the lecture itself will be described considering content, background and general settings. Subsequently, the survey instrument and methodology will be presented. In a third step, the results of the survey will be presented and discussed. Finally, further research fields will be identified.

V. Stehling (✉) · K. Schuster · A. Richert · I. Isenhardt
IMA/ZLW & IfU, RWTH Aachen University, Dennewartstr. 27,
52068 Aachen, Germany
e-mail: valerie.stehling@ima-zlw-ifu.rwth-aachen.de

Originally published in "Proceedings of the 10th International Conference on e-Learning 2015 (ICEL 2015)", © Academic Conferences 2015.
Reprint by Springer International Publishing Switzerland 2016,
DOI 10.1007/978-3-319-42620-4_37

463

Keywords Large Classes · Clicker Questions · Flipped Classroom

1 Introduction

Teaching engineering to a vast amount of students can be an extremely challenging task. Teaching these engineering students why communication and organizational aspects are extremely important for almost every field in their future careers is an even harder goal to be achieved. At the Center for Learning and Knowledge Management (ZLW) of RWTH Aachen University, the teaching method of flipping the classroom has been combined with the clicker technology in order to master this challenge. Previous research has already shown that the use of clicker questions in a traditional large lecture at the same institution with the same target group – students of mechanical engineering – is e.g. motivating the students to participate and be more attentive in class [1, 2]. The research at hand shall answer the question how the mix of the two teaching methods affects the class and how the results of the class evaluation turn out compared to previous evaluations of simply using clickers in the traditional classroom.

2 Lecture "Communication and Organization Development" ("KOE")

2.1 Content and Genesis

The lecture focused on in this paper is a lecture for first year mechanical engineering students. It is called "Communication and Organization Development" and will in the following be referred to as "KOE", from the German term "Kommunikation und OrganisationsEntwicklung". As has already been hinted at, the KOE lecture pursues one decisive goal. It aims to motivate young engineers to think interdisciplinary and learn how communication and organizational aspects are the key to success in student team work as well as professional life. To reach the targeted goal, many consecutive approaches have been introduced and tested in the lecture over the past years. In the early years, the lecture has been held as a traditional ex-cathedra lecture. It did, however, already include an extracurricular lab tutorial as well as online assessment at this stage. Shortly after, a learning management system in order to exchange data and provide online learning material has been added to the lecture. These data included slides as well as first video recordings of the lecture and served as helpful means for the students to prepare for and postprocess the lecture. In the following years it has proved as good practice to invite experts from the industry as well as alumni to the lecture. These experts were asked to give talks and enrich the lecture by combining their own experiences and practical examples with basic theoretical information about the topic that the specific class or module focused on. Altogether, the lecture consists of twelve modules [3]. Each of these modules covers one specific topic such as listed below.

1. Introduction
2. Basics of communicating in organizations
3. Basics of problem solving in teams
4. Basics of organizational development
5. Professional/cooperative processes in organizations
6. Learning and knowledge management
7. Global division of work
8. Intercultural cooperation
9. Organizational models and management approaches
10. Virtual Production
11. Innovation management
12. Change management

As a consecutive step, the lecture has been professionally videotaped and processed in small chunks of content in order to support student learning by allowing them to pause and afterwards easily reconnect and reenter the class. This adjustment of the lecture has been combined with the application of an audience response system in the lecture. The software-based system enables the students' participation and subsequently also the student-teacher interaction in class. The current and still prevailing adjustment to the lecture has been developed and introduced in 2012. It adds the method of flipping the classroom by giving an online lecture and using lecture time

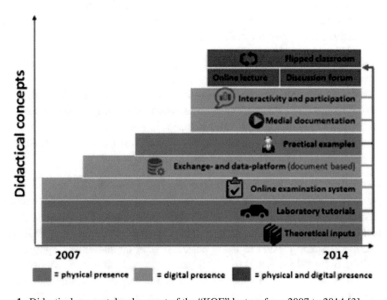

Figure 1 Didactical concept development of the "KOE" lecture from 2007 to 2014 [3]

for discussion. This large group discussion is regularly fostered by the use of the previously described audience response system and shows the current educational setting of the KOE lecture which will be described in the following Chapter.

Figure 1 shows the development and advancement of the KOE teaching approach and consecutive settings from 2007 until 2014 [3].

2.2 Current Setting of the KOE Lecture

The KOE lecture has been held for many years at RWTH Aachen University. In earlier years the lecture has been extremely challenging for both students as well as lecturers, due to the high number of students (up to 1500), a resulting high level of noise and an extremely low level of student-teacher interaction. Therefore the lecture has been gradually redesigned over the years and is now being held as a "flipped classroom". The teaching model of "flipping the classroom" is a lecture design originated in the field of blended learning. The blended learning approach combines classical classroom lectures with various e-learning designs. In the case of the flipped classroom model that means that the lecture itself is made available to the students in the form of online videos. In addition to that there are classroom sessions, where students bring in the questions aroused by watching the lecture videos. In contrast to the traditional lecture models, the classroom roles are therefore "flipped" by making the students active participants asking questions and thereby defining the content of the classroom sessions. This active participation is the one decisive element that leads to the success and popularity of the method. "Students learn best when they are active participants in the education process" [4, 5]. When talking of lecture sizes beyond a seminar size, however, this task is a special challenge to every lecturer [6, 7] as well as their students:

- Speaking in front of a large class is often frightening for students. They might fear being exposed when giving a wrong answer.
- Large classes are loud. The level of noise can quickly become very high once the attention of the students starts to decline. When answering a question in a large class the answer is often overheard because it is too loud in class.
- Even when the lecturer frequently asks questions in his/her class, he/she still cannot involve all the students at the same time [1, 2, 6, 8].

As a result, it has become common practice at many universities around the world that lecturers use technology as a means to overcome the described possible obstacles and to still gain as much feedback and insight in the student learning as possible. One technological tool that has proved to be very successful and easy to handle in the described teaching scenario is the application of an audience response system – "ARS" or often also referred to as "clickers". A software-based model of these ARS which is accessible via smartphone, laptop or any device that can connect to the internet has been acquired especially for large lecture interaction. When using technology in class there are, however, certain aspects to be carefully considered. Beatty's statement from 2004 gets to the heart of the dilemma: "Technology doesn't inherently improve learning" [9]. "Kerres also states that digital media are no 'Trojan horses' that can be brought into an organization (or situation such as a class) and unfold their effect 'overnight'. Kay and LeSage reinforce these statements by stating that '(…) the integration of an ARS into a classroom does not guarantee improved student learning' but that 'it is the implementation of pedagogical strategies in combination with the technology that ultimately influences student success'" [1, 9–11].

Along with the application of the ARS in the flipped classroom a didactical concept has been developed in order to reach the highest possible outcome of the method for both. Major benefits such as fearless interaction through anonymity, immediate feedback etc. have been put together in previous research and are shown in Table 1 below.

The didactical concept provides three different procedures depending on the poll results. It starts with the lecturer explaining specific concepts, posing a poll question afterwards and giving the students time to submit their answers. If the number of correct answers is at less than 25 %, the lecturer will explain the previously conveyed content again, pose the poll question again until it shows significant improvement and then proceed. If the number of correct answers ranges from 25 to 75 %, the lecturer will take up these answers and explain the correct answer. Given that the content is mostly understood by the students (more than 75 % correct answers), the lecturer will move on to the next topic [3].

Figure 2 shows the didactical concept of the application of the audience response system in the flipped KOE lecture.

Table 1 Benefits of using ARS in class [1]

Benefits of clickers for the student	Benefits of clickers for the lecturer
Interaction with the lecturer without fear of compromising oneself	Identification of knowledge gaps
Immediate feedback	Identification of shortcomings of the lecture [11]
Possibility to actively check their learning outcomes outside of exams	Student engagement
Be an active participant in class	Keeps students focused and involved
Anonymity	Higher attendance
Enhancement of learning	Better control of the learning progress
Classroom experience more enjoyable …	…

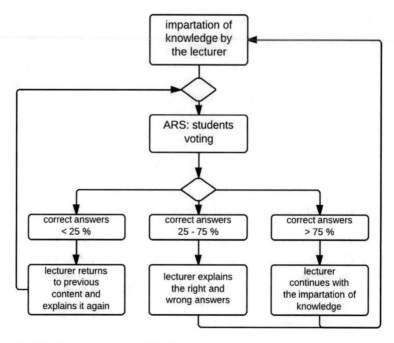

Figure 2 Didactical concept of the "KOE" lecture [3]

3 Survey Instrument and Methodology

Previous research in this particular field of study has focused on the introduction of clicker questions in a traditional but also large lecture with the same target group: first-year engineering students [1, 2]. The topic of the lecture then, however, was an introduction into computer programming for mechanical engineers. The paper at hand, as has been previously mentioned, concentrates on the evaluation of the use of clicker questions in an interdisciplinary flipped classroom setting. Of course, a comparison of evaluations of the same lecture but in two different settings would have allowed for a more intensive analysis of the one lecture itself. But despite the different topics and settings of the lectures, the questionnaire handed out to the students covered the same questions. Since the object of analysis – effects of clicker questions on student motivation and learning – was the same as before, it was possible to collect the same sort of data which makes both surveys comparable.

Worldwide evaluations show that the application of ARS in university lectures has led to e. g. higher motivation of attendance [12], more attention of the students during class and even higher knowledge acquisition than in traditional, non-interactive classes [13]. After having frequently used clicker questions from the beginning of the flipped classroom, the setting was evaluated by the students near the end of the semester in 2014. The sample covers 367 responses of first year engineering students. The questions posed were mostly closed-ended questions except for a comments section at the end. They can be divided into several sections which included questions concerning participation, impact on comprehension and content, motivational aspects, rating of the software itself and room for open ended comments. The following chapter will elaborate the results of the evaluation and – in a consecutive step – compare the most significant results with these of a previously conducted survey.

4   Results

4.1   Survey Results

The application of clicker questions in the flipped KOE lecture has – judging by the results of the survey – at large been received well by the students. Approximately 75 % of those questioned thought the application to be a reasonable approach. Around 70 % had a lot of fun answering the clicker questions and almost half of the questioned students felt that the clicker questions were motivating them to be attentive in class. The results of the survey can be summed up in five categories of the questionnaire: participation, motivation, questions, software as well as open ended questions.

4.1.1   Participation

97.8 % of the questioned students were in class when clicker questions were being posed. Of the 367 respondents, 33 % indicate to have participated in every single one of the polls. 44.7 % state that they participated only sometimes and a number of 21.5 % never entered a poll at all. To participate in a voting process, more than half of the respondents – 57.5 % – used their smart phone. Of the students who did not take part in the voting process, 34.5 % state that they do not own a suitable device for participation. A number of 25.5 % indicate that a lack of interest made them miss out on the clicker questions.

4.1.2   Comprehension and Content

85.6 % agree or somewhat agree, that the questions posed were comprehensible. 72.8 % of the respondents state that the questions posed were at a suitable or rather suitable complexity. Another 42.6 % agree or somewhat agree that the questions fostered their comprehension and were therefore helpful. 37.6 %, however, somewhat or even completely disagree. A 47.7 % stated that they would have wished for more clicker questions during the lecture and another 53.6 % even wished for more lectures with clicker questions in general. 76.3 % thought the application of clickers in class to be very useful.

4.1.3 Motivational Aspects

66.2 % of the students enjoyed or rather enjoyed participating in the polls. Around half of the respondents (54.2 %) agreed or somewhat agreed, that the application of ARS lead to a higher motivation to be attentive during the lecture. 38.5 %, however disagreed or rather disagreed.

Approximately half of the respondents (52.5 %) were motivated to participate in the polls. Another half of the questioned students (48.8 %) disagreed that the application of polls contributed to them attending class and a number of 67 % stated that the application of clicker questions enhanced the quality of the lecture.

4.1.4 Software

Around half of the questioned students stated that the software operated well while 37.9 % disagreed. The majority, a total of 85.3 % agreed that the software and terms of use were introduced and explained adequately by the lecturer. 55.9 % thought the software to be easily manageable, another 26.4 % rather agreed to that.

Figures 3 and 4 sum up the most significant results of the survey.

4.1.5 Open Ended Questions

Most comments in the open-ended comments section at the end of the survey are related to the internet connection and the ability to participate. Apparently the WiFi connection in class often broke down before all students were able to participate.

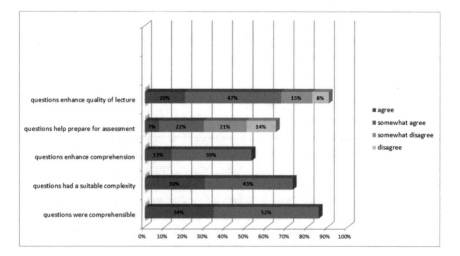

Figure 3 Results of the flipped classroom clicker evaluation (questions)

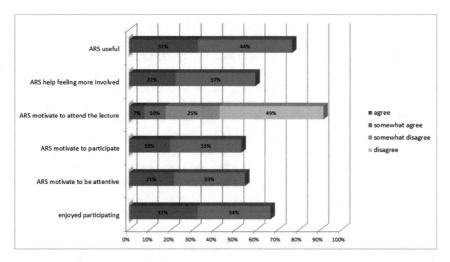

Figure 4 Results of the flipped classroom clicker evaluation (ARS)

Some students stated that the questions posed in the classroom sometimes were too easy and that they would have wished for a higher complexity factor. Other comments concern motivational state that the clicker questions motivated students to focus on the content again after having been diverted.

4.2   Flipped Classroom Versus Traditional Lecture

When comparing the results of both surveys – the first conducted in a traditional lecture and the second conducted in a flipped classroom setting – the first impression is that in both settings the application was rather successful. Only by taking a closer look at the results, smaller differences as well as some significant differences in the rating are noticeable. In order to identify similarities and differences, the most significant results elaborated in the previous chapter will be picked up again and compared to those of the earlier conducted survey. For the purpose of a higher clarity, the numbers in percentage will be rounded and the classifications "agree" and "rather agree" as well as "disagree" and "rather disagree" will be aggregated.

In the traditional lecture, 71 % of the students agree or rather agree that the clicker questions enhance the quality of the lecture. In the flipped classroom setting, the number of agreement lies slightly below at 67 %.

Some more significant differences are recorded with the following items. When asked if the polls help prepare for assessment, a total of 35 % agree in the traditional lecture and 56 % disagree. In the flipped setting, only 29 % agree but also only 35 % of the respondents disagree while the others abstain from voting. In the traditional setting, 63 % agree that the application of polls enhanced their comprehension. In the flipped setting, however, only 52 % agree to that. When queried about the complexity of the questions, 82 % of the traditional lecture respondents state that it was suitable. In the flipped lecture, only 73 % agree to that. Finally, 93 % of the questioned students from the traditional lecture indicate that the questions posed were comprehensible while in the flipped lecture a slightly smaller number of 86 % agree.

In both evaluations, the application of ARS is considered a useful approach, although the percentage of approval from the traditional lecture is slightly higher (83 %) than in the flipped setting (76 %). The following questions deal with motivational effects of the application of clickers in the classroom. 70 % of the traditional lecture state that ARS help them feel more involved in the lecture. In the flipped classroom evaluation, only 59 % agree to that. When asked if the ARS motivate students to attend the lecture, only 29 % of the respondents of the traditional lecture compared to 17 % of the flipped lecture, agree. The most significant differences between the two surveys appear in the last three items concerning motivational aspects towards participation, attention and enjoyment. In the traditional lecture, 69 % state that the use of clickers in the classroom motivates them to actively participate in the lecture. In the flipped classroom, however, only 52 % agree to that. Considering one key factor for the lecturer of the class, that ARS motivate the students to be attentive and focus on the lecture, the following results are recorded: 80 % of the traditional lecture agree while only 54 % of the flipped classroom students support this view.

The last item in focus deals with the fun factor of the polls. 81 % of the students in the traditional lecture enjoyed participating while only 66 % of the flipped lecture students agree to that.

Figures 5 and 6 sum up the most significant results of the clicker evaluation of the traditional lecture.

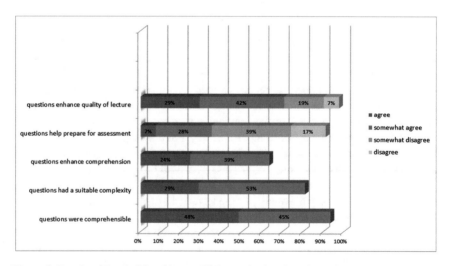

Figure 5 Results of the traditional lecture clicker evaluation (questions)

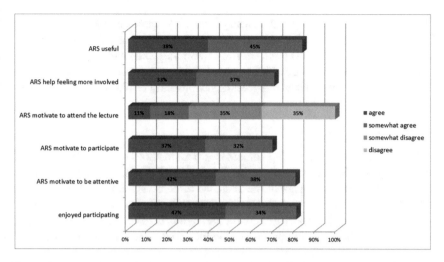

Figure 6 Results of the traditional lecture clicker evaluation (ARS)

5 Discussion and Further Research Fields

To sum up the results of the survey, audience response systems (ARS) are altogether received well with first-year engineering students. Most students appreciate the chance to actively participate in a lecture. In addition to that it also encourages them to be more attentive in order to be prepared for the questions coming up. This again helps the lecturer to stay focused and not be distracted by the high level of noise. The two major points of disagreement in both surveys were that ARS help to prepare for assessment and that they motivate to attend the lecture at all. The first result can be explained by the fact that students at the point of filling out the survey might not have been preparing for assessment yet or on the other hand that at this point they were unsure of how the test would look like and therefore not able to give an adequate statement about this just yet. In order to get an answer to this question, it should be posed after assessment.

The comparison of the two evaluations shows that the use of clicker questions in a traditional lecture has been slightly, sometimes even significantly better received than in the flipped classroom setting. At this particular point reasons for such a result can only be estimated and need to be verified by further research. Most possibly they are a mixture of many different conditions. Derived from the results of the evaluation these can be summarized under the following captions:

Complexity of/interest in the topic

One significant comment reinforces the impression that one gets when speaking to a lot of engineering first-years that attend the lecture. It reads: "mathematics is more fun". As has already been stated, engineering students in their first semester often do not understand why they should attend a lecture that at first sight does not directly connect to their original subject of study. Compared to many other courses in the curriculum of mechanical engineering, the subject of communication and organization management is felt to have a lower complexity and seems more easily intelligible to many students, which is why they often underestimate its importance. This, in turn, can lead to students feeling insufficiently challenged and sometimes even bored and frustrated.

Level of interaction

The flipped classroom setting combines several methods of interaction with poll questions just being one of many other interactive elements. This might lead to a slight reduction of the impact of the audience response system. Compared to that, in the traditional lecture, the poll questions were the only interactive element of the lecture which might have been a cause for the slightly better evaluation. The students attending the "KOE" lecture are usually in their first semester and mostly haven't attended any university lectures yet, they are rather used to traditional and obligatory courses from school. The method of flipping the classroom is yet a rather uncommon one in Germany and might also lead to an overstimulation-induced shutdown with some students. This should be subject of further research.

Technical difficulties

Apparently, the WiFi connection broke down several times during class times due to the high number of participating students in the poll. The resulting network traffic peaks led to connection difficulties or even errors. Most comments in the free comments section refer to this circumstance. Presumably this circumstance caused frustration with several students who tried to participate but couldn't.

Another interesting object of further research would be to query lecturers who use ARS in their lectures in order to find out whether or to what extent the described benefits of ARS are perceptible. Further research can be dedicated to the question if and – if yes – how the results of the polls are used for curricular optimization and didactical adjustments.

References

1. V. Stehling, U. Bach, A. Richert, S. Jeschke. Teaching professional knowledge to XL-classes with the helph of digital technologies, 2012
2. V. Stehling, U. Bach, R. Vossen, S. Jeschke. Chances and risks of using clicker software in XL engineering classes - from theory to practice, 2013
3. L. Koettgen, S. Schröder, E. Borowski, A. Richert, I. Isenhardt. Flipped classroom on top - excellent teaching through a method-mix, 2014
4. T.B. Crews, L. Ducate, J. Rathel, K. Heid, S. Bishoff, Clickers in the classroom: Transforming students into active learners. ECAR Research Bulletin (9), 2011
5. B. Davis, *Tools for Teaching*. Jossey Bass, San Francisco, 1993
6. R.J. Anderson, R. Anderson, T. Vandegrift, S. Wolfman, K. Yasuhara, Promoting interaction in large classes with computer-mediated feedback. In: *Designing for Change in Networked*

Learning Environments, ed. by B. Wasson, S. Ludvigsen, U. Hoppe, no. 2 In: Computer-Supported Collaborative Learning, Springer Netherlands, 2003, pp. 119–123

7. University of Maryland. Large classes: A teaching guide. center for teaching excellence, 2008. http://www.cte.umd.edu/library/teachingLargeClass/guide/index.html

8. B. Hasler, R. Pfeifer, A. Zbinden, P. Wyss, S. Zaugg, R. Diehl, B. Joho, Annotated lectures. student-instructor interaction in large-scale global education. Journal of Systemics, Cybernetics and Informatics **7**, 2009

9. I.D. Beatty, Transforming student learning with classroom communication systems. EDU-CAUSE, Research Bulletin **2004**, 2005

10. R.H. Kay, A. LeSage, A strategic assessment of audience response systems used in higher education. Australian Journal of Educational Technology **25** (2), 2009, pp. 235–249

11. M. Kerres, Wirkungen und wirksamkeit neuer medien in der bildung. In: *Education Quality Forum. Wirkungen und Wirksamkeit neuer Medien*, ed. by R. Keill-Slawik, M. Kerres, Waxmann, Münster, 2003

12. A. Trees, M. Jackson, The learning environment in clicker classrooms: student processes of learning and involvement in large university-level courses using student response systems. Learning, Media and Technology **32** (1), 2007, pp. 21–40

13. A.W. Nicolai Scheele, Die interaktive vorlesung in der praxis., 2004, pp. 283–294

Part III
Cognitive IT-Supported Processes for Heterogeneous and Cooperative Systems

Efficient Collision Avoidance for Industrial Manipulators with Overlapping Workspaces

Philipp Ennen, Daniel Ewert, Daniel Schilberg and Sabina Jeschke

Abstract This paper introduces an efficient collision watchdog predicting impacts between fast moving industrial robots. The presented approach considers the manipulator states in a three dimensional space. Tailored bounding volumes allow fast collision detection and distance calculations. The watchdog makes use of the internal rotary sensors of each robot to build an integrated world representation. Based on this information it is able to monitor the non-predictable behavior of all involved robots.

Keywords Collision Detection · Industrial Manipulators · Overlapping Workspaces

1 Introduction

In high-wage countries industrial production is often organized as an automated process. The setup of such a process is complex and cost-intensive, so automated processes amortize only in case of high-batch sizes. This contradicts the wishes of customers for tailored products which results in a large amount of manual work to achieve the desired flexibility [1]. Hence it becomes necessary to work on easily-adaptive and automated production systems.

An essential approach is given by self-optimizing production systems which are based on multiple axis industrial robots. Each robot within such a production system is autonomous regarding his motions, that is, it calculates its motions by itself. This may lead to a non-predictable behaviour of the ensemble of robots. If different manipulators are moving within the same workspace this possibly results in collisions.

With respect to non-predictable behaviour robot states cannot be determined in advance. On that account collision detection has to be calculated a posteriori. So

P. Ennen (✉) · D. Ewert · D. Schilberg · S. Jeschke
IMA/ZLW & IfU, RWTH Aachen University,
Dennewartstr. 27, 52068 Aachen, Germany
e-mail: philipp.ennen@ima-zlw-ifu.rwth-aachen.de

Originally published in "Procedia CIRP 2nd ICRM 2014 International
Conference on Ramp-Up Management", © Elsevier 2014.
Reprint by Springer International Publishing Switzerland 2016,
DOI 10.1007/978-3-319-42620-4_38

479

Table 1 Nomenclature

n, m	Bounding volumes/lines
r_n, r_m	Radius of bounding volume n, m
P_n, P_{n+1}	Position of origin joint n, joint $n + 1$
μ_n	Parameter of line n
d_{nm}	Minimum distance between line n and m
t_r	Response time
$\vec{v_n}, \vec{v_m}$	Maximum velocities of bounding volumes n, m
τ	Timestamp
ΔT	Time between two timestamps
κ	Numerical condition
S	Safety reserve
P_{crit}	Point of maximum velocity in a bounding volume

the basic requirements for the detection is time efficiency and a prospective decision function. The main task of the algorithm is to provide impacts inside robot ensembles. Therefore a large number of collision tests are necessary by checking all independent moving components. Possible combinations can be calculated by $(n - 1)!$ for n components. This equation is derived from following: If a robot ensemble consists of n independent components, the first component can collide with $n - 1$ others, the second one can collide with $n - 2$ additional others and the third one with $n - 3$. Finally the series $(n - 1)!$ results (Table 1).

In a nutshell: Applying an ensemble of two six axis robots with $n = 12$ independently moving components, the collision detection have to take care of impacts between $k = 39.916.800$ possible combinations of components. This slows down collision calculations. Additional robot states have to be computed by values of internal rotary sensors. Therefore joint positions have to be converted into Cartesian coordinates. This transformation also have an influence on performance. Consequently due to a high number of necessary tests, coordinate transformation and fast moving robots, the algorithm has to be very efficient in calculation and respecting movements following in near future.

2 State of the Art

In recent years many algorithms for collision detection have been developed and optimized, especially within the fields of computer graphics. As a result the classifications into *discrete* and *continuous* methods has been established [2]. For this purpose most famous algorithms for both classifications are be presented.

The continuous method (a priori) makes use of predicted trajectories of physical objects. This way the point of collision can be detected in advance. Literature here distinguishes between different types: the algebraic equation solving approach [3], the swept volume approach [4], kinetic data structures approach [5], and adaptive bisection approach [6].

The algebraic equation solving method makes use of calculating the point in time of collision. Therefore equations describing relevant trajectories have to be explicitly solved, often by using numerical techniques. The swept volumes approach is based on calculated swept volumes of all moving objects. If those volumes interference with each other or the environment, a collision is detected. The kinetic data structure approach makes use of elementary conditions. These conditions, also called certificates [5], describe criteria for object movements. If a certificate fails a collision might occur. The adaptive bisection method is based on conservative state sampling. Here, the distance, velocity and direction of movement is interpreted as the estimated time of impact between objects. If a lower bound is reached, a collision is detected.

Discrete methods (a posteriori) sample object trajectories and repeatedly apply a static interference test. In case of convex polytopes the problem can be described as a linear programming problem as follows: Two convex polytopes only overlap if no separating plane exists. A sufficient criterion for a separating plane is all vertices of the first polytopes lying in one halfspace of the plane and those vertices of the second polytopes lying in the second halfspace. Approaches for calculating the separating plane can be found in [7, 8].

In case of non-convex polytopes a preprocessing step is necessary which decomposes the polytopes into primitive convex entities [9]. This increases the number of essential pairs of objects that need to be checked for contact. Bounding volumes (BV) are established to reduce the number of pairs of objects by approximating them. In a hierarchical tree model the root BV contains all primitives of a model, each children's BVs contains separated primitives enclosed by the parent and the leaf usual contains one single primitive. Common used BV primitives are spheres [10, 11], axis-aligned bounding boxes [12–14] and orientated bounding-boxes [15–17]. Another approach for collision detection and minimum separation distance calculation is based on the Minkowski sum of two objects A and B. Cameron and Culley [18] have shown the minimum separation distance is equal to the minimum distance of the origin of the Minkowski sum $A \otimes - B$ to the surface. An algorithm solving this is provided in [19], but it has to be considered that the Minkowski sum of two convex polytopes can have $O(n^2)$ features [20], making the algorithm inefficient for such cases.

3 Collision Avoidance

The continuous methods presented in Section 2 are based on the equations of movement of every component involved. However, most proprietary robotic systems do not provide this information. Therefore the collision detection must be based on discrete methods: A watchdog is monitoring the robotic movements and triggers an emergency stop if it becomes necessary. Basically, the following guiding principle applies: For industrial manipulators a fast and safe algorithm for collision detection and avoidance is more important than exact analytical results.

Therefore it is feasible to approximate the actual robot geometry with simplified structures which completely enclose the robot. In the present case this is realized

Figure 1 Description of a
single robot component

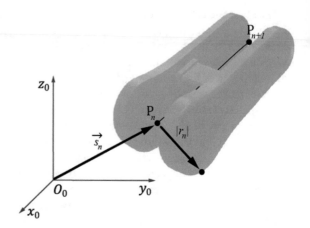

as follows: For each axis a finite line $\overrightarrow{s_n}$ is spanned through the two adjacent joints P_n and P_{n+1}. Also, a radius r_n is introduced which describes the maximum distance between the line and the surface of the related robot component (see Figure 1). The describing equation for the line $\overrightarrow{s_n}$ can be found in Equation 1.

$$\overrightarrow{s_n}(\mu) = P_n + \mu_n \cdot (P_{n+1} - P_n) \tag{1}$$

The point $\mathbf{P_n}$ represents the origin of a joint and is related to the world coordinate system O_o. Therefore a coordinate transformation is required. This can be done using the Denavit-Hartenberg representation [21].

Once the robot components are described, following conclusion is allowed: a collision exists if and only if the minimal distance between robot lines $\overrightarrow{s_n}$ and $\overrightarrow{s_m}$ is smaller than the sum of the related radiuses r_n and r_m. For fast moving robots we have to consider the whole systems response time. Therefore Equation 2 is introduced, with n and m as tested components, d_{nm} the minimal distance between those components, $\mathbf{V_n}$ and $\mathbf{V_m}$ as respective maximum velocities, t_r as response time, S as a safety reserve and τ as the timestamp.

$$d_{nm}(\tau) - 2t_r \cdot \| \overrightarrow{V}_n(\tau - 1, \tau) - \overrightarrow{V}_m(\tau - 1, \tau) \|_2 - s < r_n + r_m \tag{2}$$

Figure 2 shows the main signal flow with all functionalities affecting the response time of the system. These include sensors and the communication units, where signal flow is managed between all involved entities, the necessary coordinate transformations, and subsequently the actual collision detection. In consequence of an imminent collision, a suitable reaction is invoked and send to the actors.

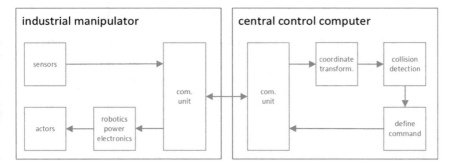

Figure 2 Signal flow

3.1 Maximum Velocity Calculation

Given a point $\mathbf{P_{n,crit}}$ at two timestamps τ and $\tau - 1$ with ΔT as relating time difference, the velocity in-between $\mathbf{P_{n,crit|\tau}}$ and $\mathbf{P_{n,crit|\tau-1}}$ can be linearly approximated as seen in Equation 3.

$$\overrightarrow{V}_n = \frac{\mathbf{P}_{n,crit|\tau} - \mathbf{P}_{n,crit|\tau-1}}{\Delta T} \tag{3}$$

For a safe estimation of imminent collision only maximum velocities are taken into account. Considering a linear relationship between velocity and distance to the point of initial movements the point $\mathbf{P_{n,crit}}$ have to be both located at maximum distance to the origin joint and inside the introduced bounding volume. So in case of the robot description model presented in Figure 1, the critical point is calculable with Equation 4.

$$\mathbf{P}_{n,crit} = \mathbf{P}_{n+1} + r_n \cdot \frac{\mathbf{P}_{n+1} - \mathbf{P}_n}{\| \mathbf{P}_{n+1} - \mathbf{P}_n \|_2} \tag{4}$$

3.2 Minimum Distance Calculation

This section will address how to calculate the minimum distance d_{nm} (τ) (see Equation 2). The calculation procedure thereby depends on the relative position to each other. Therefore, configuration tests are needed which allow to select the proper procedure. Figure 3 shows the projection of three general configurations.

The minimum distance in case (A) corresponds to the shortest distance between the lines, in case (B) it is found as the shortest distance between a point and a line and in case (C) the minimum distance is equal to the shortest distance between two points.

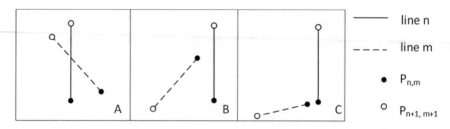

Figure 3 Different test configurations in general

For case (A) the procedure (4) is valid [22], provided that $(\mathbf{P}_{n+1} - \mathbf{P}_n) \times (\mathbf{P}_{m+1} - \mathbf{P}_m) \neq \vec{0}$, in other words the two lines have to be skew. However, this limitation is insignificant because in case of *moving* robots exact parallel lines can only occur within one single timestamp. Hence the response time t_r in Equation 2 is doubled.

$$d_{nm,A} = \frac{\| \, det[\mathbf{P}_n - \mathbf{P}_m \ \ \mathbf{P}_{n+1} - \mathbf{P}_n \ \ \mathbf{P}_{m+1} - \mathbf{P}_m] \| \|_2}{\| (\mathbf{P}_{n+1} - \mathbf{P}_n) \times (\mathbf{P}_{m+1} - \mathbf{P}_m) \|_2} \tag{5}$$

The minimum distance in case (B) is calculated by (6) [22].

$$d_{nm,B} = \frac{\| (\mathbf{P}_n - \mathbf{P}_m) \times (\mathbf{P}_{n+1} - \mathbf{P}_n) \|_2}{\| \mathbf{P}_{n+1} - \mathbf{P}_n \|_2} \tag{6}$$

And the minimum distance in case (C) is calculated by (7) [22].

$$d_{nm,C} = \| \mathbf{P}_n - \mathbf{P}_m \|_2 \tag{7}$$

3.3 Identifying Test Configuration

In total, there are nine possible configurations (shown in Figure 4), depending on their relative orientation. For correct distance calculation it is necessary to identify the case of configuration.

Assuming case (1) the value of both line-parameters $\mu_{n,m}$ are necessary, which describe the points of applied minimum distance. If they are in between [0, 1], case (1) is acknowledged, if both are higher than 1, case (9) is detected. Figure 5 shows all existing combinations in a decision tree.

For calculating μ_n and μ_m a temporary plane is necessary. In case of μ_n the plane is spanned by $\mathbf{P}_{m+1} - \mathbf{P}_m$ and $(\mathbf{P}_{m+1} - \mathbf{P}_m) \times (\mathbf{P}_{n+1} - \mathbf{P}_n)$ (see Equation 8).

$$\vec{q} = \mathbf{P}_m + \alpha \cdot (\mathbf{P}_{m+1} - \mathbf{P}_m) + \beta \cdot (\mathbf{P}_{m+1} - \mathbf{P}_m) \times (\mathbf{P}_{n+1} - \mathbf{P}_n) \tag{8}$$

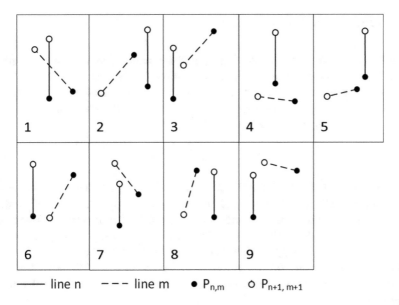

Figure 4 Overview of all test configurations

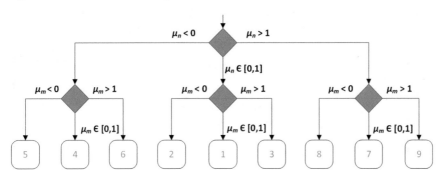

Figure 5 Decision tree for determining current test configuration

The intersection of the temporary plane with the line n gives μ_n. With $\vec{\chi}_n = (\alpha, \beta, \mu_n)^T$ the value of μ_n can be calculated by solving the system of equations in (9).

$$\vec{x}_n = A^{-1}[P_n - P_m],$$

with $A = [P_{m+1} - P_m \ (P_{m+1} - P_m) \times (P_{n+1} - P_n) \ P_n - P_{n+1}]$ (9)

3.4   Numerical Effects

If both lines n and m are approaching a parallel configuration, the inverse matrix in Equation 9 and the distance calculation (Equation 5) cannot be reliably determined. Therefore a watchdog for the numerical condition of the problem is necessary. Since the matrix A (see Equation 9) describes the whole geometrical situation, the condition can be calculated by solving Equation 10 [23].

$$\kappa(A)_\infty = \| \mathbf{A}^{-1} \|_\infty \| \mathbf{A} \|_\infty \tag{10}$$

If the condition becomes infinity, n and m are exactly parallel, otherwise Equations 5 and 9 can be solved. However, if they are approaching a parallel configuration, numerical errors during distance calculation can occur. Therefore the numerical condition should be a small value. If the value is exceeding an upper limit, the collision calculation has to be skipped. However, this does not affect the collision detection in practice: Experiments on two industrial robots (ABB IRB 120) have shown that extremely bad conditioned robot states can only occur within one timestamp (see Figure 6).

So in case of a doubled response time (see Equation 3) the skipped timestamp is considered by approximating current robot state in a previous timestamp.

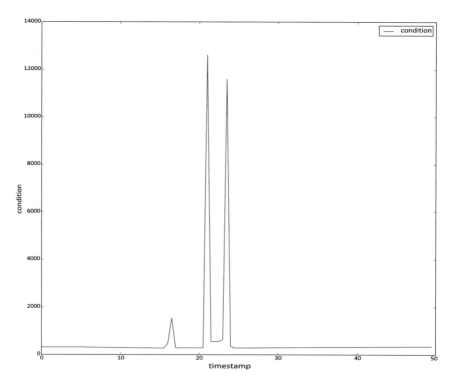

Figure 6 Numeric condition of moving robot components

3.5   Relevant Test Candidates

Not every combination of lines needs to be tested. Directly adjacent lines, linked with a joint, are overlapping all the time. So as a minimum condition, n and m are not allowed to be connected by a joint. Also depending on robot location there can be geometrical reasons which eliminate possible collisions. By considering relevant test candidates only, the time efficiency of the collision watchdog can be dramatically increased.

4   Evaluation

The presented approach was evaluated using two six axis robots (ABB IRB 120, see Figure 7). A socket communication based on TCP/IP is used for sending sensor information from the robots to a central control computer. Here the necessary transformations are calculated using the ROS framework [24]. A python based program is working as a watchdog and conducts collision detection. If the robots get into impact, the watchdog triggers an emergency stop. Considering the whole system a response time of $t_r = 0.5$ s is measured. Therefore in this use case a forward looking collision detection is mandatory for safe movements.

Test runs were based on the decision tree in Figure 5. They were successfully performed for a complete statement coverage of this decision tree.

Figure 7 Robotics cell based on two ABB IRB 120

5 Summary

We presented a hybrid approach of a both continuous and discrete collision detection. Considering current velocities allows to approximate future collision states within following timestamp. By doing this the algorithm provides collision detection for fast moving robots in case of non-negligible response time. Relevant test candidates for collision are verified by checking them for interference.

For choosing the right calculation rule an analysis of current test configuration is necessary. Therefore a decision tree is compiled. Depending on the result, the appropriate distance calculation is performed. A concluding interference test considering response time, speed and current minimal distance gives a final statement about collision. This algorithm is successfully tested on a robot cell consisting of two industrial manipulators.

6 Outlook

An optimal approach for collision detection will prevent collisions by calculating collision-free trajectories a priori. Therefore the presented approach can be integrated into numerical calculation of inverse kinematics whereby impacts could be predicted in advance. However, for non-predictable behaviour this approach cannot be realised. For a safe algorithm the approximation of future robot states have to be credible and a good estimation.

Therefore adding constraints like robotics kinematic can result in more precise position approximations. Additional while the presented approach detects collisions, it will not resolve the collision state. On that account an evasive action is necessary. Considering the contact point of imminent collision the initial evasive direction should be normal to this collision point. Using this as a boundary condition a new path can be calculated.

References

1. C. Brecher, F. Klocke, R. Schmitt, G. Schuh, *Excellence in Production*. Apprimus, Aachen, 2007
2. S. Redon, Y. Kim, M. Lin, D. Manocha, Fast continuous collision detection for articulated models. Journal of Computing and Information Science in Engineering **5** (2), 2005, pp. 126–137
3. S. Redon, A. Kheddar, S. Coquillart, *An Algebraic Solution to the Problem of Collision Detection for Rigid Polyhedral Objects*. INRIA - Rocquencourt, San Francisco, 2000
4. K. Abdel-Malek, J. Yang, D. Blackmore, K. Joy, Swept volumes: Fundation, perspectives, and applications. International Journal of Shape Modeling **12** (1), 2006, pp. 87–127
5. J. Basch, J. Erickson, L. Guibas, J. Hershberger, L. Zhang, *Kinetic Collision Detection Between Two Simple Polygons*. SiRF Technology Inc., 2003

6. F. Schwarzer, M. Saha, J. Latombe, *Adaptive Dynamic Collision Checking for Single and Multiple Articulated Robots in Complex Environments.* 2005
7. R. Seidel, ed., *Linear programming and convex hulls made easy.* ACM, 1990. 211–215
8. K. Chung, W. Wang, Quick elimination of non-interference polytopes in virtual environments. In: *Virtual Environments and Scientific Visualization '96*, Springer, 1996, pp. 64–73
9. B. Chazelle, Convex partitions of polyhedra: a lower bound and worst-case optimal algorithm. SIAM Journal on Computing **13** (3), 1984, pp. 488–507
10. P. Hubbard, Approximating polyhedra with spheres for time-critical collision detection. ACM Transactions on Graphics (TOG) **15** (3), 1996, pp. 179–210
11. S. Quinlan, ed., *Efficient distance computation between non-convex objects.* IEEE, 1994
12. N. Beckmann, H. Kriegel, R. Schneider, B. Seeger, *The R*-tree: an efficient and robust access method for points and rectangles.* ACM, 1990
13. M. Held, J. Klosowski, J. Mitchell, eds., *Evaluation of collision detection methods for virtual reality fly-throughs.* Citeseer, 1995
14. M. Ponamgi, D. Monacho, M. Lin, eds., *Incremental algorithms for collision detection between solid models.* ACM, 1995
15. S. Gottschalk, M. Lin, D. Manocha, eds., *OBBTree: A hierarchical structure for rapid interference detection.* ACM, 1996. 171–180
16. G. Barequet, B. Chazelle, L. Guibas, J. Mitchell, A. Tal, eds., *Boxtree: A hierarchical representation for surfaces in 3D.* Wiley Online Library, 1996. 387–396
17. S. Gottschalk, *Collision queries using oriented bounding boxes.* 2000
18. S. Cameron, R. Culley, eds., *Determining the minimum translational distance between two convex polyhedra.* IEEE, 1986. 591–596
19. E. Gilbert, K.S. Johnson, D. W., A fast procedure for computing the distance between complex objects in three-dimensional space. Robotics and Automation **4** (2), 1988, pp. 193–203. IEEE Journal
20. D. Dobkin, J. Hershberger, D. Kirkpatrick, S. Suri, Computing the intersection-depth of polyhedra. Algorithmica **9** (6), 1993, pp. 518–533
21. S. Niku, *Introduction to robotics: analysis, systems, applications.* Prentice Hall New Jersey, 2001
22. I. Bronshtein, K. Semendyayev, G. Musiol, H. Muehlig, *Handbook of mathematics.* Springer, 2007
23. W. Dahmen, A. Reusken, *Numerik für Ingenieure und Naturwissenschaftler.* Springer, 2008
24. M. Quigley, B. Gerkey, K. Conley, J. Faust, T. Foote, J. Leibs, E. Berger, R. Wheeler, A. Ng, eds., *ROS: an open-source Robot Operating System.* 2009

Auf dem Weg zu einer „neuen KI": Verteilte intelligente Systeme

Sabina Jeschke

Zusammenfassung Cyber Physical Systems und Internet of Things entwickeln sich rasant vorwärts und bringen gleichzeitig die Entwicklung einer neuen künstlichen Intelligenz voran, der der „verteilten künstlichen Intelligenz". Hervorstechende Merkmale sind dabei die räumliche Verteiltheit und Heterogenität ihrer Subkomponenten. Das Konzept einer verteilten Intelligenz ist dabei nicht neu, kann man doch für die Tierwelt feststellen, dass hochentwickelte Formen von Intelligenz gerade mit einer höheren Verteiltheit ihrer „intelligenz-erzeugenden Komponenten" einhergehen. Neuartig ist, dass die räumliche Verteilung der Komponenten nicht beschränkt und die Zusammensetzung der Systeme hochgradig wandelbar ist. Ein „intelligenter Agent" ist dabei durch das Vorhandensein von Sensorik, Kognition und Aktuatorik gekennzeichnet. Aufbauend auf die Theorien der neuen verteilten künstlichen Intelligenz entstehen derzeit neue Modelle hybrider Teams, bei denen Gruppen aus mehreren autonomen Robotern, Web-Agenten und Menschen miteinander kooperieren. Beispiele für die Anwendung intelligent kooperierender Teilkomponenten sind die industrielle Produktion und die Intralogistik, aber auch vermehrt Bereiche des Alltags wie etwa autonome Fahrzeuge.

Schlüsselwörter Embodiment-Theorie · Verteilte Künstliche Intelligenz · Cyber-Physical Systems · Multi-Agenten-Modell · Autonome Fahrzeuge

1 Abstract

Zusammen mit einigen bemerkenswerten Durchbrüchen wie Watson oder dem Google Car in den vergangenen Monaten stehen wir vor einer neuen Ära in der Künstlichen Intelligenz. Eine besondere Rolle nimmt dabei die „Verteilte Künstliche Intelligenz" ein, Cyber Physical Systems und Internet of Things "boomen". Im Kern handelt es sich bei ersteren um einen Verbund mehrheitlich technischer Subkomponenten, die über eine internetbasierte Dateninfrastruktur miteinander

S. Jeschke (✉)
IMA/ZLW & IfU, RWTH Aachen University, Dennewartstr. 27, 52068 Aachen, Germany
e-mail: sabina.jeschke@ima-zlw-ifu.rwth-aachen.de

Originally published in "Informatik Spektrum",   491
© Springer 2015. Reprint by Springer International Publishing Switzerland 2016,
DOI 10.1007/978-3-319-42620-4_39

kommunizieren, bei letzterem um die Ausweitung des „Beteiligungskonzepts" des
Internets: Teilnehmer sind nicht mehr ausschließlich Menschen, sondern auch
„Dinge" – wie etwa die Sensorik eines Autos, Klimadatenstationen, Prozessdaten-
rechner der Produktionstechnik und andere informationstragende und/oder mit der
Umwelt unmittelbar interagierende Systeme.

2 Cyber Physical Systems und das Internet of Things

In beiden Konzepten entsteht ein „Graph" aus Knoten und Kanten, bei dem die
Knoten die Informationseinheiten repräsentieren, die Kanten die Kommunikationswege. Die eine Sichtweise fokussiert auf die Komponenten, die zweite auf das Netzwerk – in gegenseitiger vollständiger Anerkennung, dass das eine ohne das andere
wenig nützlich wäre. Im Ergebnis bilden Cyber Physical Systems und Internet of
Things zwei Sichten auf dasselbe Phänomen, nämlich, wie es einer der führenden
Wissenschaftler von Bosch, Dr. Stefan Ferber, in seiner Keynote 2012 in Wuxi/China
zusammenfasste, „...the outlook of connecting 50 billions devices by 2015". „Connecting them with each other and with 6 billion people", hätte er noch hinzufügen
können.

3 Analogie zur menschlichen Intelligenz

Die Bedeutung dieser Konzepte geht in vielfacher Hinsicht weit über die Entwicklung
einer „normalen" technischen Neuerung hinaus. Ein Aspekt lässt dabei besonders
aufhorchen: Hier entsteht eine neuartige Form einer „Intelligenz", einer nämlich, die
auf der Vielzahl, der räumlichen Verteiltheit und der Heterogenität ihrer Subkomponenten basiert – eine „Intelligenz verteilter Systeme". Diese Intelligenz gilt es
zu untersuchen – schon allein weil die hier entstehenden Systeme unmittelbare und
umfassende Auswirkungen auf unseren Alltag haben und in der Zukunft noch viel
mehr haben werden, wie wir an Beispielen wie smart grids in der Energietechnik,
kooperativen Robotern in der Produktionstechnik, vernetzten medizinischen Systemen zur Früherkennung oder bei der car2Infrastructur Communication im Verkehr
sofort erkennen.

Als Wissbegierige, die wir aber auch immer unseren eigenen Aufbau und unsere
Funktionsweise versuchen zu verstehen, unseren Körper und unseren Geist und
insbesondere deren Wechselwirkungen, eröffnet sich hier noch eine ganz andere
Verbindung: Es mehren sich Anzeichen dafür, dass das Verständnis dieses hier entstehenden Intelligenzbegriffes auch geeignet sein könnte, das Verständnis für unsere
eigene, menschliche Intelligenz zu schärfen. So haben etwa Studien von Thompson
und Swanson 2010 (das theoretische Modell geht zurück auf Arbeiten aus 2003)
klare Anhaltspunkte dafür geliefert, dass die Vernetzungsstruktur des Gehirns nicht
vollständig hierarchisch ist wie einst angenommen. Vielmehr erinnert es in Teilen an

die dezentrale Netzwerkstruktur des Internets. Insbesondere ist die Fähigkeit einer "graceful degradation", die Eigenschaft also, dass ein Teilausfall des Systems nicht zum Totalausfall führt, eine unmittelbare Konsequenz nicht-hierarchischer Strukturen, und kennzeichnend für beide, für moderne verteilte Computersysteme ebenso wie für die Funktionsweise des menschlichen Gehirns [2]. Vor diesem Hintergrund könnte hier auch ein – lange erhoffter – Durchbruch auf dem Gebiet der „Starken KI" erzielt werden – jenem Teilgebiet der künstlichen Intelligenz also, das sich mit der Entwicklung tatsächlich und intrinsisch intelligenter Systeme befasst und nicht nur solcher mit „intelligentem Erscheinungsbild".

4 Neuartigkeit verteilter Intelligenz?

Ist das Konzept einer „verteilten Intelligenz" tatsächlich neu? Schließlich ist auch bei Menschen, ebenso bei anderen Säugetieren, der „Sitz" der Intelligenz nicht ausschließlich das Gehirn selbst: Intelligente Vorverarbeitungen finden in „biologischen Sensoren", den Sinnesorganen, statt. Entsprechendes gilt für „biologische Aktuatoren", die Muskeln, durch zuständige Bereiche im Rückenmark. Erst diese Vorverarbeitungen ermöglichen hochqualifizierte und vor allem „echtzeitfähige" Reaktionen: Ohne sie wären wir nicht einmal in der Lage, eine Fliege zu erlegen, gar nicht zu reden von einer Flucht vor einem Säbelzahntiger!

Auch hier ist die Intelligenz also zu einem gewissen Grade – wenn auch in anderer Weise – verteilt. Als Tendenz zeichnet sich in der Tierwelt ab, dass hochentwickelte Formen von Intelligenz gerade mit einer höheren Verteiltheit ihrer „intelligenzerzeugenden Komponenten" einhergehen. Die Vermutung liegt daher nahe, dass – wenngleich nicht ausnahmslos – zunehmende Verteiltheit nicht nur zu Intelligenz beiträgt, sondern möglicherweise sogar eine Voraussetzung einer hochentwickelten Intelligenz sein könnte.

5 Das Neue: Globale und variable Entitäten

Tatsächlich ist nicht die Verteiltheit an sich „das Neuartige" im „Zoo der Systeme". Das Neue liegt zum einen in der Art der Verteilung, also darin, dass Systeme wie Cyber Physical Systems weder in einem strikten noch in einem schwachen Sinn „lokalisiert" sind. Während ein Mensch, ein Tier, ein einzelner Roboter oder ein isolierter Computer in erster Näherung als eine „lokalisierte Entität" angenommen werden können, mit „kleiner" Ausdehnung und einen entsprechend beschränkten Wahrnehmungs- und Wirkungsradius also, ist ein Cyber Physical System räumlich nicht beschränkt, sondern kann aus Komponenten bestehen, die sich im Extremfall über die gesamte Welt verteilen. Ein solches System kann Informationen von weit entfernten Orten – bis auf physikalisch bedingte Latenzzeiten – synchron zusammenführen und ebenso an all diesen Orten mit seinen global agierenden Entitäten gleichzeitig fast synchron mit der Umwelt interagieren.

Die zweite wesentliche Neuerung betrifft Gestalt und Wandelbarkeit solcher Systeme: Während Mensch und Tier i. d. R. aus einer festen Anzahl von Komponenten, etwa Gliedmaßen, Sinnesorganen etc., bestehen, induziert die Perspektive des Internet of Things Konstrukte hochgradiger Dynamik und Variabilität: Ihre mit einer Teilautonomie ausgestatteten Komponenten bilden eine Art „Community", deren Mitglieder „kommen und gehen" wie sie wollen bzw. wie es die Gegebenheiten erlauben.

Eine Art zentraler „Zugangs- und Anwesenheitskontrolle" kann zwar technisch realisiert werden, ist aber nicht unmittelbarer Bestandteil des Konzepts. Vorgesehen ist hier zunächst nur eine zentrale Kenntnis über die Mitgliederstruktur zu jedem festen Zeitpunkt, eine Art „Anwesenheitsprotokoll". Darüber hinaus können einzelne Komponenten gleichzeitig zu mehreren, also verschiedenen Cyber Physical Systems gehören!

6 Embodiment-Theorie: Keine Intelligenz ohne Körper

Intelligente CPS haben damit Fähigkeiten der „Eigengestaltung" und darauf basierender Optimierung in einem bis dato in der Technik unbekannten Maße. Gestalt ist hier die entscheidende Vokabel: Eine der vielversprechendsten Theorien zur Erklärung von Entstehung und Entwicklung von Intelligenz ist die Embodiment-Theorie [6]. Hervorgegangen aus den Kognitionswissenschaften, wurde sie in den achtziger Jahren von Rodney Brooks in die künstliche Intelligenz eingebracht [4] und seitdem intensiv weiterentwickelt, unter anderem durch Hans Moravez („Moravez Paradoxon") und in Deutschland insbesondere durch Rolf Pfeiffer. Eindrucksvolle jüngere Arbeiten kamen insbesondere von Josh Bongard [5], der als erster die Lernfähigkeit von Robotern auf der Basis selbsterlernter Körpermodelle demonstrierte (Abbildung 1).

So naheliegend der Ansatz heute retrospektiv erscheinen mag – in seiner Entstehung stand er diametral auf den klassischen Interpretationen, die Intelligenz und Bewusstsein als rein interne Prozesse verstanden. Mit der Embodiment-Theorie trat ein Ansatz ins Rampenlicht, der den dahinterliegenden Prozess als einen sensomotorischen Prozess auffasste, der also auf der Koordination von Sensoren und motorischen Fähigkeiten beruht. In der künstlichen Intelligenz löste die Idee die top-down orientierten symbolischen Ansätze (GOFAI, for "Good old-fashioned AI") ab bzw. ergänzte sie um bottom-up-basierte subsymbolische, „neuronale" künstliche Intelligenz und die Subsumption-Theorie, die getrieben sind von der physikalischen Interaktion eines Agenten mit seiner Umwelt.

Nach der Embodiment-Theorie entsteht Intelligenz aus dem Wechselspiel zwischen dem Körper und seiner Umwelt. Nicht nur setzt damit die Ausbildung von Intelligenz die grundsätzliche Existenz eines Körpers voraus, – die Robotik hat durch die Embodiment-Theorie innerhalb der KI einen enormen Schub erfahren – mehr

Abbildung 1   4-beiniger
Roboter mit
„Körperbewusstsein" von J.
Bongard 2006

noch: Die Ausbildung der Intelligenz hängt von der konkreten Gestalt eines Körpers ab, weil unterschiedlich gestaltete Körper unterschiedliche physikalische Interaktion erfahren. Wenn die Embodiment-Theorie sich als zutreffend erweist – und dafür spricht vieles: Was für ein Intelligenzbegriff entsteht dann bei einem „Wesen" mit praktisch unbegrenzter Reichweite und dynamischem Austausch seiner Komponenten? Was bedeutet der Intelligenz-Begriff in einem Cyber Physical System?

7   Intelligenzkonzepte

Dazu stellt sich zunächst einmal die Frage nach der zugrunde gelegten Definition von Intelligenz. Die Konzeption dessen was Intelligenz ist – oder was sie nicht ist – füllt Bücher. Extreme liegen zwischen dem sogenannten "Biological chauvinism" (verkürzt: „nur biologische Gehirne sind intelligent", angelehnt an C. Sagan in den sechziger Jahren) und dem "Liberal functionalism" (verkürzt: „jedes verhaltensfähige System ist intelligent"; Jackendorf 1987, Putnam 1967). Leicht überzeichnet ergibt sich eine Spanne von „nur Menschen sind intelligent" bis hin zu „auch ein Toaster ist schlau".

Während das erste Extrem verhältnismäßig schnell zu den Akten gelegt werden kann, hat der zweite Ansatz intellektuell einen gewissen Charme, weil er immerhin keine „willkürlichen Grenzen" zieht. Jedoch: In einem Verständnis, dass „fast alles" intelligent ist, lassen sich wiederum kaum Schlussfolgerungen über Entstehung und Funktionsweise hochentwickelter Intelligenz gewinnen.

Im Großen und Ganzen besteht heute durchaus eine gewisse Übereinstimmung in der wissenschaftlichen Community, auch über die unterschiedlichen Fachdisziplinen und Schulen hinweg, darüber, dass ein "intelligenter Agent" typischerweise durch drei zentrale Komponenten gekennzeichnet ist:

1. Die Fähigkeit zur Wahrnehmung der Umgebung und ihrer Veränderungen, also der Besitz sensorischer Komponenten zur Wahrnehmung externer Stimuli – SEN-SORIK
2. Die Fähigkeit zur Prozessverarbeitung, also das Prozessieren der externen Daten, deren Analyse und schließlich die Anpassung des eigenen Verhaltens an die Umwelt – KOGNITION
3. Die Fähigkeit zur Reaktion, also die Möglichkeit zur unmittelbaren physikalischen Interaktion mit der Umgebung – AKTUATORIK.

8 Aufbruch in eine neue Robotik

In einem so verhältnismäßig allgemein gehaltenen Ansatz der Intelligenz als einem „Dreisprung" im oben skizzierten Sinne haben etwa heutige Industrieroboter durchaus eine gewisse Intelligenz, jedenfalls die neuester Generationen. Dabei ist klar, dass diese Intelligenz in keiner Weise einer humanen Intelligenz nahekommt. Ein zentraler Unterschied liegt insbesondere darin, dass solche Systeme weder „ziel-basiert" und als soziales Team agieren: Sie wissen kaum, wer sie sind, wer ihre Nachbarn sind, was ihre Aufgabe ist, wie sie zusammenwirken, was die Konse-quenzen ihrer Fehler sein können, welche alternativen Strategien es geben könnte usw. Die Intelligenz heutiger Industrierobotik ist mehrheitlich beschränkt auf die – sehr präzise! – Durchführung von Basisfunktionalitäten wie das Handhaben spezi-fischer Tools, Navigation, Kollisionsvermeidung, und zentrale Prüffunktionen (z. B. die integrierte Messung der Breite einer Schweißnaht). Wobei bereits das letztge-nannte keine durchgängige Eigenschaft mehr ist, so erkennen etwa Roboter in einer Lackierstraße im Automobilbereich nicht, dass sie aufgrund verstopfter Düsen ein ungleichmäßiges Lackbild produzieren.

Genau hier allerdings setzen heute moderne Robotikkonzepte an: Eine Vielzahl von Forschungsprojekten adressiert die Thematik, Roboter als Team agieren zu lassen. Der Hintergrund ist evident: Werden Menschen mit einer komplizierten Auf-gabe konfrontiert, so lösen sie sie in Teamarbeit, und dies mit möglichst heterogenen Teammitgliedern, um die wechselseitigen Kompetenzen nutzen zu können. Genau dieser Schritt steht in der Robotik derzeit an – die aktuellen Produktionssysteme umfassen zwar oft viele Roboter, aber diese agieren nur zeitlich synchron, nicht als sich abstimmendes Team. Die neuen Erkenntnisse der Theorie verteilter Systeme bilden die Grundlage für neue Modelle.

9 Roboterteams mit dezentraler Intelligenz

Dazu sind aber weitere erhebliche Steigerungen der „Intelligenz" dieser Systeme notwendig: Um kooperieren zu können ist ein Verständnis der Gesamtaufgabe notwendig, ein Verständnis der einzelnen Entität für seine Rolle im Team, ein Verständnis der Rolle anderer, eine Übersicht über im Team verfügbare Kompetenzen, Spielregeln für die „decision making processes", die Fähigkeit der Kommunikation von Intentionen für die anderen Teammitglieder, …

Letztes wird gerne unterschätzt: Naiv sollte man annehmen, dass sich Roboter aufgrund ihrer Möglichkeit drahtloser Verbindungen eher schneller und unmittelbarer austauschen können sollten als Menschen. Das gilt aber nur dann, wenn Roboter über „gleiche Protokolle" verfügen, d. h. „gleiche Sprachen" sprechen. Das wiederum ist in heterogenen Teams – heterogen durch verschiedene Hersteller – aber gar nicht gegeben. Hier müssen dann Ansätze aus einem ganz anderen Bereich der KI greifen – dem Bereich der Computer Linguistic nämlich, der in den vergangenen ein bis zwei Dekaden eine stürmische Erfolgsgeschichte zu verzeichnen hat. Ein Babelfish für Roboter!

Im Juli diesen Jahres wurde unser Team CAROLOGISTICS (eine Kooperation von drei Lehrstühlen in Aachen[1]) Weltmeister im TeamRobotik-Wettbewerb „Logistics League" auf dem RoboCup 2014, Brasilien (Abbildung 2). In der Logistics League des RoboCups treten zwei Teams von je drei Transportrobotern der Firma Festo, die „Robotinos", gegeneinander an. Auf einem eine Produktionshalle simulierenden Spielfeld ist ein effizienter Warenfluss zwischen einzelnen Fertigungsmaschinen zu realisieren. „Rohstoffe" müssen in flexiblen Reihenfolgen zu verschiedenen Bearbeitungsstationen transportiert werden, um schließlich als „fertiges Produkt" in der Auslieferungsstation zu landen. Der Schwerpunkt des Wettbewerbs liegt in der intelligenten Kooperation der vollständig autonomen Roboter als ein Team, das in der Lage ist, flexibel auf verschiedenste Ereignisse und Störungen zu reagieren. Der gesamte Ansatz des hier verwendeten Algorithmus basiert auf einer praktisch ausschließlich dezentralen Intelligenz, einem hoch-kommunikativen Ansatz zwischen den Robotern, und ist realisiert durch ein dreischichtiges, sehr robustes Multi-Agenten-Modell (Abbildung 3). Dass das Team in diesem Turnier inkl. aller Vor- und Zwischenrunden ungeschlagen blieb, möge hier bitte nicht als Angeberei verstanden werden – sondern als Beleg für die Robustheit eines solchen verteilten dezentralen Ansatzes.

---

[1] Seit nun mehr drei Jahren engagiert sich das Institutsclusters IMA/ZLW & IfU der RWTH Aachen (S. Jeschke, Maschinenbau) gemeinsam mit Kollegen des Lehr- und Forschungsgebiet für Wissensbasierte Systeme der RWTH Aachen (Prof. G. Lakemeyer, Informatik) und der Arbeitsgruppe Robotik (A. Ferrein, Elektrotechnik, FH Aachen) im Roboterteam Carologistics. Im Rahmen der Kooperation integrieren die Wissenschaftler dort die Ergebnisse ihrer Forschung im Bereich der Mobilen Robotik und dezentralen Automatisierungstechnik auf einer gemeinsamen Roboterplattform.

Abbildung 2 Robocup Logistics League, © Robocup Federation

10 Vernetzte Roboter erobern den Alltag

Eine weitere große Entwicklung in der Robotik zeichnet sich ab: Während sich die Mehrheit aktueller robotischer Systeme in spezialisierten industriellen Umgebungen befindet, sind zunehmend mehr und mehr Roboter dabei, eben dieses zu verlassen. Roboter erobern den Alltag. Dabei ist nicht nur von Staubsaug-, Rasenmäh- und Fensterputzrobotern die Rede, die sich zunehmend größerer Beliebtheit erfreuen – die in Kontext auf hohe Intelligenz aber eine untergeordnete Rolle spielen.

Eine absolut zentrale Entwicklung liegt vielmehr im Bereich autonomer Fahrzeuge. Inzwischen haben verschiedene Hersteller autonome Autos oder Trucks vorgestellt, Google 2012, das deutsche Konvoi-Projekt 2009, Daimler Bertha-Benz-Fahrt 2014, Daimler Future Truck 2014, Volvo DriveMe 2013, ... Ihr flächendeckender Einsatz ist – und zwar gerade vor dem Hintergrund erhöhter Verkehrssicherheit! – nur noch eine Frage der Zeit. Wenn die wesentlichen Mobilitätsträger Roboter sein werden, dann umgeben uns Roboter in unserem Lebensraum überall.

Anhand dieses Beispiels zurück zur verteilten Intelligenz: Auch wenn solche Fahrzeuge im Grundsatz in der Lage sein müssen, ihre Fahrleistung komplett

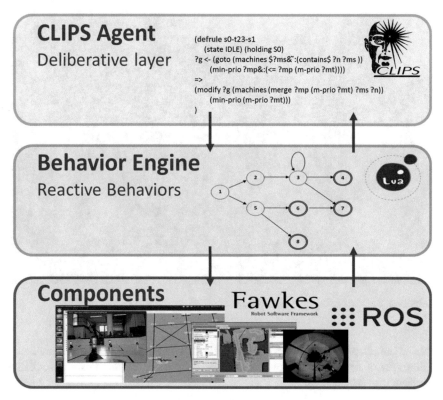

Abbildung 3 Hybride 3-Layer Architektur für die Robocup Logistics League, Team Carologistics Aachen

autonom, d.h. ohne Hilfe ihrer „Roboterkollegen", zu meistern – die Zukunft liegt auf der Vernetzung dieser Entitäten miteinander, und mit der Verkehrsinfrastruktur, um auf dieser Basis zu einem noch stabileren weil durch mehr Perspektiven bereicherten Fahrverhalten zu gelangen. Es ist offensichtlich, dass ein System, das „um die Ecke schauen" kann und Kenntnisse über seine eigene Zukunft hat, über eine deutlich höhere Robustheit verfügen wird als das effizienteste Einzelsystem. Das Stichwort lautet „car2car" bzw. „vehicle2infrastructure communication", das unterliegende Protokoll ist das Internet (Abbildung 4). Hier entsteht ein System kooperierender Roboter, das gleichzeitig ein Cyber Physical System im Sinne des Internet of Things darstellt. Wenn damit die eingangs formulierte Hypothese stimmt, dass die Intelligenz von Cyber Physical Systems starke Bezüge zu unserer eigenen, humanen Intelligenz

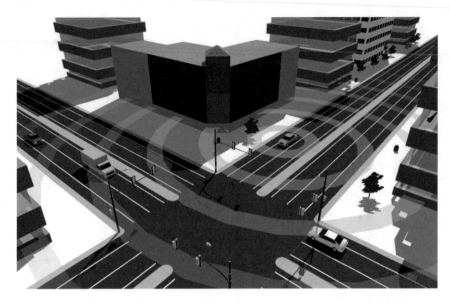

Abbildung 4 CAR2CAR, 2011 und ConnectSafe, 2011

hat – dann umgeben uns in der nahen Zukunft Systeme in unserem unmittelbaren Alltag, deren Analyse eine völlig neue Basis für unser Verständnis menschlicher Intelligenz eröffnet.

Literaturverzeichnis

1. A. Chella, R. Manzotti, Machine consciousness: A manifesto for robotics. International Journal of Machine Consciousness **1** (01), 2009, pp. 33–51
2. B. Goertzel, R. Lian, I. Arel, H. de Garis, S. Chen, A world survey of artificial brain projects, part II: Biologically inspired cognitive architectures. Neurocomputing **74** (1–3), 2010, pp. 30–49
3. M. Minsky, *The Emotion Machine: Commonsense Thinking, Artificial Intelligence, and the Future of the Human Mind*. Simon & Schuster, New York, 2006
4. R.A. Brooks, Intelligence without reason. In: *Proceedings of the 12th International Joint Conference on Artificial Intelligence - Volume 1*. Morgan Kaufmann Publishers Inc., San Francisco, CA, USA, 1991, pp. 569–595
5. R. Pfeifer, J. Bongard, *How the Body Shapes the Way We Think: A New View of Intelligence*. Bradford Books, New York, 2006
6. M. Schilling, H. Cruse, The evolution of cognition — from first order to second order embodiment. In: *Modeling Communication with Robots and Virtual Humans*, ed. by I. Wachsmuth, G. Knoblich, no. 4930 In: Lecture Notes in Computer Science, Springer Berlin Heidelberg, 2008, pp. 77–108

A Causal Foundation for Consciousness in Biological and Artificial Agents

Riccardo Manzotti and Sabina Jeschke

Abstract Traditional approaches model consciousness as the outcome either of internal computational processes or of cognitive structures. We advance an alternative hypothesis – consciousness is the hallmark of a fundamental way to organise causal interactions between an agent and its environment. Thus consciousness is not a special property or an addition to the cognitive processes, but rather the way in which the causal structure of the body of the agent is causally entangled with a world of physical causes. The advantage of this hypothesis is that it suggests how to exploit causal coupling to envisage tentative guidelines for designing conscious artificial agents. In this paper, we outline the key characteristics of these causal building blocks and then a set of standard technologies that may take advantage of such an approach. Consciousness is modelled as a kind of cognitive middle ground and experience is not an internal by-product of cognitive processes but the external world that is carved out by means of causal interaction. Thus, consciousness is not the penthouse on top of a 50 stores cognitive skyscraper, but the way in which the steel girders snap together from bottom to top.

Keywords Machine Consciousness · Consciousness · Cognitive Architecture · Externalism · Situated Cognition

1 Introduction

Is it possible to devise a general architectural principle that might lead an artificial agent to exploit what is called consciousness in human beings and various animals? In principle, there is no reason why such a feat should not be accomplished – at least if consciousness is a natural feature of certain biological beings as we believe it to

R. Manzotti (✉)
Department of Linguistics and Philosophy, Massachusetts Institute of Technology, Cambridge, MA, USA
e-mail: manzotti@mit.edu

S. Jeschke
IMA/ZLW & IfU, RWTH Aachen University, Dennewartstr. 27, 52068 Aachen, Germany

Originally published in "Cognitive Systems Research",
© Elsevier 2016. Reprint by Springer International Publishing Switzerland 2016,
DOI 10.1007/978-3-319-42620-4_40

be. However, it is still unclear what consciousness is and what the necessary and sufficient conditions for its occurrence are. Yet, consciousness appears to be a universal feature of all biological agents above a certain level of cognitive development. This is neither a sure proof that consciousness is a mandatory feature of any superior cognitive system nor that it is replicable by machine. However, it is a very strong evidence in favour of its usefulness and it physical nature. In this paper, we have three goals. First, we address a novel hypothesis about the physical conditions that lead to consciousness. Second, we show tentatively how such a condition can be exploited in artificial beings. Third, we discuss theoretical and technical consequences of the suggested approach.

In a scientific and technological context, whatever formulation of consciousness one adopts, *consciousness should not come at the end of the day as an addition to an otherwise working description of a physical system*. Since natural selection has singled out conscious agents capable of coping with the world, consciousness must make a not negligible physical difference. Thus, it must be identical to some physical structure or process. Of course, consciousness may be either the outcome of a process or may be embodied from the beginning as a consequence of some fundamental structural principle. While most approaches to machine consciousness privilege the former option, here we take into consideration the latter. The problem, thus, is *how to devise a model of the agent in which consciousness is not a nuisance but a natural and efficient way to organise its structure*. To do so, we believe, it is necessary to outline a novel theoretical model of the causal coupling between agent and environment. Consciousness is not some unexpected internal (and somehow magical) property but rather the expression of the physical structure of the agent and the way in which it couples with the external world. In a nutshell, in this paper, we put forward a hypothesis as to what consciousness is in terms of simple causal structure, rather than suggesting that consciousness is the outcome of complex emergent neural/cognitive/ computational processes. The key question is whether it is possible to single out structural principles that could be replicated in artificial beings so that they could show something akin to human and animal consciousness. Coherently with such a goal, it is fair to maintain that a huge amount of empirical evidence suggests that consciousness does not seem to be a local capacity – such as language or face recognition – but rather the expression of a general structural principle shared by all neural areas (and not necessarily contained by them). While it is practically possible to switch off many specific mental capabilities without compromising consciousness, there is no evidence that it is possible to switch off consciousness without shutting down all other functions. When consciousness is severely impaired the subject is usually able to behave in an automatic and stereotyped mode such as during certain kind of epileptic seizures or during the anecdotal night drive [1]. However, the subject is unable to exploit its capability to perform any new task. While human beings perform amazingly in terms of sensory-motor capability and specific skills such as pattern recognition, discrimination, and planning, so far consciousness has remained elusive. There is no doubt that human beings – and likely many animals both mammals and not mammals [2, 3] – exhibit and exploit a sort of integrated cognitive capability that is closely tied to the capability of having a rich experience of the world and one's perceptual states. However, it is still far from clear how such a capability arises in biological systems and to what extent it may be replicated and exploited by an artificial system. Consciousness is a notorious mongrel concept [4] insofar as it refers to various aspects of the mind such as phenomenal experience, unity of action, information integration, symbol grounding, autonomy, and development of intrinsic goals [5–9]. Yet, there are reasons – both of theoretical and empirical nature – that suggest that such a cluster of apparently loosely related mental feature is the outcome of a single architectural principle [10–14]. So far, there has been no machine able to show this kind of unified mental integration that a conscious being – be it a dog or a man – shows so uncompromisingly and smoothly. As an example of such astonishing capability to self-organise a unity of experiences one may consider the cases of severe brain damage, drug intoxication, sensori-motor impairment, neural forced rewiring, and genetic mutations [15–21]. In all these cases, as long as the patient survives, the brain is able to self-adapt to conditions that could not have been part of its genetic blueprint. This is not to say that the brain is always able to cope with any damage to the extent that the subject is still operational. Often the damage overcomes the brain's resilience. Yet, it cannot be ignored that the brain, no matter what and as long as it has some residual capability, seems to be able to self-organize so that conscious experience occurs. Here, we will focus on two aspects ubiquitous in conscious agents – namely the fact that conscious experience is always the experience of something and the fact that consciousness is always an integrated experience. Thus, content and unity. Is it possible to have a machine characterised by these two elements? And, finally, is their joint occurrence the hallmark of some structural architectural principle? As regards these two aspects, the innovative rationale behind this paper is that, so far, in the still young literature about consciousness and even machine consciousness, two approaches have polarised most of the attention – namely computational approaches and action-oriented approaches. The former have focused their attention on special computations performed inside the agent [9, 22–25] – they believe that consciousness is some emergent property popping out of the computational processes inside the agent. The latter have focused on the sensori-motor coupling between an agent and its environment [26–29]. Surely both approaches have their merit [24, 30]. Yet, it is possible that the nature of consciousness is the result of something that is physically more fundamental than the distinction between the inside and the outside of the agent. Consciousness might be the result of a very basic causal coupling between the body of the agent and the environment. It might be a basic principle *on the basis of which cognitive architectures develops*. Consciousness might thus require to overcome the traditional distinction between inner computational processes and *outer* external stimuli. The goal of this paper is to address such a conceptual shift and to check whether it can be exploited to devise the basic requirements for a conscious architecture.

2 Materials and Methods

So far, many approaches to the issue of consciousness have had a weakness – if one would not know that certain physical activities are correlated with consciousness, the theory would have had no reason to predict its occurrence. [9, 31–34].[1] Suppose to have a theory that postulates that, whenever a system performs a certain computation, consciousness occurs. What if consciousness did not occur? Would there be any difference in the outcome of the system? Of course not. In such theories, consciousness is added as a surprising and largely unexpected final effect, a bewildering *coup de theatre*. Such approaches are unlikely to provide a satisfying explanation of what consciousness is. Successful explanations in physics have a different structure. Consider a classic example. Thermodynamics is able to predict an increase in entropy and thus a variation of temperature given the initial conditions. Once you accept the premises of the theory, which can be checked experimentally, certain consequences follow. These consequences may thus be checked and their verification supports the theory. In contrast, consider now a theory that suggests that consciousness is the result of a certain internal computation, for instance Tononi's theory of information integration [9]. This is an interesting hypothesis. But the point is that, if one starts assuming that consciousness is somewhat different from the physical activity taking place inside agents, consciousness cannot make any relevant difference afterwards. In other words, a computational process does not compute differently if it is labelled as "conscious". Likewise it does not make any difference whether such a computational process is correlated with a conscious process either running in parallel or emerging at the end of the computation. In technical terms, one may say that the simpler level (in this case the computational level) drains all the causal powers of additional levels (in this case consciousness) [35, 36]. So much for approaches that look for the neural/functional/computational correlates of consciousness.

Likewise, consider an action-oriented form of explanation, for instance any stripe of enactivism [28]. One considers an agent as a physical structure situated in a certain environment and the actions that such an agent performs. Once again, if one were able to provide a satisfactory theory of all the sensori-motor contingencies – that is, actions – between the agent and the world, why should one add anything like consciousness? The behaviour of the agent would be enough and the addition of consciousness would not and could not modify what is taking place. In both cases, one could be reasonably tempted to get rid of the issue of consciousness altogether. However, this explanatory strategy (some form of eliminativism) would be unwise because (1) human beings seem to have something more than either sheer computation or sheer behaviour and (2) no purely computational agent or purely behaviour-based

---

[1] A necessary caveat: In this paper, whenever we criticize other approaches it is only to the extent that they are presented as an explanation of consciousness. Thus, for example, we have nothing either against Baars' theory of the global workspace as an excellent cognitive model, or Tononi's notion of integrated information as a mean to achieve unified representations. However, at the present state of research, we do not understand how these models could justify the emergence of consciousness at the end of a computational process or as the output of a cognitive module.

agent seems to have the degree of autonomy and adaptability that human and animal beings show. It seems undeniable that human beings cope with the most unexpected events by means of conscious reflection. Finally, they are extremely sensitive to anything remotely resembling feelings in other agents. In sum, consciousness appears to be a non-negotiable aspect of highly developed autonomous agents. The practical advantages that could result from its replication within an artificial being cannot be underestimated [37–40]. Thus, it is wise to reconsider the current state of research and consider alternative hypotheses about the basic requirements for consciousness. Thus our goal is to outline a different starting point for the whole issue and to see whether such a different starting point could be exploited along the whole conceptual path that should have, as desirable outcome, either a prediction about the occurrence of consciousness in biological agents or the design of conscious artificial agents. In short, what is the shift in perspective that we want to put forward? Basically, a theoretical framework that does not require the familiar distinction between the agent and the environment, between the internal cognitive processes and the external brute stimuli, between the internal states and the external causes, and so forth. Such distinctions, however useful, are suspicious because they presuppose a difference – inner versus outer, inside versus outside, computational versus physical, input versus output, process versus action – that is the offshoot of assuming the existence of an agent. For instance, consider the notion of action, key for enactivists, which seems to set aside any vestigial form of dualism. Yet, is it possible to define actions irrespective of the existence of an agent? Is the shift between mere causal processes and action not akin to the gap that divides objects from agents? Is the notion of agent-less actions sound? Thus, an alternative and radical hypothesis is put forward. Consciousness is not *something produced by an agent (no matter whether inside in terms of computational processes or outside in terms of actions), but rather it is the kind of basic causal structure between the body of the agent and the environment.* Our goal is to outline such elementary but powerful causal structure and then, on the one hand, to show how it is exploited by existing biological beings and, on the other hand, to show how artificial agents can take advantage of it.

2.1   The Approach

In this section, the objective is to provide an elementary causal building block of the interaction between the agent and the body that allows to single out experience as a kind of interaction. The resulting new concept of consciousness is directed to overcome the gap between 'experience' and 'physical world'. One may ask why a causal building block and not either a computational, an information-related, or a functional building block? There are many possible reasons to avoid these descriptive levels of reality. We would like to name two of them. On the one hand, these levels run the risk to provide question-begging circular explanations of agenthood. It is dubious whether an agent-independent definition of computation is available – pace Chalmers and Tononi [25, 41, 42]. On the other hand, in a physical world the most fundamental kind of relation is a causal process – no matter how elusive it may be to logic modelling [43–46]. Thus, if we want to build our architecture on top of fundamental notions, we better focus on something natural and physical such as causal processes. The goal is to single out a physical and elementary building block that could be exploited to model the development of a (conscious) agent in a physical environment. Such a building block ought to be neutral with respect both to all mentalistic notions (such as computation, function, representation, and the like) and to the distinction between internal and external. Our bet is that by adopting such a neutral stance, the ensuing understanding of the agent structure will be greatly simplified. First, however, we would like to give the reader the chance to understand *how* such an approach could work in practice. The key idea is that the body and the world are just two pieces of the same physical system and that what we call consciousness is an external cause that requires both pieces to take place. Body and world are interlocked. Let us go through the different steps carefully as shown in Figure 1:

- In the outside[2] world, there are physical scattered events (Figure 1 top). Let us call them A, B, C. They are located in time and space. A, B, and C do not have anything in common – each takes place on its own. For simplicity, assume they take place at time t_1.

  - For instance, A may be a certain light ray with a certain frequency emitted at a certain time and location and B and C surrounding colour spectra. Or A, B, C may be different words in a sentence, or three separate features of someone's face.

- In a nearby human body and after a finite amount of time (Figure 1 middle), a healthy brain, through the causal connection provided by sensorial paths, is affected by the three events A, B, and C. Such an effect takes place at a time t_2 because necessarily the finite speed of causal processes requires a delay between external events and neural activity always occurs. This delay is due partially to the medium and partially to the complexity of the neural activity. The latter is usually a greater than the former. This is not an obvious step. What now happens is that the neural structure – thanks to various neural learning mechanisms – allows the fusion of A, B, and C to create the causal circumstances[3] that allow the fusion itself to act as a cause for further interaction. A whole is born. We call P the fusion of A, B, and C.
- N is the joint effect of P and takes place inside the brain (Figure 1 bottom).

---

[2]We use outside and inside to refer to physical events inside and outside one's body.

[3]In causal terms, we may distinguish between 'the cause and that without which the cause would not be a cause' ([47], p. 119). The former may be taken to be an event actually occurring while the latter may be just a state of affairs. The latter may be formalized in terms of conditions G such that $P \wedge G \; \Box \rightarrow \; E$ which may unfolded in three conditionals $P \wedge \neg G \rightarrow \neg E$, $\neg C \wedge G \rightarrow \neg E$, and $C \wedge \neg G \rightarrow \neg E$.

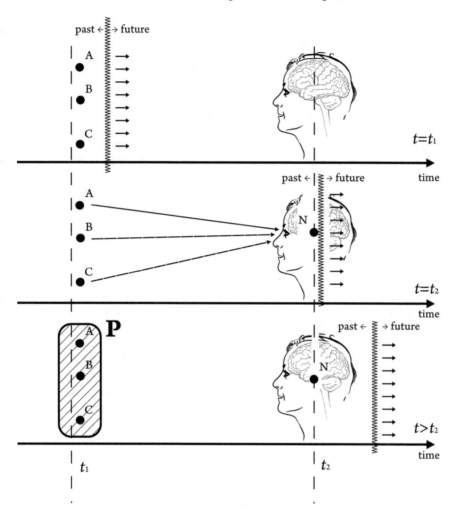

Figure 1 A three-steps model of the proposed causal structure

- As a simple analogy consider the 'distributed key' of an atomic rocket launch system: to launch it, two different 'keys' have to be turned at the same time. However, considering the matter more carefully, there are not two keys, there is just one key in two pieces. The two pieces are not keys since, if alone, they do not unlock anything.

- Because the fusion P causes N, the three originally independent events (A, B, and C) 'become' a whole. Thus, things are defined in causal and not abstract terms. P does not exist until E occurs. N is inside the body, while P remains outside of it.

- Going back to the Rocket launcher analogy, before being put into the 'lock', both 'key pieces' are as unrelated as A, B, C. After being inserted, they, together with the lock, from a system and form a 'lock-and-key principle' which is a construction partially internal and partially external. Their separation is only conceptual – physically and causally they are a whole.

- Finally, the hardest bullet to swallow, namely the temporal order. N takes place at a time t_2, while P occurs at a time t_1, with $t_1 < t_2$, meaning that the neural effect N in the brain happens after P comes into being even though P has not existed until N occurred. Does that mean that N changes its past? In some sense yes. At least it changes the causal role of the past. But a better way to put it is the following. P remains unsaturated until N takes place. It is a bit like lottery tickets. The winning ticket is a winner only after it has been extracted. However, after it has been extracted, the lucky customer had bought the winning ticket from the beginning. So, macrophysical causes may be conceived as unsaturated functions that become saturated and thus complete only when their effects takes place.

The last point merits further considerations. There is no need to invoke any kind of retro-causation moving backward in time. Rather, the case shows that physical phenomena are extended in time. This means that they get to completion within time. Therefore, when something begins to unfold, its nature is not wholly defined until it reaches some natural ending in causal terms. Nothing goes backward in time – the past is, of course, past. However, what the past was may well be defined by what happens now. Using the above time indexes, there is no need to suppose that anything goes backward from t_2 to t_1. However, there is no harm in considering that the world at time t_1 (that is, A, B, and C) changed after t_2 (that is, the fusion P takes place). If one considers a physical phenomenon as something extended in time, then it may well be the case that what happens along such time span redefines the structure of the phenomenon from the beginning. As a further example, consider the conscious perception of colour. If we apply the approach just proposed, a colour is a collection of scattered and otherwise separate physical properties until they produce a joint effect in one's brain (N).[4] When they do so, the scattered wavelengths can be considered as a set of external phenomena (A, B, C, . . .), say Tuscan red. What is hard to grasp is how to step from scattered wavelengths to the impression of a colour. The answer is that it happens in the same way as we come from a bundle of 'whatsoever-pieces' to a key – the components themselves do not constitute a whole (in respect to colour) until they produce a joint effect. The pieces merge into one key if and only if they have the opportunity to do so in causal and actual terms, meaning that there is a 'suitable lock' around (certain interactions with the eye-brain system). Then the whole may take place – without that lock, nothing can happen. By doing so, the scattered events allow Tuscan red to take place. In this account, the colour red is the causal fusion of the set of incoming wavelengths. It is neither an internal impression nor a mental

---

[4]These properties may be a inhomogeneous set of actual physical properties such as the reflected colour spectrum, the percentage of certain components, or the contrastive ratios among different areas. For the sake of the example, consider just a set of wavelengths.

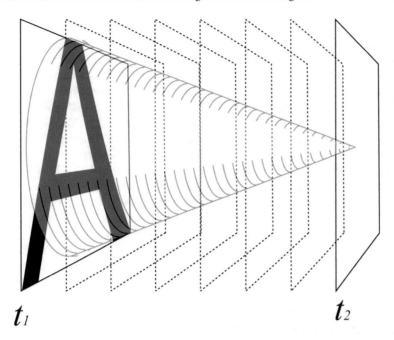

t_1 t_2

Figure 2 Where is located in time an actual cause of a contingent process such as those relevant for everyday conscious perception? Is it located in a particular instant or rather it is spread in time between the occurrence of the components of the cause and the joint effect in one's brain?

ink. Red is an external whole whose occurrence is possible thanks to causal coupling with the neural event (the joint effect N). A colour-blind person would not have the 'suitable lock' – and therefore would not be conscious of the phenomenon that standard sighted subjects call colour. If there were only a colour-blind person in a certain environment, the combination of physical phenomena getting to an end as a colour would never be able to produce a causal joint effect. In a strong physical sense, where is the macroscopical object, which is one's experience (Figure 2)? If a criterion of existence based on actual causation – namely to exist is to be the cause of some effect – is taken seriously, the object does exist only when it is the actual cause of an effect. By "actual" here we emphasise that the occurrence of the effect is mandatory. The effect happens thanks to the presence of the proper structure AND to many other contingent factors. For instance, one might be struck by a seizure and therefore be unable to complete the neural activity at the very last moment. Or, less dramatically, other processes may interfere and prevents neural processes from reaching completion as it is probably the case in many top down forms of blindness such as inattentional blindness and motion induced blindness [48, 49].

The new view is so different from the currently standard view that it is worth to make a quick comparison. The standard view in AI as well as in cognitive science is roughly the following: (1) a body capable of sensori-motor interaction with the environment; (2) external objects and their properties; (3) an input–output flow of information back and from such objects; (4) internal representations inside the body that – due to unknown reasons – are capable of producing the experience of external objects; (5) the brain as a kind of internal interpreter that gives meaning and "colour" to processed data; (6) consciousness as a property of the internal activity that is useful to control functional relations with the environment. The current orthodoxy is based on the so-called 'internalistic models' – namely those models that take the mind to be a property of what takes place inside the neural system [25, 50–52][5]. This view has its strength: putting some kind of 'interpretation layer' between the individual and the outside world allows explaining why humans – all of them built alike – tend to be rather different in their behaviours, reactions and 'feelings'. However, there are also serious issues. First, to believe that 'the internal interpreter', and 'neurons will do it "somehow"', is doubtful. In the last couple of decades, scientists from all areas have invested a lot of energy in the quest for a neural mechanism capable of producing everyday conscious experience. Up to now, there is no known law of nature predicting that neural activity should result in one's experience. Second, in such a model, consciousness neither fits the physical world nor its properties. To carry it to the extremes, that means that we constantly ignore the 'real world' by overwriting it with some internal 'fantasies', a sort of environment-driven virtual reality. Of course, this could be the case – but it sounds at least pretty counterintuitive: why should nature take this kind of detour? Third: The discrepancy between our immediate experience and the 'world' is more than just 'somewhat regrettable'. If everything we experience – from pain to colour, from pictures to music – is nothing more than a product of our neurons, then a logical problem ensues: Why should it be easier for neurons to transmute neural firings into music – than for a cello to shape airwaves into music? If the physical world is devoid of qualitative features, why should the brain – which is part of the physical world – be any better in this respect? Why should the brain create meaningful things, but not a cello? Or, to use an even catchier picture: 'If colours could not pop out of strawberries, how would they pop out of neurons?' According to such a view, then consciousness is causally superfluous and the world is invisible. Furthermore, many explanatory terms seems to be circularly dependent on the notion of agent (and thus on that of consciousness). One may say that in the standard view, the homunculus in the brain (the ghost in the machine) has been dismissed because the whole brain has become the homunculus. Yet, the brain interprets, experiences, perceives, understands, wants, and so forth. Conveniently, the brain has become a kind of new homunculus that does everything that was once done either by the soul or by the immaterial Cartesian mind. To overcome these obstacles and to achieve a foundation for an architecture that might be exploited by biological and artificial beings alike, our proposal dwells on a neutral ontology:

- The gap between experience and world . . . is gone since external events and internal perceptions become a whole.

---

[5]Ofcourse, these models do not rule out the importance of the external environment to guide the development of internal structures. Indeed, they consider necessary for a healthy brain to develop by means of continuous interactions with the environment. However, once the required neural connections are in place, the mind is taken to be an internal phenomenon.

- The human-centred view . . . is gone. Experience is driven partly internal and partly external to a physical body and it is constituted by physical events. The experience is internal to the physical system that underpins it, and it is external to one's body. The body is, of course, nothing but as a subset of a larger physical superset of processes taking place in time.
- The internal interpreter . . . is gone or at least no longer necessary. It was enough to relocate the physical underpinnings of consciousness. The consciousness of 'seeing red' (instead of seeing several scattered wavelengths) is the result of the fitting between key-parts and lock. Red is not a meaning associated to some internal representation – red is a physical phenomenon in one's physical environment.
- Consciousness 'is' the world we live in: to make an example, to see red is to be united with an external collection of physical phenomena, since experience takes places as a temporally and causally extended phenomenon which requires internal and external components. Red is external to one's body.

In short, the new approach may be recapped as follows.

- External events and a neural event form a key-locksystem that is neither internal nor external.
- The external events produce a certain neural activity whose goal is to allow a causal process between the body and the environment to take place.
- Because of the causal process between the neural activity and the environment, an actual external cause takes place.
- The external cause is both one's conscious experience and the external object/property.

In this way, because of the existence of an agent, the environment is partitioned so as to be made of a series of objects that are at the same time the experience and the object one is experiencing at any given moment.

2.2 Theoretical Advantages

For a moment, before raising the inevitable objections, let us consider this view as a tentative scientific hypothesis about the physical nature of consciousness – a scientific hypothesis insofar as it puts forward a falsifiable hypothesis as to what the mind is. If this hypothesis has any merit, a few conceptual advantages will ensue:

- First, the hard problem of consciousness Chalmers [5] addresses the problem of explaining how and why we have qualia and phenomenal experiences such as pain, colours, taste etc. (incl. 'Why does awareness of sensory information exist at all?' 'And why is there a subjective component to experience?'). With the presented approach, the core of the hard problem of consciousness is swept away. The mind and the world are no longer two incommensurable and indeed autonomous domains; they are the same one seen from two different perspectives.

- Second, the mind–body-problem. Overt and covert dualism is finally upturned. Dualism is not only the straw man of the traditional substance dualism contrasting either matter and soul or body and mind. There is also a form of dualism that suggests a juxtaposition between cognition and the brain [22, 53, 54], sometimes dubbed Cartesian Materialism [55, 56]. There is no longer the need to differentiate the way in which things look to subjects and the way in which things are. There is just a flow of physical phenomena causally interconnected.

- Third, exclusiveness. Being conscious of something is a 'private' event – but in contrast to the traditional interpretation, the privacy is no longer created by an internal individual interpreter. It is no longer an exclusive and unbridgeable metaphysical privacy. Rather, it is the kind of privacy that prevents two individuals from eating the same piece of cake. The exclusiveness follows from the fact that the pieces fuse into one key only if there is a 'suitable lock' – a suitable brain and a working body – around. The causal interaction between internal and external world links the observed object and its observer. Of course, in presence of two similar groups of events, two similar brains let similar fusions occur.

- Fourth, location of consciousness. It is possible to physically locate the (conscious) mind in the physical world. The location is not inside the neural system though. However, it is possible to pinpoint a certain physical cause and consider whether such a cause is identical to one's own experience of, say, a red patch. It is thus possible to resurrect the theory of identity in terms of broader physical processes and not just in terms of neural processes. The fact that consciousness might take place outside the body is not in contrast with the impression one may have to be located inside one's own body. Nothing in our experience points to where our experience takes place, only to what our experience is. If we cut a finger, we do not feel a pain inside the brain; rather we feel a pain in the finger. By the same token, it is not necessary that the physical phenomenon that is our conscious experience is located inside our body.

- Fifth, the misperception issue[6] – namely, the fact that apparently we may experience things that are not physically present, as it happens in the case of hallucinations or dreams e.g. – has to be dealt with differently. They are no longer the result of a somewhat 'hyper-creative' internal interpreter, but of unusual connections with real features in one's environment. It is important to realise that our dreams are just 'boring' recombinations of the basic components of our past, albeit reshuffled

---

[6]Whenever it was necessary to point to the autonomy of the mental with respect to the physical domain, the issue of misperception has been the battering ram of both philosophers and scientists. Dream and hallucinations appear as formidable evidence in favour of an inner world. However, this approach promises to locate a physical cause for any experience in the physical surrounding. All cases of conscious experience ought to be revisited as cases of (admittedly unusual) perception. The approach presented here honestly stands or falls on whether it will succeed to show that – perhaps surprisingly – whenever there is consciousness, there is a physical phenomenon, which is the content of one's experience. We cannot do justice here to the problem of misperception. However, we flesh out a template of the strategy – namely to address each purported case of misperception and to revise them in terms of perception. (One of the authors is actually working on such an account for most cases of misperception, from hallucination to illusions, from aftereffects to direct brain stimulations).

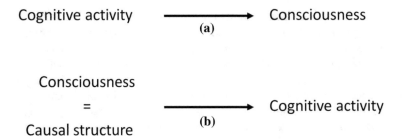

Figure 3 The direction of the explanation in most standard cognitive approaches to consciousness (**a**) and in this case (**b**). Consciousness is at the root of the cognitive architecture

in possibly original ways, they are chimeric but not innovative [57–60]. It is also important to realise that all perceptions require a temporal lag between the object and the ensuing neural activity, due to the velocity of information transportation. Combining these two insights leads to a possible and fairly simple explanation, namely, that dreams and hallucinations might be cases of very long and reshuffled perception of one's world. So, tentatively, this approach suggests that the stuff dreams (and consciousness) are made of is the same stuff the world is made of.

A way to visualise one of the main differences between this approach and other ones is given by the direction of the explanation (Figure 3). In many approaches, first a theory of cognition (or something akin to it) is fleshed out and then a hypothesis is put forward about how a set of unconscious computation/processes/cognitive modules will produce consciousness. In contrast here a different stance is adopted. The key hypothesis is that consciousness is not the outcome (or an emergent property) of some internal activity. Consciousness is a way to describe the causal connection that holds between the external world and the agent's body. Thus, consciousness is not something that is concocted inside, but rather the way in which the interaction between the body (with all its neural structures) and the environment carves out a set of entities.

3 Results and Discussions

Now, can a machine gain consciousness – i.e., real consciousness?[7] To address this question we take advantage of the suggested fundamental causal structure and consider a series of promising cases and technologies. In the following, we are not

---

[7]The distinction between strong and weak machine consciousness mirrors that between strong and weak AI [38]. Weak machine consciousness considers whether it is possible to build machines that behave as if they were consciousness. Strong machine consciousness ventures to consider the possibility of real conscious machines. We believe that skipping the 'hard problem' is not a viable option in the business of making conscious machines [61].

outlining a strong theoretical formulation. Also, we are not capable at this point to give 'a full proof' (in a strict sense). Rather, we want to show the inherent potential of the suggested causal model. As long as consciousness is interpreted as an 'internalist' concept, there would be no change in modelling it as an 'internal interpreter, e.g. transforming about 10.000 light waves of 700 nm into "red"'. Nobody knows why and how, except from that it happens. The internalist interpretation may be true, but this would not help us to come closer to any understanding of the concept behind. However, if consciousness is interpreted in the sense we have been proposing (halfway between 'internalist' and 'externalist'), then it could be realised (at least as a toy model) as we are going to explain. Thus, we can start to understand it, we can try to run tests on it, and so on. So by proposing this possible solution, we will sketch a 'lab scenario'. Here, we consider promising off-the-shelf technologies that may fit the bill if deployed in the proper way. It is worth to stress that such a neutral causal account of consciousness is coherent with a notion of consciousness as cognitive middle ground – namely consciousness is not a special ingredients or the outcome of a special computational process but a basic causal structure underpinning the agents. Thus, to use a visual metaphor, consciousness is not the penthouse on top of a 50 stores cognitive skyscraper, but it is the way in which the steel girders snap together from bottom to top. In fact, machine consciousness lies in the promising middle ground between the extremes of biological chauvinism (i.e., only brains are conscious) and liberal functionalism (i.e., any behaviourally equivalent functional systems is conscious) [62]. One of the most central concepts behind 'intelligence' and perhaps the most difficult aspect to grasp is clearly not restricted to humans. From that it follows quite naturally that the goal of building a technological system with a somewhat 'authentic intelligence' requires consciousness to be part of the game, i.e., phenomenal consciousness. It remains to be seen whether new concepts will lead to insights into other components of consciousness such as self-consciousness or forms of higher order consciousness.

3.1 Tentative Guidelines for a Conscious Artificial Architecture

What are the ideal features that a cognitive architecture should have in order to adapt to a partially unknown body and environment? On the basis of the available literature and the empirical evidence a series of key features and their justification can be listed:

- The architecture must be based on a very limited number of kinds of basic building blocks – each kind exploiting the same common structure. Thus, the description length of the architecture must be kept to a minimum.
- The basic module might be freely replicated in order to cope with multiple sensor modalities and demanding incoming stimuli. This should ensure scalability.
- The basic module has to be able to develop its own goals and to use them both for its own development and for interacting with other modules. This should allow to develop intentionality and a tight environment architecture coupling.

- In principle, adding further modules (constrained only by the system resources) should lead to an increase in performances. Once again, this is important for scalability.

An architecture with the above features should be able to adapt to unknown situations and with a minimum of predesign. Rather than specifying all the algorithms and their mutual relationships, the above approach suggests a recipe to build a cognitive architecture given a body and an environment. Such a recipe is a lot less demanding in terms of description and a priori knowledge than a detailed plan. Furthermore, a recipe of such a universal scope offers many more advantages in terms of adaptability and flexibility. Thus, the architecture we are willing to implement must satisfy the requirement of being both scalable and adaptable. Furthermore, the architecture has to take into account the whole history of the system and it must be coherent to the current understanding of the biological structure of a mammalian brain. It might have a limited number of more specialised versions of the same elementary block (for fine tuning, better performance, and optimisation), but it must not rely on explicit algorithms. These requirements are definitely compatible with the neuroscientific evidence collected from human beings and non-humans mammals. Surprisingly, these requirements are not met by most artificial architectures and AI agents. In fact, many robotic setups and architectures are the result of careful programming since designers aim to solve specific sensorimotor, relational, or logic issues. A classic example is offered by robotic feats like the Robocup[8] where teams of robots exploit algorithms devised by their designers to compete together in a soccer match. Although their behaviours may be very clever, it is not the result of real adaptation on a high-cognition level. Of course, there are some robots capable of learning new skills and to adapt to novel situation, at least to a certain degree. However, explicit attempts at integrating consciousness into a robots' intelligence are rare, and so far no model has been exceedingly convincing. Compared to current robotic agents, biological agents like mammals and humans show a totally different kind of adaptability to novel stimuli. Mammals are capable of dealing with totally unexpected environmental challenges for which they could not possibly have any kind of inborn solution. Furthermore, it is a fair bet to assume that the complexity of their neural structure largely exceeds their genetic blueprint. Most mammals are capable not only of learning how to achieve goals but also of learning what goals have to be pursued [63, 64] – which is an important issue in respect to consciousness. As it has been observed [65, 66], the cortex shows an almost universal capability of autonomously adapting to novel kind of stimuli: 'The fact that humans can learn and adapt to problems that did not exist when the initial model (the neocortex) was created is proof of the generic nature of the mechanisms used by the human brain.' [11]. Thus, it makes sense to look for very general approaches capable, albeit with possible shortcomings, to model a unified and common approach to all aspects of cognition. Empirical evidence shows that mammals exhibit a very high degree of neural plasticity and cognitive autonomy [67–69] to the extent that it is fair to sup-

---

[8]http://www.robocup.org/.

pose that any part of the cortex might develop almost any cognitive skill. If this supposition were true, it would mean that the neocortex, and possibly the thalamo-cortical system, exploit some kind of rather general architectural principle, mainly independent of the kind of incoming data. There have been various attempts in the past to devise a general cognitive architecture [11, 65, 66]. In this paper, we make yet another attempt. This time we want to take advantage of a rather simple idea: true autonomy entails teleological openness. By teleologically openness we mean that the system is capable of developing new goals autonomously on the basis of environmental conditions and stimuli [8, 64].

3.2 Not Reinventing the Wheel: Combining Multi Agent Systems and Genetic Algorithms

A tentative approach might be to realise a robot's brain as a multi-agent system (MAS) once such an endeavour may find support by some additional key hypothesis about the physical foundations of consciousness. MAS have been discussed already as a possible model to realise artificial brains, or as a model to explain the function of a brain (e.g. [70]). They have also been discussed as a possible extension of cognitive architectures – e.g. within the hybrid design of CLARION. In computer sciences, MAS have become a very popular instrument during the last years when modelling complex heterogeneous distributed systems, which are organised 'bottom up'. Taking the suggested approach to consciousness as guideline, in such a MAS each software agent would represent one 'conscious-lock' to a certain key, an external phenomenon. Thus, the resulting robotic brain would be conscious of the external events. It has the appropriate locks for it, and the mechanism of building this consciousness would be exactly the same as for the human brain. So, the tentative idea is that MAS could offer the necessary architectural backbone for a conscious mind and that, once tuned to satisfy to some specific requirement; it may be indeed the workable tool to design a new kind of cognition.

At least three questions arise immediately:

- *Complexity*: One may argue that by this approach, only a small number of locks can be realised due to the enormous programming effort needed otherwise.
- *Specification*: An even stronger objection might be that in this way, the programmer may tend to mainly 'imitate' the human consciousness but does not develop one which is appropriate for the given robot with a certain form, function and so on.
- *Proof*: A third difficult point is the answer to the question about how we would like to prove that a certain robot really has a consciousness in a strong sense.

To tackle all three problems with one approach, optimisation algorithms have to be integrated, allowing to improve the MAS during runtime. Here, due to their 'closeness' to the underlying problem (a developing brain), genetic algorithms might form a natural choice: The 'consciousness-locks' have to be specialised to species, their mode of living, and the challenges presented to them.[9] Their special characteristics are probably not the result of some kind of 'biological master plan' for all living beings, but the result of a species-exclusive evolutionary process, which over millions of years has favoured individuals which are better adapted to their environment than others. In this understanding, the consciousness-lock (realised through multi agents) would be subject to the same evolutionary process, which has driven the whole design of a certain species, including the body shapes, motor skills, brain structure and the like. Likely, genetic algorithms reproduces this kind of development. The idea of using genetic algorithms to build a conscious brain is also one of the central design principles behind the cognitive architecture. Genetic algorithms are a part of evolutionary computing, which is a rapidly growing area of artificial intelligence. Today, they play an important role in many complex optimisation problems and form an important concept for machine learning approaches. Genetic algorithms use mechanisms inspired by biological evolution, such as reproduction, mutation, recombination, and selection. Over several generations, systems are optimised: Pairs of first generation solutions are taken and recombined. The 'fittest' solutions of this match are selected for the next generation. Mutations are used to enhance the genetic variety and thus, the overall solution space. The optimisation goal – in nature given through environment and the corresponding challenges – is realised through a so-called fitness function which determines the quality of the solutions. Lately, combining multi agent systems with genetic algorithms has become popular in certain field as e.g. automated testing scenarios. Assuming that consciousness is a capability of higher life forms, the following digest gives a first impression of the number of genetic iterations which are necessary to produce this kind of complex structures: About approx. 3.5 billion years ago, the first life forms developed, monads with a very limited range of functionalities. The first plants and simple animals came up about 700 million and mammals about 200 million years ago. Humanoid life forms started to develop 70 million years ago, and we, the homo sapiens species, are only 150,000 years old. Even if it is difficult to tell from which stage in evolution consciousness has first entered the scene, the reader might understand why we consider genetic algorithms to do the optimisation job! How do we proceed 'practically', meaning: how exactly are we going to use genetic algorithms to re-build the evolution of a robotic brain? Regarding the development of consciousness, one would start with a couple of given perceptions, each of them realised through a single agent, say regarding colours, temperature and the like – a 'basic set' of conscious perceptions. This is

---

[9]Consider the following example: In the literature, we find that cats are somewhat colour-blind concerning the colour red, they see it as a shade of grey (whereas they have a perfect colour vision concerning e.g. green and blue). Well, the first finding is that we cannot be really sure about that, since we can only predict that from their eye anatomy – but what kind of 'consciousness' cats really have concerning the colour red is a totally different topic because we don't know anything about the design of the key-lock-structure at this point. It may be totally different from ours. The second – and much more important – insight is that it might be less important for a cat to be capable of seeing red than for example for a bear. Cats – being carnivores – do not have to differentiate between ripe and unripe apples since they would not eat them anyway. For a bear on the other hand – being omnivores consuming a large portion of fruits daily – the situation may show itself quite differently.

'easy' – and would form the 'first generation solution'. Now, to make the system learn new conscious elements, the second preparative step is to place the robot in a certain challenging environment – meaning that certain tasks have to be given to him – in order to challenge his 'consciousness enhancement'. Next, genetic algorithms come into play in order to produce variants of the robot's 'brain structure': the single agents will be altered, multiplied and become more complex. Some of the 'new' solutions will not survive as they do not particularly contribute to the robot's task. Others will survive as they enhance the robot's capabilities to deal with its tasks. This, the resulting brain structure, will turn out having consciousness-locks which are complex and adapted to the individual needs of the specific kind of robot and its environment. From that, there are two possibilities to infer that the robotic brain has really developed consciousness by using genetic algorithms. First, the direct inspection would address the source code itself. Starting with a 'basis set' of agents in a MAS, the resulting system would consist of old and new software agents, the latter representing new conscious capabilities. The new code can be investigated, varied and different test cases could be designed and analysed. Second, a more 'indirect' inspection would be to use a test scenario that is designed in such a way that consciousness for a certain perception area would definitely be necessary to solve a certain task. This particular conscious perception would not be part of the basic conscious skills the system it started with. Now, if – after of a couple of 'genetic rounds' – the robotic brain would come up with new solutions for the given task, which definitely requires the enhancement of its consciousness, this would be a strong signal that it has developed a new perception in a certain area. So, the combination of multi agent systems with genetic algorithms allows overcoming the aforementioned problems:

- *Complexity*: Starting from a small number of locks, their expansion is realised by genetic algorithms which enhance the number of locks in order to optimise the system's behaviour.
- *Specification*: Since the optimisation takes places in relation to a certain environment including specific challenges and particular tasks, the robotic brain develops the right causal structure, which is adapted to its own needs.
- *'Proof'*: The proof of whether a consciousness has been developed is not complete. However, on the more direct side, the investigation of the auto-generated source-code will deliver new insights. From the perspective of an indirect proof, it would address the development of a new conscious aspect rather than its existence. If a robot can adapt to a certain situation and if and only if it is in the right causal structure then that will be a strong hint that consciousness has been developed.

4 Conclusions

The just outlined causal structure, no matter how simple, suggests a kind of causal structure that may be exploited both by artificial and biological agents. As a matter of fact, there is a huge empirical evidence showing that biological beings take advantage of this kind of causal connection with their environment during their development [71–73]. Such a causal structure offers a physical basis to address the issue of consciousness since it suggests that experience might be the way in which the external object takes place thanks to the interaction with the body of the agent. In this regards this approach offers the following advantages:

1. Experience is no longer something that the agent must concoct inside.
2. Experience of an object and the object are, to a great extent, the same thing.
3. Experience (and thus the mind) is no longer the outcome of an internal process but the very process taking place between the external object and the body of the agent.
4. Experience is no longer the outcome of some problematic level of reality (information/computation/ function).
5. Experience might be modelled into an artificial agent by controlling the way in which the artificial agent interacts with the world.
6. The abovementioned way to interact with the environment might be the right one to achieve the necessary autonomy and adaptability exhibited by conscious biological agents.

These elements are, in our opinion, the preconditions for devising an architecture capable of consciousness. They suggest the causal structure of consciousness and thus something that may help in singling out relevant architectures in an artificial agent. For instance, the approach suggests that being conscious is not a matter of having the right internal code [74, 75], or using a central global dashboard [22, 76], or processing information in a certain way [9]. The advantage is that this proposal provides precise indications about how the causal coupling between the environment and the agent ought to be realised. Of course, by itself, the approach does not provide a complete picture of how to implement an intelligent agent. Many other aspects – often already addressed and partially implemented in AI and robotics – must flank what is here suggested. In sum, the suggested approach does not aim to be alternative to other approaches in AI or in robotics, rather it aims to tune them in a way that should be productive for consciousness. Another argument may be raised against the temptation of 'weak artificial consciousness via the easy way'. In nature, the development of consciousness goes along with increased intelligence. Most animals are exhibiting behavioural signs at least of phenomenological consciousness, human beings have a phenomenological consciousness. 'Evolutionary optimisation' is the most powerful optimisation known so far (even if it takes its time). Thus, it seems to be highly unlikely that natural selection took such a long way to provide us with consciousness if there was a way to get all the advantages of a conscious being without actually producing it. Of course, this does not mean 'proof' – but we cannot help but sincerely doubt it. So far, practical attempts to devise and implement conscious artificial beings have been hampered both by technical difficulties and by theoretical uncertainties. While the first issue will likely be overcome in time and is shared by other equally formidable feats in AI, developing consciousness in an artificial agent has often appeared to be an impossible endeavour (a preposterous idea). It is true that a handful of researchers have nonetheless considered the possibility, yet there is still no consensus as to what should be achieved. Based on the presented new approach to consciousness lying between internalism and externalism, a possible technological design for a conscious machine has been sketched addressing the abovementioned goals. The approach is taking advantage of an architecture exploiting selfdevelopment of new goals, intrinsic motivations, and situated cognition. From a technological point of view, multi agent systems are used to model independent conscious perceptions. Genetic algorithms – as a subgroup of evolutionary algorithms – come into play to mimic the biological evolution of the brain's structure, thus allowing in general for adaptivity and scalability, and assuring some coherence to what we know about the biological structure of brains of higher developed animals as e.g. mammals. The architecture does not pretend to be either conclusive or experimentally satisfying. In the future, this rather sketchy outline of a cognitive architecture will be enhanced to a satisfying and more comprehensive architectural model. Then, we will also integrate components of a cognitive architecture which has been partially implemented in previous setups [64, 77–79]. The goal of the full architecture model is the implementation of the kind of development and environmental coupling through consciousness we have described in the previous sections. However, the presented approach suggests a way in which a cognitive architecture may exploit the same kind of causal entanglement that in biological beings might be identical with consciousness – an agent is conscious of a given environment to the extent that its cognitive structures are the result of the proper causal coupling with that environment – namely that one's consciousness of an object might be nothing but the external object tightly coupled to the agent's body. The experience of something would be literally identical to that something. Consciousness would be situated in the environment in a very strong sense and it would also be a very physical phenomenon.

References

1. D.M. Armstrong, What is Consciousness? In: *The Nature of Mind*, ed. by J. Heil, Cornell University Press, 1981, pp. 55–77
2. D.B. Edelman, B.J. Baars, A.K. Seth, Identifying hallmarks of consciousness in non-mammalian species. Consciousness and Cognition **14** (1), 2005, pp. 169–187. doi:10.1016/j. concog.2004.09.001. http://www.sciencedirect.com/science/article/pii/S1053810004000935
3. A.K. Seth, B.J. Baars, D.B. Edelman, Criteria for consciousness in humans and other mammals. Consciousness and Cognition **14** (1), 2005, pp. 119–139. doi:10.1016/j.concog.2004.08.006. http://www.sciencedirect.com/science/article/pii/S1053810004000893
4. N. Block, On a confusion about a function of consciousness. Behavioral and Brain Sciences **18** (2), 1995, pp. 227–247. doi:10.1017/S0140525X00038188
5. D.J. Chalmers, In: *The Conscious Mind: In Search of a Fundamental Theory*, Oxford University Press, U.S.A., 1996, pp. xvii, 414
6. D. Gamez, Information integration based predictions about the conscious states of a spiking neural network. Consciousness and Cognition **19** (1), 2010, pp. 294–310. doi:10.1016/j.concog. 2009.11.001. http://www.sciencedirect.com/science/article/pii/S1053810009001767
7. S. Harnad, The Symbol Grounding Problem. Physica **D** (42), 1990, pp. 335–346

8. R. Manzotti, A Process-oriented Framework for Goals and Motivations in Biological and Artificial Agents. In: *Causality and Motivation*, ed. by R. Poli, Ontos-Verlag, Frankfurt, 2010, pp. 105–134

9. G. Tononi, An information integration theory of consciousness. BMC Neuroscience **5** (42), 2004, pp. 1–22. doi:10.1186/1471-2202-5-42. http://www.biomedcentral.com/1471-2202/5/42/abstract

10. E. Bullmore, O. Sporns, The economy of brain network organization. Nature Reviews Neuroscience **13** (5), 2012, pp. 336–349. doi:10.1038/nrn3214. http://www.nature.com/nrn/journal/v13/n5/full/nrn3214.html

11. D. George, How the Brain Might Work: A Hierarchical and Temporal Model for Learning and Recognition. Ph.D. thesis, Stanford University, Stanford, CA, USA, 2008. AAI3313576

12. R. Manzotti, A. Chella, Physical integration: A causal account for consciousness. Journal of Integrative Neuroscience **13** (2), 2014, pp. 403–427. doi:10.1142/S0219635214400044. http://www.worldscientific.com/doi/abs/10.1142/S0219635214400044

13. V.B. Mountcastle, The columnar organization of the neocortex. Brain **120** (4), 1997, pp. 701–722. doi:10.1093/brain/120.4.701. http://brain.oxfordjournals.org/content/120/4/701

14. G. Tononi, G.M. Edelman, O. Sporns, Complexity and coherency: integrating information in the brain. Trends in Cognitive Sciences **2** (12), 1998, pp. 474–484. doi:10.1016/S1364-6613(98)01259-5. http://www.sciencedirect.com/science/article/pii/S1364661398012595

15. M. Sur, A. Angelucci, J. Sharma, Rewiring cortex: the role of patterned activity in development and plasticity of neocortical circuits. Journal of Neurobiology **41** (1), 1999, pp. 33–43

16. M. Sur, J.L.R. Rubenstein, Patterning and Plasticity of the Cerebral Cortex. Science **310** (5749), 2005, pp. 805–810. doi:10.1126/science.1112070. http://www.sciencemag.org/content/310/5749/805

17. O. Sacks, *The Man Who Mistook His Wife For A Hat: And Other Clinical Tales*. Berkeley, 1970

18. V.S. Ramachandran, W. Hirstein, The perception of phantom limbs. Brain **121** (9), 1998, pp. 1603–1630. doi:10.1093/brain/121.9.1603. http://brain.oxfordjournals.org/content/121/9/1603

19. B. Merker, Consciousness without a cerebral cortex: a challenge for neuroscience and medicine. The Behavioral and Brain Sciences **30** (1), 2007, pp. 63–81; discussion 81–134. doi:10.1017/S0140525X07000891

20. D.M. Kahn, L. Krubitzer, Massive cross-modal cortical plasticity and the emergence of a new cortical area in developmentally blind mammals. Proceedings of the National Academy of Sciences **99** (17), 2002, pp. 11,429–11,434. doi:10.1073/pnas.162342799. http://www.pnas.org/content/99/17/11429

21. J.A. Hobson, *The Dream Drugstore: Chemically Altered States of Consciousness*. MIT Press, 2002

22. B.J. Baars, In the Theatre of Consciousness: Global Workspace Theory, a Rigorous Scientific Theory of Consciousness. Journal of Consciousness Studies **4** (4), 1997, pp. 292–309

23. C. Koch, G. Tononi, Can Machines Be Conscious? IEEE Spectrum **45** (6), 2008, pp. 47–51

24. M. Shanahan, B. Baars, Applying global workspace theory to the frame problem. Cognition **98** (2), 2005, pp. 157–176. doi:10.1016/j.cognition.2004.11.007

25. G. Tononi, C. Koch, The neural correlates of consciousness: an update. Annals of the New York Academy of Sciences **1124**, 2008, pp. 239–261. doi:10.1196/annals.1440.004

26. A. Clark, Whatever next? Predictive brains, situated agents, and the future of cognitive science. The Behavioral and Brain Sciences **36** (3), 2013, pp. 181–204. doi:10.1017/S0140525X12000477

27. O. Holland, The Future of Embodied Artificial Intelligence: Machine Consciousness? In: *Embodied Artificial Intelligence*, ed. by F. Iida, R. Pfeifer, L. Steels, Y. Kuniyoshi, no. 3139 In: Lecture Notes in Computer Science, Springer Berlin Heidelberg, 2004, pp. 37–53. http://link.springer.com/chapter/10.1007/978-3-540-27833-7_3

28. A. Noë, *Action in Perception*. MIT Press, 2004

29. J.K. O'Regan, How to Build a Robot that is Conscious and Feels. Minds and Machines **22** (2), 2012, pp. 117–136. doi:10.1007/s11023-012-9279-x. http://link.springer.com/article/10.1007/s11023-012-9279-x

30. R. Chrisley, Embodied Artificial Intelligence. Artifcial Intelligence **149** (1), 2003, pp. 131–150. doi:10.1016/S0004-3702(03)00055-9. http://dx.doi.org/10.1016/S0004-3702(03)00055-9
31. S. Dehaene, J.P. Changeux, L. Naccache, J. Sackur, C. Sergent, Conscious, preconscious, and subliminal processing: a testable taxonomy. Trends in Cognitive Sciences **10** (5), 2006, pp. 204–211. doi:10.1016/j.tics.2006.03.007
32. C. Koch, *The Quest for Consciousness: A Neurobiological Approach*. Roberts & Company Publishers, 2004
33. B. Libet, *Mind Time: The Temporal Factor in Consciousness*. Harvard University Press, Cambridge, Mass, 2005
34. S. Zeki, A. Bartels, Toward a theory of visual consciousness. Consciousness and Cognition **8** (2), 1999, pp. 225–259. doi:10.1006/ccog.1999.0390
35. J. Kim, *Mind in a Physical World*. MIT Press, Cambridge, Mass, 1998
36. J. Kim, Blocking Causal Drainage and Other Maintenance Chores with Mental Causation. Philosophy and Phenomenological Research **67** (1), 2003, pp. 151–176. http://www.jstor.org/stable/20140586
37. D. Gamez, Progress in machine consciousness. Consciousness and Cognition **17** (3), 2008, pp. 887–910. doi:10.1016/j.concog.2007.04.005. http://www.sciencedirect.com/science/article/pii/S1053810007000347
38. O. Holland, ed., *Machine Consciousness*. Imprint Academic, 2003
39. P. Yourgrau, *A World Without Time: The Forgotten Legacy of Godel and Einstein*. Penguin Books, New York, 2005
40. C. Adami, What Do Robots Dream Of? Science **314** (5802), 2006, pp. 1093–1094. doi:10.1126/science.1135929. http://science.sciencemag.org/content/314/5802/1093
41. D.J. Chalmers, A Computational Foundation for the Study of Cognition. Journal of Cognitive Science (Seoul) **12** (4), 2011, pp. 323–357
42. R. Manzotti, The Computational Stance Is Unfit for Consciousness. International Journal of Machine Consciousness **4** (2), 2012, pp. 401–420. doi:10.1142/S1793843012400239. http://www.worldscientific.com/doi/abs/10.1142/S1793843012400239
43. P. Dowe, Causal Processes. In: *The Stanford Encyclopedia of Philosophy*, ed. by E.N. Zalta, 2007. http://plato.stanford.edu/archives/fall2008/entries/causation-process/
44. L.A. Paul, N. Hall, *Causation: A User's Guide*. Oxford University Press, New York, 2013
45. J. Schaffer, The Metaphysics of Causation. In: *The Stanford Encyclopedia of Philosophy*, ed. by E.N. Zalta, 2014. http://plato.stanford.edu/archives/sum2014/entries/causation-metaphysics/
46. H. Reichenbach, *The Philosophy of Space and Time*. Dover, 1958
47. S. Yablo, Advertisement for a Sketch of an Outline of a Prototheory of Causation. In: *Causation and Counterfactuals*, ed. by J. Collins, N. Hall, L.A. Paul, MIT Press, Cambridge, Mass, 2004, pp. 119–137
48. Y.S. Bonneh, A. Cooperman, D. Sagi, Motion-induced blindness in normal observers. Nature **411** (6839), 2001, pp. 798–801. doi:10.1038/35081073. http://www.nature.com/nature/journal/v411/n6839/full/411798a0.html
49. D.J. Simons, C.F. Chabris, Gorillas in our midst: sustained inattentional blindness for dynamic events. Perception **28** (9), 1999, pp. 1059–1074
50. F. Crick, *Astonishing Hypothesis: The Scientific Search for the Soul*. Touchstone, New York, 1994
51. A. Revonsuo, *Inner Presence: Consciousness as a Biological Phenomenon*. MIT Press, Cambridge, Mass, 2006
52. D. Marr, *Vision*. Freeman, San Francisco, 1982
53. M.R. Bennett, P. Hacker, *Philosophical Foundations of Neuroscience*. Blackwell, Malden, Mass, 2003
54. R. Manzotti, P. Moderato, Is Neuroscience Adequate As The Forthcoming "Mindscience"? Behavior and Philosophy **38**, 2010, pp. 1–29. http://dx.doi.org/10.1142/S1793843011000765
55. J. Dewey, *Experience And Nature*. Open Court, Chicago, 1925
56. W.T. Rockwell, *Neither Brain nor Ghost*. MIT Press, Cambridge, Mass, 2005

57. N.H. Kerr, G.W. Domhoff, Do the Blind Literally "See" in Their Dreams? A Critique of a Recent Claim That They Do. Dreaming **14** (4), 2004, pp. 230–233. DREAMING 8(3) Abstracts - The Journal of the Association for the Study of Dreams

58. A. Revonsuo, C. Salmivalli, A content analysis of bizarre elements in dreams. Dreaming **5** (3), 1995, pp. 169–187. doi:10.1037/h0094433

59. E. Schwitzgebel, C. Huang, Y. Zhou, Do we dream in color? Cultural variations and skepticism. Dreaming **16** (1), 2006, pp. 36–42. doi:10.1037/1053-0797.16.1.36

60. C.S. Hurovitz, S. Dunn, G.W. Domhoff, H. Fiss, The Dreams of Blind Men and Women: A Replication and Extension of Previous Findings. Dreaming **9** (2-3), 1999, pp. 183–193. doi:10.1023/A:1021397817164. http://link.springer.com/article/10.1023/A%3A1021397817164

61. R. Manzotti, Is Consciousness Just Conscious Behavior? International Journal of Machine Consciousness **3** (2), 2011, pp. 353–360. doi:10.1142/S1793843011000765. http://www.worldscientific.com/doi/abs/10.1142/S1793843011000765

62. R. Jackendoff, *Consciousness and the Computational Mind*. MIT Press, Cambridge, Mass, 1987

63. R. Manzotti, F. Mutti, G. Gini, S.Y. Lee, Cognitive Integration through Goal-Generation in a Robotic Setup. In: *Biologically Inspired Cognitive Architectures 2012*, ed. by A. Chella, R. Pirrone, R. Sorbello, K.R. Jóhannsdóttir, no. 196 In: Advances in Intelligent Systems and Computing, Springer, 2013, pp. 225–231. http://link.springer.com/chapter/10.1007/978-3-642-34274-5_40

64. R. Manzotti, V. Tagliasco, From behaviour-based robots to motivation-based robots. Robotics and Autonomous Systems **51** (2–3), 2005, pp. 175–190. doi:10.1016/j.robot.2004.10.004. http://www.sciencedirect.com/science/article/pii/S0921889004001861

65. D. George, J. Hawkins, Towards a Mathematical Theory of Cortical Micro-circuits. PLoS Comput Biol **5** (10), 2009, p. e1000532. doi:10.1371/journal.pcbi.1000532. http://dx.doi.org/10.1371/journal.pcbi.1000532

66. J. Hawkins, S. Blakeslee, *On Intelligence*. Times Books, New York, 2004

67. J. Sharma, A. Angelucci, M. Sur, Induction of visual orientation modules in auditory cortex. Nature **404** (6780), 2000, pp. 841–847. doi:10.1038/35009043. http://www.nature.com/nature/journal/v404/n6780/abs/404841a0.html

68. J. Sharma, V. Dragoi, J.B. Tenenbaum, E.K. Miller, M. Sur, V1 Neurons Signal Acquisition of an Internal Representation of Stimulus Location. Science **300** (5626), 2003, pp. 1758–1763. doi:10.1126/science.1081721. http://www.sciencemag.org/content/300/5626/1758

69. M. Sur, P.E. Garraghty, A.W. Roe, Experimentally induced visual projections into auditory thalamus and cortex. Science **242** (4884), 1988, pp. 1437–1441

70. E.R. Kandel, L.R. Squire, Neuroscience: Breaking Down Scientific Barriers to the Study of Brain and Mind. Science **290** (5494), 2000, pp. 1113–1120. doi:10.1126/science.290.5494.1113. http://www.sciencemag.org/content/290/5494/1113

71. F. Aboitiz, D. Morales, J. Montiel, The evolutionary origin of the mammalian isocortex: towards an integrated developmental and functional approach. The Behavioral and Brain Sciences **26** (5), 2003, pp. 535–552; discussion 552–585

72. E. Thelen, Grounded in the World: Developmental Origins of the Embodied Mind. Infancy **1** (1), 2000, pp. 3–28. doi:10.1207/S15327078IN0101_02. http://onlinelibrary.wiley.com/doi/10.1207/S15327078IN0101_02/abstract

73. U. Neisser, In: *Concepts and Conceptual Development: Ecological and Intellectual Factors in Categorization*, Cambridge University Press, 1987, pp. x, 317

74. F. Tong, M.S. Pratte, Decoding patterns of human brain activity. Annual Review of Psychology **63**, 2012, pp. 483–509. doi:10.1146/annurev-psych-120710-100412

75. P.S. Churchland, T.J. Sejnowski, Neural Representation and Neural Computation. Philosophical Perspectives **4**, 1990, pp. 343–382

76. M. Shanahan, *Embodiment and the inner life: Cognition and Consciousness in the Space of Possible Minds*. Oxford University Press, USA, 2010

77. R. Manzotti, A Process-Based Architecture for an Artificial Conscious Being. In: *Process Theories: Crossdisciplinary Studies in Dynamic Categories*, ed. by J. Seibt, Springer Netherlands, 2003, pp. 285–312. http://link.springer.com/chapter/10.1007/978-94-007-1044-3_12

78. R. Manzotti, Machine Free Will: Is Free Will a Necessary Ingredient of Machine Consciousness? In: *From Brains to Systems*, ed. by C. Hernández, R. Sanz, J. Gómez-Ramirez, L.S. Smith, A. Hussain, A. Chella, I. Aleksander, no. 718 In: Advances in Experimental Medicine and Biology, Springer New York, 2011, pp. 181–191. http://link.springer.com/chapter/10.1007/978-1-4614-0164-3_15
79. R. Manzotti, L. Papi, S.Y. Lee, Does radical externalism suggest how to implement machine consciousness? In: *Biologically Inspired Cognitive Architectures 2011*, ed. by A. Samsonovich, K. Jóhannsdóttir, IOS Press, Amsterdam, pp. 232–240

From the Perspective of Artificial Intelligence: A New Approach to the Nature of Consciousness

Riccardo Manzotti and Sabina Jeschke

Abstract Consciousness is not only a philosophical but also a technological issue, since a conscious agent has evolutionary advantages. Thus, to replicate a biological level of intelligence in a machine, concepts of machine consciousness have to be considered. The widespread internalistic assumption that humans do not experience the world as it is, but through an internal '3D virtual reality model', hinders this construction. To overcome this obstacle for machine consciousness a new theoretical approach to consciousness is sketched between internalism and externalism to address the gap between experience and physical world. The 'internal interpreter concept' is replaced by a 'key-lock approach'. Here, consciousness is not an image of the external world but the world itself. A possible technological design for a conscious machine is drafted taking advantage of an architecture exploiting self-development of new goals, intrinsic motivation, and situated cognition. The proposed cognitive architecture does not pretend to be conclusive or experimentally satisfying but rather forms the theoretical the first step to a full architecture model on which the authors currently work on, which will enable conscious agents e. g. for robotics or software applications.

Keywords Consciousness · Machine Consciousness · Multi Agent System · Genetic Algorithms · Externalism

1 Introduction

Even if consciousness is not exactly a 'well-defined' term and generations of philosophers and other scientists have discussed its complex features at length, there is a certain common understanding about its central meaning: Consciousness describes the unique capability of having experiences in terms of perceptions, thoughts, feelings

R. Manzotti (✉)
Department of Linguistics and Philosophy, Massachusetts Institute of Technology,
Cambridge, MA, USA
e-mail: manzotti@mit.edu

S. Jeschke
IMA/ZLW & IfU, RWTH Aachen University, Dennewartstr. 27, 52068 Aachen, Germany

Originally published in "International Journal of Advanced Research in Artificial Intelligence (IJARAI)",
© The Science and Information (SAI) Organization 2014. Reprint by Springer
International Publishing Switzerland 2016,
DOI 10.1007/978-3-319-42620-4_41

and awareness.[1] Obviously, consciousness requires the awareness of the external world. What is still fairly mysterious is the nature of this experience. Although this capability is still very poorly understood and indeed is considered a sort of challenge for the standard picture of the world, it is a plain fact that the conscious human being is one of the outcomes of natural selection. Likewise, it seems undeniable that human beings cope with the most unexpected events by means of conscious reflection. Finally, they are extremely sensitive to anything remotely resembling the capability of feeling in other agents. In sum, consciousness appears to be a not negotiable aspect of a highly developed autonomous agents and it cannot be underestimated that the practical advantages may result from its replication within an artificial being [1–5]. Here, the problem of the physical underpinnings of consciousness rather than the problem of the self is addressed and thus the nature of what it is like to have a certain experience [6–8] rather than the problem of how the different cognitive processes combine together to form a self. This paper considers how experience may be the result of a physical system interaction with the world experience rather than the self is the goal of this proposal.

During the recent decades, one got familiar with the conception that one never gets acquainted to the world as it is, but only to a '3D virtual reality model' of the outside world that one's brain switches on as soon as one wakes up. This internal model is taken to be the inner world of consciousness how the world appears to humans and not what the world really is. To give an example, colour in the external world may be defined, albeit with some simplification, by two physical parameters: wavelength and intensity. For a human being however, light is not just the detection of a certain light frequency on the retina, but a certain experience when detecting that light frequency (say, perceiving red). In his excellent textbook on vision, Stephen Palmer claims: 'Color is a psychological property of our visual experiences when we look at objects and lights, not a physical property of those objects or lights' [9]. Nevertheless, this psychological property is without comprehensive explanation so far. It does not fit to any obvious physical property.

In a nutshell, the current main line of explanation goes as follows:

- An external event
- Goes through some kind of internal interpreter (in the human brain)
- And internally produces a certain result (within the human).

This current interpretation can be allocated to the so-called 'internalistic models' – namely those models that take the mind to be a property of what takes place inside the neural system [10–12].[2] This view has its strength: putting some kind of 'interpretation layer' between the individual and the outside world allows explaining

---

[1] Consciousness is an 'umbrella term' encompassing a variety of distinct meanings. For this purpose it is important to differentiate between consciousness and self-consciousness. In this paper, to be conscious it is only necessary to be aware of the external world, whereas self-consciousness is an acute sense of self-awareness. In Section 4.1 this topic is discussed in more detail.

[2] Of course, these models do not rule out the importance of the external environment to allow the development of internal structures. Indeed, they consider it as necessary for a healthy brain development to continuously interact with the environment. However, once the required neural

why humans – all of them build alike – tend to be rather different in their behaviours, reactions and 'feelings'. This argument can be extended to non-human species as well: if the same physical reality 'shows itself' differently to diverse entities, a tentative explanation for the heterogeneous behaviours of these unequal species living in the same environment may be put forward. It also allows an explanation of why humans seem to be capable of consciousness in the absence of obvious external stimuli – in the case of dreams, hallucinations, and afterimages.

However, there are still many open ends, some riddles concerning the conception of consciousness that cannot be answered with the abovementioned picture:

- First, to believe that 'the internal interpreter', and 'neurons will do it "somehow"', is interesting but doubtful. In the last couple of decades, scientists from all areas have invested a lot of energy in the quest for a neural mechanism capable of producing our everyday conscious experience. Up to now, there is no known law of nature predicting that neural activity should result in one's experience.
- Second, in the current model, consciousness neither fits the physical world nor its properties. To carry it to the extremes, that means that one constantly ignores the 'real world' by overwriting it with some internal 'fantasies'. Of course, this could be the case but it sounds at least pretty counterintuitive: why should nature take this kind of detour?
- Third: The discrepancy between our immediate experience and the 'world' is more than just 'somewhat regrettable'. If everything one experiences – from pain to colour, from pictures to music – is nothing more than a product of human neurons, then a logical problem occurs: Why should it be easier for neurons to transmute neural firings into music – than for a cello to shape airwaves into music? If the physical world is devoid of qualitative features, why should the brain – which is part of the physical world – be any better in this respect? Why should the brain create meaningful things, but a cello does not? Or, to use an even catchier picture: 'If colours cannot pop out of strawberries, how can they pop out of neurons?'

As anticipated at the beginning of this section, consciousness is not only a scientific conundrum but a practical goal, too. From the Artificial Intelligence community (and the authors admit that they belong to that community), another thought comes up: Whatever consciousness is in detail, it seems to form an important part at least of a human-like intelligence [13–19]. Therefore, to build artificial systems with certain intelligence, it might be necessary to give them some kind to consciousness too – even if this artificial consciousness might differ very much from the human one, or from other mammals or biological systems. Now, however consciousness might work in biological systems, one may envisage implementing consciousness in totally different ways as part of forthcoming technological systems. Therefore, alternative models to explain consciousness are of the utmost interest, either to explain the 'true nature of human consciousness', or, to allow different approaches for building an artificial/technological agent.

(Footnote 2 continued)
connections are in place, the mind is taken to be an internal phenomenon. Dreams and hallucinations are constantly quoted as obvious cases.

This paper proposes a new hypothesis concerning the nature of conscious experience, to overcome a conceptual war between externalism and internalism (Section 2). In Section 3 the consequences of this new perspective are discussed. Section 4 relates the change of perspective to the field of artificial intelligence. Finally, Section 5 summarizes the paper and gives an outlook on the next research steps to be undertaken.

2 Towards a New Conceptualization of Consciousness

2.1 The Approach

This chapter's goal is to flesh out a new concept of consciousness that is directed to overcome the gap between 'experience' and 'physical world'. To reach that goal, one has to undergo several changes in the standard mindset. Below the hardest nuts to crack are mentioned:

1. *To bridge the gap between experience and physical world, the external surrounding environment in which a brain and a body are situated have to be more prominent in our concept of consciousness (since the other way around seems even more radical).*
2. *If the 'externality' becomes more important, the next domino falling is the giving-up of the underlying 'full-bodied' human-centered view of being necessarily located fully and totally inside its body. This assumption is a subliminal driver of the current theories, but it is not based on any empirical evidence – it is something that may be true or false.*
3. *Also, if 'experience' and 'external world' come closer together, the need for some kind of internal interpreter is dramatically reduced. The transformation of the outside world into an internal representation or a virtual model is getting more and more obsolete.*
4. *The closer 'experience' and 'external world' get, the less their difference can be. This is not at all a trivial statement. On the contrary, this leads to the most difficult point to grasp: namely, that what people call consciousness 'is' the world people live in. It is not how the world appears, but what the world is.*

In Section 3 the consequences of this new model will be discussed. First, however, the authors would like to give the reader the chance to understand HOW such an approach could work in practice with a construction sketch The key idea is that the body and the world are just two pieces of the same physical system and that what the authors call the mind is a physical process that requires both pieces to take place. Body and world are interlocked gear wheels and the consciousness turns them. A schematic description of how the coupling between body and world works is sketched in Figures 1, 2 and 3.

Below the different steps are explained in detail:

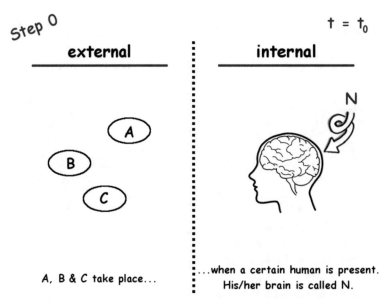

Figure 1 Step 0 – before perception, there is no external object, as one perceives it. There are smaller and scattered physical phenomena, which are not the target of any normal experience

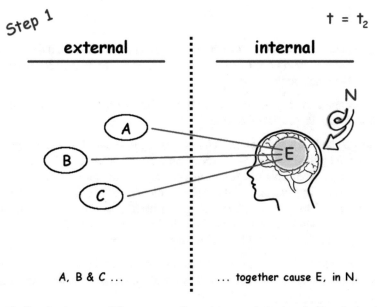

Figure 2 Step 1 – because of the presence of a certain neural structure inside a body with the proper sensor apparatus, the scattered external phenomena produces a joint effect inside one's brain

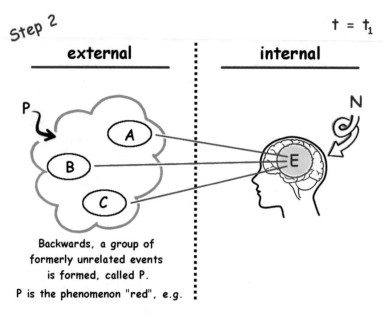

Figure 3 Step 2 – events that are responsible for the occurrence of a joint effect are a joint cause – they become a whole. One's experience is the process P which takes place thanks to both: to one's brain and by A, B, and C

- In the outside[3] world, there are physical scattered events. Let us call them A, B, C. They are located in time and space. A, B, and C do not have anything in common – each take place on their own.

  - For instance, A may be a certain light ray with a certain frequency emitted at a certain time in a certain location.

- In a nearby body, a healthy brain (call it N), through the causal connection provided by sensorial paths, is affected by these three events A, B, and C. This is not an obvious step. What now happens is that the neural structure – thanks to various neural learning mechanisms – allows the fusion of A, B, and C to create the causal circumstances[4] that would allow the fusion itself to act as a cause for further interaction. A whole is born where P is called the fusion of A, B, and C.
- E is the joint effect of P and takes place inside the brain N.

  - As a simple analogy consider the 'distributed key' of an atomic rocket launch system: to launch it, two different 'keys' have to be turned at the same time (in

---

[3] Outside and inside are used to refer to physical events inside and outside one's body.

[4] In causal terms, one may distinguish between 'the cause and that without which the cause would not be a cause' ([20], p. 119). The former may be taken to be an event actually occurring while the latter may be just a state of affairs. The latter may be formalized in terms of conditions G such that $P \wedge G \rightarrow E$ (which may unfolded in three conditionals $P \wedge \neg G \rightarrow \neg E$, $\neg C \wedge G \rightarrow \neg E$, and $\neg C \wedge \neg G \rightarrow \neg E$).

movies, usually one of them always refuses to work!). However, considering the matter more carefully, there are not two keys, there is just one key in two pieces, and to be even more precise, the two pieces are not even keys since, when alone they do not unlock anything.

- Because the fusion P of A, B, and C causes E, these three originally independent events 'become' a whole in any practical and sense – things are defined in causal terms and not in ideal terms. P has not existed until E occurred. E is inside the body, while P remains outside of it.

    – Going back to the Rocket launcher analogy, before being put into the 'lock', both 'key pieces' are as unrelated as A, B, C. After being inserted, they, together with the lock, from a group and form a 'lock-and-key principle' which is a construction partially internal and partially external. The separation between them becomes purely conceptual – physically and causally they are a whole.

- Finally, the hardest bullet to swallow, namely the temporal order. No, the authors have NOT confused the indices: E takes place at a time t_2, while P occurred at a time t_1, with $t_1 < t_2$. Why that – and does that mean that E is changing its past? In some sense yes. At least it is changing the causal role of the past.

The last point merits some further consideration. It is not the intention to invoke some sort of retro-causation that moves backward in time. Rather, the point shows something that should not come as a surprise: Physical phenomena are extended in time. This means that they get to completion within time. Therefore, when something begins to unfold, its nature is not wholly defined until it reaches some natural ending in causal terms. Nothing goes backward in time – the past is, of course, past.

However, what the past was may well be defined by what happens in the present. Using the above time indexes, there is no need to suppose that anything is going from t_2 to t_1. However, there is no harm either in considering that what the world at time t_1 was (that is, A, B, and C) changed after t_2 (that is, the fusion P takes place). If one considers a physical phenomenon as something extended in time, then it may well be the case that what happens along such an extension redefines the structure of the phenomenon.

To understand the temporal structure of the proposed causal sequence one has to take a closer look. At time $t_1 < t_2$, the actual cause P has not yet happened. Until t_2, P has not yet produced any effect. Thus, from a physical standpoint, P has not yet existed. One may put the situation in these terms – until t_2 P's existence is not causally any different from P's absence. Then E takes place at t_2. Things have changed. P is now the actual cause of an event E that might have not happened. Yet E happened and thus P was its actual cause. Has this temporal sequence any effect to the temporal order of perceived events [21, 22]? Not necessarily, as a matter of fact, the subjective temporal order of events depends on how subsequent cognitive processes exploit them. Furthermore, here the crucial issue is the internal physical and causal structure of a perceptual act rather than how the temporal order of different perceptual acts is experienced. Just to dispel any possible misunderstanding, neither the time t_2 is not the subjective time, nor the time t_1 is the objective time. Both t_1 and t_2 are physical times and they refer to when certain causal processes take place. The interval t_1-t_2 has no mandatory impact as to which order P is in temporal relation with other perceived events.

As a further example, let's get back to the case that was mentioned at the onset – namely conscious perception[5] of colour. If one applies the approach just sketched, a colour is a collection of scattered and otherwise separate physical properties until they produce a joint effect in one's brain (E).[6] When they do so, the scattered wavelengths can be considered as a set of external phenomena (A, B, C, ...), namely the colour red. What is hard to grasp is how to step from some scattered wavelengths to the impression of a colour. The answer is that it happens in exactly the same way as one comes from a bundle of 'whatsoever-pieces' to a key – the components themselves do not constitute a whole (in respect to colour) but the sum of them does. The pieces merge into one key IF AND ONLY IF they have the opportunity to do so in causal and actual terms, meaning that there is a 'suitable lock' around (certain interactions with the eye-brain system). Then the whole may take place – without that lock, nothing may happen. By doing so, the scattered events make the colour red happen. In this account, the colour red is the causal fusion of the set of incoming wavelengths. It is neither an internal impression nor a mental ink. Red is an external whole whose occurrence is made possible by causal coupling with the neural event (the joint effect E). A colour-blind person would not have the 'suitable lock' – and therefore would not be conscious of the phenomenon that normally sighted subjects call colour. In physical terms, if there was only a colour-blind person in a certain environment, the combination of physical phenomena getting to an end in a normally-sighted subject would never be able to produce a causal joint effect.

To recap, a causal notion of fusion may thus be put forward – any group of events X_i is fused if and only if there are the causal conditions for a further event E to take place. The idea is that a fusion takes place only if it is the actual cause of some event. Thus, a fusion is not an abstract entity, but a physical occurrence with its own causal efficacy. For any group of events, there is a fusion, whenever they produce together an effect. The structure of neural networks embedded into one's nervous and sensor systems are ideal in this respect – they are the causal circumstances that allow complex events in the environment to be the actual causes of some bodily events, hereafter integrated into the agent's behaviour. Thus, the fusion takes place because of some neural event, which is the effect of external groups of events. What is fused, though, is not inside the body, but it is the group of external events. The actual cause, too, remains outside of the body.

Coming back to the steps 1–4 noted in Section 2 the main changes with respect to the standard view are recapped:

---

[5] As the distinguished neuroscientist Semir Zeki once said, there is no such a thing as unconscious perception. However, in this case the authors prefer to be redundant rather than misunderstood [23].

[6] These properties may be quite an inhomogeneous set of actual physical properties such as the reflected colour spectrum, percentage of certain components, contrastive ratios among different areas, and so forth. For the sake of the example, just a set of wavelengths is considered.

- The gap between experience and world ...is gone since in the upper model, the external events and the internal perceptions become one.
- The human-centred view ...is gone, since in this model, experience is driven partly internal und partly external to a physical body and it is constituted by physical events. The experience is internal to the physical system that underpins it, and it is external to one's body. The body is, of course, nothing but as a subset of a larger physical superset of processes taking place in time.
- Here, the word external does not mean being "projected" but just being physical outside of one's neural structure. Such a notion very strongly suggests that the physical underpinnings of mental states are made of physical events taking place outside of one's body – a standpoint sometimes dubbed phenomenal externalism or externalism about phenomenal content. The goal is to single out a physical event identical with ones experience without having to resort to any mentalistic notion such as content, character, interpretation, projection, reference, and so forth.
- The internal interpreter ...is gone or at least not necessary any longer. It was enough to relocate our insights as to what the physical underpinnings of one's mind are. The consciousness of 'seeing red' (instead of seeing several scattered wavelengths) is the result of the fitting between key-parts and lock. Red is not a meaning associated to some internal representation – red is a physical phenomenon in one's physical environment.
- Consciousness 'is' the world people live in: to make an example, to see red is to be united with an external collection of physical phenomena, since experience takes places as a temporally and causally extended phenomenon that includes internal and external components.

In comparison to the traditional view previously mentioned, a new approach is fleshed out (key differences in italics):

- External events & a neural event

  - *Form a key-lock-system,*
  - Which is, therefore, *partly internal – partly external.*

- The external events produce a certain neural activity,

  - An event inside the human body

- and, as a result,

  - *An external actual cause has occurred.*

2.2 The Consequences Overall

For a moment, before raising the inevitable objections, consider this view as a tentative scientific hypothesis as to the physical nature of consciousness – a scientific hypothesis insofar as it puts forward a falsifiable hypothesis as to what the mind is. If this hypothesis had any merit, a few conceptual advantages are immediately obvious:

- First, the **hard problem of consciousness**: The hard problem of consciousness, introduced by David Chalmers [6], addresses the problem of explaining how and why one has qualia and phenomenal experiences such as pain, colours, taste etc. (incl. 'Why does awareness of sensory information exist at all?' 'And why is there a subjective component to experience?'). The presented approach sweeps away the premises on which the hard problem of consciousness is based on (and thus the hard problem itself). In short, the hard problem is based on the dustbin model of the conscious mind [24, 25] in which a set of features is relocated that have been eschewed from the physical world. The idea is that – according to Chalmers – the hard problem is not an unavoidable chasm in the structure of nature, but a false issue created by assuming wrong premises. The approach presented here addresses such premises and indeed suggests a different picture. The mind and the world would no longer be two incommensurable and indeed autonomous domains, but the same one under two different perspectives.
- Second, the **mind-body-problem**: overt and covert dualism would finally be over-turned. Dualism is not just the straw man often depicted as the traditional substance dualism contrasting matter and soul or body and mind. There are also forms of dualism that suggest a juxtaposition between cognition and the brain [26, 27], sometimes dubbed Cartesian Materialism [28, 29]. There is no longer the need to differentiate the way in which things look to subjects and the way in which things are. There is just a flow of physical phenomena causally interconnected.
- Third, **exclusiveness**: being conscious of something is a 'private' event – but in contrast to the traditional interpretation, the privacy is no longer created by an internal individual interpreter. It is no longer an exclusive and unbridgeable metaphysical privacy. Rather, it is the kind of privacy that prevents two individuals from eating the same piece of cake. It is a notion akin to that considered by the philosopher Mark Johnston who considers privateness as the impossible to receive the same anti-flu shot [30]. The exclusiveness follows from the fact that the pieces fuse into one key only if there is a 'suitable lock' – the suitable brain of a conscious agent - around. The causal interaction between internality and external world links the observed object and its observer. Of course, in presence of two similar groups of events, two similar brains let similar fusions occur.
- Fourth, **location of consciousness**: it is possible to physically locate the (conscious) mind into the physical world. The location is not some inside neural activity though. However, it is possible to pinpoint a certain physical process and consider whether such a process is identical to ones own experience of, say, a red patch. It is thus possible to resurrect the theory of identity in terms of broader physical processes and not just in terms of neural processes. The fact that consciousness takes place partially outside the body is not in contrast with the impression one may have to be located inside the own body. Nothing in our experience points to where our experience takes place, only to what our experience is. If someone cuts

a finger, he or she does not feel a pain inside the brain, but rather a pain in the finger. By the same token, it is not necessary that the process has to be located within our body.

- Fifth, *the misperception issue*[7] – namely, the fact that apparently one may experience things that are not physically present, as it happens in the case of hallucinations or dreams e. g. has to be dealt with differently. They are no longer the result of a somewhat hyper-creative internal interpreter, but of an unusual connection with real features in one's environment since this component has been removed from the picture. First, it is important to realize that our dreams are just 'boring' recombinations of the basic components of our past, albeit reshuffled in possibly original ways, they are chimeric but not innovative [31–34]. Second, it is important to realize that ALL perceptions require a temporal lag between the object and the neural activity, due to the velocity of information transportation. Combining these two insights leads to a possible and fairly simple explanation approach, namely, that dreams and hallucinations may just happen to be cases of very long and reshuffled perception of ones world. So, tentatively, this approach suggests that the stuff dreams (and consciousness) are made of is the same stuff the world is made of.
- Sixth, **tabula rasa**. According to this view there is no mental content distinct from a physical event (that may be part of one's body, of course). Thus, one may experience a red apple or an itch in the elbow, but one cannot experience a pure mental content that one may experience. This is a very physical view that rules out any immaterial or purely mental content. By the same token, at the very beginning, organisms cannot have any experience since, by definition, they have not yet been in contact with any physical phenomenon. This does not prevent, of course, that either new born infants or foetuses may have consciousness as long as (1) they have a working neural system and (2) they perceive external events through parts of their bodies (or their mothers' bodies). However, the approach presented here rules out any innate or purely mental content of experience.

---

[7]Whenever it was necessary to point to the autonomy of the mental with respect to the physical domain, the issue of misperception has been the battering ram of both philosophers and scientists. Dream and hallucinations appear as formidable evidence in favour of an inner world. However, this approach promises to locate in the physical surrounding a physical cause for ANY experience. All cases of conscious experience ought to be revisited as cases of (admittedly unusual) perception. The approach presented here honestly stands or falls on whether it will succeed to show that – perhaps surprisingly – whenever there is consciousness there is a physical phenomenon, which is the content of one's experience. The authors cannot do justice here to the problem of misperception by and large. However, one can flesh out a template of the strategy – namely to address each purported case of misperception and to revise it in terms of actual perception. (One of the authors is actually working on such an account for most cases of misperception, from hallucination to illusions, from aftereffects to direct brain stimulations).

3 Consciousness Opening New Perspectives for Artificial Intelligence

Can the outlined approach help in shaping and devising an architecture capable of consciousness? The authors believe that it can, because it suggests a causal structure of consciousness and thus something that may help in singling out relevant architectures in an artificial agent. For instance, the approach suggests that being conscious is not a matter of either having the right internal code [35, 36], or having a central global dashboard [37, 38], or processing information in a certain way [39]. The advantage of this proposal is that it allows for rather precise indications as to why the causal coupling between the environment and the agent ought to be realized. Of course, by itself, the approach does not provide a complete picture of how to implement an intelligent agent. Many other aspects – often already addressed and partially implemented in AI and robotics – must flank what is here suggested. In sum, the suggested approach to consciousness does not aim to be alternative to other approaches in AI or in robotics, rather it aims to tune them in a way that should be productive for consciousness.

3.1 From Machine Intelligence to Machine Consciousness

It may be useful to make a comparison between current attempts to implement intelligence and consciousness. The understanding of what intelligence is – or what it is not – fills book. The notion derives from the Latin verb 'intellegere' ('understanding', more literally 'to have the choice between', 'to read between' [40, 41]). As a scientific term originated in psychology, the concept of intelligence addresses the cognitive capabilities of an entity, usually a human being. In these general terms, the notion partially overlaps with the psychological notion of consciousness. This is also partially due to the fact that both notions (consciousness and intelligence) are mongrel concepts that encompass several vague and not entirely coherent aspects.

By and large, an agent with intelligence is often considered as divided into three central parts following each other:

1. *Recognition of external changes*

   • Having **sensory** components in order to receive stimuli from the external environment

2. *Information processing*

   • Being capable of processing the sensory data together with internal knowledge in order to adapt behaviour, **cognition**

3. *Reaction*

   • Having the capability to interact with external environment, realized by **actuators**

Hereby, the Latin verb 'cognoscere' translates into 'conceptualize, recognize'. Cognition comprises the processes of information processing within an intelligent actor (the sensory input is transformed, reduced, elaborated, stored, recovered, and used). Cognitive processes are divided into conscious and unconscious ones, e. g., by far not all learning processes are conscious. From this argumentation chain – from intelligence to cognition to consciousness – it follows that consciousness plays an important role in the understanding of intelligence.

Even if the majority of research done in the field of intelligence is directed towards human intelligence, the upper description states clearly that intelligence is NOT a primacy of humans. Obviously, many animals have a certain form of intelligence – proof is already given by observing your pet cat – even if it may differ from the human. Interestingly, the scientific status of 'consciousness' in animals continues to be hotly debated even if it is obvious that most animals have a phenomenal consciousness including a sense of pain, colour recognition, temperature etc. As mentioned above, the confusion is partly due to the variety of conceptions of consciousness. Researchers from different fields include very different aspects into the concept: (a) phenomenal consciousness, (b) the capability of thinking (thinking, remembering, planning, expecting), (c) self-consciousness (awareness of oneself), (d) consciousness of uniqueness (of oneself and of others) etc. Whereas phenomenal consciousness is probably part of most animals, it is still unclear if at least highly developed animals as mammals dispose of additional types of consciousness [5]. So the research space may be unfolded according to two broad criteria; one related to the kind of agents (animal, human or machine) and the other related to the kind of cognition involved (sensori-motor skill, symbolic capability aka traditional intelligence, linguistic capability, consciousness).

So which interim conclusions can be deduced? Machine consciousness lies in the promising middle ground between the extremes of biological chauvinism (i. e., only brains are conscious) and liberal functionalism (i. e., any behaviourally equivalent functional systems is conscious) [42]. One of the most central concepts behind 'intelligence' and perhaps the most difficult aspect to grasp is clearly not restricted to humans. From that it follows quite naturally that when building a technological system with a somewhat 'authentic intelligence', consciousness will have to play its part. Phenomenal consciousness – that is. It remains to be seen whether new concepts to realize this aspect will lead to insights into other components of consciousness.

3.2 Weak Versus Strong Machine Consciousness

The traditional and historically outdated distinction between weak AI and strong AI results from two different requirements: on the one hand, it follows from researchers focusing on different goals (more 'practical' vs. more 'principal'). On the other hand, a comprehensive philosophical debate on the nature of intelligence is driving the debate, including its exclusiveness or non-exclusiveness for humans (or other biological systems), ethical aspects, and the general possibility of reconstructing real intelligence, just to mention a few important aspects.

Weak AI addresses the position of artificial intelligence in philosophy that machines can demonstrate human-like intelligence, but do not necessarily have a mind, mental states or consciousness. Contrarily, strong AI supposes that some forms of artificial intelligence can reason and solve problems[8] as opposed to just making the humans feel that the machines are intelligent. In short: a weak AI-capable agent seems to be intelligent whereas a strong AI-capable agent is intelligent.

Obviously, the philosophical question behind this distinction is strongly related to the problem of consciousness. From that, it is not surprising that some authors suggested the possibility to distinguish between weak and strong artificial consciousness [3, 5]. In analogy to the weak vs. strong AI debate, weak artificial consciousness aims to deal with agents that behave as if they were conscious, at least in some respects, whereas strong artificial consciousness tries to address the design and construction of 'truly' conscious machines. Thus, the distinction between weak and strong artificial consciousness mirrors the dichotomy between true conscious agents and 'as if' conscious agents.

Although the distinction between weak and strong artificial consciousness sets a temporary working ground [5], it suggests a misleading view in so far as it suggests that a concept for a 'weak artificial consciousness' will help to gain a 'first understanding' on what consciousness might be and how it could be realized. Since it misses indispensables for the understanding of cognition – namely experience, i.e. phenomenal consciousness – the concept will not be adequate to overcome 'the riddle': Skipping the 'hard problem' is not a viable option in the business of making conscious machines [43].

Another argument may be raised against the temptation of 'weak artificial consciousness via the easy way': in nature, the development of consciousness goes along with increased intelligence. Most animals are exhibiting behavioural signs at least of phenomenological consciousness, human beings have a phenomenological consciousness and 'above'. 'Evolutionary optimization' is the most powerful optimization known so far (even if it takes its time). Thus, it seems to be highly unlikely that natural selection took such a long way to provide us with consciousness if there was a way to get all the advantages of a conscious being without actually producing it. Of course, this does not mean 'proof' – but the authors cannot help but to sincerely doubt it.

4   Concepts for Building a Conscious Machine

Now, can a machine gain consciousness – that is, strong artificial consciousness as described in the previous section? Why – or why not? Is consciousness not a property of natural system that may thus be, at least in principle, realized by another physical

---

[8]Sometimes, the term "artificial general intelligence" ("AGI") is used to address strong AI, in particular by science fiction writers and within the community of futurists.

system? And if so, how can that be done? Armed with the absence of a theoretical reason to reject the practical possibility, this paper addresses this issue.

In the following, the authors are not outlining a strong theoretical formulation. Also, they are not capable – at this point – to give 'a full proof' (in a strict sense). Rather, it is the intention to show the inherent potential in the given interpretation of consciousness (Section 2): As long as consciousness is interpreted as an 'internalistic' concept, there would be no change in modelling it: It remains to be something like 'internal interpreter, e. g. transforming 10.000 × 700 nm into "red"'. Nobody knows why and how, except from that it happens. The internalistic interpretation may be true but this would not help oneself to come closer to any understanding of the concept behind it. However, if consciousness is interpreted in the sense as the authors proposed (halfway between 'internalistic' and 'externalistic'), then it could be realized (at least as a toy model) as will be explained in the following. Thus, one can start to understand it and try to run tests on it, and so on. So by proposing this possible solution, the authors will sketch a 'lab scenario'. Here, promising off-the-shelf technologies are considered that may fill the bill if deployed in the proper way.

4.1 Preparations for a Tentative Architecture for a Conscious Agent

Currently, many robotic setups and architectures are the result of careful programming since designers aim to solve specific sensorimotor, relational, or logic issues. A classic example is offered by robotic feats like Robocup[9] where teams of robots exploit algorithms devised by their designers to compete together in a soccer match. Although their behaviours may be very clever it is not the result of real adaptation on a high-cognition level. Of course, there are some robots capable of learning new skills and to adapt to novel situation, at least to a certain degree. However, explicit attempts of integrating consciousness into a robots' intelligence are rare, and so far no model has been exceedingly convincing.

Compared to current robotic agents, biological agents like mammals and humans show a totally different kind of adaptability to novel stimuli. Mammals are capable of dealing with totally unexpected environmental challenges for which they could not possibly have any kind of inborn solution. Furthermore, it is a fair bet to assume that the complexity of their neural structure largely exceeds their genetic blueprint. Most mammals are capable not only of learning how to achieve goals but also of learning what goals have to be pursued [44, 45] – which is an important issue in respect to consciousness. As it has been observed [46, 47], the cortex shows an almost universal capability of autonomously adapting to novel kind of stimuli: 'The fact that humans can learn and adapt to problems that did not exist when the initial model (the neocortex) was created is proof of the generic nature of the mechanisms used by the human brain.' [48]. Thus, it makes sense to look for very general approaches

---

[9]http://www.robocup.org/.

capable, albeit with possible shortcomings, to model a unified and common approach to all aspects of cognition.

Empirical evidence shows that mammals exhibit a very high degree of neural plasticity and cognitive autonomy [49–51] to the extent that it is fair to suppose that any part of the cortex might develop almost any cognitive skill. If this supposition were true, it would mean that the neocortex, and possibly the thalamocortical system, exploit some kind of rather general architectural principle, mainly independent of the kind of incoming data.

There have been various attempts in the past to devise a general cognitive architecture [46–48]. This paper makes yet another attempt and takes advantage of a rather simple idea: true autonomy entails teleological openness. By being teleologically open the authors mean that the system is capable of developing new goals autonomously on the basis of environmental conditions and stimuli [45].

4.2 Objectives and Motivations of the Architecture

What are the ideal features that a cognitive architecture should have in order to adapt to a partially unknown body and environment? On the basis of the available literature and the empirical evidence a series of key features and their justification may be listed:

- The architecture must be based on a very limited number of kinds for basic building blocks – each kind exploiting the same common structure. Thus, the description length of the architecture must be kept to a minimum.
- This basic module might be freely replicated in order to cope with multiple sensor modalities and demanding incoming stimuli. This should ensure scalability.
- The basic module has to be able to develop its own goals and to use them both for its own development and for interacting with other modules. This should allow developing intentionality and a tight environment-architecture coupling.
- In principle, adding further modules (constrained only by the system resources) should lead to an increase in performances. Once again, this is important for scalability.

An architecture with the above features should be able to adapt to unknown situations and with a minimum of predesign. Rather than specifying all the algorithms and their mutual relationships, the above approach suggests a recipe to build a cognitive architecture given a body and an environment. Such a recipe is a lot less demanding in terms of description and a priori knowledge than a detailed plan. Furthermore, a recipe of such a universal scope offers many more advantages in terms of adaptability and flexibility.

Thus, the architecture the authors are willing to implement must satisfy the following requirements:

- Structure:

  - It must be scalable
  - It must be adaptable
  - It must take advantage of memory more than speed
  - It must be hierarchical

- Capabilities:

  - It must take into account the whole history of the system
  - It must develop fine grained new goals
  - It must develop overarching goals emerging out of the finer structure

- Additional Do's and Don'ts:

  - It must not rely on explicit algorithms
  - It might have a limited number of more specialized versions of the same elementary block (for fine tuning, better performance, and optimization)
  - It must be coherent to what one knows about the biological structure of a mammalian brain

4.3   Combining Multi Agent Systems with Genetic Algorithms

A tentative approach might be to realize a robot's brain as a multi-agent system (MAS) once such an endeavour may find support by some additional key hypothesis as to the physical foundations of consciousness. MAS have been discussed already as a possible model to realize artificial brains, or as a model to explain the function of a brain (e. g. [52]). They have also been discussed as a possible extension of cognitive architectures as e. g. within the hybrid design oft CLARION (e. g. [53]). In computer sciences, MAS have become a very popular instrument during the last years when modelling complex heterogeneous distributed systems, which are organized 'bottom up'. Their strength lies in predicting appearance of complex phenomena. The single agents have a certain degree of autonomy, they represent local views (in general, no agent has a full global view of the system, due to the complexity and the number of dynamically changing external dependencies), and they work decentralized ('no master brain'). Topics where multi-agent systems are used include in particular the modelling of social and/or cooperative structures. Multi-agent system may be one of the key architectural principles necessary for a conscious mind.

Taking the new approach to consciousness as described in Section 2 as basis, in such a MAS each software agent would represent one 'conscious-lock' to a certain key, an external phenomenon. Thus, the resulting robotical brain would be conscious of the external events it has the appropriate locks for it, and the mechanism of building this consciousness would be exactly the same as for the human brain. So, the tentative idea is that MAS could offer the necessary architectural backbone for a conscious mind and that, once tuned to satisfy to some specific requirement, may be indeed the workable tool to begin designing a new kind of cognition.

At least three questions pose themselves immediately:

- **Complexity**: One may argue that by this approach, only a small number of locks can be realized due to the enormous programming effort needed otherwise.
- **Specification**: An even harder objection might be that in this way, the programmer may tend to mainly 'imitate' the human consciousness but does not develop one which is appropriate for the given robot with a certain form, function and so on.
- **Proof**: A third difficult point is the answer to the question as how one would like to prove that a certain robot really has a consciousness in a strong sense (compare Section 4.2).

To tackle all three problems with one approach, optimization algorithms have to be integrated, allowing to improve the multi agent system during runtime. Here, due to their 'closeness' to the underlying problem (a developing brain), genetic algorithms might form a natural choice: The 'consciousness-locks' have to be specialized to species, their mode of living, and the challenges presented to them.[10] Their special characteristics are probably not the result of some kind of 'biological master plan' for all living beings, but the result of a species-exclusive evolutionary process, which over millions of years has favoured individuals which are better adapted to their environment than others. In this understanding, the consciousness-lock (realized through multi agents) would be subject to the same evolutionary process, which has driven the whole design of a certain species, including the body shapes, motor skills, brain structure and the like. Genetic algorithms are precisely reproducing this kind of development.

The idea of using genetic algorithms to build a conscious brain is also one of the central design principles behind the cognitive architecture [54]. Genetic algorithms are a part of evolutionary computing, which is a rapidly growing area of artificial intelligence. They are inspired by Darwin's theory about evolution. The idea was introduced in the 1960 s by Ingo Rechenberg in his work 'Evolution strategies'. His idea has been extended by many other researchers over the last decades. Today, they play an important role in many complex optimization problems and form an important concept for machine learning approaches. Genetic algorithms use mechanisms inspired by biological evolution, such as reproduction, mutation, recombination, and selection. Over several generations, systems are optimized: Pairs of first generation solutions are taken and recombined. The 'fittest' solutions of this match are selected for the next generation. Mutations are used to enhance the genetic variety and thus, the

---

[10]Consider the following example: Literature states that cats are somewhat colour-blind concerning the colour red, they see it as a shade of grey (whereas they have a perfect colour vision concerning e. g. green and blue). Well, the first finding is that one cannot be really sure about that, since one can only predict that from their eye anatomy – but what kind of 'consciousness' cats really have concerning the colour red is a totally different topic because at this point the design of key-lock-structure is unknown. It may be totally different to the human one. The second – and much more important – insight is that it might be less important for a cat to be capable of seeing red than for example for a bear: cats – being carnivores – do not have to differentiate between ripe and unripe apples since they would not eat them anyway. For a bear on the other hand – being omnivores consuming a large portion of fruits daily – the situation may show itself quite differently.

overall solution space. The optimization goal – in nature given through environment and the corresponding challenges – is realized through a so-called fitness function which determines the quality of the solutions. Lately, combining multi agent systems with genetic algorithms has become popular in certain field as e. g. automated testing scenarios.

Assuming that consciousness is a capability of higher development of life forms, the following digest gives a first impression of the number of genetical iterations which are necessary to produce this kind of complex structures: About 3.5 billion years ago, the first life forms developed, monads with a very limited range of functionalities. Based on the development of genetical heredity through DNA molecules, advancements and progresses could be passed over to the next generation leading to first plants and simple animals which arose about 700 million years ago. 200 million years ago, mammals started to populate the earth. Humanoid life forms developed 70 million years ago and the homo sapiens species is only 500.000 years old. Even if it is difficult to tell from which stage in evolution consciousness has first entered the scene, referring to the current state of the art its development is part of a growing and more and more complex brain (ibidem). From this analysis, the reader might understand why the others consider genetic algorithms for the optimization job!

4.4   The Practical Side

How to proceed 'practically', meaning: how exactly are the genetic algorithms used to re-build the evolution of a robotical brain?

- Regarding the development of consciousness, one would start with a couple of given perceptions, each of them realized through a single agent, say regarding colours, temperature and the like which seem to play an important role for all living beings – a 'basic set' of conscious perceptions, so to say. This is 'easy' – and would form the 'first generation solution'.
- Now, to make the system learn new conscious elements, the second preparative step is to place the robot in a certain challenging environment – meaning that certain tasks have to be given to him – in order to challenge his 'consciousness enhancement'.
- Next, genetic algorithms come into play in order to produce variants of the robot's 'brain structure': the single agents will be multiplied and altered through the means of the genetic algorithms. They will become multiplied, more complex and more varying. Some of the 'new' solutions will not survive as they do not particularly contribute to the tasks the robot is given. Others will survive as they enhance the robot's capabilities to deal with its tasks. This, the resulting brain structure, will turn out having consciousness-locks which are complex and adapted to the individual needs of the specific kind of robot and its environment.

From that, there are two possibilities to infer that the robotic brain is really developed something like a consciousness by using genetic algorithms:

- First, the direct inspection would address the source code itself. Starting from a 'basis set' of agents in a MAS, the resulting system would consist of old and new software agents, the latter representing new conscious capabilities. The new code can be investigated, varied and different tests cases could be designed and analysed.
- A more 'indirect' inspection would be: A test scenario could be designed where consciousness for a certain perception area would definitely be necessary to solve a certain task. By design, this particular conscious perception would not be part of the basic conscious skills the system is starting with.[11] Now, if – after of a couple of some (more) 'genetic rounds' – the robotic brain would come up with new solutions for the given task, which definitely requires the enhancement of its consciousness, this would be a strong signal that it has developed a new perception in a certain area.

So, the combination of multi agent systems with genetic algorithms allows overcoming the upper mentioned problems:

- **Complexity**: Starting from a small number of locks, their expansion is realized by genetic algorithms which enhance the number of locks in order to optimize the systems behaviour
- **Specification**: Since the optimization takes places in relation to a certain environment including specific challenges and particular tasks, the robotical brain develops a consciousness which is adapted to its own needs.
- **'Proof'**: The proof of whether a consciousness has been developed is not complete. However, on the more direct side, the investigation of the auto-generated sourcecode will deliver new insights. From the perspective of an indirect proof, it would address the development of a new conscious aspect rather than its existence. If a robot can adapt to a certain situation IF and only IF it develops a conscious perception for something that will be a strong hint that consciousness has been developed.

---

[11]Consider the following example: assume having a robot with colour consciousness as one of the basic components. This robot is part of a cooperative structure with other robots and humans, working together in a production line. Due to long geographical distances within factory, it would be absolutely necessary to be capable to 'visualize temperature', meaning to have a visual perception for extended areas within wavelengths between 700 nm and 1 mm (infrared). Here, humans would not be able to 'see' the wavelengths, since they have no consciousness for this wavelength area. However, the robot (using genetic algorithms on the multi agent system which is forming the 'conscious part' of the robotical brain) could develop a perception for this wavelength. By that, the robot might be capable to solve the task – finding a heat leak in an extended machinery – opposed to the human. Due to the fact that the brain is built in the upper described key-lock system that would mean that some kind of 'new' consciousness has been developed.

5   Summary and Outlook

Putting it all together: According to [55], there are three motivations to pursue artificial consciousness [56–58]:

1. *Implementing and designing machines resembling human beings (cognitive robotics);*
2. *Understanding the nature of consciousness (cognitive science);*
3. *Implementing and designing more efficient control systems.*

Based on the presented new approach to consciousness lying between internalism and externalism, a possible technological design for a conscious machine has been sketched addressing the upper mentioned goals. The approach is taking advantage of an architecture exploiting self-development of new goals, intrinsic motivation, and situated cognition. From a technological point of view, multi agent systems are used to model independent conscious perceptions. Genetic algorithms – as a subgroup of evolutionary algorithms – come into play to mimic the biological evolution of the brain's structure, thus allowing in general for adaptivity and scalability, and assuring some coherence to what humans know about the biological structure of brains of higher developed animals as e. g. mammals.

The architecture does not pretend to be either conclusive or experimentally satisfying. In the future, this rather sketchy outline of a cognitive architecture will be enhanced to a satisfying and more comprehensive architectural model. At this point, the authors will also integrate components of a cognitive architecture that has been partially implemented in previous setups [59–61]. The goal of the full architecture model is the implementation of the kind of development and environmental coupling through consciousness which was described in the previous sections.

On the other hand, up-to-date examples of highly distributed systems will be analysed in respect to their decision making processes (e. g., IBM's Watson which is operating on very distributed resources originally). These Systems show a new quality of artificial intelligence from which can be learned from: If high-developed intelligence includes consciousness, and if these big data oriented approached do produce results with a certain intelligence, than the interesting question arises whether these systems MUST have developed a certain consciousness, as part of their intelligence. If there is any merit in that, one could observe the emergence and the 'building' of consciousness in artificial system. Just by watching and interpreting, one could avoid arguing on the basis of biases and presumptions, bringing the whole debate back into the laboratories of natural sciences.

References

1. C. Adami, What do robots dream of? Science **314** (5802), 2006, pp. 1093–1094
2. G. Buttazzo, Can a machine ever become self-aware? In: *Artificial Humans*, ed. by R. Aurich, W. Jacobsen, G. Jatho, Goethe Institut, Los Angeles, 2000, pp. 45–49

3. A. Chella, R. Manzotti, eds., *Artificial Consciousness*. Imprint Academic, 2007
4. D. Gamez, Progress in machine consciousness. Consciousness and Cognition **17** (3), 2008, pp. 887–910
5. O. Holland, ed., *Machine Consciousness*. Imprint Academic, 2003
6. D.J. Chalmers, The conscious mind: In search of a fundamental theory. Oxford University Press, U.S.A., 1996, pp. xvii, 414
7. J. Kim, *Mind in a Physical World*. MIT Press, Cambridge, Mass, 1998
8. T. Nagel, What is it like to be a bat? Philosophical Review **83**, 1974, pp. 435–50
9. S.E. Palmer, *Vision science: photons to phenomenology*. MIT Press, Cambridge, Mass., 1999
10. F. Crick, *Astonishing Hypophdthesis: The Scientific Search for the Soul*. Touchstone, New York, 1994
11. C. Koch, G. Tononi, Can machines be conscious? IEEE Spectrum **45** (6), 2008, pp. 47–51
12. A. Revonsuo, *Inner Presence: Consciousness as a Biological Phenomenon*. MIT Press, Cambridge, Mass, 2006
13. Editors of Scientific American Magazine, *The Scientific American Book of the Brain*. Lyons Press, 1999
14. A. Damasio, *Feeling of What Happens: Body and Emotion in the Making of Consciousness*. Harcourt Brace & Company, New York, 1999
15. G.M. Edelman, G. Tononi, *A Universe Of Consciousness: How Matter Becomes Imagination*. Allen Lane, London, 2000
16. W. James, *The Principles of Psychology*. Henry Holt and Company, New York, 1890
17. B. Kuipers, Drinking from the firehose of experience. Journal of Artificial Intelligence in Medicine **44** (2), 2008, pp. 155–170
18. R. Kurzweil, *How to Create a Mind: The Secret of Human Thought Revealed*. Viking Adult, New York, 2012
19. J.R. Searle, *Minds, Brains, and Science*. Harvard University Press, 1984
20. S. Yablo, Advertisement for a sketch of an outline of a prototheory of causation. In: *Causation and Counterfactuals*, ed. by J. Collins, N. Hall, L.A. Paul, MIT Press, Cambridge, Mass, 2004, pp. 119–137
21. D.C. Dennett, M. Kinsbourne, Time and the observer: The where and when of consciousness in the brain. Behavioral and Brain Sciences **15** (2), 1992, pp. 183–201
22. R. Flach, P. Haggard, The cutaneous rabbit revisited. Journal of Experimental Psychology. Human Perception and Performance **32** (3), 2006, pp. 717–732
23. S. Zeki, A. Bartels, Toward a theory of visual consciousness. Consciousness and Cognition **8** (2), 1999, pp. 225–259
24. M. Cook, Descartes and the dustbin of the mind. History of Philosophy Quarterly **13** (1), 1996, pp. 17–33
25. S. Shoemaker, Qualities and qualia: What's in the mind? Philosophy and Phenomenological Research **50**, 1990, pp. 109–131
26. M.R. Bennett, P. Hacker, *Philosophical Foundations of Neuroscience*. Blackwell, Malden, Mass, 2003
27. R. Manzotti, Moderato, Is neuroscience adequate as the forthcoming "mindscience"? Behavior and Philosophy **38**, 2010, pp. 1–29
28. J. Dewey, *Experience And Nature*. Open Court, Chicago, 1925
29. W.T. Rockwell, *Neither Brain nor Ghost*. MIT Press, Cambridge, Mass, 2005
30. M. Johnston, Appearance and reality. In: *The Manifest*, Princeton University Press, Princeton, NJ, 2002
31. C.S. Hurovitz, S. Dunn, G.W. Domhoff, H. Fiss, The dreams of blind men and women: A replication and extension of previous findings. Dreaming **9** (2), 1999, pp. 183–193
32. N.H. Kerr, G.W. Domhoff, Do the blind literally "see" in their dreams? a critique of a recent claim that they do. Dreaming **14** (4), 2004, pp. 230–233
33. A. Revonsuo, C. Salmivalli, A content analysis of bizarre elements in dreams. Dreaming **5** (3), 1995, pp. 169–187

34. E. Schwitzgebel, C. Huang, Y. Zhou, Do we dream in color? cultural variations and skepticism. Dreaming **16** (1), 2006, pp. 36–42
35. P.S. Churchland, T.J. Sejnowski, Neural representation and neural computation. Philosophical Perspectives **4**, 1990, pp. 343–382
36. F. Tong, M.S. Pratte, Decoding patterns of human brain activity. Annual Review of Psychology **63**, 2012, pp. 483–509
37. B.J. Baars, In the theatre of consciousness: Global workspace theory, a rigorous scientific theory of consciousness. Journal of Consciousness Studies **4** (4), 1997, pp. 292–309
38. M. Shanahan, *Embodiment and the inner life: Cognition and Consciousness in the Space of Possible Minds*. Oxford University Press, USA, 2010
39. G. Tononi, An information integration theory of consciousness. BMC Neuroscience **5** (42), 2004, pp. 1–22
40. D.J. Chalmers, Facing up to the problem of consciousness. Journal of Consciousness Studies **2** (3), 1995, pp. 200–219
41. R. Pfeifer, C. Scheier, *Understanding Intelligence*. MIT Press, Cambridge, MA, USA, 1999
42. R. Jackendoff, *Consciousness and the Computational Mind*. MIT Press, Cambridge, Mass, 1987
43. R. Manzotti, Is consciousness just conscious behavior? International Journal of Machine Consciousness **3** (2), 2011, pp. 353–360
44. R. Manzotti, F. Mutti, G. Gini, S.Y. Lee, Cognitive integration through goal-generation in a robotic setup. In: *Biologically Inspired Cognitive Architectures 2012*, ed. by A. Chella, R. Pirrone, R. Sorbello, K.R. Jóhannsdóttir, no. 196 In: Advances in Intelligent Systems and Computing, Springer, 2013, pp. 225–231
45. R. Manzotti, V. Tagliasco, From behaviour-based robots to motivation-based robots. Robotics and Autonomous Systems **51** (2), 2005, pp. 175–190
46. D. George, J. Hawkins, Towards a mathematical theory of cortical micro-circuits. PLoS Comput Biol **5** (10), 2009, p. e1000532
47. J. Hawkins, S. Blakeslee, *On Intelligence*. Times Books, New York, 2004
48. D. George, How the brain might work: A hierarchical and temporal model for learning and recognition. phdphdthesis, Stanford, CA, USA, 2008
49. J. Sharma, A. Angelucci, M. Sur, Induction of visual orientation modules in auditory cortex. Nature **404** (6780), 2000, pp. 841–847
50. J. Sharma, V. Dragoi, J.B. Tenenbaum, E.K. Miller, M. Sur, V1 neurons signal acquisition of an internal representation of stimulus location. Science **300** (5626), 2003, pp. 1758–1763
51. M. Sur, P.E. Garraghty, A.W. Roe, Experimentally induced visual projections into auditory thalamus and cortex. Science **242** (4884), 1988, pp. 1437–1441
52. E.R. Kandel, J.H. Schwartz, T.M. Jessell, S.A. Siegelbaum, A.J. Hudspeth, *Principles of Neural Science, Fifth Edition*, 5th edn. McGraw-Hill Medical, New York, 2012
53. R. Sun, The CLARION cognitive architecture: Extending cognitive modeling to social simulation. In: *Cognition and Multi-Agent Interaction*, ed. by R. Sun, Cambridge University Press, 2005, pp. 79–100
54. B. Goertzel, D. Duong, OpenCog NS: A deeply-interactive hybrid neural-symbolic cognitive architecture designed for global/local memory synergy. In: *AAAI Fall Symposium: Biologically Inspired Cognitive Architectures'09*, vol. FS-09-01. 2009, vol. FS-09-01
55. R. Sanz, C. Hernández, Towards architectural foundations for cognitive self-aware systems. In: *Biologically Inspired Cognitive Architectures 2012*, ed. by A. Chella, R. Pirrone, R. Sorbello, K.R. Jóhannsdóttir, no. 196 In: Advances in Intelligent Systems and Computing, Springer, 2013, pp. 53–53
56. J. Bongard, V. Zykov, H. Lipson, Resilient machines through continuous self-modeling. Science **314** (5802), 2006, pp. 1118–1121
57. R. Pfeifer, J. Bongard, *How the Body Shapes the Way We Think: A New View of Intelligence*. Bradford Books, New York, 2006
58. R. Sanz, I. López, M. Rodríguez, C. Hernández, Principles for consciousness in integrated cognitive control. Neural Networks **20** (9), 2007, pp. 938–946

59. R. Manzotti, L. Papi, S.Y. Lee, Does radical externalism suggest how to implement machine consciousness? In: *Biologically Inspired Cognitive Architectures 2011*, ed. by A. Samsonovich, K. Jóhannsdóttir, IOS Press, Amsterdam, pp. 232–240

60. R. Manzotti, A process-based architecture for an artificial conscious being. In: *Process Theories: Crossdisciplinary Studies in Dynamic Categories*, ed. by J. Seibt, Springer Netherlands, 2003, pp. 285–312

61. R. Manzotti, Machine free will: Is free will a necessary ingredient of machine consciousness? In: *From Brains to Systems*, ed. by C. Hernández, R. Sanz, J. Gómez-Ramirez, L.S. Smith, A. Hussain, A. Chella, I. Aleksander, no. 718 In: Advances in Experimental Medicine and Biology, Springer New York, 2011, pp. 181–191

TIDAQL: A Query Language Enabling On-line Analytical Processing of Time Interval Data

Philipp Meisen, Diane Keng, Tobias Meisen, Marco Recchioni and Sabina Jeschke

Abstract Nowadays, time interval data is ubiquitous. The requirement of analyzing such data using known techniques like on-line analytical processing arises more and more frequently. Nevertheless, the usage of approved multidimensional models and established systems is not sufficient, because of modeling, querying and processing limitations. Even though recent research and requests from various types of industry indicate that the handling and analyzing of time interval data is an important task, a definition of a query language to enable on-line analytical processing and a suitable implementation are, to the best of our knowledge, neither introduced nor realized. In this paper, we present a query language based on requirements stated by business analysts from different domains that enables the analysis of time interval data in an on-line analytical manner. In addition, we introduce our query processing, established using a bitmap-based implementation. Finally, we present a performance analysis and discuss the language, the processing as well as the results critically.

Keywords Time Interval Data · Query Language · On-line Analytical Processing · Distributed Query Processing

1 Introduction

Nowadays, time interval data is recorded, collected and generated in various situations and different areas. Some examples are the resource utilization in production

P. Meisen (✉) · T. Meisen · S. Jeschke
IMA/ZLW &IfU, RWTH Aachen University,
Dennewartstr. 27, 52068 Aachen, Germany
e-mail: philipp.meisen@ima-zlw-ifu.rwth-aachen.de

D. Keng
School of Engineering, Santa Clara University, Santa Clara, USA

M. Recchioni
Airport Devision, Inform GmbH Aachen, Aachen, Germany
e-mail: marco.recchioni@inform-software.de

Originally published in "Proceedings of the 17th International Conference on Enterprise Information Systems (ICEIS 2015)",
© ScitePress 2015. Reprint by Springer International Publishing Switzerland 2016,
DOI 10.1007/978-3-319-42620-4_42

environments, deployment of personnel in service sectors, or courses of diseases in healthcare. Thereby, time interval data is used to represent observations, utilizations or measures over a period of time. Put in simple terms, time interval data is defined by two time values (i. e. start and end), as well as descriptive values associated to the interval: like labels, numbers, or more complex data structures. Figure 1 illustrates a sample database of five records.

For several years, business intelligence and analytical tools have been used by managers and business analysts, inter alia, for data-driven decision support on a tactical and strategic level. An important technology used within this field, is on-line analytical processing (OLAP). OLAP enables the user to interact with the stored data by querying for answers. This is achieved by selecting dimensions, applying different operations to selections (e. g. roll-up, drill-down, or drill-across), or comparing results. The heart of every OLAP system is a multidimensional data model (MDM), which defines the different dimensions, hierarchies, levels, and members [1].

The need of handling and analyzing time interval data using established, reliable, and proven technologies like OLAP is desirable in this respect and an essential acceptance factor. Nevertheless, the MDM needed to model time interval data has to be based on many-to-many relationships which have been shown to lead to summarizability problems. Several solutions solving these problems on different modeling levels have been introduced over the last years, leading to increased integration effort, enormous storage needs, almost always inacceptable query performances, memory issues, and often complex multidimensional expressions [2, 3]. Additionally, these solutions are, considering real-world scenarios, only applicable to many-to-many relationships having a small cardinality which is mostly not the case when dealing with time interval data. As a result, the usage of MDM and available OLAP systems is not sufficient, even though the operations (e. g. roll-up, drill-down, slice, or dice) available through such systems are desired.

Enabling such OLAP like operations in the context of time interval data, requires the provision of extended filtering and grouping capabilities. The former is achieved by matching descriptive values against known filter criteria logically connected using operators like *and*, *or*, or *not*, as well as a support of temporal relations like *starts-with*, *during*, *overlapping*, or *within* [4]. The latter is applied by known aggregation operators like *max*, *min*, *sum*, or *count*, as well as temporal aggregation operators like *count started* or *count finished* [5].

key	resources	type	location	start	end
2285954	3	cleaning	POS F6	2015/01/01 16:21	2015/01/01 17:13
2285965	5	maintenance	POS F5	2015/01/01 16:25	2015/01/01 17:10
2285971	1	maintenance	POS F5	2015/01/01 17:02	2015/01/01 17:17
2285972	3	room service	POS F5	2015/01/01 16:42	2015/01/01 16:55
2285990	4	miscellaneous	POS F6	2015/01/01 16:20	2015/01/01 17:05

Figure 1 A sample time interval database with intervals defined by [start, end), an id, and three descriptive values

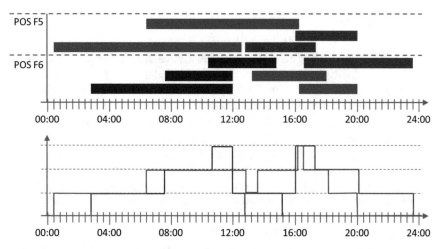

Figure 2 On top the time interval data (10 records) shown in a Gantt-Chart, on the bottom the aggregated time-series

The application of the *count* aggregation operator for time interval data is exemplified in Figure 2. The color code identifies the different types of a time interval (e. g. cleaning, maintenance, room service, miscellaneous). Furthermore, the swim-lanes show the location. The figure illustrates the count of intervals for each type over one day across all locations (e. g. POS F5 and POS F6) using a granularity of minutes (i. e. 1,440 aggregations are calculated).

In this paper, we present a query language allowing to analyze time interval data in an OLAP manner. Our query language includes a data definition (DDL), a data control (DCL), and a data manipulation language (DML). The former is based on the time interval data model introduced by Meisen et al. [6], whereby the latter supports the two-step aggregation technique mentioned in Meisen et al. [5]. Furthermore, we outline our query processing which is based on a bitmap-based implementation and supports distributed computing.

This paper is organized as follows: In Section 2, we discuss related work done in the field of time interval data, in particular this section provides a concise overview of research dealing with the analyses of time interval data. We provide an overview of time interval models, discuss related work done in the field of OLAP, and present query languages. In Section 3, we introduce our query language and processing. The section presents among other things how a model is defined and loaded, how temporal operators are applied, how the two-step aggregation is supported, how groups are defined, and how filters are used. We introduce implementation issues and empirically evaluate the performance regarding the query processing in Section 4. We conclude with a summary and directions for future work in Section 5.

2 Related Work

When defining a query language, it is important to have an underlying model, defining
the foundation for the language (e. g. the relational model for SQL, different interval-
based models for e. g. IXSQL or TSQL2, the multidimensional model for MDX, or
the graph model for Cypher). Over the last years several models have been introduced
in the field of time intervals, e. g. for temporal databases [7], sequential pattern mining
[8, 9], association rule mining [10], or matching [11].

Chen et al. [12] introduced the problem of mining time interval sequential patterns.
The defined model is based on events used to derive time intervals, whereby a time
interval is determined by the time between two successive time-points of events. The
definition is based on the sequential pattern mining problem introduced by Agrawal
and Srikant [13]. The model does not include any dimensional definitions, nor does
it address the labeling of time intervals with descriptive values.

Papapetrou et al. [14] presented a solution for the problem of "discovering frequent
arrangements of temporal intervals". An e-sequence is an ordered set of events. An
event is defined by a start value, an end value and a label. Additionally, an e-sequence
database is defined as a set of e sequences. The definition of an event given by
Papapetrou et al. is close to the underlying definition within this paper (cf. Figure 1).
Nevertheless, facts, descriptive values, and dimensions are not considered.

Mörchen [15] introduced the TSKR model defining tones, chords, and phrases
for time intervals. Roughly speaking, the tones represent the duration of intervals,
the chords the temporal coincidence of tones, and the phrases represent the partial
order of chords. The main purpose of the model presented by Mörchen is to overcome
limitations of Allen's [4] temporal model considering robustness and ambiguousness
when performing sequential pattern mining. The model neither defines dimensions,
considers multiple labels, nor recognizes facts.

Summarized, models presented in the field of sequential pattern mining, association rule mining or matching do generally not define dimensions and are focused on
generalized interval data, or support only non-labelled data. Thus, these models are
not suitable considering OLAP of time interval data, but are a guidance to the right
direction.

Within the research community of temporal databases different interval-based
models have been defined [7]. The provided definitions can be categorized in weak
and strong models. A weak model is one, in which the intervals are used to group
time-points, whereas the intervals of the latter carry semantic meaning. Thus, a
weak interval-based model is not of further interest from an analytical point of view,
because it can be easily transformed into a point-based model. Nevertheless, a strong
model and the involved meaning of the different operators – especially aggregation
operators – are of high interest from an analytical view. Strong interval-based models
presented in the field of temporal databases lack to define dimensions, but present
important preliminary work.

In the field of OLAP, several systems capable of analyzing sequences of data
have been introduced over the last years. Chui et al. [16] introduced S-OLAP for

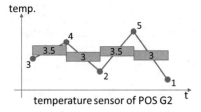

Figure 3 Illustration of the model introduced by Koncilia et al. [19]. The intervals (rectangles) are created for each two consecutive events (dots). The facts are calculated using the average function as the *compute value function*

analyzing sequence data. Liu and Rundensteiner [17] analyzed event sequences using hierarchical patterns, enabling OLAP on data streams of time point events. Bebel et al. [18] presented an OLAP like system enabling time point-based sequential data to be analyzed. Nevertheless, the system neither support time intervals, nor temporal operators.

Recently, Koncilia et al. [19] presented I OLAP, an OLAP system to analyze interval data. They claim to be the first proposing a model for processing interval data. The definition is based on the interval definition of Chen et al. [12] which defines the intervals as the gap between sequential events. However, Koncilia et al. assume that the intervals of a specific event-type (e. g. temperature) for a set of specific descriptive values (e. g. POS G2) are non-overlapping and consecutive. Considering the sample data shown in Figure 1, the assumption of non-overlapping intervals is not valid in general (cf. record 2,285,965 and 2,285,971). Figure 3 illustrates the model of Koncilia et al. showing five temperature events for POS G2 and the intervals determined for the events. Koncillia et al. mention the support of dimensions, hierarchies, levels, and members, but lack to specify what types of hierarchies are supported and how e. g. non-strict relations are handled.

Also recently, Meisen et al. [6] introduced the TIDAMODEL "enabling the usage of time interval data for data-driven decision support". The presented model is defined by a 5-tuple (P, Σ, τ, M, δ) in which P denotes the time interval database, Σ the set of descriptors, τ the time axis, M the set of measures, and δ the set of dimensions. The time interval database P contains the raw time interval data records and a schema definition of the contained data. The schema associates each field of the record (which might contain complex data structures) to one of the following categories: temporal, descriptive, or bulk. Each descriptor of the set Σ is defined by its values (more specific its value type), a mapping- and a fact-function. The mapping-function is used to map the descriptive values of the raw record to one or multiple descriptor values. The mapping to multiple descriptor values allows the definition of non-strict fact-dimension relationships. Additionally, the model defines the time axis to be finite and discrete, i. e. it has a start, an end, and a specified granularity (e. g. minutes). The set of dimensions δ can contain a time dimension (using a rooted plane tree for the definition of each hierarchy) and a dimension for each descriptor (using a directed acyclic graph for a hierarchy's definition). Figure 4 illustrates the modeled sample

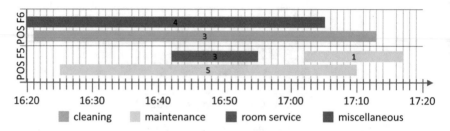

Figure 4 Data of the sample database shown in Figure 1 modeled using the TIDAMODEL [6]

database of Figure 1 using the TIDAMODEL. The figure shows the five intervals, as well as the values of the descriptors location (cf. swim-lane) and type (cf. legend). Dimensions are not shown. The used mapping function for all descriptors is the identity function. The used granularity for the time dimension is minutes.

Another important aspect when dealing with time interval data in the context of OLAP, is the aggregation of data and the provision of temporal aggregation operators. Kline and Snodgrass [20] introduced temporal aggregates, for which several enhanced algorithms were presented over the past years. Nevertheless, the solutions are focused on one specific aggregation operator (e. g. SUM), do not support multiple filter criteria, or do not consider data gaps. Koncilia et al. [19] address shortly how aggregations are performed using the introduced *compute value functions* and *fact creating functions*. Temporal operators are neither defined nor mentioned. Koncilia et al. point out that some queries need special attention when aggregating the values along time, but a more precise problem statement is not given. Meisen et al. [5] introduce a two-step aggregation technique for time interval data. The first one aggregates the facts along the intervals of a time granule and the second one aggregates the values of the first step depending on the selected hierarchy level of the time dimension. Figure 5 illustrates the two-step aggregation technique. In the illustration, the technique is used to determine the needed resources within the interval [16:30, 16:34]. Within the first step, the sum of the resources for each granule is determined and within the second step the maximum of the determined values is calculated, i. e. 14. Additionally, they introduce temporal aggregation operators like *started* or *finished count*.

The definition of a query language based on a model and operators (i. e. like aggregations), is common practice. Regarding time-series, multiple query languages and enhancements of those have been introduced [21]. In the field of temporal databases time interval-based query languages like IXSQL, TSQL2, or ATSQL have been defined [7] and within the analytical field, MDX [22] is a widely used language to query MDMs. Considering models dealing with time interval data in the context of analytics, Koncilia et al. [19] published the only work the authors are aware of that mentions a query language. Nevertheless, the query language is neither formally defined nor further introduced.

Summarized, it can be stated that recent research and requests from industry indicate that the handling of time interval data in an analytical context is an important

2. step: max →14
16:32: 10   [16:30, 16:34]
16:31: 14↑ 8 :16:33
16:30: 14↑ | | 8 :16:34

start	end	res.	loc.
16:21	16:31	4	POS1
16:21	16:38	3	POS1
16:22	16:32	2	POS2
16:25	16:38	5	POS2

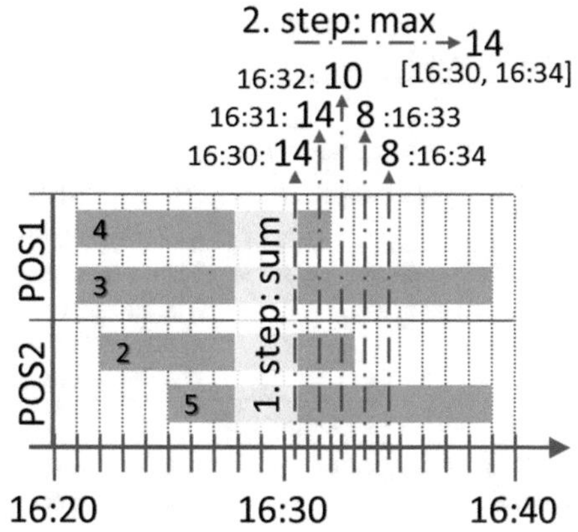

16:20   16:30   16:40

Figure 5 Two-step aggregation technique presented by Meisen et al. [5]

task. Thus, a query language is required capable of covering the arising requirements. Koncilia et al. [19] and Meisen et al. [5, 6] introduced two different models useful for OLAP of time interval data. Different temporal aggregation operators, as well as standard aggregation operators, are also presented by Meisen [5]. Nevertheless, a definition of a query language useful for OLAP and an implementation of the processing are, to the best of our knowledge, not formally introduced.

3 The Tida Query Language

In this section, we introduce our time interval data analysis query language (TIDAQL). The language was designed for a specific purpose; to query time interval data from an analytical point of view. The language is based on aspects of the previously discussed TIDAMODEL. Nevertheless, the language should be applicable to any time interval database system which is capable of analyzing time interval data. Nevertheless, some adaptions might be necessary or some features might not be supported by any system.

3.1 Requirements

The requirements concerning the query language and its processing were specified during several workshops with over 70 international business analysts from different domains (i.e. aviation industry, logistics providers, service providers, as well as language and gesture research). We aligned the results of the workshop with an extended literature research. Table 1 summarizes selected results.

Table 1 Summary of the requirements concerning the time interval analysis query language (selected results)

Requirement	Description
Data Control Language (DCL)	
[DCL1]: Authorization aspects	It is expected that the language encompasses authorization features, e. g. user deletion, role creation, granting and revoking permissions
[DCL2]: Permissions grantable on global and model level	Permissions must be grantable on a model and a global level. It is expected that the user can have the permission to add data to one model but not to another. For simplicity, it should be possible to grant or revoke several permissions at once
Data Definition Language (DDL)	
[DDL1]: Loading and unloading	The language has to offer a construct to load new and unload models. The newly loaded model has to be available without any restart of the system. An unloaded model has to be unavailable after the query is processed. However, queries currently in process must still be executed
[DDL2]: Non-onto, non-covering, non-strict hierarchies	Each descriptor dimension must support hierarchies which might be non-onto, non-covering, and/or non-strict [23]
[DDL3]: Raster levels	A raster level is a level of the time dimension. For example: the *5-minute raster*-level defines members like [00:00, 00:05) …[23:55, 00:00). Several raster levels can form a hierarchy (e. g. 5-min → 30-min → 60-min → half-day → day)
Data Manipulation Language (DML)	
[DML1]: Raw data records	The language must provide a construct to select the raw time interval data records
[DML2]: Time series by time-windows	The language must support the specification of a time-window for which time-series of different measures can be retrieved
[DML3]: Temporal operators	It must be possible to use temporal operators for filtering as e. g. defined by Allen [4]. Depending on the type of selection (i. e. raw records or time-series) the available temporal operators may differ
[DML4]: The two-step aggregation technique	Meisen et al. [5] present a two-step aggregation technique which has to be supported by the language. Both aggregation operators (see Figure 5) must be specified by a query selecting time-series, no predefined measure should be necessary
[DML5]: Complete time series	A time-series is selected by specifying a time-window (e. g. [01.01.2015, 02.01.2015) and a level (e. g. minutes). The resulting time-series must contain a value for each member of the selected level, even if no time interval covers the specified member. The value might be N/A or null to indicate missing information
[DML6]: Insert, update and delete	The language must offer constructs to insert, update and delete time interval data records
[DML7]: Open, half-open, or closed intervals	The system should be capable of interpreting intervals defined as open, e. g. (0, 5), closed, e. g. [0, 5], or half-opened, e. g. (0, 5]
[DML8]: Meta-information	It is desired that the language supports a construct to receive meta-information from the system, e. g. actual version, available users, or loaded models
[DML9]: Bulk load	It is desired, that the language provides a construct to enable a type of bulk load, i. e. increased insert performance

3.2   Data Control Language

The definition of the DCL is straight forward to the DCL known from other query languages e. g. SQL. As defined by requirement [DCL1], the language must encompass authorization features. Hence, the language contains commands like **ADD, DROP, MODIFY, GRANT, REVOKE, ASSIGN** and **REMOVE**. In our implementation, the execution of a DCL command always issues a direct commit, i. e. a roll back is not supported. Figure 6 shows the syntax diagram of the commands. Because of simplicity, a value is not further specified and might be a permission, a username, a password, or a role.

To fulfill the [DCL2] requirement, we define a permission that consists of a scope-prefix and the permission itself. We define two permission-scopes GLOBAL and MODEL. Thus, a permission of the GLOBAL scope is defined by

```
GLOBAL.<permission>
```

(e. g. GLOBAL.manageUser). Instead, a permission of the MODEL scope is defined by

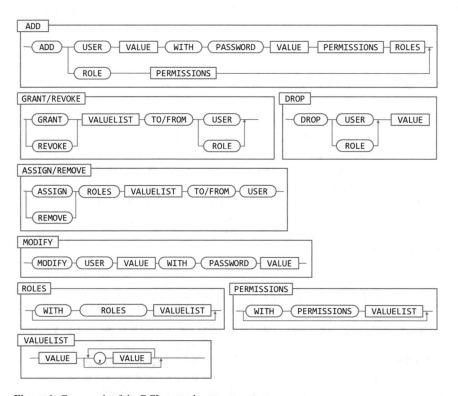

Figure 6 Commands of the DCL query language

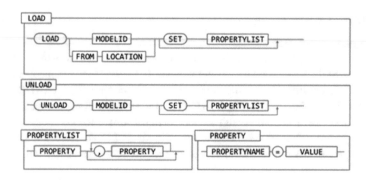

Figure 7 Commands of the DDL query language

MODEL.<model>.<permission>

(e. g. MODEL.myModel.query).

For query processing, we use the Apache Shiro authentication framework (http://
shiro.apache.org/). Shiro offers annotation driven access control. Thus, the permission to e. g. execute a DML query is performed by annotating the processing query
method.

3.3 Data Definition Language

The DDL is used to define, add, or remove the models known by the system. [DDL1]
requires a command within the DDL which enables the user to load or unload a
model. The syntax diagram of the **LOAD** and **UNLOAD** command is shown in
Figure 7. A model can be loaded by using a model identifier already known to the
system (e. g. if the model was unloaded), or by specifying a location from which the
system can retrieve a model definition to be loaded. Additionally, properties can be
defined (e. g. the *autoload* property can be set, to automatically load a model when
the system is started). In the following subsection, we present an XML used to define
a TIDAMODEL.

3.3.1 The XML TIDAMODEL Definition

As mentioned in Section 2, the TIDAMODEL is defined by a 5-tuple (P, Σ, τ, M, δ).
The time interval database P contains the raw record inserted using the API or the
INSERT command introduced later in Section 3.4.1. From a modelling perspective
it is important for the system to retrieve the descriptive and temporal values from the
raw record. Thus, it is essential to define the descriptors Σ and the time axis τ within

the XML definition. Below, an excerpt of an XML file defining the descriptors of our sample database shown in Figure 1 is presented:

```
<model id="myModel">
 <descriptors>
  <string id="LOC" name="location" />
  <string id="TYPE" name="type" />
  <int id="RES" null="true" />
 </descriptors>
</model>
```

The excerpts shows that a descriptor is defined by a tag specifying the type (i.e. the descriptor implementation to be used), an id-attribute, and an optional name-attribute. Additionally, it is possible to define if the descriptor allows *null* values (default) or not. To support more complex data structures (and one's own mapping functions), it is possible to specify one's own descriptor-implementations:

```
<descriptors>
 <ownImpl:list id="D4" />
</descriptors>
```

Our implementation scans the class-path automatically, looking for descriptor implementations. An added implementation must provide an XSLT file, placed into the same package and named as the concrete implementation of the descriptor-class. The XSLT file is used to create the instance of the own implementation using a Spring Bean configuration (http://spring.io/).

```
<!- File: my/own/desc/List.xslt ->
<xsl:template match="ownImpl:list">
 <xsl:call-template name="beanDesc">
  <xsl:with-param name="class">
   my.own.desc.List
  </xsl:with-param>
 </xsl:call-template>
</xsl:template>
```

The time axis of the TIDAMODEL is defined by:

```
<model id="myModel">
 <time>
  <timeline start="20.01.1981"
            end="20.01.2061"
            granularity="MINUTE" />
 </time>
</model>
```

The time axis may also be defined using integers, i. e. [0, 1000]. Our implementation includes two default mappers applicable to map different types of temporal
raw record value to a defined time axis. Nevertheless, sometimes it is necessary to
use different time-mappers (e. g. if the raw data contains proprietary temporal values) which can be achieved using the same mechanism as described previously for
descriptors.

Due to the explicit time semantics, the measures M defined within the
TIDAMODEL are different than the ones typically known from an OLAP definition. The model defines three categories for measures, i. e. *implicit time measures*,
descriptor bound measures, and *complex measures*. The categories determine when
which data is provided during the calculation process of the measures. Our implementation offers several aggregation operators useful to specify a measure, i. e. *count*,
average, min, max, sum, mean, median, or *mode*. In addition, we implemented two
temporal aggregation operators *started count* and *finished count*, as suggested by
Meisen et al. [5]. We introduce the definition and usage of measures in Section 3.4.2.

The TIDAMODEL also defines the set of dimensions δ. The definition differs
between descriptor dimensions and a time dimension, whereby every dimension
consists of hierarchies, levels, and members. It should be mentioned that, from a
modelling point of view, each descriptor dimension fulfills the requirements formalized in [DDL2] and that the time dimension supports raster-levels as requested in
[DDL3]. The definition of a dimension for a specific descriptor or the time dimension
can be placed within the XML definition of a model using:

```
<model id="myModel">
 <dimensions>
  <dimension id="DIMLOC" descId="LOC">
   <hierarchy id="LOC">
    <level id="HOTEL">
     <member id="DREAM" rollUp="*" />
     <member id="STAR" rollUp="*" />
     <member id="ADV" reg="TENT"
             rollUp="*" />
    </level>
    <level id="ROOMS">
     <member id="POSF" reg="POS F"
             rollUp="DREAM" />
     <member id="POSG" reg="POS G"
             rollUp="DREAM" />
    </level>
    <level id="STARROOMS">
     <member id="POSA" reg="POS A"
             rollUp="STAR" />
    </level>
   </hierarchy>
  </dimension>
 </dimensions>
</model>
```

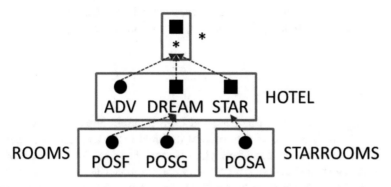

Figure 8 Illustration of the dimension created with our web-based dimension-modeler as defined by the XML

Figure 8 illustrates the descriptor dimension defined by the previously shown XML excerpt. The circled nodes are leaves which are associated with de-scriptor values known by the model (using regular expressions). Additionally, it is possible to add dimensions for analytical processes to an already defined model, i.e. to use it only for a specific session or query. The used mechanism to achieve that is similar to the loading of a model and will not further be introduced.

The definition of a time dimension is straight forward to the one of a descriptor dimension. Nevertheless, we added some features in order to ease the definition. Thus, it is possible to define a hierarchy by using pre-defined levels (e.g. templates like 5 min raster, day, or year) and by defining the level to roll up to, regarding the hierarchy. The following XML excerpt exemplifies the definition:

```
<model id="myModel">
 <dimensions>
  <timedimension id="DIMTIME">
   <hierarchy id="TIME5TOYEAR">
    <level id="YEAR" template="YEAR"
           rollUp="*" />
    <level id="DAY" template="DAY"
           rollUp="YEAR" />
    <level id="60R" template="60RASTER"
rollUp="DAY" />
    <level id="5R" template="5RASTER"
           rollUp="60R" />
    <level id="LG" template="LOWGRAN"
           rollUp="5R" />
   </hierarchy>
  </timedimension>
 </dimensions>
</model>
```

A defined model is published to the server using the **LOAD** command. The following subsection introduces the command, focusing on the loading of a model from a specified location.

3.3.2 Processing the LOAD Command

The loading of a model can be triggered from different applications, drivers, or platforms. Thus, it is necessary to support different loaders to resolve a specified location. In the following, some examples illustrate the issue. When firing a **LOAD** query from a web-application, it is necessary that the model definition was uploaded to the server, prior to executing the query. While running on an application server, it might be required to load the model from a database instead of loading it from the file-system. Thus, we added a resource-loader which can be specified for each context of a query. Within a servlet, the loader resolves the specified location against the upload-directory, whereby our JDBC driver implementation is capable of sending a client's file to the server using the data stream of the active connection. After retrieving and validating the resource, the implementation uses a model-handler to bind and instantiate the defined model. As already mentioned, the bitmap-based implementation presented by Meisen et al. [5] is used. The implementation instantiates several indexes and bitmaps for the defined model. After the instantiation, the model is marked to be up and running by the model-handler and accepts DML queries. Figure 9 exemplifies the initialized bitmap-based indexes filled with the data from the database of Figure 1.

3.4 Data Manipulation Language

Considering the requirements, it can be stated that the DML must contain commands to **INSERT**, **UPDATE**, and **DELETE** records. In addition, it is necessary to provide **SELECT** commands to retrieve the time interval data records, as well as results retrieved from aggregation (i. e. time-series). Furthermore, a **GET** command to retrieve meta-information of the system is needed.

3.4.1 INSERT, UPDATE, and DELETE

Figure 10 illustrated the three commands **INSERT**, **UPDATE**, and **DELETE** using syntax diagrams which fulfill the requirement [DML6]. The **INSERT** command adds one or several time interval data records to the system. First, it parses the structure of the data to be inserted. The query-parser validates the correctness of the structure, i. e. the structure must contain exactly one field marked as *start* and exactly one field marked as *end* even though the syntax diagram suggest differently. Additionally, the parser verifies if a descriptor (referred by its id) really exists within the model.

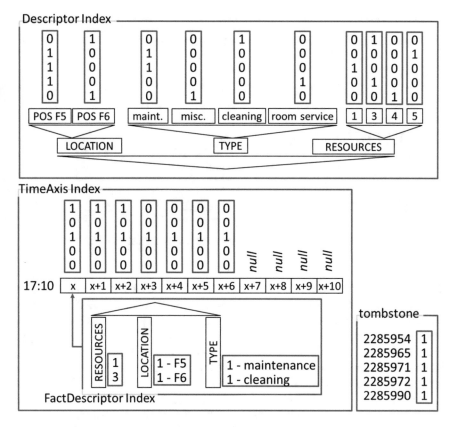

Figure 9 Example of a loaded model [5] filled with the data shown in Figure 1

Finally, it reads the values and invokes the processor by passing the structure, as well as the values. The processor iterates over the defined values, validates those against the defined structure, uses the mapping functions of the descriptors to receive the descriptor values, and calls the mapping function of the time-axis. The result is a so-called *processed record* which is used to update the indexes. The persistence layer of the implementation ensures that the raw record and the indexes get persisted. Finally, the tombstone bitmap is updated which ensures that the data is available within the system.

A deletion is performed by setting the tombstone bitmap for the specified id to 0. This indicates that the data of the record is not valid and thus the data will not be considered by any query processors anymore. The internally scheduled clean-up process removes the deleted records and releases the space.

An update is performed by deleting the record with the specified identifier and inserting the record as described above.

To support bulk load, as desired by [DML9], an additional statement is introduced. The statement **SET BULK TRUE** is used to enable the bulk load, whereby **SET**

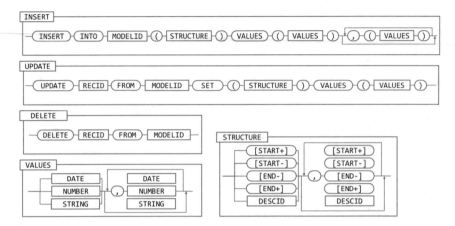

Figure 10 Syntax diagrams of the commands **INSERT**, **UPDATE** and **DELETE**

BULK FALSE stops the bulk loading process. When enabling the bulk load, the system waits until all currently running **INSERT**, **UPDATE**, or **DELETE** queries of other sessions are performed. New queries of that type are rejected across all sessions during the waiting and processing phase. When all queries are handled, the system responds to the bulk-enabling query and expects an insert-like statement, whereby the system directly starts to parse the incoming data stream. As soon as the structure is known, all incoming values are inserted. The indexes are only updated in memory. If and only if the memory capacity reaches a specified threshold, the persistence-layer is triggered. In this case, the current data in memory is flushed and persisted using the configured persistence-layer (e. g. using the file-system, a relational database, or any other NoSQL database). The memory is also flushed and persisted whenever a bulk load is finished.

3.4.2   SELECT Raw Records and Time-Series

The **SELECT** command is addressed by the requirements [DML1], [DML2], [DML3], [DML4], [DML5], and [DML7]. Figure 11 illustrates the select statements to select records and time-series. Because of space limitations, we removed more detailed syntax diagrams for the **LOGICAL** and **GROUP EXPRESSION**. The non-terminal **MEASURES** is specified later in this subsection when introducing the **SELECT TIMESERIES** in detail.

As illustrated, the intervals can be defined as open, half-open or closed (cf. [DML7]). The processing of the intervals is possible, thanks to the discrete time-axis used by the model. Using a discrete time-axis with a specific granularity makes it easy to determine the previous or following granule. Thus, every half-open or open interval can be transformed into a closed interval using the previous or following

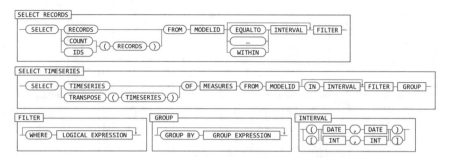

Figure 11 Syntax diagrams of the **SELECT RECORDS** and **SELECT TIMESERIES** commands

granule. Hence, the result of the parsing always contains a closed interval which is used during further query processing.

As illustrated in Figure 11, the **SELECT RECORDS** statement allows to retrieve records satisfying a logical expression based on descriptor values (e. g. **LOC = "POS F5" OR (TYPE = "cleaning" AND DIMLOC.LOC.HOTEL = "DREAM"))** and/or fulfilling a temporal relation (cf. [DML3]). Our query language supports ten different temporal relations following Allen [4]: **EQUALTO, BEFORE, AFTER, MEETING, DURING, CONTAINING, STARTINGWITH, FINISHINGWITH, OVERLAPPING**, and **WITHIN**. The interested reader may notice that Allen introduced thirteen temporal relationships. We removed some inverse relationships (i. e. inverse of meet, overlaps, starts, and finishes). When using a temporal relation-ship within a query, the user is capable of defining one of the intervals used for comparison. Thus, the removed inverse relationships are not needed, instead the user just modifies the self-defined interval. In addition, we added the **WITHIN** relationship which is a combination of several relationships and allows an easy selection of all records within the user-defined interval (i. e. at least one time-granule is contained within the user-defined interval).

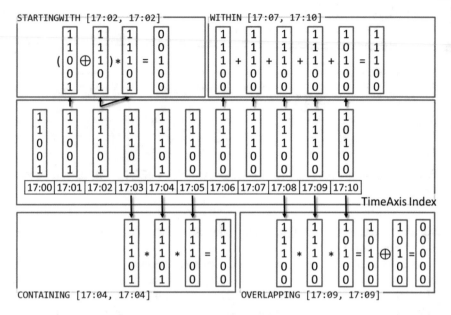

Figure 12 Examples of the processing of temporal relationships using bitmaps (and the sample database of Figure 1)

When processing a **SELECT RECORD** query, the processor initially evaluates the filter expression and retrieves a single bitmap specifying all records fulfilling the filter's logic [5]. In a second phase, the implementation determines a bitmap of records satisfying the specified temporal relationship. The two bitmaps are combined using the and-operator to retrieve the resulting records. Depending on the requested information (i. e. count, identifiers, or raw records (cf. [DML1])), the implementation creates the response using bitmap-based operations (i. e. count and identifiers) or retrieving the raw records from the persistence layer. Figure 12 depicts the evaluation of selected temporal relationships using bitmaps and the database shown in Figure 1.

The **SELECT TIMESERIES** statement specifies a logical expression equal to the one exemplified in the **SELECT RECORDS** statement. In addition, the statement specifies a **GROUP EXPRESSION** which defines the groups to create the time-series for (e. g. **GROUP BY DIMLOC.LOC.ROOMS**). Furthermore, the measures to be calculated for the time-series and the time-window (cf. [DML2]) are specified. It is also possible to specify several comma-separated measures. Figure 13 shows the syntax used to specify measures (cf. **MEASURES** in Figure 11).

A simple (considering the measures) example of a **SELECT TIMESERIES** query is as follows:

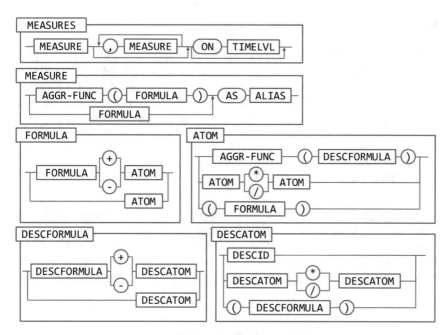

Figure 13 Syntax diagrams of the MEASURES definition

```
SELECT TRANSPOSE(TIMESERIES)
OF MAX(SUM(RESOURCES)) AS "needed Res"
ON DIMTIME.TIME5TOYEAR.5RASTER
FROM myModel
IN [01.01.2015, 02.01.2015)
WHERE DIMLOC.LOC.HOTEL="DREAM"
GROUP BY TYPE
```

As required by [DML4], a measure can be defined using the two-step aggregation technique. The first aggregation (in the example SUM) is specified for a specific descriptor and the second optional aggregation function (in the example MAX) aggregates the values across the stated level of the time-dimension.

When processing the query, the system retrieves the bitmaps for the filtering and the grouping conditions. The system iterates over the bitmaps of the specified groups and the bitmaps of the granules of the selected time-window. For each iteration, the implementation combines the filter-bitmap, group-bitmap, and the time-granule-bitmap and applies the first aggregation function. The second aggregation function is applied whenever all values of a member of the specified time-level are determined by the first step (cf. Figure 6). This processing technique ensures that for each time-granule a value is calculated, even if no interval covers the granule (cf. [DML5]).

3.4.3 GET Meta-Information

[DML8] demands the existence of a command which enables the user to retrieve meta-information, like the version of the system. This requirement is fulfilled by adding a **GET** command to the query language. A statement like **GET VERSION**, **GET USERS**, or **GET MODELS** enables the user to retrieve information provided from the system. Filtering is currently not required and thus, not supported.

4 Implementation Issues

This section introduces selected implementation aspects of the language and its query processing. First, we introduce processing implementations for the most frequently used query-type **SELECT TIMESERIES** and show performance results for the different algorithms. In addition, we present considerations of analysts using the language to analyze time interval data and address possible enhancements.

4.1 SELECT TIMESERIES Processing

In Section 3, we outlined the query processing based on the TIDAMODEL and its bitmap-based implementation (cf. Sections 3.3.2 and 3.4.2). For a detailed description of the bitmap-based implementation we refer to Meisen et al. [5]. In this section, we introduce three additional algorithms which are capable to process the most frequently used **SELECT TIMESERIES** queries, introduced in Section 3.4.2.

Prior to explaining the algorithms, it should be stated, that we did not implement any algorithm based on AGGREGATIONTREEs [20], MERGESORT, or other related aggregation algorithms defined within the research field of temporal databases. Such algorithms are optimized to handle single aggregation operators (e.g. count, sum, min, or max). Thus, the implementation would not be a generic solution usable for any query. Nevertheless, such algorithms might be useful to increase query performance for specific, often used measures. It might be reasonable to add a language feature, which allows to define a special handling (e.g. using an AGGREGATIONTREE) for a specific measure.

Next, we introduce our naive implementation. All three presented algorithm do not support queries using group by, multiple measures, nor multi-threading scenarios. To support these features, commonly used techniques (e.g. iterations and locks) could be used.

```
01  TimeSeries naive(Query q, Set r) {
02  TimeSeries ts = new TimeSeries(q);
03  // filter time def. by IN [a, b]
04  r = filter(r, q.time());
05  // filter records def. by WHERE
06  r = filter(r, q.where());
07  // it. ranges def. by IN and ON
08  for (TimeRange i : q.time()) {
09  // filter records for the range
10  r' = filter(r, i);
11  // det. measures def. by OF
12  ts.set(i, calc(i, r', q.meas()));
13  }
14  return ts;
15  }
```

The algorithm filters the records of the database, which fulfill the defined criteria of the **IN** (row 04) and **WHERE** clause (row 06). Next, it calculates the measure for each defined range (row 10). The calculation of each measure depends mainly on its type (i. e. measure of lowest granularity (e. g. query #1 in Table 2), measure of a level (e. g. query #2), or two-step measure (e. g. query #3)). Because of space limitations, we state the complexity of the **calc**-method instead of presenting it. The complexity is $O(k \cdot n)$, with k being the number of granules covered by the **TimeRange** and n being the number of records.

The other algorithms we implemented are based on INTERVALTREEs (INTTREE) as introduced by Kriegel [24]. The first one (A) – of the two INTTREE – based implementations – uses the tree to retrieve the relevant records considering the **IN**-clause (row 05 of the naive algorithm). Further, the algorithm proceeds as the naive algorithm.

The second implementation (B) differs from the first one, by created a new INTTREE for every query.

```
01  TimeSeries iTreeB(Query q, Set r) {
02  TimeSeries ts = new TimeSeries(q);
03  // filter records def. by WHERE
04  IntervalTree iTree =
05  createAndFilter(r, q.in(),
06  q.where());
07  // it. ranges def. by IN and ON
08  for (TimeRange i : q.time()) {
09  // use iTree to filter by i
10  r' = filter(iTree, i);
11  // det. measures def. by OF
12  ts.set(i, calc(i, r', q.meas()));
13  }
14  return ts;
15  }
```

As shown, the algorithm filters the records according to the **IN**- and **WHERE**-clause and creates an INTTREE for the filtered records (row 04). When iterating over the defined ranges, the created iTree is used to retrieve the relevant records for each range (row 08).

4.2 Performance

We ran several tests on an Intel Core i7-4810MQ with a CPU clock rate of 2.80 GHz, 32 GB of main memory, an SSD, and running 64-bit Windows 8.1 Pro. As Java implementation, we used a 64-bit JRE 1.6.45, with XMX 4,096 MB and XMS 512 MB. We tested the parser (implemented using ANTLR v4) and processing considering correctness. In addition, we measured the runtime performance of the processor for the three introduced algorithms (cf. Section 4.2), whereby the data and structures of all algorithms were held in memory to obtain CPU time comparability. We used a real-world data set containing 1,122,097 records collected over one year. The records have an average interval length of 48 minutes and three descriptive values: person (cardinality: 713), task-type (cardinality: 4), and work area (cardinality: 31). The used time-granule was minutes (i.e. time cardinality: 525,600). We tested the performance using the **SELECT TIMESERIES** queries shown in Table 2. Each query specifies a different type of query (i.e. different measure, usage of groups, or filters) and was fired 100 times against differently sized sub-sets of the real-world data set (i.e. 10, 100, 1,000, 10,000, 100,000, and 1,000,000 records).

The results of the runtime performance tests are shown in Figure 14. As illustrated, the bitmap-based implementation performs better than the naive and INTTREE algorithms when processing query #1 and #3. Regarding query #2 the INTTREE-based implementations perform best. As stated in Table 3, the most important criterion to determine the performance is the selectivity. Regarding a low selectivity the INTTREE-based algorithm (B) performs best.

Nevertheless, considering persistency and reading of records from disc the algorithm might perform worse. We would also like to state briefly, that other factors (e.g. kind of aggregation operators used) influence the performance of the bitmap

Table 2 The shortened queries used for testing

#	Query
1	**OF** COUNT(TASKTYPE **IN** [01.JAN, 01.FEB) **WHERE WA**.LOC.TYPE='Gate'
2	**OF** SUM(TASKTYPE) **ON** TIME.DEF.DAY **IN** [01.JAN, 01.FEB) **WHERE** WORKAREA='SEN13'
3	**OF** MAX(COUNT(WORKAREA)) **ON** TIME.DEF.DAY **IN** [01.JAN, 01.FEB) **WHERE** TASKTYPE='short'

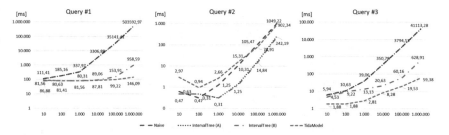

Figure 14 The measured average CPU-time performance (out of 100 runs per query)

Table 3 Statistics of the test results

In DB	Number of records			Selectivity		
	Selected by query			Selected/in DB		
	#1	#2	#3	#1	#2	#3
10^1	1	0	0	0.1000	0.0000	0.0000
10^2	5	0	7	0.0500	0.0000	0.0700
10^3	12	2	46	0.0120	0.0020	0.0460
10^4	147	9	480	0.0147	0.0009	0.0480
10^5	1.489	121	5.148	0.0149	0.0012	0.0515
10^6	15.378	1.261	51.584	0.0154	0.0013	0.0516

algorithm, so that it outperforms the INTTREE-based implementation, even if a low selectivity is given.

4.3 Considerations

The query language and processing introduced in this paper, is currently used within different projects by analysts and non-experts of different domains to analyze time-interval data. In the majority of cases, the introduced language and the processing is capable of satisfying the user's needs. Nevertheless, there are limitations, issues, and preferable enhancements. In the following, we introduce selected requests/improvements:

1. The presented query language and its processing do not support any type of transactions. A record inserted, updated, or deleted is processed by the system as an atomic operation. Nevertheless, roll-backs needed after several operations have to be performed manually. This generally increases implementation effort on the client-side.
2. The presented XML definition of dimensions (cf. Section 3.3.1) uses regular expressions to associate a member of a level to a descriptor value. Regular

expressions are sometimes difficult to be formalized (especially for number ranges). An alternative, more user-friendly expression language is desired.

3. The **UPDATE** and **DELETE** commands (cf. Section 3.4.1) need the user to specify a record identifier. The identifier can be retrieved from the resultset of an **INSERT**-statement or using the **SELECT RECORDS** command. Nevertheless, users requested to update or delete records by specifying criteria based on the records' descriptive values.

4. When a model is modified, it has to be loaded to the system as new, the data of the old model has to be inserted and the old model has to be deleted. Users desire a language extension, allowing to update models. Nevertheless, the implications of such a model update could be enormous.

5 Conclusion

In this paper, we presented a query language useful to analyze time interval data in an on-line analytical manner. The language covers the requirements formalized by several business analyst from different domains, dealing with time interval data on a daily basis. We also introduced four different implementations useful to process the most frequently used type of query (i.e. **SELECT TIMESERIES**).

An important task for future studies is to confirm, or define new models and present novel implementations solving the problem of analyzing time interval data. In addition, future work should focus on distributed and incremental query processing (e.g. when rolling-up a level). The mentioned considerations (cf. Section 4.3) of our introduced language and its implementation should be investigated. Another interesting area considering time-interval data is on-line analytical mining (OLAM). Future work should study the possibilities of analyzing aggregated time series to discover knowledge about the underlying intervals. Finally, an enhancement of the processing of the two-step aggregation technique should be considered. Depending on the selected aggregations an optimized processing strategy might be reasonable.

Acknowledgments The approaches presented in this paper are supported by the German Research Foundation (DFG) within the Cluster of Excellence "Integrative Production Technologies for High-Wage Countries" and the project "ELLI – Excellent Teaching and Learning in Engineering Sciences" as part of the Excellence Initiative at the RWTH Aachen University.

References

1. E. Codd, S. Codd, C. Salley, *Providing OLAP (On-Line Analytical Processing) to User-Analysts: An IT Mandate*. 1993. E. F. Codd and Associates (sponsored by Arbor Software Corp.)

2. J. Mazón, J. Lichtenbörger, T. J., Solving summarizability problems in fact-dimension relationships for multidimensional models. In: *11th Int. Workshop on Data Warehousing and OLAP (DOLAP '08). Napa Valley, California, USA, 26.–30. October*. 2008, pp. 57–64

3. R. Kimball, M. Ross, *The data warehouse toolkit: The definitive guide to dimensional modeling*, 3rd edn. Wiley Computer Publishing, 2013
4. J. Allen, Maintaining knowledge about temporal intervals. Communication ACM **26** (11), 1983, pp. 832–843
5. P. Meisen, D. Keng, T. Meisen, M. Recchioni, S. Jeschke, Bitmap-based on-line analytical processing of time interval data. In: *12th Int. Conf. on Information Technology. Las Vegas, Nevada, USA, 13.–15. April.* 2015
6. P. Meisen, T. Meisen, M. Recchioni, D. Schilberg, S. Jeschke, Modeling and processing of time interval data for data-driven decision support. In: *IEEE Int. Conf. on Systems, Man, and Cybernetics, San Diego, California, USA, 04.–08. October.* 2014
7. M. Böhlen, B. R., J. C. S., Point-versus interval-based temporal data models. In: *14th Int. Conf. on Data Engineering, Orlando, Florida, USA, 23.–27. Feburary.* 1998, pp. 192–200
8. P. Papapetrou, G. Kollios, S. S., G. D., Mining frequent arrangements of temporal intervals, knowledge and information systems **21** (2), 2009, pp. 133–171
9. F. Mörchen, Temporal pattern mining in symbolic time point and time interval data. In: *IEEE Symp. on Computational Intelligence and Data Mining (CIDM 2009), Nashville, Tennessee, USA, 30. March-2. April.* 2009
10. F. Höppner, F. Klawonn, Finding informative rules in interval sequences. In: *IDA2001. LNCS*, vol. 2189, ed. by F. Hoffmann, N. Adams, D. Fisher, G. Guimarães, D. Hand, Springer, Heidelberg, 2001, pp. 123–132
11. A. Kotsifakos, P. Papapetrou, V. Athitsos, Ibsm: Interval-based sequence matching, 13th siam int. conf. on data mining (sdm13), austin, texas, usa, 02.–04. may. 2013
12. Y. Chen, M. Chiang, M. Ko, Discovering time-interval sequential patterns in sequence databases. Expert Systems with Applications **25** (3), 2003, pp. 343–354
13. R. Agrawal, R. Srikant, Mining sequential patterns. In: *Int. Conf. Data Engineering, Taipei, Taiwan.* 1995, pp. 3–14
14. P. Papapetrou, G. Kollios, S. S., D. Gunopulos, Discovering frequent arrangements of temporal intervals. In: *5th IEEE Int. Conf. on Data Mining (ICDM'05), IEEE Press.* 2005, pp. 354–361
15. F. Mörchen, A better tool than allen's relations for expressing temporal knowledge in interval data. In: *12th ACM SIGKDD Int. Conf. on Knowledge Discovery and Data Mining, Philadelphia, Pennsylvania, USA.* 2006
16. C. Chui, B. Kao, E. Lo, D. Cheung, S-olap: An olap system for analyzing sequence data. In: *ACM SIGMOD International Conference on Man-agement of Data, Indianapolis, Indiana, USA.* 2010
17. M. Liu, E. Rundensteiner, K. Greenfield, C. Gupta, S. Wang, I. Ari, A. Mehta, E-cube: multidimensional event sequence analysis using hierarchical pattern query sharing. In: *ACM SIGMOD International Conference on Management of Data, Athens, Greece.* 2011
18. B. Bebel, M. Morzy, T. Morzy, Z. Królikowski, R. Wrembel, Olap-like analysis of time point-based sequential data. In: *Advances in Conceptual Modeling*, ed. by S. Castano, P. Vassiliadis, L. Lakshmanan, M. Lee, 2012. 978-3-642-33998-1
19. C. Koncilia, T. Morzy, R. Wrembel, E. J., *Interval OLAP: Analyzing Interval Data, Data Warehousing and Knowledge Discovery (DaWaK 2014)*, vol. 8646. Springer Int., 2014
20. N. Kline, R. Snodgrass, Computing temporal aggregates. In: *11th Int. Conf. on Data Engineering (ICDE 1995), Taipei, China, 06.–10. March.* 1995, pp. 222–231
21. D. Rafiei, A. Mendelzon, Querying time series data based on similarity. IEEE Transactions on Knowledge and Data Engineering **12** (5), 2000
22. G. Spofford, S. Harinath, C. Webb, D.H. Huang, F. Civardi, *MDX-Solutions: With Microsoft SQL Server Analysis Services 2005 and Hyperion Essbase.* John Wiley & Sons, 2006
23. T. Pedersen, Aspects of data modeling and query processing for complex multidimensional data. Ph.D. thesis, Aalborg Universitetsforlag, Aalborg, Department of Computer Science, Aalborg Univ., 2000. No. 4
24. H. Kriegel, M. Pötke, T. Seidl, Object-relational indexing for general interval relationships. In: *7th Int. Symposium on Spatial and Temporal Databases (SSTD 2001), Los Angeles, California, 12.–15. July.* 2001, pp. 522–542

Decisive Factors for the Success of the Carologistics RoboCup Team in the RoboCup Logistics League 2014

Tim Niemueller, Sebastian Reuter, Daniel Ewert, Alexander Ferrein, Sabina Jeschke and Gerhard Lakemeyer

Abstract The RoboCup Logistics League is one of the youngest application- and industry-oriented leagues. Even so, the complexity and level of difficulty has increased over the years. We describe decisive technical and organizational aspects of our hardware and software systems and (human) team structure that made winning the RoboCup and German Open competitions possible in 2014.

Keywords RoboCup Logistics League · Robotino · Festo

1 Introduction

The Carologistics RoboCup Team is a cooperation of the Knowledge-based Systems Group, the IMA/ZLW & IfU Institute Cluster (both RWTH Aachen University), and the FH Aachen University of Applied Sciences initiated in 2012. Doctoral, master, and bachelor students of all three partners participate in the project and bring in their specific strengths tackling the various aspects of the RoboCup Logistics League sponsored by Festo (LLSF): designing hardware modifications, developing functional software components, system integration, and high-level control of a group of mobile robots.

Our approach to the league's challenges is based on a distributed system where robots are individual autonomous agents that coordinate themselves by communicating information about the environment as well as their intended actions. In this

T. Niemueller (✉) · G. Lakemeyer
Knowledge-based Systems Group, RWTH Aachen University, Ahornstr. 55, 52056 Aachen, Germany
e-mail: niemueller@kbsg.rwth-aachen.de

S. Reuter · D. Ewert · S. Jeschke
IMA/ZLW & IfU, RWTH Aachen University, Dennewartstr. 27, 52068 Aachen, Germany

A. Ferrein
Electrical Engineering Department, FH Aachen, Eupener Str. 70, 52066 Aachen, Germany

Originally published in "RoboCup 2014: Robot World Cup XVIII",
© Springer 2015. Reprint by Springer International Publishing Switzerland 2016,
DOI 10.1007/978-3-319-42620-4_43

Figure 1 Carologistics (three Robotino 2 with laptops on top) and BavarianBendingUnits (two larger Robotino 3) during the LLSF finals at RoboCup 2014

paper we outline decisive factors for our successes, like building on and extending proven methods and components and hosting events to attract new students to cope with the challenge of students leaving the team.

Our team has participated in RoboCup 2012, 2013, and 2014 and the RoboCup German Open (GO) 2013 and 2014. We were able to win the GO 2014 (cf. Figure 1) as well as the RoboCup 2014 in particular demonstrating a flexible task coordination scheme, and robust collision avoidance and self-localization.

In the following Section 2 we give an overview of the LLSF. Then we describe our team's robots, software components (Section 3), and aspects of our task coordination and simulation (Section 4). We detail our involvement in the league's organization and outreach events in Section 5, before concluding in Section 6.

2   The RoboCup Logistics League

RoboCup [1] is an international initiative to foster research in the field of robotics and artificial intelligence. The basic idea of RoboCup is to set a common testbed for comparing research results in the robotics field. RoboCup is particularly well-known for its various soccer leagues. In the past few years focus application-oriented leagues such as urban search and rescue or domestic service robotics received more and more attention. In 2012, the new industry-oriented Logistics League Sponsored by Festo (LLSF) was founded to tackle the problem of production logistics. Groups of up to three robots have to plan, execute, and optimize the materialflow in a factory automation scenario and deliver products according to dynamic orders. Therefore, the challenge consists of creating and adjusting a production plan and coordinate the group of robots [2]. The LLSF has attracted an increasing number of teams since established (8 teams in 2014).

The LLSF competition takes place on a field of 11.2m × 5.6m surrounded by walls (Figure 1). Two teams are playing at the same time competing for points, (travel) space and time. Each team has an exclusive input storage (blue areas) and delivery zone (green area). Machines are represented by RFID-readers with signal lights on top. The lights indicate the current status of a machine, such as "ready", "producing" and "out-of-order". There are three delivery gates, one recycling machine, and twelve production machines per team. Material is represented by orange pucks with an RFID tag. At the beginning all pucks have the raw material state S_0 and are in the input storage, and can be refined through several stages to final products using the production machines. These machines are assigned a type randomly at the start of a match which determines what inputs are required and what output will be produced, and how long this conversion will take [3]. Finished products must then be taken to the active gate in the delivery zone.

The game is controlled by the referee box (refbox), a software component which keeps track of puck states, instructs the light signals, and posts orders to the teams [4]. After the game is started, no manual interference is allowed, robots receive instructions only from the refbox. Teams are awarded with points for delivering ordered products, producing complex products, and recycling. The refbox can be seen as a higher-level production planning entity as used in industry, e.g. ERP or MES-Systems.

3 The Carologistics Platform

The standard robot platform of this league is the Robotino by Festo Didactic [5]. The Robotino was developed for research and education and features omni-directional locomotion, a gyroscope and webcam, infrared distance sensors, and bumpers. It is also equipped with a static puck holder to move pucks. The teams may equip the robot with additional sensors and computation devices.

3.1 Hardware System

The robot system currently in use is still based on the Robotino 2 which is now replaced by the new version 3. The modified Robotino depicted in Figure 2a used by the Carologistics RoboCup team features two additional webcams and a Sick laser range finder. One of the webcams is used for recognizing the signal lights of the production machines, the other to detect pucks in front of the robot. The former omni-directional camera is no longer used as it was prone to distortion and time-intensive calibration. Tracking pucks especially during rotational movement presented another challenge while the benefit of detecting pucks anywhere next to

(a)   (b)

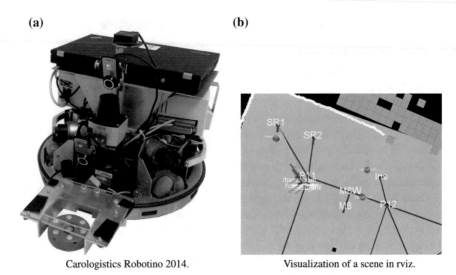

Carologistics Robotino 2014.   Visualization of a scene in rviz.

Figure 2 Carologistics Robotino, sensor processing, and visualization

the robot was minimal. The webcams are mounted with serrated locking plates for a firm adjustment to defined angles. The Sick TiM551 laser scanner is used for collision avoidance and self-localization. In comparison to the Hokuyo laser scanner with a scanning range of 4 meters we used last year, the Sick TiM551 has a maximal scanning of 10 meters. An additional laptop on the robot increases the computation power and allows for more elaborate methods for self-localization, computer vision, and navigation. A custom-made passive guidance device is mounted to the front of the robots to allow for proper control of the pucks. Optical sensors mounted to the guidance device are used to measure the longitudinal distance for approaching the signal lights.

Next year we intend to migrate to the new Robotino 3. This is a requirement to cope with the new field stations coming in 2015 [4]. Preliminary experiments indicated a smooth migration of our control system to the new platform.

3.2 Architecture and Middleware

The software system of the Carologistics robots combines two different middlewares, Fawkes [6] and ROS [7]. This allows us to use software components from both systems. The overall system, however, is integrated using Fawkes. Adapter plugins connect the systems, for example to use ROS' 3D visualization capabilities (cf. Figure 2b). The overall software structure is inspired by the three-layer architecture paradigm [8]. It consists of a deliberative layer for high-level reasoning, a reactive execution layer for breaking down high-level commands and monitoring their

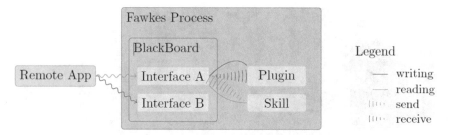

Figure 3 Components communicate state data via interfaces stored in the blackboard. Commands and instructions are send as messages. Communication is universally shared among functional plugins and behavioral components

execution, and a feedback control layer for hardware access and functional components. The lowest layer is described in Section 3.3. The upper two layers are detailed in Section 4. The communication between single components – implemented as *plugins* – is realized by a hybrid blackboard and messaging approach [6]. This allows for information exchange between arbitrary components. As shown in Figure 3, information is written to or read from *interfaces*, each carrying certain information, e. g. sensor data or motor control, but also more abstract information like the position of an object. The information flow is somewhat restricted – by design – in so far as only one component can write to an interface. Reading, however, is possible for an arbitrary number of components. This approach has proven to avoid race conditions when for example different components try to instruct another component at the same time. The principle is that the interface is used by a component to provide state information. Instructions and commands are sent as messages. Then, multiple conflicting commands can be detected or they can be executed in sequence or in parallel, depending on the nature of the commands.

3.3 Functional Software Components

A plethora of different software components is required for a multi-robot system. Here, we discuss the lowest layer in our architecture which contains functional modules and hardware drivers. All functional components are implemented in Fawkes. Drivers have been implemented based on publicly available protocol documentation, e. g. for our laser range finders or webcams. To access the Robotino base platform hardware we make use of a minimal subset of OpenRobotino, a software system provided by the manufacturer.

Localization is based on Adaptive Monte Carlo Localization which was ported from ROS and then extended. For locomotion, we integrated the collision avoidance module [9] which is also used by the AllemaniACs[1] RoboCup@Home robot.

---

[1] See the AllemaniACs website at http://robocup.rwth-aachen.de.

Figure 4 Vision-based machine detection providing 3D pose information to approach a machine

A computer vision component for robust detection of the light signal state on the field has been developed specifically for this domain. A new such component we developed is a vision-based machine detection module. It will allow to detect and approach the machines more precisely as it yields a 3D pose. Figure 4 shows the visualization of the extracted features.

4 High-Level Decision Making and Task Coordination

Task coordination is performed using an incremental reasoning approach [10]. In the following we describe the behavior components, and the reasoning process in two particular situations from the rules in 2014. For computational and energy efficiency, the behavior components need also to coordinate activation and deactivation of the lower level components to solve computing resource conflicts.

4.1 Behavior Components for the LLSF

Tasks that the high-level reasoning component of the robot must fulfill in the LLSF are:

Exploration: Gather information about the machine types by sensing and reasoning to gain more knowledge, e. g. the signal lights' response to certain types of pucks.
Production: Complete the production chains as often as possible dealing with incomplete knowledge.
Execution Monitoring: Instruct and monitor the reactive mid-level Lua-based Behavior Engine.

A group of three robots perform these steps cooperatively, that is, they communicate information about their current intentions, acquire exclusive control over resources like machines, and share their beliefs about the current state of the environment. This continuous updating of information suggests an incremental reasoning approach. As facts become known, the robot needs to adjust its plan.

4.2 Lua-Based Behavior Engine

In previous work we have developed the Lua-based Behavior Engine (BE) [11]. It mandates a separation of the behavior in three layers, as depicted in Figure 5: the low-level processing for perception and actuation, a mid-level reactive layer, and a high-level reasoning layer. The layers are combined following an adapted hybrid deliberative-reactive coordination paradigm with the BE serving as the reactive layer to interface between the low- and high-level systems.

The BE is based on hybrid state machines (HSM). They can be depicted as a directed graph with nodes representing states for action execution, and/or monitoring

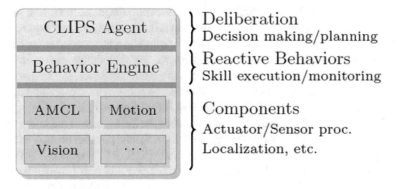

Figure 5 Behavior Layer Separation

of actuation, perception, and internal state. Edges denote jump conditions implemented as Boolean functions. For the active state of a state machine, all outgoing conditions are evaluated, typically at about 15 Hz. If a condition fires, the active state is changed to the target node of the edge. A table of variables holds information like the world model, for example storing numeric values for object positions. It remedies typical problems of state machines like fast growing number of states or variable data passing from one state to another. Skills are implemented using the light-weight, extensible scripting language Lua.

4.3 Incremental Reasoning Agent

The problem at hand with its intertwined world model updating and execution naturally lends itself to a representation as a fact base with update rules for triggering behavior for certain beliefs. We have chosen the CLIPS rules engine [12], because using incremental reasoning the robot can take the next best action at any point in time whenever the robot is idle. This avoids costly re-planning (as with approaches using classical planners) and it allows us to cope with incomplete knowledge about the world. Additionally, it is computationally inexpensive.

The CLIPS rules are roughly structured using a fact to denote the current overall state that determines which subset of the rules is applicable at any given time. For example, the robot can be idle and ready to start a new sub-task, or it may be busy moving to another location. Rules involved with physical interaction typically depend on this state, while world model belief updates often do not. The state is also required to commit to a certain action and avoid switching to another one if new information, e.g., contributed by other robots on the field, becomes available. While it may be better in the current situation to pursue another goal, aborting or changing an action usually incurs much higher costs.

The rules explained in the following demonstrate what we mean by incremental reasoning. The robot does not create a full-edged plan at a certain point in time and then executes it until this fails. Rather, when idle it commits to the 'then-best' action. As soon as the action is completed and based on its knowledge, the next best action is chosen. The rule base is structured in six areas: exploration, production step decision, coordination with other robots, process execution, world modeling, and utilities.

In Figure 6 we show a simplified rule for the production process. The game is in the production phase, the robot is currently idle and holds a raw material puck S_0 or no puck: **(phase PRODUCTION) (state IDLE)(holding NONE|S0)**. Furthermore there is a $T5$-machine, whose team-color matches the team-color of the robot, which has no produced puck, is not already loaded with an S_0, and no other robot is currently bringing an S_0. If these conditions are satisfied and ***PRIORITY-LOAD-T5-WITH-S0*** is the highest priority of currently active rules, the rule fires proposing to load the machine with the name ?name with an puck in state S_0. It also switches the state.

```
(defrule load-T5-with-S0
  (declare (salience ?*PRIORITY-LOAD-T5-WITH-S0*))
  (phase PRODUCTION)
  ?s <- (state IDLE)
  (holding NONE|S0)
  (team-color ?team-color)
  (machine (mtype T5) (loaded-with $?l&~:(member$ S0 ?l))
    (incoming $?i&~:(member$ BRING_S0 ?i)) (name ?name)
    (produced-puck NONE) (team ?team-color))
  =>
  (printout t "PROD: Loading T5 " ?name " with S0" crlf)
  (assert (proposed-task (name load-with-S0) (args (create$ ?name))))
  (retract ?s)
  (assert (state TASK-PROPOSED))
)
```

Figure 6 CLIPS Production Process Rules

```
(defrule goto-proc-complete
  (declare (salience ?*PRIORITY-WM*))
  (state GOTO-FINAL)
  (goto-target ?name)
  ?h <- (holding ?)
  (lights GREEN-ON YELLOW-OFF RED-OFF)
  (machine (name ?name) (output ?output) (loaded-with $?lw) (junk ?jn))
  =>
  (printout t "Production completed at " ?name "|" ?mtype crlf)
  (retract ?h)
  (assert (holding ?output))
  (foreach ?puck ?lw
    (assert (worldmodel-change (machine ?name)
                (change REMOVE_LOADED_WITH) (value ?puck))))
  (assert (worldmodel-change (machine ?name) (change SET_NUM_C0)
                (amount (+ ?jn (length$ ?lw)))))
)
```

Figure 7 CLIPS World Model Update Rules

There is a set of such production rules with their conditions and priorities determining what the robot does in a certain situation, or – in other terms – based on a certain belief about the world in the fact base. This simplifies adding new decision rules. The decisions can be made more granular by adding rules with more restrictive conditions and a higher priority.

After a proposed task was chosen, the coordination rules of the agent cause communication with the other robots to announce the intention and ensure that there are no conflicts. If the coordination rules accept the proposed task, process execution rules perform the steps of the task (e. g. getting an S_0 from the input storage and bringing it to the machine). Here, the agent calls the Behavior Engine to execute the actual skills like driving to the input storage and loading a puck.

The *world model* holds facts about the machines and their state, what kind of puck the robot is currently holding (if any) and the state of the robot. A simplified examples for a world model update is shown in Figure 7. The world model update rule is invoked after a task or sub-task from the production rule presented above was successfully completed, i.e. an S_0 puck was taken to a machine of the type T_5. The rule shows the inference of the output puck type given a machine's reaction. The conditions **(state GOTO-FINAL) (goto-target ?name)** denote that the robot finished locomotion and production at the target machine ?name. Furthermore, the robot sees only a green light at the machine, which indicates that the machine successfully finished the production. If all these conditions hold, the rule updates the world model about what kind of puck the robot is holding. Additionally it assumes all pucks removed that were loaded in the machine and increases the amount of consumed pucks. The world model is synchronized with other robots with another set of rules.

In comparison to 2013, the agent evolved to enable a tighter cooperation of the three agents. This required smaller atomic tasks, which are performed by the agents, a coordination mechanism to ensure the robots perform no redundant actions, more fine-grained production rules, and synchronization of the world model. The latter allows for dynamically adding or removing robots without interference to the overall production process. Furthermore, the agent became more robust against failure of behavior execution and wrong perception by adding a set of more distinctive world model update rules.

4.4 Multi-robot Simulation in Gazebo

The character of the LLSF game emphasizes research and application of methods for efficient planning, scheduling, and reasoning on the optimal work order of production processes handled by a group of robots. An aspect that distinctly separates this league from others is that the environment itself acts as an agent by posting orders and controlling the machines' reactions. This is what we call *environment agency*. Naturally, dynamic scenarios for autonomous mobile robots are complex challenges in general, and in particular if multiple competing agents are involved. In the LLSF, the large playing field and material costs are prohibitive for teams to set up a complete scenario for testing, let alone to have two teams of robots. Additionally, members of related communities like planning and reasoning might not want to deal with the full software and system complexity. Still they often welcome relevant scenarios to test and present their research. Therefore, we have created an *open simulation environment* [3] to support research and development. There are three core aspects in this context: (1) The simulation should be a turn-key solution with simple interfaces, (2) the world must react as close to the real world as possible, including in particular the machine responses and signals, and (3) various levels of abstraction are desirable depending on the focus of the user, e.g. whether to simulate laser data to run a self-localization component or to simply provide the position (possibly with some noise).

In recent work [3], we provide such an environment (Figure 8). It is based on the well-known Gazebo simulator addressing these issues: (1) its wide-spread use and

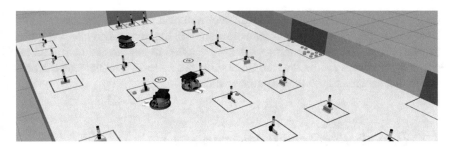

Figure 8 The simulation of the LLSF in Gazebo. The circles above the robots indicate their localization and robot number

open interfaces already adapted to several software frameworks in combination with our models and adapters provides an easy to use solution; (2) we have connected the simulation directly to the referee box, the semi-autonomous game controller of the LLSF, so that it provides precisely the reactions and *environment agency* of a real-world game; (3) we have implemented *multi-level abstraction* that allows to run full-system tests including self-localization and perception or to focus on high-level control reducing uncertainties by replacing some lower-level components using simulator ground truth data. This allows to develop an idealized strategy first, and only then increase uncertainty and enforce robustness by failure detection and recovery. More information, media, and the software itself are available at http://www.fawkesrobotics.org/projects/llsf-sim/.

In the LLSF, the large playing field and material costs are prohibitive for teams to set up a complete scenario, let alone to have two teams of robots. Additionally, members of related communities like planning and reasoning might not want to deal with the full software and system complexity. Still they often welcome relevant scenarios to test and present their research. Therefore, we propose a new simulation sub-league for the LLSF based on our simulation [3].

5 Continued Involvement and Outreach

We have been active members of the Technical and Organizational Committees and proposed various ground-breaking changes for the league like merging the two playing fields or using physical processing machines in 2015 [2, 4]. Additionally we introduced and currently maintain the autonomous referee box for the competition as explained in the next section.

5.1 LLSF Referee Box

The Carologistics team has developed the autonomous referee box (refbox) for the LLSF which was deployed in 2013 [2]. It strives for full autonomy on the game controller, i. e. it tracks and monitors all puck and machine states, creates (randomized) game scenarios, handles communication with the robots, and interacts with a human referee. In 2014 the refbox has been adapted to the merged fields and two opposing teams on the field at the same time. We have also implemented a basic encryption scheme for secured communications.

5.2 Organizational and Didactic Aspects

One of the primary goals of the RoboCup Logistics League lies in providing a test bed for cyber-physical systems [13]. The competition also serves as an excellent educational tool to give students a hands-on experience in dealing with robotics in industrial applications and future challenges for industry and research.

To improve the outreach of the league and involve a larger group of students the team and its associated institutes host *Hackathons*: students are invited to delve into the teams robotic system and develop a solution for a specific simplified challenge in a night's time. This year, the task was to explore an arena and find color-coded objects. A particular complication was that the arena was dark, i. e. without external lighting (see Figure 9). The robots were equipped with headlights to enable perception with a considerably smaller field of view.

Hackathons also serves as recruiting platform for new team members, which are always needed due to the natural fluctuation, e. g. due to students leaving the university. Until now, Hackathons have been held each year since 2012, attracting up to 50 attendees for each event. The majority of the current student team members entered through one of the Hackathons.

(a)   (b)

Field with illuminated obstacles where the robot has to find objects in the dark with just a headlight.

View from the robot's front camera with detected pucks in the light cone of the headlight mounted on the robot.

Figure 9 The 2014 Carologistics Hackathon challenge

The LLSF is also a formidable teaching platform. The KBSG offers regular lab courses where students are introduced to the robot platform and have to work on a specific problem. Example topics are inter-robot communication[2] or a new task coordination component[3] based on Procedural Reasoning Systems [14].

5.3 Research

The LLSF provides an excellent domain for research, in particular for a focus on task coordination for multi-robot systems. Compared to other RoboCup leagues, the problem is less dynamic (compared to soccer) and less cluttered and unorganized (compared to urban search and rescue or domestic service robots). That does not mean that it is easy to compete in the league but it does provide an interesting balance for researchers from related fields like planning or knowledge representation and reasoning to apply their results in a robotic environment.

In our context, creating a new central planning component is slated for inclusion as a part for the next phase of an ongoing research project on hybrid reasoning.[4] There, we want to explore the possibility to have a globally optimizing reasoner that offers suggestions for the robot group to maximize the overall achievable score with the given resources. This inclusion in largerscale research projects is crucial to advance the state of the art in a team or a league.

6 Conclusion

In this paper we have outlined several decisive factors for winning all competitions and technical challenges in 2014 without loosing a single game. It is important to have a joint team from different areas to cope with the large variety of challenges, from hardware modification to software integration. Our robust and proven base system that has been developed and used for and in other RoboCup leagues allowed us to focus on the domain-specific challenges like flexible and efficient task coordination. In particular our focus on behavioral and multi-robot coordination components and the availability of our Gazebo-based simulation environment were crucial advantages. Even though the more elaborate approach meant a disadvantage through higher complexity for the simpler rules in 2012 and 2013, it was worthwhile since we were able to cope with the increasing level of difficulty of the league more quickly.

Finally, our outreach program organizing large yearly Hackathons to attract and recruit new students is crucial to keep the team vivid and compensate leaving team members.

---

[2]http://kbsg.rwth-aachen.de/teaching/WS2012/LabRoCoCo.

[3]http://kbsg.rwth-aachen.de/teaching/WS2014/LabPRoGrAMR.

[4]http://www.hybrid-reasoning.org.

The website of the Carologistics RoboCup Team with further information and media can be found at http://www.carologistics.org.

Acknowledgments The team members in 2014 are: Daniel Ewert, Alexander Ferrein, Sabina Jeschke, Nicolas Limpert, Gerhard Lakemeyer, Matthias Löbach, Randolph Maaßen, Victor Mataré, Tobias Neumann, Tim Niemueller, Florian Nolden, Sebastian Reuter, Johannes Rothe, and Frederik Zwilling.
We gratefully acknowledge the financial support of RWTH Aachen University and FH Aachen for participation at RoboCup 2014 in João Pessoa, Brazil. We thank the Bonding student organization for co-organizing and providing food, caffeinated drinks, and support for the Hackathons 2013 and 2014.
F. Zwilling and T. Niemueller were supported by the German National Science Foundation (DFG) research unit *FOR 1513* on Hybrid Reasoning for Intelligent Systems (http://www.hybrid-reasoning.org).

References

1. H. Kitano, M. Asada, Y. Kuniyoshi, I. Noda, E. Osawa, RoboCup: The Robot World Cup Initiative. In: *Proc. 1st Int. Conf. on Autonomous Agents*. 1997
2. T. Niemueller, D. Ewert, S. Reuter, A. Ferrein, S. Jeschke, G. Lakemeyer, RoboCup Logistics League Sponsored by Festo: A Competitive Factory Automation Benchmark. In: *RoboCup Symposium 2013*. 2013
3. F. Zwilling, T. Niemueller, G. Lakemeyer, Simulation for the RoboCup Logistics League with Real-World Environment Agency and Multi-level Abstraction. In: *RoboCup Symposium*. 2014
4. T. Niemueller, G. Lakemeyer, A. Ferrein, S. Reuter, D. Ewert, S. Jeschke, D. Pensky, U. Karras, Proposal for Advancements to the LLSF in 2014 and beyond. In: *ICAR – 1st Workshop on Developments in RoboCup Leagues*. 2013
5. U. Karras, D. Pensky, O. Rojas, Mobile Robotics in Education and Research of Logistics. In: *IROS 2011 – Workshop on Metrics and Methodologies for Autonomous Robot Teams in Logistics*. 2011
6. T. Niemueller, A. Ferrein, D. Beck, G. Lakemeyer, Design Principles of the Component-Based Robot Software Framework Fawkes. In: *Int. Conference on Simulation, Modeling, and Programming for Autonomous Robots (SIMPAR)*. 2010
7. M. Quigley, K. Conley, B.P. Gerkey, J. Faust, T. Foote, J. Leibs, R. Wheeler, A.Y. Ng, ROS: an open-source Robot Operating System. In: *ICRA Workshop on Open Source Software*. 2009
8. E. Gat, Three-layer architectures. In: *Artificial Intelligence and Mobile Robots*, ed. by D. Kortenkamp, R.P. Bonasso, R. Murphy, MIT Press, 1998, pp. 195–210
9. S. Jacobs, A. Ferrein, S. Schiffer, D. Beck, G. Lakemeyer, Robust collision avoidance in unknown domestic environments. In: *RoboCup Symposium 2009*, ed. by J. Baltes, M.G. Lagoudakis, T. Naruse, S.S. Ghidary. 2009
10. T. Niemueller, G. Lakemeyer, A. Ferrein, Incremental Task-level Reasoning in a Competitive Factory Automation Scenario. In: *Proc. of AAAI Spring Symposium 2013 - Designing Intelligent Robots: Reintegrating AI*. 2013
11. T. Niemueller, A. Ferrein, G. Lakemeyer, A Lua-based Behavior Engine for Controlling the Humanoid Robot Nao. In: *RoboCup Symposium 2009*. 2009
12. R.M. Wygant, CLIPS: A powerful development and delivery expert system tool. Computers & Industrial Engineering **17** (1–4), 1989

13. T. Niemueller, D. Ewert, S. Reuter, U. Karras, A. Ferrein, S. Jeschke, G. Lakemeyer, Towards Benchmarking Cyber-Physical Systems in Factory Automation Scenarios. In: *KI 2013: Advances in Artificial Intelligence*, Springer, 2013
14. R. Alami, R. Chatila, S. Fleury, M. Ghallab, F. Ingrand, An architecture for autonomy. The International Journal of Robotics Research **17** (4), 1998

Evaluation of the RoboCup Logistics League and Derived Criteria for Future Competitions

Tim Niemueller, Sebastian Reuter, Alexander Ferrein,
Sabina Jeschke and Gerhard Lakemeyer

Abstract In the RoboCup Logistics League (RCLL), games are governed by a semi-autonomous referee box. It also records tremendous amounts of data about state changes of the game or communication with the robots. In this paper, we analyze the data of the 2014 competition by means of Key Performance Indicators (KPI). KPIs are used in industrial environments to evaluate the performance of production systems. Applying adapted KPIs to the RCLL provides interesting insights about the strategies of the robot teams. When aiming for more realistic industrial properties with a 24/7 production, where teams perform shifts (without intermediate environment reset), KPIs could be a means to score the game. This could be tried first in a simulation sub-league.

Keywords CPS · CPPS · Autonomous Logistic · Mobile Robots · Benchmarking CPS

1 Introduction

Benchmarking of autonomous mobile robots and industrial scenarios alike are difficult due to many dynamic factors. The scenarios might be too diverse to compare or the environment is not observable (enough). This makes it problematic to evaluate such domains objectively. The *RoboCup Logistics League* (RCLL) is a medium complex domain inspired by actual challenges in industrial applications – in particular that of intra-logistics in a smart factory environment, that is, moving goods in a factory among a number of machines for processing. When developing the league,

T. Niemueller (✉) · G. Lakemeyer
Knowledge-based Systems Group, RWTH Aachen University, Aachen, Germany
e-mail: niemueller@kbsg.rwth-aachen.de

S. Reuter · S. Jeschke
IMA/ZLW & IfU, RWTH Aachen University,
Dennewartstr. 27, 52068 Aachen, Germany

A. Ferrein
MASCOR Institute, Aachen University of Applied Sciences, Aachen, Germany

Originally published in "RoboCup 2015: Robot World Cup XIX",
© Springer 2015. Reprint by Springer International Publishing Switzerland 2016,
DOI 10.1007/978-3-319-42620-4_44

591

it was ensured that the domain remained partially observable – enough, so that one could autonomously judge the game.

In an industrial setting, companies strive to improve in terms of *Key Performance Indicators (KPI)*. KPIs are, for example, the time required to move a part through its production process along several machines, or how many products are currently worked on (work in progress) at a time.

Our goal is to *make KPI applicable in the RCLL* in a meaningful way. As a first step, we have analyzed games of the RCLL competition in 2014 focusing on the two top performing teams Carologistics and BBUnits. We provide an evaluation in terms of KPIs mapped to the RCLL game. This is possible, because the referee box, a program that controls and monitors the game, also records relevant data like game state changes and robot communication. The KPIs adapted for the RCLL provide the *performance metrics* by which we can analyze this data. Based on this analysis we give possible explanations on the differences in performance seen from the two teams. The information gained also allows for improving the RCLL as a testbed for industrial applications.

Additionally, on the road to a more realistic industrial setting it is conceivable to aim for a 24/7 production where teams take over shifts without an intermediate environment reset. That would allow for better judging of system robustness and flexibility of the task-level coordination of a team. However, this requires new metrics to score the game, which the adapted KPIs might provide. The RCLL simulation [1] might be a suitable basis to try this in a reasonable way.

In the following Section 2, we introduce the RCLL in more detail. In Section 3, we give an overview of related work regarding robotic competitions and benchmarks. KPIs and their adaptation to the RCLL is presented in Section 4, before applying them for analyzing the RCLL 2014 finale in Section 5. We conclude in Section 6.

2   RoboCup Logistics League

RoboCup [2] is an international initiative to foster research in the field of robotics and artificial intelligence. The basic idea of RoboCup is to set a common testbed for comparing research results in the robotics field. RoboCup is particularly well-known for its various soccer leagues. In the past few years, application-oriented leagues received increasing attention. In 2012, the new industry-oriented RoboCup Logistics League (RCLL, previously LLSF), was founded to tackle the problem of production logistics. Groups of up to three robots have to plan, execute, and optimize the material flow in a smart factory scenario and deliver products according to dynamic orders. Therefore, the challenge consists of creating and adjusting a production schedule and coordinate the group of robots. In the following, we describe the rules of 2014, that we used for our evaluation.

The RCLL competition takes place on a field of 11.2 m × 5.6 m (Figure 1). Two teams are playing at the same time competing for points, (travel) space and time. Each team has an exclusive input storage (blue areas) and delivery zone (green area in Figure 1). Machines are represented by RFID readers with signal lights on top

Figure 1 Carologistics (three Robotino 2 with laptops on top) and BavarianBendingUnits (two larger Robotino 3) during the RCLL finals at RoboCup 2014

indicating the machine state. At the beginning all pucks (representing the products) have the raw material state, are in the input storage, and can be refined (through several stages) to final products using the production machines. These machines are assigned a type randomly at the start of a match which determines what inputs are required and what output will be produced, and how long this conversion will take [1]. Finished products must then be taken to the active gate in the delivery zone. The game is controlled by the referee box (refbox), a software component which instructs and monitors the game [3]. It posts orders dynamically that state the product type (required final puck state), how many items are requested, and a time window when the order must be delivered. Pucks are identified by a unique ID stored on an RFID tag to maintain the puck's virtual state. After the game is started, no manual interference is allowed, robots receive instructions only from the refbox. Teams receive points for producing complex products, delivering ordered products, and recycling. The RCLL is also very interesting from a planning and scheduling point of view [4].

2.1   The Referee Box

Overseeing the game requires tracking of more than 40 pucks and their respective states, watching machine areas of 24 machines to detect pucks that are moved out of bounds, checking for the completion of production steps along the production chain awarding points and keeping a score. This can easily overwhelm a human referee and make the competition hard to understand for the audience. Therefore, we introduced a (semi-)autonomous referee box (refbox) in 2013. It controls and monitors all machines on the field, tracks the score, and provides information for visualization to the audience. The interface for the human referees (e.g., to start or pause the game) is shown in Figure 2. The refbox communicates with all robots on the field. Some core aspects are listed in the following.

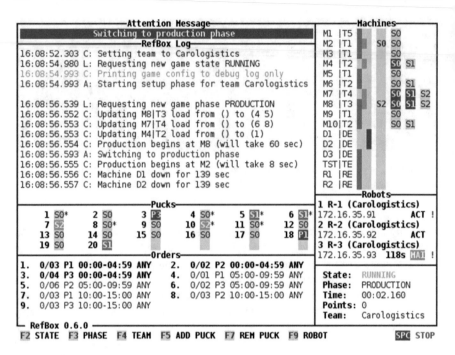

Figure 2 The Referee Box UI

Control. The refbox must oversee the game implementing the rules defined in the rule book.[1] For this very purpose it uses the rule-based system CLIPS [5]. This part is responsible for awarding points if the robots accomplished a (partial) task.

Communication. It must communicate with the robots on the field to provide information, send orders, and receive reports.

Representation. A textual or graphical application is required to visualize the current state of the game and to receive command input from the human referees.

Interfacing. The referee box needs to communicate with the programmable logic controller (PLC) which is used to set the light signals and read the RFID sensors.

Data Recording. The refbox records each and any message received or sent over the network, all state changes of the internal fact base that is used to control the game, and comprehensive game reports. This is crucial for this work.

3 Related Work: Competitions and Benchmarks

Competitions and benchmarking through competitions have become very popular for many research fields from the AI planning and scheduling community (e. g. [6, 7])

---

[1]The current rules can be found at http://www.robocup-logistics.org/rules.

leading ultimately to the development of PDDL and its extensions over SAT solvers [8] to game-based benchmark for learning algorithms [9] and robotics research. Since its beginnings in the 90's (see [10]), a large number of robotics competitions were launched in all fields of robot applications from autonomous driving (e.g. DARPA Grand Challenges, http://www.darpa/mil/grandchallenge) to disaster response (for instance, European Land Robot Trial, http://www.elrob.org) to landmine disposal (e.g., Minesweeper, http://www.landminefree.org). The motivations for running a competition are manifold. There are aspects to promote or compare research output and approaches. For exchanging ideas and experiences, symposia or user-group meetings are often organized together with a competition to foster the open exchange of solutions and ideas. Additionally, competitions are very motivating and can, in particular, activate students to be part of a competition team.

Among the established robotics competitions, the RoboCup competition [2] is a very successful example. While one of the frequently mentioned motivations of RoboCup is to compare approaches that work well in practice, the comparison of different approaches is nonetheless difficult. A reason is, in part, that robots systems are highly integrated and it is, in general, not possible to exchange software modules or test functionalities in isolation easily. In [11], the authors argue that competition challenges should lead to better algorithms and systems by a continual development process. Anderson et al. [12] critically review the contributions of a number of competitions. Proper benchmarks are not simply given and defined by performing a robotic competition. The organizers of a competition have to define determining factors in order to develop a robotic competition into a benchmark. Many competitions work toward this goal. Under the roof of the RoboCup Federation, in particular, the RoboCup Rescue [13] and RoboCup@Home [14] competitions have to be mentioned. In the RoboCup Rescue competition, for instance, benchmarks for assessing the quality of generated environment maps are established (see e.g. [15]). In RoboCup@Home, the rules change from year to year and an innovative scoring system helps to define a benchmark for fully integrated domestic service robots. Other approaches focus more on certain components such as motion algorithms [16]. The recent RoCKIn project (http://rockinrobotchallenge.eu/) aims at setting up a robot competition that increases the scientific an technological knowledge [17, 18].

In the next section, we will define the key performance indicators for production systems. These performance indicators can be used in order to judge the performance of a team. Analyzing the data recorded by the referee box using KPIs, the RCLL could indeed define a logistic benchmark in the future.

4 Key Performance Indicators

The traditional goal of production systems (in the sense of systems producing goods, not rule-based production systems) is to maximize production output while minimizing production costs. In the context of increasing market competition, product delivery times and reliability gain importance as buying criteria alongside price and quality of the product [19]. High delivery reliability and short delivery times of products demand for short throughput times of all required intermediate parts and high schedule reliability of all sub-processes within the logistic system [20]. The demand for short throughput time (time span for an order to be created) and high schedule reliability (extent to which planned orders are finished in time) conflicts with the minimization of costs which calls for a high utilization of production resources [21]. Furthermore, the minimization of throughput time and the maximization of output rate contradict each other: As a maximization of output depends on a high level of work in progress (WIP, production orders that are processed in parallel), short throughput times can only be achieved by a low level of WIP [20].

For example, a high utilization of production entities implies a high level of WIP to prevent shortages within the material flow. But it will also slow down the throughput time, because it requires a lot of transport resources. Hence, high machine utilization and short throughput times cannot be achieved together [22].

This conflict among the objectives *logistic performance* and *logistic costs* is called the scheduling dilemma of logistics [22]. Figure 3 shows Key Performance Indicators as measures for *logistic performance* and *logistic costs* [21]. KPIs are used in industry to make the efficiency of logistic systems assessable.

The logistic performance can be described by the measures throughput time, delivery reliability and delivery lateness of orders. The *throughput time TTP* for an operation is defined as start of the order processing ($T_{\text{operation start}}$) till the end of the order processing ($T_{\text{operation end}}$) [21]: $\text{TTP} = T_{\text{operation end}} - T_{\text{operation start}}$. An exemplary throughput of a product of type P2 is shown in Figure 4. The production of a product P2 is consists of a manufacture of a intermediate product S1 and S2. The critical path – the minimal throughput time of a product P2 – is formed by the throughput time of the intermediate product S2 and the final assembly. The manufacturing of an intermediate product S2 consists of two operations on the machines T1 and T2 as well as the time span needed for transportation of the intermediate products (S1 and S2) and the waiting times.

The *delivery lateness DEL* is a measure for the deviation of the actual ($T_{\text{actual delivery date}}$) and the planned delivery date ($T_{\text{planned delivery date}}$) [21]: $\text{DEL} = T_{\text{actual delivery date}} - T_{\text{planned delivery date}}$. As the actual delivery of an order can be before and after the specified delivery date, a positive lateness describes an order

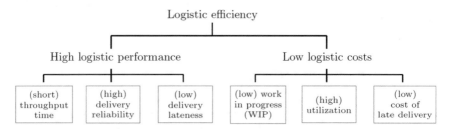

Figure 3 Key Performance Indicator (KPI) within Production Logistics

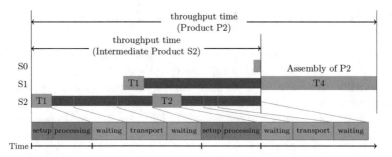

Figure 4 Throughput Time Components

that was delivered too late and a negative lateness describes an order that was delivered too early. The lateness of an order has a negative impact on the overall delivery reliability of the production system.

The *delivery reliability DERE* is an indicator to measure if a production systems sticks to scheduled delivery times. It describes the percentage of orders that are delivered within a defined delivery reliability tolerance. The number of in-time deliveries refers to all production orders that are completed within the specified tolerance band of permissible delivery lateness. The number of orders (NO) are all posted orders within the observation period. The *delivery reliability DERE* can be expressed as [21]: $DERE = \frac{DEL}{NO} * 100\,\%$.

The *logistic costs* influence the effectiveness of a logistics system just as the logistic performance does. As the logistic cost increase, the product price increases and decreases the customers willingness to buy the product. Measure for the logistic costs are work in progress, utilization and cost of late delivery.

The *work in progress WIP* describes the amount of orders that are started within a production system but are not yet completed. It can be calculated by subtracting the system output from the system input. For discretization, the period of observation can be split into equidistant time slots such as standard hours. Thus, the development of the WIP can be tracked.

The *utilization U* describes the ratio of idle and working time of production resources such as production machines or transportation entities. In terms of a production machine the utilization describes the amount of time the machine is processing an item ($T_{\text{operation}}$) in relation to the duration of a reference period ($T_{\text{duration of reference period}}$) [21]: $U = \frac{\sum T_{\text{operations}}}{T_{\text{duration of reference period}}} * 100\,\%$.

The *cost of late delivery COLD* are expenses due to a delay of an order delivery. Cost could be due to increased cost for express shipment or default penalties, or the cost of late delivery can be expressed as a lack of customer trust.

4.1 KPIs Applied in the RoboCup Logistics League

The RCLL aims to simulate a realistic, yet simplified, production environment. With the given resources (stationary production machines and mobile robots for transportation) the teams have to maximize the production output with respect to a certain set of products. In this section, we map KPIs to the RCLL.

The *throughput time TTP* in our scenario is defined as the time from the insertion of the first input product (of any accepted type) for a machine until all required inputs have been provided and the processing has been completed. For example, in Figure 5 in the second line for M2, the Busy-Blocked-Busy cycle is the TTP for a production of 84 s. The *delivery lateness DEL* is directly applicable given that orders have a delivery time window stating a latest time for delivery. The *delivery reliability DERE* can by calculated by dividing the number of delivered products by the total number of products ordered. In the RCLL, *work in progress (WIP)* can be interpreted as machines currently being blocked for an order. This contains machines blocked for

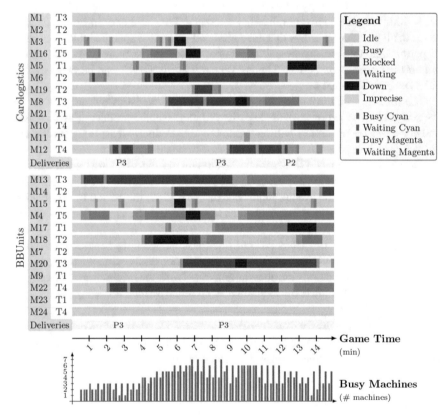

Figure 5 Machine states over the course of the final game at RoboCup 2014. The lower graph shows the occupied machines per 20 s time block

the production of intermediate as well as final products, i. e. by the green and orange blocks in Figure 5. The *utilization U* of a machine is calculated by dividing the actual busy time by the overall game time, i. e. in Figure 5 all bright green areas. The *cost of late delivery COLD* are expressed in the scoring scheme of the RCLL. A delivery in the requested time window is awarded with 10 points, while a late delivery only scores 1 point, setting COLD to 9 points. Furthermore the RCLL punishes over-production by also reducing the score from 10 to 1 point.

The teams have to balance logistic performance and logistic costs. On the one hand, the teams aim to maximize the logistic performance by short throughput times and low delivery lateness leading to a high output rate and high delivery reliability. On the other hand, the teams have to take care of high WIP which is a prerequisite for high resource utilization, but has a negative effect on throughput time and delivery reliability.

5   Analysis of the RCLL 2013 and 2014

For the presented data analysis we have used recordings of the RoboCup competition 2014. The data comprises about 75 GB of refbox data of communication, the state changes of the internal knowledge-based system, text logs, and comprehensive reports of all games played. The data is organized using MongoDB which provides fast and efficient access [23].

The basic analysis was performed using aggregation and map-reduce features of the database as well as retrieval and analysis scripts written in Python. While we have records for all games of the competitions, for brevity we focus on the two top performers in 2014, the Carologistics and BBUnits teams.

5.1   *Exemplary Application of KPI to the RoboCup 2014 Finale*

We will exemplary apply the KPIs for the RoboCup 2014 final of the RCLL between the Carologistics (cyan) and the BBUnits (magenta) which ended with a score of 165 to 124.[2] We base our analysis on Figures 5 and 6 for this game.

---

[2]Video of the final is available at https://youtu.be/_iesqH6bNsY.

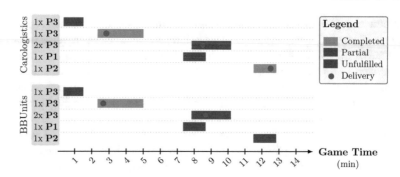

Figure 6 Adherence to delivery schedule (finals RoboCup 2014). Each row represents an order for the indicated team on the left. The blocks denote their respective delivery windows in the game time represented on the Y-axis. Green boxes mean (partially) completed orders, red unfulfilled ones. Red dots indicate the time of delivery

Figure 5 shows the machines (M1–M24) grouped per team above the time axis. Each row expresses the machine's state over the course of the game. Gray means it is currently unused (idle). Green means that it is actively processing (busy) or blocked while waiting for the next input to be fed to the machine. After a work order has been completed, the machine is waiting for the product to be picked up (orange). The machine can be down for maintenance for a limited time (dark red). Sometimes the machine is used imprecisely, that is, the product is not placed properly under the RFID device. The row 'Deliveries' shows products that are delivered at a specific time. Below the time axis, Figure 5 shows the busy machines over time. Each entry consists of a cyan and magenta column and represents a 20 s period. The height of each column shows the number of machines that are producing in that period (bright team color) or waiting for the product to be retrieved (dark team color).

Figure 6 shows the orders grouped by teams, which product type was requested and how many were requested. In each row, the colored box denotes the delivery time window in which the product has to be delivered. If the box is green, the order was fulfilled (partial fulfillment means that a smaller number of products was delivered than requested), if it is red no product was delivered in time. The red circles mark the time of delivery. Both teams were able to fulfill the second and third P3 order (partially), but only cyan managed to deliver a P2 product.

The *throughput time TTP* of an order within a machine is denoted by the green (light green and dark green) boxes in Figure 5. Cyan generally retrieves (partially) finished work orders faster, while magenta often leaves machines blocked for considerable time (dark green areas). The *delivery lateness DEL* can be best seen in Figure 6. The delivered orders (green boxes) would result in a DEL of zero, while unfulfilled orders would result the maximum DEL of the full game time (in seconds). In some games, orders were delivered after the delivery time window and therefore would get a smaller positive DEL. In the given game, the *delivery reliability DERE* of cyan is 50 %, that of magenta is 33 %. The *work in progress WIP* machines are shown as busy machines in Figure 5. As we can see, the WIP was generally equal or higher for magenta, having more machines in use. Looking at the machine states however, most of this time is blocked time in which a machine waits for the next input. If combining with the machines waiting for removal of the finished product, magenta has more machines unusable for new productions on average. The typical *machine utilization U* is currently low in the RCLL due to an emphasis on the logistics aspect that causes long travel times. In the finals, the overall utilization of all machines was about 2.3 % by cyan and 1.8 % by magenta (thus cyan has utilized the machines more than 25 % better). If there had been a late delivery (which did occur in other games), the *cost of late delivery COLD* would be severe (9 points). What we do see is that some orders were missed completely (resulting in a maximum DEL of losing the full 10 delivery points). In particular, no team managed to fulfill the P1 order. Only cyan managed to complete the work order at all (cf. Figure 5, row for T3 machine).

KPI Discussion. It seems that especially the lower throughput time TTP of cyan contributed to their success. Machines can be used again much faster. For example, only this made it possible to match the delivery time of the P2 order that magenta missed due to very long blocking times. Considering the waiting times makes this even more severe. The cyan team followed the strategy to store products finished before a matching order was received. This meant that the involved machines could be used again much faster (the waiting times of cyan are much shorter). Even with more work in progress machines of magenta, the cyan team used more T1 machines (3 instead of 2). Magenta even left M17 with a finished puck untouched for half of the game. A contributing factor here could be that magenta lost a robot during the game due to a software problem. It seems that the other robots could not recover the state of M17 (instead they later produced at another T1 machine). Concluding it seems that the cyan strategy focused more on low TTP and high throughput, while magenta's strategy was to maximize the overall machine usage and the WIP.

While BBUnits lost a robot in this game, similar statements can be made about a play-offs game between the teams a day earlier that ended 158 to 122 for the Carologistics where both teams had all robots running continuously.[3]

5.2 Overall Tournament Evaluation

Analyzing the data of all games at RoboCup2014 (within the Round-Robin Phase, Playoffs and Finals) in terms of machine state graphs (Figure 5) and adherence to delivery schedules (Figure 6) as well as using KPIs as statistical queries yields insights for the development of the competition as a whole.

---

[3] Video of the play-offs game is available at https://youtu.be/77V-7LzMBY8.

A key insight is that the current dynamic order scheme parameterization is unsuitable for the given resources (robots and machines). Even the best teams at most delivered 3 of 6 ordered products in any game. This seems to be, in particular, because the order time windows are too short. Especially with the modified game in 2015 with vastly more product variants this must be taken into account, since opportunistic production is virtually impossible.

A possible solution would be to considerably increase the time of a game. This would give the robot teams more time to work on the orders and we could gather more data to valuate the KPIs for a game. It also increases demands for system robustness, a crucial factor for industrial applications. The increased time could be tried first in a simulation league. Work is currently underway to create a common and open simulation for the RCLL based on [1] by the Carologistics and BBUnits team,[4] which could provide the basis for the project.

6 Conclusion

In recent years we have developed the RCLL as a domain of medium complexity towards being a testbed for industry-inspired robotic applications. The domain is partially observable by the referee box which allows to record detailed data about the course of the game. This data combined with Key Performance Indicators known from industrial environments allow for analyzing games objectively. We can also use this analysis combined with statistical evaluation to optimize the competition to be more balanced and to improve it as a testbed for industrial robotic applications in smart factory environments.

In an example analysis of the finals in 2014 we have determined some factors based on KPI that may explain the outcome of the game, i.e. that the winning team Carologistics' strategy was focused on short throughput times rather than a high number of machines busy at the same time as the competitor BBUnits did. While we have seen that the order schedule should be tuned to better fit the given resources for more interesting games, teams also need to investigate better scheduling strategies that allow to use the given resources more effectively. KPIs can be one aspect of determining the utility in this regard.

To aim for a more realistic scenario, it is conceivable to develop the RCLL towards a long-time evaluation in the sense of a *24/7 robot competition*. Each team gets assigned a shift in which it has to realize the material flow in the production system without a reset of the environment. Within this scenario a more complex grading scheme is needed as the state of the production system is changing in terms of the amount of work in progress, blocked machines and orders that are currently selected for production. The introduced KPIs are a possible approach to adapt the *grading scheme* to this scenario. It will also require that the teams take different initial states into account and that they provide accurate information to the refbox during

---

[4]The project is available at https://github.com/robocup-logistics.

a handover to the next team. Especially the development of a *simulation league* can help to facilitate this in a shorter time frame. It would allow teams to adapt more gently. Work in this direction is on-going as described in Section 5.2.

More information, the recorded data as well as the evaluation scripts are available at http://www.fawkesrobotics.org/p/llsf2014-eval.

References

1. F. Zwilling, T. Niemueller, G. Lakemeyer, Simulation for the RoboCup Logistics League with Real-World Environment Agency and Multi-level Abstraction. In: *RoboCup Symposium*. 2014
2. H. Kitano, M. Asada, Y. Kuniyoshi, I. Noda, E. Osawa, Robocup: The Robot World Cup Initiative. In: *1st Int. Conference on Autonomous Agents*. 1997
3. T. Niemueller, G. Lakemeyer, A. Ferrein, S. Reuter, D. Ewert, S. Jeschke, D. Pensky, U. Karras, Proposal for Advancements to the LLSF in 2014 and beyond. In: *ICAR – 1st Workshop on Developments in RoboCup Leagues*. 2013
4. T. Niemueller, G. Lakemeyer, A. Ferrein, The RoboCup Logistics League as a Benchmark for Planning in Robotics. In: *WS on Planning and Robotics (PlanRob) at Int. Conf. on Aut. Planning and Scheduling (ICAPS)*. 2015. (to appear)
5. R.M. Wygant, CLIPS: A powerful development and delivery expert system tool. Computers & Industrial Engineering **17** (1–4), 1989
6. D. Long, H. Kautz, B. Selman, B. Bonet, H. Geffner, J. Koehler, M. Brenner, J. Hoffmann, F. Rittinger, C.R. Anderson, et al., The AIPS-98 planning competition. AI magazine **21** (2), 2000
7. A.E. Howe, E. Dahlman, A critical assessment of benchmark comparison in planning. Journal of Artificial Intelligence Research **17** (1), 2002, pp. 1–33
8. A. Balint, A. Belov, M. Järvisalo, C. Sinz, Sat challenge 2012 random sat track: Description of benchmark generation. SAT CHALLENGE 2012, 2012, p. 72
9. S. Karakovskiy, J. Togelius, The mario ai benchmark and competitions. Computational Intelligence and AI in Games, IEEE Transactions on **4** (1), 2012, pp. 55–67
10. T.R. Balch, H.A. Yanco, Ten years of the AAAI mobile robot competition and exhibition. AI Magazine **23** (1), 2002
11. M. Anderson, O. Jenkins, S. Osentoski, Recasting robotics challenges as experiments [competitions]. Robotics Automation Magazine, IEEE **18** (2), 2011
12. J. Anderson, J. Baltes, C.T. Cheng, Robotics competitions as benchmarks for AI research. The Knowledge Engineering Review **26** (01), 2011
13. H. Kitano, S. Tadokoro, Robocup rescue: A grand challenge for multiagent and intelligent systems. AI magazine **22** (1), 2001
14. T. Wisspeintner, T. Van Der Zant, L. Iocchi, S. Schiffer, Robocup@home: Scientific competition and benchmarking for domestic service robots. Interaction Studies **10** (3), 2009
15. R. Madhavan, R. Lakaemper, T. Kalmár-Nagy, Benchmarking and standardization of intelligent robotic systems. In: *Int. Conf. on Adv. Rob. (ICAR)*. 2009
16. D. Calisi, L. Iocchi, D. Nardi, A unified benchmark framework for autonomous mobile robots and vehicles motion algorithms (movema benchmarks). In: *WS on experimental methodology and benchmarking in robotics research at RSS*. 2008
17. A. Ahmad, I. Awaad, F. Amigoni, J. Berghofer, R. Bischoff, A. Bonarini, R. Dwiputra, G. Fontana, F. Hegger, N. Hochgeschwender, et al., Specification of general features of scenarios and robots for benchmarking through competitions. RoCKIn Deliverable D **1**, 2013
18. F. Amigoni, A. Bonarini, G. Fontana, M. Matteucci, V. Schiaffonati, Benchmarking through competitions. In: *European Robotics Forum–Workshop on Robot Competitions: Benchmarking, Technology Transfer, and Education*. 2013

19. P. Nyhuis, H.P. Wiendahl, *Fundamentals of production logistics: theory, tools and applications.* Springer Science & Business Media, 2008
20. F. Wriggers, T. Busse, P. Nyhuis, Modeling and deriving strategic logistic measures. In: *IEEE Int. Conf. on Ind. Engineering and Engineering Mgmt.* 2007
21. H. Lödding, *Handbook of manufacturing control: Fundamentals, description, configuration.* Springer Science & Business Media, 2012
22. P. Nyhuis, H.P. Wiendahl, Logistic production operating curves–basic model of the theory of logistic operating curves. CIRP Annals-Manufact. Tech. **55** (1), 2006
23. T. Niemueller, G. Lakemeyer, S.S. Srinivasa, A Generic Robot Database and its Application in Fault Analysis and Performance Evaluation. In: *Int. Conf. on Intelligent Robots and Systems (IROS).* 2012

RoboCup Logistics League Sponsored by Festo: A Competitive Factory Automation Testbed

Tim Niemueller, Daniel Ewert, Sebastian Reuter, Alexander Ferrein,
Sabina Jeschke and Gerhard Lakemeyer

Abstract A new trend in automation is to deploy so-called cyber-physical systems (CPS) which combine computation with physical processes. The novel RoboCup Logistics League Sponsored by Festo (LLSF) aims at a such CPS logistic scenarios in an automation setting. A team of robots has to produce products from a number of semi-finished products which they have to machine during the game. Different production plans are possible and the robots need to recycle scrap byproducts. This way, the LLSF is a very interesting league offering a number of challenging research questions for planning, coordination, or communication in an application-driven scenario. In this paper, we outline the objectives of the LLSF and present steps for developing the league further towards a benchmark for logistics scenarios for CPS. As a major milestone we present the new automated referee system which helps in governing the game play as well as keeping track of the scored points in a very complex factory scenario.

Keywords RoboCup Logistics League · Robotino · Festo

1 Introduction

A new trend in automation is to deploy so-called cyber-physical systems (CPS) to larger extents. These systems combine computation with physical processes. They include embedded computers and networks which monitor and control the physical processes and have a wide range of applications in assisted living, advanced automotive systems, energy conservation, environmental and critical infrastructure control,

T. Niemueller (✉) · G. Lakemeyer
Knowledge-based Systems Group, RWTH Aachen University, Aachen, Germany
e-mail: niemueller@kbsg.rwth-aachen.de

D. Ewert · S. Reuter · S. Jeschke
IMA/ZLW & IfU, RWTH Aachen University, Dennewartstr. 27,
52068 Aachen, Germany

A. Ferrein
MASCOR Institute, Aachen University of Applied Sciences, Aachen, Germany

Originally published in "RoboCup 2013: Robot World Cup XVII",
© Springer 2013. Reprint by Springer International Publishing Switzerland 2016,
DOI 10.1007/978-3-319-42620-4_45

or manufacturing [1]. One application area of CPS are logistics scenarios in automation settings. As production is going to move away from mass production towards customized products, the challenges for the automation process will increase. This will open the floor to mobile robots in order to help with the manufacturing process.

In particular, mobile robots will be deployed for transportation tasks, where they have to get semi-finished products in place to be machined in time. This is right where the novel *RoboCup Logistics League Sponsored by Festo* (LLSF) starts. Teams of robots have to transport semi-finished products from machine to machine in order to produce some final product according to some production plan. Machines can break down, products may have inferior quality, additional important orders come in and need to be machined at a higher priority. Due to increasing demands for flexibility production facilities will become dynamic environments, where shop floor layouts and the number, location and type of the engaged machinery change constantly. The robots therefore need to be able to identify these machines either visually or by direct communication. For the LLSF, a team consisting of up to three robots starts in the game area of about 5.6 m × 5.6 m. A number of semi-finished products is represented by RFID-tagged *pucks*. Each is in a particular state, from raw material through intermediate steps to being a final product. The state cannot be read by the robot but must be tracked and communicated among the robots of a team. On the playing field are *machines*, RFID devices with a signal light indicating their processing status. When placed on a proper machine type, a puck changes its state according to the machine specification which is communicated via broadcast messages. The outcome and machine state is indicated by particular light signals. During the game a number of different semi-finished products need to be produced with ten machines on the playing field. Orders are posted to the robots requesting particular final products to be delivered to specific delivery gates and in specified time slots. All teams use the same robot base, a Festo Robotino which may be equipped with additional sensor devices and computing power, but only within certain limits (Figure 1).

The LLSF has a number of very interesting research questions to be addressed. The robots need to be autonomous, detect the pucks, detect and identify the light signals from the machines, know where they are and where to deliver the final product to. These are the basic robotics problems such as localization, navigation, collision

Figure 1 A LLSF
competition during the
RoboCup 2012 in Mexico
City

avoidance, computer vision for the pucks and the light signals. Of course, all these modules have to be integrated into an overall robust software architecture. On top of that, the teams need to schedule their production plan. This opens the field to various concepts from full supply chain optimization approaches to simpler job market approaches where a master robot gives sub-tasks to other robots. If the robots are able to prioritize the production of certain goods they can maximize the points they can achieve. In order to avoid resource conflicts (there can only be one robot at a machine at a time), to update other robots about the current states of machines and pucks, and to delegate (partial) jobs of the production process the robots must communicate. The only means allowed is via a wireless network. Since at RoboCup events there are so many wireless networks, connectivity cannot be guaranteed and is even likely to be interrupted every now and then. Hence robots must have both, useful cooperative behavior, but also sensible single-robot fallbacks. Robots that can recognize each other can have a decisive advantage. In the first phase exploration and roaming concepts must be implemented. A lot of these tasks present cutting edge research problems for the fields of factory automation and CPS, especially in the logistics sector. We intend to emphasize these problems in future tournaments to improve the attractiveness of the league as a benchmark for research and development in these fields.

The LLSF competition started in 2012 for the first time awarding a first RoboCup champion. In order to develop the league further, a number of rule changes have been proposed and implemented such as allowing more computing power on- and off-field, more diverse production plans or the introduction of a Technical Challenge. Another major novelty is the introduction of the referee box which, as in the other leagues, governs the automated game play. A particular benefit in the LLSF is the fact that the game is easy to observe automatically, and hence a large autonomy of the game play can be achieved. To foster the acceptance and implementation of the connection by the teams—while still being backward compatible for this year—certain aspects of the rules have been added and modified to grant bonus points for robots that can communicate with the refbox. For example, during the exploration phase, teams must announce their findings to the refbox, only then they can score.

This paper presents the LLSF as a competitive RoboCup league and details the objectives as well as the ensuing research questions. We outline the leagues' road-map for the next years which we intend to gradually increase in complexity, presenting the automated referee system as a major milestone. The rest of the paper is organized as follows. In the next section, we review cyber-physical systems and their connection to logistics problems arising in automation processes. In Section 3, we discuss the current version of the rules and detail the game objectives. In Section 4, we present the automated referee box (refbox). We first sketch the CLIPS rule base which the refbox is based on, before we describe the tasks and the implementation of the refbox in detail. After proposing possible future advancements of the league in Section 5 we conclude in Section 6.

2   CPS and Logistics Challenges

Digital devices have infused and changed the private and industrial world. Almost every process is supported, controlled or monitored by a digital device. Taking this a step further leads to the development of so-called cyber physical systems (CPS) [1]. Here, every entity of a system is equipped with suitable hardware and software for carrying relevant information and for communicating autonomously with its environment to exchange these information [2]. In this way, the entity is transformed from a passive element into an acting agent, which directly interacts with and manipulates its surrounding. This allows for decentralized control as well as increased flexibility and robustness of the affected processes. Examples for possible applications include a.o. the fields of advanced automotive systems, process control, manufacturing, and distributed robotics [3]. Especially logistic scenarios will benefit from CPS by promoting transparency and robustness due to decentralized control. Shipping goods or material can carry address information, handling instructions, or even directly instruct operating facilities. Additionally, due to throughout interconnections and built-in sensors, real-time tracking and tracing becomes possible in so far unrivaled accuracy and quality. However, to allow for large-scaled, reliable, and flexible CPS, research must still find answers to a number of challenges. According to [4] these include finding solutions for

- Situation recognition,
- Planning and anticipatory, partially or completely autonomous behavior,
- Cooperation and negotiation,
- Strategies of self-organization and adaptation.

For the application of CPS in manufacturing systems, the requirements for communication within the smart factory are detailed in [5]: CPS must allow for horizontal and vertical communication within the automation pyramid. *Vertical communication* refers to information exchange between the different levels of manufacturing management and control. Nowadays, these are associated to organization-wide enterprise resource planning systems (ERP), manufacturing execution systems (MES) for detailed production planning across machines and the underlying supervisory control and data acquisition (SCADA). While the boundaries between the levels are not fixed, in the long run CPS are expected to completely obliterate them. CPS must therefore be able to exchange information on different levels of abstraction. Thus, non-hierarchical communication structures will evolve which shift the production from a centralized production control to a production controlled by the products which are to be manufactured [6]. *Horizontal communication* refers to data exchange on the same level along the production chain. Thereby, information is handed to downstream production processes which will adapt their production parameters to the individual characteristics of the currently manufactured product. As we will show in the following sections, these challenges are similar to the ones faced in the LLSF.

(a)   **(b)**

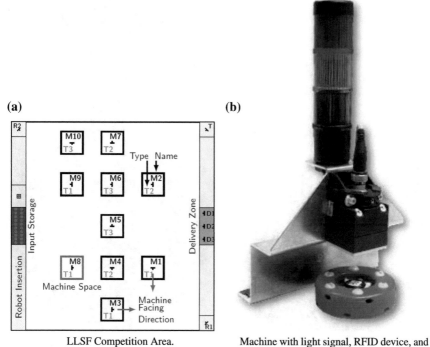

LLSF Competition Area.   Machine with light signal, RFID device, and game puck.

Figure 2 LLSF Competition Area and field machine. **a** LLSF Competition Area. **b** Machine with light signal, RFID device, and game puck

3 Game Rules and Objectives

As presented in Section 2, the deployment of CPS in the logistics of production sites opens a new field of challenges for factory automation. To recommend the LLSF as a suitable testbed for research and education regarding CPS in logistics, in Section 3 we will present the ideas of the LLSF in more detail.

In 2012 the Logistics League Sponsored by Festo (LLSF) was officially founded. The general intention is to create a simplified and abstracted *factory automation scenario* with an emphasis on logistics applications. Teams of up to three robots operate in a fenced area of about 5.6 m × 5.6 m as shown in Figure 2a. Figure 3 shows the original Festo Robotino and a modified version of the Carologistics RoboCup team.[1] The task is to complete a production chain by carrying a (semi-finished) product (a puck in the game) along different machines (signal lights on the playing field as shown in Figure 2b). Points are awarded for intermediate and completed products.

---

[1] http://www.carologistics.org.

(a)   (b)

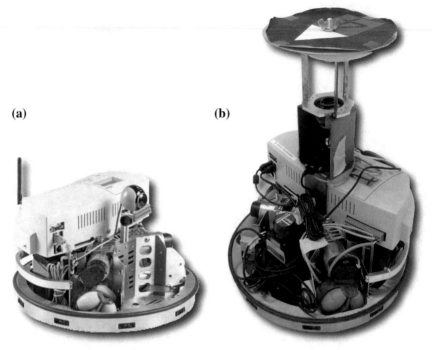

Standard Festo Robotino.   Modified Robotino of the Carologistics Ro-
                                             boCup team.

Figure 3 Standard Festo and modified Carologistics Robotino. **a** Standard Festo Robotino.
b Modified Robotino of the Carologistics RoboCup team

On the field two margin areas on opposite sides contain the puck input storage, delivery zone, and several machines that deal as delivery gates for the final products or as recycling stations. Each puck has a programmable radio frequency identification (RFID) chip with which the different product states S_0, S_1, S_2, and P_1, P_2, P_3 are distinguished. Initially, all pucks are in state S_0. In the enclosed inner field, ten signals equipped with an RFID device mounted on its front represent production machines. Each machine is assigned a random but defined type out of the types T_1–T_5, which is initially unknown to the robots. The type determines the input and output of a machine. Pucks transition through their states by being processed through machines. The complete production tree is shown in Figure 4. Circular nodes indicate a puck's state and rectangular nodes show the respective machine type. For example, the T_1 machine in the upper branch takes an S_0 puck as input with an S_1 puck as output. If a machine, like T_2, requires multiple inputs, these can be presented to the machine in any order. However, until the machine cycle completes, all involved pucks must remain in the machine space. The last input puck will be converted to the output puck, all others become junk and must be recycled at a recycling station.

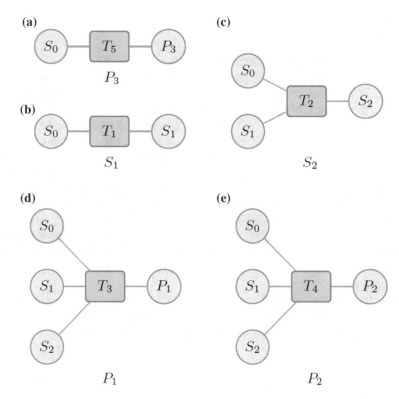

Figure 4 Production Chain Diagrams showing the machines and inputs relative to their outputs

The machines indicate their state after processing a puck using light signals. A green signal means that the particular machine production cycle has been completed, i. e., all required input products have been presented one after another, and now the puck has been transformed to the machine's respective output; for instance, after a T_1 machine transformed a puck from state S_0 to S_1, an orange light indicates partial success (more pucks are required).

Besides typical robotics tasks such as motion planning or self-localization, the robot needs to plan an efficient sequence of actions to produce as many products as possible in a fixed amount of time. Moreover, the robot has to deal with incomplete knowledge as it is not known in advance what machine has which type. Thus, the robots need to combine *sensing and reasoning* to incrementally update their belief about the world. Based on the knowledge gained, it has to find a strategy to maximize its production output, ideally minimizing costs such as travel distance.

4   Autonomous Referee Box

The LLSF game comprises a rather constrained task for now. Yet overseeing the game requires tracking of more than 20 pucks and their respective states, watching machine areas of 10 machines to detect pucks that are moved out of bounds, placing late order pucks for visual triggering at specified places at certain times, and overseeing completion of the production chain awarding points and keeping a score. This can easily overwhelm a human referee and make the competition hard to understand for a visitor. In fact, in 2012 we needed to review a camera recording of a game to award points in hindsight because a situation was overseen by the two human referees. Therefore, we strive for the implementation of a (semi-)autonomous referee box (refbox). The refbox shall control, monitor, and score the overall game. The human referees instruct the refbox (e.g., to start or pause the game), which then communicates with the robots on the field. The humans act as a second tier referee correcting for misjudgment of yet unforeseen situations. Additionally, full autonomy cannot be achieved, yet, because there is no sensor to detect pucks moved out of a machine area.

In the following we will first introduce CLIPS, a logic-based system used to implement the core functionality of the refbox and to formally describe the competition rules. We then give details about the implementation of the refbox.

4.1   The CLIPS Rule-Based Production System

CLIPS is a rule-based production system using forward chaining inference based on the Rete algorithm [7]. The CLIPS rule engine [8] has been developed and used since 1985 and is thus mature and stable. It was designed to integrate well with the C programming language,[2] which specifically helps to integrate with the refbox. Its syntax is based on LISP as it was conceived to replace LISP-based systems. The Carologistics team has already used CLIPS to implement the task coordination component [9]. Therefore it appeared suitable for the refbox implementation.

CLIPS has three building blocks [10]: a fact base or working memory, the knowledge base, and an inference engine. *Facts* are basic forms representing pieces of information which have been placed in the fact base. They are the fundamental unit of data used by rules. Facts can adhere to a specified template. It is established with a certain set of slots, properties with a specified name which can hold information. The *knowledge base* comprises heuristic knowledge in the form of condition-action rules, and procedural knowledge in the form of functions. *Rules* are a core part of the production system. They are composed of an antecedent and consequent. The antecedent is a set of conditions, typically patterns which are a set of restrictions that determine which facts satisfy the condition. If all conditions are satisfied based on the existence, non-existence, or content of facts in the fact base the rule is activated

---

[2]And C++ using clipsmm, see http://clipsmm.sf.net.

and added to the agenda. The consequent is a series of actions which are executed for the currently selected rule on the agenda, for example to modify the fact base. *Functions* carry procedural knowledge and may have side effects. They can also be implemented in C++. In our framework, we use them to utilize the underlying robot software, for instance to communicate with the reactive behavior layer described below. CLIPS' *inference engine* combines working memory and knowledge base. Fact updates, rule activation, and agenda execution are performed until stability is reached and no more rules are activated. Modifications of the fact base are evaluated if they activate (or deactivate) rules from the knowledge base. Activated rules are put onto the agenda. As there might be multiple active rules at a time, a conflict resolution strategy is required to decide which rule's actions to execute first. In our case, we order rules by their salience, a numeric value where higher value means higher priority. If rules with the same salience are active at a time, they are executed in the order of their activation, and thus in the order of their specification. The execution of the selected rule might itself trigger changes to the working memory, causing a repetition of the cycle.

4.2 Tasks of Referee Box

The refbox has several tasks it must fulfill. It must control the game, communicate with the robots, represent the current state of the game, and interface with the devices on the playing field. We will now briefly explain the four tasks before we describe some aspects of the implementation.

- *Control.* The refbox must oversee the game implementing the rules defined in the rule-book.[3] For this very purpose we use CLIPS. This part is also responsible for awarding points if the robots accomplished a (partial) task.
- *Communication.* It must communicate with the robots on the field to provide information, send orders, and receive reports.
- *Representation.* A textual or graphical application is required to visualize the current state of the game and to receive command input from the human referees. A simplified visualization can be used to explain the game to visitors.
- *Interfacing.* The referee box needs to communicate with the programmable logic controller (PLC) which is used to set the light signals and read the RFID sensors on the pucks.

4.3 Implementation

The referee box has been implemented by members of the LLSF TC. Its infrastructure is written in C++ and the game controller core in CLIPS. It uses Boost for some of its internals, for example asynchronous I/O and signal propagation.

---

[3]The current rules can be found at http://www.robocup-logistics.org/rules.

The base program creates the environment for the CLIPS core, in which the actual game controller is implemented. This core is a knowledge-based system. The facts in the working memory are used to keep track of the state of the game and to communicate within the core Rules trigger on specific conditions and events, For example the reception of a message, or the completion of a production cycle of a machine. A time fact is periodically asserted (currently at 25 Hz) to allow for time-based triggering, such as in the case of the production completion. This allows us to specify durative actions.

There are currently two interfaces to represent the game state and to accept commands. A textual shell which uses the ncurses library is used for quick operation by the human referee. It shows the most important information and accepts commands. A graphical user interface (GUI) has been implemented using the Gtkmm library. It features a visual display of the playing field and will be focused on visualization and explanation of the game to the audience in 2013.

There are two ways of communication: client-server stream connections using TCP, and peer-to-peer broadcast communication using UDP. This might be changed to multicast at a later point in time. The stream connections are used to connect the refbox tools like the shell or GUI. Broadcast communication is used to communicate with the robots. Both protocols use Google protocol buffers (protobuf) for message specification and serialization. A small framing protocol allows for transmitting messages of different types over the same connection. This is particularly important in RoboCup, where network resources are scarce and combining messages and reducing the number of connection handshakes is beneficial. Using protobuf for message specification gives a very efficient serialization in terms of message size, and allows for forward compatibility, i.e. older clients can still read the messages as new fields are added, they are simply ignored. It also allows for optional fields, further reducing the amount of data required to be sent.

The refbox has been released as Open Source software and can be downloaded at http://www.robocup-logistics.org/refbox. The page also contains links to the documentation of the refbox.

5   Gradual Advancement of the Logistic Scenarios

In this section we will propose possible advancements of the LLSF to keep the league challenging for existing teams while preserving a sufficiently low entry barrier for new teams. The basic idea for developing the league further is to gradually increase the complexity of the game. The objective is to create a benchmark for logistics tasks scenarios addressing the challenges mentioned in Section 2. Therefore, we need to push the boundaries towards more realistic scenarios in the following years. We envision challenges in the following areas.

5.1 Basic Robotic Problems

For now, only one team is in the arena at a time. Mostly it is sufficient to navigate with odometry or correct the pose estimate by the unambiguous positions of the observable landmarks. At some point, an opposing team could be in the arena at the same time. It could even try to obstruct machines and hinder the opposing team (within certain limits). With opponents, the whole setup becomes much more complicated and teams need to have solutions for increased situation awareness, collision avoidance, navigation, and self-localization.

5.2 Mobile Manipulation

A realistic logistics robot will most probably be equipped with a handling device. In the next few years, the pucks will not only be pushed around by the robots, but also have to be lifted up and put into the machine or a conveyor belt. The possibilities to extend the concept of a "machine" as we have it right now are manifold. They would need to be in such a way that a manipulator arm or a fork lift mechanism on the robots needs to be installed. The interaction with machines must not stay limited to one interacting robot. Also cooperative scenarios, where one robot moves the puck, while a second simultaneously activates the machine in a given way or tasks which have to be executed nearly in parallel on different machines are imaginable.

5.3 Multi-robot Coordination

At this time multiple robots can be advantageous if they split the serial task and complete parts of it with each robot. In the future, we could require multiple robots to cooperate to be able to complete the task at all. For example, two goods may be required to be delivered to two machines in a small time window. Also cooperation of multiple teams would be very interesting, as it would require to agree on a common communication protocol. For instance, one team could produce an item that another has to deliver, and only completing both steps by robots of different teams would score.

5.4 Logistics Management

The game setup allows for numerous interesting scenarios for logistics management. While the current task still is fixed in many aspects, this should change in the future. Production tasks, machine and product specifications could be changed dynamically, as well as a shift of the optimization goals from a purely market driven maximization of the production output to a cost driven maximization of machine utilization. Also the playfield structure should allow for changes by rotating machines as a first step towards a change of the factory layout, e. g., from a workshop-based layout to a flow-production system. While a part of these changes is communicated by the refbox (see Section 4), other changes have to be recognized by the robots. Competitive teams should have to build more autonomous robotic agents that can cope with such an environment are able to exchange relevant information and react as a team.

5.5 Robot Hardware Restrictions

Currently, the Robotino robot by Festo is the only legitimate robot platform that can be used for the game. Step-wise, hardware restrictions could be dropped to move towards more realistic scenarios. One way for the league could be to proceed towards a number of autonomous fork-lift systems in a real factory scenario.

5.6 Referee Box Development

The current referee box is the first of its kind in the RoboCup domain which has achieved such a high level of autonomy. For the future we want to further increase this autonomy. By adding an overhead camera over the field, for instance, we could automatically track the pucks and robots on the field. This would put the refbox in a position to automatically declare foul play if a puck is moved out of a machine area while it was still needed.

The rules should be implemented in a *rule tic-toc* way. This means that major changes to the rules only become implemented every two years. Major changes will be announced in one year, but only the year after this rule will come into effect. This will give all the teams enough time to tackle the new problem and work towards solutions that are required by the rule change.

Another way to push boundaries further is the introduction of Technical Challenges. In the 2013 competition, the TC announced three technical challenges where teams need to show their technical skills in the field of robotic navigation, localization and computer vision. As in other RoboCup leagues, participation in the Technical Challenge will not be mandatory, but teams will very likely use this possibility to showcase their latest development there. This way, after one particular Technical Challenge has been solved by most of the teams, it can be replaced by a more complicated task. Additionally an open challenge, where teams can show a task of their choosing, can serve as a showcase to propose ideas for future development of the league and its rules. This concept has already been applied successfully to other RoboCup leagues with a strong Technical Committee which is pushing boundaries from year to year.

6 Conclusion

In this paper we gave an overview of the new RoboCup Logistics League Sponsored by Festo (LLSF). The idea is to simulate a factory automation scenario with mobile robots. A team of robots has to "machine" different semi-finished products in order to produce some end-product and then deliver it at a certain delivery point. The team needs to identify machines, derive feasible production plans and realize them in due time. Additionally, they need to cope with failures in the production process that may occur, products need to be recycled and high priority orders may be placed arbitrarily. All this must be dealt with to optimize the outcome of the factory (and earn as many points as possible). With this league, exactly the challenges that arise in logistics setting of cyber-physical systems (CPS) are covered. The requirements of the systems are that they enjoy among others planning abilities and autonomous behavior, self-organization strategies or cooperation facilities. Therefore, the LLSF is very well suited to become a benchmark for logistics scenarios of CPS.

We outlined the rules of the league and presented a road-map which will develop the league further towards this eligibility as a benchmark. Requirements towards cooperation and communication, both within the team and with external entities will play an increasing role. A number of rule changes have already been proposed such as allowing more computing power and fostering autonomous behavior with several technical challenges. A major improvement, which we described is the introduction of the referee box, an autonomous referee for the LLSF. Different than in other leagues where referee boxes are mainly used to control the flow of the game, in LLSF the referee box keeps track of the machines types and the produced items and awards the points to the teams as the factory scenario is very complex for a human referee to follow. In the near future, we also want to extend the referee box with an automated robot and product tracking system to take another step towards the automated factory.

Acknowledgments We thank the anonymous reviewers for their helpful comments.

References

1. E. Lee, Cyber Physical Systems: Design Challenges. In: *11th IEEE International Symposium on Object Oriented Real-Time Distributed Computing (ISORC-08)*. 2008, pp. 363–369
2. P. Marwedel, *Embedded System Design*, 1st edn. Springer, 2005. http://www.amazon.com/exec/obidos/redirect?tag=citeulike07-20&path=ASIN/0387292373
3. E.A. Lee, Computing foundations and practice for cyber-physical systems: A preliminary report. University of California, Berkeley, Tech. Rep. UCB/EECS-2007-72, 2007
4. M. Broy, M. Cengarle, E. Geisberger, Cyber-physical systems: Imminent challenges. In: *Large-Scale Complex IT Systems. Development, Operation and Management, Lecture Notes in Computer Science*, vol. 7539, ed. by R. Calinescu, D. Garlan, Springer Berlin Heidelberg, 2012, pp. 1–28. doi:10.1007/978-3-642-34059-8_1. http://dx.doi.org/10.1007/978-3-642-34059-8_1

5. B. Vogel-Heuser, M. Witsch, eds., *Erhöhte Verfügbarkeit und transparente Produktion, Embedded Systems – 1 Tagungen und Berichte*, vol. 2. kassel university press, Kassel, 2011. http://nbn-resolving.org/urn/resolver.pl?urn=urn:nbn:de:0002-31799

6. R. Hensel, Industrie 4.0 revolutioniert die produktion. VDI Nachrichten, 2012

7. C.L. Forgy, Rete: A fast algorithm for the many pattern/many object pattern match problem. Artificial Intelligence **19** (1), 1982

8. R.M. Wygant, CLIPS: A powerful development and delivery expert system tool. Computers & Industrial Engineering **17** (1–4), 1989

9. T. Niemueller, G. Lakemeyer, A. Ferrein, Incremental Task-level Reasoning in a Competitive Factory Automation Scenario. In: *Proc. of AAAI Spring Symposium 2013 - Designing Intelligent Robots: Reintegrating AI.* 2013

10. J.C. Giarratano, *CLIPS Reference Manuals*, 2007. http://clipsrules.sf.net/OnlineDocs.html

The Carologistics Approach to Cope with the Increased Complexity and New Challenges of the RoboCup Logistics League 2015

Tim Niemueller, Daniel Ewert, Sebastian Reuter, Alexander Ferrein, Sabina Jeschke and Gerhard Lakemeyer

Abstract The RoboCup Logistics League (RCLL) has seen major rule changes increasing the complexity, e. g. by raising the number of product variants from 3 to almost 250, and introducing new challenges like the handling of physical processing machines. We describe various aspects of our system that allowed to improve the performance in 2015 and our efforts to advance the league as a whole.

Keywords RoboCup Logistics League · Robotino · Festo

1 Introduction

In 2015 the RoboCup Logistics League (RCLL) has changed considerably by introducing new machines on the field that require more elaborate handling, and by increasing the complexity through increasing the number of possible products from 3 to almost 250. The Carologistics team was able to adapt best to the new circumstances. This was possible, because from the beginning of the team in 2012, flexible and robust solutions were chosen and developed. Members of three partner institutions bring their individual strengths to tackle the various aspects of the RCLL: designing hardware modifications, developing functional software components, system integration, and high-level control of a group of mobile robots. Only the effective combination of these approaches explains the team's overall performance.

Our team has participated in RoboCup 2012–2015 and the RoboCup German Open (GO) 2013–2015. We were able to win the GO 2014 and 2015 (cf. Figure 1) as well as the RoboCup 2014 and 2015 in particular demonstrating flexible task coordination, and robust collision avoidance and self-localization. We have publicly

---

T. Niemueller (✉) · G. Lakemeyer
Knowledge-based Systems Group, RWTH Aachen University, Aachen, Germany
e-mail: niemueller@kbsg.rwth-aachen.de

D. Ewert · S. Reuter · S. Jeschke
IMA/ZLW & IfU, RWTH Aachen University, Dennewartstr. 27, 52068 Aachen, Germany

A. Ferrein
MASCOR Institute, Aachen University of Applied Sciences, Aachen, Germany

Originally published in "RoboCup 2015:
Robot World Cup XIX", © Springer 2015. Reprint by Springer International
Publishing Switzerland 2016, DOI 10.1007/978-3-319-42620-4_46

Figure 1 Teams Carologistics (robots with additional laptop) and Solidus (pink parts) during the RCLL finals at RoboCup 2015, Hefei, China, that ended 46 to 16

released our software stack used in 2014 in particular including our high-level reasoning components for all stages of the game[1] [1].

In the following we will describe some of the challenges originating from the new game play in 2015. In Section 3 we give an overview of the Carologistics platform and the changes that were necessary to adapt to the new game play. Some parts have been used during the German Open 2015, but several components were extended and improved afterwards. We continue highlighting our behavior components in Section 4 and our continued involvement for advancing the RCLL as a whole in Section 5 before concluding in Section 6.

1.1   Game Play Changes 2015

The goal is to maintain and optimize the material flow in a simplified Smart Factory scenario. Two competing groups of up to three robots each use a set of exclusive machines spread over a common playing field to produce and deliver products (cf. [2–4]).

In 2015, the RCLL has changed considerably by introducing actual processing machines based on the Modular Production System (MPS) platform by Festo Didactic [3] as shown in Figures 2 and 4. For more details on the new game play we refer to [5]. The new machines require to equip the robot with a gripper for product handling, adaptation to the general game play due to vastly extended production schedules, and suggest switching to the Robotino 3 platform, which is larger and supports a higher payload.

In the *exploration phase* the robots are given zones on the playing field in which they are supposed to look for machines, and – if one is found – identify and report them. In 2015, the machines can be freely positioned within the zones, in particular at any randomly chosen orientation. Additionally, the referee box can no longer

---

[1]Software stack available at http://www.fawkesrobotics.org/p/llsf2014-release.

Figure 2 Carologistics
Robotino approaching a ring
station MPS

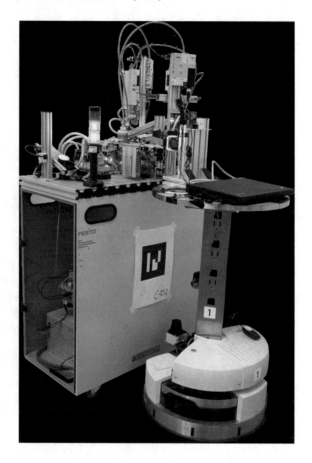

provide ground truth information about the poses of machines in the production phase, making a successful exploration a mandatory requirement. Additionally the exploration procedure per machine became more complex. The robots must first identify whether there is a machine in a zone or not, generally requiring multiple positions per zone to be checked. In our system, we sweep the zone with the laser scanner looking for machine edges and a camera looking for markers to make a quick decision whether a zone is empty or not. If a machine is identified, robots must align on the output side of a machine (frequently requiring to go around the machine which costs time) to identify the marker and light signal with high confidence.

In the *production phase*, work orders are much more diverse in 2015 increasing the number of product variants from 3 to about 250. An example production chain is shown in Figure 3. This in turn requires that the high-level reasoning component needs to make more decisions dynamically and opportunistic production is virtually impossible. It also requires to coordinate the robots more closely to achieve delivery in the desired time windows. Another challenge is that points are only awarded on successful delivery, even the points for intermediate processing steps. This requires

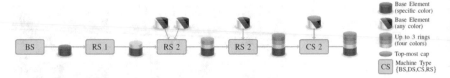

Figure 3 Refinement steps for the production of a highest complexity product in the RCLL 2015 (legend on the right)

Figure 4 Play-offs game at the German Open 2015

the robots to be much more robust and the handling to be nearly perfect since harm done by failures in the production later along the chain is much more severe than before (Figure 4).

2   The Carologistics Platform

The standard robot platform of this league is the Robotino by Festo Didactic [6]. The Robotino is developed for research and education and features omni-directional locomotion, a gyroscope and webcam, infrared distance sensors, and bumpers. The teams may equip the robot with additional sensors and computation devices as well as a gripper device for product handling.

2.1 Hardware System

The robot system currently in use is based on the Robotino 3. The modified Robotino used by the Carologistics RoboCup team is shown in Figure 5 and features three additional webcams and a Sick laser range finder. The webcam on the top is used to recognize the signal lights, the one above the number to identify machine markers, and the one below the gripper is used experimentally to recognize the conveyor belt. We have recently upgraded to the Sick TiM571 laser scanner used for collision avoidance and self-localization. It has a scanning range of 25 m at a resolution of $\frac{1}{3}$ degrees. An additional laptop increases the computation power.

Figure 5 Carologistics Robotino 2015

Several parts were custom-made for our robot platform. Most notably, a custom-made gripper based on Festo fin-ray fingers and 3D-printed parts is used for product handling. The gripper is able to adjust for lateral and height offsets.

2.2 Architecture and Middleware

The software system of the Carologistics robots combines two different middlewares, Fawkes [7] and ROS [8]. This allows us to use software components from both systems. The overall system, however, is integrated using Fawkes. Adapter plugins connect the systems, for example to use ROS' 3D visualization capabilities. The overall software structure is inspired by the three-layer architecture paradigm [9]. It consists of a deliberative layer for high-level reasoning, a reactive execution layer for breaking down high-level commands and monitoring their execution, and a feedback control layer for hardware access and functional components. The lowest layer is described in Section 3. The upper two layers are detailed in Section 4. The communication between single components – implemented as *plugins* – is realized by a hybrid blackboard and messaging approach [7]. This allows for information exchange between arbitrary components. Information is written to or read from *interfaces*, each carrying certain information, e.g. sensor data or motor control, but also more abstract information like the position of an object. The information flow is somewhat restricted – by design – in so far as only one component can write to an interface. Reading, however, is possible for an arbitrary number of components. This approach has proven to avoid race conditions when for example different components try to instruct another component at the same time. The principle is that the interface is used by a component to provide state information. Instructions and commands are sent as messages. Then, multiple conflicting commands can be detected or they can be executed in sequence or in parallel, depending on the nature of the commands.

3 Advances to Functional Software Components

A plethora of different software components is required for a multi-robot system. Here, we discuss some components and advances of particular relevance to the game as played in 2015.

3.1 Basic Components

The lowest layer in our architecture which contains functional modules and hardware drivers. All functional components are implemented in Fawkes. Drivers have been implemented based on publicly available protocol documentation, e.g. for our laser range finders or webcams. For this year, we have extended the driver of our laser range finder for the Sick TiM571 model. To access the Robotino base platform hardware we make use of a minimal subset of OpenRobotino, a software system provided by the manufacturer. For this year, we have upgraded to using the version 2 API in order to use the Robotino 3 platform. Localization is based on Adaptive Monte Carlo Localization which was ported from ROS and then extended. For locomotion, we integrated the collision avoidance module [10] which is also used by the AllemaniACs[2] RoboCup@Home robot.

3.2 Light Signal Vision

A computer vision component for robust detection of the light signal state on the field has been developed specifically for this domain. For 2015, we have improved the detection component to limit the search within the image by means of the detected position of the machine, which is recognized through a marker and with the laser range finder. This provides us with a higher robustness towards ambiguous backgrounds, for example colored T-shirts in the audience. Even if the machine cannot be detected, the vision features graceful degradation by using a geometric search heuristic to identify the signal, losing some of the robustness towards the mentioned disturbances.

3.3 Conveyor Belt Detection

The conveyor belts are rather narrow compared to the products thus require a precise handling (Figure 6). The tolerable error margin is in the range of about 3 mm. The marker on a machine allows to determine the lateral offset from the gripper to

---

[2]See the AllemaniACs website at http://robocup.rwth-aachen.de.

Figure 6 Vision-based conveyor belt detection (training images)

the conveyor belt. It gives a 3D pose of the marker with respect to the camera and thus the robot. However, this requires a precise calibration of the conveyor belt with

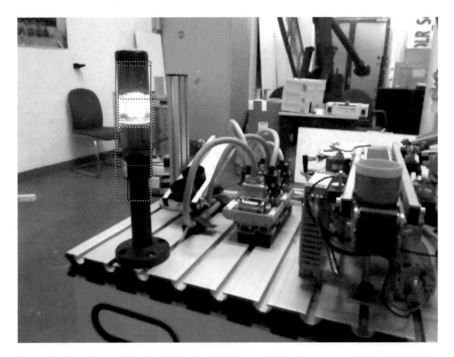

Figure 7 Vision-based light-signal detection during production (post-processed for legibility)

respect to the marker. While ideally this would be the same for each machine, in practice there is an offset which would need to be calibrated per station. Combined with the inherent noise in the marker detection this approach requires filtering and longer accumulation times. The second approach is to detect a line using the laser scanner. This yields more precise information which we hope to further improve with the higher resolution of the Sick TiM571.

The third approach we are investigating is using a dedicated vision component to detect the conveyor belt in an image. So far, we have implemented a detection method based on OpenCV's cascade classifier [11, 12]. First, videos are recorded that are split into images from which positive and negative samples are extracted. These samples are then fed into a training procedure that extracts local binary features from which the actual classifiers are built. This training requires tremendous amounts of computing power and is generally done offline. The detection, however, is swift and allows for on-line real-time detection of the conveyor belt as shown in Figure 7. We are currently evaluating the results and considering further methods to improve on these results.

4   High-Level Decision Making and Task Coordination

The behavior generating components are separated into three layers, as depicted in Figure 8: the low-level processing for perception and actuation, a mid-level reactive layer, and a high-level reasoning layer. The layers are combined following an adapted hybrid deliberative-reactive coordination paradigm.

The robot group needs to cooperate on its tasks, that is, the robots communicate information about their current intentions, acquire exclusive control over resources like machines, and share their beliefs about the current state of the environment.

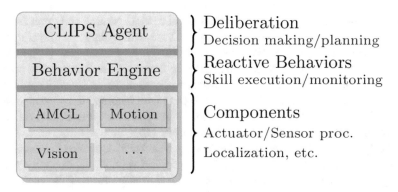

Figure 8 Behavior layer separation

Currently, we employ a distributed, local-scope, and incremental reasoning approach [5]. This means that each robot determines only its own action (local scope) to perform next (incremental) and coordinates with the others through communication (distributed), as opposed to a central instance which plans globally for all robots at the same time or for multi-step plans.

In the following we describe the reactive and deliberative layers of the behavior components. For computational and energy efficiency, the behavior components need also to coordinate activation of the lower level components.

4.1   Lua-Based Behavior Engine

In previous work we have developed the Lua-based Behavior Engine (BE) [13]. It serves as the reactive layer to interface between the low- and high-level systems. The BE is based on hybrid state machines (HSM). They can be depicted as a directed graph with nodes representing states for action execution, and/or monitoring of actuation, perception, and internal state. Edges denote jump conditions implemented as Boolean functions. For the active state of a state machine, all outgoing conditions are evaluated, typically at about 15 Hz. If a condition fires, the active state is changed to the target node of the edge. A table of variables holds information like the world model, for example storing numeric values for object positions. It remedies typical problems of state machines like fast growing number of states or variable data passing from one state to another. Skills are implemented using the light-weight, extensible scripting language Lua.

4.2   Incremental Reasoning Agent

The problem at hand with its intertwined world model updating and execution naturally lends itself to a representation as a fact base with update rules for triggering behavior for certain beliefs. We have chosen the CLIPS rules engine [14], because using incremental reasoning the robot can take the next best action at any point in time whenever the robot is idle. This avoids costly re-planning (as with approaches using classical planners) and it allows us to cope with incomplete knowledge about the world. Additionally, it is computationally inexpensive. More details about the general agent design and the CLIPS engine are in [15].

The agent for 2015 is based on the previous years [15]. One major improvement is the introduction of a task concept. Until the 2014 version, the executive part of the agent consisted of separate rules for each action to perform. This resulted in a considerable amount of rules often repeating similar conditions. While this is not a problem in terms of performance, it does make maintaining the agent code more involved. For 2015, we have introduced the concept of tasks. Tasks group several steps necessary to perform a certain behavior, for example producing a low complexity product consisting only of a base element and a cap, into a single entity. Then, generic code can execute the steps of a task. We retained the flexibility and extensibility of our approach by allowing to add execution rules for steps which need special treatment or parametrization, and monitoring rules to react to disturbances during execution, for example a product being dropped during transport.

Additionally, we are progressing on making the world model synchronization generic. So far, we have had an explicit world model consisting of a specified set of elements. Now, we are aiming for a more generic world model consisting of key-value pairs. This will allow for simpler updates to the world model as modifications do not require changing an explicit world model schema anymore.

4.3 Multi-robot Simulation in Gazebo

The character of the RCLL game emphasizes research and application of methods for efficient planning, scheduling, and reasoning on the optimal work order of production processes handled by a group of robots. An aspect that distinctly separates this league from others is that the environment itself acts as an agent by posting orders and controlling the machines' reactions. This is what we call *environment agency*. Naturally, dynamic scenarios for autonomous mobile robots are complex challenges in general, and in particular if multiple competing agents are involved. In the RCLL, the large playing costs are prohibitive for teams to set up a complete scenario for testing, let alone to have two teams of robots. Members of related communities like planning and reasoning might not want to deal with the full software and system complexity, yet they welcome relevant scenarios to test and present their research. Therefore, we have created an *open simulation environment* [15, 16] based on Gazebo.[3]

After RoboCup 2014 the BBUnits team joined the effort and we cooperatively extended the simulation for the new game as shown in Figure 9. The simulation is developed publicly[4] and we hope for more teams to join the effort. With the new game, the need for a simulation is even more pressing as the cost for the field has drastically increased. Additionally, we envision that long-term games and future changes to the game can be tried in the simulation before implementing it in the real world [17].

---

[3]More information, media, the software itself, and documentation are available at http://www.fawkesrobotics.org/projects/llsf-sim/.

[4]The project is hosted at https://github.com/robocup-logistics.

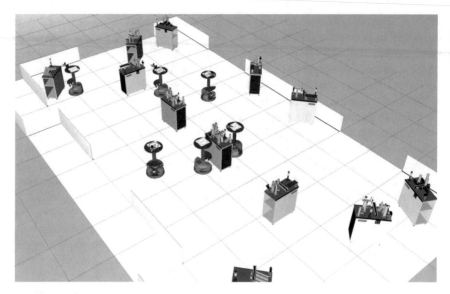

Figure 9 Simulation of the RCLL 2015 with MPS stations

5   League Advancements and Continued Involvement

We have been active members of the Technical, Organizational, and Executive Committees and proposed various ground-breaking changes for the league like merging
the two playing fields or using physical processing machines in 2015 [2, 3]. Additionally we introduced and currently maintain the autonomous referee box for the
competition and develop an open simulation environment.

5.1   RCLL Referee Box and MPS Stations

The Carologistics team has developed the autonomous referee box (refbox) for the
RCLL which was deployed since 2013 [2]. It strives for full autonomy, i. e. it tracks
and monitors machine states, creates (randomized) game scenarios, handles communication with the robots, and interacts with a human referee.

In 2015, the refbox required updates for the modified game involving MPS stations. A noteworthy difference is that in the current game, the products can no longer
be tracked as they do not have an RFID chip anymore. There are plans to remedy this
later, for example using bar codes or RFID chips again on products. This would in
particular allow to grant production points not only on delivery, but once a production
step has been performed. The new MPS stations require programs that provide sensor
data to and execute commands from the refbox. For example, all stations require a
setup step to be performed, for example to instruct the particular ring color a ring

station should mount in the next step. Festo Didactic sponsored an internship for a student who performed the development embedded in the Carologistics team. It involved developing a program for each of the four station types to reliably carry out the production steps and implementing a robust communication to the stations. As Wifi is used for communication which is inherently unreliable in particular during RoboCup events, the refbox and stations must be robust towards temporary connection losses. The basic prototype was used successfully at the GO2015 and we are continuing the development to further stabilize and improve the game.

5.2 League Evaluation

We have made an effort [17] to analyze past games based on data recorded during games by the referee box and propose new additional criteria for game evaluation

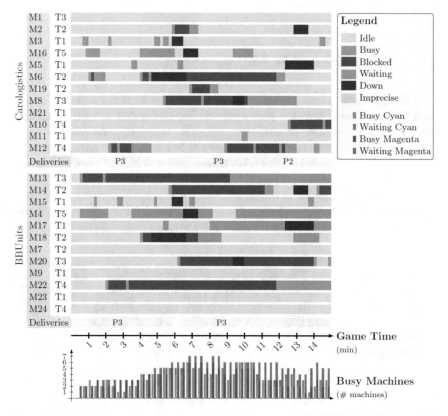

Figure 10 Machine states over the course of the final game at RoboCup 2014. The lower graph shows the occupied machines per 20 sec time block

based on Key Performance Indicators (KPI) used in industrial contexts to evaluate factory performance. For example, one such KPI is the throughput time TTP which describes the time required at a specific station to complete processing and retrieve the processed workpiece. We have analyzed several games and will focus now on the finals of RoboCup 2014 between the Carologistics (cyan) and BBUnits (magenta) which ended with a score of 165 to 124.[5] It seems that especially the lower throughput time TTP of cyan contributed to the Carologistics' success. Machines can be used again much faster. This is shown by the green times in Figure 10, which is effective utilization of the machines. The BBUnits team has much longer waiting times (orange) that resulted for one in machines to be blocked for other uses and for another meant that the production chain for a product was not advanced quickly. For more details we refer to [17]. For a generalization to mobile industrial robotics scenarios cf. [18].

We imagine that KPIs can play a role in the future of the league and could be tested in simulated tournaments before. The Technical Committee has added a technical challenge to the rules 2015 which involves performing such a simulated game in order to foster adoption and participation in the simulation efforts.

5.3 Public Release of Full Software Stack

Over the past eight years, we have developed the *Fawkes Robot Software Framework* [7] as a robust foundation to deal with the challenges of robotics applications in general, and in the context of RoboCup in particular. It has been developed and used in the Middle-Size [19] and Standard Platform [20] soccer leagues, the RoboCup@Home [21, 22] service robot league, and now in the *RoboCup Logistics League* [15] as also shown in Figure 11.

The Carologistics are the first team in the RCLL to publicly release their software stack. Teams in other leagues have made similar releases before. What makes ours unique is that it provides a complete and *ready-to-run package with the full software*

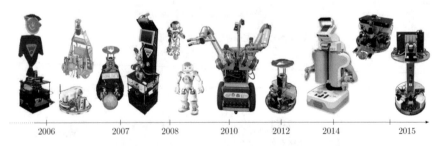

2006	2007	2008	2010	2012	2014	2015

Figure 11 Robots running (parts of) Fawkes which were or are used for the development of the framework and its components

---

[5]Video of the final is available at https://youtu.be/_iesqH6bNsY.

(and some additions and fixes) that we used in the competition in 2014. This in particular *includes* the complete *task-level executive* component, that is the strategic decision making and behavior generating software. This component was typically held back or only released in small parts in previous software releases by other teams (for any league). We are currently preparing a similar release of the software stack 2015.

5.4 Agent-Based Programming for High-Level Control

In Section 4.2, we have discussed our incremental high-level reasoning component. Furthermore, we have used the RCLL scenario as an evaluation testbed for the agent-based programming and reasoning system YAGI [23]. YAGI (Yet Another Golog Interpreter) is an approach to develop an alternative implementation to the well-known logic-based agent control language GOLOG. It combines imperative programming with reasoning to formulate the behaviors of the robot.

6 Conclusion

In 2015, we have in particular adapted to the new game. We upgraded to the Robotino 3 platform, developed new custom hardware additions like a gripper, and adapted and extended the behavior and functional components. We have also continued our contributions to the league as a whole through active participation in the league's committees, publishing papers about the RCLL and proposing new performance criteria. The development of the simulation we initiated has been transferred to a public project where other teams have joined the effort. Most notably, however, have we released the complete software stack including all components and configurations as a ready-to-run package.

The website of the Carologistics RoboCup Team with further information and media can be found at http://www.carologistics.org.

Acknowledgments The team members in 2015 are Daniel Ewert, Alexander Ferrein, Mostafa Gomaa, Sabina Jeschke, Gerhard Lakemeyer, Nicolas Limpert, Matthias Löbach, Randolph Maaßen, David Masternak, Victor Mataré, Tobias Neumann, Tim Niemueller, Sebastian Reuter, Johannes Rothe, and Frederik Zwilling. We gratefully acknowledge the financial support of RWTH Aachen University and FH Aachen University of Applied Sciences. We thank Sick AG and Adolf Hast GmbH & Co. KG for sponsoring our efforts by providing hardware and manufacturing support. F. Zwilling and T. Niemueller were supported by the German National Science Foundation (DFG) research unit *FOR 1513* on Hybrid Reasoning for Intelligent Systems (http://www.hybrid-reasoning.org).

References

1. T. Niemueller, S. Reuter, A. Ferrein, Fawkes for the robocup logistics league. In: *RoboCup Symposium 2015 – Development Track*. 2015
2. T. Niemueller, D. Ewert, S. Reuter, A. Ferrein, S. Jeschke, G. Lakemeyer, RoboCup Logistics League Sponsored by Festo: A Competitive Factory Automation Benchmark. In: *RoboCup Symposium 2013*. 2013
3. T. Niemueller, G. Lakemeyer, A. Ferrein, S. Reuter, D. Ewert, S. Jeschke, D. Pensky, U. Karras, Proposal for Advancements to the LLSF in 2014 and beyond. In: *ICAR – 1st Workshop on Developments in RoboCup Leagues*. 2013
4. T. Niemueller, G. Lakemeyer, A. Ferrein, Incremental Task-level Reasoning in a Competitive Factory Automation Scenario. In: *Proc. of AAAI Spring Symposium 2013 - Designing Intelligent Robots: Reintegrating AI*. 2013
5. T. Niemueller, G. Lakemeyer, A. Ferrein, The RoboCup Logistics League as a Benchmark for Planning in Robotics. In: *25th Int. Conference on Automated Planning and Scheduling (ICAPS) – Workshop on Planning in Robotics*. 2015
6. U. Karras, D. Pensky, O. Rojas, Mobile Robotics in Education and Research of Logistics. In: *IROS 2011 – Workshop on Metrics and Methodologies for Autonomous Robot Teams in Logistics*. 2011
7. T. Niemueller, A. Ferrein, D. Beck, G. Lakemeyer, Design Principles of the Component-Based Robot Software Framework Fawkes. In: *Int. Conference on Simulation, Modeling, and Programming for Autonomous Robots (SIMPAR)*. 2010
8. M. Quigley, K. Conley, B.P. Gerkey, J. Faust, T. Foote, J. Leibs, R. Wheeler, A.Y. Ng, ROS: an open-source Robot Operating System. In: *ICRA Workshop on Open Source Software*. 2009
9. E. Gat, Three-layer architectures. In: *Artificial Intelligence and Mobile Robots*, ed. by D. Kortenkamp, R.P. Bonasso, R. Murphy, MIT Press, 1998, pp. 195–210
10. S. Jacobs, A. Ferrein, S. Schiffer, D. Beck, G. Lakemeyer, Robust collision avoidance in unknown domestic environments. In: *RoboCup Symposium 2009*, ed. by J. Baltes, M.G. Lagoudakis, T. Naruse, S.S. Ghidary. 2009
11. P. Viola, M. Jones, Rapid object detection using a boosted cascade of simple features. In: *IEEE Conf. on Comp. Vision and Pattern Recog. (CVPR)*. 2001
12. R. Lienhart, J. Maydt, An extended set of haar-like features for rapid object detection. In: *Proc. Int. Conf on Image Processing*. 2002
13. T. Niemueller, A. Ferrein, G. Lakemeyer, A Lua-based Behavior Engine for Controlling the Humanoid Robot Nao. In: *RoboCup Symposium 2009*. 2009
14. R.M. Wygant, CLIPS: A powerful development and delivery expert system tool. Computers & Industrial Engineering **17** (1–4), 1989
15. T. Niemueller, S. Reuter, D. Ewert, A. Ferrein, S. Jeschke, G. Lakemeyer, Decisive Factors for the Success of the Carologistics RoboCup Team in the RoboCup Logistics League 2014. In: *RoboCup Symposium – Champion Teams Track*. 2014
16. F. Zwilling, T. Niemueller, G. Lakemeyer, Simulation for the RoboCup Logistics League with Real-World Environment Agency and Multi-level Abstraction. In: *RoboCup Symposium*. 2014
17. T. Niemueller, S. Reuter, A. Ferrein, S. Jeschke, G. Lakemeyer, Evaluation of the RoboCup Logistics League and Derived Criteria for Future Competitions. In: *RoboCup Symposium 2015 – Development Track*. 2015
18. T. Niemueller, G. Lakemeyer, S. Reuter, S. Jeschke, A. Ferrein, Benchmarking of cyber-physical systems in industrial robotics. In: *Cyber-Physical Systems: Foundations, Principles and Applications*, ed. by H. Song, D.B. Rawat, S. Jeschke, C. Brecher, Elsevier, Amsterdam, 2016 (in press)
19. D. Beck, T. Niemueller, AllemaniACs 2009 Team Description. Tech. rep., Knowledge-based Systems Group, RWTH Aachen University, 2009
20. A. Ferrein, G. Steinbauer, G. McPhillips, T. Niemueller, A. Potgieter, Team ZaDeAt 2009 – Team Report. Tech. rep., RWTH Aachen Univ., Graz Univ. of Technology, and Univ. of Cape Town, 2009

21. S. Schiffer, G. Lakemeyer, AllemaniACs Team Description RoboCup@Home. TDP, Knowledge-based Systems Group, RWTH Aachen University, 2011
22. A. Ferrein, T. Niemueller, S. Schiffer, G.L. and, Lessons Learnt from Developing the Embodied AI Platform Caesar for Domestic Service Robotics. In: *AAAI Spring Symposium 2013 - Designing Intelligent Robots: Reintegrating AI*. 2013
23. A. Ferrein, C. Maier, C. Mühlbacher, T. Niemueller, G. Steinbauer, S. Vassos, Controlling Logistics Robots with the Action-based Language YAGI. In: *IROS Workshop on Task Planning for Intelligent Robots in Service and Manufact.* 2015

AUDIME: Augmented Disaster Medicine

Alexander Paulus, Michael Czaplik, Frederik Hirsch, Philipp Meisen,
Tobias Meisen and Sabina Jeschke

Abstract In this positioning paper we present the AUDIME project approach in
which we plan to evaluate the usability and social acceptance of smart and wearable devices in the context of mass-casualty-incidents. AUDIME aims to provide a
platform which captures, evaluates, and provides data from various sources on an
incident site. This way, triage, patient monitoring, information sharing and communication are to be improved, which simplifies decision making at executive staff level.
In contrast to previous projects, AUDIME does not replace any low-tech or non-tech
approaches (e.g. triage cards,) but enhances the information handling by capturing
analogue information using smart or wearable devices and distributing digitalized
data to qualified recipients in real-time.

Keywords Mass-casualty-incident · Tele-medicine · Information System
· Wearable Device

1 Introduction

In the scenario of a mass-casualty incident (MCI) the first triage of casualties is an
essential part in the process of medical assistance. Due to the chaotic nature in the
area of accident, arriving (medical) forces may miss to take the correct and advised
steps when encountering casualties in the field [1]. This often leads to insufficient
treatment of patients or misleading information forwarded to the executive emergency physician (EEP). As a result of this deficit, the coordination and distribution
of available medical staff, devices, and transport capacities, as well as alerting potentially necessary additional forces gets complicated as an overview of all available

A. Paulus (✉) · P. Meisen · T. Meisen · S. Jeschke
IMA/ZLW & IfU, RWTH Aachen University, Dennewartstr. 27, 52068 Aachen, Germany
e-mail: alexander.paulus@ima-zlw-ifu.rwth-aachen.de

M. Czaplik · F. Hirsch
Department of Anaesthesiology, University Hospital Aachen, Aachen, Germany

Originally published in "Proceedings of the 17th International
Conference on E-health Networking, Application & Services (HealthCom)",
© IEEE 2015. Reprint by Springer International Publishing Switzerland 2016,
DOI 10.1007/978-3-319-42620-4_47

information is not available for the EEP's disposal, which is one of the major problems during the MCI scenario.

To mitigate these problems, we initiated the project AUgmented DIsaster MEdicine (AUDIME), which is founded by the German "Federal Ministry of Education and Research" (BMBF) under the topic of "Civil Security–Custody and Rescue in Complex Emergency Situations". AUDIME is intended to evaluate the effects of using mobile and wearable devices to form an on-scene IT support platform for medical and organisational decision making in the context of an MCI. For scientific experiments we aim to provide a prototype system, which provides three central support sub-systems for medical staff equipped with compatible hardware:

- assistive information display for first arriving medical staff regarding triaging schemata,
- gathering of information from patients by on-scene physicians, and
- provision of gathered information sets to executive forces on and off accident site.

With AUDIME, information display and gathering is done in the field using smart glasses or similar wearable devices carried by key personnel like emergency physicians. Utilizing mobile devices improves information reception and capture using state of the art hardware (like tablets, smart phones or also wearable devices). In addition it enables the display of structured data summaries to executives. This also includes information displayed from on-site patient monitoring devices, leading to the need of an integration of heterogeneous data sources to support an on-scene physician with enhanced real-time information. Furthermore, AUDIME is planned to profit from the recent availability of telemedical assistance systems (e.g. *TemRas* [2]) to encounter the shortage of medical professionals in pre-hospital emergency situations to enhance patient treatment in an MCI by using tele-medical assistance. A close integration of those systems into AUDIME is desired as we have all audiovisual equipment needed already available at site of the incident and in the ambulances. In AUDIME, major fields of application of telemedic assistance will be triage support and off-site patient monitoring.

Using high-tech components in a medical environment, which currently is dominated by low- or non-tech components is mostly unevaluated. This includes acceptance and usability by medical and supportive staff as well as patient acceptance. AUDIME will therefore also feature an evaluation of the social impacts on the on-scene personnel in an emergency situation when utilizing wearable devices for communication and information exchange.

2 Approach

During the first hour of an MCI, the arriving medical forces face a problem mainly consisting of estimating the situation and coordinating the few available medical staff members. A hierarchy among the available forces is built ad-hoc, putting the first arriving emergency physician in charge of organizing the medical care of all patients. Within the next few minutes, this emergency physician will have to acquire an overview of all triaged patients, including their triage priorities (if already known), and a sketch of the disaster area. Gathering all needed information is time-consuming and prone to error, especially as the acquired information may be outdated when arriving at the EEP from various sources. Most of these sources may be semi-qualified paramedics or non-medical personnel (e.g. fire-fighters), which are expected to apply simple triaging schemata to classify a patient into one of four categories (red, yellow, green; critical to non-critical and black for dead or without chance to survive). Recent studies in the context of mass exercises have shown those first triages to only be correct in about 80 % of all cases [1], which may lead to increased numbers of casualties caused by incorrect or insufficient treatment. To improve the correct classification of patients, AUDIME will support triages by displaying pre-defined instructions, using smart glasses, to the medical staff. If available, telemedic support can also be utilized. Furthermore, after the triage, the triage card has been filled describing the current state of the patient. This information should be automatically scanned using OCR and provided to the information system.

Enabling the use of high-tech devices in an emergency situation faces several difficulties which were discovered during previous projects attempting to establish new hardware devices to *replace* low-tech components like triage cards (cf. [3–6]). In contrast to these approaches, AUDIME aims to *enhance* the low-tech solutions by operating on top of the accustomed techniques, while still preserving the low-tech solution as a fall-back approach. Information should be captured from low-tech components by wearable devices, integrated into common data structures and dynamically shared to all recipients.

In the following we will introduce two of the main components of the AUDIME project, the used wearable devices and the information management.

2.1 Wearable Devices

The central concept of wearable device utilization in AUDIME is based on a hands-free approach to mobile or wearable devices. On the one hand, this offers several advantages compared to usual mobile hardware, but also carries some problems. One of the main reasons requiring a hands-free approach is the fact that, especially in the first time of an MCI, on-scene staff has many tasks which require one or both hands (e.g. triage, filling triage cards, moving wreckage). Using wearable devices enables additional information display and data capture without influencing the medical staff operating at high stress or influencing their handwork. On the other hand, a major drawback is reliable device handling. State-of-the-art wearable devices are commonly operated using voice commands, which may emerge infeasible due to high surrounding noise. Touching on-device buttons might be unwanted for hygienic reasons. Furthermore, there are multiple devices with different used technology available which have to be evaluated w.r.t. technical usability and social user acceptance.

As they are a central concept of AUDIME, most of the planned research focus will involve the usage of wearable devices. It is planned to use multiple state of the art technologies to explore different kinds of wearable concepts. In this context, the applicability of virtual reality concepts assisting triage processes will be tested. This also includes the manageability by the staff using the devices. The concept will be kept generic, such that no excessive special training is needed to operate wearable devices in the field. Configuration and setup prior to usage is kept as simple as possible to take as little time as possible.

Due to their multiple capabilities, wearable devices can be used to achieve different tasks which previously required multiple low-tech devices and/or personal effort:

- Assisted triage with recording and live sharing of findings,
- Detail and overview information retrieval by executive forces from on-scene physicians,
- General communication with dedicated channels for each conversation, previously done by general radio usage,
- Live data display from medical hardware (e.g. ECG),
- Previously unavailable remote diagnostics using live video analysis, and
- Support for tele-medic assistance while providing live video or data feed to the tele-physician.

If GPS signals are available, patient tracking is also possible due to geo-tagged readings of the triage tags. This allows for live overview maps of the incident scene and treatment places.

2.2   Information Management

Raw heterogeneous information material captured by smart phones, wearable devices, tablets, and medical devices (e.g. ECG) has to be stored, analysed, filtered, and integrated into existing common data models and eventually shared with interested personnel on- and off-site. The timeliness of the shared data is of utmost importance. AUDIME aims to provide a real-time data analysis and distribution using dedicated hardware to perform such steps, while mobile devices only perform simple capture, cache and display tasks. All communication and data operations are handled by the so-called Information-Integration-Layer (IIL). Running on powerful, yet mobile, hardware, it performs all necessary tasks to extract and distribute data from raw material (e.g. pictures, video clips) of mobile devices.

Thus, one of the preconditions for the development of AUDIME was the existence of an on-site communication network similar to WiFi. Although such an infrastructure cannot be guaranteed today, there will be solutions to establish adequate networks in an ad-hoc manner in the future, according to Braunstein et al. [7]. We therefore expect to be capable of transferring data from clients to servers and vice versa with little to no delay, causing only the effectiveness of the running analysis algorithms to limit the performance of our system. With only the raw server performance considered a bottleneck, one of the major technical tasks of AUDIME is to provide a server backend software architecture, which analyses data send by mobile devices and returns extracted, summarized and, if possible, enhanced information as fast as possible, preferably in real time.

To demonstrate and evaluate this concept, we will test the real-time analysis of uploaded raw data on the example of the Eulerian Video Magnification algorithm [8]. It features the magnification and extraction of subtle, non-human-readable changes in picture sequences, which can be used to analyse heart beat rate or breathing frequency from observed patients. To achieve this goal, we capture a short video sequence of the patient (e.g. his/her face), send it to the backend via WiFi, run the algorithm and return acquired vital parameters to the wearable device. If additional data regarding the current patient is available, it will be sent as well. This allows the analysis of vital parameters of yet unreachable patients. As a side effect we plan to evaluate the applicability of the algorithms with regard to bad conditions like dim light or time needed to generate reliable results. Additional approaches could include the application of machine-learning algorithms to estimate helpful parameters from the gathered data. In the context of big data, there are several approaches for efficient and distributed data handling as shown in [9].

Aside from assisting the emergency physician at the patient, all data is also stored securely in a database. This enables other on-scene personnel, e.g. the EEP or area mangers, to access data summaries or overviews. Those data overviews will be accessible using tablets or notebooks. Additional advancing personnel will also be able to request general anonymized live data from the IIL (e.g. number of patients per triage

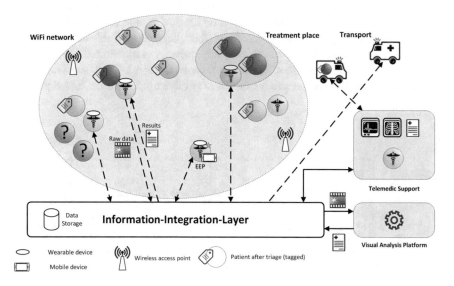

Figure 1 Architectural overview of the AUDIME information infrastructure

category or the overview map) to not arrive at the site totally unprepared. Preshared information such as a sketch of the accident surroundings or hints about blocked access ways can help arriving staff to be on site and evacuate critical patients faster than before as indispensable drive ways can be kept clear from the beginning.

Figure 1 gives a high level overview of the information technical concept and components of the AUDIME approach. Triaged patients are assigned one of the four categories and their triage tag is scanned using the physician's wearable device. If necessary, an estimation of the heart rate and/or breathing frequency is attempted using the *Video Analysis Platform* which is accessible via the IIL. The EEP as well as tele-medic support can access all stored data at any time.

3   First Perceptions

To estimate and determine the functional and non-functional requirements to our approach, we conducted surveys of emergency physicians and paramedics. Participants were asked to fill out anonymized questionnaires regarding expectations, possibilities, and possible problems when using wearable devices. A short introduction (written and oral) was given prior to the surveys to introduce the concept of wearable devices in the context of a MCI. From those surveys we were able to define a total of 17 major use cases for on-site personnel in correlation with our project and scientific research goals. The results can be seen in Figure 2. We differentiated the presented use cases into four categories: information request, information recording, communication, and some additional miscellaneous use cases. The first category contains all use cases which mainly focus on retrieving already present data from the IIL. This information is then displayed on the device until the user chooses otherwise. This also includes possible updates, which become available during the time of display. The information request is mostly done by medical staff on-site but is also intended for additional forces on the approach (UC04 + UC05). Most notably is UC06, which enables the on-site physician to get live updates of his recent patients

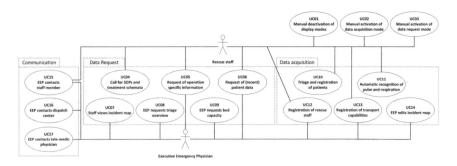

Figure 2 Simplified view of all use cases in the AUDIME project

(e.g. when triaged by another physician) to support treatment decisions. Executive staff handling different devices will obtain different views of the available data, with more focus on overview and organizational purpose (UC06–UC09). There will also be some limited interaction possibilities for this personnel, like editing a location map (UC14). All data input/generation use cases are grouped under the context of information recording, forming the second group mentioned. There are various ways of acquiring data, most frequently by scanning the triage cards using the wearable devices (UC10) and using the EVM algorithm (UC11). Furthermore, there are use cases concerning staff and transport registration to track available resources (UC12 + UC13). Use cases UC01–UC03 are used on wearable devices only to select the current action and form the forth category. This is due to energy saving purposes. UC02 and UC03 are intended to provide a menu to the user, offering different capabilities described in the request or data acquisition group.

In addition to the shown functional requirements, we identified several non-functional requirements which origin in the medical context of an MCI. Examples are hygienic considerations when operating a device while being in contact with patient blood or devices having to be operable wearing emergency equipment (e.g. various types of gloves), helmets, or masks. Due to the unpredictable nature of an MCI, several more tests will have to be conducted to evaluate the usability concept of wearables in real-world conditions.

4 Summary and Outlook

In this positioning paper, we presented the AUDIME approach featuring wearable devices for MCI management. We described the general goals of the project, namely information gathering, evaluating and distributing, which shall increase the manageability of complex and time-critical situations in the case of a disaster. AUDIME aims to set up on top of the already present low-tech solutions without replacing them and without interfering with already known procedures. As a first step, we conducted a user survey and derived the main use cases from the results.

Alongside wearable devices, AUDIME will also utilize common mobile hardware like tablets and smart phones. A "bring-your-own-device" concept is planned to harness additional personal hardware also present at the site and on the approach. Those devices could be used to acquire data from AUDIME information system and instruct medical helpers prior to arrival.

In the upcoming time, we will define a general concept for the information integration layer, as well as the specifications for wearable device software. Information interfaces need to handle different data sources and open up an efficient on-site data handling. An approach to extract medical information from video sequences will be tested as well as the social acceptance of such data acquisition methods.

Acknowledgments The authors would like to thank the members of the AUDIME consortium including the Institute of Information Management in Mechanical Engineering RWTH Aachen

University, Department of Anaesthesiology University Hospital Aachen, Centre for Security and Society Albert-Ludwig-Universität Freiburg, TECH2Go Mobile Systems GmbH and GS Elektromedizinische Geräte GmbH. This work has been developed in the project AUDIME (reference number: 13N13261), which is partly funded by the German ministry of education and research (BMBF) within the research programme "Forschung für die zivile Sicherheit 2012–2017".

References

1. N. Ellebrecht, L. Latasch, Vorsichtung durch Rettungsassistenten auf der Grossuebung SOGRO MANV 500. Notfall und Rettungsmedizin **15** (1), 2012, pp. 58–64
2. S. Thelen, M. Czaplik, P. Meisen, D. Schilberg, S. Jeschke, Using off-the-shelf medical devices for biomedical signal monitoring in a telemedicine system for emergency medical services. Biomedical and Health Informatics, IEEE Journal of **19** (1), 2015, pp. 117–123. 10.1109/JBHI.2014.2361775
3. Anton Donner, Stefan Erl, Christine Adler, Anton Metz, und Marion Krüsmann, Projekt e-Triage: Datenmanagement für die elektronische Betroffenenerfassung und Akzeptanz bei Rettungskräften
4. Projekt Speed Up. http://www.speedup-projekt.de/SpeedUp-p-9.html
5. Projekt Sofortrettung bei Grossunfall mit Massenanfall von Verletzten (SOGRO). http://edok01.tib.uni-hannover.de/edoks/e01fb13/772097763.pdf
6. Projekt Adaptive Lösungsplattform zur Aktiven technischen Unterstützung beim Retten von Menschenleben (ALARM). http://tmcc.charite.de/fileadmin/user_upload/microsites/sonstige/tmcc/ALARM/ALARM-Projekt_Praesentation_Allgemein.pdf
7. B. Braunstein, T. Trimble, R. Mishra, B.S. Manoj, R.R. Rao, L. Lenert, Feasibility of using distributed wireless mesh networks for medical emergency response. In: *AMIA 2006, American Medical Informatics Association Annual Symposium, Washington, DC, USA, November 11–15, 2006.* 2006
8. H.Y. Wu, M. Rubinstein, E. Shih, J. Guttag, F. Durand, W.T. Freeman, Eulerian video magnification for revealing subtle changes in the world. ACM Trans. Graph. (Proceedings SIGGRAPH 2012) **31** (4), 2012
9. Jesko Elsner et al, Implementating a volunteer notification system into a scalable, analytical realtime data processing environment

Fostering Interdisciplinary Integration in Engineering Management

Tobias Vaegs, Inna Zimmer, Stefan Schröder, Ingo Leisten,
René Vossen and Sabina Jeschke

Abstract Research in the challenging field of industrial engineering and engineering management often needs the expertise from more than one discipline. Various disciplinary competences have to be combined to answer research questions and to solve specific (engineering) problems at the interfaces of different professional disciplines. The disciplines being part of the research and problem solving process have to be consequently integrated to form an efficiently performing interdisciplinary consortium. Current research states that this interdisciplinary integration process has to include various dimensions. This paper introduces three sets of interdisciplinary integration tools. Together they cover all of the dimensions explained before and lead to an enhanced interdisciplinary integration. Having just rolled out a set of integration tools, measurement tools are adjusted to evaluate and optimize them continually.

Keywords Interdisciplinary Integration · Knowledge Management · Terminology · Virtual and Physical Cross Linkage

1 Motivation and Problem Statement

The challenging area of industrial engineering proves just as demanding as the management of engineering processes, as all their different facets need to be considered: Problem analysis, design of tailor-made solutions, creation of products out of these solutions and converting them into profit [1]. The complexity of the problems to be solved and research questions to be answered has constantly increased over the years [2].

Not only does the complexity of the problems increase, but more and more problems originate from the interfaces between professional disciplines which traditionally may have been perceived as investigating completely different research subjects [3]. For these problems an isolated disciplinary search for solutions can no longer

T. Vaegs (✉) · I. Zimmer · S. Schröder · I. Leisten · R. Vossen · S. Jeschke
IMA/ZLW & IfU, RWTH Aachen University, Dennewartstr. 27, 52068 Aachen, Germany
e-mail: tobias.vaegs@ima-zlw-ifu.rwth-aachen.de

Originally published in "2013 IEEE International Conference
on Industrial Engineering and Engineering Management ",
© IEEE 2013. Reprint by Springer International Publishing Switzerland 2016,
DOI 10.1007/978-3-319-42620-4_48

645

suffice [4, 5]. They have to be tackled by interdisciplinary consortia, where every discipline has to be integrated into an interdisciplinary research process. Various disciplines need to jointly work together on the same problem, each one contributing and integrating their special knowledge and experience.

This leads to another increase in the complexity of the process of finding solutions, especially when research disciplines not only cooperate with other but also with partners from the business world. Because of this fact, the needed cooperation structure has to be effectively managed to ensure a good performance of the consortium [6]. This management includes the integration of all participating disciplines. Experiences in our three research subjects, i.e. various interdisciplinary research consortia, show that this holds not only for consortia comprising disciplines from the field of engineering.

1. The *Cluster of Excellence (CoE) "Integrative Production Technology for High-Wage countries"* at RWTH Aachen University was initiated by the *German Research Foundation* (DFG)[1] and the *German Council of Science and Humanities* (WR)[2] as part of the German excellence initiative.[3] It is devoted to the resolution of the polylemma of production and serves (together with II.) as an example for a rather local interdisciplinary consortium [7]. Many engineering scientists jointly investigate ways to solve the tradeoff between scale and scope and between plan and value oriented production [8]. In this consortium the different sub-disciplines in the engineering context and their differences become apparent.

2. The *Cluster of Excellence (CoE) "Tailor-Made Fuels from Biomass"* at RWTH Aachen University adopts an interdisciplinary approach towards research of new, biomass-based synthetic fuels, in order to verify their potential, with regard to modern combustion technologies, while simultaneously reducing the dependence on fossil fuels [9]. The long-term goal is to determine the optimal combination of fuel components based on renewable materials and their production processes [10].

3. In the *funding priority "Innovative Capability in the Context of Demographic Change"* of the research and development (R&D) program "Working – Learning – Developing Skills. Potential for Innovation in a Modern Working Environment" of the German Federal Ministry of Education and Research (BMBF) several cooperative projects investigate the opportunities due to the demographic change especially in the working environment. These projects mostly offer a larger variety of different scientific disciplines than the CoEs, and the applying practice poses an incremental partner in the consortia. Moreover, the cooperative project partners are scattered all over Germany [11]. Hence, this research subject serves as an example for a more distributed interdisciplinary consortium.

---

[1] http://www.dfg.de/en/index.jsp.

[2] http://www.wissenschaftsrat.de/en/home.html.

[3] The excellence initiative was started in 2005 by the government and federal states. It is divided into three funding lines: graduate schools, clusters of excellence and future concepts for top-level research. In 2012 its second funding period started.

The management of such research consortia has to integrate the different disciplines into the consortium to realize its interdisciplinary potential [3]. However, often enough such consortia hardly evolve further than to a collection of different disciplines working next to each other [12]. This leaves interdisciplinary potential unused. Section 2 of this contribution introduces the dimensions in which the integration of the disciplines can be achieved. Afterwards, Section 3 introduces three sets of interdisciplinary integration tools to foster interdisciplinary cooperation covering all dimensions of integration. Section 4 explains how the effectiveness of the implemented tools can be investigated, before Section 5 summarizes this contribution and gives an outlook on further research.

2 Dimensions of Integration

According to the current state of research the integration of the different disciplines in several dimensions is required to obtain interdisciplinarity within a consortium. Inspired by Gibbons [13] and Bergmann [14], Jahn specifies four dimensions of integration, whose composition and conditions need to be understood in order to combine different disciplines into a consortium and generate added value in scientific research and knowledge [3]. He defines the following four dimensions: social/organizational, factual/technical, communicative and cognitive/epistemic (cf. D1-D4 in Figure 1).

2.1 Social/Organizational Dimension

The social and organizational dimension intends to define and relate both different interests and activities of involved disciplines as well as between subprojects and organizational institutions. It is about a distinction and merging of different stocks of scientific and non-scientific knowledge. Controversial research interests and topics influenced by research and funding policies need to be brought to a common denominator. Furthermore, the gap between research findings, their usefulness and the validity of research results has to be filled. The challenge is to combine controversial research approaches and results in order to bring them together aiming at compatibility of interests and activities to build a collective identity [15].

2.2 Factual/Technical Dimension

The technical dimension provides a transformation of technical solutions to create a normative embedded system, which is sustainable and functional. The use of technical solutions has to be integrated in aspects of daily life, social fabric as well as the work process and discipline-specific methods and workflows. The technical tools used to exchange ideas should integrate problem solutions and make them practicable. User-oriented technical solutions facilitate the application of solutions through technical means, which should not disturb the work flow, but support it and be integrated into the work process [3].

2.3   Communicative Dimension

The communicative dimension focuses on the distinction and linkage of different interdisciplinary linguistic expressions and communicative practices. The development of a common speech practice in everyday research is substantial for the integrative knowledge generation and common solution findings. In this way discipline-specific terminologies can be expanded and integrated into them. The aim is to achieve a basic interdisciplinary understanding despite the still important differentiation of the individual disciplines [16].

2.4   Cognitive/Epistemic Dimension

The cognitive or epistemic dimension aims at distinguishing and associating different discipline-specific knowledge as well as connecting scientific and practical knowledge [17].

The goal is to understand the methods and concepts of other disciplines, to recognize own limits and explicate common methodological and theoretical developments. Different points of view in research have to be mutually understood, which enables the consortium to arrive at joint findings. The merging of interdisciplinary knowledge leads to a solution exchange in different disciplines. An integrative consolidation of methods and research findings used in a discipline might be useful for the other discipline and thereby produce an added value for both.

All dimensions of integration are essential to foster interdisciplinary collaboration [3]. With these dimensions, different research levels are covered, so that interdisciplinary cooperation, organization, network, communication, knowledge supply and technology can be integrated into the interdisciplinary work process of a consortium [14]. Section 3 introduces three sets of tools which in combination support the integration of different disciplines into an interdisciplinary consortium covering all dimensions of integrations.

3   Interdisciplinary Integration Methods

Among various implemented measures to tackle the demands gathered during the past funding phases of our research subjects [5] three sets of interdisciplinary integration tools are explained here in detail. They were designed to achieve an integration of the different disciplines into the consortia in all dimensions mentioned in Section 2. Additionally, they serve the most prominent demands discovered in the areas of a cooperation platform, enhanced communication and an understanding of used terminology in the consortium.

Tailor-made means to foster interdisciplinary integration in the areas of physical cross linkage, virtual cross linkage and the acquisition of interdisciplinary terminology were introduced. By combining measures of physical and virtual cross linkage a personal connection between the different actors is supported while the limitations faced by widely distributed consortia can be overcome more easily. A special focus in the cross linkage activities was placed on developing a common terminology as a necessary basis for interdisciplinary research. Figure 1 illustrates how the different sets of tools relate to the dimensions of integration and Table 1 as well as the following subsections explain their impact on the different dimensions.

3.1 Physical Cross Linkage

In the area of physical cross linkage a set of means was implemented consisting of various formal and informal meetings to inform colleagues in the consortia about ones activities and findings. These meetings are also used to collaboratively develop a common understanding of the topics researched by the consortium.

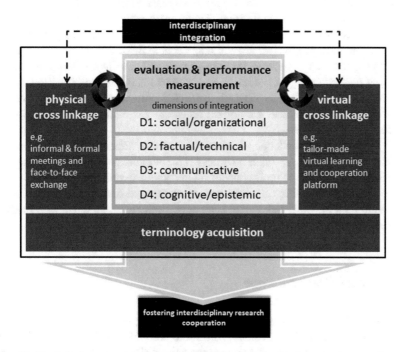

Figure 1 The holistic approach to foster interdisciplinary integration in research consortia

Table 1 Dimensions of integration covered by interdisciplinary integration methods + = weaker impact, ++ = stronger impact

Dimensions of integration	Sets of interdisciplinary integration tools		
	Physical cross linkage	Virtual cross linkage	Terminology acquisition
Social/organizational	++	+	+
Factual/technical	+	++	+
Communicative	+	+	++
Cognitive/epistemic	++	++	++

The means are especially suited to cover the social and organizational dimension, where interdisciplinary integration can be achieved through bringing together all actors and taking into account different interests and activities. Various personal networking meetings foster a personal exchange between the participants in face to face interactions and strengthen personal benefit [18].

The factual/technical dimension, however, is hardly covered by this tool set. Physical networking does not include technical aspects, because here the personal and social component dominates, which is hard to transfer via technical tools. Personal presence does not necessarily require technical solutions to exchange ideas, interests and activities, although the personal experience can be complemented by the technical tools [19].

The communicative dimension is related to social interaction and exchange, which often takes place in a personal circle. Therefore a common language integrating discipline-specific terminology is important to create a basic interdisciplinary understanding. The tool set physical cross linkage acts at these dimension and aims at a discipline-specific language transfer through personal meetings, where a transparent and equal understanding of processes and contents is discussed [20]. However, as opposed to the specific means of terminology acquisition, in the process of physical cross linkage the creation of a common language game happens in a natural and therefore unstructured way.

The exchange of knowledge via physical presence offers a more diverse use of cognitive channels than any other medium [21]. Therefore physical cross linkage is a powerful set of integration tools concerning the cognitive/epistemic dimension [22]. On the one hand scientific outcomes can be exchanged directly within a personal presentation; on the other hand there is an opportunity to get in touch with methods and concepts of other disciplines.

3.2   Virtual Cross Linkage

In the field of virtual cross linkage a tailor-made virtual learning and cooperation platform was developed for each consortium to complement the means of physical cross linkage. It fosters the exchange among the different actors and supports the joint development of solutions and their internal and external transfer. The different groups, projects, and the people involved are connected independent of time and place and networking outside physical meetings is supported. This leads to an enhanced social integration despite the challenges created by differences in time and place [19].

Virtual cross linkage heavily builds up on technical means to interconnect the different disciplines internally as well as externally. The different technical tools are designed not only to be easy to use but also to be well integrated into the every-day work of the partners. The aim is to create an added value for its users by providing ways to present findings in a more formal way as well as to exchange ideas with colleagues in a more informal manner [23].

Similarly to the means of physical cross linkage their virtual counterparts lead to an enhanced understanding of the different disciplines by fostering interdisciplinary communication processes. Due to terminology usage a basic insight is produced and accomplishes the fundament of physical and virtual meetings. However, the development of the understanding takes place in a rather arbitrary and not well managed way as opposed to the highly structured way of terminology acquisition.

Analysis of physical cross linkage shows that the personal benefit of the events decreases with an increasing number of participants. It is therefore recommended to focus on meetings within subprojects [7]. Additionally, the diversity in interdisciplinary research projects often derives from the inclusion of organizationally and geographically distributed participants [24, 25]. To exploit the advantages of physical meetings, these need to be held in the right frequency and with a manageable number of participants. They require a relatively high effort from the actors in a consortium distributed over a larger area. Therefore, virtual means are a good addition to physical cross linkage, as it enables flexible meetings with lower effort for the participants.

3.3 Terminology Acquisition

The terminology acquisition is another set of virtual integration tools comprising means of acquiring and visualizing the terminology used by different actors in the consortium. This happens mainly through an online tool, where actors in the consortium can contribute definitions, combined with the automated extraction of characteristics of the used terminology from given communication data.

For a joint research effort as well as for the social and organizational interaction within, a mutual understanding of the participants plays an important role. The terminology acquisition aims at a common understanding and a shared language among the actors in the consortium. This is needed to formulate contributions to the common research problem, define common goals and exchange information about disciplinary methods.

The terminology acquisition measures make use of technical means as well, but the focus lies in gathering and presentation of the content, which needs to be supported technically to be performed. The main purpose of the tool set – the acquisition of the terminology used – is rather non-technical [25].

Communicative integration is addressed to some extend by all the presented tool sets. However, especially the terminology acquisition contributes to a transparent and equal understanding of processes and contents. It specifically provides a platform for the actors of a consortium to become aware of and discuss about the differences in their terminology. This understanding leads to a more efficient knowledge generation and fosters a joint finding of solutions [20, 22, 26].

To a similar extend than the virtual cross linkage the terminology acquisition as well helps to illustrate, compare, exchange and advance information and knowledge about the topics of the consortium. It creates an advanced understanding of the terminologies present in the consortium as well as the awareness of the mere existence of different terminologies. This knowledge supports the actors in understanding results of their colleagues as well as participating in consortium-wide discussions about relevant topics.

4   How to Measure Interdisciplinary Integration

To foster interdisciplinary cooperation with the objective of increasing effectiveness of interdisciplinary research consortia, measurement approaches are necessarily implemented and adapted to the needs of interdisciplinary consortia as examples for publicly funded, non-profit organizations [27, 28]. Thereby the measurement of interdisciplinarity is essential, e.g. to reveal the impact of research work and/or give recommendations to improve collaborative cooperation. Beyond that, means of measurement can verify the holistic covering of the previously presented dimensions through adequate sets of interdisciplinary integration tools. Each tool is reflected against the backdrop of its performance capability with regard to all dimensions enabling conclusions about its effectiveness.

The integration and promotion of interdisciplinary cooperation and their success has to be measured iteratively [29]. Thus established performance measurement methods from the private sector are adjusted and introduced within scientific research consortia. For instance, with the help of the balanced scorecard (BSC), objectives of the research alliances are improved and the cross-hierarchical communication and cooperation can be promoted [30, 31].

The BSC comprises four perspectives: internal perspective/research cooperation, perspective on learning and development, output/client perspective and financial perspective. To measure the cluster performance, key performance indicators (KPI) were successfully identified (e.g. frequency, quality and benefits of meetings; scientific cooperation; publications [30]) within the interdisciplinary research context and gathered once a year. The analysis of this data enables the derivation of recommendations of actions for the cluster management to enhance the cluster-wide interdisciplinary cooperation. As an example for enhanced communication and cooperation measures (outlined due to BSC results) within the first funding period of the CoEs at RWTH Aachen, colloquia for research assistants, a tailor-made advanced training program for PhD-students, adaption and expansion of the cluster-internal knowledge map, initiation of informal summer fairs, thematic exchange by regularly project meetings and the organization of Summer Schools can be listed.

Public funding organizations such as the DFG and the WR express the need for developing further concepts to measure, steer and regulate interdisciplinary collaborations [32, 33]. Building on existing elaborated methods like the BSC, additional measures are launched within the current research work. At this, a performance measurement and management approach shall be developed. Bottom-up interdisciplinary cooperation shall be promoted with the vision to accompany performance measurement by performance management and to cope with the mentioned dimensions of integrations. While measurement is merely the process of assessing progress towards achieving predetermined goals, management builds on that process, adding relevant communication and action achievements against these predetermined goals [34]. Further need for research is especially focused on evolving performance measurement to foster interdisciplinarity and occupy the whole dimensions of integration.

5   Outlook

During the upcoming funding phase of the three research subjects the effectiveness of the implemented tools will be assessed regularly. Together with their participants the contained tools will be enhanced, and new demands will be incorporated.

Since we see the problems resulting from differing terminologies inside interdisciplinary research consortia as prominent, further ways of supporting the actors of the consortia in coping with problems especially from the field of terminology acquisition will be implemented.

Creating awareness of the mere fact that the disciplines inside a consortium use different terminologies and that this leads to potential problems is a first step. With the data from the terminology acquisition and the corresponding visualization, however, tools can be imagined, that directly support the actors in their daily work. An example for this would be the assessment of joint interdisciplinary publications for their potential of misunderstandings. Differing disciplines of authors as well as included terms, for which fundamentally different definitions exist, would in this case suggest a raised potential.

Acknowledgments This research has been funded by the German Research Foundation (DFG) as part of the Clusters of Excellences *'Integrative Production Technology for High-Wage Countries'* and *'Tailor-Made Fuels from Biomass'* at RWTH Aachen University. Additionally, the work was supported by the German Federal Ministry of Education and Research (BMBF) under Grant 01HH11088 and was co-financed by the European Social Funds (ESF).

References

1. G. Pahl, W. Beitz, J. Feldhusen, K.H. Grote, *Engineering design: a systematic approach.* Springer, Lonon, 2007
2. R. Anderl, M. Eigner, U. Sendler, R. Stark, *Smart engineering: interdisziplinäre Produktentstehung.* Springer, Berlin, 2012
3. T. Jahn, Transdisziplinarität in der Forschungspraxis. In: *Transdisziplinäre Forschung: integrative Forschungsprozesse verstehen und bewerten.*, Campus-Verlag, Frankfurt am Main, 2008
4. J. Bryson, Hybrid manufacturing systems and hybrid products: Services, production and industrialisation. In: *Studies for Innovation in a Modern Working Environment - International Monitoring*, ed. by K. Henning, IMA/ZLW & IfU - RWTH Aachen University, 2009
5. C. Jooß, F. Welter, A. Richert, S. Jeschke, Fostering Innovative Capability in Germany – The Role of Interdisciplinary Research Networks. In: *Enabling Innovation - Innovative Capability – German and International Views*, ed. by S. Jeschke, I. Isenhardt, F. Hees, S. Trantow, Springer, Berlin, Heidelberg, 2011, pp. 289–300
6. U. Bach, I. Leisten, *Entwicklung eines deliberativen Governance-Ansatzes für die Arbeitsforschung.* Budrich UniPress, 2013
7. C. Brecher, ed., *Integrative Produktionstechnik für Hochlohnländer.* Springer, Berlin, Heidelberg, 2011
8. Cluster of excellence 'Integrative production technology for high-wage countries' renewal proposal for a cluster of excellence - excellence initiative by the german federal and state governments to promote science and research at german universities
9. Cluster of excellence - tailor-made fuels from biomass. yearly report, 2008
10. R. Vossen, F. Welter, I. Leisten, A. Richert, I. Isenhardt, Making scientific performance measurable - experiences from a german cluster of excellence. Graz, Austria, 2011
11. M. Haarich, S. Sparschuh, C. Zettel, S. Trantow, F. Hees, Innovative capability - learning capability - transfer capability. promoting innovation systematically. In: *Enabling Innovation: Innovative Ability - German and International Views*, ed. by S. Jeschke, I. Isenhardt, F. Hees, S. Trantow, Springer, 2011
12. I. Leisten, *Transfer Engineering in transdisziplinären Forschungsprojekten.* Books on Demand, 2012
13. M. Gibbons, H. Nowotny, P. Scott, *Wissenschaft neu denken.* Velbrück, Weilerswist, 2004
14. M. Bergmann, *Transdisziplinäre Forschung: integrative Forschungsprozesse verstehen und bewerten.* Campus-Verlag, Frankfurt am Main; New York, 2008
15. R. Häußling, *Grenzen von Netzwerken.* VS Verlag für Sozialwissenschaften, Wiesbaden, 2009
16. L. Wittgenstein, *Philosophische Untersuchungen.* Suhrkamp, Frankfurt a.M., 1967
17. H.J. Bullinger, W. Bauer, P. Kern, "Innovation der Arbeit: Zukunft für Menschen durch nachhaltige Arbeitskonzepte.". In: *Arbeits- und Dienstleistungsforschung als Innovationstreiber. Bilanzen, Herausforderungen, Zukünfte*, ed. by S. D., Fraunhofer Verlag, Stuttgart, pp. p.4–14
18. R. Defila, A. Di Giulio, M. Schermann, *Forschungsverbundmanagement. Handbuch für die Gestaltung inter- und transdisziplinärer Projekte.* Vdf Hochschulverlag, Zürich, 2006
19. J. Sydow, *Management von Netzwerkorganisationen – Zum Stand der Forschung.* Gabler, Wiesbaden, 2010
20. T. Hellmuth, *Terminologiemanagement - Aspekte einer effizienten Kommunikation in der computerunterstützten Informationsverarbeitung.* Verlag für Wissenschaft und Forschung, Konstanz, 1997
21. C. Mast, S. Huck, M. Hubbard, *Unternehmenskommunikation: Ein Leitfaden.* Lucius & Lucius, Stuttgart, 2008
22. G. Budin, *Kommunikation in Netzwerken - Terminologiemanagement.* Springer, Wien, 2006
23. P. Cyganski, B. Hass, *"Potenziale sozialer Netzwerke für Unternehmen in"* Web 2.0. Springer, Berlin, Heidelberg, 2011
24. T. Vaegs, C. Jooß, I. Leisten, A. Richert, S. Jeschke, A virtual collaboration platform to enhance scientific performance within transdisciplinary research networks. In: *ICELW 2013, New York, USA*

25. T. Vaegs, F. Welter, C. Jooß, I. Leisten, A. Richert, S. Jeschke, Cluster terminologies for promoting interdisciplinary scientific cooperation in clusters of excellence. In: *INTED2013, Valencia, Spain.* URL http://library.iated.org/view/VAEGS2013CLU

26. H. Sonneveld, K. Loening, *Terminology. Applications in interdisciplinary communication.* John Benjamins Publishing Co., Amsterdam, 1993

27. F. Welter, S. Schröder, I. Leisten, A. Richert, S. Jeschke, Scientific performance indicators - empirical results from collaborative research centers and clusters of excellence in germany. In: *IRSPM 2012*

28. D. Greiling, "Trust and performance management in non-profit organizations.". The Innovation Journal: Public Sector Innovation Journal. **12** (3), 2007, p. Article 9

29. C. Jooß, R. Vossen, I. Leisten, A. Richert, S. Jeschke, Knowledge engineering in interdisciplinary research clusters. In: *IEEM 2012, Hong Kong*

30. F. Welter, *Regelung wissenschaftlicher Exzellenzcluster mittels scorecardbasierter Performancemessung.* Books on Demand, 2013

31. P. Horváth, R. Gleich, D. Voggenreiter, *Controlling umsetzen - Fallstudien, Lösungen und Basiswissen.* Schäffer-Poeschel, Stuttgart, 2001

32. DFG/WR – Deutsche Forschungsgemeinschaft und Wissenschaftsrat. Bericht der gemeinsamen Kommission zur Exzellenzinitiative an die gemeinsame Wissenschaftskonferenz, 2008. URL http://www.gwk-bonn.de/fileadmin/Papers/GWK-Bericht-Exzellenzinitiative.pdf

33. M. Sondermann, D. Simon, A. Scholz, S. Hornbostel, *Die Exzellenzinitiative: Beobachtungen aus der Implementierungsphase.* ifQ-Working Paper. Bonn, 2008

34. M. Bourne, M. Franco, J. Wilkes, Corporate performance management. Measuring Business Excellence **7** (3), 2003, pp. 15–21. 10.1108/13683040310496462. URL http://www.emeraldinsight.com/journals.htm?articleid=843751

Arbeit in der Industrie der Zukunft – Gestaltung Kooperativer Arbeitssysteme von Mensch und Technik in der Industrie 4.0

Florian Welter, Stella Schulte-Cörne, Anja Richert, Frank Hees
and Sabina Jeschke

Zusammenfassung Die Arbeit in der Industrie der Zukunft wird zunehmend unter den Bedingungen der Industrie 4.0 entwickelt und gestaltet. Hierhinter verbirgt sich eine umfassende Vernetzung und Digitalisierung der Produktion, die auf den technischen Grundlagen cyber-physischer Systeme und des Internets der Dinge basiert. Bezogen auf die industrielle Arbeit wird damit eine Entwicklung hin zu kooperativen Arbeitssystemen, bestehend aus Menschen, Maschinen/Robotern und IT-Systemen möglich. Dabei gilt es für Forschung und Entwicklung, pilotartige Demonstratoren zu entwerfen und hinsichtlich der Implikationen für den Faktor Arbeit in der Industrie der Zukunft zu untersuchen. Zu diesem Zweck sind entsprechende wissenschaftliche und industrielle Demonstratoren von Teilvorhaben zu flankieren, in denen insbesondere geeignete Organisations- und Lernformen, Fragen der Akzeptanz und Sicherheit von kooperativen Arbeitssystemen sowie wirtschaftsräumliche Implikationen dieser neuen Arbeitssysteme für den Standort Deutschland untersucht werden. Da Industrie 4.0 den Menschen als integralen Bestandteil der Produktion sowie als Erfahrungs- und Entscheidungsträger versteht, ist zur Untersuchung derselben ein ganzheitliches und verschiedene Ebenen und Disziplinen berücksichtigendes Untersuchungsdesign wesentlich – nicht zuletzt auch deswegen, um die tradierte Technikzentrierung bei der Gestaltung von Arbeitssystemen zu überwinden. Das Anliegen dieses Beitrags ist es, eine der vielen Facetten von Industrie 4.0 durch die Vorstellung von zentralen Zielen zu Arbeit in der Industrie der Zukunft zu beleuchten und letzteres der Scientific Community vorzustellen.

Schlüsselwörter Arbeitsforschung · Industrie 4.0 · Mensch-Maschine-Kooperation · Kooperatives Arbeitssystem · Organisation · Lernen

F. Welter (✉) · S. Schulte-Cörne · A. Richert · F. Hees · S. Jeschke
IMA/ZLW & IfU, RWTH Aachen University, Dennewartstr. 27,
52068 Aachen, Germany
e-mail: florian.welter@ima-zlw-ifu.rwth-aachen.de

Originally published in "VerANTWORTung für die Arbeit der Zukunft; 61.
Kongress der Gesellschaft für Arbeitswissenschaft (GfA)", © Gesellschaft für
Arbeitswissenschaft e.V. 2015. Reprint by Springer International Publishing
Switzerland 2016, DOI 10.1007/978-3-319-42620-4_49

1 Digitalisierung als Treiber einer Industrie 4.0 Entwicklung

Der Megatrend Digitalisierung durchdringt zunehmend alle Bereiche der Lebens- und Arbeitswelt [1] und ist mit seinen Chancen und Herausforderungen alltäglicher Bestandteil von Wirtschaft [2, 3] und Wissenschaft [4, 5]. Die wachsende Bedeutung wissenschaftlicher Analysen zum Themenfeld der Digitalisierung manifestiert sich gegenwärtig in der Ausrichtung diverser Aktivitäten des BMBF, wie z.B. der Bekanntmachung „Dienstleistungsinnovation durch Digitalisierung" [4] im Rahmen des Aktionsplans Dienstleistung 2010 oder der Ausrichtung des Wissenschaftsjahrs 2014 mit dem Fokus „Die digitale Gesellschaft" [6]. Hier ist hervorzuheben, dass der Begriff der Digitalisierung nicht einfach als die bloSSe Erfassung, Aufbereitung und Speicherung analoger Informationen auf digitalen Datenträgern verstanden wird. Im Kontext der Entwicklung zu einer Industrie 4.0 fungiert „Digitalisierung vielmehr als Innovationsmotor, verändert die Wertschöpfung, die Rolle der Kunden, erlaubt den Aufbau von neuartigen, eigenständigen und umfassenden Servicesystemen und fördert die Internationalisierung von Dienstleistungen" [4]. Digitalisierung vernetzt gesellschaftliche und wirtschaftliche Teilsysteme raum- und zeitunabhängig [1, 4] und wirkt sich dadurch maSSgeblich auf die Entwicklung einer Industrie 4.0 aus.

2 Industrie 4.0 – Technologischer Paradigmenwechsel: dezentrale statt zentrale Steuerung

Die aktuelle Automatisierungstechnik ist dominiert und durchdrungen von dem Paradigma einer zentralistischen Steuerung. Master-Slave-Systeme sind das dominierende Architekturmodell: Ein Zentralserver verteilt die Aufgaben, die abhängigen Einheiten erfüllen ihre Aufgaben und berichten an den zentralen Server. Daran ändert auch die Vernetzung aller Komponenten einer Produktionsanlage, als ein wichtiges Merkmal von Industrie 4.0, zunächst nichts [7]. Solche Systeme sind weder zielbasiert noch agieren sie als "soziales Team": Sie wissen kaum wer sie sind, wer ihre Nachbarn sind, was ihre Aufgabe ist, wie sie zusammenwirken, was die Konsequenzen ihrer Fehler sein können, welche alternativen Strategien es geben könnte usw. Die Art etwa, wie eine kooperative Robotik wirkt – z.B. bei einer Vielzahl von Robotern entlang einer Automobilfertigungsstrasse –, ist i.d.R. heute lediglich ein zeitlich synchrones Abarbeiten verschiedener Aufgaben durch die beteiligten Roboter [7]. Ein Ziel, das in den vergangenen Jahren bereits mehr und mehr erreicht werden konnte, ist die Flexibilität der Anlagen in Bezug auf die Fertigung mehrerer Varianten auf derselben Produktionslinie. Durch die Ausstattung der Systeme mit mehr Sensoriken und internem Prozessor wurde es möglich, dass Systeme die angelieferten Teile bzw. den Zustand des halbfertigen Produkts korrekt identifizieren und auf dieser Basis die nächsten Schritte durchführen. Im Automobilbereich beherrschen moderne Produktionslinien in Deutschland heute selbstverständlich die Fertigung mehrerer Modellvarianten. Die Zukunft der Automatisierungstechnik liegt also darin, die Einzelsysteme mit mehr Intelligenz auszustatten, zunächst einmal vor allem, um Prozess- und Produktqualität zu erhöhen [8]. Aufbauend darauf setzen moderne Robotikkonzepte beim Prinzip der "Kooperation" an: Eine Vielzahl von Forschungsprojekten in der Automatisierungstechnik adressiert die Thematik, Roboter als "Team" interagieren zu lassen. Der Hintergrund ist evident: Komplizierte Aufgaben lösen heterogene "Teammitglieder" besser, da sie die unterschiedlichen Kompetenzen nutzen können. Dieser Schritt steht derzeit, aus rein technischer Sicht, in der Robotik an. Noch einmal wesentlich weiter geht der Schritt, den Menschen als Teil und Leiter eines dann entstehenden hybriden Systems/Teams zu integrieren. Ein solcher Teamprozess unterliegt dann notwendigerweise einem dezentralen Steuerungsparadigma mit völlig neuartigen Arbeitsorganisationen, -prozessen, Führungs- und Entscheidungsoptionen sowie Verantwortlichkeiten für die beteiligten Akteure und beschreibt damit die tiefgreifenden Veränderungen, den Industrie 4.0, konsequent zu Ende gedacht, für die Arbeit in der Industrie der Zukunft bedeutet [8]. Vor diesem Hintergrund ergeben sich zahlreiche neue Herausforderungen und Ziele, die die Gestaltung hybrider, kooperativer Arbeitssysteme betreffen und eine multidimensionale Sicht (menschlicher, organisationaler und technischer Aspekte) auf das Thema Arbeit in der Industrie der Zukunft erfordern. Entsprechend sind gleichermaSSen die Interessen, mitunter auch die Bedenken, von Akteuren aus Wissenschaft, Wirtschaft, Gesellschaft und Politik zu adressieren, die jeweils die Entwicklung zu einer Industrie 4.0 beeinflussen.

3 Übergeordnete Ziele von Arbeit in der Industrie der Zukunft

Wissenschaftlich-experimenteller Demonstrator: Ein wissenschaftlich- experimenteller Demonstrator eines zukünftigen 4.0-Szenarios dient als Labor und stellt eine vollständig vernetzte Industrie 4.0 Umgebung zur Simulation und Analyse von Arbeitsprozessen und ihren Wirkungen dar. Sein Zweck ist u.a., Menschen, Maschinen (in der Montagezelle auch Lightweight Roboter, optimiert auf Mensch-Maschine-Kooperation auf engstem Raum) und IT-Systeme in einem innovativen Arbeitssystem zu integrieren, um kooperative und vernetzte Arbeitsabläufe zu demonstrieren, die in dieser Form noch nicht in der industriellen Praxis eingesetzt werden können und dürfen. Mit einer entsprechenden Demonstrationsanlage wird die Möglichkeit gegeben, neue Wege der Wissensgenerierung aufzuzeigen, weitere Anforderungen an Kompetenzen und deren Entwicklung abzuleiten sowie innovative Organisations- und Gestaltungsmodelle von Arbeit für die und in der Industrie 4.0 zu entwickeln. Ein wissenschaftlich-experimenteller Demonstrator fungiert als exploratives Untersuchungsumfeld, in dem interdisziplinäre Erkenntnisse gewonnen und darüber hinaus Implikationen für Wirtschaft, Gesellschaft und Politik abgeleitet werden.

Industrieller Demonstrator: Für die Gestaltung einer pilotartigen industriellen Demonstratoranlage, z.B. in der Fertigung eines deutschen GroSSunternehmens, ist auf Basis eines heute bestehenden Systems und konkreter Anforderungen an weniger belastende, demografiefeste Arbeitsplätze [9] zunächst ein anwendungsbezogenes Konzept eines zukünftigen hybriden, kooperativen Arbeitssystems zu simulieren. Darauf basierend lassen sich Teilsysteme einer modularen, flexiblen, hybriden Endmontage als Vordemonstrator realisieren. Ein Vordemonstrator ist insbesondere durch die Werksleitung und gemeinsam mit Berufsgenossenschaften zu evaluieren und hinsichtlich sicherheitstechnischer Komplexität, Arbeits- und Gesundheitsschutz [10] und Gefährdungspotenzial zu kategorisieren. Nach der Evaluation eines Vordemonstrators ist die Umsetzung eines produktionsnahen Gesamtdemonstrators anzustreben.

Arbeitsorganisation und Gestaltungspotenziale: Die Generierung von handlungs- und gestaltungswirksamen Ergebnissen zur zukünftigen Industriearbeit (z.B. Konzepte erweiterter Arbeitsautonomie, Konzepte situationsgerechter Unterstützung des Mitarbeiters in der Industrie 4.0, [11]) stellt ein weiteres übergeordnetes Ziel für die Arbeit in der Industrie der Zukunft dar. Im Fokus stehen dabei die Entwicklung neuer Konzepte der Arbeitsorganisation und die Beschreibung neuartiger Gestaltungpotenziale hybrider, kooperativer Arbeitssysteme durch den Menschen. Hier werden Prozesse der Arbeitsgestaltung, -teilung und -organisation, der Datenintegration [12] und des Informationsaustauschs, wie auch der Kommunikation in den Blick genommen und auf dazugehörige Flexibilisierungspotenziale hin untersucht.

Wahrnehmungs- und Wirkungsanalyse Industrie 4.0: Um für die Arbeit in der Industrie 4.0 vorbereitet zu sein und die Entwicklung dahingehend zu gestalten, ist das Herausarbeiten von Implikationen und Implementierungshilfen auf verschiedenen Ebenen (Multi-Level-Analysen) notwendig. Die Wahrnehmung von Möglichkeiten und Konsequenzen, arbeitspraktischer und -organisatorischer Chancen und Risiken fortschreitender Digitalisierung und Automatisierung aus Sicht der Gesellschaft, der räumlichen Verortung und Wandlungsfähigkeit von Industriestandorten [13] sowie von Arbeitnehmern ist zum"Enabling" einer Industrie 4.0 notwendig und weitergehend zu eruieren.

Kooperation und Sicherheit: Durch die erfolgreiche Gestaltung der hybriden Arbeitssysteme in der Industrie 4.0 wird die Verbindung von Vorteilen der Economies of Scale (sinkende Durchschnittskosten bei steigenden Stückzahlen, [14]) und der Economies of Scope (zunehmende Variantenvielfalt und Flexibilität für den Kunden [14]) ermöglicht. Modellierte Systeme einer engen und physisch unmittelbaren Kooperation von Mensch und Maschine erlauben wirtschaftliche Lösungen bei kleinsten LosgröSSen. Es stellen sich jedoch dadurch Fragen der Akzeptanz und Sicherheit kooperativer Arbeitssituationen zwischen Mensch und Maschine, die aus Sicht der Arbeitspsychologie zu beantworten sind [15]. Ziel ist es, Erfolgsfaktoren und Risiken zur Akzeptanz [16] und Verbreitung kooperativer Arbeitssysteme zu identifizieren und zu systematisieren. Die Sicherheit solcher innovativen Arbeitssysteme gilt es – u.a. mit Vertretern von Berufsgenossenschaften und weiteren Intermediären – zu diskutieren, um eine zeitnahe Überführung von Sicherheitsaspekten, die an Demonstratoren analysiert wurden, in die Praxis zu ermöglichen.

Lernen und Kompetenzentwicklung: Die Untersuchung von Lern- und Kompetenzentwicklungsprozessen [17] in kooperativen Arbeitssystemen ist ein weiterer Kern von Arbeit in der Industrie der Zukunft. Ziel ist es hier, neuartige Formen der Kommunikation und Interaktion in kooperativen Arbeitssystemen zu identifizieren wie auch Lehr- und Lernszenarien und grundlegende Rollen- und Teamentwicklungsprozesse als Voraussetzung und Möglichkeit sozialer Innovationen zu beschreiben. Darüber hinaus stehen Fragen veränderter Führungsaufgaben des Menschen, der Orchestrierung intelligenter, vernetzter Systeme und der Identifikation salutogener Potenziale der Arbeit [18] der Industrie 4.0 im Fokus.

4 Ausblick

Die übergeordneten Ziele von Arbeit in der Industrie der Zukunft verdeutlichen die Relevanz einer integrierten Arbeitsforschung im Kontext von Industrie 4.0 auf Basis eines explorativ-experimentellen Designs. Hierdurch sind sowohl Chancen als auch Risiken hinsichtlich der Entwicklung zu einer Industrie 4.0 zu betrachten, und Wege aufzuzeigen, wie insbesondere die Potenziale für die arbeitenden Menschen, die Unternehmen und den Standort Deutschland am besten genutzt werden können. Wohingegen für die arbeitenden Menschen u.a. die Frage der Akzeptanz von zukünftigen Vernetzungsformen in kooperativen Arbeitssystemen aus Mensch und Technik zu beantworten ist, stellen sich auf Ebene eines gesamten Unternehmens u.a. Fragen nach neuen Organisationsstrukturen und -prozessen sowie nach geeigneten Wegen, organisationale Veränderungen nachhaltig zu implementieren. Des Weiteren ist eine engere Vernetzung von Inhalten bestehender nationaler und internationaler Forschungs- und Entwicklungsprojekte im Kontext von Industrie 4.0 unter dem Aspekt zu forcieren, dass sich Arbeit in der Industrie der Zukunft, in Abhängigkeit von unterschiedlichen Faktoren, wie z.b. kultureller, politischer, rechtlicher, wirtschaftsstruktureller und technologischer Art, sehr divers in einer globalisierten Welt entwickeln wird.

Danksagung Das Paper basiert z. T. auf Vorarbeiten, die im Rahmen des BMBF-Förderschwerpunkts „Innovationsfähigkeit im demografischen Wandel" im BMBF-Programm „Arbeiten – Lernen – Kompetenzen entwickeln. Innovationsfähigkeit in einer modernen Arbeitswelt" (A-L-K-Programm) gefördert wurden.

Literaturverzeichnis

1. E. Wenzel, O. Dziemba, *#wir Wie die Digitalisierung unseren Alltag verändert*. Redline, München, 2014
2. D. Lucke, D. Görzig, M. Kacir, J. Volkmann, *Strukturstudie Industrie 4.0 für Baden-Württemberg*, Ministerium für Finanzen und Wirtschaft Baden-Württemberg, Fraunhofer-Institut für Produktionstechnik und Automatisierung IPA, 2014

3. Z_punkt. Megatrends 2020plus Herausforderungen und Chancen für Unternehmer, 2014. Unternehmer Positionen Nord Eine Initiative der HSH Nordbank AG (Hrsg)

4. BMBF. Bekanntmachung des Bundesministeriums für Bildung und Forschung von Richtlinien zur Förderung von Maßnahmen für "Dienstleistungsinnovation durch Digitalisierung" im Rahmen des "Aktionsplans Dienstleistung 2010", 2014. http://www.bmbf.de/foerderungen/23808.php. Stand: 27.08.2014

5. S. Winkler-Nees, Anforderungen an wissenschaftliche Informationsinfrastrukturen. Tech. Rep. 180, Working Paper Series des Rates für Sozial- und Wirtschaftsdaten, 2011

6. BMBF. Das Wissenschaftsjahr 2014 – Die digitale Gesellschaft, 2014. http://www.bmbf.de/de/23173.php. Stand: 01.10.2014

7. S. Jeschke, *Kybernetik und die Intelligenz verteilter Systeme. Nordrhein-Westfalen auf dem Weg zum digitalen Industrieland. Arbeitspapier zum Cluster Informations- und Kommunikationstechnologie.* IKT.NRW, 2014. Im Druck

8. D. Ewert, *Adaptive Ablaufplanung für die Fertigung in der Factory of the Future.* No. 830 In: VDI Reihe 10. 2014

9. S. Jeschke, ed., *Innovationsfähigkeit im demografischen Wandel. Beiträge der Demografietagung des BMBF im Wissenschaftsjahr 2013.* Campus, Frankfurt, 2013

10. K. Henning, A. Richert, F. Hees, eds. *Präventiver Arbeits- und Gesundheitsschutz 2020: Tagungsband zur Jahrestagung 2007 des BMBF-Förderschwerpunkts am 15. und 16. November in Aachen, Aachener Reihe Mensch und Technik,* vol. 59. Aachen, Mainz, 2008

11. H. Kagermann, W. Wahlster, J. Helbig, *Umsetzungsempfehlungen für das Zukunftsprojekt Industrie 4.0. Abschlussbericht des Arbeitskreises Industrie 4.0.* Forschungsunion im Stifterverband für die Deutsche Wissenschaft, Berlin, 2013

12. T. Meisen, Framework zur Kopplung numerischer Simulationen für die Fertigung von Stahlerzeugnissen. Ph.D. thesis, 2012. Dissertation

13. BMBF. Zukunftsbild "Industrie 4.0", 2014

14. C. Brecher, ed., *Integrative production technology for high-wage countries.* Springer, Berlin, 2012

15. E. Ulich, *Arbeitspsychologie.* vdf Hochschulverlag AG, 2011

16. G. Baxter, I. Sommerville, Socio-technical systems: From design methods to systems engineering. Interacting with Computers **23** (1), 2011, pp. 4–17

17. T. Mühlbradt, Was macht Arbeit lernförderlich? eine Bestandsaufnahme. MTM-Schriften Industrial Engineering **Ausgabe 1**, 2014

18. E. Ulich. Arbeitsinduziertes Lernen. Lebenslanges Lernen: Wissen und Können als Wohlstandsfaktoren, 2014

Part IV
Target Group-Adapted User Models for Innovation and Technology Development Processes

Development of a Questionnaire for the Screening of Communication Processes in Transdisciplinary Research Alliances

Wiebke Behrens, Claudia Jooß, Anja Richert and Sabina Jeschke

Abstract Transdisciplinarity is a research strategy that is increasingly employed in a multitude of fields. Communication between the actors is of importance when it comes to ensuring successful collaboration. In order to assess communication processes in a transdisciplinary research alliance, a process screening questionnaire has been developed that can be easily evaluated and thus allows timely feedback to the actors. The process screening questionnaire has been utilized repeatedly on a specific research alliance. This article describes the scientific basis for the development of the process screening questionnaire, exemplifies its application, gives a summary of the results of this specific use, and summarizes benefits and future measures of improvement.

Keywords Transdisciplinarity · Evaluation · Communication · Feedback

1 Introduction

For centuries, the traditional scientific approach has clearly distinguished between scientific knowledge and practical knowledge [1]. Even though interdisciplinarity – which "seeks coherence between the knowledges produced by different disciplines" [1] has become a widely accepted research concept, interdisciplinarity had still been limited to the scientific world. However, over the course of the past four decades, a paradigm shift has begun to occur. Since the 1970s, transdisciplinarity has been the subject of an "intensive scholarly debate" [2].

It was first considered a theoretical principle that allowed collaboration across disciplines aimed at a common purpose and was based on a set of generalized axioms as connecting principle between the disciplines [3]. However, during the last decade, the perception of transdisciplinarity has undergone a radical change. From a scientific theory it has developed into a practical research approach that has come to be known

W. Behrens (✉) · C. Jooß · A. Richert · S. Jeschke
IMA/ZLW & IfU, RWTH Aachen University, Dennewartstr. 27, 52068 Aachen, Germany
e-mail: wiebke.behrens@ima-zlw-ifu.rwth-aachen.de

Originally published in "International Journal of Innovation, Management and Technology (IJIMT)", © International Journal of Innovation, Management and Technology (IJIMT) 2015. Reprint by Springer International Publishing Switzerland 2016, DOI 10.1007/978-3-319-42620-4_50

665

as the "Zurich approach" [2] after the venue of a ground-breaking conference in which transdisciplinarity steered into a new direction. Transdiciplinarity is nowadays considered a "reflexive, integrative, method-driven scientific principle" that aims at providing solutions to "societal problems and concurrently of related scientific problems by differentiating and integrating knowledge" [4]. In other words, "transdisciplinarity cannot be an end in itself. It is meant to achieve particular aims" [5], and these aims are found in the establishment of practical answers to non-scientific problems [6]. Transdisciplinary research increasingly follows a holistic approach [7] while looking for solutions to problems that arise from the fusion of scientific and societal knowledge interests [6, 8]. While the working process focuses on a problem solving approach if considered by practice partners, the scientific perspective places more emphasis on knowledge generation [9]. The ideal transdisciplinary research identifies structures and analyzes problem areas in which the source of the problem and its further development are unknown [10]. Furthermore, the ideal setting consists of scientists from various disciplines as well as experts from the non-scientific world. According to Jahn [11, 12], there are three initial starting points of transdisciplinary research. The first approach is based on a practical problem and creates results that influence social discourse. The second approach is based on knowledge generation and follows a more scientific interest, with the purpose of gaining new insights and developing new models and theories. These results are mainly intended to influence the scientific discussion. Thirdly, Jahn [11] points out an integrated approach which is based on a common research topic and generates compatible knowledge that can be integrated into both target contexts.

These approaches point at the difficulties research partners of transdisciplinary research alliances face when tackling problems of both scientific and non-scientific demands. This paper will first point out the importance of communication for transdisciplinary research alliances. It will then describe the development of a screening questionnaire based on previous work on communication processes in transdisciplinary research alliances. Furthermore, the first application of the screening questionnaire in a research context will be described, the results will be discussed and further research needs will be indicated.

2   Communication in Transdisciplinary Research Alliances

Researchers that are involved in a transdiciplinary research team typically view a problem, its causes, consequences and solutions through the lens of their own discipline [10]. Their approaches towards non-scientific problems have been shaped by different styles of thinking. According to Leisten [8], both scientific as well as entrepreneurial interests influence the transdisciplinary research project and the collaborative knowledge generation. Leisten [8] states that knowledge transfer from one project partner to the other is only possible if a common context has been created which allows a shared understanding across system boundaries like research versus corporate practice. Misconceptions in the communication process can lead to cooperation failures or underachievement because partners talk at cross purposes.

The assessment and evaluation of transdisciplinary research usually takes place ex-post and focuses on the achievement of certain goals or the creation of solutions to common problems. Evaluation is crucial due to "the complexity and high risk of transdiciplinary research" [13]. Krott [13] also mentions three basic approaches to the assessment and evaluation of transdisciplinary research: evaluation of the project's research activities by the scientific community, scientific meta-evaluation focussing on performance, and lastly, political evaluation based on the impact of the research on non-scientific stakeholders. Additionally, the evaluation of research is increasingly being demanded by society [14]. They also notice a changing trend which requests the scientific community to give account of the expenses that have been "invested" into research by the larger society. According to Smrekar et al. [14], final evaluations also act as an incentive to align the research focus according to the evaluation criteria.

There are some possible methods to assess and evaluate the end results of transdisciplinary research projects that refer to measurable objectives and indicators [15, 16], but it is almost impossible find the underlying structures that lead to a certain measurable outcome, especially with regard to the communication patterns that shape a research alliance and can steer a project onto a successful path at best. Bergmann et al. [17] do provide some evaluation criteria for communication processes in their "Guide for the Formative Evaluation of Research Projects". However, these guidelines have been established for ex-post evaluation procedures rather than for continuous assessment and evaluation during the research process. This means that to our knowledge, there are no methods or tools for the assessment and evaluation of communication processes available that can be applied while the research is still in progress.

2.1 Questionnaire Development Based on Guiding Principles for the Evaluation of Communication Processes

Our process screening questionnaire is aimed at assessing the perception of the communication process of different project partners especially with regard to the collective understanding of research and work processes in a transdisciplinary context. It has been developed in order to enable this communication of communication processes.

The communication process becomes ascertainable through the establishment of operative guiding principles [18] that allow the identification of elements which "describe a communication process in its entirety, but at the same time point out the specific structures and special features of the process design in a social context" [8]. Luhman [19] defines operative guiding principles as "differences that control the information processing options of a theory". According to Michulitz [18], these guiding principles need to be able to "demonstrate [...] the differentiation of content as well as the modifiability of a communication process" in order to be applicable to the evaluation of communication. The process screening questionnaire has been

Table 1 Operative guiding principles of communication processes in transdisciplinary research projects [8, 18]

Operative guiding principles of the communication process within a transdisciplinary research project	Central questions regarding the operative guiding principles
P1 Individuals	Who communicates with whom?
P2 Topics	What are the reason for and the subject of the communication?
P3 Routes	How does communication develop?
P4 Location	Where does communication occur?
P5 Time	When does communication occur?
P6 Tools	Which tools are there to support communication?

developed on the basis of the following guiding principles presented in Table 1 that have been proposed by Michulitz [18] and adapted by Leisten [8].

The first guiding principle "Individuals" evaluates the influence that actors from different backgrounds exert on the communication process. Principle two ("Topics") investigates the starting point of the communication process and through this establishes the main reason of existence for the research alliance [20]. "Routes" comprises the processes and structures of communication in organizations, and "Location" takes a deeper look at the spatial organization of a communicating research project. The guiding principle "Time" signifies the importance of temporal procedures for the communication process, and "Tools" evaluates those elements that aid its support and development. The process screening questionnaire contains questions that put the guiding principles in concrete terms and can be administered while the research is still in progress.

2.2   First Application of the Process Screening Questionnaire

The process screening tool has been tested on a complex, hierarchically organized transdisciplinary research alliance with more than 80 subsidiary projects, called "Innovative Capability in Demographic Change".

With the aim of promoting Germany on its way to becoming Europe's leading innovator, the German federal government passed a so-called high tech strategy plan of action in 2006. It uses an integrative approach in order to unite the work of different ministries and departments [21, 22]. Within this political space, the Research and Development Program "Working - Learning - Developing Skills. Potential for Innovation in a Modern Working Environment (German abbreviation: A-L-K)" has been established in 2007. It promotes research into personnel and organizational development as well as skill acquisition and supports the establishment of innovation-friendly

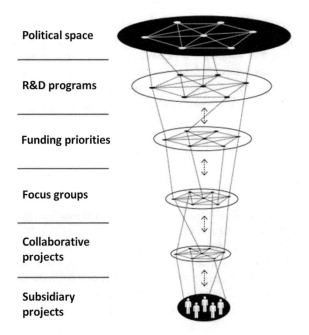

| Political space |
| R&D programs |
| Funding priorities |
| Focus groups |
| Collaborative projects |
| Subsidiary projects |

Figure 1 The hierarchical structure of the Research and Development Program "Working - Learning - Developing Skills. Potential for Innovation in a Modern Working Environment"

frameworks for the coherent cooperation between science, economy and politics. The demographic change and the challenges and chances it brings caused the German Federal Ministry of Education and Research (German abbreviation: BMBF) to establish the funding priority "Innovative Capability in Demographic Change" within the above mentioned Research and Development Program. It explores how Germany's innovation capacity may be increased through the systematic identification and utilization of innovative potential and "contributes considerably to Germany's future competitive ability" [23].

The funding priority is arranged in four hierarchical levels of recursion in order to bring together relevant actors and institutions from research and practice, which allows for a scientific exchange on a deeper level. The hierarchical structure of the funding priority is presented in Figure 1. More than 80 subsidiary projects are clustered into 27 collaborative research projects which each represent a certain research approach. Focus groups again cluster the collaborative research projects according to thematic similarities. Focus groups in this setting are defined as a measure to enable an exchange of knowledge and experiences between the smaller projects and to release synergetic effects through the transfer of research results [24].

The funding priority is able to flexibly adjust to emerging research needs and environmental changes through its creation as a (self-)learning program. This setup contains the creation of a meta-project, which takes up a unique position beyond the thematic aim of the funding priority. It not only supports the individual actors within the different layers, but vertically connects the projects across the hierarchical levels. The meta project provides support to the focus group speakers in the design of communication and cooperation processes [24]. Within its scope, the process screening questionnaire has been developed as a means of evaluating the status quo of the project's communication process so that immediate measures for the improvement of the transdisciplinary collaboration might be taken, should the need become evident.

3 Method

The aim of the process screening questionnaire is the continuous and systematic assessment of communication and cooperation structures between the different actors of research alliances. The process screening questionnaire has been developed in order to support the reflection of complex and dynamic research and development activities, as well as the synchronization of the transdisciplinary research partners towards their common goal. It has been developed in such a way that recommendations for action can easily be derived from the results of the screening. It evaluates the perception of the communication process on two different levels, namely the focus groups and the funding priority.

It has been applied on the research alliance "Innovative Capability in Demographic Change". In close collaboration, representatives of the funding priority, the project sponsor as well as members of the meta project expanded and adapted the guiding principles as proposed by Michulitz [18] and Leisten [8] to fit the specific characteristics of the interdisciplinary research alliance "Innovative Capability in Demographic Change".

The process screening questionnaire is semi-standardized and contains 47 questions that had to be rated on a four-point Likert scale. Example questions include "How is information passed on within your focus group/the funding priority" (guiding principle 3, "Routes") and "How often do personal meetings take place within your focus group"? (Guiding principle 5, "Times").

From September to November 2013 and again from July until December 2014, speakers of the collaborative projects, focus group representatives and agents of the project sponsor were invited to participate in an online survey.

In 2013, 45 people were invited for participation, of which 28 completed the questionnaire (62.2 %). In the following year, 33 of 56 invited participants completed the survey (58.9 %). The participants were not necessarily the same in 2014 as in 2013 since invitations were issued according to the individuals' positions within the funding priority network rather than on personal involvement.

4 Results

The participants answered on a four point Likert scale ranging from 1 (very good or very applicable) to 4 (bad or not applicable at all). Lower numbers equal a better rating.

Table 2 exemplarily shows the results of the guiding principle-category "Tools". The average rating per year is calculated for each question. Figure 2 allows graphical comparison of the results. The questions typically refer to two levels of the hierarchical structure (the focus groups and the funding priority), which makes it possible to evaluate differences in communication processes across the levels. All questions are listed in the appendix.

The analysis of the results for the first guiding principle "Individuals", which evaluates the questions of who communicates with whom, shows that most of the participants assess the communication process within the focus group as better than the processes on the funding priority level. The same is true for the second guiding principle. It analyses the satisfaction of the participants with the cooperation within focus group and funding priority especially with regard to publications and fostering and inhibiting factors for transdisciplinary research. The participants were more satisfied with the work of the focus groups on all aspects and at both times they were interviewed. Guiding principle 3 "Routes") contains questions regarding processes and structures of communication within focus groups and funding priority. The pattern continues – participants were more satisfied with the transparency of decision making processes and the feedback of results from collaborative and subsidiary projects within their focus group than on the level of the funding priority. The results for this guiding principle also showed that a discussion about possible deficits and their correction was more likely to take place within a focus group.

Guiding principle 4 ("Location") inquired about the quality of public relations especially with regard to online presence, print media and presence at exhibitions.

Table 2 Answers to the operative guiding principle "Tools". Lower numbers equal better ratings

	Average (2013)	Average (2014)
How would you assess the incentive for active involvement within the focus group?	2.18	2.03
How would you assess the incentive for active involvement within the funding priority?	2.5	2.45
Would you say that the communication within the focus group resulted in new project ideas with the cooperation partners of the funding priority?	2.4	2.43
Would you say that the communication within the funding priority resulted in new project ideas with cooperation partners of the funding priority?	2.48	2.5
How satisfied are you with the virtual platform demoscreen.de?	1.61	1.75

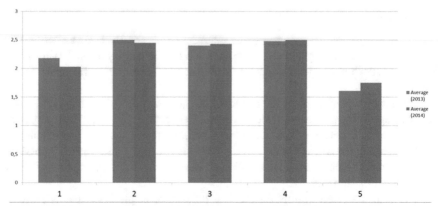

1) How would you assess the incentive for active involvement within the focus group?
2) How would you assess the incentive for active involvement within the funding priority?
3) Would you say that the communication within the focus group resulted in new project ideas with cooperation partners of the funding priority?
4) Would you say that the communication within the funding priority resulted in new project ideas with cooperation partners of the funding priority?
5) How satisfied are you with the virtual platform demoscreen.de?

Figure 2 Answers to the operative guiding principle ?Tools?. Lower numbers equal better ratings

"Location" is the only category in which the work of the funding priority continuously received better feedback then the focus groups. The following set of questions ("Time") asked participants to indicate how often members of the focus groups and the funding priority met face-to-face. For the focus groups, meetings took place mostly on a monthly basis, while the representatives of the funding priority mostly came together every six months. For the final category ("Tools"), participant were asked to assess the incentive for active involvement within the focus group and funding priority as well as the range of new project ideas that followed from communicating with partners of the funding priority. Again, the average ratings were better for the focus groups on all questions.

5 Discussion

The process screening questionnaire has been developed with the aim of screening and assessing the communication processes within a transdisciplinary research alliance. The results above exemplify the utilization of the questionnaire in the research alliance "Innovative Capability in Demographic Change". The following discussion demonstrates the analysis of the results.

The participants' answers about the communication processes within the research alliance "Innovative Capability in Demographic Change" generally indicated that on the level of the focus group, communication processes were more sufficient and effective than on the level of the funding priority. This may be true because representatives of the funding priority exclusively come from academic settings, which is why on this level, transdisciplinarity by definition cannot happen.

The results of the implementation of the process screening questionnaire on the research alliance "Innovative Capability in Demographic Change", however, need to be considered with caution, since the number of participants is not sufficient for a proper statistical analysis.

For the further use of the process screening questionnaire within the context of this funding priority, the number of participants needs to be increased. It is also advised to evaluate the communication processes on more than the two top hierarchical levels. This is especially true for the research alliance "Innovative Capability in Demographic Change", since here, the actual transdisciplinary work commonly takes place on lower levels in close collaboration of scientific institutions with partners from practice, while the representatives of the focus group and funding priority itself generally belong to a purely scientific circle.

6 Conclusion

Even though the discourse concerning transdisciplinary research has been intense during the past decades, no instrument had been devised which actually analyzed the communication processes in an active research alliance. The process screening questionnaire closes the gap and is able to provide direct feedback in order to enable immediate adjustments in the communication processes.

The process screening questionnaire provides a basis for comprehensive analyses of the different aspects of the communication processes that can either lead to a successful completion of a transdisciplinary research project or to improvable results that would be difficult to implement in a practical setting.

However, there is still demand for further research and adaption of the process screening questionnaire so that the process screening questionnaire can be utilized on other transdisciplinary research alliance that may even have a different structure.

In order to reach the participants of complex research alliances, those members primarily working on transdisciplinary levels actually need to be consulted, even if they may not be the primary decision-makers. Also, a higher number of participants need to be invited so that statistically valid conclusions can be drawn. In summary, the process screening questionnaire has been accepted very well by the participants and can increase the success of transdisciplinary communication through the direct feedback of results.

Appendix

Guiding principle 1 - Individuals

1. Are the results of the focus group/the funding priority communicated to others?

2. Are the target groups of the focus group/the funding priority being addressed?
3. Are the target groups of the focus group/the funding priority effectively being reached?

Guiding principle 2 - Topics

4. How satisfied are you with the cooperation within your focus group/the funding priority?
5. How would you assess the quality of the scientific publications within the focus group/the funding priority?
6. How would you assess the quality of the transdisciplinary publications within the focus group/the funding priority?
7. To what extent would you agree to the following statement: The initiation of inter- and transdisciplinary publications and products within the focus group/the funding priority is an aim worth striving for.
8. Which of the following factors do you personally consider as supporting factors for transdiciplinary cooperation? Multiple factors may be chosen.

   a. Common goals
   b. A common, structured approach
   c. A common agenda
   d. Communication

9. Which of the following factors do you personally consider inhibiting factors for transdisciplinary cooperation? Multiple factors may be chosen.

   a. Increase in complexity
   b. Knowledge and acceptance barriers
   c. Globalization

Guiding principle 3 - Routes

10. How is information passed on within the focus group / the funding priority?
11. How would you assess the transparency with regard to decision making within the focus group/the funding priority?
12. Are the results of the collaborative projects reported back quickly and transparently within the focus group/the funding priority?
13. How would you assess the communication procedures within the focus group/the funding priority with regard to (a) promptness and (b) effectiveness?
14. In your opinion, is there sufficient opportunity within the focus group/funding priority to talk about (a) possible deficits, (b) their correction, and (c) intended measures of improvement?
15. How would you assess the opportunities for the contribution of feedback within your focus group/the funding priority?
16. How would you assess the ratio between face-to-face and media-based communication within your focus group/the funding priority?

Guiding principle 4 - Location

17. Please rate the quality of public relations of your focus group/the funding priority with regard to (a) online presence, (b) print media, (c) presence at exhibitions.

Guiding principle 5 - Time

18. How often do members of the focus group/funding priority meet in person?

Guiding principle 6 - Tools

19. How would you assess the incentive for active involvement within the focus group/funding priority?
20. Would you say that the communication within the focus group/funding priority resulted in new project ideas with cooperation partners of the funding priority?
21. How satisfied are you with the virtual platform demoscreen.de?

References

1. K. Cronin, *Transdisciplinary research (TDR) and sustainability*. 2008
2. T. Jahn, M. Bergmann, F. Keil, Transdisciplinarity: Between mainstreaming and marginalization. Ecol. Econ. **79**, 2012, pp. 1–10
3. E. Jantsch, Towards Interdisciplinarity and Transdisciplinarity in Education and Innovation. In: *Interdisciplinarity. Problems of Teaching and Research in Universities*, ed. by CERN, Paris, 1972, pp. 97–121
4. D. Lang, A. Wiek, M. Bergmann, M. Stauffacher, P. Martens, P. Moll, M. Swilling, C. Thomas, Transdisciplinary research in sustainability science: practice, principles, and challenges. Sustain. Sci. **7** (1), 2012, pp. 25–43
5. K. Hollaender, M. Loibl, Wilts, *Management of transdisciplinary research*, EOLSS Publisher Co., Oxford, 2002
6. J. Mittelstraß, Methodische Transdisziplinarität. Tech. Theor. Prax. **14** (2), 2005, pp. 18–23
7. R. Kötter, P. Balsinger, Interdisciplinarity and Transdisciplinarity: A Constant Challenge To The Sciences. Issues Integr. Stud., 1999
8. I. Leisten, *Transfer Engineering in transdisziplinären Forschungsprojekten*. RWTH Aachen University, Books on Demand, Norderstedt, 2012
9. R. Scholz, D. Lang, A. Wiek, A. Walter, M. Stauffacher, Transdisciplinary Case Studies as a Means of Sustainability Learning: Historical Framework and Theory. Int. J. Sustain. High. Educ. **7** (3), 2006, pp. 226–251
10. C. Pohl, G. Hadorn, Methodenentwicklung in der transdisziplinären Forschung. In: *Transdisziplinäre Forschung. Integrative Forschungsprozesse verstehen und bewerten*, ed. by M. Bergmann, E. Schramm, Campus Verlag, Frankfurt, New York, 2008, pp. 69–93
11. T. Jahn, Soziale Ökologie, kognitive Integration und Transdisziplinarität. Tech. Theor. Prax. **14** (2), 2005, pp. 32–38
12. T. Jahn, Transdisziplinarität in der Forschungspraxis. In: *Transdisziplinäre Forschung. Integrative Forschungsprozesse verstehen und bewerten*, ed. by M. Bergmann, E. Schramm, Campus Verlag, Frankfurt, New York, 2008, pp. 21–37
13. M. Krott, Evaluation of transdisciplinary research. In: *Evaluation of transdisciplinary research*, ed. by UNESCO, EOLSS Publisher Co., Oxford, 2002, pp. 1010–1030
14. O. Smrekar, C. Pohl, S. Stoll-Kleemann, Evaluation: Humanökologie und Nachhaltigkeitsforschung auf dem Prüfstand. GAIA - Ecol. Perspect. Sci. Soc. **14** (1), 2005, pp. 73–76
15. K. Huutoniemi, Evaluating interdisciplinary research. In: *The Oxford Handbook of Interdisciplinarity*, ed. by R. Frodeman, J. Klein, C. Mitcham, Oup, Oxford, 2010, pp. 309–320

16. C. Pohl, P. Perrig-Chiello, B. Butz, G. Hirsch Hadorn, D. Joye, R. Lawrence, M. Nentwich, T. Paulsen, M. Rossini, B. Truffer, D. Wastl-Walter, U. Wiesmann, J. Zinsstag, *Questions to evaluate inter- and transdisciplinary research proposals*. td-net for Transdisciplinary Research, Berne, 2011. Working paper
17. M. Bergmann, B. Brohmann, E. Hoffmann, M. Loibl, R. Rehaag, E. Schramm, J. Voß, *Qualitätskriterien transdisziplinärer Forschung. Ein Leitfaden für die formative Evaluation von Forschungsprojekten, ISOE-Studientexte*, vol. 13. 2005
18. C. Michulitz, *Kommunikationsprozessanalyse - Ein interdisziplinärer Beitrag zur Analyse der Kommunikation in Organisationen*, 1st edn. Shaker, Aachen, 2005
19. N. Luhmann, *Soziale Systeme. Grundriß einer allgemeinen Theorie*. Suhrkamp, Frankfurt am Main, 1984
20. N. Luhmann, Die Paradoxie des Entscheidens. In: *Organisation und Entscheidung*, VS Verlag für Sozialwissenschaften, 2000, pp. 123–151
21. BMBF, *Ideen. Innovation. Wachstum. Hightech-Strategie 2020 für Deutschland*. Referat Innovationspolitische Querschnittsfragen, Bonn, 2010
22. Deutsche Bundesregierung. Leitbild eines innovativen Deutschlands – Hightech-Strategie der Bundesregierung, 2015. http://www.hightech-strategie.de/de/Leitbild-13.php. Accessed: 21-Apr-2015
23. M. Haarich, I. Leisten, A. Richert, S. Jeschke, Learning Processes In Research Associations Through Communication And Cooperation Development - The Meta Project Demoscreen. In: *EDULEARN11 Proc.* 2011, pp. 1259–1267
24. C. Jooß, S. Kadlubek, A. Richert, S. Jeschke, Gestaltung von Kommunikations- und Kooperationsprozessen im Förderschwerpunkt "Innovationsfähigkeit im demografischen Wandel". In: *Exploring Demographics*, ed. by S. Jeschke, A. Richert, F. Hees, C. Jooß, Springer Fachmedien, Wiesbaden, 2015, pp. 11–23

AutoHD – Automated Handling and Draping of Reinforcing Textiles

Burkhard Corves, Jan Brinker, Isabel Prause, Mathias Hüsing, Bahoz Abbas, Helga Krieger and Philipp Kosse

Abstract In almost all industrial sectors handling processes are automated through the use of robotic systems. However, in the manufacture of fiber-reinforced structures with complex geometries, the handling of dry, pre-impregnated semi-finished textiles is still performed mainly manually resulting in long processing times, low reproducibility and high manufacturing costs. A previous AiF research project "Auto-Preforms" aimed at the automation of the entire production process of components with uniaxial curvature. The scope of this AiF research project "AutoHD" is to fully automate the draping and handling process of complex, three-dimensional fiber composite structures with high degrees of deformation and multiaxial curvature (e.g. car wings). Based on a draping simulation wrinkles can already be recognized during the draping process and counteracted by the developed mechanical structure. This is achieved by the utilization of the bending stiffness of textile semi-finished products, a flexible end-effector and a built-in optical quality assurance process. In this paper the main aspects of preforming processes are described revealing the challenges of the project. With examples of currently existing systems, the objective and innovative contribution of the project are described. The paper serves as initial presentation of the project and its solution approaches.

Keywords Automated Handling · Draping · Preforming Process · Mechanism Design · Composite Materials

B. Corves (✉) · J. Brinker · I. Prause · M. Hüsing
Department of Mechanism Theory and Dynamics of Machines (IGM),
RWTH Aachen University, Kackertstr. 16-18, 52072 Aachen, Germany
e-mail: corves@igm.rwth-aachen.de

B. Abbas
IMA/ZLW & IfU, RWTH Aachen University, Dennewartstr. 27, 52068 Aachen, Germany

H. Krieger
Institute of Textile Technology (ITA), RWTH Aachen University,
Otto-Blumenthal-Str. 1, 52074 Aachen, Germany

P. Kosse
Laboratory for Machine Tools and Production Engineering (WZL),
RWTH Aachen University, 52056 Aachen, Germany

Originally published in "Mechanisms, Transmissions and Applications.
Proceedings of the Third MeTrApp Conference 2015", ? Springer 2015.
Reprint by Springer International Publishing Switzerland 2016,
DOI 10.1007/978-3-319-42620-4_51

677

1 Introduction

For medium-sized suppliers of automotive and engineering industries, the production of technically sophisticated fiber-reinforced plastics (FRP) is an important and economically rewarding field of activity. However, further development of FRP so far fails due to the low degree of automation of the production process; particularly the automated handling. Although in almost all industrial sectors handling processes are automated through the use of robot and gripper systems, the handling and draping of dry and pre-impregnated semi-finished textiles into complex geometries is still mostly performed manually. The reasons for this are high demands on technology and lack of knowledge about the behavior of flexible semi-finished products [1, 2]. This in turn results in long processing times, low reproducibility and high manufacturing costs.

First approaches for the efficient handling of textile reinforced structures (preforms) for FRP components are presented in a joint AiF project "AutoPreforms" [3]. The research project "AutoPreforms" aimed at the automation of the entire production process of components with low geometric complexity like uniaxial curvatures. Complex, three-dimensional parts with high degrees of deformation, such as car wings, impose high requirements particular to the handling and quality assurance (Figure 1).

The aim of the research project "AutoHD" is to fully automate the draping and handling process during preforming of composite structures with complex multiaxial geometries for the first time. This is achieved by the utilization of the bending stiffness of textile semi-finished products, a flexible end-effector and a built-in optical quality assurance process. On the basis of draping simulation an optimal draping strategy and process control is determined in order to, for example, minimize the occurrence of wrinkling and other drape defects. Additional rules are derived for wrinkle elimination in order to correctively adjust the draping process. The project solely focuses on the manufacturing steps of handling and draping, since automated

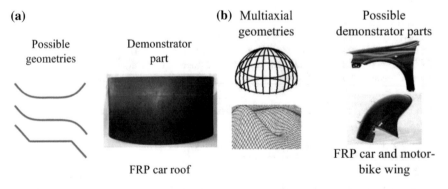

(a)   **(b)** Multiaxial   Possible
                                                   geometries   demonstrator parts

Possible   Demonstrator
geometries   part

                                                                                   FRP car and motor-
                    FRP car roof   bike wing

Figure 1 Geometries realized within the project "AutoPreforms" (a) and examples of complex multiaxial geometries within "AutoHD" (b), (image ref. upon request)

solutions for the remaining process steps (e.g. cutting, consolidation) have already been developed [4–6].

2 Preform Processes

In the following paragraph, aspects of preform processes are described revealing the main challenges of the project. First, the general manufacturing process of a preform is introduced. Next, the draping and handling in textile preforming processes are presented. Finally, the need of sensor systems in order to avoid wrinkling is described.

A preform is a non-impregnated, textile, multi-layer reinforcement structure. In the preforming process, complex, three-dimensional, near net shaped textile reinforcement structures can be realized. These structures have a high potential for the automated mass production of high performance fiber-reinforced plastics [4]. Initial approaches for a semi-automated production have already been developed [7–9]. However, these approaches consider single steps of a process chain such as the investigations of sewing parameters for the automated sewing of fiber composites. The economical use of preforming technologies, e.g. within the medium-sized automotive industry, is only possible if the entire process chain of multiaxial handling is automated for mass production. The automation of the three-dimensional placing of the reinforcing textiles thereby represents a key role.

2.1 Draping of Fiber-Reinforced Plastics

Forming or draping of textile structures is an essential component in the manufacturing process of textile preforms. The deformation behavior of flexible textile semi-finished products is very complex. Therefore, the material selection of the semi-finished textile is typically performed for a given component geometry using a trial-and-error process. Draping is performed manually by experienced professionals correcting the resulting draping defects (e.g. wrinkles, gaps and loops). This procedure is technically and economically deficient with respect to a large-scale production of FRP components. The time and cost-intensive process of textile selection by trial-and-error and manual production can be improved by material modeling and process simulation. The objective of the simulation is to predict the deformation behavior of the reinforcing textile during preforming.

There are several computer-based methods for the so-called draping simulation. The available methods differ in their computational complexity and level of detail. Purely kinematic or on the finite element method (FEM) based models can be applied [10]. If external process forces (e.g. the motion and gripping of the robot) that act on the textile should be considered, FEM-based models must be used. As input for the simulation, the mechanical material properties of the semi-finished textile (e.g.

Young's modulus, bending stiffness, friction coefficients between the reinforcing fibers as well as between the reinforcing fibers and the tool) are required.

2.2 Handling in Textile Preform Processes

Previously developed semi-automated systems provide solutions for specific handling tasks (cf. among others [10–14]). The forming of the reinforcing textiles by these systems is partly three-dimensional, but the individual transformations are realized only uniaxial (i.e. there are no spherical deformations (e.g. hemisphere)). The placing of multiaxial reinforcing textiles in curved geometries with existing systems is, if at all, only possible at very low degrees of deformation.

When selecting a suitable handling system for textile semi-finished products, different criteria must be considered (e.g. the geometry of the gripping and placing position, the dimensions of the workpiece as well as its material properties [15]). An important criterion is the reliability of the gripping process. According to [16] the reliability depends on a variety of factors, such as the material flexibility, operating speed, gripping technology and the environment.

2.3 Process Monitoring by Means of Sensor Systems

For a continuous process-relevant protection of significant quality features only non-destructive, manufacturing-integrated audit procedures (referred to as inline inspection systems [17]) are of importance. For testing of multiaxially curved preforms – and in particular for the detection of wrinkles – it is essential to capture the 3D geometry. The recognition of wrinkling must take place early during the draping process in order to be able to counteract without loss of time. The strict requirements under industrial conditions (e.g. short cycle times, robustness, and reliability) can solely be met by integrating the sensors directly into the end-effector. A simultaneous quality control and feedback of the test results to the corrective intervention process (process control) is not achieved by currently existing methods.

3 Handling Devices in Textile Preform Processes

An adaptive multi-functional end-effector (Figure 2), which was developed as part of a previous joint research project called "AutoPreforms", is able to handle different geometries (such as flat, bent, concave, convex or a combination thereof (see also Figure 1)). The component geometries treated by this end-effector can be divided into three groups (plane components with curved sides, curved components with constant radius of curvature and bent components). The gripping elements of the

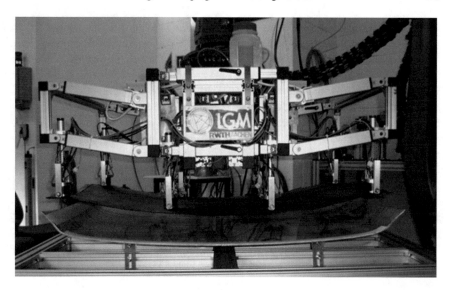

Figure 2 Adaptive multi-functional end-effector [20]

end-effector can be moved relatively to each other, where the distances between the gripping elements remain constant, so that the reinforcing textile is not damaged. The movement of the gripping elements is active and controlled by the robot controller. In addition, the gripping elements are passively pivotable and displaceably mounted in height. So they can orient themselves when placed towards the surface of the mold and in addition to the actively shaped curvature [3, 18–20].

A prototype for the automated handling of clothing textiles was developed within the scope of the joint research project "Integrated 3D sewing system" funded by the Federal Ministry of Education and Research (BMBF) [14]. Monitoring the gripping operation was carried out with the help of a camera system attached to a stationary system of linear axes [21]. For the conformal gripping of reinforcing textiles, a large-area vacuum gripping system whose gripping surface can be changed by sliding chamber walls was realized at the Bremer Werk for assembly systems [22]. At the University of Louisiana at Lafayette, USA a gripping system with suction surfaces attached to the movable kinematic units was developed. The system serves both the separation and handling of limp materials [23]. The Brötje-Automation GmbH, Wiefelstede developed a handling system for similar applications that is based on selectively controllable suckers. These suckers are arranged in a matrix form. Thus, an adaptation of the gripping surface is possible [11]. Within the Fraunhofer Innovation Cluster "KITe hyLITE" a handling unit was designed and developed with which unidirectional fiber semi-finished products can be formed and positioned in a mold [13]. Another example of a rigidly constructed end-effector is the "CFKTex" system with a selectively activatable gripping surface [12]. The Institute for Machine Tools and Industrial Management of the Technical University of Munich realized a system that enables the automated production of three-dimensional preforms, wherein also fixing of the reinforcing textiles is possible by binder activation. In a current research project of the German Aerospace Center called "EVo", a continuous process chain for the automated manufacture of FRP components for automotive and aerospace applications is realized. The developed preform unit also includes an end-effector with 90 separately switchable vacuum grippers and is quality assured to handle and stack blanks. The transformation of the resulting layer stacks is done in a consolidation process, whereby an on-line correction of the deformation is not possible [24]. The mentioned systems, if at all, allow draping only with very low degrees of deformation. The quality of the conversion can be detected comprehensively, but is not used as a basis for on-line correction of the process. Regarding the flexibility of the production of automotive components, this represents a limitation that need to be addressed. Examples of such complex, multiaxial geometries are shown in Figure 1 (right). The aforementioned sub-goal of an inline-wrinkle detection and correction has been integrated into any forming unit of the projects mentioned. Extensive reviews of literature about mechanisms for automated handling and draping of multiaxial reinforcing textiles are presented by [12, 25].

4 Objective and Innovative Contribution

The aim of the recently started joint research project "AutoHD" is to develop an automated handling and draping technology for fiber-reinforced semi-finished products for the series production of multiaxially curved structural components. The main task of the approach lies in the specific combination of draping, material-appropriate utilization of bending stiffness of reinforcing textiles, an end-effector for multiaxial draping and process-integrated quality control. The complete sequence of this process is shown in Figure 3. First, a planar semi-finished textile is grasped by means of a suitable gripper. The gripper is located on an end-effector which in turn is mounted on an industrial robot for transport and coarse positioning of the textile. Using external cameras, the process is monitored, so that the actual position is directly adjusted in accordance to variations from the nominal position. After successful coarse positioning, fine positioning is carried out through the end-effector by reshaping and draping the semi-finished textile so that it can be placed in the multiaxially curved mold – preferably without wrinkling.

An optical sensor system monitoring the draping is also integrated within this process step. A second control loop, the draping control loop, can intervene directly in the handling process once deviations between the data of the sensor system and the data of draping simulation and especially wrinkles occur. The described process flow ensures a highly automated draping and handling process. The on-line quality assurance provides a zero defect production. This results in increased output rates and the saving of expensive reinforcing textiles.

The automated handling of textile preforms aims at the economical production of FRP parts of different geometries. The innovative aspect of the approach is to automate the current manual handling and multiaxial draping by a suitable robot

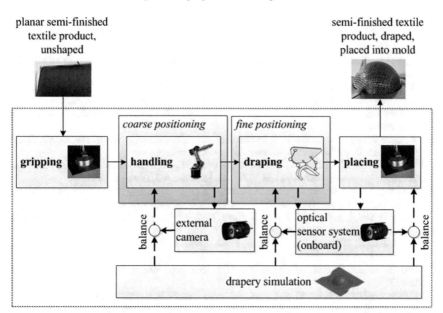

planar semi-finished textile product, unshaped

semi-finished textile product, draped, placed into mold

coarse positioning — fine positioning

gripping → handling → draping → placing

external camera

optical sensor system (onboard)

balance

drapery simulation

Figure 3 Automated drapery and handling process (image ref. upon request)

end-effector system with integrated gripping elements. By integrating a sensor system, changes of the position of the textiles or the emergence of wrinkles when placed into the mold are registered. The connection of the sensor system to the control loop of draping allows for a direct intervention of the end-effector on these errors. A tool for process monitoring, without the need of separate sensors within additional process steps, ensures integrated optical inspection of wrinkles, correct positioning as well as correct fiber orientation of the placed semi-finished products. Thereby, potential errors can already be detected in the running process resulting in increased process quality and reduction of discard. For the first time, the measured 3D fiber orientation allows the balance between the real and the simulated draped preforms. The degree of automation in the production of FRP parts is thus increased significantly and represents an important contribution to the establishment of FRP as a material for high volume applications.

Acknowledgments The authors thank the German Federation of Industrial Research Associations "Otto von Guericke" eV (AiF) for the financial support of the research project "Automated handling and draping of reinforcing textiles for multiaxially curved composite structures – AutoHD" Project No 18264 N / 4 (cf. www.auto-hd.de).

References

1. R. Berger, *VDMA: Serienproduktion hochfester Faserverbundbauteile: Perspektiven für den deutschen Maschinen- und Anlagenbau.* 2012
2. A. Schnabel, Entwicklung von lokal angepassten textilen albzeugen für die Großserienfertigung von faserverstärkten Kunststoffen. Ph.D. thesis, RWTH Aachen, 2013. Diss. Shaker
3. K. Henning, T. Gries, P. Flachskampf, Autopreforms: Gemeinsamer abschlussbericht. Tech. rep., Aachen, 2008
4. T. Grundmann, Automatisiertes Preforming für schalenförmige komplexe Faserverbundbauteile. Ph.D. thesis, RWTH Aachen, 2009. Diss. Shaker
5. W. Hufenbach, ed., *Textile Verbundbauweisen und Fertigungstechnologien für Leichtbaukonstr. des Maschinen- und Fahrzeugbaus.* SDV – Die Medien AG, Dresden, 2007
6. T. Gries, T. Grundmann, C. Greb, F. Kruse, J. Stüve, Technologies for series production of near-net-shape textile preforms. In: *Use of Textile Composites Technology for Safer Vehicles,* ed. by D. Hui, P. Loboda, X. Chen, Y. Wang. National Technical University of Ukraine, Kiev Polytechnic Institute, Kiev, 2009
7. A. Geßler, K. Gliesche, R. Keilmann, E. Laourine, J. Kröber, A. Pickett, Textile Integrationstechniken zur Herstellung vorkonfektionierter Verstärkungsstrukturen für fvk "intex". BMBF-Projekt, Förderkennzeichen 03n3060. Tech. rep., EADS, Ottobrunn, 2002. Schlussbericht
8. C. Weimer, P. Mitschang, M. Neitzel, Continuous manufacturing of tailored reinforcements for liquid infusion processes based on stitching technologies. In: *6. International Conference on Flow Processes in Composite Materials, Auckland, New Zealand, 15.-16.06.2002.* 2002. Paper No. FPCM6-DE-2
9. T. Gries, T. Grundmann, C. Greb, F. Kruse, J. Stüve, Technologies for series production of near-net-shape textile preforms. World Journal of Engineering **7** (H. 1), 2010, pp. 67–74
10. L. Girdauskaite, G. Haasemann, S. KrzywinskiIn, Modellierung und simulation. In: *Textile Werkstoffe für den Leichtbau,* ed. by C. Cherif, Springer, 2011, pp. 573–636
11. Brötje-Automation GmbH, Automated handling system for fibre composites. JEC Compo-sites Magazine **47** (57), 2010, p. 38
12. C. Ehinger, Automatisierte Montage von Faserverbund-Vorformlingen. Ph.D. thesis, Technical University Munich, 2012. Diss., Herbert Utz Verlag
13. G. Köhler, A. Ochs, M. Schneider, Automatisierung in der Leichtbauproduktion. wt Werkstatttechnik online **99** (9), 2009, pp. 614–617
14. K. Zöll, Nähtechnik zur Fertigung textiler Hüllen. Ph.D. thesis, RWTH Aachen, 2002. Diss. Shaker
15. D. Biermann, W. Hufenbach, G. Seliger, Serientaugliche Bearbeitung und Handhabung faserverstärkter Hochleistungswerkstoffe. Untersuchungsbericht zum Fchungs- und Handlungsbedarf. Tech. rep., TU Dresden, 2008
16. J. Stephan, Beitrag zum Greifen von Textilien. Ph.D. thesis, IPK, Techn. Univ., Berlin, 2001. Diss.
17. R. Schmitt, A. Orth, B. Damm, A knowledge based method to evaluate image processing algorithms - a case sample with texture analysis. In: *EOS Conference on Industrial Imaging and Machine Vision, European Optic Society (EOS), World of Photonics Congress 2005, Munich, June.* 2005
18. B. Corves, M. Kordi, M. Hüsing, P. Flachskampf, Economic manu-facturing of fibre reinforced composite structures using robot-supported automation. In: *Proceedings of MUSME 2008, the International Symposium on Multibody Systems and Mechatronics, 8-12 April 2008.* 2008. Paper n. 03
19. M. Kordi, M. Hüsing, B. Corves, Development of a multifunctional robot endeffector system for automated manufacture of textile preforms. In: *IEEE/ASME International Conference on Advanced Intelligent Mechatronics.* 2008
20. M. Kordi, Entwicklung von Roboter-Endeffektoren zur automatisierten Herstellung textiler Preforms für Faserverbundbauteile. Ph.D. thesis, RWTH Aachen, 2009. Diss. Shaker

21. J. Feldhoff, Texturbasierte Bildsegmentierung am Beispiel der Steuerung von Handhabungsprozessen für das automatisierte Nähen. Berichte aus der Produktionstechnik. Ph.D. thesis, RWTH Aachen, 2001. Bd. 2001, 24, Diss. Shaker

22. Patent DE102005047645B4. Bremer Werk für Montagesysteme GmbH, I. Gebauer, C. Dörsch, F. Eckner. Vorrichtung zum aufnehmen und ablegen flächiger Teile, 2010

23. R. Kolluru, K. Valavanis, S. Smith, N. Tsourveloudis, Design fundamentals of a reconfigurable robotic gripper system. In: *IEEE Transactions on System, Man and Cybernetics - Part A: Systems and Humans 30*, 2000

24. S. Torstrick, F. Kruse, M. Wiedemann. Rtm processing for net-shaped parts in high quantities. http://www.cfk-convention.com/uploads/media/130521_CFK-CONV_DLR_Torstrick. pdf. 7. CFK-Valley Stade Convention, Stade; last accessed: 12.06.2013

25. G. Straßer, Greiftechnologie für die automatisierte Handhabung von technischen Textilien in der Faserverbundfertigung. Ph.D. thesis, Technical University Munich, 2012. Diss. H. Utz Verlag

New Intermodal Loading Units in the European Transport Market

Alexia Fenollar Solvay, Max Haberstroh, Sebastian Thelen,
Daniel Schilberg and Sabina Jeschke

Abstract The increasing transport demand and the need for higher cargo volume, together with the need to standardize dimensions and handling processes are characteristics of the current intermodal freight transport. To develop sustainable and resilient transport system, a modal shift and a strengthening of intermodal transport is necessary. Therefore appropriate loading units have to be developed. This includes not only combining the advantages of existing loading units, complying with current European legislation, but also satisfying the needs of the market. This paper presents a new intermodal transport system designed within the European project TelliSys, including variations of a new loading unit as well as an innovative super low deck tractor and a compatible chassis. Additionally a comparison of usage and technical and economical characteristics between utilized intermodal loading units and the new loading units will be given.

Keywords Intermodal Loading Units · MegaSwapBox · TelliSys · Super Low Deck Tractor

1 Introduction

Statistics and studies foresee a growing transport in terms of transport demand and cargo weight. Just in the last decade the gross weight of goods in containers handled in European ports has increased by 30 % [1]. This increase of freight transport and the steady growth in emission of greenhouse gases particularly challenge the European road traffic [2]. Many roads are already overloaded and the surroundings often do not offer enough room for sufficient technical improvement [3]. This is particularly problematic, since over three quarters of the total freight volume (76.4 %) in Europe (EU-27) is carried on roads and only 17.1 %, respectively 6.5 % of the goods are transported by rail and waterways [4]. To develop sustainable and environmentally

A. Fenollar Solvay (✉) · M. Haberstroh · S. Thelen · D. Schilberg · S. Jeschke
IMA/ZLW & IfU, RWTH Aachen University, Dennewartstr. 27, 52068 Aachen, Germany
e-mail: alexia.fenollar@ima-zlw-ifu.rwth-aachen.de

Originally published in "Proceedings of the 13th International Symposium on Heavy
Vehicle Transport Technology (HVTT 2014), San Luis, Argentina, 27-31 October 2014",
© The National Academies of Sciences, Engineering and Medicine 2014. Reprint by Springer
International Publishing Switzerland 2016, DOI 10.1007/978-3-319-42620-4_52

friendly transport, it is necessary to increase the share of inherently more resource-friendly modes of transport, which means shifting parts of the road transport to other modes, i.e. rail and short sea [5, 6].

The European Union supports and promotes the development of new transportation systems which contribute to relieve road traffic and enable an efficient combined and intermodal transport [5]. To enhance multimodal transportation the obstacles between transport modes have to be eliminated and the handling processes have to be designed more efficiently. The development of appropriate loading units is an important step to provide smooth interaction between modes of transport and setting an efficient transport system [7].

2 Overview of the European Intermodal Transport

The European intermodal transport is evolving and adapting to new demands. An analysis of the current use of loading units for all transport modes and their handling processes has revealed a trend towards the use of bigger loading units and the standardization of dimensions to increase the efficiency of the handling processes [8]. The diversity of loading units currently being used in the European market, leads to different handling processes and various equipment, making thereby transhipment operations inefficient in time, energy and costs [2].

Current intermodal loading units can be divided in three groups: swap-bodies, containers and semi-trailers. All of them have a main domain of application, where they are efficiently transported but none of them were developed with the focus of intermodal transportation [9]. Table 1 summarize the main characteristics of current loading units and describe their main usage.

The standard worldwide loading unit that is used in intermodal transport is the ISO 1A container which dominates the maritime transport. Standardization of overseas vessels and railway wagons strengthen the use of these loading units. However, their use is limited by their volume capacity and their flexibility during the loading processes. On another hand 45 ft containers (UNIT 45) represent only 3 % of total containers handled [4]. They are more often used in hinterland traffic than ISO container because of a higher volume capacity on road, but the standard dimension in ships and wagons restricts their use. Both container variants cannot be efficiently used in road transport due to their high tare weight.

For this reason, Swap-bodies and Mega-trailers have the biggest share in road transport. Freight forwarders are looking for high volume capacities and flexible loading processes. The use of craneable Mega-Trailers is increasing since they allow combined intermodal road-rail transport. Despite of this increasing trend, the system is not efficient in terms of wagons standardization, handling processes and the transportation of dead weight (i.e. tires) [10].

The development of a new loading unit for the intermodal transport domain becomes necessary, since current loading units are only partly suitable for combined and trimodal transport. This new loading unit has to be developed for its application

Table 1 Basic dimensions and main characteristics of intermodal loading units

		ISO 1A container	HC 45 ft container (UNIT 45)	Swap body (series A)	Semi-trailer (Mega-trailer)
External dimensions (mm)	Length	12192	13716	13192	13620
	Width	2438	2500	2500	2550
	Height	2438	2896	2675	>4000
Volume (m^3)		66	89	77	75
Payload (t)		27	29.7	34	25
Loading of europallets	No	25	33	30	34
	%*	85.6	95.6	97	96
Usage		Overseas transport	Inland waterways and hinterland transport	Combined transport (road/rail)	Road and combined transport

*% (Pallet area/total area), the remaining area is used for manoeuvrability

by road, rail and water, while bringing together the advantages from other loading units. Therefore, the aim is to comply with transport standards (handling and manoeuvrability), to adhering current European legislation and, especially, to satisfy the needs of the market by considering customer requirements.

3   TelliSys Project

The project "Intelligent Transport System for Innovative Intermodal Freight Transport" (TelliSys) offers a family of MegaSwapBox (MSB) especially developed for intermodal transport. TelliSys aims to optimize the performance of intermodal logistic chains by developing a transport system that enables the transport of goods by road, rail and short see. The objective is to provide cooperative interactions between the different modes of transport and to promote efficient intermodal transport.

Within a period of three years, starting on December 2012, the EU funded project TelliSys develops a new transport system and a modular family of volume-optimized loading units. The new loading units, called MegaSwapBoxes, have been designed for intermodal transport and tailored for specific customer segments. To enable the transport by road, the complete system consists of an adaptable chassis for the loading units and an innovative super low deck tractor. With a maximized internal boxs height, the complete system has a total height of maximum 4 m for road transportation.

Figure 1 TelliBox

A first prototype of the MSB has already been build and demonstrated within the project TelliBox, a direct predecessor of TelliSys. Aim of the project was to integrate the advantages of containers, swap-bodies and mega-trailers in a new loading unit, reaching 3 m internal height (cf. Figure 1) [11].

The development of all TelliSys's products is subdivided into six essential phases: analysis, concept, evaluation, decision, design and construction phase [12]. Currently all TelliSys partners are working on the construction phase, which will lead to the construction of a prototype of each essential component. These prototypes will be tested extensively to demonstrate their functionality and an optimization loop will be integrated in order to allow additional improvements to the developed system, if required. Thus, the project TelliSys will not only deliver ready to use prototypes but also a proof of concept.

4   Requirements of a New Intermodal Loading Unit

In the context of the TelliSys project an analysis of the current European transport market was carried out. In order to collect the important requirements for a new load-ing unit, statistics research and extensive interviews with main transport actors, i. e.

freight forwarding companies, manufacturers, transport operators, inland terminals and ports were performed during more than six months [13]. The collected results were worked out and in the following the main requirements are listed:

- The need to adhere always with the current legislations for intermodal transport in the EU, especially focusing on weight and height.
- Use of standard handling devices, like corner castings and/or grappler pockets for efficient and easy trans-shipments.
- A minimum payload of 24 t to compete with other loading units and to assure a flexible transport of goods.
- Maximization of internal height with the aim to transport up to three stacked pallets (optimal internal height between 2.95 and 2.98 m).
- The ability to stack at least three times (2+1) the loaded unit in order to save space during storage and to allow efficient transport on water ways.
- Theft proof to reduce economic damages for the freight forwarders.
- Openable long sides to improve the efficiency of the loading and unloading processes and increase the possibility to open new markets.
- Cargo security to ensure secure transport of goods.

Meeting all these requirements would generate a loading unit that is too expensive for the customer. In addition, not all requirements are relevant for every customer. Therefore the development of a modular family of intermodal loading units becomes necessary, where each family member has to be defined by different use cases. For example, a length of 45 ft. is a must for the western European market, while the eastern European market requires the 40 ft. standard length which also offers advantages in oversea transport.

To embed this family of new intermodal loading units in the current market, not only the unit itself has to be improved, but a whole transport system has to be developed. Only this system will guarantee the highest increase in efficiency for all modes of transport.

5 TelliSys Designs

Based on the lead user requirements, prototypes for all TelliSys components have been designed. The new MSB is based on unique selling propositions like stackability, openable sides, maximal loading height, trimodality, euro pallet wide and cargo security. The idea to build a modular family of intermodal loading units was developed to cover different use cases. Thereby each MSB family member is addressed to a concrete market and/or for specific applications. This highly user oriented approach of the development is continuously supported by the lead users.

Two main models of MSB have been defined to cover the market, both varying in size (40 or 45 ft) (cf. Table 2). The Continental MSB is a loading unit designed to be used along west Europe and for short sea transport. The Intercontinental MSB addresses the necessities in over sea and long distance transport and fulfils the requirements of the east European market.

In order to include use cases that do not require stackable loading units, a special model of the Continental MSB has been designed. This MSB offers higher internal height and two openable long sides, thus allowing the transport of more cargo volume and flexible loading processes.

The new super low deck tractor is designed to provide an extra low fifth wheel height, enabling a total external height of the loading unit of 3150 mm while complying with European legislation, i.e. having a total vehicle height under 4 m (cf. Table 3) [14]. To design and construct this new super low deck tractor, special tires

Table 2 Main characteristics of the MSB family

Characteristics	Continental MSB		Intercontinental MSB
Size	40 ft/45 ft	45 ft	40 ft/45 ft
Height (internal)	2940 mm	2970 mm	2970 mm
Wide (internal)	Euro-pallet	Euro-pallet	Euro-pallet
Long sides	One open	Two open	Closed
Pay load	Min. 24 t	Min. 24 t	Min. 24 t
Roof	Hinged	Hinged	Hinged
Handling	Top corner casting	Corner castings Grappler pockets	Corner castings Grappler pockets
Stackability	3 times loaded containers (2+1)	Not stackable	3 times loaded containers (2+1)

Table 3 Characteristics of the TelliSys concept developed for the low deck tractor and chassis

Characteristics	Tractor
Type	6 × 2 truck
Tank capacity	700 l
Tyres	Wheelbase 3300 mm
Axle	11 t driver axle pressure
Coupling height	850 mm (loaded and unloaded)
Regulation	Euro 6
Weight	Max. 8 t
	Chassis
Size	45 ft
Weight	Target weight max. 4 t
Hitch	Gooseneck

with a smaller diameter have been developed. Despite the smaller diameter, these tires are able to bear the same weight on the drive axle as the regular tires.

Finally a chassis has been designed, being as light as possible and flexible to transport conventional loading units as well as the new MSBs (cf. Table 3).

6 Comparison Between Intermodal Loading Units

6.1 Comparison of Technical Characteristics

Current intermodal loading units present different technical characteristics. These characteristics have been collected and compared with the mentioned lead user requirements and the MSB characteristics. Figure 2 represents in a mesh diagram main requirements desired by the costumers and shows which requirements are met by each loading unit. The characteristics are valuated with the following scale: 1 meaning no correspondence, 2 for purpose-built and 3 signifying an intrinsic characteristic.

The results reveal that the new intermodal loading unit (MegaSwapBox) meets most of the customers' requirements while the swap body misses the most of the listed specifications compared to the other intermodal loading units. Mega-trailer and the 45 ft. HC container have a similar rate but stay behind the opportunities of the MSB.

6.2 Comparison of Usage of Different Loading Units

Each loading units has its distinctive advantages and disadvantages depending on the usage or main utilization, i. e. type of goods to be transported, origin and destination or mode of transport involved. The following table (Table 4) summarizes the

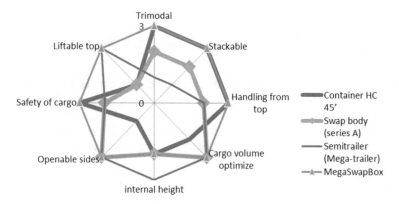

Figure 2 Comparison of technical and technological parameters of intermodal loading units

Table 4 Advantages and disadvantages of intermodal loading units

	Advantages	Disadvantages
ISO 1A container	High availability, good interoperability, high stability, safety of cargo and stackability	Insufficient use of loading area with euro pallets, lower volume, lower internal height, loading of goods only from back, only vertical handling
45 ft HC container	Increased capacity compared to ISO 1A containers, high stability, safety of cargo and stackability	Loading of goods only from back, only vertical handling and exceeded loading gauge (on standard wagons)
Swap body (series A)	Good interoperability (road – rail), possible horizontal handling (without external equipment) and good loading/unloading process	Only box SBs are stackable, safety of cargo (not in case of box SB), not applicable for shipping and no optimised handling process
Semi-trailer (mega-trailer)	Good loading area utilisation with euro pallets, availability, no need of terminals, easy loading/unloading process, flexibility	Not stackable, need of special wagons and transportation of empty weight by rail
MSBs	High cargo volume, stackable, 3 m internal height, liftable top and easy loading/unloading process	Special road chassis, railway wagons with low platform on C45 lines

comparative analysis performed for current intermodal loading units and the MSBs by its implementation.

The current loading units are only efficient for a specific case of implementation or mode of transport. In this way swap bodies for example are efficient for combined transport (road-rail) while mega-trailers have advantages in road transport. This limitation is overcome by the MSBs offering new possibilities for the intermodal traffic by being more efficient in trimodal transport.

6.3 Comparison of Economic Costs and Benefits

In order to evaluate TelliSys costs and benefits, the authors have developed a realistic transport scenario. The aim was to calculate and compare the complete costs of a transport scenario that is currently performed by mega-trailers with the TelliSys system. The 45 ft. continental MSB is the one considered for this transport scenario and it is assumed that mega-trailer and MSB have the same volume capacity. The

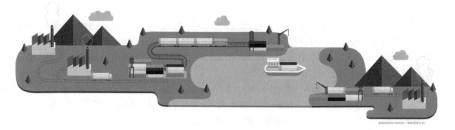

Figure 3 Intermodal scenario

amount of goods being transported per year from point A to B through Europe was defined and an implementation period of 5 years chosen. This realistic scenario had the following characteristics depending on the transport system:

- Mega-trailer: 85 % of the transport distance is covered by road and 15 % by short see.
- TelliSys: 20 % of the transport distance is covered by road for pre and post carriage, 65 % by train and 15 % by inland waterways and/or short sea (cf. Figure 3).

The calculation of this specific scenario shows that switching the current transport (mega-trailer) in favour of the TelliSys system would save up to a 15 % of the complete transport costs. Besides direct costs, indirect extra benefits of TelliSys like the ability to be stacked and save storage place or be easily transported by waterways, have not been directly valued. Also, the possibility to be more environmental friendly by performing the transport with electric trains or to release traffic from roads and therefore being more secure are characteristics that can bring an extra benefit.

Nevertheless the decision on which loading unit is to be used for a specific transport depends on the characteristics of the transport, e. g. trimodal or combined transport, the need of a stackable loading unit for storage or transport, type of loading process and other characteristics.

7 Conclusion

The current European freight transport system offers room for improvement in terms of costs, time and money. Harmonizing handling characteristics of loading units could raise the efficiency of transhipment operations. As a solution the project TelliSys offers a new transport system (family of MSB, tractor and chassis) that will potentially lead to a further standardisation of the equipment while maximizing the transportable volume.

The viability and successful implementation of the product in the market depends on the customers' acceptance and the required economical investment. To avoid new equipment and structural investment the MSB uses standard container handling equipment and has a modular design to be adaptable to different use cases and to be tailored for a vast variety of goods.

The MSBs offer a good chance to enhance and support intermodal transport in Europe, offering a flexible, sustainable and efficient deployment of different modes of transport. The specially developed super low deck tractor and chassis enable the transport of higher loading units, while transporting up to 44 t and complying with the last environmental standards.

Compared to existing loading units, TelliSys meets more user requirements, like maximizing transportable volume, and has the advantages of being stackable and openable from a long side, while being stable and cargo securing. Besides, the economic benefit compared to a direct competitor strongly supports its successful implementation in the market.

References

1. Eurostat. Energy, transport and environment indicators Publications Office of the European Union, Luxembourg, 2013
2. European Commission. Intermodal Loading Units. Directorate-General for Energy and Transport, Brussels, 2004
3. D. Stead. European Journal of Transport and Infrastructure Research. Mid-term review of the Europeans Commison's 2001 Transport White Paper, 2006
4. Eurostat, *Key figures on Europe. Online Eurostat yearbook (op. 2013)*. Publications Office of the European Union (Key figures on Europe), Luxembourg, 2013
5. European Commision. White Paper. Roadmap to a Single European Transport Area – Towards a competitive and resource efficient transport system, 2011
6. S. Jeschke, Global Trends in Transport Routes and Goods Transport: Influence on Future International Loading Units. In: *16th ACEA Scientific Advisory Group (SAG) meeting, Bruxelles, June 2011*. 2011
7. H. Silborn, W. Debauche, C. Orosz, J. Poutchy-Tixier, M. Ruesch, *Measures promoting alternatives to the road and intermodal terminals*. World Road Association, 2007
8. A. Carrillo Zanuy, M. Kendra, J. Čamaj, J. Mašek, S. Stolz, P. Márton. VEL-Wagon. Versatile, Efficient and Longer Wagon for European Transportation. Deliverable 2.1 Intermodal application of VEL-Wagon, 2011
9. H. Sennewald, M. Klingender, M. Haberstroh, A. Fenollar Solvay. Intelligent Transport System for Innovative Intermodal Freight Transport, 2013. TelliSys Deliverable no. D1.1 Report on market analysis
10. M. Burkhardt, Road-Rail Combined Transport: new developments and best practices. In: *55. session of the UNECE Working Party on Intermodal Transport and Logistics*. Geneva, 2012
11. M. Klingender, S. Jursch, Enchancing intermodal freight transport by means of an innovative loading unit. In: *International Conference on Engineering Design, ICED11, Kopenhagen, August 2011*. 2011
12. G. Pahl, F.J. Beitz, W. and, K. Grote, *Konstruktionslehre (p. 20)*. Springer-Verlag, Berlin, Heidelberg, 2007

13. M. Haberstroh, M. Klingender, A. Fenollar Solvay, D. Schilberg, S. Jeschke, An Intelligent Transport System for Intermodal Freight Transport. In: *IST World Congress*. 2013
14. The European Parliament and the Council. Proposal for a Directive of the European Parliament and of the Council amending Directive 96/53/EC of 25 July 1996 laying down for certain road vehicles circulating within the Community the maximum authorised dimensions in national and international traffic and the maximum authorised weights in international traffic, 2013

In-Line Quality Control System for the Industrial Production of Multiaxial Non-crimp Fabrics

Marcel Haeske, Bahoz Abbas, Tobias Fuertjes and Thomas Gries

Abstract A new approach is presented in which existing systems for error detection are used to build up a closed loop control for quality optimization in the production process of NCF. Only with this online system, productivity and product quality can be increased. The fibre material is screened for fibre disorientation and gaps and deviations in areal weight at different positions alongside the production line using state of the art equipment for error detection. Additionally, the machine-parameters are monitored and the yarn tension of both reinforcement fibres and stitching yarn are measured continuously. An automated and self-learning control is developed that automatically adjusts the settings of the production machine. By this, a minimum of errors in the textile is achieved. In addition, the quantitative description generated in the process is used to develop an error map for each batch of material produced. This error map allows for production waste to be minimized, as the produced material can be automatically sorted into different quality grades, meeting the demands of different customers. Furthermore, the material cutting can be adjusted that only low quality areas are removed. By this, only high quality material will be used for the component part production. An entire roll of non-crimp fabric would not be declared as waste. The quality control is realized by an automatic control device. By a pressure roll in front of the warp knitting unit the rovings are homogenized.

Keywords Non-Crimp Fabric · Quality · Production · Pressure Rol

M. Haeske (✉) · T. Gries
ITA, RWTH Aachen University, Otto-Blumenthal-Str. 1, 52074 Aachen, Germany
e-mail: marcel.haeske@ita.rwth-aachen.de

B. Abbas
IMA/ZLW & IfU, RWTH Aachen University, Dennewartstr. 27, 52068 Aachen, Germany

T. Fuertjes
WZL, RWTH Aachen University, Steinbachstr. 19, 52074 Aachen, Germany

Originally published in "Proceedings of the 20th International Conference on Composite Materials (ICCM 2015), Copenhagen, Denmark, 19-24 July 2015", © International Committee on Composite Materials (ICCM) 2015. Reprint by Springer International Publishing Switzerland 2016, DOI 10.1007/978-3-319-42620-4_53

1 Introduction

The use of fibre reinforced composites is getting a new standard in lightweight con-
structions. Due to their great mechanical properties and low density of the materials
in comparison to metal materials, composites find their way into more and more
application fields. In the automotive and aerospace industry as well as for wind
power plants, both glass and carbon fibre reinforced composites are the state of the
art solution regarding weight reduction and performance improvement [1]. Textiles
used for the composite production exist in various forms of textile architecture. This
can be for instance: nonwoven, braids, woven fabric or non-crimp fabrics (NCF).
Commonly, multiaxial non-crimp fabrics are used as semi-finished fabrics for fibre
reinforcement plastics. In NCF reinforced composites, the fibres are not crimped and
have therefore substantial in-plane orientation. This fact results in high mechanical
properties of the textile in various directions depending on the numbers of layers
and the angle of the fibre lay-up. NCF production takes place on a warp knitting
machine with multi-axial weft insertion [2]. For the production of multiaxial NCFs
the straight reinforcement fibres (rovings or spread tows) are laid parallel in a defined
angle. Several layers are laid on top of each other. Directly after the fibre placement
all layers are joined with a warp knitted technology with a standard Polyester yarn.
By this, the so-produced fabric can be wound up and handled during the upcoming
preform steps. In Figure 1, a non-crimp fabric is illustrated.

 Due to the increasing amount of application of fibre reinforced composites, the
quality of the produced fabrics takes on more and more prominence these days.
The industry requires to process only high quality material to acquire the highest
possible mechanical properties of the composite part. Low quality material is declared
as waste or is only used for low performance application. The quality impairment
occurs directly during the NCF production. Due to a non-uniform fibre tension, gaps
or thick spots appear between single fibre rovings. As this error happens irregularly,
various areas of the NCF have very poor quality and are not used as reinforcement.

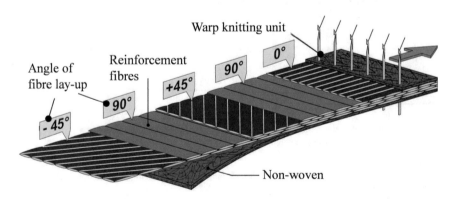

Figure 1 Non-crimp fabric [3]

Since the industry needs to know which areas of the produced fabric are of high quality, the usage of quality control systems is inevitable [4].

To investigate the properties of produced fabrics, various solutions exist for monitoring the quality of multiaxial non-crimp, braided and woven fabrics as well as performs. Those technologies are limited to the singular inspection of their area of application, such as breakage of weft or knitting yarns [5, 6], discontinuous measurements [7], off-line measurement [7–9] and qualitative instead of quantitative acquisition of defects [6, 10, 11] as well as focus on the inspected materials (e.g. only carbon fibres) [12, 13]. Special sensors for the acquisition of the area density were developed for this purpose [14]. But so far combinations of the stand-alone solutions to acquire and document all defects as a whole are non-existent. In addition to that the gained data is not used for quality management of the running process.

Existing measuring procedures to acquire defects in interplies are limited to carbon fibre fabrics and show low resolutions [12, 13] or are only used off-line [7–9]. Image analyzing systems are currently solely able to inspect the top layers. Defects like gaps are mainly caused by the knitting process and therefore only occur in the interplies afterwards. Bi-axial non-crimp fabrics are therefore attractive for mass production as no interplies have to be investigated [15].

All these monitoring systems allow a detection of the fabrics offline and are stand-alone solutions. Nowadays, no inline-system exists that allows a continuous and permanent quality monitoring including thorough quality documentation [4]. A quantitative acquisition and documentation of defects (gaps between reinforcement fibres, thick spots, angle deviation of the reinforcement fibres and voids) of multiaxial non-crimp fabrics is the key for a high productivity. An overall system that gathers material- and process parameters as well as defects has still to be developed.

This overall system development is performed in the AiF Auto-NCF project funded by the Federal Ministry for Economic Affairs and Energy (BMWi), Germany, with three project partners: ITA and WZL of RWTH Aachen University and the IfU as Assoc. Institute of RWTH Aachen University, Germany.

In this project an in-line quality control system is developed. By the use of different sensors, the process parameters are tracked during the manufacturing process. Areal weight is detected by an X-ray sensor while fibre misalignments, gaps and thick spots are detected with a camera system (see Section 2). In Figure 2 the overall concept of the in-line quality control system is shown. The goal is a real-time control system. With this system it becomes possible to record material and process parameters as well as defects. The result is an error map in which all defects are recorded with type, location and category. Each produced fabric roll is delivered to the customer in the downstream process including its unique error map. By the use of this error map, all areas of each produced textile are defined and the industry can process the best suited areas. As industrial companies have different requirements regarding the textile quality, the amount of waste material and the quality costs can be reduced tremendously.

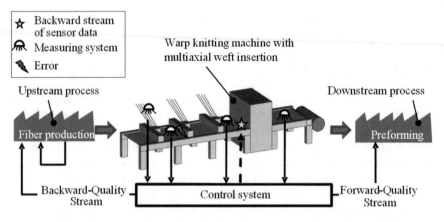

Figure 2 In-line quality control system on a warp knitting machine with multiaxial weft insertion

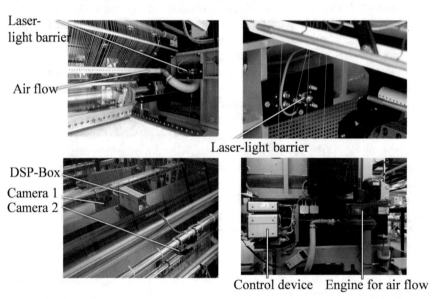

Figure 3 Integration of the light-barrier, air blow unit and camera system in the machine

2 Machine Modifications

In this chapter, the realized machine modifications are described. In Figure 3 the integration of different components in the LIBA COPCENTRA MAX 3 CNC, Liba, Naila, Germany, machine at the ITA, Aachen, Germany is shown. The components are installed at the machine by the company Protechna Herbst GmbH & Co. KG, Ottobrunn, Germany. The fibre material is screened for fiber disorientation and gaps at different positions alongside the production line. Additionally, the machine parameters

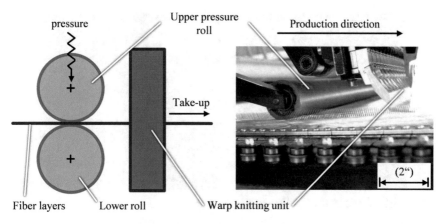

Figure 4 Integration of the pressure roll in the machine set-up

are monitored and the yarn tension of both reinforcement fibres and warp knitting yarns are measured continuously. Furthermore, a yarn breakage detection system is installed.

The yarn breakage detection is realized by an air blow unit. This unit is installed next to the warp knitting yarns. Fibre breakage is detected with a visible red light laser with 660 nm. As soon as a fibre breaks, that fibre is blown into the light-barrier which stops the machine.

Furthermore, a camera system is installed to realize the fibre screening regarding fibre misalignments and gaps. This system includes a DSP-Box and two cameras (one in front and one in the back of the warp knitting unit).

Moreover, a pressure roll is integrated right in front of the warp knitting unit of the machine. In Figure 4 this set-up is illustrated. By the use of this pressure roll it becomes possible to homogenize the roving layers. The pressure applied on the roll is adjusted with springs on each side. The set pressure influences the formation of gaps. A uniform pressure across the width of the fabric realizes the reduction of gaps and fibre misalignments.

Furthermore, a real-time machine vision system for areal weight detection and fibre orientation are to be installed in the next steps. By the integration of all these components the overall model based closed loop control regarding a self-configuration pressure roll will be realized.

The images taken by the camera system are compared before and after the warp knitting process step. If the cameras detect a defect, the pressure roll pressure is adjusted according to the recorded defect. Appropriate controlling strategies are developed and validated with a functional model so that a closed loop control is realized. Through this the desired target values, e.g. gap frequency, are obtained automatically and adjusted by the pressure roll.

At the same time the camera installed after the warp knitting unit is recording all defect information and safes them in the error map.

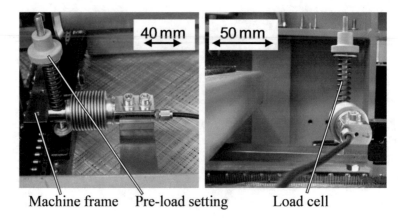

Machine frame   Pre-load setting   Load cell

Figure 5 Installation of load cells

3 Methods

To investigate the influence of the pressure roll on the propagation of gaps, various trials have been performed. Load cells to measure pre-load and operation loads have been installed at the machine (see Figure 5). The implementation of a pressure measurement film is used to determine the load distribution. This pressure measurement film indicates applied pressure differences along the width of the measurement area. The film was placed below the pressure roll covering the entire width.

Furthermore, the homogenizing effect of the compression roll for different settings is investigated. Various machine speeds from 30 m/h (low speed) up to 150 m/h (high speed) are combined with different spring loads varying from 50 to 100 N. In addition two different areal weights are investigated: 300 and 600 g/m² per layer. PPG Hybon glass fibres with 600 tex have been used for all trials.

4 Results

The examinations with the pressure measurement film show a curved, irregular characteristic of the force along the pressure rolls. As shown in Figure 6, the pressure indicating film is suitable for the localization of gaps and thick spots. Clearly the pressure peaks depict the thick spots and the throughs depict the gaps in the fabric. The edges of the gaps display higher peaks than anywhere else in the fabric because the missing rovings accumulate at the edges of the gaps and create thick spots which result in a high pressure on the pressure-measurement-sheet. Conclusions about the height and width of thick spots can be drawn from the resulting force distribution.

In Figure 7, the results of the use of the pressure roll is shown. For 300 g/m² areal weight per layer and low production speed, the number of gaps can be reduced

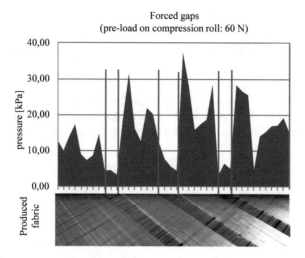

Figure 6 Forced gaps in a produced fabric

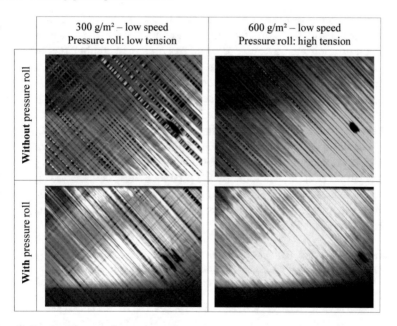

Figure 7 Result of the use of a pressure roll

significantly if a pressure roll is used. For low areal weights a low tension shows the best results. For $600/m^2$ areal weight per layer and low production speed, a high tension applied on the fabric shows good results regarding gap distribution and homogenization.

The best fabric appearance for a biaxial non-crimp fabric with an area weight of $300 g/m^2$ per layer is achieved with a high roving tension, a low production speed and a spring force of 70–80 N each spring. This equals a compression force of 171.7–187.5 N. The trials with a low thread tension did not result in adequate fabric appearances.

5 Conclusion

It can be pointed out that the pressure roll system increases significantly the homogenization of the fabric. By this, the next level of quality in the NCF production can be reached.

To compensate gaps different compression forces can be set in the machine control panel depending on roving tension, areal density and machine speed. Thus the integration of a self-configuration pressure roll system in combination with a camera system, which is independent from the machine control panel and adjusts the compression force automatically with a change in layer structure, textile- or process parameters as well with the occurrence of defects, is reasonable. A spring force setting between 70 and 80 N shows the best results regarding fibre misalignments. Based on that value the force is adjusted at the detection of a gap. An automatic control enables a defined setting of the compression force. Through automatically controllable pressure rolls the same fabric with similar characteristic is producible after any period of time. Additionally the product quality is increased significantly through the reduction of thick spots and gaps. In the next steps the further optimization of the pressure roll would therefore result in an enhancement of the composite part characteristics.

Acknowledgments This IGF-project "Auto-NCF" 494 ZN of the Forschungsvereinigung Forschungskuratorium Textil e.V., Reinhardtstraße 12–14, 10117 Berlin is funded via the AiF by the Federal Ministry for Economic Affairs and Energy (BMWi) according to a decision of the German Federal Parliament.
We would like to thank all project partners and companies listed on the official project homepage (http://auto-ncf.de) for their support and financial aid which leads to the great success of this project.

References

1. R. Lässig, M. Eisenhut, A. Mathias, R. Schulte, F. Peters, T. Kühmann, T. Waldmann, W. Begemann, Serienproduktion von hochfesten Faserverbundbauteilen. Perspektiven für den deutschen Maschinen- und Anlagenbau. Tech. rep., Roland Berger Strategy Consultants, 2012. Studie

2. S. Lomov, *Non-Crimp Fabric Composites: Manufacturing, Properties and Applications*. CRC Press
3. LIBA Maschinenfabrik GmbH. Kettenwirkautomat mit multiaxialem Schußeintrag COPCEN-TRA MAX 3, 2003. Booklet, Naila
4. C. Mersmann, Industrialisierende Maschine-Vision-Integration im Faserverbundleichtbau. Ergebnisse aus der Produktionstechnik **8/2012**, 2012
5. BTSR International S.P.A., SMART MATRIX WARP – Production control system. Tech. rep., Italy, 2010. BTSR International S.P.A., Firmenschrift
6. Protechna Herbst GmbH & Co. KG, Ottobrunn. http://www.protechna.de. 09.04.2015
7. U. Hassler, M. Rehak, R. Hanke, Carbon Fibre Preform Inspection by Circular X-ray Tomosynthesis, IEEE Nuclear Science Symposium, Dresden. 2008
8. U. Hassler, S. Schloetzer, R. Hanke, Computed tomography for analysis of fiber distribution in carbon fiber preforms. In: *International Symposium on Digital Industrial Radiology and Computed Tomography (DIR), Lyon, France*. 2007
9. S. Mohr, S. Gayetskyy, U. Haßler, R. Hanke, Konturangepasste Lagenextraktion aus Computertomographie-Daten von Faserverbundwerkstoffen. DGZfP-Jahrestagung 2009. In: *Zerstörungsfreie Materialprüfung, ZfP in Forschung, Entwicklung und Anwendung, Münster, Germany, Mai 18-20.* 2009
10. Elbit Vision System Ltd. IQ-TEX 4 Trademark – Vision Empowered Process Monitoring. http://www.evs.co.il. 05.06.2012
11. KARL MAYER Textilmaschinenfabrik GmbH, Multiaxial und Biaxial, Protechna ProCam System is watching you. Kettenwirk-Praxis 1/2010 , 2010
12. M. Schulze, H. Heuer, Zerstörungsfreie Prüfung von CFK-Gelegen mit Hochfrequenz-Wirbelstrom. Darmstadt, 2011
13. M. Schulze, M. Küttner, N. Heuer, H. andMeyendor, Mehrfrequenz-Wirbelstromprüfverfahren zur Qualitätskontrolle bei der Produktion von Kohlefaser-Multiaxialgelegen. In: *DGZfP-Jahrestagung 2009. Zerstörungsfreie Materialprüfung, ZfP in Forschung, Entwicklung und Anwendung, Münster, Germany, Mai 18-20.* 2009
14. ZAP Systemkomponenten GmbH + Co. KG, *Echtzeitmessung: Flächengewicht, Warendicke, Dichte*. ZAP Systemkomponenten GmbH + Co. KG, Straubing, 2010
15. C. Greb, A. Schnabel, J. Haring, T. Gries, Evaluation of novel preforming technologies for large-scale composite production. In: *Emerging Oppoertunities: Materials and Process Solutions, Sampe 2012 Conference & Exhibition, May 21-24, 2012, Baltimore, Maryland*, ed. by S.W. In: Beckwidth, M. Maher, M. Newaz, G. Reyes-Villaneuav. 2012

Exploring Demographics – Transdisziplinäre Perspektiven zur Innovationsfähigkeit im demografischen Wandel

Claudia Jooß, Anja Richert, Frank Hees and Sabina Jeschke

Zusammenfassung Als Gegenkonzept zu einer rein akademisch und disziplinär geprägten Wissenserzeugung kann die Produktion von Wissen im Anwendungskontext gesehen werden. Dabei werden die Interessen gesellschaftlicher, wirtschaftlicher und politischer Akteure in den Forschungsprozess integriert (vgl. T. Jahn, Transdisziplinarität in der Forschungspraxis. In: M. Bergmann (Hrsg.): Transziplinäre Forschung – Integrative Forschungsprozesse verstehen und bewerten. Campus, Frankfurt am Main, 2008, S. 7, 22). Solchen transdisziplinären Perspektiven wird ein besonders positiver Einfluss auf die Wissensproduktion und Innovation unterstellt (vgl. F. J. van Rijnsoever, L. K. Hessels, Factors associated with disciplinary and interdisciplinary research collaboration. In: Research Policy, 40 (3), 2011, S. 463). Dieser Beitrag diskutiert ein grundlegendes Verständnis von Transdisziplinarität und beleuchtet die Vielfalt transdisziplinärer Konstellationen im Förderschwerpunkt „Innovationsfähigkeit im demografischen Wandel" des Bundesministeriums für Bildung und Forschung (BMBF). Zu Beginn wird das Verständnis von Innovationsfähigkeit im Kontext des demografischen Wandels aufgezeigt. Abschließend werden die Ziele und die Struktur des Sammelbandes erläutert.

Schlüsselwörter Transdisziplinarität · Wissensproduktion · Innovationsfähigkeit · Demografischer Wandel · Integration · Förderschwerpunkt

1 Innovationsfähigkeit im Kontext des demografischen Wandels

Der demografische Wandel ist in Deutschland bereits an vielen Stellen bemerkbar: Durch die sich verändernde Altersstruktur, die niedrigere Geburtenraten im Allgemeinen und einen steigenden Anteil der älteren Bevölkerung zur Konsequenz hat, verkleinert sich der Anteil erwerbstätiger Personen in Deutschland. Dies ist neben der Integration von Zuwanderern in das Bildungs- und Beschäftigungssystem

C. Jooß (✉) · A. Richert · F. Hees · S. Jeschke
IMA/ZLW & IfU, RWTH Aachen University, Dennewartstr. 27, 52068 Aachen, Germany
e-mail: claudia.jooss@ima-zlw-ifu.rwth-aachen.de

Originally published in "Exploring Demographics ", © Springer 2015.
Reprint by Springer International Publishing Switzerland 2016,
DOI 10.1007/978-3-319-42620-4_54

und dem Fachkräftemangel eine der tiefgreifenden Herausforderungen des demografischen Wandels ([1]: 2). Eine direkte Konfrontation und Auseinandersetzung der Gesellschaft, Wirtschaft und Wissenschaft mit diesem Thema ist daher unumgänglich.

Das Ziel des Förderschwerpunktes liegt darin, den genannten Herausforderungen des demografischen Wandels aktiv zu begegnen und die damit einhergehenden Chancen für unsere Gesellschaft nutzbar zu machen [2]: 1).

Vor diesem Hintergrund ist es notwendig, ein ganzheitliches Innovationsverständnis zugrunde zu legen. Daher ist es nach Trantow et al. [3] von Bedeutung, den Begriff des Nutzens[1] über wirtschaftliche Aspekte hinaus um soziale und ökologische Aspekte zu ergänzen und das Kriterium der Veränderungskraft miteinzubeziehen:

> „Innovation ist die Realisierung neuartiger Ideen, die zu nachhaltigen Veränderungen beitragen. Innovationsfähigkeit von Unternehmen umfasst das komplexe Zusammenspiel der menschlichen, organisationalen und technischen Voraussetzungen zur kontinuierlichen Hervorbringung und Realisierung neuartiger Ideen, die zu nachhaltigen Veränderungen beitragen."
> ([3]: 5)

Vor dem Hintergrund dieses ganzheitlichen Verständnisses werden im Förderschwerpunkt die Innovationspotenziale sowohl jüngerer als auch älterer Menschen systematisch erschlossen, um diese für die Steigerung der Wettbewerbsfähigkeit des Standortes Deutschlands zu nutzen. Die Ergebnisse der Identifikation dieser Potenziale fließen in eine demografieorientierte Personalentwicklung und Organisationsgestaltung in Unternehmen, um den vielfältigen Implikationen der sich ändernden Verhältnisse in der Bevölkerungs- und Altersstruktur entgegenzutreten. Für diese Entwicklungen müssen Instrumente und Methoden geprüft und bedarfsgerecht weiterentwickelt werden [1]: 2).

Die exemplarisch skizzierten Auswirkungen des demografischen Wandels sind an vielen Stellen unserer Gesellschaft bemerkbar, was eine aktive und gemeinsame Gestaltung von Wissenschaft und Wirtschaft erfordert. Damit lassen sich diese als komplexe und wissenschaftsübergreifende (also auch disziplinenübergreifende) Herausforderungen bezeichnen, die ihren Ursprung im außerwissenschaftlichen Bereich (bspw. in der Gesellschaft) haben und deren Lösungen als dringlich und von der Öffentlichkeit als relevant eingestuft werden. Über institutionelle Wege (bspw. Politik und Forschungsförderung) werden diese Herausforderungen an die Wissenschaft herangetragen, um diese gemeinsam zu lösen ([4]: 185). Solchen als transdisziplinär bezeichneten Forschungsaktivitäten wird ein besonders positiver Einfluss auf die Wissensproduktion und Innovation bzw. die Innovationsfähigkeit zugeschrieben ([5]: 463).

Woher rührt diese Zuschreibung? Welche Merkmale machen Transdisziplinarität so besonders? Und was bedeutet dies für die kooperativen Prozesse und Strukturen? Diese Fragen greift das folgende Kapitel auf.

---

[1] Vgl. hierzu die Definition von Innovation nach Bullinger und Schlick als „nutzenstiftende Problemlösung [...]. Sie umfasst den gesamten Prozess von der Idee über Entwicklung und Produktion bis hin zur Markteinführung bzw. Realisierung" (Bullinger und Schlick 2002:16, zitiert nach Trantow et. al [3]: 5).

# 2	Transdisziplinarität – vom Forschen für die Praxis zum Forschen mit der Praxis

Ähnlich zu dem Begriff der Interdisziplinarität finden sich bei dem Begriff der Transdisziplinarität verschiedene Definitionen mit nicht unbedingt deckungsgleichen Bedeutungen ([4]: 174). Jantsch stellte bei seiner Einführung des Begriffs einen, 'common purpose' in den Mittelpunkt seines Verständnisses ([6]: 21).[2] Im Gegensatz zur Crossdisziplinarität verbindet nicht eine gemeinsame Basis die Disziplinen, sondern ein Ziel bzw. ein Zweck, gemeinsam Lösungen zu erarbeiten. Mit der Einschätzung von Mittelstrass, dass Inter- und Transdisziplinarität weder neu noch originell, aber auch nicht die normale Form wissenschaftlicher Forschung sind ([7]: 19), unterstreicht er zwei wesentliche Aspekte der Diskussion, die zu dem Begriff der Transdisziplinarität geführt wird:

Zum einen verdeutlicht Mittelstrass die weit zurückliegenden Ursprünge dieser Thematik. Diese zeigt eine terminologische Ambiguität der Begrifflichkeiten, die mit unterschiedlichen Verständnissen der jeweiligen Kooperationsstrukturen[3] einhergehen ([8]: 10ff.). Bezüglich der Transdisziplinarität unterscheidet Bergers in ihrer Arbeit zwei definitorische Schwerpunkte, eine wissenschaftstheoretische und eine wissenschaftssoziologische/akteursbezogene Annäherung ([9]: 18f).[4] In der wissenschaftstheoretischen Definition wird Transdisziplinarität von Mittelstrass als eine durch Integration ausgezeichnete Form der Interdisziplinarität, und damit als die wirkliche Interdisziplinarität sowie als Wissenschaftsprinzip bestimmt[5]:

> „[Mit] Transdisziplinarität [ist] gemeint, dass Kooperation zu einer andauernden, die fachlichen und disziplinären Orientierungen selbst verändernden wissenschaftssystematischen Ordnung führt. Dabei stellt sich Transdisziplinarität sowohl als eine Forschungs- und Arbeitsform der Wissenschaft dar, wo es darum geht, außerwissenschaftliche Probleme [...] zu lösen, als auch als ein innerwissenschaftliches, die Ordnung des wissenschaftlichen Wissens und der wissenschaftlichen Forschung selbst betreffendes Problem. In beiden Fällen ist Transdisziplinarität ein Forschungs- und Wissenschaftsprinzip, das dort wirksam wird, wo eine allein fachliche oder disziplinäre Definition von Problemlagen und Problemlösungen nicht möglich ist bzw. über derartige Definitionen hinausgeführt wird." ([12]: 9f.)

Zum anderen wird in der wissenschaftssoziologischen bzw. akteursbezogenen Definition Transdisziplinarität als interdisziplinäre Forschung bestimmt, die gemeinsam mit der außerwissenschaftlichen Praxis entwickelt und durchgeführt wird. Bergmann, als ein Vertreter des akteursbezogenen Verständnisses, stimmt Mittelstrass in seiner Bestimmung zu, dass „Transdisziplinarität [...] ein Forschungs- und

---

[2]Der Begriff wurde nach Jahn erstmals 1970 von Jantsch auf einer OECD–Konferenz (Organisation for Economic Co-operation and Development) eingeführt.

[3]In einer umfassenden Diskursanalyse werden Merkmale zur Beschreibung der unterschiedlichen Begrifflichkeiten (Disziplinarität, Multi-, Cross-, Inter- und Transdisziplinarität) und den damit einhergehenden unterschiedlichen Kooperationsstrukturen herausgearbeitet.

[4]Diese findet sich ebenso im englischsprachigen Diskurs wieder ([10]: 116; Hirsch Hadorn et al. 2010 in [11]: 433–436).

[5]Neben Mittelstrass sind weitere Vertreter eines wissenschaftstheoretischen Zugangs u. Jantsch 1972 und Piaget 1972.

Wissenschaftsprinzip, aber keine Methode [sei]" ([13]: 9), bestimmt den Begriff allerdings in der Auseinandersetzung von Wissenschaft und Praxis. In einem transdisziplinären Forschungsprojekt greifen seiner Ansicht nach zwei Prozesse ineinander: ein wissenschaftlicher Erkenntnis- und ein lebensweltlicher Problemlösungsprozess. Diese Art des Forschens wird in der Literatur in der sog. Modusdebatte geführt. Diese thematisiert unterschiedliche Formen disziplinenübergreifender Wissenschaftspraxis sowie die Art und Weise, wie ein „Forschen für die Praxis" konzipiert sein sollte ([14]: 6). Zentrales Moment ist dabei der sogenannte „mode 2" oder „new mode" ([15]: VII; [14]: 6; [16]: 7), der als Gegenkonzept zur rein disziplinären und akademisch isolierten Wissensproduktion („mode 1"; [15]: VII; [14]: 6; [16]: 7; [6]: 22ff) bezeichnet wird:

> „Mode-2 knowledge production, where transdisciplinarity is achieved by focusing on research problems as they emerge in contexts of application and where the heterogeneity of knowledge producers introduces additional criteria of assessment apart from scientific quality." ([17]: 223).[6]

Mittelstrass verdeutlicht zudem die weit zurückliegenden Ursprünge dieser Thematik ([8]: 2f.) Für den Begriff der Transdisziplinarität lassen sich zwei definitorische Schwerpunkte, eine wissenschaftstheoretische und eine wissenschaftssoziologische/akteursbezogene, identifizieren.[7] Vom wissenschaftstheoretischen Standpunkt kann Transdisziplinarität als Wissenschaftsprinzip verstanden werden, da die „Kooperation zu einer andauernden, die fachlichen und disziplinären Orientierungen selbst verändernden wissenschaftssystematischen Ordnung führt" ([12]: 9). Andererseits kann eine Annäherung an Transdisziplinarität ebenso von einem wissenschaftssoziologischen bzw. akteursbezogenen Standpunkt erfolgen. In diesem Kontext stellt Transdisziplinarität eine Form der interdisziplinären Forschung dar, die die außerwissenschaftliche Praxis einbindet. Dem wissenschaftssoziologischen bzw. akteursbezogenen Verständnis folgend lassen sich unterschiedliche Merkmale[8] zur Abgrenzung transdisziplinärer Forschungsprozesse zusammenfassen ([8]: 19).

Zur Erreichung transdisziplinärer Lösungen ist es notwendig, verschiedene Methoden der in die Kooperation eingebundenen Akteure zusammenzuführen, um der Komplexität der zu bewältigenden Probleme angemessen begegnen zu können. Des Weiteren liegt im Falle von Transdisziplinarität ein gemeinsames Erkenntnisinteresse von Wissenschaft und Praxis vor, welches sich durch die wissenschaftliche Bearbeitung gesellschaftlicher Problemstellungen ergibt. Die Kooperationsstruktur ist dementsprechend nicht nur als disziplinen-, sondern ebenso als domänenübergreifende Zusammenarbeit zu beschreiben ([8]: 19ff). Ausgehend von einer Konstitution eines gemeinsamen Forschungs-gegenstandes liegt das Ziel transdisziplinärer

---

[6]Bruce et al. ergänzen, dass es sich bei diesem Verständnis von Transdisziplinarität um eine domänenübergreifende Kooperationsstruktur handelt ([18]: 459). Auch Balsiger macht daran seine Unterscheidung zu interdisziplinären Forschungsprozessen fest ([4]: 185).

[7][9]: 18f. Diese findet sich ebenso im englischsprachigen Diskurs wieder. ([10]: S. 116 und Hirsch Hadorn et al. 2010 in [11]: 433–436.

[8]Gegenstandsbereich, Forschungsmethoden, Art des Erkenntnisinteresses und Kooperationsstruktur ([8]: 10ff.).

Forschungsprozesse darin, neues und anschlussfähiges Wissen gemeinsam zu generieren und in die jeweiligen Kontexte (vgl. Abbildung 1 „Forschung" und „Unternehmen") zu integrieren. Das entscheidende Element transdisziplinärer Forschung liegt in dem sogenannten integrativen Zugang ([6]: 30; [14]: 9). Dieser setzt einen gemeinsamen Aushandlungsprozess voraus, indem beispielsweise Fragen, Verständnisse, Zuordnung von Problemen etc. von den beteiligten Akteuren thematisiert werden. Der Prozess der gemeinsamen Wissensproduktion beginnt dabei bereits schon vor dem eigentlichen Projektstart. Dieser ist aktuellen Studien folgend besonders herausfordernd, was sich anhand der Abbildung 1 erläutern lässt.

In der gemeinsamen Antragsformulierung müssen bereits Herausforderungen gemeistert werden, indem die Problemstellungen formuliert werden, die sowohl dem Komplexitätsniveau und der Individualität der praktischen Probleme entsprechen als auch dem wissenschaftlichen Anspruch an generischem Erkenntnisgewinn und den methodischen Gütekriterien gerecht werden. Gleichzeitig ist aus kommunikationswissenschaftlicher Perspektive zu bedenken, dass ein „nachrichtentechnikorientierter" Wissenstransfer nicht möglich ist: Es geht nicht um die Frage, wie wissenschaftliche Erkenntnisse bestmöglich für die Praxis „verpackt" werden können, sondern darum, wie ein gemeinsames Verständnis für die jeweiligen Bedarfe über die gesamte Zusammenarbeit entwickelt werden kann. Hier wird plakativ von der Entwicklung eines gemeinsamen „Sprachspiels"[9] innerhalb des transdisziplinären Forschungsprojektes gesprochen ([14]: 14ff.).

Nach dieser wissenschaftstheoretischen Einführung transdisziplinärer Forschungsprozesse wird nun beleuchtet, wie diese kooperativen Prozesse und Strukturen im Förderschwerpunkt verankert sind. Darüber hinaus werden der Entstehungskontext sowie der Aufbau und die Ziele des Sammelbandes erläutert.

3   Transdisziplinäre Perspektiven im Förderschwerpunkt: Aufbau und Ziele des Sammelbandes

Die Entstehung dieses Buch geht zurück auf das Forschungs- und Entwicklungsprogramm „Arbeiten – Lernen – Kompetenzen entwickeln. Innovationsfähigkeit in einer modernen Arbeitswelt" (A-L-K), das einen zentralen Bestandteil der Hightech-Strategie der Bundesregierung darstellt. Das Bundesministerium für Bildung und Forschung (BMBF) hat in diesem Rahmen u. a. den Förderschwerpunkt „Innovationsfähigkeit im demografischen Wandel" ins Leben gerufen [20].

Dieser Sammelband verfolgt das Ziel, die Vielfalt der entwickelten Ansätze, Konzepte, Instrumente und Methoden aus den transdisziplinären Verbundprojekten

---

[9]Der Begriff „Sprachspiel" wurde vor allem durch Ludwig Wittgensteins *Philosophische Untersuchungen* (1967) geprägt. Leisten greift auf diese zurück, indem er die Idee der Sprachspiele als einen Erklärungspfad bezeichnet, „indem angenommen werden kann, dass innerhalb einer Sprachgemeinschaft prinzipiell alle Teilnehmer auf die gleiche Menge sprachlicher Signifikanzen zurückgreifen können, deren Bedeutung sich jedoch durch spezifische Verwendungen in spezialisierten gesellschaftlichen Subsystemen ausdifferenziert hat" ([14]: 14).

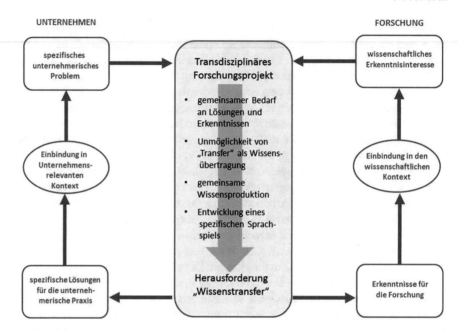

Abbildung 1 Herausforderung „Wissenstransfer" in transdisziplinären Forschungsprojekten ([14]: 15)

aufzuzeigen (vgl. Teile II, III und IV). Zudem werden die Erkenntnisse und Diskussionen aus der übergreifenden Verbundprojektarbeit abgebildet und gebündelt, um zukünftige Handlungsfelder zu identifizieren und Impulse für die Gestaltung weiterer Forschungsprogramme aufzuzeigen (vgl. Teil I).

Um den skizzierten komplexen Herausforderungen des demografischen Wandels sowie den Zielen der Demografie- und der Hightech-Strategie [2] angemessen zu begegnen, haben sich neben den transdisziplinären Verbundprojektstrukturen im Rahmen des Forschungs- und Entwicklungsprogramms A-L-K weitere Strukturen etabliert. Um sowohl die Strukturen mit Ihren Inhalten, Zielen und Aufgaben darzustellen und eine Einordnung in den relevanten politischen Raum vorzunehmen, wird den einzelnen Teilen eine **Einführung in diesen Sammelband** vorangestellt, die sich in fünf Beiträge gliedert:

Als Regierungsdirektor des Ressorts *Forschung für Produktion, Dienstleistung und Arbeit* im Bundesministerium für Bildung und Forschung BMBF obliegt *Rudolf Leisen* das Vorwort dieser Veröffentlichung. Zur Einordnung des Förderschwerpunktes in den politischen Raum beschreibt *Ilona Kopp*, wie Politikgestaltung durch Forschungsförderung realisiert werden kann. Darin werden sowohl Leitlinien, Strukturen als auch die Wirkung von Förderstrukturen am Beispiel des Förderschwerpunktes Innovationsfähigkeit im demografischen Wandel skizziert. In ihrem Beitrag zur Gestaltung von Kommunikations- und Kooperationsprozessen im Förderschwerpunkt konkretisieren *Claudia Jooß et al.* bereits etablierte als auch neu eingeführte

Strukturelemente. Diese werden durch exemplarische Aufgaben konkretisiert und in den förderstrukturellen Rahmen des A-L-K Programms, das als ein lernendes System beschrieben wird, eingeordnet. Nachdem hiermit sowohl der forschungspolitische als auch der förderstrukturinterne Rahmen aufgezeigt ist, schlagen *Oleg Cernavin et al.* schließlich eine Brücke zu einer förderschwerpunktübergreifenden Übersicht bereits bestehender Transferstrukturen. Diese werden im Kontext des demografischen Wandels erläutert. Darüber hinaus werden Potenziale, Herausforderungen und Handlungsempfehlungen zum Transfer transdisziplinärer Forschungsprozesse aufgezeigt.

Anschließend bündelt Teil I die Inhalte, Diskussionen und Ergebnisse der Förderschwerpunkt-Tagung „Chancen durch Demografie – Konzepte und Lösungen für den Wandel". Darüber hinaus werden die Inhalte und die Beschreibung der einzelnen Prozessschritte des Memorandums gerahmt. Das Memorandum zeigt Impulse für die Ausgestaltung zukünftiger Forschungsprogramme auf und fasst weitere Forschungsbedarfe aus Perspektive der Förderschwerpunktakteure zusammen. Im gemeinsamen Einführungsbeitrag des Teams der FokusgruppensprecherInnen sowie des Metaprojektes DemoScreen wird dieser umfassende Erstellungsprozess erläutert und visualisiert – von der Idee bis zur Erstellung des Memorandums.

Die Beiträge in Teil I beinhalten fünf Handlungsfelder, die eine Diskussionsgrundlage des Memorandums darstellen [19]. In fünf parallelen Sessions wurde auf der Förderschwerpunkt-Tagung diskutiert, welche Herausforderungen, Gestaltungsaufgaben und Forschungsbedarfe aus Sicht der Förderschwerpunktakteure mit diesen Handlungsfeldern einhergehen. In einem letzten Prozessschritt hat das Team der FokusgruppensprecherInnen ein Memorandum formuliert. Dieses bündelt die Inhalte der sechs Beiträge in diesem Teil, zeigt Impulse für die Ausgestaltung zukünftiger Forschungsprogramme auf und fasst weitere Forschungsbedarfe zusammen. Das Memorandum des Förderschwerpunktes Innovationsfähigkeit im demografischen Wandel fungiert in Ergänzung zu den „Eschborner Thesen zur Arbeitsforschung"[10] als forschungspolitisches Perspektivenpapier.

Als Sammelband dieses Förderschwerpunktes werden in den Teilen II, III und IV aus den insgesamt 27 transdisziplinären Verbundprojekten gemeinsam erarbeitete Lösungen zur Gestaltung des demografischen Wandels präsentiert. Die Besonderheit der Verbundprojekte liegt darin, dass unterschiedliche, relevante und betroffene Akteure sowie Institutionen gemeinsam in den oben beschriebenen Aushandlungsprozess treten. Die Strukturen des Forschungs- und Entwicklungsprogramms A-L-K ermöglichen in diesem Förderschwerpunkt damit einen integrativen Zugang, der insgesamt über 80 bundesweit beteiligte Institutionen aus Wissenschaft, Wirtschaft und intermediären Organisationen beinhaltet. Diese Vielfalt spiegelt sich in den insgesamt 52 Beiträgen, die sowohl theoretische Forschungs- als auch praktische Handlungsfelder sowie „Good Practice-Beiträge" umfassen. Um den Erfahrungsaustausch und die Vernetzung zwischen den einzelnen Verbundprojekten zu unterstützen, wurden im Rahmen der Aachener Förderschwerpunkt-Tagung „Chancen durch Demografie

---

[10]http://www.rkw-kompetenzzentrum.de/nc/publikationen/details/rkw/publikationen/
eschborner-thesen-zur-arbeitsforschung-697/.

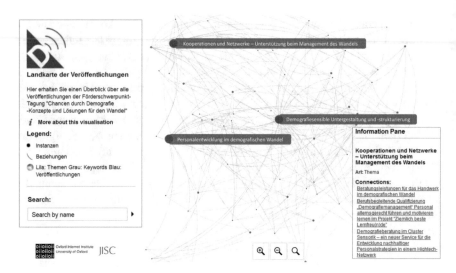

Abbildung 2 Visualisierung (Auszug) des Publikationsmarktplatzes auf der Aachener Förderschwerpunkt-Tagung

– Konzepte und Lösungen für den Wandel" die eingereichten Abstracts zu diesem Sammelband auf einem Publikationsmarktplatz visualisiert (vgl. Abbildung 2).

Das Ziel lag darin, Autorengruppen mit thematischen Übereinstimmungen in den Themengebieten zu identifizieren, um Schnittstellen aufzuzeigen und nach Bedarf verbundprojektübergreifende Publikationen zu ermöglichen. Insgesamt sind hieraus 48 Beiträge entstanden, die in den folgenden Themenfeldern gebündelt sind:

Das Themenfeld **Demografiesensible Unternehmensgestaltung und -strukturierung** des **Teils II** bündelt insgesamt 17 Beiträge. Darin werden Konzepte, Untersuchungen und Lösungen beschrieben, die ein erfolgreiches Kompetenzmanagement im demografischen Wandel ermöglichen und so unter anderem eine langfristige Erwerbsfähigkeit oder einen nachhaltigen Wissenstransfer in einem Unternehmen unterstützen.

Teil III widmet sich mit 21 Beiträgen dem Thema **Personalentwicklung im demografischen Wandel**. Trainings, Instrumente und Konzepte werden vorgestellt, mittels derer die Innovations- und Beschäftigungsfähigkeit von Mitarbeiterteams gefördert werden, so dass eine Unterstützung des Personalmanagements hinsichtlich der Herausforderungen des demografischen Wandels erfolgen kann.

Das Themenfeld **Kooperationen und Netzwerke – Unterstützung beim Management des Wandels** (vgl. **Teil IV**) umfasst insgesamt zehn Beiträge. Die darin adressierten Cluster, Netzwerke und Arbeitsgruppen verfolgen das Ziel, die heterogenen Akteursgruppen im Themenfeld Demografie und Arbeitsforschung zu unterstützen. Anhand der hier aufgezeigten Beispiele mit häufig regionalem Bezug wird der Frage nachgegangen, inwieweit bspw. Infrastruktur und Mitarbeitervernetzung solcher Netzwerke eingeschätzt und Potentiale dieser Netzwerke weiter ausgeschöpft werden können. Nachdem in diesem Artikel sowohl das Verständnis von Innovationsfähigkeit im demografischen Wandel als auch von Transdisziplinarität beleuchtet wurde, wünschen wir allen Leserinnen und Lesern viele Denkanstöße bei der Erkundung von demografieorientierten Ergebnissen, zukünftigen Impulsen, Forschungsbedarfen und forschungspolitischen Perspektiven.

Literaturverzeichnis

1. S. Jeschke, R. Vossen, I. Leisten, C. Jooß, T. Vaegs, *Arbeit im demografischen Wandel: Strategien für das Arbeitsleben der Zukunft*. Technische Hochschule Aachen Zentrum f. Lern- u. Wissensmanagement, Aachen, 2013
2. I. Kopp, Politikgestaltung durch Forschungsförderung am Beispiel des Förderschwerpunktes. In: *Exploring Demographics*, ed. by S. Jeschke, A. Richert, F. Hees, C. Jooß, Springer Fachmedien Wiesbaden, 2015, pp. 3–10
3. S. Trantow, Die fähigkeit zur innovation einleitung in den sammelband. In: *Enabling Innovation. Innovationsfähigkeit deutsche und internationale Perspektiven*, ed. by S. Jeschke, I. Isenhardt, F. Hees, S. Trantow, Springer, Berlin, Heidelberg, 2011
4. P.W. Balsiger, *Transdisziplinarität*, 1st edn. Fink (Wilhelm), München, 2005
5. F.J. van Rijnsoever, L.K. Hessels, Factors associated with disciplinary and interdisciplinary research collaboration. Research Policy **40** (3), 2011, pp. 463–472
6. T. Jahn, Transdisziplinarität in der Forschungspraxis. In: *Transdisziplinäre Forschung: integrative Forschungsprozesse verstehen und bewerten.*, Campus-Verlag, Frankfurt am Main, 2008
7. J. Mittelstraß, Methodische Transdisziplinarität. Technikfolgenabschätzung. Theorie und Praxis **14** (2), 2005, pp. 18–23
8. C. Jooß, *Gestaltung von Kooperationsprozessen interdisziplinärer Forschungsnetzwerke*. BoD Books on Demand, 2014
9. M. Bergers, *Multi-, Trans- und Interdisziplinarität. Eine exemplarische Anwendung der sozialen Netzwerkanalyse. Bachelorarbeit am Institut für Soziologie und am IMA/ZLW der RWTH Aachen*. 2012
10. L.R. Lattuca, *Creating Interdisciplinarity: Interdisciplinary Research and Teaching Among College and University Faculty*. Vanderbilt University Press, 2001
11. R. Frodeman, *The Oxford handbook of interdisciplinarity*. Oxford University Press, Oxford, New York, 2010
12. J. Mittelstraß, *Transdisziplinarität : wissenschaftliche Zukunft und institutionelle Wirklichkeit*. UVK, Univ.-Verl., Konstanz, 2003
13. M. Bergmann, *Methoden transdisziplinärer Forschung. Ein Überblick mit Anwendungsbeispielen*. Campus, Frankfurt am Main, 2010
14. I. Leisten, Transfer Engineering in transdisziplinären Forschungsprojekten. Ph.D. thesis, Books on Demand, Norderstedt, 2012
15. M. Gibbons, C. Limoges, H. Nowotny, S. Schwartzman, P. Scott, M. Trow, *The new production of knowledge : the dynamics of science and research in contemporary societies*. Sage Publications, London, 1994
16. M. Bergmann, *Transdisziplinäre Forschung: integrative Forschungsprozesse verstehen und bewerten*. Campus-Verlag, Frankfurt am Main; New York, 2008
17. H. Nowotny, P. Scott, M. Gibbons, *Re-thinking science: knowledge and the public in an age of uncertainty*. Polity, Cambridge, UK, 2001
18. A. Bruce, C. Lyall, J. Tait, R. Williams, Interdisciplinary integration in Europe: the case of the Fifth Framework programme. Futures **36**(4), 2004, pp. 457–470

19. P.D.G. Becke, D.P. Bleses, O. Cernavin, P.D.A. Ducki, D.C. Jooß, D.R. Klatt, P.D.T. Langhoff, P.D.F.W. Nerdinger, Einführungsbeitrag: Von den Handlungsfeldern der Förderschwerpunkt-Tagung zum Memorandum. In: *Exploring Demographics*, ed. by S. Jeschke, A. Richert, F. Hees, C. Jooß, Springer Fachmedien Wiesbaden, 2015, pp. 55–66
20. BMBF. Innovationsfähigkeit im demografischen Wandel, 2014. http://pt-ad.pt-dlr.de/de/279.php. Zuletzt aufgerufen: 08.09.2014

Gestaltung von Kommunikations- und Kooperationsprozessen im Förderschwerpunkt „Innovationsfähigkeit im demografischen Wandel"

Claudia Jooß, Sabine Kadlubek, Anja Richert and Sabina Jeschke

Zusammenfassung Über die Laufzeit des Forschungs- und Entwicklungsprogramms „Arbeiten – Lernen – Kompetenzen entwickeln. Innovationsfähigkeit in einer modernen Arbeitswelt" haben sich in den einzelnen Förderschwerpunkten unterschiedliche Strukturen etabliert. Ziel dieses Beitrags ist es, diese in einem Überblick darzustellen, zu erläutern und anhand exemplarischer Aufgaben zu konkretisieren. Somit wird eine Grundlage geschaffen, Kommunikations- und Kooperationsprozesse zwischen und über diese Strukturelemente hinweg zu gestalten und für die (Weiter-) Entwicklung künftiger Förderschwerpunkte zu nutzen.

Schlüsselwörter Metaprojekt DemoScreen · Lernendes Programm · Gestaltung · Kooperation

1 Einleitung

Im Rahmen des Forschungs- und Entwicklungsprogramms (FuE-Programm) „Arbeiten – Lernen – Kompetenzen entwickeln. Innovationsfähigkeit in einer modernen Arbeitswelt" (A-L-K) des Bundesministeriums für Bildung und Forschung (BMBF) werden seit 2007 neue Ansätze der Personal-, Organisations- und Kompetenzentwicklung gefördert, um innovationsfreundliche Rahmenbedingungen zu gestalten und kohärente Strategien zwischen Wissenschaft, Wirtschaft und Politik zu etablieren ([1]: 7). Um diese Ansätze kontinuierlich an die sich wandelnden sozioökonomischen Bedingungen anpassen zu können, muss das FuE-Programm so konzipiert sein, dass es selbst lern- und innovationsfähig ist ([1]: 9). Die dafür erforderliche Flexibilität wird durch „offene Programmstrukturen" ([1]: 26) sowie integrierte Lern- und Strukturinstrumente gewährleistet. Um zu erläutern, wie sich diese im Förderschwerpunkt widerspiegeln, wird dieser in einem ersten Schritt in den förderstrukturellen Rahmen des A-L-K Programms eingeordnet. Nachfolgend gibt Kapitel 2 einen Überblick über das A-L-K Programm als lernendes System

C. Jooß (✉) · S. Kadlubek · A. Richert · S. Jeschke
IMA/ZLW & IfU, RWTH Aachen University, Dennewartstr. 27, 52068 Aachen, Germany
e-mail: claudia.jooss@ima-zlw-ifu.rwth-aachen.de

Originally published in "Exploring Demographics ", © Springer 2015.
Reprint by Springer International Publishing Switzerland 2016,
DOI 10.1007/978-3-319-42620-4_55

samt seiner Struktur, die sowohl die Etablierung als auch die Konstituierung der Lernprozesse unterstützt.

Anschließend wird in Kapitel 3 die Weiterentwicklung des lernenden Programms am Beispiel des Förderschwerpunktes „Innovationsfähigkeit im demografischen Wandel" erläutert. Dabei steht der Beitrag des Metaprojekts „DemoScreen – Kommunikation, Kooperation und Innovationsfähigkeit im demografischen Wandel" zur kontinuierlichen Steigerung und Fortentwicklung der Lernfähigkeit des Programms im Fokus. Da ein zentrales Ziel des Metaprojektes darin liegt, eine nachhaltige Zusammenarbeit der einzelnen Förderschwerpunktakteure zu etablieren, wird die Rekursionsebene der Fokusgruppe, das bereits etablierte Strukturelement der FokusgruppensprecherInnen sowie das neu eingeführte Strukturelement des Teams der FokusgruppensprecherInnen erläutert und mit exemplarischen Aufgaben konkretisiert. Somit werden Kommunikations- und Kooperationsprozesse aufgezeigt. In Kapitel 4 wird ein abschließendes Fazit gezogen.

2 Einordung des Förderschwerpunktes in den förderstrukturellen Rahmen

Im Folgenden wird das lernende A-L-K Programm aus einer systemtheoretisch-kybernetischen Perspektive als lernendes System betrachtet. Anschließend werden die offenen Programmstrukturen des A-L-K Programms samt der Elemente zur Programmsteuerung und -gestaltung, welche das Programm zur kontinuierlichen Lern- und Transferfähigkeit befähigen, detaillierter skizziert.

2.1 FuE-Programm A-L-K als lernendes System

Um den Veränderungen in der Arbeitswelt gerecht zu werden, ist das A-L-K Programm des BMBF als lernendes Programm konzipiert. Nach Trantow spielen Lernprozesse eine besondere Rolle, „da sie eine kontinuierliche Anpassung der Programmaktivitäten an aktuelle Problemstellungen und Forschungsbedarfe erlauben" ([2]: 27).

Die dem FuE-Programm A-L-K immanente Lernfähigkeit wird u.a. durch die Vernetzung aller Rekursionsebenen (vgl. Abbildung 1) und den damit einhergehenden Reflexionsprozessen zwischen verschiedenen strukturellen Elementen (vgl. Kapitel 2.2.2) gewährleistet und gefördert. „Dies schließt ein, dass die künftig durch das Programm erzielten Ergebnisse auch in die Entscheidungsprozesse zur weiteren Umsetzung des Programms einfließen und die Vertiefung und Erweiterung von Forschungs- und Entwicklungsaufgaben anleiten" ([1]: 9). Jedoch wird einer lernorientierten Programmausrichtung nicht genügend Rechnung getragen, wenn ausschließlich die innerhalb des Programms generierten Forschungsergebnisse berücksichtigt und verwertet werden. Es ist daher unabdingbar, auch die im nationalen und internationalen Forschungskontext erzielten Ergebnisse und Erkenntnisse zu rezipieren. Überdies findet ebenfalls ein Transfer der Programmergebnisse in den programmexternen Forschungskontext statt. Somit ist das A-L-K Programm und dessen Umwelt durch Kopplungen miteinander verbunden.

Mithilfe der modernen Systemtheorie, die in erster Linie als System-Umwelt-Theorie bezeichnet werden kann ([3]: 2), lässt sich das A-L-K Programm allgemein als System beschreiben, welches sich von seiner Umwelt abgrenzt. Nach Strunk und Schiepek wird „unter einem System eine von der Umwelt abgegrenzte funktional geschlossene Entität verstanden, die aus Elementen besteht, die miteinander in Wechselwirkung stehen. Systeme können offen sein für Austauschprozesse mit ihrer Umwelt" ([4]: 8). Im konkreten Fall des A-L-K Programms handelt es sich um solch ein offenes System. Ein offenes System nimmt zunächst Informationen aus der Umwelt auf (Input). Systemimmanente Transformationsprozesse verarbeiten diese Informationen anschließend zu einem Output, welcher wiederum in das System zurückgespielt wird. Im Sinne der Kybernetik wird ein System erst durch solche Rückkopplungsprozesse zu einem adaptiven und lernfähigen System ([5]: 13ff.).

Da die Rückkopplungen im Fall des A-L-K Programms auf unterschiedlichen Rekursionsebenen des Programms erfolgen, erscheint es an dieser Stelle notwendig, im folgenden Verlauf die Struktur des A-L-K Programms zu skizzieren.

2.2 Struktur des FuE-Programms A-L-K

Das A-L-K Programm ist aufgrund seiner lernförderlichen Programmstruktur und Strukturelemente der Programmsteuerung speziell darauf ausgelegt, sich flexibel auf neue Forschungsbedarfe und Umweltveränderungen anpassen zu können, kurz, lernfähig zu sein. Im Folgenden werden die unterschiedlichen Rekursionsebenen sowie Strukturelemente vorgestellt, welche in die Programmkonzeption integriert sind (vgl. Abbildung 1).

2.2.1 Rekursionsebenen des FuE-Programms A-L-K

Bevor im weiteren Verlauf näher auf die Struktur des FuE-Programms A-L-K eingegangen wird, lohnt sich zunächst ein Blick auf die Förderaktivitäten der Bundesregierung, um so das Programm in den politischen Kontext der Bundesregierung einbetten zu können. Im Rahmen der Hightech-Strategie möchte die Bundesregierung alle Forschungsaktivitäten bündeln und damit „eine inhaltliche Klammer zu innovationspolitischen Themen über die Ressorts der Bundesregierung hinweg [bilden]" ([6]: 4). Mit der Hightech-Strategie wurden fünf Bedarfsfelder identifiziert, die zukünftige globale Bedarfe und Herausforderungen adressieren: Klima/Energie, Gesundheit/Ernährung, Mobilität, Sicherheit und Kommunikation ([6]: 5). Ziel dieser Strategie ist es, „die enormen Potenziale Deutschlands in Wissenschaft und

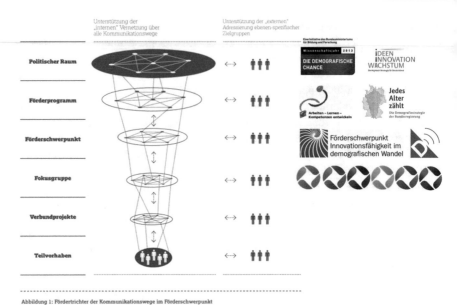

Abbildung 1: Fördertrichter der Kommunikationswege im Förderschwerpunkt

Abbildung 1 Lernförderliche Förderstruktur des Programms „Arbeiten – Lernen – Kompetenzen entwickeln. Innovationsfähigkeit in einer modernen Arbeitswelt"

Wirtschaft gezielt zu aktivieren und Lösungen für die globalen und nationalen Herausforderungen bereitzustellen" ([6]: 5). Folglich ist das FuE-Programm A-L-K eines der vielen unterschiedlichen Förderaktivitäten der Bundesregierung mit dem speziellen Ziel, „Innovationsfähigkeit aus der Verknüpfung von Personal, Organisations- und Kompetenzentwicklung in einer modernen Arbeitswelt [zu] stärken" ([1]: 7).

Gegenwärtig besteht das A-L-K Programm aus den folgenden fünf Förderschwerpunkten: „Innovationsstrategien jenseits traditionellen Managements", „Balance von Flexibilität und Stabilität in einer modernen Arbeitswelt", „Präventiver Arbeits- und Gesundheitsschutz", „Innovationsfähigkeit im demografischen Wandel" und als jüngster Schwerpunkt „Betriebliches Kompetenzmanagement im demografischen Wandel" (vgl. Abbildung 2). Damit geben die Förderschwerpunkte den inhaltlichen Rahmen für die unterschiedlichen Projekte vor und spiegeln gleichzeitig eingeleitete fachliche Themenfelder des Programms wider.

Um einen tiefen wissenschaftlichen Austausch zu ermöglichen sowie aufgrund der Quantität und Diversität der Forschungsprojekte, ist jeder Förderschwerpunkt im A-L-K Programm in *Fokusgruppen* untergliedert (vgl. [7]: 129). Fokusgruppen fungieren als thematisch fokussierte Untergruppen zu den Förderschwerpunkten. Sie wurden zum Wissens- und Erfahrungsaustausch gegründet und ermöglichen damit eine fachliche Kommunikation auf einer höheren Aggregationsebene. Im Förderschwerpunkt „Innovationsfähigkeit im demografischen Wandel" wurden beispielsweise insgesamt sechs Fokusgruppen mit den folgenden Schwerpunkten

 Förderschwerpunkt
Präventiver Arbeits-
und Gesundheitsschutz

 Förderschwerpunkt
Innovationsstrategien jenseits
traditionellen Managements

 Förderschwerpunkt
Balance von
Flexibilität und Stabilität

 Förderschwerpunkt
Innovationsfähigkeit im
demografischen Wandel

 Förderschwerpunkt
Betriebliches
Kompetenzmanagement
im demografischen Wandel

Abbildung 2 Förderschwerpunkte des A-L-K Programms

eingerichtet: „Altersheterogene Innovations-teams als Erfolgsfaktor des demografi-schen Wandels", „Demografiemanagement und Vernetzung", „Erwerbsbiografien als Innovationstreiber im demografischen Wandel", „Kompetenzentwicklung und Quali-fizierung für den demografischen Wandel", „Messung von Innovationspotenzialen vor dem Hintergrund des demografischen Wandels" sowie „Regionale Aspekte des demografischen Wandels".

Fokusgruppen bündeln zudem mehrere transdisziplinäre *Verbundprojekte*, die sich aus Wissenschaft, Wirtschaft und Intermediären zusammensetzen. Ziel dieser Bündelung und Zusammenführung ist es, Synergieeffekte durch den Austausch von Forschungsergebnissen zu nutzen sowie Forschungs- und Handlungsempfehlungen innerhalb der Fokusgruppe zu generieren (vgl. [7]: 130). Zu diesem Zweck bringt jede Fokusgruppe jeweils mehrere, thematisch ähnliche Verbundprojekte zusammen. Im Rahmen gemeinsamer Fokusgruppentreffen wird den einzelnen Verbundprojek-ten die Möglichkeit geboten, sich mit anderen Verbundprojekten zu inhaltlich ähn-lichen Themenfeldern auszutauschen. Damit wird ein kontinuierlicher Wissensaus-tausch und -transfer gewährleistet. Im Förderschwerpunkt „Innovationsfähigkeit im demografischen Wandel" gibt es insgesamt 27 transdisziplinäre Verbundpro-jekte. Diese bilden einen thematischen Nukleus, unter dem wiederum mehrere Teil-vorhaben gebündelt sind. Die Teilvorhaben sind ebenfalls transdisziplinär aufgestellt.

2.2.2 Strukturelemente des FuE-Programms A-L-K

Neben den zuvor ausführlich skizzierten Rekursionsebenen, die unterschiedliche Aggregationsgrade im Hinblick auf den Wissensaustausch abdecken und so Bün-delung und Transfer ermöglichen, wurden im Rahmen des lernenden A-L-K Pro-gramms Strukturelemente wie das Monitoringprojekt und Metaprojekt integriert. Ein *Monitoringprojekt* ist ein programmbegleitendes Instrument, dessen Aufgabe darin besteht, Forschungsergebnisse aller Förderschwerpunkte des Programms iter-ativ zu erfassen, zu reflektieren und Handlungsbedarfe sowie Erkenntnislücken für die weitere Programmentwicklung zu identifizieren ([8]: 453). Ziel des Monitorings ist es somit, organisationale Lernprozesse zu etablieren und zu institutionalisieren ([2]: 51).

Ein weiteres Strukturelement im A-L-K Programm stellt das *Metaprojekt* dar. Metaprojekte übernehmen eine zentrale Funktion bei der inhaltlichen

Weiterentwicklung der Förderschwerpunkte, der Bündelung und Synthese verschiedener Forschungsergebnisse sowie ihrer Dissemination. Aktuell werden vier der oben genannten Förderschwerpunkte von jeweils einem Metaprojekt begleitet: „Strategischer Transfer im Arbeits- und Gesundheitsschutz" (StArG), „Innovationsfähigkeit als Managementaufgabe, Synthese, Transfer und Begleitung von Forschungs- und Entwicklungsvorhaben" (MANTRA), „BALANCE – Flexibilität und Stabilität in der Forschungswelt" (BALANCE) und „Kommunikation, Kooperation und Innovationsfähigkeit im demografischen Wandel" (DemoScreen). Als Metaprojekt im Förderschwerpunkt „Innovationsfähigkeit im demografischen Wandel" hat DemoScreen eine Moderatorenrolle inne und stellt sicher, dass ein nachhaltiger Forschungs- und Wissenstransfer in die Gesellschaft, Wirtschaft, Forschung und Politik realisiert wird. Dabei erfüllt DemoScreen sowohl Dienstleistungs- als auch Forschungsaufgaben für den Förderschwerpunkt.

3 Weiterentwicklung des lernenden Programms am Bespiel des Förderschwerpunktes „Innovationsfähigkeit im demografischen Wandel"

Im Folgenden steht der Beitrag, den das Metaprojekt DemoScreen im Förderschwerpunkt „Innovationsfähigkeit im demografischen Wandel" zur Weiterentwicklung des lernenden A-L-K Programms leistet, im Fokus der Betrachtung. Somit wird im weiteren Verlauf des Beitrags das Metaprojekt DemoScreen detaillierter vorgestellt. Anschließend wird die Rekursionsebene der Fokusgruppe im förderstrukturellen Rahmen visualisiert und definiert sowie das Team der FokusgruppensprecherInnen näher beschrieben.

3.1 Beitrag des Metaprojektes DemoScreen im Förderschwerpunkt

Mit der internen Vernetzung über alle Rekursionsebenen hinweg und der externen Adressierung ebenenspezifischer Zielgruppen leistet DemoScreen einen Beitrag zur kontinuierlichen Verbesserung und Weiterentwicklung des FuE-Programms.
   Ziele des Metaprojektes DemoScreen sind,

- Die nachhaltige Zusammenarbeit der Förderschwerpunktakteure zu unterstützen und zu entwickeln,
- Die Forschungs- und Projektergebnisse sowohl förderschwerpunktintern (Förderprogramm, Förderschwerpunkt, Fokusgruppen, Projekte) als auch förderschwerpunktextern (Wirtschaft, Forschung, Gesellschaft, Politik, Intermediäre) sichtbar zu machen und damit
- Die komplexe Aufgabe des Transfers erfolgreich und nachhaltig zu unterstützen.

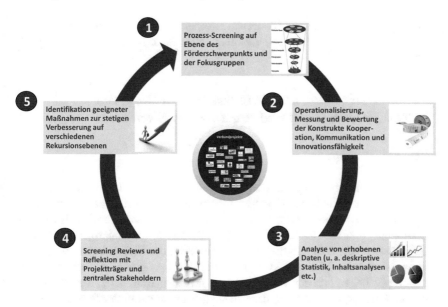

Abbildung 3 Konzeption des Prozess-Screenings anhand von wissenschaftlichen Arbeiten zu inter- und transdisziplinären Forschungsprojekten

Die Ziele werden in den vier Handlungsfeldern Vernetzung und Allianzbildung, Befähigung und Dissemination, Methodenintegration und Zielgruppenadaption sowie Operationalisierung und Bewertung bearbeitet. Das Metaprojekt Demo-Screen wird als Instrument der Forschungsförderung definiert, das sowohl Dienstleistungsaufgaben für den Förderschwerpunkt übernimmt als auch Forschungsaufgaben für die Kommunikation, Kooperation und Innovationsfähigkeit transdisziplinärer Forschungsverbünde eruiert. Im Laufe seiner Begleitforschung entwickelt DemoScreen Instrumente, wie die Workshop-Reihe, die Systematisierung und Bündelung sowie das Prozess-Screening (siehe Abbildung 3). Das letztgenannte Instrument wurde bereits prozessbegleitend implementiert, um die Kommunikation und Kooperation im Förderschwerpunkt zu erfassen, Optimierungsmöglichkeiten zu identifizieren und die Übertragbarkeit auf weitere Metaprojekte und Forschungsprogramme zu diskutieren. Somit stößt das Prozess-Screening organisationale Lernprozesse innerhalb des Förderschwerpunktes an. Alle von DemoScreen entwickelten Instrumente ermöglichen im Sinne des lernenden Programms u. a. die Reflexion der komplexen und dynamischen Forschungs- und Entwicklungsaktivitäten sowie die Ableitung von Handlungsempfehlungen bereits während der Projektlaufzeit. Im Selbstbild von DemoScreen wird *Meta* nicht als Strukturbeschreibung, sondern als eine Aufgabe im Sinne eines *Enabling* verstanden.

Ferner unterstützt DemoScreen das Team der Fokusgruppen und Fokusgruppen-sprecherInnen bei der Gestaltung der Kommunikations- und Kooperationsprozesse (virtuell und real), regt zu dessen Reflexion an und erarbeitet Maßnahmen zu dessen Optimierung. Eine besondere Herausforderung stellt dabei die Vernetzung der heterogenen Projekte und Akteure über alle Rekursionsebenen hinweg dar (Abbildung 4).

3.2 Konkretisierung und Aufgabenbeschreibung von Rekursionsebenen und strukturellen Elementen

3.2.1 Fokusgruppe

Im Förderschwerpunkt „Innovationsfähigkeit im demografischen Wandel" arbeiten sechs Verbundprojekte in einer Fokusgruppe zusammen und etablieren eine Kooperation, die über die Förderphase hinaus reicht. Eine Fokusgruppe wird als zielgruppenbezogene, thematische Bündelung von geförderten Verbundprojekten in einem Förderschwerpunkt definiert, die sich aus Vertretern der Verbundprojekte, des Projektträgers und des Metaprojektes zusammensetzt. Ausgewählte Aufgaben einer Fokusgruppe liegen u.in der Vernetzung und im Austausch, in der Verstärkung der Außenwirkung, in der Stärkung der Nachhaltigkeit sowie in der Entwicklung von gemeinsamen Positionen, Handlungs- und Forschungsempfehlungen.

Neben der Ebene der Verbundprojekte wird die Fokusgruppe als strukturelles Element der Forschungsförderung dargelegt und individuell durch die Akteure gestaltet. Durch die Aushandlung gemeinsamer Ziele, Inhalte und Rahmenbedingungen der Zusammenarbeit etabliert sich eine Kooperationsform, deren Fokus in der internen Vernetzung über Verbundprojekt-Grenzen hinweg sowie in der Adressierung förderschwerpunktexterner Zielgruppen (insbesondere im Bereich von Intermediären) liegt.

Um eine nachhaltige Außenwirkung des gesamten Förderschwerpunktes „Innovationsfähigkeit im demografischen Wandel" auf den spezifischen Zielgruppenebenen zu erreichen, gilt es, die Vielfalt der heterogenen Projekte und Akteure über alle Rekursionsebenen hinweg zu vernetzen. Daher existieren neben den in Abbildung 1 visualisierten Rekursionsebenen weitere strukturelle Elemente:

- FokusgruppensprecherIn
- FokusgruppensprecherInnen-Team

- (Be-)Förderung der Kommunikation- und Kooperationsprozesse im Förderschwerpunkt
- Wissenschaftliche Begleitforschung zu den Inhalten „Kommunikation und Kooperation" und „Verwertung"
- Aufdeckung und Nutzung der Synergien zwischen den Verbundprojekten
- Sprachrohr des Förderschwerpunktes zu verschiedenen Zielgruppen

Abbildung 4 Exemplarische Aufgaben des Metaprojektes DemoScreen

3.2.2 FokusgruppensprecherIn

Um die Etablierung der oben genannten Kooperationsform in der Fokusgruppe zu unterstützen und diese gemeinsam mit den Fokusgruppenakteuren individuell zu gestalten, wird in jeder Fokusgruppe ein/eine Fokusgruppensprecher/-in gewählt. Das Selbstverständnis eines Fokusgruppensprechers/einer Fokusgruppensprecherin lässt sich wie folgt konkretisieren: Der/Die Fokusgruppensprecher/-in sieht sich als AktivatorIn und ModeratorIn, um gemeinsame Ziele, Inhalte und Rahmenbedingungen der Zusammenarbeit auszuhandeln. Dabei ist es von Bedeutung, die zu etablierende Kooperationsform vorzuleben. Den inhaltlichen Austausch (z. B. die Erarbeitung inhaltlicher Linien) und die Vernetzung (z. B. gemeinsame Veranstaltungen, Veröffentlichungen, Präsentation auf DemoScreen.de) über Verbundprojekt-Grenzen hinweg sowie die Identifikation, Adressierung und Einbindung förderschwerpunkt externer Akteure gilt es anzuregen und zu organisieren. Um gemeinsam mit der Fokusgruppe die interne Vernetzung und die externe Adressierung zu gestalten, ist eine Identifikation der FokusgruppensprecherInnen mit den Projektinhalten sowie die Bearbeitung des Begriffsfelds „Demografie" notwendig (Abbildung 5).

3.2.3 FokusgruppensprecherInnen-Team

Das Team der FokusgruppensprecherInnen lässt sich als strukturelles Element des Förderschwerpunktes zwischen der Rekursionsebene Förderschwerpunkt und der Rekursionsebene Fokusgruppe zuordnen (vgl. Abbildung 1). Dies bedeutet neben der Vernetzung der Fokusgruppen, dass das Team die inhaltlichen Linien und Schwerpunkte des Förderschwerpunktes aktiv mitgestaltet. Durch die Bündelung der Inhalte, Kernbotschaften, Konzepte und Instrumente der Fokusgruppen werden neben den relevanten Rekursionsebenen (Förderschwerpunkt, Förderprogramm und politischer Raum) auch die jeweiligen externen Zielgruppenebenen beispielsweise durch die Formulierung gemeinsamer Handlungsempfehlungen, das Aufzeigen von Forschungsperspektiven, gemeinsame Veröffentlichungen und Veranstaltungen (z. B. Ergebnisband, Abschlussveranstaltungen, Memorandum, etc.) adressiert. Eine Transparenz bezüglich der politischen Institutionen, Akteure, Zielgruppen sowie die Komptabilität mit Inhalten des Arbeitskreises „Zukunft der Arbeitsforschung" sind wünschenswert (Abbildung 6).

- Vernetzung der Akteure und inhaltlicher Diskurs von Forschungsinhalten
- Profilschärfung der Aktivitäten des gesamten Förderschwerpunktes
- Verstärkung der Außenwirkung
- Entwicklung von gemeinsamen Positionspapieren, Handlungs- und For-
  schungsempfehlungen

Abbildung 5 Exemplarische Aufgaben einer Fokusgruppe

- AktivatorIn und ModeratorIn, um gemeinsame Ziele, Inhalte und Rahmenbedingungen der Zusammenarbeit auszuhandeln
- Regt den inhaltlichen Austausch und die Vernetzung über Verbundprojekt-Grenzen hinweg an
- Gestaltung der internen Vernetzung und externen Adressierung
- Organisatorische Aufgaben für die Fokusgruppe, wie z.B. die Ausgestaltung der Fokusgruppentreffen oder die Koordination von Publikationen und (Messe-) Auftritten

Abbildung 6 Exemplarische Aufgaben einer Fokusgruppensprecherin/eines Fokusgruppensprechers

4 Fazit

Die Diversität und Transdisziplinarität der Forschungsprojekte des FuE Programms A-L-K sowie deren mannigfaltige Forschungsergebnisse bedürfen einer Programmstruktur, die Lernräume für beteiligte Forschungsakteure unterstützen und fördern. Diesen Bedarf greift das Metaprojekt DemoScreen auf und richtet sein Augenmerk auf die Gestaltung der Kommunikations- und Kooperationsprozesse. Nach Jooß 2014 liegt eine wesentliche Voraussetzung für die Gestaltbarkeit von Kooperationsprozessen darin, dass diese bei den beteiligten Akteuren ansetzt und während des gesamten Kooperationsprozesses stattfindet ([9]: 78). Gestaltung wird dabei als eine reflexive Unterstützung der Kooperationen verstanden, die mit dem Ziel einhergeht, Lernprozesse zu initiieren ([9]: 78) (Abbildung 7).

Dieses Verständnis zeigt sich zum einen in dem neu implementierten Lerninstrument des Prozess-Screenings, welches bereits während der Förderlaufzeit und nicht lediglich ex post stattfindet. Darüber hinaus ist eine Zusammenarbeit der beteiligten Akteure von Bedeutung. Im Rahmen dieses Beitrags wurden durch die intensive Zusammenarbeit des Metaprojektes, der VertreterInnen des Projektträgers sowie durch das Team der FokusgruppensprecherInnen die einzelnen Rekursionsebenen und Strukturelemente beschrieben und mit Beispielen hinterlegt (vgl. Kapitel 3.2), mit dem Ziel, die weitere Zusammenarbeit zu konkretisieren und zu gestalten. Im Sinne eines lernenden Programms dienen diese als Grundlage zur (Weiter-)

- Aktive Mitgestaltung inhaltlicher Linien und Schwerpunkte des Förderschwerpunktes
- Unterstützung bei der Entwicklung von gemeinsamen (Forschungs-) Leitlinien
- Setzt Impulse für die Ausgestaltung zukünftiger Forschungsprogramme

Abbildung 7 Aufgaben des FokusgruppensprecherInnen-Teams

Entwicklung künftiger Förderschwerpunkte und können auf Grund flexibler, sich noch ergebender Bedarfe erweitert und konkretisiert werden.

Literaturverzeichnis

1. Bundesministerium für Bildung und Forschung, Arbeiten – Lernen – Kompetenzen entwickeln. Innovationsfähigkeit in einer modernen Arbeitswelt. Bericht im Rahmen des BMBF-Forschungs- und Entwicklungsprogramms. Tech. rep., Berlin, 2007
2. S. Trantow, *Ein kybernetisches Modell für das Internationale Monitoring von F&E-Programmen im Bereich der Arbeitsforschung*. Books on Demand, Norderstedt, 2012
3. H. Willke, *Systemtheorie: Systemtheorie 3. Steuerungstheorie: Grundzüge einer Theorie der Steuerung komplexer Sozialsysteme: III*, auflage: 3., bearb. aufl. edn. UTB, Stuttgart, 2001
4. G. Strunk, G. Schiepek, *Systemische Psychologie: Eine Einführung in die komplexen Grundlagen menschlichen Verhaltens*, auflage: 2006 edn. Spektrum Akademischer Verlag, 2006
5. N. Wiener, Kybernetik. In: *Futurum Exactum. Ausgewählte Schriften zur Kybernetik und Kommunikationstheorie*, ed. by B. Dotzler, Springer, Wien, New York, 1948, pp. 13–29. 2000
6. Bundesministerium für Bildung und Forschung, Ideen. Innovation. Wachstum. Hightech-Strategie 2020 für Deutschland. Tech. rep., Berlin, 2010
7. U. Bach, *Deliberative Governance in der Arbeitsforschung: Ein Ansatz zur Demokratisierung von Forschungsprozessen in der anwendungsorientierten Forschung*. Budrich UniPress, Opladen, 2013
8. M. Haarich, S. Sparschuh, C. Zettel, S. Trantow, F. Hees, Innovationsfähigkeit – lernfähigkeit – transferfähigkeit. innovationen systematisch fördern. In: *Enabling Innovation – Innovationsfähigkeit – deutsche und internationale Perspektiven*, ed. by S. Jeschke, I. Isenhardt, F. Hees, S. Trantow, Springer, Berlin, Heidelberg, 2011
9. C. Jooß, *Gestaltung von Kooperationsprozessen interdisziplinärer Forschungsnetzwerke*. BoD – Books on Demand, 2014
10. Bundesministerium für Bildung und Forschung, Wohlstand durch Forschung. Bilanz und Perspektiven der Hightech-Strategie für Deutschland. Tech. rep., Berlin, 2013
11. M. Harich, Innovationsfähigkeit – Lernfähigkeit – Transferfähigkeit. Innovationen systematisch fördern. In: *Enabling Innovation*, ed. by S. Jeschke, I. Isenhardt, F. Hees, S. Trantow, Springer, Berlin, Heidelberg, 2011

New Challenges in Innovation-Process-Management. A Criticism and Expansion of Unidirectional Innovation-Process-Models

Markus Kowalski, Florian Welter, Stella Schulte-Cörne, Claudia Jooß, Anja Richert and Sabina Jeschke

Abstract Innovation-Process-Models have become an increasingly important issue of research. The main point of criticism on process-models is that innovation-processes are not rigid and phases do not run compulsorily linear, but rather in recursive loops and with disruptions. Therefore, existing unidirectional innovation-process-models need to be modified. The aim of this paper is to develop a cybernetic innovation-process-model, which is based on the famous concepts of Wheelwright/Clark (*Innovation-Funnel*) and Cooper (*Stage-Gate-Process*). Suitable measuring-instruments are required to gain the central cognition of cybernetics that viable systems have an invariant structure. This paper sheds light on shortcomings of already existing unidirectional innovation-process-models and as current studies promulgate, such as Vanhaverbeke already 2013 pointed out in his work "Rethinking Open Innovation beyond the Innovation Funnel", it is time for a new step forward by enriching and broadening *Open Innovation*. A possible next step thus can be the implementation of a cybernetic innovation-process-model.

Keywords Innovation Management · Innovation-Process-Model · Cybernetics · Open Innovation · Closed innovation · Innovation-Funnel · Stage-Gate-Process

1 Introduction and Problem Statement

"Innovation is more than having new ideas: it includes the process of successfully introducing them or making things happen in a new way" (John Adair). The design of innovation-process-models (IPM) is as manifold as the use. It depends on the intention of the user and is often molded in *own-principles*, which disembogue in unidirectional IPMs.

In today's globalized world, continuous creation of innovation is one of the major goals that businesses with research activities (hereinafter *business*) have to

M. Kowalski (✉) · F. Welter · S. Schulte-Cörne · C. Jooß · A. Richert · S. Jeschke
IMA/ZLW & IfU, RWTH Aachen University, Dennewartstr. 27,
52068 Aachen, Germany
e-mail: markus.kowalski@ima-zlw-ifu.rwth-aachen.de

Originally published in "The XXV ISPIM Conference – Innovation for Sustainable Economy & Society", © ISPIM 2014. Reprint by Springer International Publishing Switzerland 2016, DOI 10.1007/978-3-319-42620-4_56

accomplish ([1]: 37). As Schumpeter ([2]: 81) already knew, knowledge and innovative ideas are widespread so that a new perspective is needed. In this vein, Chesbrough [3] characterizes the term *Open Innovation* to evidence the need of businesses to use external knowledge systematically to increase innovation performance ([4]: 1). They combine knowledge generated through external search with knowledge developed inside the business [5].

Open innovation-process-models have become an increasingly important issue of research [6]. The management of innovations has to be done holistically at normative, strategic and operational level ([7]: 7). The main point of criticism on *Process-Models* [8] is that innovation-processes are not rigid and phases do not run compulsorily linear, but rather in recursive loops and with disruptions. Reasonable extensions of unidirectional approaches offer cybernetic principles (e.g. emergence or recursion) that support an iterative management of innovation-processes and thus enable a holistic process-contemplation ([9]: 76). Therefore, existing unidirectional IPM of product, process and business-model innovations need to be modified and suitable measuring-instruments are required to gain the central cognition of cybernetics that viable systems have an invariant structure ([9]: 80).

The aim of this paper is to develop a cybernetic IPM, which is based on the concepts *Innovation-Funnel* [10] and *Stage-Gate-Process* [11]. The central research question is: What modifications are necessary, to convert unidirectional IPMs into a cybernetic IPM?

2 Unidirectional Innovation-Process-Models – A Need for a New Paradigm

"We are bound no longer by the straightjacket of the past" (Douglas MacArthur). Is there a need to rethink open innovation and existing unidirectional IPMs? An innovation-process is a complex composite of activities that in general last a longer period of time. There is a need for a contingency approach regarding the management of innovations in a process ([10]: 187).

2.1 A Historical Overview

In literature on innovation-management, a significant IPM must be balanced between reduction of complexity and specialization ([8]: 6). Historically, in the Anglo-Saxon literature, different generations of process-models are revealed. The first generation of *Stage-Review-Process* ([12]: 62), which divides the innovation-process into discrete phases to standardize activities that would otherwise be accomplish ad hoc, was followed by a second generation of *Stage-Gate-Process* ([13]: 46) in which the activities do no longer run strictly sequential. Henceforth, imbrications are possible to accelerate the innovation-process. The third generation of a modified *Stage-Gate-Process* ([14]: 479) and *Value-Proposition-Cycle* ([12]: 93) bases on the assumption that business boundaries are systematically open and permeable to outside knowledge.

Newer developments in research, due to complex and heterogeneous incidents, support the change from third generation *Stage-Gate-Model* to *Flexible-Models* ([15]: 7). Enkel/Gassmann [16] refer to *outside-in* and *inside-out* processes in contrast to Chesbrough/Growther [4] who distinguish between *inbound* (integration of external inputs into a business's new-product-development) and *outbound* (licensing or spinning off startups) innovation.

2.2 *Closed Versus Open Innovation*

Innovation has often been recognized as one critical driving force in enhancing social welfare. As in former times and today, the accessibility of knowledge is one of the most acquainted prerequisite for business' innovation activities and success [17].

The use of external sources has been termed an open innovation model as opposed to a closed innovation model. In a closed approach, businesses internalize their business-specific research and development activities and commercialize them through internal development ([3], [4]: 1). According to Chesbrough ([4]: 1), the open approach is defined as: "[…] the use of purposive inflows and outflows of knowledge to accelerate internal innovation and broaden the markets for external use of innovation". This definition is supported by the quote, that: "[…] not all of the smart people work for us, so we must find and tap into knowledge and expertise of bright individuals outside our company" ([18]: 38).

In the closed innovation world, businesses peripheries are systematically kept as impermeable as possible. In contrast, boundaries in open innovation are systematically permeable to outside knowledge influx. However, with changing external business conditions, both models are no longer suitable ([19]: 37). For example

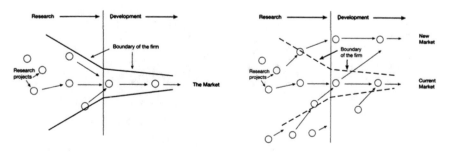

Figure 1 Closed and open innovation model ([3]: XXII)

mobility, abundance of venture capital and increase of labor availability stirred this development [3] (Figure 1).

Open innovations include the integration of customers [20], involvement in external networks [21] and community-based innovations [22]. These patterns have in common that external knowledge is utilized for businesses' internal innovation-processes and aims at enhancing innovation or business performance. However, following Lichtenthaler/Lichtenthaler ([5]: 1315) open innovation constitutes "[...] firms' interorganizational knowledge transactions to extend their existing knowledge base".

2.3   Process-Models in the Anglo-Saxon Literature

The design of IPMs is as manifold as the use and depends on the intention of the user. Different process-models share three overarching phases: 1. Idea generation and selection; 2. Idea realization; 3. Idea commercialization ([23]: 51). Those phases can be identified in practical applications. Two of the most disseminated process-models in practice are the Innovation-Funnel and the Stage-Gate-Process.

The Stage-Gate-Process ([14]: 479) contains five phases: scoping, build business case, development, testing and validation, full production and market launch (cf. Figure 2). Moving from one stage to another necessitates that a specific set of requirements have to be performed ([24]: 215). Stages and gates are not static but should be customized to the specific innovation-process. Cooper ([24]: 223) characterizes them as fluid and overlapping gates, which have fuzzy "go" [24] conditions.

The Innovation-Funnel is a framework for the development of strategies in businesses and ensures a concentration on significant ideas ([10]: 187). Chesbrough ([3]: XXII) uses this approach in the closed and open innovation model.

Due to different objectives, all of these models have their legitimation, but also limitations' regarding their linearity becomes obvious. There is a need to change the

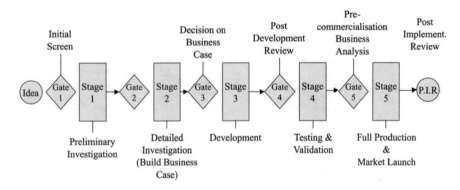

Figure 2 Third generation of Stage-Gate-Process ([14]: 479)

theoretical open innovation framework due to the linearity of existing IPMs. The enlargement of the open innovation framework connotes that the Innovation-Funnel should be complemented by a cybernetic IPM with recursive loops and the possibility of disruptions ([25]: 8).

3   Enriching and Broadening Open Innovation by Implementation of a Cybernetic Innovation-Process-Model

At the beginning of an innovation-process there are many new ideas. A few projects with good perspectives are stepwise selected and only the best product-innovations are brought to market ([26]: 146). These phases of the innovation-process do not run compulsorily linear. A repetition of some steps and a reusability of ideas must always be possible, if improvements or changes are needed ([27]: 222).

A first suggestive expansion of the Innovation-Funnel is given by Andrew Gaule's [28] approach he offered in the booklet entitled "Open Innovation in Action". The funnel-model, he suggests, is also widening at the end and not only at the beginning, he calls it "double lens". In the management of innovation-processes the technology-oriented perspective, with the basic requirement to anticipate new technologies and innovations, is enhanced by potential markets and customers ([29]: 150).

We go one step further in the expansion of unidirectional IPMs, using cybernetic principles which support a dynamic and systematic management of innovation-processes ([30]: 30). For promoting businesses with a cybernetic innovation-process-management both, organic (adaptable) and mechanic (prestructured) elements, have to be combined along the entire process. As an example, the mode of the Innovation-Funnel contains cybernetic elements such as early ideas are recirculated into the system for possible future actions. Therefore, we developed a cybernetic IPM, transferring a traditional linear operating business into a vibrant business (e.g. recursive knowledge loops). This model is genuine sustainable, because it does not simply guide ideas to the market or shelves them ad acta. It assesses the early ideas concerning their potential of innovation to either forward them to [new] markets and customers or to let ideas be stored and revitalized at a later point of time.

The design of the innovation-process-model is comparable to a human nervous-system, in which the Innovation-Funnel and the Open-Innovation-Model were integrated. Thus they were enhanced by a *pipe-system* that is used *in* the innovation-funnel. This implemented *pipe-system*, which consists of several *regulation-pipes*, is necessary to control and steer the broad range of early ideas in the exploration-phase but also to select and bundle research-projects at later stages in the innovation-process. Therefore, ideas are assessed and selected *in* the *regulation-pipes* regarding potential research-projects and *"smartly"* returned back into the innovation-funnel. Ideas that at a certain point are not feasible or unnecessary are either completely rejected (*exit-pipes*) or remain as floating ideas in the system with a

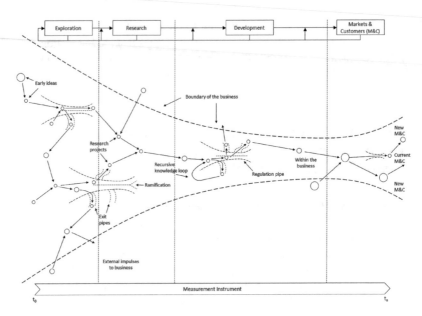

Figure 3 Cybernetic innovation-process-model (own representation)

potential of revitalization. There are also exit-pipes for ideas, which implicate a high potential for innovation at different points of time in the innovation-process. Furthermore both, the already developed project-ideas and also ideas with value-adding potential are accompanied by measures which encourage innovations.

All ideas of the *internal* and *external* technological base are managed in a dynamic and permeable *pipe-system* in four interconnected innovation-process-phases: exploration, research, development, markets/customers (cf. Figure 3).

A necessary requirement of a successful IPM is a systematic control of all activities ([31]: 165). Therefore, a suitable measurement-instrument of innovation-processes is Kaplan and Norton's ([32]) Balanced Scorecard. For a prototypical trial of the developed model in practice, the classic Scorecard was adapted to a particular use-case, here two interdisciplinary research-cluster at RWTH Aachen University (*Integrative Production Technology for High-Wage-Countries and Tailor-Made Fuels for Biomass*). These interdisciplinary research-cluster, due to their characteristic form of a network, enable the possibility to incorporate internal and external impulses in order to generate innovative solutions for social challenges.

As part of a first application of this cybernetic IPM in practice, it is already evident, that configurations (e. g. flanked by communication-promoting-measures) of innovation-processes thus gets more targeted and output-oriented.

4   Conclusion

There is a major gap in the open innovation literature and this gap has been impeding the progress of open innovation as a viable system. This paper sheds light on short-comings of already existing unidirectional IMPs and as current work and studies promulgate [25, 28] it is time for a new step forward by enriching and broadening open innovation. A possible proposal for this new mode of thought is the implementation of a cybernetic IPM.

This paper contributes to the development of theory in innovation-management, which is based on contemporary literature about IPMs and which is already partially proved by the implementation of a prototypical Scorecard in a use-case.

The cybernetic IPM is genuine sustainable, because it does not simply guide ideas to the market or shelves them ad acta. It assesses the early ideas concerning their potential for innovation to either forward them to [new] markets and customers or to let ideas be stored and revitalized at a later point of time. There is a challenge in IPM, to make strategic sense of networks, lead-users and ecosystems, and their ramification for competitive advantage ([33]: 73). Therefore, the developed cybernetic IPM is a first approach which uses the complexity of innovations and innovation-pro-cesses but also the business-environment. Future research should examine the short-comings of existing unidirectional IPMs and discuss practical implications concerning our cybernetic expansion.

Acknowledgments This work was performed as part of the Cluster of Excellence Integrative Production Technology for High-Wage Countries, which is funded by the excellence initiative by the German federal and state governments to promote science and research at German universities.

References

1. F. Drucker, P. *The Practice of Management*. Harper & Row, New York, 1954
2. A. Schumpeter, J. *Capitalism, socialism, and democracy*. Harper & Row, New York, 1942
3. W. Chesbrough, H. *Open Innovation: The New Imperative for Creating and Profiting from Technology*. Harvard-Business-School Press, Boston, Massachusetts, 2003
4. W. Chesbrough, H. Open innovation: A new paradigm for understanding industrial innovation. In: *Open innovation: Researching a new paradigm*, ed. by W. Chesbrough, H. W. Vanhaverbeke, J. West, Oxford University Press, Oxford, 2006, pp. 1–12
5. U. Lichtenthaler, E. Lichtenthaler, A capability-based framework for open innovation: Complementing absorptive capacity. Journal of Management Studies **46**, 2009, pp. 1315–1338
6. U. Lichtenthaler, The evolution of technology licensing management: Identifying five strategic approaches. R&D Management **41**, 2011, pp. 173–189
7. O. Gassmann, P. Sutter, *Praxiswissen Innovationsmanagement: Von der Idee zum Markterfolg*. Carl Hanser Verlag, München, 2011
8. G. Cooper, R. A process model for industrial new product development. IEEE Transactions on Engineering Management (30), 1983, pp. 2–11
9. F. Malik, *Strategie des Managements komplexer Systeme. Ein Beitrag zu Management-Kybernetik evolutionärer Systeme*, 4th edn. Bern, Stuttgart, Wien, 1992

10. C. Wheelwright, S. B. Clark, K. *Revolutionizing product development: quantum leaps in speed, efficiency, and quality.* The Free Press, New York, 1992
11. G. Cooper, R. *Product Leadership: Creating and Launching Superior New Products.* 1998
12. D. Hughes, G. C. Chafin, D. Turning new product development into a continuous learning process. Journal of Product Innovation Management **13**, 1996, pp. 89–104
13. G. Cooper, R. *New Products: The Key Factors in Success.* 1990
14. E. Kleinschmidt, H. Geschka, G. Cooper, R. *Erfolgsfaktor Markt. Kundenorientierte Produktinnovationen.* Springer, Berlin, 1996
15. R. Rothwell, Towards the fifth-generation innovation process. International Marketing Review **11**, 1994, pp. 7–31
16. O. Gassmann, E. Enkel, Open innovation: Externe hebeleffekte in der innovation erzielen. Zeitschrift Führung + Organisation (3), 2006, pp. 132–138
17. A. Schumpeter, J. *Business Cycles. A Theoretical, Historical and Statistical Analysis of the Capitalist Process.* New York, London, 1939
18. W. Chesbrough, H. The era of open innovation. MIT Sloan Management Review (44), 2003, pp. 35–41
19. S. Brunswicker, W. Vanhaverbeke. Beyond innovation in large enterprises: How do small and medium-sized enterprises (smes) open up to external innovation sources?. http://ssrn.com/abstract=1925185. Available Online at SSRN
20. T. Piller, F. Toolkit for idea competitions: a novel method to integrate users in new product development. R&Dmanagement **36**, 2006, pp. 307–318
21. F. Dittrich. The credit rating industry: Competition and regulation, 2007. http://ssrn.com/abstract=991821. Available Online at SSRN
22. E. von Hippel, R. Lakhani, K. How open yource software works: "free" user-to-user assistace. Research Policy **32**, 2003, pp. 923–943
23. J. Gerpott, T. *Strategisches Technologie- und Innovationsmanagement*, 2nd edn. Stuttgart, 2005
24. G. Cooper, R. Perspective: The stage-gate idea-to-launch process – update, what's new, and nexgen systems. Journal of Product Innovation Management (JPIM) (25), 2008, pp. 213–232
25. W. Vanhaverbeke, Rethinking open innovation beyond the innovation funnel. Technology Innovation Management Review **April 2013**, 2013, pp. 6–10
26. R. Reichwald, T. Piller, F. *Interaktive Wertschöpfung: Open Innovation, Individualisierung und neue Formen der Arbeitsteilung.* Gabler, Wiesbaden, 2009
27. E. Porter, M. *Wettbewerbsstrategie: Methoden zur Analyse von Branchen und Konkurrenten.* Campus Verlag, New York/Frankfurt, 1999
28. A. Gaule, *Open Innovation in Action: How to Be Strategic in the Search for New Sources of Value.* H-I Network, 2006
29. W. Vanhaverbeke, Business models in open innovation and commercialization – a dynamic approach. In: *Innovationsführerschaft durch Open Innovation: Chancen für die Telekommunikations-, IT- und Medienindustrie*, ed. by A. Picot, S. Doeblin, Springer-Verlag, Berlin, 2009, pp. 147–154
30. S. Beer, *Brain of the Firm: A Development in Management Cybernetics.* Herder and Herder, 1972
31. P. Horváth, R. Gleich, D. Voggenreiter, *Controlling umsetzen: Fallstudien, Lösungen und Basiswissen.* Schäffer-Poeschel, 2012
32. S. Kaplan, R. P. Norton, D. The balanced scorecard: Measures that drive performance. Harvard-Business-Review **January-February**, 1992, pp. 71–79
33. W. Chesbrough, H. M. Appleyard, M. Open innovation and strategy. California Management Review (50), 2007, pp. 57–76

Neue und flexible Formen der Kompetenzentwicklung

Thomas Langhoff, Friedemann W. Nerdinger, Stefan Schröder, Freya Willicks and Stephanie Winter

Zusammenfassung Im Kontext des demografischen Wandels („Älter, Weniger, Bunter") und damit einhergehender veränderter Erwerbsbiografien, neuer und kontinuierlicher Qualifizierungsmöglichkeiten und -pfade, der zunehmend digitalisierten Arbeitswelt, generationenübergreifender sowie interkultureller Zusammenarbeit und vielen weiteren Faktoren, werden neue und flexible Formen der Kompetenzentwicklung notwendig. Letztlich besteht auf dem Gebiet der neuen und flexiblen Formen der Kompetenzentwicklung der Bedarf von Gestaltungskonzepten, die u. a. den Wissensaustausch, die Kompetenzentwicklung und Kollegialitätsbeziehungen heterogener Individuen und Organisationen fördern. Im Rahmen dieses Beitrages, Resultat einer zweieinhalbstündigen Workshop-Session im Rahmen der Förderschwerpunkt-Tagung 2014 des BMBF-Förderschwerpunktes „Innovationsfähigkeit im demografischen Wandel", wird neben einer theoretischen Einführung in die Thematik ein besonderer Fokus auf die erzielten Ergebnisse des Workshops gelegt. Dabei werden die bearbeiteten Handlungsfelder vorgestellt, deren Ergebnisse gelistet und abschließend die Ergebnisrückführung in das Memorandum des Förderschwerpunktes vorgestellt (vgl. Becke, G., Bleses, P., Cernavin, O., Ducki, A., Langhoff, T., Klatt, R., Nerdinger, F. W. (2015): Memorandum: Förderschwerpunkt Innovationsfähigkeit im demografischen Wandel. In: Jeschke, S., Richert, A., Hees, F., Jooß, C.: Exploring Demographics. Transdisziplinäre Perspektiven zur Innovationsfähigkeiten im demografischen Wandel, S. 119-121.).

Schlüsselwörter Kompetenzen · Kompetenzentwicklung · Kompetenzanalyse · Anreizsysteme

T. Langhoff (✉)
Prospektiv Gesellschaft für betriebliche Zukunftsgestaltungen mbH, Kleppingstr. 20, 44135 Dortmund, Germany
e-mail: langhoff@prospektiv.de

F.W. Nerdinger
Institut für Betriebswirtschaftslehre, Universität Rostock, 18051 Rostock, Germany

S. Schröder · F. Willicks · S. Winter
IMA/ZLW & IfU, RWTH Aachen University, Dennewartstr. 27, 52068 Aachen, Germany

Originally published in "Exploring Demographics",
© Springer 2015. Reprint by Springer International Publishing Switzerland 2016,
DOI 10.1007/978-3-319-42620-4_57

1 Einführung

Kompetenzen sowie deren Entwicklung sind bereits seit Jahrzehnten fester Bestandteil der (geistes-)wissenschaftlichen Theorien und Forschungen. Diese sind so vielfältig wie die Diskussionen um eine Begriffsdefinition von Kompetenzen. Eine umfassende Definition von Kompetenzen stellt der Pädagoge Weinert vor. Dieser definiert Kompetenzen als

> „bei Individuen verfügbare oder durch sie erlernbare, kognitive Fähigkeiten und Fertigkeiten, um bestimmte Probleme zu lösen, sowie die damit verbundenen motivationalen, volitionalen und sozialen Bereitschaften und Fähigkeiten, um die Problemlösungen in variablen Situationen erfolgreich und verantwortungsvoll nutzen zu können" ([1]: 27).

Es handelt sich also um eine im Individuum verankerte spezifische Problem-lösefähigkeit, die sich während des Lebens (weiter) entwickeln kann. Dabei spielen sowohl biologische Faktoren als auch unterschiedliche Lebensbedingungen, -umstände und Erfahrungen eine wichtige Rolle. Insbesondere in Bezug auf das Altern scheinen sich die Voraussetzungen der Kompetenzentwicklung zu ändern. Basierend auf dieser Beobachtung, wurde das Kompetenzmodell des Alterns entwickelt ([2]: 14ff.). Dieses geht entgegen der Annahme einer kontinuierlichen Abnahme der Leistungsfähigkeit im Alter von der Möglichkeit des Entgegenwirkens dieser Vorhersage durch Anpassungs- und Änderungsfähigkeiten an altersbedingte Beeinträchtigungen und sich verändernde Situationen und Voraussetzungen aus ([2]: 14ff.). Das Modell grenzt sich somit klar von dem seit den 70er Jahren herrschenden Defizitmodell des Alters ab und beleuchtet die Unterschiede der Kompetenzgewinne und -verluste in den verschiedenen kognitiven und physischen Bereichen (vgl. Abbildung 1: Kompetenzmodell des Alterns). Gerade das hohe Potenzial älterer Beschäftigter, Schlüsselkompetenzen und Erfahrung aufzubauen bzw. bereits aufgebaut zu haben, nimmt in den aktuellen gesellschaftlichen Entwicklungen einen immer höheren Stellenwert ein und zeigt die Bedeutung einer gezielten Kompetenzentwicklung auch im Alter.

Maßgeblichen Einfluss hat die Kompetenzentwicklung auf den Erhalt und Ausbau von Innovationsfähigkeit verschiedener Organisationen bzw. Unternehmen. Besonders vor dem Hintergrund des demografischen Wandels ist die Entwicklung als eine der zentralen Herausforderungen für Unternehmen zu sehen, um in einem zunehmend globaler werdenden Wettbewerbsumfeld bestehen zu können.

Unter dem Begriff der Innovationsfähigkeit sind nicht ausschließlich technologische Entwicklungen zu verstehen, sondern vielmehr die gezielte Verknüpfung technologischer Innovationen mit Aspekten der Personal-, Organisations- und Kompetenzentwicklung.

Defizitmodell des Alterns	Kompetenzmodell des Alterns
▪ Bis 70er Jahre des 20. Jahrhunderts	▪ Von 90er Jahre des 20. Jahrhunderts bis heute
▪ Altern als defizitärer Verlauf	▪ Altern als ein Abschnitt des lebenslangen Entwicklungsprozesses
▪ Gekennzeichnet durch Verluste, Mängel und Defizite	▪ Gekennzeichnet durch ein Wechselspiel von Gewinnen und Verlusten
▪ Biologischen Faktoren	▪ Multifaktoriell (biologische, intra- & interindividuelle, soziale und kulturelle Faktoren)
▪ Verallgemeinerung der Ergebnisse	▪ Differenzierung der Ergebnisse

Abbildung 1 Kompetenzmodell des Alterns

2 Kompetenzentwicklung im Kontext des Förderschwerpunktes „Innovationsfähigkeit im demografischen Wandel"

In diesem Zusammenhang ergeben sich verschiedene Fragestellungen, die sich u. a. mit der Entwicklung von Kompetenzen in der unmittelbaren und mittelbaren Zukunft beschäftigen, sowie einen Zusammenhang zwischen Arbeitsformen, Kompetenzentwicklung und Kompetenzeinsatz diskutieren. Im Kontext der Tagung des Förderschwerpunktes „Innovationsfähigkeit im demografischen Wandel" des Bundesministeriums für Bildung und Forschung (BMBF), einer der Förderschwerpunkte des Förderprogramms „Arbeiten, Lernen, Kompetenzen entwickeln (A-L-K)", wurden „neue und flexible Formen der Kompetenzentwicklung" vorgestellt, weiterentwickelt und diskutiert.

Neben Grundproblematiken wurden zusätzlich Herausforderungen und Anknüpfungspunkte, Gestaltungsaufgaben und schließlich Forschungsfragen für die Wissenschaft und den betrieblichen und wirtschaftlichen Einsatz entwickelt. Herausforderungen und Anknüpfungspunkte stellen neben der ökonomischen „Betrachtungsweise" von Kompetenzentwicklung auch veränderte Verantwortlichkeiten und Lernformen, Widerstand und Bereitschaft der Mitarbeiter oder die Frage, wann eine Kompetenzanalyse überhaupt sinnvoll für verschiedene Zielgruppen ist (Personalmanagement, Führungskräfte, Politik, Unternehmen, Individuum), dar. Diese Grundproblematiken und Herausforderungen bzw. Anknüpfungspunkte leiten zu Gestaltungsaufgaben der Wissenschaft über. Diese reichen von der Nachhaltigkeit und der Motivation zur Weiterqualifizierung der Arbeitnehmer bis hin zur Gestaltung von Kompetenzmanagementsystemen. Die sich daraus ergebenden Forschungsfragen sind so vielfältig wie das Gebiet der Kompetenzen. Diskussionsgrundlage

Tabelle 1 Auszug Memorandum: Neue und flexible Formen der Kompetenzentwicklung [3]

#1	Konzepte und Modelle für die Qualifizierung und Kompetenzentwicklung sind erforderlich, die sich zukünftig stärker auf vielfältige betriebliche und persönliche Bedarfslagen ausrichten, und diese angemessen berücksichtigen
#2	Neue Formen der organisationalen Lernfähigkeit sind zu entwickeln, die die Dynamik der digitalisierten Arbeitswelt und des demografischen Wandels mit der zunehmenden Vielfalt der Personengruppen berücksichtigen. Hier stellen sich insbesondere auch Fragen der Gestaltung generationenübergreifender sowie interkultureller Zusammenarbeit
#3	Zunehmend ist mehr Flexibilität beim Erwerb von Formalqualifikationen als auch eine schnelle Kompetenzentwicklung in den Organisationen verlangt, für die es oft keine formalen Wege gibt. Hier sind Gestaltungskonzepte gerade für kleinere und mittlere Unternehmen erforderlich
#4	Modelle zur Qualifizierung für den Umstieg zwischen verschiedenen Tätigkeitsfeldern, den Wiedereinstieg und Neueinstieg sowie die ‚horizontale Laufbahngestaltung' sind erforderlich. Kompetenzentwicklung und Innovationen basieren zukünftig zunehmend auf der Basis der Reflexivität von Personen und Organisationen. Es sind Modelle und Konzepte zu entwickeln, wie diese Reflexivität gefördert wird
#5	Es bedarf Gestaltungskonzepte, die Wissensaustausch, Kompetenzentwicklung und Kollegialitätsbeziehungen in der Vielfalt ermöglichen und Innovationsfähigkeit fördern

bildeten die folgenden fünf Thesen (siehe Tabelle 1: Auszug Memorandum: Neue und flexible Formen der Kompetenzentwicklung) des Memorandums [3].

In diesem Kontext zeigt sich die Bedeutung der Konzept- und Modellentwicklung für die Qualifizierung und Kompetenzentwicklung aller Mitarbeiter (insbesondere aber auch der älteren Beschäftigten). Zu diesem Zweck sind neue Formen der organisationalen Lernfähigkeit zu schaffen, die die Dynamik der digitalisierten Arbeitswelt und des demografischen Wandels mit der zunehmenden Vielfalt der Personengruppen („Älter, Weniger, Bunter"), berücksichtigen und Fragen der Gestaltung generationenübergreifender sowie interkultureller Zusammenarbeit einbeziehen. Den Erfordernissen einer zunehmenden Flexibilität beim Erwerb von Formalqualifikationen sowie der schnelleren Kompetenzentwicklung, insbesondere in kleineren und mittleren Unternehmen, muss mithilfe von Gestaltungskonzepten Rechnung getragen werden. Des Weiteren sind Modelle zur Qualifizierung für den Umstieg zwischen verschiedenen Tätigkeitsfeldern, den Wiedereinstieg und Neueinstieg sowie die „horizontale Laufbahngestaltung" erforderlich. Dabei spielt auch die Reflexivitätsfähigkeit von Personen und Organisationen eine immer wichtigere Rolle, die entsprechend gefördert werden muss. Letztlich besteht auf dem Gebiet der neuen und flexiblen Formen der Kompetenzentwicklung der Bedarf von Gestaltungskonzepten, die den Wissensaustausch, die Kompetenzentwicklung und Kollegialitätsbeziehungen heterogener Individuen und Organisationen fördern.

Diese Forderungen spiegeln einige der Voraussetzungen wider, die zur nachhaltigen Sicherung der Innovationsfähigkeit deutscher Unternehmen beitragen. Denn wie das FuE-Programm „Arbeiten, Lernen, Kompetenzen entwickeln. Innovationsfähigkeit in einer modernen Arbeitswelt" [4] verdeutlicht, ist Innovationsfähigkeit mehr als ein technisch ausgerichteter Begriff. Vielmehr ist es die Verknüpfung von technologischen Entwicklungen mit Personal-, Organisations- und Kompetenzentwicklungen, die es gezielt zu gestalten gilt.

In einer zweieinhalbstündigen, moderierten Workshop-Session tauschten sich ExpertInnen aus den Bereichen Wirtschaft, Wissenschaft und Politik zum Themenfeld aus. Deren Ergebnisse finden sich im Folgenden wieder und werden zusammenfassend aufgeführt. Ferner dienen diese als Grundlage zur Weiterentwicklung bzw. Ausformulierung des Memorandums des Förderschwerpunktes [5].

3 Workshopergebnisse

3.1 Impulsvortrag

Mithilfe eines Impulsvortrages wurde die Kompetenzentwicklung vor dem Hintergrund des demografischen Wandel kontextualisiert. Auf Grundlage der wirtschaftspolitischen Standortbestimmung „Wir werden weniger, bunter, älter" ergeben sich verschiedene neue Anforderungen, unter anderem eine neue Form der Reflektion von Kompetenzentwicklung und -erfassung. Ferner findet ein Wandel innerhalb klassischer Karriereverläufen (z. B. Managementkarriere, Aufstieg in Unternehmenshierarchien, Fachkarriere, Expertenlaufbahn etc.) statt. Ein verändertes Karriereverständnis bzw. Personalmanagement betrifft hierbei beispielsweise die vorherrschenden Meinungen wie „ununterbrochene Berufstätigkeit", „Es darf nur aufwärts gehen", „Wer bis 40 nicht Führungskraft ist, wird es auch nicht mehr", „Spätestens mit 50 gibt es kaum mehr Jobrotation, externe Wechsel oder Aufstieg", „Weiterbildungsbeteiligung nimmt mit dem Alter ab", „Angebot von Altersteilzeit und vorgezogenem Berufsausstieg" ([6]: 64). Hierbei werden häufig die Lebensziele der Mitarbeiter, die demografische Entwicklung, veränderte Altersstrukturen, der Fachkräftemangel, der Wettbewerb um Talente sowie erhöhte Flexibilität nicht adäquat berücksichtigt. Heute ist Karriere nicht mehr ausschließlich mit Aufstieg gleichzusetzen. Die Zunahme und Flexibilisierung von Karrieremodellen führt zu Führungs-, Fach-, Projekt- und Übergangskarrieren, wie beispielsweise Bogen- oder Patchworkkarrieren.

Aus diesem Grund wird einer systematischen Karriereplanung vermehrt Aufmerksamkeit zu Teil. Fragestellungen können diesbezüglich lauten: Welche Diversität an Tätigkeiten gibt es? Wo liegen Stärken und Interessen der Belegschaft? Methoden, um hierzu Antworten liefern zu können, werden im Rahmen des Förderschwerpunktes entwickelt und erprobt. Als beispielhaft sind das ursprünglich innerbetrieblich ablaufende Job-Profiling als Grundlage interner und externer Rekrutierung, die Arbeitsplatzkarte, der Kompetenzpass und die Qualifikationsmatrix zu nennen. Ebenfalls muss die zunehmende Individualisierung der Vermarktung von Kompetenzentwicklung im Blick behalten werden. Arbeit läuft immer mehr projektbezogen ab und die Nachfrage nach überfachlicher Qualifikation ist groß, obwohl diese selten an z. B. Universitäten gelehrt wird.

Fragen, die sich u. a. hieraus ergeben, lauten: Was bedeutet dies nun für die Kompetenzentwicklung und systematische Karriereplanung? Wie sieht generell die Zukunft der Kompetenzentwicklung aus? Welche Grundproblematiken, Herausforderungen, Anknüpfungspunkte, Gestaltungsaufgaben sowie For-schungsfragen ergeben sich daraus?

3.2 Bearbeitung der Handlungsfelder – Leitfrage 1: Grundproblematik

Auf Grundlage dieses Impulsvortrags wurden die sich aus dem demografischen Wandel herleitenden Grundproblematiken in Bezug auf neue und flexible Kompetenzentwicklung gesammelt und diskutiert. Diese zeichnen sich durch ihre Vielfältigkeit aus und sind nachfolgend aufgeführt:

• Was sind die zu entwickelnden Kompetenzen von morgen?
• Wie entwickelt man diese Kompetenzen?
• Wie unterscheiden sich branchenspezifische von übergreifenden Kompetenzen?
• Was macht individualisierte Kompetenzentwicklung in standardisierten und/oder komplexen Systemen aus?
• Welcher Zusammenhang besteht zwischen individueller und kollektiver Kompetenzentwicklung?

3.3 Bearbeitung der Handlungsfelder – Leitfrage 2: Herausforderungen und Anknüpfungspunkte

Anhand der thematisierten Grundproblematiken lassen sich Herausforderungen und Anknüpfungspunkte formulieren. Dabei zeigt sich, dass sowohl die Voraussetzungen als auch die Umsetzung und Wahrnehmung von Kompetenzentwicklung in diesem Kontext genannt wurden und in der folgenden Aufzählung ihre Spezifizierung finden:

• Entwicklungsmöglichkeiten müssen gegeben sein (besonders in kleinen und mittelständischen Betrieben).
• Die Managementausbildung an Hochschulen muss angepasst werden.
• Wie wirken sich Incentives für Kompetenzentwicklung aus und für welche Akteure sind diese geeignet?
• Die Anerkennung von informell erworbenen Kompetenzen gestaltet sich als anspruchsvoll.
• Veränderte Lernformen müssen berücksichtigt werden.
• „Atypische"/diskontinuierliche Erwerbsbiografien nehmen zu.

3.4 Bearbeitung der Handlungsfelder – Leitfrage 3: Gestaltungsaufgaben

Die Grundproblematiken und Herausforderungen bieten Gestaltungsräume für die Wissenschaft. In der Arbeitssession konnten folgende Gestaltungsaufgaben identifiziert werden:

- Wie kann man die Mitarbeiter zur Kompetenzentwicklung motivieren? Anreizsysteme?
- Neue Lernformen müssen entwickelt und zielgruppenadäquat getestet/eingesetzt werden (Lerntypen definieren und beachten).
- Kompetenzmanagement (Messung, Entwicklung, …) muss in Unternehmen integriert werden.
- Die Nutzung und Entwicklung von Kompetenzen in unternehmensübergreifenden, vernetzten Gruppen (z. B. in Projekten) muss erfolgen.
- Abhängige Beschäftigte und Selbstständige müssen gemischt werden.
- Den Kompetenzbegriff gilt es neu zu gestalten.
- Überfachliche Kompetenzen müssen gelehrt werden.
- Betriebsübergreifende/interkulturelle Projektarbeit muss gefordert und gefördert werden.
- Generationenunterschieden muss begegnet werden.

3.5 Bearbeitung der Handlungsfelder – Leitfrage 4: Forschungsfragen

Letztlich wurden Forschungsfragen hergeleitet, die u. a. die Komplexität dieses Forschungsgebietes aufgreifen und verdeutlichen:

- Wer legt den Soll-Wert von nicht-fachlichen Kompetenzen fest (für einen bestimmten Tätigkeitsbereich) und auf welche Weise geschieht dies?
- Welche Konsequenzen folgen daraus?
- Was sind die Voraussetzungen überbetrieblicher und institutionsübergreifender Kompetenzentwicklungsangebote (z. B. in Form von Rahmenbedingungen, die geschaffen werden müssen)?
- Lernen unabhängig vom Alter: Was gibt es für Lerntypen? Was ist die Motivation? Welche Lernarrangements bestehen und in welcher Form werden diese bzw. müssen diese aufbereitet werden?
- Wie gestaltet sich die Entwicklung von Arbeit und Kompetenzen in globalen Wertschöpfungsketten?
- Wie lässt sich eine Willkommenskultur etablieren (z. B. Kompetenzentwicklung bei unterschiedlichem ethnischem Hintergrund)?
- Welche Unterschiede gibt es in der Selbst- und Fremdbewertung (z. B. Unterschiede zwischen Männern und Frauen)?

- Wie lassen sich Altersbilder und -stereotype abbauen (z. B. Jüngere über Ältere, Ältere über Jüngere etc.)?
- Wie kann Kompetenzentwicklung im Rahmen von Projektarbeiten und für Freelancer erfolgen?
- Wie kann man atypischen/diskontinuierlichen Erwerbsbiografien von Beschäftigten Rechnung tragen?
- Welche strukturellen Veränderungen zur „Übertragbarkeit des Gelernten" ergeben sich?
- Wie lassen sich die betrieblichen Altersgruppen für die Kompetenzentwicklung motivieren?
- Welche Voraussetzungen müssen für Kompetenzerweiterung in überbetrieblichen Kooperationen, um beispielsweise Ressourcen zu schonen, gelten?

3.6 Überarbeitung des Memorandums auf Grundlage der Workshopergebnisse

Im Nachgang der Tagung flossen die Ergebnisse des Workshops in die Überarbeitung des Memorandums ein. Hierzu wurde der Teil „neue und flexible Formen der Kompetenzentwicklung" angepasst und überarbeitet [5]. Tabelle 2 visualisiert die Ergebnisse.

Tabelle 2 Auszug Memorandum: Neue Orientierungs- und Entwicklungspfade im Humanressourcen-Management [3]

#1	Neue Formen der organisationalen Lernfähigkeit sind zu entwickeln, die die Dynamik der digitalisierten Arbeitswelt und des demografischen Wandels mit der zunehmenden Vielfalt der Personengruppen berücksichtigen. Hier stellen sich insbesondere auch Fragen der Gestaltung generationenübergreifender sowie interkultureller Zusammenarbeit
#2	Zunehmend ist mehr Flexibilität beim Erwerb formaler Qualifikationen als auch eine schnelle Kompetenzentwicklung in den Organisationen gefordert, für die es oft keine formalen Wege gibt. Hier sind Gestaltungskonzepte gerade für kleinere und mittlere Unternehmen notwendig
#3	Konzepte und Modelle für Wissenstransfer, Qualifizierung und Kompetenzentwicklung sind erforderlich, die sich zukünftig stärker auf vielfältige betriebliche und persönliche Bedarfslagen ausrichten, und diese angemessen berücksichtigen
#4	Modelle zur Qualifizierung für den Umstieg zwischen verschiedenen Tätigkeitsfeldern, den Wiedereinstieg und Neueinstieg sowie die ‚horizontale Laufbahngestaltung' sind erforderlich. Kompetenzentwicklung und Innovationen basieren zukünftig zunehmend auf der Basis der Reflexivität von Personen und Organisationen. Es sind Modelle und Konzepte zu entwickeln, wie diese Reflexivität gefördert wird

4 Fazit

Insgesamt lässt sich konstatieren, dass die Ergebnisse des Workshops gewinnbringend in den Dialog der FokusgruppensprecherInnen und in den förderschwerpunktinternen Diskurs eingebracht werden konnten. Im Speziellen ließen sich aus den Ergebnissen neue Impulse und Forschungsfragen für beispielsweise die Orientierungs- und Entwicklungspfade im Humanressourcen-Management ableiten. Diese beschäftigen sich unter anderem mit veränderter Erwerbsarbeit durch den Einfluss der sogenannten „Digital natives", sowie der Integration von fortschreitender technologischer Wissensentwicklung in zum Teil veralteten Arbeitsorganisationen und Qualifikationsmuster. Ferner sind neue Formen der organisationalen Lernfähigkeit zu entwickeln sowie zunehmende Flexibilität beim Erwerb von Qualifikationen als auch eine schnelle Kompetenzentwicklung in Organisationen gefordert. Die finale Version des Memorandums, in welche die Ergebnisse einflossen [5], sowie das ausführliche Thesenpapier sind in diesem Band zu finden [3].

Literaturverzeichnis

1. F. Weinert, Vergleichende Leistungsmessung in Schulen - eine umstrittene Selbstverständlichkeit. In: *Leistungsmessungen in Schulen*, ed. by F. Weinert, 2nd edn., Beltz, Weinheim, Basel, 2001, pp. 17–31
2. P. Baltes, M. Baltes, Gerontologie: Begriff, Herausforderung und Brennpunkte. In: *Zukunft des Alterns und gesellschaftliche Entwicklung*, ed. by P. Baltes, J. Mittelstraß, U. Staudinger, de Gruyter, Berlin, 1994, pp. 1–34
3. G. Becke, P. Bleses, O. Cernavin, A. Ducki, C. Jooß, R. Klatt, T. Langhoff, F.W. Nerdinger, Einführungsbeitrag: Von den Handlungsfeldern der Förderschwerpunkt-Tagung zum Memorandum. In: *Exploring Demographics*, ed. by S. Jeschke, A. Richert, F. Hees, C. Jooß, Springer Fachmedien Wiesbaden, 2015, pp. 55–66
4. Bundesministerium für Bildung und Forschung, Arbeiten – Lernen – Kompetenzen entwickeln. Innovationsfähigkeit in einer modernen Arbeitswelt. Bericht im Rahmen des BMBF-Forschungs- und Entwicklungsprogramms. Tech. rep., Bundesministerium für Bildung und Forschung, Berlin, 2007
5. G. Becke, P. Bleses, O. Cernavin, A. Ducki, R. Klatt, T. Langhoff, F.W. Nerdinger, Memorandum: Förderschwerpunkt Innovationsfähigkeit im demografischen Wandel. In: *Exploring Demographics*, ed. by S. Jeschke, A. Richert, F. Hees, C. Jooß, Springer Fachmedien Wiesbaden, 2015, pp. 119–121
6. E. Regnet, Neue Karrieremodelle in einem veränderten wirtschaftlichen Umfeld. In: *Personalentwicklung bei längerer Lebensarbeitszeit – ältere Mitarbeiter von heute und morgen entwickeln*, ed. by D. e.V., Bertelsmann, Bielefeld, 2012, pp. 64–77

Long Term Examination of the Profitability Estimation Focused on Benefits

Stephan Printz, Kristina Lahl, René Vossen and Sabina Jeschke

Abstract Strategic investment decisions are characterized by high innovation potential and long-term effects on the competitiveness of enterprises. Due to the uncertainty and risks involved in this complex decision making process, the need arises for well-structured support activities. A method that considers cost and the long-term added value is the cost-benefit effectiveness estimation. One of those methods is the "profitability estimation focused on benefits – PEFB"-method developed at the Institute of Management Cybernetics at RWTH Aachen University. The method copes with the challenges associated with strategic investment decisions by integrating long-term non-monetary aspects whilst also mapping the chronological sequence of an investment within the organization's target system. Thus, this method is characterized as a holistic approach for the evaluation of costs and benefits of an investment. This participation-oriented method was applied to business environments in many workshops. The results of the workshops are a library of more than 96 cost aspects, as well as 122 benefit aspects. These aspects are preprocessed and comparatively analyzed with regards to their alignment to a series of risk levels. For the first time, an accumulation and a distribution of cost and benefit aspects regarding their impact and probability of occurrence are given. The results give evidence that the PEFB-method combines precise measures of financial accounting with the incorporation of benefits. Finally, the results constitute the basics for using information technology and data science for decision support when applying within the PEFB-method.

Keywords Cost-benefit Analysis · Multi-criteria Decision · Profitability Estimation Focused on Benefits · Risk and Uncertainty Analysis

S. Printz (✉) · K. Lahl · R. Vossen · S. Jeschke
IMA/ZLW & IfU, RWTH Aachen University, Dennewartstr. 27, 52068 Aachen, Germany
e-mail: stephan.printz@ima-zlw-ifu.rwth-aachen.de

Originally published in "International Science Index:
Online special Journal Issues", © World Academy of Science Engineering
and Technology 2015. Reprint by Springer International Publishing
Switzerland 2016, DOI 10.1007/978-3-319-42620-4_58

1 Introduction

Strategic investment decisions (SID) are defined as "substantial investments that involve high levels of risk, produce hard-to-quantify (or intangible) outcomes and have a significant long term impact on corporate performance [1]". Hence, the process of strategic decision making (SDM) has emerged as an important research field over the past decade [2].

Controlling the complexity and uncertainty surrounding SID presents particular challenges for the management [3]. In addition, SID have an effect on the whole competitiveness of the organization [4, 5]. For this reason, efficient information search and evaluations are necessary [6]. The evaluation of accounting for SID has to pay more attention to scenario-based techniques [7, 8].

Field studies give evidence that traditional profitability analysis assessing SID is supplanted by sophisticated techniques in terms of linking qualitative and (quantitative) financial aspects [1, 9–11]. While the quantification and assignment of cost is examined extensively, there are fewer methods for the assignment of long-term benefits [12]. In fact, the evaluation of utilizing quantitative and qualitative criteria in decision making is a challenge when implementing effective decisions [13]. However, involving teams in the decision making process (DMP) improves the quality of the decision [14] and allows for alternative evaluations in the problem solving process [15]. Hence, much of the DMP in companies is decided as a team [16, 17].

The PEFB-method [18, 19] faces challenges with SID. However, since its development there has been no evaluation of the method itself regarding its applicability. Hence, a review of requirements and a comparative analysis of the gathered cost and benefit aspects are required. With these results, the applicability of the PEFB-method is confirmed and the baseline for future research is set up.

2 State of the Art

Research has aimed to answer the questions surrounding which analyses are being used to assess SID [20]. In fact, financial accounting information assists managers to give a quantitative overview of the current company situation and prepare for future decisions [21]. Hence, SID are usually based on economic criteria, often without considering qualitative issues [22]. Even if qualitative criteria is incorporated in the SID, there is a lack of structured and validated methodologies [23, 24]. However, scientists argue that an organization's philosophy itself and organizational context vary across circumstantial settings [25, 26].

The DMP is characterized by different attitudes and different knowledge of uncertainties arising as a result of imprecisions and vagueness of information [27]. In particular, SDM is involved with questions affecting the long-term success of the company, the allocation of significant resources and the trade-off in ambiguous situations as a result of insufficient information [28, 29].

Table 1 Requirements for new SID approach

Requirement [R]	Description of the requirement
R1	Precise measure of financial accounting
R2	Incorporation of qualitative criteria
R3	Structured and valid methodology
R4	Treating of insufficient information
R5	No remarkable skills to use required
R6	Team decision

In general, involving teams in decision making (DM) improves the solution quality and generates a wider variety of problem solutions [17]. The DM can be supported by group decision making frameworks. Most of the proposed methodology frameworks within the business environment are related to mathematical decision support frameworks [30, 31]. These frameworks are related to Analytical Hierarchy Processes [27] and Fuzzy preference relations [32, 33]. However, the implementation of mathematical models needs experienced customers and sometimes fails due to a lack of skills and its complexity [30, 34]. As a result of these uncertainties and the limitations of skills and abilities, there is a search for new management approaches [3]. The requirements of the new approach are summarized in Table 1. This method should provide accuracy of financial accounting decisions (R1) through the incorporation of qualitative criteria (R2). The SID has to be performed in a structured series of steps and provide valid results. In particular, so as to prioritize the sequence of investment, a timeline is required (R3). Furthermore, the treatment of insufficient information and information quality has to be integrated (R4). Finally, the method has to be non-complex and easy to use so as to reduce time and effort expenditure (R5). Finally, the approach has to integrate research and evaluations by financial experts, just like all experts affected within the company (R6).

3 The PEFB-Method

The PEFB-method is a participation-oriented, cybernetic approach for the evaluation of a SID. Based upon the utility analysis and the profitability analysis of IT-investment [19, 35, 36] both measurable monetary factors and non-monetary aspects are considered and thus requirements (R2) and (R6) are met. Figure 1 shows an overview of the PEFB-method according to the problem solving process [37]. Hence, the PEFB-method meets the demands of (R3). Moreover, to perform the method, no remarkable skills are required (R5). To demonstrate compliance with the given evidence, a short description of the PEFB-method is shown below. Examination of the requirements of (R1) and (R4)is part of the research design and a realization of the long-term evaluation.

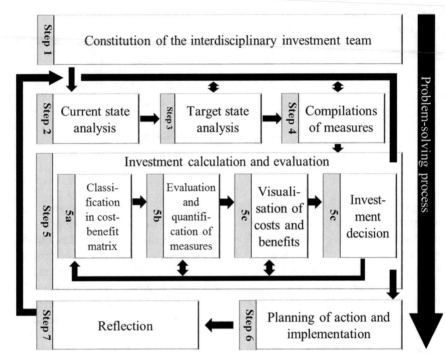

Figure 1 Steps of the PEFB-method

The following description shows the stages of the PEFB-method [38]. The PEFB-methods consists of seven steps:

1. Constitution of the interdisciplinary investment team,
2. Current situation analysis,
3. Target situation analysis,
4. Compilation of measures,
5. Investment evaluation,
6. Planning of actions and
7. Reflection.

Step 1 consists of building a representative interdisciplinary investment team containing executives and employees involved and affected. Hence, not only are the concerned departments represented, but also the various levels of hierarchy. This team will be responsible for conducting the evaluation, implementation and reflection process.

By building up an interdisciplinary investment team and through the support of a facilitator, a holistic view on the SID is ensured. Within the current state analysis (step 2), skills and competencies inside the enterprise are identified. Furthermore, by means of the target state analysis (step 3), the strategically, tactical and operational objectives of the project are identified and noted on a specification sheet. Step 4

Table 2 Cost-Portfolio

	High	Medium	Low
Direct	9	7	4
Indirect	8	5	2
Difficult to ascertain	6	3	1

Table 3 Benefit-Portfolio

	High	Medium	Low
Direct	1	3	6
Indirect	2	5	8
Difficult to ascertain	4	7	9

involves a compilation of concrete measures so as to achieve the target level of enterprise.

In step 5, the investment calculation and evaluation is carried out through 4 sub tasks. It encompasses the classification of costs and benefits, the evaluation and quantification of measures, visualization of cost and benefits as well as the investment decision. The method uses two different portfolios for the classification of costs and benefits (cf. Table 2, cf. Table 3). Costs and benefits of the investment are defined and assigned in terms of their impact on the project as "direct", "indirect" or "difficult to ascertain" aspects.

Direct costs or benefits are those, which are related to the investment and enable the impact to be measured directly, e.g. acquisition costs or increase of productivity. Accordingly, indirect costs or benefits are a derivate from direct effects, for instance maintenance costs or increased quality. Finally, "difficult to ascertain" costs or benefits contain effects which can only be presumed, like demotivation of employees or improving the image of the company. Moreover, the measures are classified regarding their probabilities of occurrence into the classes high, medium, and low.

Corresponding to the introduced impact classes, the measures are assigned in a 3×3-matrix. Each cell of the matrix contains a so called risk level, reaching from 1 to 9. The two portfolios differ in the arrangement of the risk level. Meanwhile, direct costs with high probability are assigned to the risk level 9 (cf. Table 2), benefit risk levels are designated contrarily. Direct benefits with high probability of occurrence refer to risk level 1 (cf. Table 3).

Within the framework of visualization, the levels define a ranking scale of measures. After the classification and quantification of the aspects, the filled cells of the matrix are aligned in two numerical series. The overall costs (Cm) for each risk level (j) are calculated from the individual cost aspects (cj) (refer to (1)). The modality for the computation of the overall benefits (Bn) each benefit (bj) is done similarly (see (2)).

$$C_m = \sum_{j=1}^{(9-m)+1} c_j \tag{1}$$

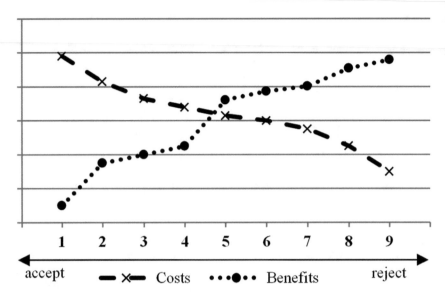

accept   ━✕━ Costs   •••●•• Benefits   reject

Figure 2 PEFB-chart

$$B_n = \sum\nolimits_{j=1}^{9} b_j \qquad (2)$$

The value of each risk level is recorded in a risk oriented PEFB-chart (see Figure 2). In relation to the possible intersection situations, four general cases can be distinguished. On the one hand, when the cost function is beneath the benefit function in all nine levels, the investment is economically evaluated as recommendable without restrictions. On the other hand, if the cost function is always above the benefit function, the investment is evaluated as definitely uneconomical. Finally, in the case of an intersection of both curves, an interpretation of the risk level is required. Investment scenarios with an intersection of both graphs in risk level 1 are the most economically reasonable. In terms of the descriptive interpretation, the overall costs are exceeded by direct and highly probable benefits. Contrarily, the worst economically reasonable investment scenario is at the location of the intersection point at the risk level 9. In this intersection point, direct and highly probable costs exceed all possible benefits.

In step 6, the next stage of the investment decision is determined. Depending on the results, this step relates to whether a plan of action or plan implementation is necessary. In particular, the advice regarding the evaluation of strategies is directly incorporated into the transfer process. Finally, the interdisciplinary investment team appraises the experiences gathered during the process (step 7). Accordingly, in the reflection phase a review of the whole investment evaluation process is carried out and appropriate recommendations are made. However, there is need to prove the compliance of the PEFB-method with (R1) and (R4). Thus, further research with regard to the fulfillment of the requirements is needed.

4 Research Design

The PEFB-method was applied in research projects as well as industrial projects. To evaluate the method's compliance with the precise measure of financial accounting (R1) and the treatment of insufficient information (R4), a chronological overview is shown in Table 4. Due to non-disclosure agreements, only the results of 14 independent workshops are allowed to be used for public evaluation. Besides the 14 applied PEFB-methods introduced over the last 10 years, the date of application and a short description of the assessed subject are given. In addition, a categorization of the projects is also provided. The categories display the areas of assessment divided into technology and methodology evaluation. Among the 14 displayed application fields, 7 belong to the category of technology assessment, whereas 4 projects belong to the methodology assessment. In 3 cases a mixture of the methodology approach and a launch of a new technology were

In order to perform a baseline study of correlating data, aspects aligned into different impact classes were excluded. If aspects are aligned within the same impact class, but dedicated to other probabilities of occurrence, they were adapted manually. These aspects are marked in the cost and benefit library. The adaption was performed

Table 4 Applied areas of the PEFB-method

Year	Project description	Area of assessment	
		Technology	Methodology
2006	Launch of a data processing service	X	X
2006	IT outsourcing solutions	X	
2006	Construction of a parking garage		X
2008	Semantic-based knowledge flow system for the European home textiles industry	X	X
2011	Intelligent Mega-Swap-Boxes for advanced intermodal freight transport	X	
2012	Lead-user method for innovation search		X
2012	Broadcast search for innovations		X
2012	Ideas competition for innovation search		X
2012	Safety technology for firefighters	X	
2013	Mass produced textile preforms by automated handling and online quality assurance	X	
2013	IT-support during the development of engineering standardization	X	X
2014	Online quality assurance for hot edge/hot air welding	X	
2014	Intelligent transport system for innovative intermodal freight transport	X	
2015	Automation of tricot machines	X	

either by majority or the lower risk level. Due to non-total order of the different risk levels the average risk level is not calculated.

5 Cost Aspects

Regarding these different types of applications, more than 96 different cost aspects were gathered. Among those, 13 aspects were inconsistent within their aligned impact classes. Hence, these values were canceled for the evaluation. Moreover, 5 aspects were adapted manually (marked). Table 5 gives an overview of the remaining 83 cost aspects in their original terms. Besides the aspects itself, the number of nomination [n] is illustrated just like the risk level. The total number of the aggregated costs aspects is 125, thus the average number of cost aspects for one PEFB-application is 8.93. Besides some common aspects like capital costs, the reduction of staff, or the demotivation of the employees, most of the aspects are individual for each project. The average number of nomination is 1.51. The distribution of all cost aspects within the impact classes are presented in Figure 3. The impact class of direct costs has a share of 57.83 %. This impact class consists of 75.00 % of high probability costs (risk level 9). This makes up the majority in this class followed by medium probabilities of 18.75 % (risk level 7). In the end direct costs with low probability gain a share of 6.25 % (risk level 4). Compared to impact class of direct costs, indirect costs have an overall share of 25.30 %. The class consists of a share 38.10 % for high probability costs (risk level 8), 33.33 % for medium costs (risk level 5) and 28.57 % for low probabilities (risk levels 2).

In particular, the impact class of difficult to ascertain costs has a share of 16.87 %. The majority within this impact class is formed by low probabilities with a share of 57.14 % (risk level 1). Meanwhile, medium probabilities have a share of 28.57 % (risk level 3). Finally, high probabilities (risk level 6) gain a share of 14.29 %.

6 Benefit Aspects

In addition, the preparation of the benefit aspects was done in the same way. By data adjustment, 30 benefit aspects with inconsistent impact class alignments were deleted from the overall number of 122 benefit aspects. The remaining 92 benefit aspects are shown in Table 6. The 6 benefit aspects which are manually adjusted are marked.

Collectively, the total number of aggregated benefit aspects is 131. Like the cost aspects, most of the benefit aspects are mentioned once, thus the average number of nominations is 1.42. The overall average number of benefits aspects for one

Table 5 Library of cost aspects

Cost aspects	Number [n]	Risk level
Aboriginal costs	1	9
Acceptance by machine operator	1	1
Additional charges to ensure IT-security while using IT-based standardization	1	5
Adoption of the vehicles (GPS)	1	9
Annual granting costs	1	9
Assembly	2	8
Bad declaration of performance relationships	1	8
Bought-in parts	1	9
Calibration of the equipment	1	4
Capital costs	3	1
Changeover costs	1	1
Choice of suitable intermediary	1	9
Communication effort with solver	1	9
Contract costs	2	9
Contract negotiation	1	9
Costs for sample picture creation	1	9
Debriefing costs	1	5
Demotivation	4	1
Denoting trucks	1	9
Destination charges	3	8
Developing a standard	1	5
Development of IT-platform	1	9
Development prototype	1	7
Digitalization of samples	1	9
Editing workshop results	1	9
Empty running	1	1
Expenses for data utilization (customer)	1	8
Finance costs	3	9
Flexibility boundary	1	4
Flexibility for just in time	1	3
Formulating a problem	3	9
Formulation of tender	1	9
Garment (jackets, gloves)	1	9
Handling and washing procedure for equipment	1	9
Hardware costs	1	9
Helmet (camera + vision, communication device)	1	9

(continued)

Table 5 (continued)

Cost aspects	Number [n]	Risk level
Hidden costs (generous conditions at contract closing/high debts at later change requests)	1	2
High sill, height of chassis and cam distance	1	2
Higher risk of injuries	2	1
Higher risk of know-how theft through intensive exchange with potential competitors	3	3
Higher risk of standardization employee's distraction	1	1
Higher space requirement	1	9
Higher system complexity	1	9
Image loss	3	3
Insurance for the technical equipment	1	7
Integration existing equipment	1	9
Investment costs	4	9
Lead-User identification und recruitment	1	9
Legal conflict	3	2
Legal counsel	3	9
Less control	1	6
Less flexibility (backload)	1	5
Less staff	8	8
License software	2	9
Limited service offer by contractor	1	2
Low sill	2	1
Maintenance by users	1	8
More staff	2	7
New specialist jobs	1	8
No accurate service provision by the outsourcing contractor	1	3
Oncosts through downtime and maintenance time	2	5
Premium	2	9
Production	2	5
Provide culture for acceptance of external knowledge	3	5
Psychological context monitoring	1	9
Recalls for fixing issues	1	2
Recurrent expenses through supplier change	1	4
Re-using the equipment	1	8
Rewarding lead-users	1	7
Run workshop	1	9
Sensors	1	9
Service	2	7
Service provision	1	9
Set-up time	1	7
Shuttle costs per year	1	7

Table 5 (continued)

Cost aspects	Number [n]	Risk level
Shuttle trains leasing costs per year	1	7
Smart life line	1	9
Spare parts	1	2
Starting up maintenance	1	6
Supply of data	1	9
Terminal costs when driver changes	1	7
Trend analysis	1	9
Work of the expert jury	2	9

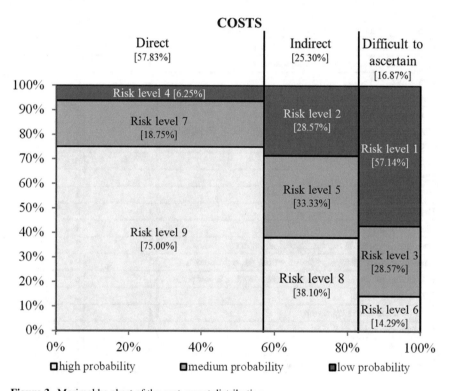

Figure 3 Marimekko chart of the cost aspect distribution

application is 9.36. Common benefit aspects are, e.g. documentation, image gain or increased flexibility through implementation of technology.

Collectively, the total number of aggregated benefit aspects is 131. Like the cost aspects, most of the benefit aspects are mentioned once, thus the average number of nominations is 1.42. The overall average number of benefits aspects for one application is 9.36. Common benefit aspects are, e.g. documentation, image gain or increased

Table 6 Library of benefit aspects

Benefit aspects	Number [n]	Risk level
Acceptance by employee	1	4
Amortization period	3	1
Availability secured	1	8
Basis for argumentation	1	1
Better control of action by team leader	1	7
Better procurement conditions	1	3
Better selling effect on Point of sale	1	3
Better view over entire situation	1	1
Central headquarters	1	5
Constructing production line	1	6
Contest's participation rate as competitions indicator	1	5
Cost reduction supply chain management	3	1
Cross-linkage of convoy with headquarters	1	8
Deal closure	2	3
Decrease damage on equipment	1	2
Decrease number of and damages on victims	1	2
Decrease of dependency on single employees with specialist know-how	1	8
Digital ordering pricing	2	3
Documentation	4	8
Door opening conception	1	1
Driver as dispatcher	1	8
Early error detection	1	1
Early involvement in the creation of standards	1	2
Ease of internal changes, reorganization, fusions/takeovers	1	9
Economies of scale on contractor's side causing lower prices	1	6
Efficiency increases	1	3
Employees receive inspiration	1	7
Error management	2	2
Fees	2	1
Focus on core business, core competency	1	7
Grasp frange	3	1
Higher arctic truck height through gooseneck	1	4
Higher machine workload	2	1
Idea goes into new product	2	3
Image gain/improvement	4	7
Increase of employee satisfaction	1	7
Increase of professionalism (no more self-made solutions)	1	1
Increased acceptance of respective standardization	1	5
Increased customer identification	2	7

(continued)

Table 6 (continued)

Benefit aspects	Number [n]	Risk level
Increased efficiency/success rate of rescue/intervention missions	1	3
Increased flexibility	4	4
Increased willingness to pay	1	5
Innovations through maintenance and updates	1	5
Inspiration for solving similar problems	1	2
Insurance possible	1	8
Integration/Interface coverage	1	8
Inter-company networking of experts	1	9
Inter-functionality of equipment	1	4
Learning aptitude	1	3
Less controls of incoming goods	1	5
Less damage by hail, birds, trees etc	1	5
Less risks for accidents on rail	1	2
Less room costs	1	4
Less substandard goods	2	1
Less trucks on road	1	4
Loading height	3	1
Lower transport costs	6	1
Maintenance on demand	1	3
Marketing tool	3	7
Mechanic lifting device	1	2
More flexible network	1	7
Multilayer field of application	1	5
New product	1	1
Newest level of data security	1	5
No IT-worries, no time exposure	1	2
One open side	1	1
Only one basic material	1	5
Potential through interdisciplinary approach	2	2
Process reliability	2	1
Production working capital	1	8
Productivity gain of already active standardization employees	1	1
Productivity gain of new standardization employees	1	1
Qualified consulting through outsourcing contractor	1	2
Quality of contact increases with higher number of participants	1	7
Real life experience of sample	2	3
Reduction of market analysis lost	1	3
Reduction of sample cost	1	3
Reduction of stock cost	1	3
Reduction of working capital	1	3

Table 6 (continued)

Benefit aspects	Number [n]	Risk level
Reduction stock cost	1	8
Vehicle conditioning	1	5
Risk of oil price increase	1	3
Road taxes independency	1	2
Saving of development work	2	1
Savings by low insurance	2	5
Support in tactical decisions	1	4
Tactical support for resource management at scene of accident	1	4
Tailgate	1	2
Total benefits	1	1
Transfer of debts to contractor	1	3
Tri-Modal	3	1
Vehicle store place	1	1

flexibility through implementation of technology. In contrast to the cost aspect distribution, the allocation of the benefit matrix is shown in Figure 4. The direct benefits gain a share of 43.48 %. Within the impact class, high probability benefits (risk level 1) have a share of 55.00 % followed by 40.00 % with medium probability (risk level 3). Direct benefits with a low probability of occurrence have a share of 5.00 % (risk level 6). With respect to the impact class of indirect benefits the share of 35.87 %, consist of 36.36 % for high probability aspects (risk level 2) as well as 36.36 % of medium probability aspects (risk level 5). Aspects with low probabilities have a share of 27.28 % (risk level 8). On top of that, the impact class of difficult to ascertain benefits has an overall share of 20.65 %. The proportion within the impact class consists of 42.11 % for high probabilities (risk level 4), 47.37 % for medium probabilities (risk level 7) and 10.52 % for low probabilities.

7   Discussion of the Results

The distribution of high probability cost aspects decreases throughout the impact classes (cf. Figure 3). In contrast, the share of low probability aspects increases with the opportunity to use qualitative aspects. Based on this fact, the increasing uncertainty among the impact classes leads to a risk-averse assessment. Furthermore, the share of the impact classes decreases from direct costs to costs that are difficult to ascertain. One possible reason for the major share of direct costs might stem from an accounting department.

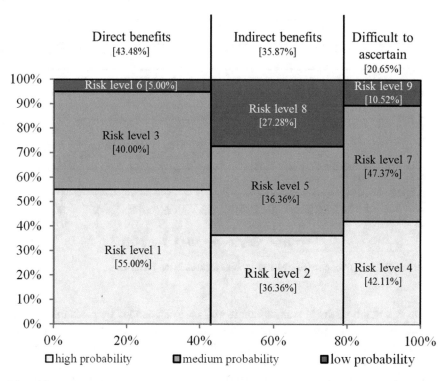

Figure 4 Marimekko chart of the benefit aspect distribution

In particular, a share of 75.00 % for risk level 9 and 18.75 % for risk level 7 represents the precise recording of these cost factors. Analyzing the impact class of indirect costs (25.30 %) highlights a change in the distribution of probability share. All probabilities roughly gain a share of around 30.00 %. This result represents the transitional period from the quantitative to the qualitative aspects, where cost aspects are not easily provided by the accounting department. Despite that, difficult to ascertain costs gain a share of 16.87 %. Actually, this result provides evidence for literary research by describing that cost estimation could be done precisely by accounting departments. Interpretation of the benefit evaluation clearly shows that the relevant requirement is met, including long-term effects on SID Thus, the PEFB-method meets the demand of the precise measure of financial accounting (R1). In contrast, the benefit aspects differ in their share of impact classes and probabilities (cf. Figure 4). The share of the direct impact class of benefits is around 14 % lower than the share of the direct costs. Indeed, the share of the indirect impact class increases around 13 % compared to costs. With a difference of nearly 4 %, the impact class of difficult to ascertain benefits is in a similar situation to the costs class. For every class there is a significant difference regarding the distribution of the probabilities of occurrence within the impact classes. Moreover, there is an increase in the share of all probabilities in the impact class of difficult to ascertain aspects. It is likely therefore that this constitutes

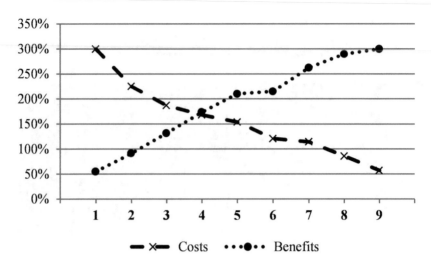

Figure 5 Relative frequency of cost and benefits aspects in the PEFB-chart

to the fact that benefits in comparison to costs are surrounded by more uncertainty within the evaluated projects. The benefit distribution reveals that a resolution of insufficient information is ensured (R4). The overall purpose of the PEFB-method is to gain knowledge regarding the investment decision. Due to reflecting potential impacts, future development becomes more certain. With respect to the challenge of SID, the use of the different impact classes represents qualitative aspects just like quantitative aspects. Hence, the results of the overall distribution of cost and benefit aspects are different. Regarding the impact of cost and benefit distribution, Figure 5 shows the relative frequency of the introduced cost and benefit aspects in the PEFB-chart. A sample of one assessed project reveals that the average value of the cost and benefit aspects fluctuates at around ±10 %, hence, the representation of the qualitative analysis is possible. The general intersection point of the evaluated projects is stated at the risk level 3.92.

According to the classification of the risk portfolio, there are two interpretations. The SID depends on the probability of benefits with high probability of occurrence and direct benefits with medium probability of occurrence. Indeed, SID are functions from which it is particularly difficult to ascertain benefit aspects with a high probability of occurrence (risk level 4). Secondly, the run of the cost curve illustrates the SID to be a function of all costs aspects with high probability of occurrence just like direct and indirect cost aspects with medium probability of occurrence. In general, the comparative analysis indicates that the PEFB-method is able to provide a precise measure of cost accounting besides the compilation of long-term effects. Hence, the PEFB-method could be stated as one possible method to assess SID. However, the presented evaluation is only one step to refine the PEFB-method. The purpose is to define a catalogue with aggregated cost and benefit aspects with predefined risk levels. Therefore, the company has to choose the aspect itself and put it in the organizational context. Referring to the results of the conducted study, it is possible to illustrate for the first time the relationship between benefit and cost aspects as a function of temporal sequences. The results illustrate the importance of indirect benefits in decision making processes. However, these aspects are surrounded by uncertainty, thus there is a need for more information. Modern information technology provides approaches like data science, predictive analysis and big data methods to gather more information. The combination of a participation oriented decision making process and data science is an interesting field of research for the advancement of the PEFB method.

8 Future Research

In general, progresses in information technology and participation oriented methods need to be linked to future research. Especially, methods of information search like data science are becoming promising approaches increasing information quality and the validity of SID. In order to use data science for information search within the PEFB-method, a methodology to assess information quality has to be developed. The challenge is to identify information sources and their ranking. In addition, future research based upon this evaluation should focus on several aspects. In particular, the visualization of the PEFB-results needs a refining. The interpretation of the intersection point gives no statement regarding profitability. Hence, the challenge is to combine visualization of the results and a sensitivity analysis of the quantified aspects. Moreover, the evaluation of the risk level alignment, the distribution of the quantified aspects and real monetary values should be evaluated, too. In addition, a review of the success rate of implemented technologies or applied methodologies is required. With regard to the correlating data preparation, 43 aspects were excluded from the evaluation because of inconsistency in the alignment into impact classes. The split of the excluded data is 13 cost aspects and 30 benefit aspects. In fact, the number of benefits is more than 2.3 times higher than the elimination rate of the cost aspects. Hence, the examination of these aspects and the reason for the inconsistency is required. A promising approach to explain the inconsistency might be the constitution of the interdisciplinary investment team. In particular, the role of the interdisciplinary investment team with regards to risk preference, personal affection and moderating team effects should be examined. Furthermore, the subject of the PEFB-method has to be reviewed in terms of the surrounding uncertainty expressed by the desired target situation.

References

1. F. Alkaraan, D. Northcott, Strategic capital investment decision-making: A role for emergent analysis tools? Br. Account. Rev. **38**(2), 2006, pp. 149–173

2. V. Papadakis, Do CEOs shape the process of making strategic decisions? Evidence from Greece. Manag. Decis. **44**(3), 2006, pp. 367–394
3. I. Gorzeń-Mitka, M. Okręglicka, Improving Decision Making in Complexity Environment. Procedia Econ. Finance **16**, 2014, pp. 402–409
4. C. Carr, C. Tomkins, Context, culture and the role of the finance function in strategic decisions. a comparative analysis of Britain, Germany, the U.S.A. and Japan. Manag. Account. Res. **9**(2), 1998, pp. 213–239
5. C. Carr, Strategic investment decisions: the importance of SCM. a comparative analysis of 51 case studies in U.K., U.S. and German companies. Manag. Account. Res. **7**(2), 1996, pp. 199–217
6. F. Rothärmel, *Strategic management*, 2nd edn. McGraw-Hill Education, New York, NY, 2015
7. T. Alessandri, D. Ford, D. Lander, K. Leggio, M. Taylor, Managing risk and uncertainty in complex capital projects. Q. Rev. Econ. Finance **44**(5), 2004, pp. 751–767
8. K. Miller, H. Waller, Scenarios, Real Options and Integrated Risk Management. Long Range Plann. **36**(1), 2003, pp. 93–107
9. R. Slagmulder, W. Bruggeman, L. van Wassenhove, An empirical study of capital budgeting practices for strategic investments in CIM technologies. PROECO Int. J. Prod. Econ. **40**(2), 1995, pp. 121–152
10. R. Adler, Strategic investment decision appraisal techniques: The old and the new. BUSHOR Bus. Horiz. **43**(6), 2000, pp. 15–22
11. H. Siegwart, U. Singer, Neues Verfahren für die Wirtschaftlichkeitsbeurteilung von Investitionen in neue Produktionstechnologien. Kostenrechnungspraxis Krp **2**, 1991, pp. 63–70
12. M. Schönheit, *Wirtschaftliche Prozessgestaltung: Entwicklung Fertigung Auftragsabwicklung.* Springer, 2013
13. O. Itanyi, U. Ewurum, W. Ukpere, Evaluation of decision making criteria with special reference to quantitative and qualitative paradigms. Afr. J. Bus. Manag. **6**(44), 2012, pp. 11,110–11,117
14. P. Drucker, *The Effective Executive: Effektivität und Handlungsfähigkeit in der Führungsrolle gewinnen.* Vahlen, München, 2014
15. J. DuBrin, *Leadership: research findings, practice, and skills.* Houghton Mifflin, Boston, 2001
16. J. Bonito, *Interaction and influence in small group decision making.* 2014
17. O. Negulescu, E. Doval, The Quality of Decision Making Process Related to Organizations' Effectiveness. Procedia Econ. Finance **15**, 2014, pp. 858–863
18. E. Savelsberg, *Innovation in European freight transportation basics, methodology and case studies for the European markets.* Springer, Berlin, London, 2008
19. D. Weydandt, *Beteiligungsorientierte wirtschaftliche Bewertung von technischen Investitionen für prozessorientierte Fertigungsinseln.* Shaker, Aachen, 2000
20. G. Sandahl, S. Sjögren, Capital budgeting methods among Sweden's largest groups of companies. the state of the art and a comparison with earlier studies. Int. J. Prod. Econ. **84**(1), 2003, pp. 51–69
21. A. Socea, Managerial Decision-Making and Financial Accounting Information. Procedia - Soc. Behav. Sci. **58**, 2012, pp. 47–55
22. R. Dekkers, Impact of strategic decision making for outsourcing on managing manufacturing. Int. J. Oper. Prod. Manag. **31**(9), 2011, pp. 935–965
23. T. Baines, An integrated process for forming manufacturing technology acquisition decisions. Int. J. Oper. Prod. Manag. **24**(5), 2004, pp. 447–467
24. G. Frank, D. de Souza, J. Ribeiro, M. Echeveste, A framework for decision-making in investment alternatives selection. Int. J. Prod. Res. **51**(19), 2013, pp. 5866–5883
25. P. Miller, T. O'Leary, *Capital budgeting, co-ordination and strategy: A field study of inter-firm and intra-firm mechanisms.* University Press Oxford, Oxford, 2005
26. F. Verbeeten, Do organizations adopt sophisticated capital budgeting practices to deal with uncertainty in the investment decision? Manag. Account. Res. **17**(1), 2006, pp. 106–120
27. N. Bryson, Group decision-making and the analytic hierarchy process: Exploring the consensus-relevant information content. Comput. Oper. Res. **23**(1), 1996, pp. 27–35

28. E. Wallace, R. Rijamampianina, Strategic decision making with corporate emotional intelligence. Probl. Perspect. Manag. **3**, 2005, pp. 83–91
29. H. Dincer, G. Gencer, N. Orhan, K. Sahinbas, The Significance of Emotional Intelligence on the Innovative Work Behavior of Managers as Strategic Decision-Makers. Procedia - Soc. Behav. Sci. **24**, 2011, pp. 909–919
30. A. Kengpol, P. Boonkanit, The decision support framework for developing Ecodesign at conceptual phase based upon ISO/TR 14062. Int. J. Prod. Econ. **131**(1), 2011, pp. 4–14
31. D. Golmohammadi, M. Mellat-Parast, Developing a grey-based decision-making model for supplier selection. Int. J. Prod. Econ. **137**(2), 2012, pp. 191–200
32. F. Herrera, L. Martínez, P. Sánchez, Managing non-homogeneous information in group decision making. Eur. J. Oper. Res. **166**(1), 2005, pp. 115–132
33. I. Palomares, R. Rodriguez, L. Martinez, An attitude-driven web consensus support system for heterogeneous group decision making. Expert Syst. Appl. **40**(1), 2013, pp. 139–149
34. M. Barfod, K. Salling, S. Leleur, Composite decision support by combining cost-benefit and multi-criteria decision analysis. Decis. Support Syst. **51**(1), 2011, pp. 167–175
35. K. Nagel, *Nutzen der Informationsverarbeitung: Methoden zur Bewertung von strategischen Wettbewerbsvorteilen, Produktivitätsverbesserungen und Kosteneinsparungen.* Oldenbourg, München, 1988
36. H. Ott, Wirtschaftlichkeitsanalyse von EDV-Investitionen mit dem WARS-Modell am Beispiel der Einführung von CASE. Wirtschaftsinformatik **35**(6), 1993, pp. 522–531
37. R. Sell, R. Schimwegm, *Probleme lösen: in komplexen Zusammenhängen denken; mit 19 Tabellen.* Springer, Berlin, 2002
38. R. Printz, S. Vossen, S. Jeschke, *Adaption of the profitability estimation focused on benefits due to personal affection.* presented at the Workshop on Assessment methodologies - energy, mobility and other real world applications, Coimbra, to be published 2016

Real-Time Machine-Vision-System for an Automated Quality Monitoring in Mass Production of Multiaxial Non-crimp Fabrics

Robert Schmitt, Tobias Fürtjes, Bahoz Abbas, Philipp Abel, Walter Kimmelmann, Philipp Kosse and Andrea Buratti

Abstract Fiber-reinforced plastics (FRPs) are used in an increasing number of applications due to their advanced light-weight properties. Beside classical deployments in high-value industries like aerospace and medical engineering, FRP materials are pushed towards mass production by the automotive industry. To mass-produce FRP products, textile structures are commonly used as semifinished products, such as multiaxial non-crimp fabrics (NCFs). However, poor repeatability and missing textile defect detection in the automated manufacturing of FRP components are major cost factors and challenge economically the series production. Reduction of these cost factors is not yet possible due to the lack of closed-loop control systems. There is currently no real-time quality monitoring system capable of ensuring quality in NCF production. The purpose of this study is to develop tools and concepts for real-time quality control of non-crimp fabrics. Therefore, a real-time machine-vision system has been developed with the purpose of detecting relevant quality features in a textile sample in deterministic time conditions. The embedded system ensures the execution of all process steps, i.e. image acquisition, processing, and evaluation, under real-time conditions. The main focus of this work is laid on the real-time algorithms for an accurate and robust detection of the fiber orientation under industrial conditions. The developed real-time system has been tested on a textile sample and an assessment of the measurement uncertainty has been performed. Results show that the proposed system can successfully assess common textile quality features.

R. Schmitt (✉) · W. Kimmelmann
Chair of Production Metrology and Quality Management (WZL),
RWTH Aachen University, Manfred-Weck-Haus 219-225, 52074 Aachen, Germany
e-mail: R.Schmitt@wzl.rwth-aachen.de

T. Fürtjes · P. Kosse · A. Buratti
WZL, RWTH Aachen University, ADITEC-Gebäude 100-105, 52074 Aachen, Germany

B. Abbas
IMA/ZLW & IfU, RWTH Aachen University, Dennewartstr. 27,
52068 Aachen, Germany

P. Abel
Institute of Textile Technology (ITA), RWTH Aachen University,
Otto-Blumenthal-Str. 1, 52074 Aachen, Germany

Originally published in ScienceDirect, IFAC-PapersOnline, Vol. 48
no. 3, © IFAC 2015, © Science Direct 2015.
Reprint by Springer International Publishing Switzerland 2016,
DOI 10.1007/978-3-319-42620-4_59

Keywords Automation · Real-Time Systems · Quality Control · Manufacturing
Processes · Sensor Fusion

1 Introduction

Fiber-reinforced plastics (FRPs) have a large potential as a way to reduce vehicle
and airplane weight. Therefore, FRPs, and especially carbon fiber-reinforced plastics
(CFRPs), are subject of interest for aerospace industries, automotive industries and
wind energy industries for its excellent mechanical properties and light weight [1].

Commonly, fiber textile products as multiaxial non-crimp fabrics (Figure 1, abbreviated: NCFs) are used as semifinished products for FRP. Multiaxial NCFs are textile
structures with high mechanical properties, which generally consist of stretched and
parallel layers of reinforcing fibers (rovings) or bands (fiber bundle). In order to
combine these layers together, it is possible to utilize an additional material, such as
mesh fiber. NCF production takes place on a warp knitting machine with multi-axial
weft insertion, ensuring an economically advantageous industrialization of the manufacturing processes. Alongside the effective manufacturing, the main advantages of
those textiles are high flexibility in terms of layer structure parameters (number of
layers, fiber orientation, mass per unit area and fiber material) and textile parameters
(bond type, seam length and thread inlet) [1].

Due to the heterogeneous and anisotropic nature of composite materials, the
defects that can occur in CFRP components are different to those found in metal.
Many defects occur at the interfaces between different layers, such as gas bubbles
(porosities) between fiber and matrix, angular deviation of fibers (fiber misalign-
ment), inclusions, voids, and thermal or shrinkage cracks. These defects can drastically alter local mechanical properties of components and can result in cracks, which
can propagate to mechanical failure of the component. Thus, it is important to monitor the manufacturing process in order to detect such defects and discard flawed
components [2].

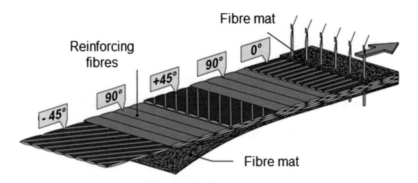

Figure 1 Manufacturing of multiaxial non-crimp fabrics

Even though the aforementioned production process can be applied to mass manufacturing, the high costs of production present an obstacle to the exploitation of the full potential. This is partly due to the fact that it is currently impossible to check product quality on the process line [3]. Indeed, there is no inline system able to perform continuous quality control of the process.

The purpose of this work is to present concepts and tools for developing an automated quality monitoring system capable of supervising the production of multiaxial non-crimp fabrics and ensuring their quality in real-time.

2 State of the Art

2.1 Quality Control in CFRP Production

The monitoring of production of multiaxial NCFs can be achieved by adopting non-destructive testing methods in the process line. Non-destructive testing methods (also called NDT methods) consist of a broad group of analysis techniques to estimate properties of materials, components or systems without permanently altering or damaging them. They are important in industry and science, as they can provide significant savings in time and money for quality testing [2].

Several methods are currently used for carbon-fiber composite inspection, such as ultrasonic testing, laser shearography, eddy current testing, thermography techniques, X-radiography, acoustic testing, white light interferometry, microwave, infrared and electron radiation. Each method can detect different material properties or defects and is suitable for a specific task [2, 3]. The vast majority (95 %) of non-destructive inspection tests in the production of fibre-reinforced composites are based on ultrasound [2, 4–6]. Indeed, these methods are suitable for detecting air voids, inclusions, delaminations, local resin inhomogeneities, fiber cracks and glue failures in components [2, 7]. Nevertheless, ultrasound testing methods are suitable to inspect just finished products during late stages of production, since semi-finished products cannot be tested in wet conditions. Air-coupled ultrasound inspection is not applicable to textile semi-finished products or textile fabrics because of sound attenuation through air gaps. Similarly, other methods based on different electromagnetic radiations (e.g. infrared or microwave) can be used to obtain the tomography of a work piece [8]. However, these methods present several problems when applied to CFRP products, as they feature a restricted measurement range, need extensive sample preparation or can damage the component. Image processing techniques are also used in microstructural analysis through scanning electron microscopy (SEM), transmission electron microscopy (TEM) or atomic force microscopy (AFM) [6]. X-ray detectors can output received data to a computer, allowing online data processing. As composites materials are usually highly transparent to X-rays, it is appropriate to use low-energy X-rays in order not to get saturated images. X-ray computed tomography has a limited measurement range and its industrial development is thwarted by high investment costs and security issues regarding protection from radiation [6, 9].

Thermography is a non-destructive testing method based on infrared radiation that can be used to detect inner defects of components [10, 11]. It can provide just blurred images of a reference surface with just few measuring points and for this reason it is suitable for qualitative analysis. As for ultrasound testing, thermography is applicable on finished products and not suitable for inspection during early stages of production, e.g. for monitoring non-crimp fabrics production.

2.2 Optical Systems for Quality Control

Optical sensor systems are well-established tools for industrial quality assurance tasks. The advantages of these systems are their relative low price and their fast and reliable functioning. Regarding quality monitoring of CFRP products, a method based on a laser sensor and image processing is described in [12]. It can detect fiber orientation and inclusions which occur during the production of preforms. Another example involving an optical system was developed by the company Profactor [13]. In this approach, an optical scanner performs eight image acquisitions under different lighting conditions for generating a reliable picture of the structure. An image processing software highlights the presence of angular deviations in the fiber orientation. Nevertheless, this system has been never used for industrial applications, as satisfying results have been reached just under laboratory conditions.

An optical quality monitoring for 3D-parts requires multisensor data fusion approaches. Multisensor data fusion is the process of combining data and related information from disparate sensors so that the achieved inferences are better than those which would be achieved by using the same sensors individually [14, 15]. Multisensor data fusion has gained a key role in production metrology. The purpose of combining sensor data is to improve significantly quality and accuracy of measurements in a production process. Regarding the lightweight industry, an important application developed at the Laboratory for Tool Machines (WZL) of RWTH Aachen University is capable of performing inline inspection of textile structures [16]. This automated system carries out the placement and alignment of fiber-reinforced composite structures. It consists of an optical sensor which detects the local fiber orientation and of a laser light-section sensor that identifies the position of the contour of textile preforms. Whereas the optical sensor coarsely detects the contour of each layer, the laser light-section source exploits this information to follow a scanning path. Subsequently, the light section sensor can provide more accurate measurements of the contour. Finally, the joint information is used to determine position and alignment for each textile layer. This system is also used to control the 3D-layer deformation that occurs in the drapery process.

Another multisensor machine vision application for large 3D- carbon-fiber reinforced structures has been developed at the WZL of RWTH Aachen University [17]. It combines light section sensors and image processing cameras with a robot system to perform quality control tasks. Data fusion occurs by integrating measuring information from these different sensors and thus reconstructing the 3D-component geometry.

Current commercial available optical measurement systems are not able to provide their results in deterministic time intervals. Despite this, an optimal integration of an automated quality monitoring system in a time-critical production process requires that the exact calculation time is known. For this reason, it is necessary to develop a suitable optical measurement system which can operate in real-time conditions and inspect NCFs.

2.3   Real-Time Conditions for Quality Control Systems

Real-time systems are commonly used for measurement and quality control tasks of technical applications. In production engineering such systems are typically used for cycle-dependent and automated component inspection, as well as for production process control. A real-time system is defined as a system that "controls an environment by receiving data, processing them, and returning the results sufficiently and reliable quickly to affect the environment at that time" [18].

The most important property of a real-time system is the maintenance of strictly predetermined time conditions. Therefore, real-time systems are classified according to the consequence of missing a deadline in:

- Hard – missing a deadline leads to a total system failure
- Form – missing a deadline is tolerable but the result may not used
- Soft – missing a deadline is tolerable but causes worse system performances

Timeliness is another important characteristic of real-time systems. This quantity can be mathematically expressed as follows:

$$A \equiv r + \Delta e \leq d \tag{1}$$

where the time condition A has to hold true in any case, r is the starting point of a new task, d represents the time by which the task has to be completed and Δe the time space in which the task needs to operate. To ensure the desired time-behavior of a real-time system, special hardware (e.g. FPGA), operating systems (e.g. Real-Time LabView) and interfaces (e.g. CANopen) are required.

Typical image processing applications for real-time systems are inline-quality assurance tasks or advanced driver assistance systems in vehicles [19]. Nevertheless, there are presently no optical real-time systems for process-control in CFRP production.

3 Concepts and Methods

Even if the aforementioned sensors can efficiently perform specific tasks (i.e. to identify or quantify given defect types), they are not able to perform a complete inspection of multiaxial non-crimp fabrics in real-time. For ensuring quality in multiaxial NCF production, it is necessary to develop a quality control system which guarantees a deterministic sensor feedback to close time-critical production control loops. The most important steps are: the development of a reliable image processing algorithm and of real-time hardware concepts, the optimization of the algorithm, testing and evaluation of the complete system.

3.1 Quality Control in CFRP Production

The inspection method is based on the detection of the local position and orientation of the fibers from the upper textile layer of the fabric. By using image processing techniques, fiber orientation can be measured with high accuracy and represented in a map, thus enabling to assess quality of multiaxial non-crimp fabrics (Figure 2).

NCFs feature a highly-reflective and non-uniform surface which is difficult to acquire with optical systems. By means of a diffuse illumination device, a camera with high dynamical contrast and an enhanced orientation algorithm, those challenges can be addressed (Figure 2). As regards the algorithms, the structure tensor method [20] proved to be the most robust technique and capable of measuring the orientation of fibers [16].

By performing an analysis of the structure tensor of small image regions based on the eigenvector matrix concept, it is possible to determine the local pixel orientation. The structure tensor defines the orientation of a region, exploiting the fact that gray

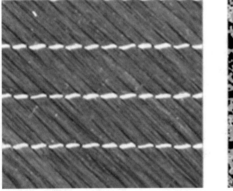

Figure 2 *Orientation Estimation* (*original*, final orientation map)

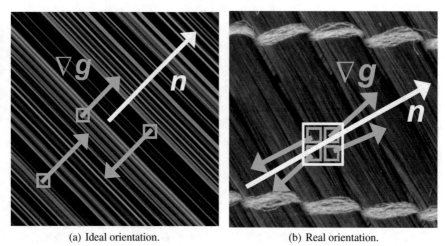

(a) Ideal orientation.   (b) Real orientation.

Figure 3 Fiber orientation estimation model

value variation in the direction perpendicular to fibers is higher (ideally maximum) than that in the fiber direction (ideally zero). The squared scalar product of the gradient vector ∇g and the normal to the local orientation n is a suitable measure for the direction of the maximum gray value variation:

$$(\nabla g^T n)^2 = |\nabla g|^2 \cos^2(\measuredangle(\nabla g, n)) \tag{2}$$

Figure 3a shows the model of an ideally-oriented neighborhood where the gradients have the same orientation as the texture. Since real fibers are not perfectly placed, the normal vector n is modelled as the result of the eigensolution for a certain neighborhood.

Maximizing the integral (Equation 3) for a W-dimensional local neighborhood leads to a tensor whose eigenvector related to the maximum eigenvalue gives the local orientation. The window function w determines size and shape of the neighborhood around the point x.

$$\int w(x - x')(\nabla g(x')^T n)^2 d^W x' \rightarrow MAX \tag{3}$$

Applying this algorithm to the image and filtering seams information, the local fiber orientation can be calculated (Figure 4). The main fiber orientation for a region of interest is calculated through a histogram analysis of its local fiber orientations. This result can be used to segment the different textile areas in an image, to analyze local deviations of the fiber orientation and to provide reliable information for the production control loop.

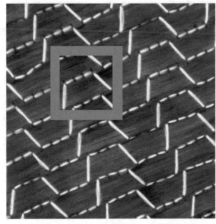 

(a) Region of Interest (ROI, e. g. 32x32 Px).   (b) Local fiber orientations (high significance).

Figure 4   Fiber orientation measurement

3.2   Real-Time Sensor Concept

A real-time machine vision prototype was built at the WZL for the inline inspection in the manufacturing of multiaxial non-crimp fabrics. The system components are summarized in Figure 5.

The main components of the real-time machine vision system are:

• National Instruments PXI rack with a LabVIEW RT operating system (Quad-Core, 2,3 GHz)
• Diffuse dome illumination device
• Basler high-resolution industrial camera with Camera Link interface for high frame rates
• National Instruments FlexRio FPGA-Module for the image processing (100 MHz)

The detection of flaws in the textile structure as well as an intelligent recognition of fabric material and type and their classification are further important features of the real-time machine vision system. Additionally, a laser light-section sensor, which is able to measure the 3D-placement of the fabric layers, can be added in future.

3.3   Pipelining and Optimizing

The LabVIEW RT operating system provides symmetric multiprocessing (SMP) for optimizing the image processing speed. Thus, the algorithm can be divided into several pipelines, which can be run parallel on different CPU- or FPGA-cores.

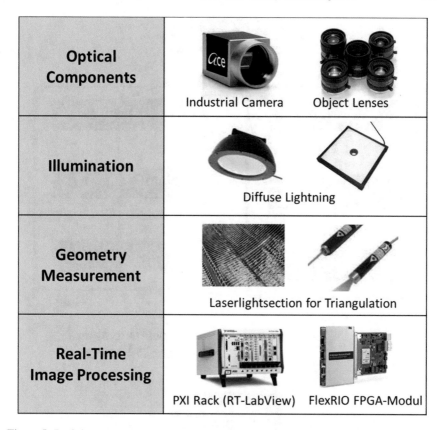

Optical Components	Industrial Camera Object Lenses
Illumination	Diffuse Lightning
Geometry Measurement	Laserlightsection for Triangulation
Real-Time Image Processing	PXI Rack (RT-LabView) FlexRIO FPGA-Modul

Figure 5 Real-time sensor-concept

In this case the selected algorithm – which is supposed to measure the global orientation of the fibers – is divided into two parallel and independent pipelines (Figure 6). In the first pipeline the mask for the segmentation is calculated. In the other pipeline the structure tensor method for the determination of the local fiber orientation is executed. In the last step the results of the two pipelines are merged to assess the global orientation of the fibers.

In contrast to conventional sensor systems, the developed real-time algorithm works pixel by pixel. This means that for every system-cycle the local orientation of one pixel can be measured. The advantage of this procedure is a much shorter computation time in total.

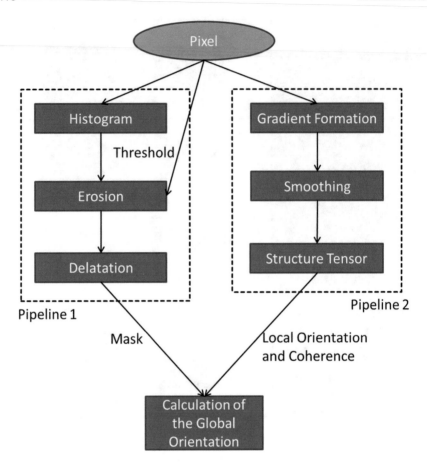

Figure 6 Pipelining of the image processing method

3.4 Evaluation Methods

For the evaluation of the real-time sensor system three main points were investigated. First, what the reachable deterministic frame rate of the system is. Second, how much the measurement uncertainty of the real-time system is. And finally, which factors have a significant influence on the measurement uncertainty.

To evaluate the real-time behavior and the frame-rate of the system, the analysis functions of the embedded system were used. These analysis functions provide an exact report about the time period of every process step.

The uncertainty in measurement and the influences of different parameters to the uncertainty, was determined according to the "Guide to the Expression of Uncertainty in Measurement" [21]. With the real-time system prototype the algorithms could be validated on the shop floor to proof its functionality and also robustness against outer

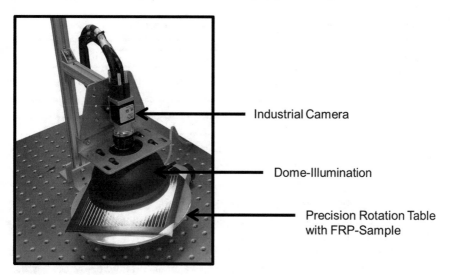

Figure 7 Real-time sensor system and the rotation table

noise (e.g. changing illumination, vibrations, etc.). To create an appropriate database for the analysis, 2 different carbon fiber probes with different optical and geometrical properties were investigated by using 5 different filter algorithms, 3 different camera resolutions, 2 different apertures, and 3 different exposure times.

The specimens were put on a high-precision rotation table (Figure 7) with a resolution of $0.598 \, \mu\text{rad}$, which is utilized as a measurement reference value for comparing the results of the real-time system and testing its repeatability. The rotation table is adjusted perpendicular to the optical axis underneath the real-time machine-vison system. The textile material sample is fixed on the rotation table (Figure 7). The reference rotation table is rotated in discrete steps $\Delta\Theta\text{REF} = \Delta\pi = 15°$ (range: $0° \leq \Theta\text{REF} < 180°$) to analyse the repeatability of the orientation measurement. This procedure is repeated 25 times for the given range.

Since the maximum positioning uncertainty of the rotation table ($0.598 \, \mu\text{rad}$) is small compared to the expected uncertainty of the measurement, the rotation table is a suitable reference to approximate the 'true' orientation and thus, its systematic error and its uncertainty can be set to zero.

4 Results

The FPGA run with a frequency of 100 MHz. Because of the pixel-wise algorithm, the theoretically reachable frame rate only depends on the camera resolution. For a camera resolution of 2048 * 2048 pixel the maximum frame rate is 23.841 frames per second. For the developed real-time system the actual frame rate is 22.72 fps.

Table 1 Influence of different parameters on the measurement uncertainty

Parameter	Influence on systematic error	Influence on random error
Gradient	+	+
Integration	0	0
Resolution	+	+
Material	+	++
Exposure time	0	0
Aperture	+	0

The reasons for this lower value are the necessity of a buffer of the pixel rows for the filter algorithm and the use of trigonometric functions, which requires more than one FPGA cycle for each computation.

For all measurements the coverage factor $kp = 2$ was chosen for a confidence level of 95 %. All investigated combinations have an uncertainty below $0.3°$, which is defined as the maximum measurement uncertainty for industrial usage [3]. The influence of the system configuration on the uncertainty was also investigated. The best configuration has an uncertainty of $0.0119°$ and the worst configuration an uncertainty of $0.1314°$. The results of the measurement uncertainty analysis of the developed real-time system show that a comparison between the measured orientation and the specification of the multiaxial fabrics can provide information about the production quality.

Finally, a full-factorial design of experiments was performed to analyze the influences of the different factors on the measurement uncertainty. Results have shown that the material type is the most critical factor which affects the random component of the uncertainty. On the other hand, resolution, aperture, filter algorithm, and also material type have a significant influence on the systematic error (Table 1).

5 Conclusion

Fiber-reinforced plastics, and especially non-crimp fabrics, are an important material group for manufacturing lightweight products. However, the lack of adequate manufacturing systems capable of ensuring quality of textile-based composite structures presents an obstacle to their mass production.

The proposed real-time machine-vision system aims to automate process control of fiber-reinforced plastics production. The system can inspect quality of each textile structure during the manufacturing process. This information can be used to establish a quality assurance system, thereby closing the quality control loop. A major challenge was to guarantee that the control system provides quality information in real time. Therefore, the image processing algorithms have been split into different pipelines to shorten the computation time without affecting the measurement uncertainty.

The proposed methods for measuring the fiber orientation trigger a robust automated manufacturing process of textile preforms for composite parts, improving the process quality and reducing the scrap quota. The next step of our research will be to integrate the real-time vision system directly into a production process.

Furthermore, the real-time machine vision system can be applied to several other stages of the production chain, such as cutting the textile layers or assembling preforms, so as to further the development of an economically feasible FRP production.

References

1. G.W. Ehrenstein, *Faserverbund-Kunststoffe. Werkstoffe, Verarbeitung, Eigenschaften*. Hanser, München, 2006
2. P. Vaara, J. Leinonen, *Technology Survey on NDT of Carbon-fiber Composites*. Kemi-Tornio University of Applied Sciences, 2012
3. C. Mersmann, *Industrialisierende Machine-Vision-Integration im Faserverbundleichtbau, Ergebnisse aus der Produktionstechnik*, vol. 8/2012. Apprimus Verlag, Aachen, 2012
4. NetComposites. Interactive Knowledge Base on NDE of Composites. http://www.netcomposites.com. Accessed 10/2013
5. H. Gerhard, S. Predak, G. Busse, Zerstörungsfreie Prüfung zur Verfolgung des Schädigungsablaufs in faserverstärkten Polymeren. In: *Tagungsband zur DGM-Tagung: Verbundwerkstoffe und Werkstoffverbunde*. Wiley-VCH, 2001, pp. 23–29
6. S. Predak, S. Lütze, T. Zweschper, R. Stössel, G. Busse, Vergleichende zerstörungsfreie Charakterisierung. Materialprüfung **44**(1-2), 2002, pp. 14–18
7. W.o. Michaeli, *Einführung in die Technologie der Faserverbundwerkstoffe*. Hanser, München, 1990
8. T. Pfeifer, M. Benz, Qualitätssicherung für Verbundkunststoffe durch Ultraschallrückstreuung. Gummi, Fasern, Kunststoffe **53**(11), 2000, pp. 764–768
9. E. Hering, M. Stohrer, R. Martin, *Physik für Ingenieure*. Springer-Lehrbuch. Springer, Berlin, 2007
10. K. Hazra, S. Blake, K. Potter, M. Wisnom, In-process fiber orientation measurement and foreign object damage (FOD) prevention for complex preform or prepreg lay-ups. SAMPE Journal 2012 **48**(1), 2012. Januar/Februar
11. A. Dillenz, G. Busse, Ultraschall-Burst-Phasen-Thermografie. Materialprüfung **43**(1/2), 2001, pp. 30–34
12. T. Ullmann, T. Schmidt, Qualitätssicherung im Flugzeugbau: Nichts wird dem Zufall überlassen. DLR Magazin **126**, 2010, pp. 27–29
13. P. GmbH. FilaInspect mit mit Richtungsbeleuchtung, 2013. http://www.profactor.at/fileadmin/13_Downloads/MV/101109_PROFACTOR_mv_fila_inspect.pdf
14. W. Palfinger, S. Thumfart, C. Eitzinger, Photometric stereo on carbon fiber surfaces. In: *Presented at the 35th Workshop of the Austrian Association for Pattern Recognition, Graz, May 26–27 (2011)*. 2011
15. M.E. Liggins, D.L. Hall, J. LLINAS, *Handbook of multisensor data fusion. Theory and practice*. The electrical engineering and applied signal processing series. CRC Press, Boca Raton, FL, 2009
16. R. Schmitt, T. Pfeiffer, C. Mersmann, A. Orth, A method for the automated positioning and alignment of fibre-reinforced plastic structures based on machine vision. In: *Proceedings of the 58th CIRP General Assembly, Manchester, England, August 28th, 2008*

17. R. Schmitt, C. Mersmann, A. Schönberg, Machine-Vision zur Messung der Faserorientierung bei der textilbasierten FVK-Herstellung. Technisches Messen **77**(4), 2012, pp. 237–242

18. M. James, *Programming Real-time Computer Systems*. Prentice-Hall Inc., Englewood Cliffs, NJ, 1965. P. 4. ISBN 013-730507-9

19. E. Shang, J. Li, X. An, J. He, *Lane Detection using Steerable Filters and FPGA-based Implementation*. IEEE, 2011

20. B. Jaehne, H. Haussecker, P. Geissler, *Handbook of Computer Vision and Applications*. Academic Press, New York, 1999

21. I.S. Organisation, ISO/IEC Guide 98-3:2008: Guide to the expression of uncertainty in measurement. Tech. rep., ISO, Geneva, 2008

Diving In? How Users Experience Virtual Environments Using the Virtual Theatre

Katharina Schuster, Max Hoffmann, Ursula Bach, Anja Richert
and Sabina Jeschke

Abstract Simulations are used in various fields of education. One approach of improving learning with simulations is the development of natural user interfaces, e.g. driving or flight simulators. The Virtual Theatre enables unrestricted movement through a virtual environment by a Head Mounted Display and an omnidirectional floor. In the experimental study presented (n = 38), the effects of objective hardware characteristics were being tested in two groups. The task was the same: Remembering positions of objects after spotting them in a maze. One group fulfilled the task in the Virtual Theatre, the other group on a laptop. Personal characteristics (gaming experience, locus of control) and perception measures for immersion (spatial presence, flow) were also assessed. Analyses show that the Virtual Theatre indeed leads to more spatial presence and flow, but has a negative effect on the task performance. This contradicts the common assumption that immersion leads to better learning.

Keywords Immersion · Spatial Presence · Flow · Learning · Simulators · Natural User Interfaces

1 Introduction

Simulations are used in various fields of education. By imitating real-world processes, personnel skills can be developed, increased or maintained. Especially if the learning process requires expensive equipment or usually would take place in a hazardous environment, the use of simulations is not only beneficial but absolutely necessary [1, 2]. Apart from the software, the user interfaces of the technological systems applied in the simulation environment can affect the learning process [3]. One approach of improving learning with simulations is the development of natural user interfaces. A common example is the use of flight simulators including authentic

K. Schuster (✉) · M. Hoffmann · U. Bach · A. Richert · S. Jeschke
IMA/ZLW & IfU, RWTH Aachen University, Dennewartstr. 27,
52068 Aachen, Germany
e-mail: katharina.schuster@ima-zlw-ifu.rwth-aachen.de

Originally published in "Design, User Experience, and Usability. User Experience
Design for Diverse Interaction Platforms and Environments", © Springer 2014.
Reprint by Springer International Publishing Switzerland 2016,
DOI 10.1007/978-3-319-42620-4_60

user interfaces within pilot training instead of using just the simulation on a regular desktop computer. According to the classical memory theory, if the context in which we use our knowledge i.e. in which we have to transfer it to new situations resembles the context in which we learned the information in the first place, our memory works better. Moreover, how well we can retrieve knowledge from our long term memory depends on the quality of how well we encoded the information [4]. In the case of computer-aided learning, encoding information can be considered as a task, which is partitioned in at least two parallel sub-tasks: Dealing with the content and controlling the learning environment with the respected user interfaces [5]. Therefore a lot of research and development activities follow the assumption that if the user can interface with the system in a natural way, more focus can be used for training than for the control itself [6]. However, to assume that hardware or software characteristics automatically lead to better learning outcomes is risky. Not every new approach which is technically feasible improves learning in the sense of task performance. The danger of de-signing complex and expensive virtual learning environments without having a positive impact on learning outcomes is obvious. However, judging the value of a virtual environment simply by its effect on task performance misses out on other factors which support learning. Boosting the students' motivation to deal longer, more steady or more effectively with the given content is also an important goal of virtual learning environments [7, 8]. Apart from learning outcome and motivation, a peak to a different domain reveals a third intended effect of virtual environments. According to the entertainment sector, the extent to which a game or in general a virtual environment can "draw you in" functions as a quality seal [1]. This phenomenon is often referred to as immersion. A figurative definition is given by Murray [9]: "Immersion is a metaphorical term derived from the physical experience of being submerged in water. We seek the same feeling from a psychologically immersive experience that we do from a plunge in the ocean or swimming pool: the sensation of being surrounded by a completely other reality, as different as water is from air that takes over all of our attention our whole perceptual apparatus" [9].

Enabling natural movement as the most basic form of interaction is considered an important hardware quality to create immersion [10]. Manufacturers of hardware that are supposed to enhance immersion claim that "Moving naturally in virtual reality creates an unprecedented sense of immersion that cannot be experienced sitting down" [11]. Almost 20 years ago, this could already be confirmed by Slater et al. [10]. Another basic assumption in the context of virtual learning environments and natural user interfaces is that greater immersion means better learning and potentially higher training transfer [3, 6]. This suggests that immersion would be the precondition for better learning, caused by the qualities of the user interfaces. However, if virtual environments are used in educational contexts, those assumptions need to be confirmed by empirical evidence. The presented study therefore focuses on the following questions:

- Do natural user interfaces create a higher sense of immersion?
- Do natural user interfaces lead to better learning?
- Is immersion a necessary precondition for learning with natural user interfaces in virtual environments?

If assessed in an experimental setting, the construct of immersion needs to be specified. Spatial presence and flow are considered key constructs to explain immersive experiences. In general, flow describes the involvement in an activity [12, 13], whilst spatial presence refers to the spatial sense in a mediated environment [10, 14]. Spatial presence, as indicated in the name of the construct, refers to the spatial component of being immersed, i.e. the spatial relation of oneself to the surrounding environment. If we experience spatial presence in a mediated environment, we shift our primary reference frame from physical to virtual reality [14].

2   Method

2.1   Experimental Setting

The study presented in this paper assesses the relationship between personal characteristics, objective hardware characteristics, subjective experiences and task performance. Their expected relationship is visualized in Figure 1.

All participants had to solve the same task in the same virtual environment, which was a large-scaled maze in a factory building. Within the maze, 11 different objects were located. The first task for the participants was to navigate through the maze and to imprint the positions of the objects to their memory. For that, they were given eight minutes of time. The second task was to recognize the objects seen before in the maze. The third task was to locate the positions of the objects on a map of the maze. This was done on a self-programmed application on a tablet (Nexus 10) with a drag-and-drop control mode. The view of the maze in the first and second task is pictured in Figure 2.

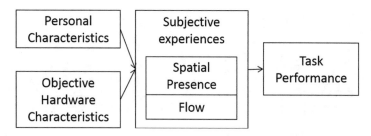

Figure 1 Expected relationship between personal characteristics, hardware characteristics, subjective experiences and task performance

Figure 2 View of the virtual environment used in the study in the first and in the third task

For both groups, the participants were given the chance to explore a test scenario (an Italian piazza) freely for about three minutes before the actual task started. This was in order to get used to the respected control mode. All experimenters who conducted the experiments were trained in advance by experienced researchers. First they were being trained the functions of the hardware. In a second step, they took the observing position in a test run, and thirdly they conducted a test run on their own with the experienced researcher being the observer and giving feedback afterwards. Two groups of test persons were compared, having to use hardware which differed from each other regarding the following characteristics:

- Control mode of the field of view,
- Control mode of locomotion,
- Display and
- Body posture of the user.

Due to the composition of the simulator which was applied in the study, the hardware characteristics could only be tested in a certain combination and could not be isolated any further. The whole experiment took one hour. The complete procedure is visualized in Figure 3.

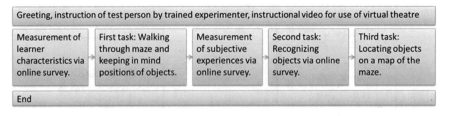

Figure 3 Procedure of the experiments

2.2 Measurements

In this study, spatial presence was measured with elements of the MEC Spatial Presence Questionnaire of Vorderer et al. Several studies conducted by the authors strengthened the postulate of spatial presence being best explained as a two-level Model. This includes process factors (attention allocation, spatial situation model, self location, possible actions), variables referring to states and actions (higher cognitive involvement, suspension of disbelief), and variables addressing enduring personality factors (i.e. the trait-like constructs domain specific interest, visual spatial imagery, and absorption) [15]. Suspension of disbelief refers to the extent of how much a person pays attention to technical and content-related inconsistencies. The more a person can fade out the action of "looking for errors", the higher the feeling of spatial presence will be according to the theory. In our study, instead of the subscales attention allocation and absorption, we used the Flow-Shortscale of Rheinberg. Flow is the mental state of operation in which a person performing an activity is fully immersed in a feeling of energized focus, full involvement, and enjoyment in the process of the activity. In essence, flow is characterized by complete absorption in what one does, as well as the feeling of smooth and automatic running of all task-relevant thoughts [12, 13].

The perception of a learning situation is highly likely not to be influenced just by objective criteria such as the technical configuration of the learning environment. The strength of spatial presence experienced in a VE is supposed to vary both as a function of individual differences and the characteristics of the VE [3]. A general interest in the topic appeals to a person's curiosity and the motivation to learn something new. If chances to learn or experience something new are low, the motivation to learn decreases [7, 16]. However, we believe that not only interest in a topic but also in the way of presenting it can influence subjective experiences during the learning situation as well as learning outcome. The subscale domain specific interest of the MEC-SPQ refers to the topic of the medium, in our case the virtual environment. Because of the given considerations mentioned above and since interest in mazes didn't seem like a helpful operationalization for domain specific interest, we adapted it to interest in digital games. Additionally, participants were asked to state their gaming frequency, i.e. how many hours they played digital games per week. How well a person learns with the assistance of technology depends not least on whether a person has any experience with the respective technology and if not, feels capable of learning it quickly. The construct "locus of control regarding the use of technology" describes the extent to which a person believes that he or she can control technical devices in everyday life. It is a technology-related personality trait of human-computer interaction and was measured with the short scale from Beier [17]. Based on all theoretical considerations, the general hypothesis of the study was that natural user interfaces should have a positive effect on subjective experiences during the learning situation as well as on learning outcome, in our case operationalized in task performance.

The set of hardware characteristics functioned as the first independent variable in the presented study. Furthermore, the test persons' locus of control regarding the use of technology (second independent variable), interest in digital games (third independent variable) and gaming frequency (fourth independent variable) were measured before the first task. As dependent variables, spatial presence and flow were measured after the first task which had to be fulfilled either in the Virtual Theatre or at the laptop. As dependent measures of task performance, three different variables were analyzed: The number of objects that were correctly recognized in the second task, the third task reaction time and the accuracy of locating the objects on the map in the third task.

2.3 Hardware

Laptop. In the presented study, learning in a Virtual Theatre was compared to a somehow conventional learning with a laptop. The type being used was a Fujitsu Lifebook S761 with a 13, 3 inch display and a 1366×768 display resolution. The field of view was controlled with a mouse. Locomotion was controlled by WASD-keys, where W/S keys controlled forward and backward while A/D keys controlled left and right. The hardware usually results in a sitting body posture while using the device.

Virtual Theatre. The Virtual Theatre is an innovative platform which enables unrestricted movement through a virtual environment and therefore is used in an upright body posture. The control mode of locomotion is walking naturally. The components of the Virtual Theatre which came to use in the study are pictured in Figure 4 and moreover described in the following sections. For a more detailed and complete description of the technical system see Ewert et al. [1] and Johansson [6, 18].

Figure 4 Head Mounted Display and omnidirectional floor of the Virtual Theatre

Head Mounted Display. In the Virtual Theatre, the field of view is controlled by natural head movement. It presents the user with a seamless 3d visualization of a virtual environment. All head movement is instantly reproduced within the simulation, so the user can look around freely. Visual feedback is received via a zSight [19] Head Mounted Display (HMD), providing a 70° stereoscopic field of view with SXGA resolution for each eye. The HMD weighs 450 grams. It is powered by rechargeable batteries which are located in a vest which the user wears while using the Virtual Theatre. The HMD can be adjusted to the shape of the user's head as well as to his or her eye distance. On a rough scale, the lenses inside the display can be adjusted to short-sightedness and farsightedness.

Omnidirectional floor. The user can move around within the environment by just walking in the desired direction. The omnidirectional floor consists of polygons of rigid rollers with increasing circumferences. Each polygon constitutes 16 rollers and together all polygons form 16 sections with an angle of 22.5°. Each section consists of one roller from each polygon, which means from cylinder origo towards the periphery, the rollers are parallel and increasing in length. Rollers are driven from below by a belt drive assembly to minimize distance between sections. In the central part of the floor there is a circular static area where the user can move without enabling the floor. Floor movements here would only cause the feet of the user to be drawn together. As the user moves outside the static boundary, the floor begins to move according to a control algorithm [18].

Tracking System. To track the movements of a user, the virtual theatre is equipped with 10 infrared cameras. They record the position of designated infrared markers attached to the HMD and an additional hand tracer. The position of the HMD serves on the one hand as an input for controlling the omnidirectional floor: The inward speed of the rollers increases linearly with the measured distance of the HMD to the floor center. On the other hand it is used to direct the position and line of vision of the user within the virtual environment [6, 18]. The position of the hand tracer serves for triggering emergency shutdown: As soon as the hand drops below 0.5m, e.g. in the event of a user falling down, all movement of the omnidirectional floor is immediately stopped. It should be noted at this point that throughout all studies conducted so far with the Virtual Theatre, this never happened.

2.4 Participants

A total of 38 students between 20 and 33 years (M = 24.71; SD = 3.06; n = 13 female) volunteered to take part in the study. The sample therefore represents a potential user group of virtual environments in higher education. They responded to a call for participation which was hung out at bulletin boards throughout the university but also posted on the front page of the virtual learning platform of the university and on several research and learning related blogs, social mediaplatforms and news feeds.

As an incentive and as a sign of appreciation, all participants took part in a drawing for a cordless screwdriver. All participants were healthy and highly interested in participating in the study. They did not report suffering from any physical or mental disorders. To rule out effects due to ametropia, participants were asked in advance to bring their corrective lenses just in case. If participants had been assigned to the Virtual Theatre group, they were asked to wear sturdy shoes.

3 Results

3.1 Correctional Approach

To gain further information regarding the relationships of the characteristics of the person, the subjective experiences, and the task performance an explorative approach was taken. Thus we calculated correlations between relevant constructs. Results are displayed in Figure 5.

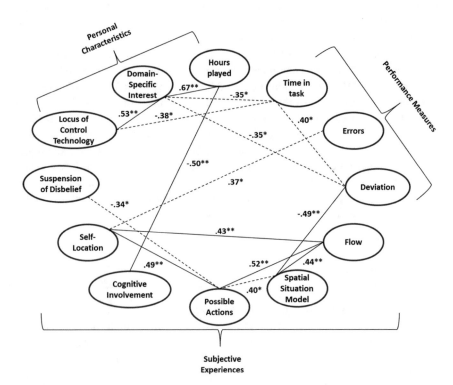

Figure 5 Correlations between personal characteristics, subjective experiences and task performance

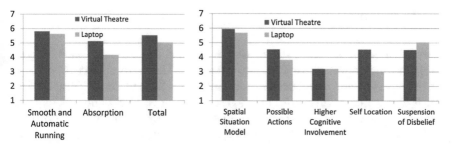

Figure 6 Effects of objective hardware characteristics on flow and spatial presence

3.2 Comparing Hardware Conditions

Furthermore hypotheses regarding influences of hardware conditions on subjective experiences and task performance measures were tested with ANOVAs. With regard to the effects of the Virtual Theatre and the laptop on flow, significant differences were found ($F(1, 36) = 4.18$; $p < 0.05$). Thus more flow has been experienced in the Virtual Theatre (see Figure 6). Taking a closer look on subscales there is a highly significant difference between conditions in self-reported absorption ($F(1, 36) = 10.63$; $p < 0.01$), but not in smooth and automatic running. There are also effects of hardware conditions on spatial presence. Self location in the Virtual Theatre was rated significantly higher ($F(1, 36) = 15.79$; $p < 0.001$), which refers to the feeling of actually being in the virtual environment. Similarly, participants in the Virtual Theatre show higher scores on the possible actions subscale of spatial presence ($F(1, 36) = 4.90$; $p < 0.05$). There were no further significant effects regarding spatial presence (see Figure 6).

In addition to that we calculated the effects of hardware on task performance measures. In the recognition task of the objects from the virtual environment, participants in the Virtual Theatre condition made significantly more errors ($F(1, 36) = 10.93$; $p < 0.01$), which opposes our hypothesis (see Figure 7). There were no differences regarding time on task and deviation between the two treatment conditions.

Figure 7 Effects of objective hardware characteristics on task performance, here: Recognition of previously presented objects

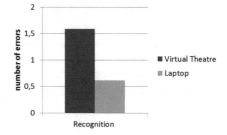

4 Discussion

Two different types of hardware as well as personal characteristics were analyzed regarding their effect on subjective experiences and task performance. Concerning the effects of natural user interfaces, the results show that the Virtual Theatre indeed leads to more flow. This is due to a higher self-reported level of absorption in the Virtual Theatre group. Although flow is an activity related construct, this result is in line with the theoretical assumptions that hardware which allows natural walking can support the feeling of "diving" into the virtual environment, which in general terms is often referred to as immersion.

Next, the effects of the hardware on spatial presence are analyzed in more detail. Students who used the Virtual Theatre reported a higher self-location in the virtual environment which indicates that they had shifted their primary reference frame from the physical to the virtual world. Although the given task didn't require any further nonmental actions but navigating through the virtual environment, students in the Virtual Theatre reported higher on the subscale of possible actions. However, for the other subscales, no differences were measured. Their role will be discussed when looking at the results of the explorative correlations.

According to the results of this study, immersion is not the precondition for better learning in virtual environments with natural user interfaces. Thus, the underlying model of the study (see Figure 1) needs to be adjusted for further research. The only effect of the Virtual Theatre on task performance was a negative influence on recognition. This result is contradictory to the assumption that immersion leads to better learning. It seems that controlling the hardware was less intuitive than expected. This probably lead to the typical situation for learning with virtual environments: Dividing the available cognitive resources on the two parallel sub-tasks of dealing with the content and controlling the learning environment with the respected user interfaces [5]. Moreover, the combination of an HMD and real physical locomotion could lead to cognitive dissonance. When wearing the HMD, the user can see, where he or she walks in virtuality, but not in physical reality. Therefore the user takes a risk and has to trust in the technology in order to continue his or her actions. Last but not least, walking on the omnidirectional floor is a new experience for users and therefore could result in the fear of falling. All interpretations for the given results are going to be addressed in a follow-up study, where previous participants of the study will be interviewed on their experiences.

When we look at the correlations of the different constructs which were under study we see a slightly different picture: Although not all results are discussed in detail, we want to put the most relevant ones into focus. The positive correlation of self location and errors is in line with the previously discussed results, but therefore also contradictory to the assumption that immersion leads to better learning. However, the negative correlation of the self-reported spatial situation model and deviation supports this assumption. Domain specific interest i.e. interest in digital games correlated negatively with deviation and with time, just like locus of control regarding the use of technology and time. Therefore, in this experimental setting, the experience with virtual environments and the belief of being able to control technology had a bigger influence on those two performance measures than flow and spatial presence, the two indicators for immersive experience. The fact that flow correlated highly significant with the self-reported spatial situation model, possible actions and self location indicates that the constructs are two related facets of immersive experiences. The only correlation between personal characteristics and subjective experiences was between gaming frequency and higher cognitive involvement. In other words, experienced gamers were not challenged enough in this experimental setting.

5 Limitations of the Present Study and Future Research

Finally, some limitations of the present study are considered that should be pursued for future research. One limitation refers to the type of hardware examined. Since the different technical characteristics of the Virtual Theatre can only be tested in a set, it is not possible to isolate single effects. Moreover, the relationship between spatial presence and learning is not clear yet, since one subscale (self location) correlates with worse performance (making more errors) whilst another one (spatial situation model) correlates with better performance (less deviation when locating objects on a map). The other aspect concerns the task chosen for this experiment. Low levels of cognitive involvement in both groups indicate that the whole sample might not have been challenged enough. Since challenge is an important precondition for the motivation for learning, more challenging tasks are going to be tested in the future.

This first exploratory study on the effects of the Virtual Theatre on subjective experiences and learning confirmed a few theoretical assumptions but also contradicted others. In a next step, interviews with participants from both groups are going to be conducted. A deeper insight on the participants' experiences will allow a more differentiated view on the subject of our research. With a constant validation of the relationship between personal characteristics and hardware, it is possible to predict the level of subjective experiences to a certain extent. Thus, a profile could be developed, for whom the Virtual Theatre would have the most additional benefit. Moreover, different scenarios could be developed for different user types.

Acknowledgments The present work is supported by the Federal Ministry of Education and Research within the project "Excellent teaching and learning in engineering sciences (ELLI)". The authors would like to thank Prof. Martina Ziefle for constant advice on the presented work. Special thanks to Anna Claus and Johanna Purschke for recruiting participants and carrying out the experiments as well as to Matthias Heinrichs for technical support.

References

1. D. Ewert, K. Schuster, D. Johansson, D. Schilberg, S. Jeschke, Intensifying learner's experience by incorporating the virtual theatre into engineering education. In: *Proceedings of the 2013 IEEE Global Engineering Education Conference.* Berlin, 2013, pp. 207–212
2. S. Malkawi, O. Al-Araidah, Students' assessment of interactive distance experimentation in nuclear reactor physics laboratory education. European Journal of Engineering Education **38**(5), 2013, pp. 512–518
3. B. Witmer, M. Singer, Measuring Presence in Virtual Environments: A Presence Questionnaire. Presence: Teleoperators and Virtual Environments **7**(3), 1998, pp. 225–240
4. P. Zimbardo, R. Gerrig, *Psychologie*, 7th edn. Springer, Heidelberg, 2003
5. D. Baacke, "Medienkompetenz": Theoretisch erschließend und praktisch folgenreich. Merz – Medien und Erziehung **43**(1), 1999, pp. 7–12
6. D. Johansson, *Convergence in Mixed Reality-Virtuality Environments. Facilitating Natural User Behaviour.* No. 53 In: Örebro Studies in Technology. Örebro University, Örebro, Sweden, 2012
7. A. Hebbel-Seeger, Motiv: Motivation?!– Warum Lernen in virtuellen Welten (trotzdem) funktionieren kann. Zeitschrift für E-Learning – Lernkultur und Bildungstechnologie **7**(1), 2012, pp. 23–35
8. F. Müller, Interesse und Lernen. Report – Zeitschrift für Weiterbildungsforschung **29**(1), 2006, pp. 48–62
9. J. Murray, *Hamlet on the Holodeck: The Future of Narrative in Cyberspace.* MIT Press, Cambridge, 1998
10. M. Slater, M. Usoh, A. Steed, Taking Steps: The Influence of a Walking Technique on Presence in Virtual Reality. ACM Transactions on Computer-Human Interaction **2**(3), 1995, pp. 201–219
11. Virtuix Technologies, 2014. http://www.virtuix.com/
12. M. Csikszentmihalyi, J. Lefevre, Optimal experience in work and leisure. Journal of Personality and Social Psychology **56**(5), 1989, pp. 815–822
13. F. Rheinberg, S. Engeser, R. Vollmeyer, Measuring Components of Flow: the Flow-Shot-Scale. In: *Proceedings of the 1st International Positive Psychology Summit.* Washington DC, USA, 2002
14. M. Slater, Measuring Presence: A Response to the Witmer and Singer Presence Questionnaire. Presence: Teleoperators and Virtual Environments **8**(5), 1999, pp. 560–565
15. P. Vorderer, W. Wirth, F. Gouveia, F. Biocca, T. Saari, F. Jäncke, S. Böcking, H. Schramm, A. Gysbers, T. Hartmann, C. Klimmt, J. Laarni, N. Ravaja, A. Sacau, T. Baumgartner, P. Jäncke. Mec spatial presence questionnaire (mec-spq): Short documentation and instructions for application. report to the european community, project presence: Mec (ist-2001-37661), 2004. http://www.ijk.hmt-hannover.de/presence
16. W. Edelmann, *Lernpsychologie*, 5th edn. Beltz PVU, Weinheim, 1996

17. G. Beier, Kontrollüberzeugung im Umgang mit Technik. Report Psychologie **9**, 1999, pp. 684–693
18. D. Johansson, L. de Vin, Towards Convergence in a Virtual Environment: Omnidirectional Movement, Physical Feedback, Social Interaction and Vision. Mechatronic Systems Journal **2**(1), 2012, pp. 11–22
19. Sensics. http://sensics.com/products/head-mounted-displays/zsight-integrated-sxga-hmd/specifications/

Using Off-the-Shelf Medical Devices for Biomedical Signal Monitoring in a Telemedicine System for Emergency Medical Services

Sebastian Thelen, Michael Czaplik, Philipp Meisen, Daniel Schilberg and Sabina Jeschke

Abstract In order to study new methods of telemedicine usage in the context of emergency medical services, researchers need to prototype integrated telemedicine systems. To conduct a one year trial phase – intended to study a new application of telemedicine in German emergency medical services – , we used off-the-shelf medical devices and software to realize real-time patient monitoring within an integrated telemedicine system prototype. We demonstrate its feasibility by presenting the integrated real-time patient monitoring solution, by studying signal delay and transmission robustness regarding changing communication channel characteristics, and by evaluating issues reported by the physicians during the trial phase. Where standards like HL7 and the IEEE 11073 family are intended to enable interoperability of product grade medical devices, we show that research prototypes benefit from the use of web technologies and simple device interfaces, as they simplify product development for a manufacturer and ease integration efforts for research teams. Embracing this approach for the development of new medical devices eases the constraint to use off-the-shelf products for research trials investigating innovative use of telemedicine.

Keywords Biomedical Telemetry · Emergency Medical Services (EMS) · Telemedicine

1 Introduction

The continued progress in mobile communication technology development is a driving force behind applications and research for telemedicine usage in Emergency Medical Services (EMS). Many research projects have reported success in building

S. Thelen (✉) · P. Meisen · D. Schilberg · S. Jeschke
IMA/ZLW & IfU, RWTH Aachen University, Dennewartstr. 27,
52068 Aachen, Germany
e-mail: sebastian.thelen@ima-zlw-ifu.rwth-aachen.de

M. Czaplik
MedIT, RWTH Aachen University, Pauwelsstr. 20, 52074 Aachen, Germany

Originally published in "IEEE Journal of Biomedical and Health Informatics ",
© IEEE 2014. Reprint by Springer International Publishing Switzerland 2016,
DOI 10.1007/978-3-319-42620-4_61

telemedicine systems using different technologies to enable on-scene EMS staff to consult with specialists in a remote location. Some examples are [1–4]. In our pilot studies, the successful implementation of telemedicine systems for EMS, offering synchronous – often referred to as real-time – telemedical consultation functionality, has been demonstrated [5, 6].

The telemedicine model applied in these two pilot studies has its origin in the German, physician based, EMS system: two paramedics (an ambulance's crew in most German EMS) establish, either from a scene or from inside the ambulance, a telemedicine session with a specially trained EMS physician (hereafter called tele-EMS physician), who is located in a remote location (hereafter called teleconsultation center). During the telemedicine session, bidirectional real-time voice communication is the main information channel. Additionally, the tele-EMS physician receives information from a multitude of sources from the scene of incident and from within the ambulance: amongst others, the biomedical signals acquired by the patient monitoring device. In the teleconsultation center, two tele-EMS physicians are on call for involved ambulances; a tele-EMS physician supervises only a single ambulance team at once.

To collect evidence on the medical and organizational impact resulting from the use of this telemedicine model, the declared goal of our pilot studies was to conduct them in regular EMS operations. Consequently, the telemedicine system must not hinder patient treatment, making usability a major concern; besides, application of telemedicine must not prolong on-scene time intervals significantly, which was already analyzed in former studies [7]. In addition to usability, organizational aspects and statutory regulations of medical devices are of major importance to the technical system realization [8–10]. Philips Healthcare (Boeblingen, Germany) joined the research project and supported it by providing off-the-shelf medical devices for patient monitoring (Philips MRx monitoring/defibrillator device) and accompanying software. From a technical perspective, the challenge to incorporate these devices into an integrated telemedicine system – together with additional communication technology and custom software – remained.

We have already reported general results from a one month test phase of this integrated telemedicine system in [11], where we discuss overall requirements and the performance we achieve with the overall system architecture. Still, we have not yet reported on the solution to the aforementioned integration challenge, which is a major enabler of our study's trail phase and as such may provide valuable insight for future telemedicine projects in EMS.

During our work we realized that the most challenging system integration aspect is the biomedical real-time signal transmission for continuous patient monitoring and diagnostics. Current literature leaves central questions unanswered regarding this topic: How can this integration be achieved, what is the impact of existing interoperability standards, and what performance regarding usability, robustness, and real-time aspects are achievable? To address these open issues, the paper is organized as follows. In Section 2, we summarize related work. In Section 3, we present the requirements that the integrated biomedical signal monitoring has to fulfill. In Section 4, we present our system design that integrates the biomedical signal monitoring. In

Section 5, we present the performance evaluation. In Section 6, we discuss the results of our performance evaluation and the impact of relevant interoperability standards. In section seven, we draw the conclusion of our work presented in this paper.

2 Related Work

Research in mobile telemedicine systems has been broadly surveyed by others; for an overview we refer to [12] and instead focus on the works most relevant for this article, i.e., those that contribute important details to biomedical signal transmission for patient monitoring in EMS telemedicine systems. Lin [12] offers a broad technology-focused survey of telemedicine research; he discusses application areas, communication technology, and the kind of data that is transmitted. From the projects he describes, two are of concrete relevance for our work: Anogianakis et al. [13] describe the early maritime telemedicine system MERMAID which enables tele-consultation centers to provide medical assistance for patient treatment on board of ships using satellite communication. Kyriacou et al. [14] report on a real-time telemedicine system for use in EMS, the general concept being close to our use of telemedicine; however, the integration of biomedical signal transmission differs. They access monitoring data directly at the patient monitoring devices using a RS-232 interface and custom transmission algorithms and viewing applications. Thereby they effectively implement software that has to be classified as active medical device, if intended for patient diagnostics [10].

Alesanco and Garcia [15] analyze the transmission of clinical real-time electrocardiogram (ECG) via 3G mobile networks using simulations. They configure their proposed transmission algorithm – real-time/reliable ECG transmission protocol – in such a way that it maintains a signal delay below 3–4 s. According to their clinical trials, 4 s is the maximum acceptable delay. Trigo et al. [16] present a prototype implementation of real-time ECG transmission using the IEEE 11073 standards family, while harmonization for the ECG device specialization was still ongoing. This prototype is not clinically evaluated, but demonstrates general feasibility. The IEEE 11073 ECG device specialization standard was published in 2012 [17].

LifeBot DREAMS is a commercially advertised, mobile EMS telemedicine system in the US [18, 19]; technical details and findings for use by other telemedicine studies are not reported. Similarly, Viewcare is a Danish telemedicine initiative that utilizes telemedicine in ambulance vehicles [20]; detailed information is scarce as well.

3 System Requirements

In a multidisciplinary team of emergency physicians – experienced in telemedicine usage from a former study – , engineers, and organizational experts, we conducted a thorough requirements analysis before starting to design the system. To document the requirements in a structured and comprehensive way, we used use-case descriptions. An iterative design process ensured that the requirements, user expectations and system development remained aligned. In [21], we have given a general overview of the system requirements, but because of space limitations, we neglected specific details necessary for the discussion in this paper; the functional requirements that the biomedical signal monitoring integration has to fulfill are:

1. Minimal need for additional system interaction on-scene, after a teleconsultation has been successfully initiated, e.g., data transmission either happens automatically or is controlled by the tele-EMS physician, not by the on-scene paramedics.
2. To protect patient and staff privacy, no data is transmitted without an active teleconsultation session.
3. To minimize potential for confusion during a teleconsultation session, only data from the active case may be displayed to the tele-EMS physician.
4. The tele-EMS physician must be able to switch the currently active, i.e., displayed, case by selecting the target ambulance with a single user interface interaction, e.g., a single button press and potentially a confirmation.
5. Continuous transmission of at least two ECG leads and pulse oximetry (SpO_2) waveforms as well as numeric values of heart-rate, SpO_2, pulse, and blood-pressure with minimal achievable delay, ideally not above three seconds.
6. Transmission of twelve-lead ECG recordings.
7. Periodic transmission of numeric trend values for heart-rate, SpO2, and pulse with a one minute interval, in case the network conditions on-scene are too bad to support requirement 5.
8. All data acquired during a teleconsultation session, such as the biomedical signals from requirements 5–7, must be available in the application that the tele-EMS physicians uses to create the documentation of the teleconsultation session.

4 System Design of Integrated Biomedical Signal Monitoring

In [11, 21], we already introduced the general system architecture, but did not discuss the design of the biomedical signal transmission for patient monitoring, its real-time performance, nor our findings in regard to its applicability for future telemedicine projects. The – commonly used – combined patient monitor and defibrillator Heart-Start MRx (model M3535A, Philips Healthcare, Andover, MA, US) is the device we chose for integration into the system; the biomedical signals that this device acquires and that are relevant to our work are listed in Table 1, together with the communication interfaces used to access them. In addition to the HeartStart MRx, we tested our integration approach with the multi parameter monitor IntelliVue X2 (Philips Healthcare, Boeblingen, Germany). The HeartStart MRx can be equipped with different configuration options that provide different interfaces and communi-

Table 1 Relevant biomedical signal transmission capabilities offered by HeartStart MRx

Signal	Transmission periodicity	Communication interface
Waveforms: 2-lead ECG, pulse oximetry (SpO$_2$)	Continuous, real-time	Internet Protocol (IP) via Ethernet
Numeric values: heart-rate, SpO$_2$, Pulse, blood-pressure	On-charge	IP via Ethernet
Numeric values: heart-rate, SpO$_2$, Pulse, blood-pressure	Periodically; once per minute	Bluetooth File Transfer Profile
Twelve-lead ECG (3.3 s recording)	On-capturing/on-demand	Bluetooth File Transfer Profile

cation capabilities. Relevant for our system integration are the options *IntelliVue* Net (continuous real-time waveform transmission in a Local Area Network (LAN)), *12-LTx Bluetooth* (transmission of twelve-lead ECG recordings via Bluetooth), and *Per Data Tx* (periodic transmission of biomedical signal's numeric values via Bluetooth).

The *IntelliVue* Net option, which provides access to a patient's biomedical real-time vital signals, usually utilizing a direct LAN connection to the proprietary, combined software/hardware appliance IntelliVue Information Center (hereafter called IIC; Philips Healthcare, Andover, MA, US) for signal transmission [22], whereas the other two options wirelessly push data via Bluetooth to a connected device. The latter data can be automatically imported into and displayed by the proprietary software HeartStart Telemedicine System (hereafter called HTS; Philips Healthcare, Andover, MA, US); alternatively, the HeartStart DataSDK (Philips Healthcare, Andover, MA, US) – a software development kit – may be used to implement import adapters that enable access to this data from custom applications [23].

Both the IIC and the HTS are certified medicine products and their intended use covers patient diagnostics [22, 23]. In the integrated telemedicine system, the use of periodic data transmission is – according to requirement 7 – solely considered as a fallback option. Whilst no single integration option fulfills the requirements mentioned above, in combination they cover the requirements 5–7: *IntelliVue Net* option for req. 5, *12-LTx Bluetooth* option for req. 6, and *Per Data-Tx* for req. 7. Because of their different underlying interface technologies and differing proprietary display applications, we had to implement two distinct integration strategies as shown in Figure 1. For the continuous real-time waveforms and numeric values of requirement 5 we used networking technologies and for the twelve-lead ECG and periodic data transmission of requirement 6 and 7 we used Bluetooth and file transfer services. In order to fulfill the remaining requirements 1–5, we had to integrate both IIC and HTS into the tele-EMS physician's workplace and automate the process of switching the currently displayed monitoring devices as well as the activation and deactivation of data transmission from the scene. Figure 2 shows the various components and their connections that implement the integration strategies. In the remainder of this section, we provide further detail to the system integration.

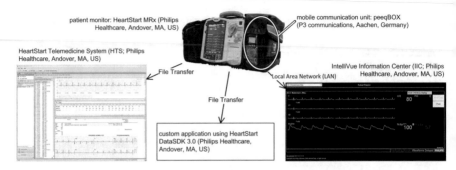

patient monitor: HeartStart MRx (Philips Healthcare, Andover, MA, US)

mobile communication unit: peeqBOX (P3 communications, Aachen, Germany)

HeartStart Telemedicine System (HTS; Philips Healthcare, Andover, MA, US)

File Transfer

Local Area Network (LAN)

IntelliVue Information Center (IIC; Philips Healthcare, Andover, MA, US)

File Transfer

custom application using HeartStart DataSDK 3.0 (Philips Healthcare, Andover, MA, US)

Figure 1 Device and application overview for the integrated biomedical signal monitoring system

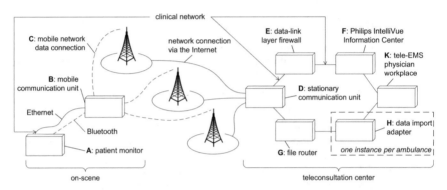

clinical network

C: mobile network data connection

network connection via the Internet

E: data-link layer firewall

F: Philips IntelliVue Information Center

K: tele-EMS physician workplace

B: mobile communication unit

D: stationary communication unit

Ethernet

Bluetooth

H: data import adapter

A: patient monitor

G: file router

one instance per ambulance

on-scene

teleconsultation center

Figure 2 Components and communication links in the integrated biomedical signal monitoring system

4.1 Mobile Communication Unit

The Mobile Communication Unit (B in Figure 2) securely connects all on-scene devices to the teleconsultation center; it simultaneously uses three dedicated mobile network data connections (C) to create a single, encrypted Internet Protocol (IP) connection to the Stationary Communication Unit (D) via the Internet. This connection forms a secured Virtual Private Network (VPN), containing the on-scene and the teleconsultation center IP networks.

The Mobile Communication Unit's three mobile network modems support both 2G (GPRS and EDGE) and 3G (UMTS, HSDPA and HSUPA) data connections, individually choosing the fastest network technology that is available [24]. In order to ensure on-scene usability, the Mobile Communication Unit is housed inside the patient monitor's bag, as shown in Figure 1, that was modified to compensate for the space required by the Mobile Communication Unit.

4.2 Continuous, Near Real-Time Signal Transmission

The transmission of continuous, near real-time waveform signals is realized by connecting the patient monitor (A) via Ethernet to the Mobile Communication Unit. The Mobile Communication Unit connects the patient monitor to the clinical network of the IIC (F) using an additional VPN from Mobile Communication Unit to Stationary Communication Unit to create an Ethernet bridge on the data-link layer. The data-link layer firewall (E) generally blocks communication between connected patient monitors and the IIC. In this default setting, the patient monitor is unable to establish a connection to the IIC (requirement 2). To enable the connection and transmission of the biomedical signals, the tele-EMS physician selectively activates each patient monitor's connection (requirement 1). This setup prevents unintended data transmission and resulting privacy concerns, because the monitoring device would otherwise begin to transmit signals to the IIC as soon as it is turned on, even without an active teleconsultation session.

On the tele-EMS physician workspace (K) the data transmitted from the on-scene patient monitor is accessible through the IIC's intranet patient viewer web-application. This web-application's monitoring widget – displaying a patient's biomedical signal waveforms and numeric values – is embedded into a wrapper web-application. The wrapper web-application selects the patient to be viewed when loading the embedded monitoring widget without additional interaction by the tele-EMS physician (requirements 3 and 4). Direct access to the numerical values and trend data for the waveform signals is enabled via the IIC's HL7 unsolicited messages interface that publishes data in a configurable interval. This data is then forwarded by a custom import adapter into the case documentation (requirement 8).

4.3 Periodic Signal Transmission

For transmission of twelve-lead ECG recordings and periodic trend values of a patient's biomedical signals, the patient monitor (A) connects to the Mobile Communication Unit (B) via Bluetooth using the Bluetooth File Transfer Profile. For each data transmission, i.e., a single twelve-lead ECG record or a set of biomedical signal trend values, the patient monitor transmits three files to the Mobile Communication Unit via this Bluetooth connection.

The Mobile Communication Unit forwards these received files using File Transfer Protocol (FTP) to the Stationary Communication Unit (D), which in turn forwards these files again via FTP to the file router (G). Here, these files are delivered, again via FTP, to two distinct data directories on the data import adapter instance (H) that is assigned to the originating ambulance. Both data directories of a data import adapter instance are watched by a separate data import service. One data import service is the HTS, which imports all incoming data into its case database, the other is a custom data import application implemented using the DataSDK, which forwards the data to the case documentation (requirement 8). The way that the HTS displays transmissions – grouped to cases from all connected patient monitors in a single list – contradicts requirement 3. Our solution is to use a separate HTS instance for each ambulance in the system, and thus one data import adapter (H) per ambulance, in order to fulfill requirement 3.

On the tele-EMS physician workspace (K) the HeartStart Telemedicine Viewer (hereafter called HTS Viewer; Philips Healthcare, Andover, MA, USA) – a propri- etary, dedicated application that connects to one HTS at a time and allows remote access to the HTS with an identical user interface – is connected to the HTS instance associated to the current ambulance. By utilizing a dedicated HTS instance per ambu- lance, the list of transmissions that the HTS Viewer presents to the tele-EMS physi- cian contains only one active case. We were unable to automate the selection of this active case, thus the tele-EMS physician must manually select the sole active case, partially violating requirement 3. In order to display the transmissions from another ambulance, the HTS Viewer must be connected to the HTS associated with the desired ambulance, a task we were able to automate using the user interface automation tool AutoIt (AutoIt Consulting, GB), fulfilling requirement 4. To our knowledge, there is no programming interface available to control the HTS Viewer.

5   Evaluation Methods and Results

In this paper, we focus on two major aspects of our evaluation of the presented bio- medical signal monitoring system: the system's real-time performance in Section 5.1 and the usability and robustness we achieve with our system integration in Section 5.2. Outside of the scope of this paper is to report more broadly about the one year trail phase that we used to evaluate the complete telemedicine system and its organiza- tional and medical procedures, which ended in July 2013: six ambulance teams in five different EMS departments were using the system during their regular EMS opera- tions; two tele-EMS physicians staffed the teleconsultation center during this time. A one month field test before starting the trail phase ensured that the system technically performed as necessary. The physicians carrying out the field test evaluation (age 33–44, assigned as EMS physician for 3–14 years) rated the system performance as satisfactory, thus the trial phase could start, as we reported in [11].

Table 2 Threshold values of communication channel characteristics for stable ECG and SpO$_2$, waveform transmissions

Channel parameter	Threshold value
Delay	3.15 s
Jitter (with 1 s base delay)	0.5 s
Data loss	27 %
Data corruption	24 %

5.1 Continuous, Near Real-Time Signal Monitoring

Responsible for the near real-time biomedical signal monitoring is the IIC and its connection to the patient monitor. To establish baseline parameters about this connection's robustness against failures in the underlying communication channel, i.e., delay, jitter, data loss, and data corruption, we connected the patient monitor via Ethernet to the IIC's clinical network and injected a network emulator, [25], into this Ethernet connection. The three parameters were then individually increased, starting at zero, with the other parameters held at zero; the threshold value, above that at least one of the ECG or the SpO$_2$ waveforms started showing the slightest form of interruptions, i.e., missing data points, is recorded (Table 2).

The end-to-end, i.e., patient monitor to tele-EMS physician display, signal delay for transmission of continuous, near real-time ECG and SpO$_2$ waveforms was measured by a test engineer, who directly observed the waveforms displayed on the patient monitor as well as on the tele-EMS physician's workplace display; while transmitting regular signals, either from a patient-simulation device (ECG) or a test subject's finger (SpO$_2$), a clearly perceivable change in the signal was triggered; the time difference was measured from onset of this change on the patient monitor's display to its occurrence on the tele-EMS physician's workplace display. For each signal type (ECG and SpO$_2$), the delay was independently measured ten times when the Mobile Communication Unit simultaneously used three 3G data connections (signal strengths at 77–87 %/61–74 %/35–45 %) and ten times, when it simultaneously used three 2G data connections (signal strengths at 93–96 %/70–77 %/58 %).

The grouped histogram of all 40 samples is shown in Figure 3. The signal in the teleconsultation center is delayed with a median of 4.9 s (range 3.4–7.1 s); no correlation with the signal type (ECG and SpO$_2$) or the data connection's technology

Figure 3 Grouped histogram of ECG and SpO$_2$ signal's waveform end-to-end delay between occurrence on the patient monitor display and the graphical user interface of the teleconsultation center

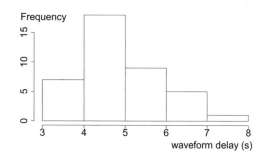

(2G and 3G) exists. Due to this range of measurements and with an additional security factor applied, tele-EMS physicians were instructed to expect a maximum signal delay of 10s, when instructing patient treatment.

5.2   Usability and System Integration

During the one year trial phase, the tele-EMS physicians and the technical support team gathered quantitative data regarding technical system performance in a ticket system to track the occurrence and elimination of technical performance issues. Out of the 377 issues reported as defects, 17 were related to the biomedical signal monitoring. These 17 issues (Figure 4) can be categorized into three groups: general connectivity (7 issues), continuous waveforms (6 issues), and periodic signals (4 issues). The general connectivity issues describe unacceptable amount or length of dropouts for continuous waveforms and unacceptable delay of transmitted twelve-lead ECGs, all caused by poor mobile data connectivity from on-scene to the teleconsultation center.

The continuous waveform issues were largely caused by unexpected behavior of the embedded IIC display component (4 out of 6 issues, i.e., 66.7 %), for which no causes could be identified: the color-coded waveforms and numerical values were suddenly all displayed white, but could still be read (independently reported twice); the component's display froze and it became unusable; no waveforms, only the numerical values were displayed. The remaining issues were caused by errors in the custom integration software, failing to handle unexpected situations (2 issues).

Most issues (3 out of 4 issues, i.e., 75 %) related to the periodic signals were caused by the custom integration software failing to control the HTS Viewer on a tele-EMS physician's workplace to connect to the correct HTS; the remaining issue was caused by the HTS Viewer staying in an unusable state for about one minute when connecting to a HTS with many stored cases.

Figure 4 Histogram of defect issues concerning the biomedical signal monitoring as reported during the one-year trial phase, grouped by issue category

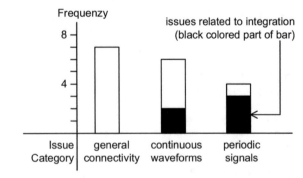

6　Discussion

Using off-the-shelf medical devices and software to build an integrated telemedicine system proved successful. The presented integration approach fulfills the functional requirements that a multidisciplinary team of experts from within the domain of telemedicine in EMS specified prior. The completion of a one year trial phase in regular EMS, using this system, validates both the technical system and the requirements upon which it is built. This trial enabled medical researchers to gather data to assess the effectiveness of the underlying model for telemedicine and its organizational impact on EMS. The medical trial evaluation is not yet finished but initial analysis is promising [6, 11].

Comparing the doubled maximum channel delay tolerated by the patient monitor to IIC connection (6.3 s) to the maximum round trip time expected in GPRS (mean below 2 s, 95 %-percentile can increase up to 4 s) and UMTS (mean at about 0.5 s, 95 %-percentile at about 1 s) networks [26], you find evidence that a GPRS network is sufficient for near real-time biomedical signal transmission by using the presented setup in most circumstances, given an adequate signal strength and link quality. In the fully integrated telemedicine system, the negative effects of data loss and data corruption are prevented by the VPN tunnel using a reliable transport protocol; this in turn, adds to the channel's delay and causes more jitter. Considering our findings from [11], some rural areas exist that do not provide sufficient network connectivity to support adequate real-time monitoring using this setup and that require the backup mechanism provided by the HTS integration.

The measured end-to-end delay for near real-time biomedical signal waveforms exceeded the otherwise suggested maximum delay of 4 s in 82.5 % of the samples, with the median exceeding this suggested delay by 0.9 s [15]. Regarding the 10 s expected delay that was used to instruct the tele-EMS physicians for the trial phase, it is unclear, if a generalized requirement can be upheld. However, in diverse clinical circumstances a delay of 10 s seems to be rather inacceptable, e.g., when ECG suddenly presents an asystole that actually occurred already 10 s ago. Furthermore, communication between the tele-EMS physician and the paramedics concerning alterations in real-time ECG curve (e.g. extra systoles) can be hindered.

The technical integration issues reported during the trail phase fall into two categories: the issues that can be resolved by improving the custom integration software and the issues that result from insufficient means of integration offered by the proprietary application. The two integration issues reported for the IIC fall into the former category and can be eliminated in a future system, the three integration issues reported for the HTS fall into the latter category. The IIC's web technology based interface allows a more stable integration, because of its individually addressable and embeddable user interface components. Contrary, the attempt to integrate the native, stand-alone HTS Viewer application using user interface automation proved to be an unreliable method.

All reported issues for the patient monitoring but the two IIC issues with missing color, severely impact the diagnostic capabilities that the telemedicine system offers to the tele-EMS physician. The issue of HTS staying in an unusable state could be solved by backing up old cases and removing them from the main database. Nevertheless, the tele-EMS physician's most important source of information is the bidirectional voice communication to the paramedics on-scene, which was reported as at least adequate even under bad conditions [11]. Hence, a defect in the patient monitoring is not a life threatening situation for a patient in this telemedicine setting.

7 Conclusion

We described how to use a commercially available off-the-shelf monitor/defibrillator (HearStart MRx, Philips Healthcare, Andover, MA, USA) and related software to create an integrated telemedicine system prototype, sufficiently fulfilling requirements regarding statutory regulations, usability, and functionality, in order to enable research of telemedicine in regular EMS. Despite numerous existing medical data and interface standards, only HL7 messaging was of importance for the integration; otherwise, web and common computer networking technology offered good integration methods. Whilst our work here focuses on integrating one device, our general approach of integrating off-the-shelf components instead of building a complete system from scratch can speed up the creation of prototypes for research in clinical trials.

Standards such as HL7 and IEEE 11073 promise full interoperability between medical devices and external software systems; with our work we have shown that, in order to easily create an integrated telemedicine system prototype for clinical research, a full implementation of these standards is not necessary.

Building user interfaces with standard web technologies that offer individually addressable display components vastly eases the work of researchers, who aim to evaluate new methods for utilizing existing technology. On the data interface level, using common computer standards and application protocols – in our case Bluetooth File Transfer Profile together with messages stored in files – offers an integration point for non-real-time data that is very simple to access. A device manufacturer searching for a simple device interface, unwilling or unable to use IEEE 11073, should consider Bluetooth File Transfer Profile instead of using a proprietary protocol implemented on top of other Bluetooth Profiles.

At this point, very little work has been done to systematically identify which biomedical signals (and signal qualities) actually are necessary for specific telemedicine procedures in EMS. The aspect of signal delay in this regard has been briefly touched by our work, but clearly, more research directed to this special aspect is necessary. Resulting knowledge would drastically improve the support a telemedicine system can offer to a tele-EMS physician who has to decide for a treatment method that heavily relies on the availability of biomedical signals within a certain maximum delay; this very specific maximum delay then defines real-time for the case at hand, which can automatically be assessed by the telemedicine system.

References

1. S. Pavlopoulos, E. Kyriacou, A. Berler, S. Dembeyiotis, D. Koutsouris, A novel emergency telemedicine system based on wireless communication technology—AMBULANCE. IEEE Transactions on Information Technology in Biomedicine 2(4), 1998, pp. 261–267
2. F. Chiarugi, D. Trypakis, V. Kontogiannis, P. Lees, C. Chronaki, M. Zeaki, N. Giannakoudakis, D. Vourvahakis, M. Tsiknakis, S. Orphanoudakis, Continuous ECG monitoring in the management of pre-hospital health emergencies. In: *Computers in Cardiology*. IEEE, 2003, pp. 205–208
3. C.S. Chang, T.H. Tan, Y.F. Chen, Y.F. Huang, M.H. Lee, J.C. Hsu, H.C. Chen, Development of a Ubiquitous Emergency Medical Service System Based on Zigbee and 3.5G Wireless Communication Technologies. In: *Medical Biometrics, Lecture Notes in Computer Science*, vol. 6165, ed. by D. Zhang, M. Sonka, Springer Berlin Heidelberg, 2010, pp. 201–208
4. K. Shimizu, Telemedicine by mobile communication. IEEE Engineering in Medicine and Biology Magazine 18(4), 1999, pp. 32–44
5. S. Bergrath, D. Rörtgen, R. Rossaint, S.K. Beckers, H. Fischermann, J.C. Brokmann, M. Czaplik, M. Felzen, M.T. Schneiders, M. Skorning, Technical and organisational feasibility of a multifunctional telemedicine system in an emergency medical service - an observational study. Journal of Telemedicine and Telecare 17(7), 2011, pp. 371–377
6. S. Bergrath, M. Czaplik, R. Rossaint, F. Hirsch, S. Beckers, B. Valentin, D. Wielpütz, M.T. Schneiders, J. Brokmann, Implementation phase of a multicentre prehospital telemedicine system to support paramedics: feasibility and possible limitations. Scandinavian Journal of Trauma, Resuscitation and Emergency Medicine 21(54), 2013, pp. 1–10
7. M. Skorning, S. Bergrath, D. Rörtgen, S.K. Beckers, J.C. Brokmann, B. Gillmann, J. Herding, M. Protogerakis, C. Mutscher, R. Rossaint, Teleconsultation Does Not Prolong Time Intervals in a Simulated Prehospital Cardiac Emergency Scenario. Circulation **120, Supplement 1**, 2009, p. S1482
8. M.T. Schneiders, D. Schilberg, S. Jeschke, A Joint Organizational and Technical Development of a Telematic Rescue Assistance System for German Emergency Medical Services. In: *Proceedings of eTELEMED 2011: The Third International Conference on eHealth, Telemedicine, and Social Medicine*, 2011, pp. 150–155
9. F. Hulstaert, M. Neyt, I. Vinck, S. Stordeur, M. Huic, S. Sauerland, M.R. Kuijpers, P. Abrishami, H. Vondeling, H. van Brabandt, The pre-market clinical evaluation of innovative high-risk medical devices: Health Service Research (HSR). Brussels, 2011
10. E. French-Mowat, J. Burnett, How are medical devices regulated in the European Union? Journal of the Royal Society of Medicine **105 Suppl 1**, 2012, pp. S22–8
11. M. Czaplik, S. Bergrath, R. Rossaint, S. Thelen, T. Brodziak, B. Valentin, F. Hirsch, S.K. Beckers, J.C. Brokmann, Employment of Telemedicine in Emergency Medicine. Methods of information in medicine 53(2), 2014, pp. 99–107
12. C.F. Lin, Mobile telemedicine: a survey study. Journal of medical systems 36(2), 2012, pp. 511–520
13. G. Anogianakis, S. Maglavera, A. Pomportsis, S. Bountzioukas, F. Beltrame, G. Orsi, Medical emergency aid through telematics: design, implementation guidelines and analysis of user requirements for the MERMAID project. International Journal of Medical Informatics 52(1-3), 1998, pp. 93–103
14. E. Kyriacou, S. Pavlopoulos, A. Berler, M. Neophytou, A. Bourka, A. Georgoulas, A. Anagnostaki, D. Karayiannis, C. Schizas, C. Pattichis, A. Andreou, D. Koutsouris, Multi-purpose HealthCare Telemedicine Systems with mobile communication link support. BioMedical Engineering OnLine 2(7), 2003
15. A. Alesanco, J. García, Clinical Assessment of Wireless ECG Transmission in Real-Time Cardiac Telemonitoring. IEEE Transactions on Information Technology in Biomedicine 14(5), 2010, pp. 1144–1152

16. J.D. Trigo, F. Chiarugi, A. Alesanco, M. Martínez-Espronceda, C.E. Chronaki, J. Escayola, I. Martínez, J. García, Standard-compliant real-time transmission of ECGs: harmonization of ISO/IEEE 11073-PHD and SCP-ECG. In: *Engineering in Medicine and Biology Society, Annual International Conference of the*. IEEE, 2009, pp. 4635–4638

17. Health informatics – Personal health device communication – Part 10406: Device specialization – Basic electrocardiograph (ECG) (1- to 3-lead ECG). Tech. Rep. 11073-10406, 2012. ISO/IEEE

18. LifeBot. LifeBot DREAMS: The Most Advanced Telemedicine Systems, 2013. http://www.lifebot.us.com/dreams/. (September 27, 2012)

19. R.L. Heath. Resuscitation and life support system, method and apparatus, 2010. 7,672,720

20. Viewcare. Viewcare Pre-Hospital Telemedicine System, 2013. http://viewcare.com/index.php/da/praehospital. (October 23, 2013)

21. S. Thelen, M.T. Schneiders, D. Schilberg, S. Jeschke, A Multifunctional Telemedicine System for Pre-hospital Emergency Medical Services. In: *eTELEMED 2013, The Fifth International Conference on eHealth, Telemedicine, and Social Medicine*, ed. by L. Van Gemert-Pijnen, H.C. Ossebaard. IARIA XPS Press, 2013, pp. 53–58

22. Philips Healthcare. IntelliVue Information Center System Release L.0: Installation and Service Manual, 2009. http://incenter.medical.philips.com/doclib/getDoc.aspx?func=ll&objId=8673388&objAction=Open (March 3, 2014)

23. Philips Healthcare. HeartStart Telemedicine System User Guide, 2009. http://incenter.medical.philips.com/doclib/enc/fetch/2000/4504/577242/577243/577245/577817/577818/HeartStart_Telemedicine_System_User_Guide.pdf

24. T. Brodziak, I. Forkel, P. Irla, P. Kornatowski, P. Seidenberg. Secure and Reliable Communication for Telemedical Applications in Emergency Medical Services, 2013. http://www.comnets.uni-bremen.de/cewit-tzi-workshop-2013/PDF/papers/Brodziak.pdf. (April 21, 2013)

25. deba81, mknambiar, pravinpardeshi. Wide Area Network Emulator: WANem Virtual Appliance, 2011. May 27, 2011

26. F. Vacirca, F. Ricciato, R. Pilz, Large-Scale RTT Measurements from an Operational UMTS/GPRS Network. In: *Wireless Internet (WICON'05), First International Conference on*. 2005, pp. 190–197

Part V
Semantic Networks and Ontologies for Complex Value Chains and Virtual Environments

Improving Factory Planning by Analyzing Process Dependencies

Christian Büscher, Hanno Voet, Tobias Meisen, Moritz Krunke, Kai
Kreisköther, Achim Kampker, Daniel Schilberg and Sabina Jeschke

Abstract Production companies in high-wage countries face growing complexity in their production conditions due to increasing variance and shorter product lifecycles. To enable the needed flexibility in production with respect to short-term changes, factory planning has to be transparent in such a way that the effects on production are traceable. Therefore, a modular planning approach combined with a continuous information management is necessary. The combination of the approaches of Condition Based Factory Planning and Virtual Production Intelligence provides the basis for an analysis of process dependencies during factory planning projects. This analysis is supposed to increase transparency of information flows and to reach traceability.

Keywords Condition Based Factory Planning · Virtual Production Intelligence · Integrative Information Model · Key Performance Indicators

1 Introduction

Due to globalized markets and worldwide activities, nowadays, companies have to deal with higher market dynamics in a highly volatile production environment [1, 2]. Since companies from high-wage countries cannot compete with companies from low-wage countries from a cost perspective, they try to meet the customers' demands by offering highly individualized and customized products [1]. A core competence for this strategy is the handling of short product and technology lifecycles, which are necessary to surpass competitors. Nevertheless, such a shortening changes the way of factory planning. A constant optimization of production processes as well as a higher flexibility regarding short-term changes and a consequent value orientation in the planning process become necessary [2, 3].

C. Büscher (✉) · T. Meisen · D. Schilberg · S. Jeschke
IMA/ZLW & IfU, RWTH Aachen University, Dennewartstr. 27,
52068 Aachen, Germany
e-mail: christian.buescher@ima-zlw-ifu.rwth-aachen.de

H. Voet · M. Krunke · K. Kreisköther · A. Kampker
WZL, RWTH Aachen University, Steinbachstr. 19, 52074 Aachen, Germany

Originally published in "Proceedings of the 47th CIRP Conference
on Manufacturing Systems", © Elsevier 2014. Reprint by Springer
International Publishing Switzerland 2016, DOI 10.1007/978-3-319-42620-4_62

These requirements result in multiple challenges. To ensure a high flexibility for short-term changes in factory planning projects, the planning process has to be analyzed more precisely. Dependencies between process steps have to be more transparent than they are today in order to increase traceability of changes within the planning process and their effects on production. Such an analysis of process dependencies has to take into account that knowledge within factory planning processes is dispersed among different experts and is used in independently developed methods [4]. Additionally, for the quantitative evaluation of the planning success and a systematic decision support during the planning process, enhanced information processing techniques are needed.

Solutions have been developed within the Cluster of Excellence "Integrative Production Technology for High-Wage Countries" at RWTH Aachen University. An advanced approach to guarantee a high transparency during factory planning processes is the concept of "Condition Based Factory Planning" (CBFP). CBFP has been employed to facilitate the factory planning process without restricting its flexibility. To achieve this aim, the process is decomposed into standardized planning modules [5]. This modularity enables a detailed analysis of the different planning tasks.

Additionally, an integrative concept called "Virtual Production Intelligence" (VPI) has been developed [6]. It indicates a concept that enables product, factory and machine planners to plan products and the production processes collaboratively and holistically [6]. The concept comprises methods and procedures to consolidate and to propagate data that is generated from historical and simulated data in real and virtual production. Furthermore, it includes visualization and interaction techniques to analyze and to explore the stored information. The concept focusses on the setup of a domain specific integrative information model which provides the possibility of an integrated analysis of the process characteristics and an improved decision support.

This paper introduces a way to combine both approaches, CBFP and VPI, to improve factory planning. The integration of an integrative information model into the underlying IT-system of the VPI enables the technical system to interact with the user efficiently by using the domain specific models and semantics of the CBFP. Thus, several challenges have to be approached. At first, a robust and extendable information model is necessary to facilitate the required analysis of process dependencies during runtime. Furthermore, provided data has to be mapped to the formalized information model. Here, adaptive information integration is used so that data is annotated with its semantic concepts during the integration process and, as a result, can be interpreted by the system [7]. On this basis, key performance indicators (KPI) for dependency analysis in factory planning are provided and implemented in a software demonstrator along with the analysis algorithms.

To reach these objectives, the following questions will be answered in the present paper:

- How can the planning modules of the CBFP be supported by an integrative information model?

- Which algorithms are relevant when analyzing process dependencies within factory planning processes?
- Which relevant KPI summarize the gain of information and support factory planners with regard to their decision making?

2 Related Work

Existing factory planning approaches often have a deterministic and analytical point of view. They divide the factory planning process in several discrete, consecutive phases [8]. For these different phases, standard workflows are defined based on an ideal information flow. These standard workflows induce the desired output of a phase in a minimum amount of time. The factory is typically divided into different elements such as the building, production facilities or logistics, which are planned by different experts who are skilled in their particular trade and who therefore have particular knowledge [9]. In reality, the knowledge exchange between the different experts is often not conducted adequately. As a consequence, every discipline mainly works in its specific field of expertise.

Due to the strong temporal orientation of these existing approaches, a weak spot consists in the fact that the process adaptability and flexibility are extremely limited when it comes to changes during factory planning projects. If e.g. production quantities change during the project, the existing models provide no solution to deal with the changed planning situation. Due to the low transparency with regard to changed information in classical approaches, the whole workflow has to be passed through again to secure that no effect caused by the changes is ignored. This leads to massive time losses [5].

Another problem is that classical approaches do not consider the specific requirements of factory planning projects. By offering one standard workflow for all factory planning projects, it is impossible to cover all individual characteristics of a specific project. E.g. in a reconfiguration project of an already existing plant, some planning steps as the plant structure planning might not be necessary as, in this case, the plant structure already exists. On the contrary, the same planning steps become relevant when a completely new plant is planned. Thus, the standard workflow leads to an inefficient planning process as the whole factory planning process is not tailor-made but over-engineered.

Factory planning processes are mainly influenced by the knowledge of experts and are therefore significantly shaped through experience values. Based on historical production and planning data, experts adapt information to the requirements of the current planning process to gain a qualitative evaluation of different scenarios [4]. During this process, the planning is gradually enhanced by taking into account additional constraints and boundary conditions in order to face the requirements of the real factory [10]. Thus, one aim of the current research consists in systematizing the knowledge of experts, which is a core element in factory planning.

In the past, different approaches to use knowledge and experience values within planning processes have been researched and developed in several fields of production and business management such as Decision Support Systems (DSS) [11] and Business Intelligence (BI) systems [12, 13]. In terms of factory planning, such systems are employed to process both historical data generated during former planning processes as well as data determined in simulation applications used during the process. This is expected to guarantee the propagation of well-structured data to people involved in the planning process in an appropriate form [14]. In this context, numerous software solutions exist providing IT support within the planning phase. However, most of the existing systems are standalone solutions that focus on one aspect of the planning task. Therefore, these heterogeneous systems are insufficient with regard to the evaluation of the overall planning process [15].

In order to apply the idea of BI on the different IT solutions of a manufacturing enterprise – and therefore following the idea of the Digital Factory [16] – the term has to be extended on the fields of management and on the production level at the same time. According to [7], an integrative solution to serve the desired interoperability between heterogeneous, distributed systems was presented with the adaptive information integration, which is the basis for the VPI approach presented in this paper.

Based on this approach, an integrative information analysis and the evaluation of an entire process become possible. To present the gained information, key performance indicators can be used. KPI are mainly employed for planning, management, analysis and monitoring purposes [17], but can generally be used for any process evaluation and decision support. KPI achieve significance particularly in conjunction with the process goals and by considering the entire process [18]. This requires a consistent calculation of KPI, their consistency, the demarcation of the areas of application and the correct understanding of their information content.

3 Methods

3.1 Condition Based Factory Planning (CBFP)

As an answer to the weaknesses of existing factory planning approaches described in Section 2, the concept of Condition Based Factory Planning has been developed within the Cluster of Excellence [5]. The main idea of CBFP is to structure the different planning tasks not with regard to a temporal chronology but with regard to their contents. In CBFP, the single planning tasks are encapsulated in different planning modules with defined input and output information (see Figure 1). Due to the resulting dependencies between the modules, the planning process is determined.

The planning process within the single modules is standardized, which leads to significant time savings in factory planning projects. The required information of the single modules is linked by particular dependencies. The visualization and

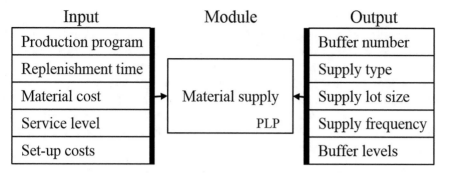

Input	Module	Output
Production program		Buffer number
Replenishment time		Supply type
Material cost	→ Material supply ←	Supply lot size
Service level	PLP	Supply frequency
Set-up costs		Buffer levels

Figure 1 Planning module material supply [5]

analysis of these dependencies lead to a higher transparency within the project. E.g. if changes occur, the project team can directly trace the changes and effects on the specific planning modules by analyzing the dependencies between input and output information of different modules.

Within the single modules, the transformation of input into output information takes place. For this transformation, in CBFP, different matrices within the specific modules were developed, which indicate the input information that is needed in order to generate the output information. In the module *Layout planning*, a specific list of required machines including their characteristics as dimensions or connections for power supply is needed to generate a detailed layout.

For single modules, it has also been investigated whether specific pieces of output information can be generated automatically without being in need of a human planner using his experience and intuition to derive and evaluate his planning results. For the different modules, software tools have been collected, which are useful and applicable for the specific planning tasks.

As all factory planning tasks have been divided into several modules with all of them using specific input and output information, it is obvious that the management of these data and the pieces of information within a factory planning project becomes very complex. The CBFP approach provides a theoretical construct for the information flow, but does not contain an integrative information model handling all data along the process: this model is realized through the coupling of the CBFP approach with the VPI.

3.2 Virtual Production Intelligence (VPI)

"Virtual Production Intelligence" (VPI) focuses on the integrated handling and analysis of information generated in the context of virtual production. It follows the idea of Luhn, who coined the term "business intelligence system" in 1958 [12]. Herein, he describes principles and operations of a fully-automatized system that facilitates the processing and its propagation to the responsible departments within an enterprise. Here, intelligence is defined as "the ability to apprehend the interrelationships of presented facts in such a way as to guide action towards a desired goal" [12]. Nowadays, business intelligence is used as an umbrella term for applications of every description that facilitate the access to and the analysis of information with the aim to improve and to optimize decision, overall efficiency and performance [14].

VPI refers to the mentioned concept of an integrated handling and analysis of information generated in the context of virtual production. The concept pursues three main objectives, which are [6]:

- **Holistic**: Addressing all sub processes of product development, factory and production plan etc.
- **Integrated**: Supporting the usage and the combination of already existent approaches instead of creating new and further standards.
- **Collaborative**: Considering roles, which are part of the planning process, as well as their communication and delivery processes.

Hence, the VPI represents a contribution to and a necessary major step towards the realization of the digital factory addressing the major challenges interoperability, user interaction as well as visual analysis and simulation [16]. The IT system that provides functionality and visualization capabilities to offer, in turn, solutions to the users to overcome these challenges is called VPI platform. The main goal of the VPI platform comprises the reduction of planning efforts and the increase of planning efficiency by providing an integrative analysis and its presentation in terms of a cockpit [19].

This requires the definition of the underlying integrated information model. Using the already mentioned adaptive information integration, data can automatically be transferred between the sources and the data sinks. Furthermore, data can be consolidated for analysis [20]. Such an information- centered approach facilitates a structured information management that makes an integrated and efficient data access possible, e.g. by supporting structured query languages. Methods and tools for data analyses and evaluation, such as Data Warehousing and OLAP (On-Line Analytical Processing), facilitate various possibilities to manipulate and to maintain existing data [21].

The platform serves for planning and support concerns by providing an integrated and explorative analysis in various fields of application. Figure 2 illustrates how the platform is used by various user groups in these fields of application.

4   Application Scenario

The integration of the domain specific models and semantics of the CBFP into an integrative information model, which serves as a basis for the underlying VPI platform, enables the factory planner to get a holistic and transparent view of the whole

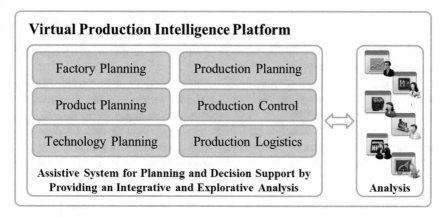

Figure 2 VPI platform [6]

planning process. The objective of the application scenario presented within this paper consists of an IT based analysis of process dependencies and its presentation to the planner. Due to the analysis and the presentation, the planner is enabled to react shortly to unforeseen changes in the planning process. As a consequence, he can adapt further planning steps more efficiently.

To consolidate the approaches aforementioned, in a first step, a small application scenario for factory planning projects has been developed. To reduce the complexity of the planning situation, only seven modules were analyzed in this scenario. These seven modules represent the most important factory planning modules, which are relevant in almost every factory planning project from reconfiguration up to new plant projects. The different information and planning data of these modules have to be managed and relations have to be analyzed intuitively and dynamically over time by making use of the VPI platform. The modules and the rough relations between them are illustrated in Figure 3.

The selected modules show a high diversity and heterogeneity with regard to the required data and also in their structure (c.f. Figure 1). Whereas the module *Profitability calculation* uses data e.g. taken from accounting, in the module *Product analysis*, very detailed information concerning the design department is analyzed.

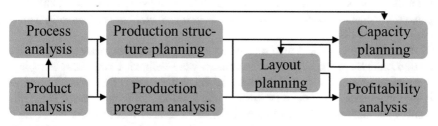

Figure 3 Planning modules in the application scenario

Additionally, there is a strong interdependence between these different modules so that aspects like the traceability of changed information or the transparency in data management can be tested in-depth. Consequently, these modules form a basis for a "prototype" within the application scenario.

Using this application scenario, a software demonstrator will be developed as part of the VPI platform that supports factory planners with regard to the information management in planning projects. The mentioned advantages as transparency and a content-based project structure of CBFP must be opened up and disadvantages, especially the complex data management, must be prevented. In this connection, the software demonstrator will continuously be developed and successively be completed with other modules from CBFP.

5 Results

5.1 Information Model, Consistency Checking and Analysis

Following the concept of VPI, the information model of the regarded domain has been formulated. The information model is directly generated from results of expert interviews and data sources used in the context of the analysis. It is modeled as an ontology and is therefore used as an explicit specification of the vocabulary and the valid constraints of the specific domain. Furthermore, description logic based reasoning becomes possible, so that constraint violations or unspecified but implicitly valid information can be extracted.

In the following example, the dependencies between module parameters are regarded (see Figure 4). The information model used for this restricted example consists of two base concepts, namely *Module* and *Parameter*. A module is further described by some meta-attributes like its name. Furthermore, the relation between *Module* and *Parameter* is specified. A module consists of parameters, whereby a parameter is related to one or more modules as either input or output. Besides, parameters are related, if an input parameter affects an output parameter when being changed. The first inter-relation is described by introducing the relation *isParameterOf* and its inverse form *hasParameter*. Similarly, the second inter-relation is described by the relations *isDependentOf* and *isRelatedTo*. Using these relations, the additionally defined concepts *ModuleParameter*, *InputParameter* and *OutputParameter* can be derived for the domain.

After having generated the information model, the mappings between the information model and the concrete data sources were defined. Using these mappings, integration services were implemented by using the ADIuS framework [7], which is an implementation of the mentioned adaptive information integration. The implemented services provide two functionalities: First, they facilitate the autonomous extraction, transformation and loading of data from the data source into a database. Second, the data is automatically described using concepts of the information model.

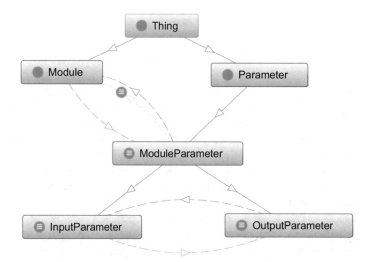

Figure 4 Excerpt of information model regarding module parameter dependencies

In the example suggested above, an integration service has been established to extract the data from a Microsoft Excel document into a MySQL database.

The integration process is finished by a consistency check to verify, for example, data quality. Therefore, the provided and implicit derived information is forwarded to a consistency checking module. The rules to validate the consistency are directly extracted from the information model and are based upon the formulated vocabulary. E.g. by using the rule

```
when
      $p: Parameter ( )
      not Module( hasParameter( $p ) )
then
      protocol("Inconsistent parameter: " + $p )
end
```

it is checked, whether each defined parameter is at least related to one module.

The integration process is fully automated. In case of inconsistent data or other unexpected problems, the user is informed via the user interface. In case of a successful integration, the user is enabled to further evaluate the data using visual analysis or provided key performance indicators. The underlying user interface is presented in Section 5.2.

5.2 VPI Platform for Factory Planning

The VPI platform is fully accessible via a provided web interface. It gives access to different analysis methods and visualizations. In the following, the exemplified analysis of module parameter dependencies is presented. After having finished the integration process, the user can directly access the integrated data and evaluate it. For the evaluation, different views, currently the module viewer and the parameter analysis, are provided. The module viewer (see Figure 5) provides a simple data visualization of the integrated and derived data. On the left side, the different available modules are shown, whereas on the right side, the input and the output parameters (parameters) of the selected module as well as the relations between these parameters (matrix) are displayed.

The parameter dependencies can be analyzed by making use of the parameter analysis (see Figure 6). First, the user selects the module and the input parameter that has to be analyzed more detailed. Second, the user triggers the analysis.

The server-side evaluation uses a breadth-first search algorithm to identify parameters that are affected when the given input parameter is changed during the planning process. In each step, the algorithm uses the provided and derived parameter relations for the identification of changed output parameters. In the first cycle, these parameters are used to identify modules whose input parameters are changed. They

Figure 5 Display of integrated data within the module viewer

PRO	Parameters	Analyze
	Product variant	
	Product structure (Parts list)	
	Customer's order history	
	Forecast data	
	Project task and delimination	
	Market analysis	
	Product strategy	

Figure 6 Module and parameter selection within parameter analysis

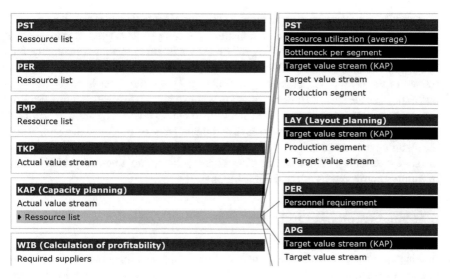

Figure 7 Display of evaluation result

define the first set of parameters that is further evaluated in the next cycle. Each such result set is pruned by parameters that have been evaluated in a previous cycle. Such a pruned set defines the next set of parameters that is evaluated. The evaluation process is stopped when no more changed parameters can be identified. As the set of input parameters is limited and within each cycle, at least one parameter is removed from this set, the algorithm terminates.

The evaluation result is visualized in the parameter analysis view (see Figure 7). The view shows the changed input parameters of each module, whereby the level of dependence gets lower from left to right meaning that the directly affected input parameters are shown at the left-most position. Furthermore, the user can highlight the interrelation between the parameters (as shown in Figure 7).

5.3   KPI for Dependency Evaluation

Besides the pure analysis and visualization of the process dependencies, KPI have been developed to evaluate the dependencies. In a first step, the number of direct dependencies of a parameter is calculated and displayed in the evaluation results. This represents the complexity of the modifications' impact of one parameter. Based on these direct dependencies, correlation matrices for all parameters are developed, which have not been implemented, yet.

Furthermore, a KPI for dependency evaluation has been derived:

$$K_{dep,i} = \sum_{j=1}^{n} x_{i,j} \tag{1}$$

This KPI considers for each parameter i all dependencies found within the analysis cycles. In addition, it rates each parameter according to its degree of dependence to a parameter j with a factor of $x_{i,j}$. This factor is not determined manually, but automatically rule-based. On the one hand, the factor has a static part, which is related to the level of dependence (cf. Section 5.2). On the other hand, a dynamic part expresses the specific process situation within a precise factory planning project. If e.g. a parameter j has already been determined but depends on a parameter i which has to be changed, the factor increases as parameter j has to be calculated again. Thus, the planning process slows down. This information helps increasing the transparency of factory planning. (The dynamic aspect has not been implemented yet but is part of future improvements of the demonstrator.)

6 Conclusion and Outlook

In this paper, the approaches of Condition Based Factory Planning and Virtual Production Intelligence have been focused. The described application scenario, in which both approaches are consolidated, helps to increase transparency within factory planning processes by providing analysis algorithms of process dependencies. Parameter information of selected CBFP modules can be integrated into the presented demonstrator as a part of the VPI platform. Thus, dependencies between the process parameters can automatically be analyzed and visualized, which has to be done manually so far.

In the next steps, the demonstrator will be extended. Templates and standardized data formats for the planning modules are generated so that not only the theoretical parameters but even the precise values within a planning project can be integrated. By using these standardized data formats, the completeness and a high quality of the data are ensured. These standard templates are also the precondition for an automation of the transformation within the planning modules. Furthermore, additional modules of the CBFP will be integrated into the demonstrator to model any kinds of factory planning projects. At the same time, the analysis algorithms will be improved, more KPI such as a measure of the planning stability will be developed and the visualization in terms of a cockpit will be enhanced. Finally, the demonstrator will be used in actual factory planning projects to evaluate the improvement in decision support.

Acknowledgments The approaches presented in this paper are supported by the German Research Foundation (DFG) within the Cluster of Excellence "Integrative Production Technologies for High-Wage Countries" at RWTH Aachen University.

References

1. C. Brecher, S. Jeschke, G. Schuh, S. Aghassi, J. Arnoscht, F. Bauhoff, S. Fuchs, C. Jooß, W.O. Karmann, S. Kozielski, S. Orilski, A. Richert, A. Roderburg, M. Schiffer, J. Schubert, S. Stiller, S. Tönissen, F. Welter, Integrative production technology for high-wage countries.

In: *Integrative production technology for high-wage countries*, ed. by C. Brecher, Springer, Heidelberg, 2011, pp. 17–81

2. A. Kampker, K. Kreisköther, P. Burggräf, A. Meckelnborg, M. Krunke, S. Jeschke, M. Hoffmann, Value-oriented layout planning using the virtual production intelligence (vpi). POMS 2013 Proceedings, 24th Annual Conference POMS (Production and Operations Management Society), Denver, Colorado, USA, 2013

3. P. Burggräf, Value-oriented factory planning. Ph.D. thesis, Apprimus Verlag, RWTH Aachen, Aachen, 2012

4. J.A. Tompkins, J. White, Y. Bozer, J. Tanchoco, *Facilities planning*, 4th edn. Wiley & Sons, Hoboken and NJ, 2010

5. G. Schuh, A. Kampker, C. Wesch-Potente, Condition based factory planning. Production Engineering **5** (1), 2011, pp. 89–94. doi:10.1007/s11740-010-0281-y

6. R. Reinhard, C. Büscher, T. Meisen, D. Schilberg, S. Jeschke, Virtual production intelligence – a contribution to the digital factory. Proceedings of the 5th International Conference on Intelligent Robotics and Applications (ICIRA 2012) (1), 2012, pp. 706–715

7. T. Meisen, P. Meisen, D. Schilberg, S. Jeschke, Adaptive information integration: Bridging the semantic gap between numerical simulations. In: *Enterprise Information Systems*, vol. 102, ed. by Z.Z.F.J.C.J. Zhang R, Zhang J, Springer, Berlin/Heidelberg, 2012, pp. 51–65

8. VDI guideline 5200 part 1. Factory planning – planning procedures, 2011

9. A. Kampker, A. Meckelnborg, P. Burggräf, T. Netz, Factory planning scrum: Integrative factory planning with agile project management. Proceedings of the International Conference on Competitive Manufacturing COMA' 13, 30 January – 1 February 2013, Stellenbosch, South Africa, 2013, pp. 345–350

10. P. Nyhuis, H.P. Wiendahl, *Fundamentals of Production Logistics: Theory, Tools and Applications*. Springer, Berlin and Heidelberg, 2009

11. D. Arnott, G. Pervan, A critical analysis of decision support systems research. Journal of Information Technology **20** (2), 2005, pp. 67–87. doi:10.1057/palgrave.jit.2000035

12. H.P. Luhn, A business intelligence system. IBM Journal of Research and Development **2** (4), 1958, pp. 314–319

13. F. Melchert, R. Winter, M. Klesse, Aligning process automation and business intelligence to support corporate performance management. In: *Proceedings of the Tenth Americas Conference on Information Systems*, New York, 2004, pp. 4053–4063

14. H. Baars, H. Lasi, H.G. Kemper, M. Koch, Manufacturing execution systems and business intelligence for production environments. AMCIS 2010 Proceedings, 2010

15. R. Anderl, M. Rezaei, Federative factory data management an approach based upon service oriented architecture (soa). In: *Digital Enterprise Technology*, ed. by P.F. Cunha, P.G. Maropoulos, Springer Science+Business Media, LLC, New York, 2007, pp. 67–74. doi:10.1007/978-0-387-49864-5_7

16. VDI guideline 4499 part 1. Digital factory – fundamentals, 2008

17. V. Jovan, S. Zorzut, Use of key performance indicators in production management. In: *2nd IEEE International Conference on Cybernetics & Intelligent Systems*, IEEE, Piscataway, 2006, pp. 1–6

18. T. Reichmann, *Controlling: Concepts of management control, controllership, and ratios*. Springer, Berlin and New York, 1997

19. C. Büscher, M. Hoffmann, R. Reinhard, D. Schilberg, S. Jeschke, Performance indicators for factory planning on the basis of the virtual production intelligence approach. Proceedings of the 22nd International Conference on Production Research (ICPR 22), July 28 – August 1, Igussu Falls, Brasilien, 2013

20. T. Meisen, P. Meisen, D. Schilberg, S. Jeschke, Application integration of simulation tools considering domain specific knowledge. Proceedings of the 13th International Conference on Enterprise Information Systems (1), 2011, pp. 42–53

21. W.H. Inmon, *Building the data warehouse*, 4th edn. Wiley technology publishing Timely, practical, reliable. Wiley, Indianapolis and Ind, 2005

Ontologiebasiertes Informationsmanagement für die Fabrikplanung

Christian Büscher, Tobias Meisen and Sabina Jeschke

Zusammenfassung Trotz der aktuellen Bestrebungen zur vernetzten Planung und Steuerung von Produktionsprozessen im Zuge der Virtuellen Produktion und Digitalen Fabrik unterstützen Planungs- und Simulationsanwendungen häufig immer noch lediglich die Analyse und Optimierung einzelner Planungsaspekte. Um den gegenwärtigen Herausforderungen von immer komplexeren und sich schnell änderenden Produktionsumgebungen zu begegnen, benötigen Fabrikplaner hingegen Werkzeuge, die eine integrierte Sicht und damit die Möglichkeit der frühzeitigen und validen Bewertung eines ganzen Planungsszenarios bieten. Dafür müssen sich Informationssysteme auf ein einheitliches Verständnis der domänenspezifischen Konzeptualisierung stützen, was zu einer der zentralen Herausforderungen der Fabrikplanung führt: eine konsistente und kohärente Informationsmodellierung entlang des gesamten Planungsprozesses ausgehend von den angedeuteten heterogenen Datenquellen. Der Ansatz der Virtual Production Intelligence (VPI) liefert ein grundlegendes Konzept für ein derartiges Informationssystem, mit dem Daten entlang von Planungsprozessen integriert, analysiert und bewertet werden können. Der VPI-Ansatz stellt dabei die semantische Interoperabilität zwischen verschiedenen Planungs- und Simulationsanwendungen sowie deren individuellen Datenmodellen und -strukturen her. In diesem Beitrag wird die Informationsmodellierung der VPI anhand der Modellierung von Ontologien für die Fabrikplanung präsentiert. Zudem wird dargestellt, wie diese Ontologien in den VPI-Ansatz eingebettet und innerhalb des webbasierten Informationsproduktes, der VPI-Plattform, implementiert sind. Die so erzeugte konsolidierte Informationsgrundlage befähigt Fabrikplaner zu neuartigen Analysen und Darstellungen der Planungsergebnisse und bietet somit eine verbesserte Entscheidungsunterstützung.

Schlüsselwörter Integratives Informationssystem · Ontologien · Fabrikplanung · Virtual Production Intelligence

C. Büscher (✉) · T. Meisen · S. Jeschke
IMA/ZLW & IfU, RWTH Aachen University,
Dennewartstr. 27, 52068 Aachen, Germany
e-mail: christian.buescher@ima-zlw-ifu.rwth-aachen.de

Originally published in ?Tagungsband der 18. IFF-Wissenschaftstage, Magdeburg, Germany, 24-25 June 2015?, © Fraunhofer IFF 2015.
Reprint by Springer International Publishing Switzerland 2016,
DOI 10.1007/978-3-319-42620-4_63

1 Einleitung

Heutzutage sind vor allem Unternehmen in Hochlohnländern mit der Herausforderung konfrontiert, maßgeschneiderte und qualitativ hochwertige Produkte trotz der zunehmenden Komplexität des Produktionsprozesses und der großen Anzahl an Parametern zu wettbewerbsfähigen Kosten anzubieten [1]. Insbesondere die Fabrikplanung nimmt dabei eine entscheidende Rolle ein, da sie in vielerlei Hinsicht die Grundlage für die spätere Produktion legt. Methoden einer kosten- und ressourceneffizienten Planung und Produktionsentwicklung ermöglichen es, die erforderlichen Planungsprozesse maßgeschneidert und modular zu gestalten. Darüber hinaus sind die fortlaufende Optimierung von Fabrikplanungsprozessen, flexible Anpassungen an kurzfristige Veränderungen sowie eine konsequente Wertorientierung unabkömmlich [2].

Aktuelle Fabrikplanungsansätze basierenmeist auf den Erfahrungen von Planungsexperten aus unterschiedlichen Domänen, um die Vielzahl der verschiedenen Herausforderungen und Bedürfnissen des Planungsprozesses zu meistern. Hinsichtlich eines effizienten und ergebnisorientierten Entwicklungsprozesses müssen sie eng zusammenarbeiten und laufend Informationen miteinander austauschen. Dabei werden sie durch verschiedene Planungswerkzeuge unterstützt, die für spezielle Aufgaben hergestellt und entwickelt wurden [3]. Den unterschiedlichen Datenformaten und -modellen der Anwendungen, die aus historischen Gründen unabhängig voneinander entstanden sind, ist es geschuldet, dass eine große Zahl an sogenannten Insellösungen in der Praxis existiert. Die Herausforderung besteht somit darin, Daten aus heterogenen Planungs- und Simulationsanwendungen innerhalb der Fabrik- und Produktionsplanung in eine konsolidierte Informationsbasis zu integrieren. Dies ermöglicht eine effizientere Informationsbeschaffung und -aufbereitung, darauf aufbauend eine ganzheitliche Bewertung von Planungsszenarien und gibt Planern die notwendige Grundlage für kreatives und effektives Planen [4].

Im Zuge der Digitalen Fabrik werden diese Bemühungen unter dem Konzept der Virtuellen Produktion zusammengefasst [5]. Die grundlegende Idee der Virtuellen Produktion besteht darin, eine Interoperabilität der genannten heterogenen IT-Anwendungen zu erreichen. Dies ist bislang noch nicht ausreichend realisiert [6]. Neben dem Paradigma der Standardisierung werden Ansätze erforscht, um die Interoperabilität von verteilten, heterogenen Planung- und Simulationsanwendungen mittels Integrationstechniken zu erreichen. Im Bereich der semantischen Technologie, besonders des Semantic Web, wurden vielversprechende Ansätze entwickelt, die auf einer Informations- und Anwendungsintegration sowie auf der Verwendung von Ontologien beruhen [7].

Auf dieser Grundlage wurde das Konzept der Virtual Production Intelligence (VPI) zusammen mit der zentralen Komponente der adaptiven Informationsintegration und der technischen Umsetzung der VPI-Plattform erarbeitet [8, 9]. Die VPI überträgt grundlegende IntelligenceKonzepte der Integration und Analyse von Daten aus dem Umfeld von Geschäftsprozessen auf die Virtuelle Produktion. Dabei sind insbesondere die Informationsmodellierung und die explizite Formalisierung der relevanten Begriffe von entscheidender Bedeutung, um Domänenwissen in ein technisches System zu integrieren. Innerhalb der Domäne der Fabrikplanung wurde bereits eine Anwendung präsentiert, die ein solches kontinuierliches und semantisches Informationsmanagement für das Aachener Fabrikplanungsvorgehen (engl. Condition Based Factory Planning, CBFP) zum Ziel hat [10].

In diesem Beitrag werden die grundlegenden Prozesse der VPI, die Informationsmodellierung und -integration, basierend auf einer semantischen Annotation von Planungsdaten anhand von Domänenontologien präsentiert. Des Weiteren wird die Umsetzung innerhalb der genannten Anwendungen vorgestellt, bei denen Daten aus heterogenen Planungs- und Simulationsanwendungen integriert werden. Das Konzept fokussiert die Erstellung eines domänenspezifischen, integrativen Informationsmodells mit dem Ziel einer integrierten Analyse der Prozesseigenschaften und einer verbesserten Entscheidungsunterstützung.

Um diese Zielsetzungen zu erreichen, werden in diesem Beitrag die folgenden Fragen beantwortet:

- Wie können die relevanten Konzepte der Fabrikplanung sowie deren Wechselbeziehung explizit formalisiert werden?
- Was sind die relevanten Schritte der Modellierung und Integration, um semantische Interoperabilität von Planungs- und Simulationsanwendungen in der Fabrikplanung zu erzielen?
- Was sind die primären Begriffe aus dem Bereich der Fabrikplanung, die für die Ontologie-Erstellung in Betracht gezogen werden müssen?

2 Stand der Technik und Forschung

Die Informationsmodellierung bildet die Grundlage jedes Informationssystems für die Bereitstellung technischer Funktionalitäten, um spezifische Informations bedürfnisse von Aufgaben bzw. Nutzern gerecht zu werden. Dieser Prozess kann durch eine Anpassung des von KRCMAR definierten Lebenszyklusmodell der Informationswirtschaft [11] in den Gesamtkontext des Informationsmanagement eingebettet werden. Abbildung 1 zeigt die Hauptprinzipien des Informationsmanagementzyklus, um eine neue Domäne zu implementieren oder eine bereits existierende zu erweitern.

Ausgehend von der Identifikation des Informationsnutzers werden die Informationsbedürfnisse identifiziert und gesammelt, denen in der gegenwärtigen Informationsinfrastruktur nicht Rechnung getragen wird. Anschließend werden mögliche Datenquellen identifiziert und in die Informationsstruktur integriert. Dies wird durch die Bereitstellung als Informationsressourcen verwirklicht. Darüber hinaus werden domänenspezifische Analyse- und Bewertungsmethoden genutzt, um bestehende Informationen anzureichern. Diese werden dem Nutzer mittels eines Informationsproduktes bereitgestellt. In diesem Beitrag liegt der Fokus auf dem Prozess der Integration.

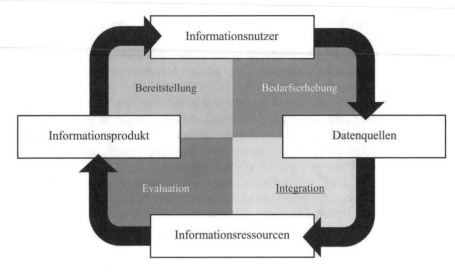

Abbildung 1 Informationsmanagementzyklus

In den vergangenen Jahren wurde auf den Gebieten der Informations- und der Anwendungsintegration eine Vielzahl an Lösungen, wie z. B. das Enterprise Application Integration (EAI) entwickelt. Diese Technologien sind insbesondere für die Fabrikplanung von Bedeutung, in der relevanten Datenquellen historisch und individuell gewachsen sind. Die Konsolidierung von Informationen aus verschiedenen Datenquellen mit unterschiedlichen Datenstrukturen wird in der Regel als Informationsintegration bezeichnet, die Konsolidierung aller IT-Lösungen entlang von Geschäftsprozessen als Anwendungsintegration [7]. Die VPI basiert auf der Informationsintegration und besonders auf die materialisierten Integration – im Gegensatz zur virtuellen Integration – bei der Daten aus Quellen geladen, bereinigt und zentral gespeichert werden [20]. Das Ziel besteht darin, eine konsolidierte Informationsbasis für einen effizienten und effektiven Zugang zu Informationen zu erzeugen, der hinsichtlich eines direkten Zugangs zu den ursprünglichen Datenquellen nur bedingt möglich ist [7].

Der Datenaustausch basiert hierbei auf verschiedenen Datenformaten und -strukturen. In erster Linie dienen computergestützte Technologien (CAx) und Enterprise Resource Planning (ERP) Systeme mit unterschiedlichen offenen und proprietären Datenformaten als Datenquelle. In der Praxis sind zudem Office-Lösungen, wie Excel, Word, oder PowerPoint als Erzeuger, Analyse- und Visualisierungssysteme in Produkt- und Produktionsentwicklung weit verbreitet [4].

Um die beschriebenen Informationsintegration zu realisieren, ist eine erweiterte Betrachtung der Informationsmodellierung, die wie folgt verstanden wird, notwendig.

"Information modeling is the cornerstone of information systems analysis and design. [...] Information modeling is the process of formally documenting the problem domain for the purpose of understanding and communication among the stakeholders." ([12]: 44).

Das Informationsmodell als Produkt des Modellierungsprozesses, bietet die Grundlage für eine explizite Formalisierung der relevanten Begriffe einer Domäne sowie deren Zusammenhänge und Abhängigkeiten. Durch klassische Modellierungssprachen wie das Entity-Relationship-Model (ERM) oder die Unified Modeling Language (UML) werden Begriffe und Beziehungen identifiziert und visualisiert. Um das Informationsmodell nicht nur den Anwendern zwecks eines gemeinsames Verständnisses sondern auch dem technische System zugänglich zu machen, muss das Model in maschinenlesbare Formate überführt werden. Somit kann das Informationsmodell als zentrale Instanz in das Informationssystem integriert werden.

Das Ziel der semantischen Technologien, besonders des Semantic Web, besteht in der formalen Repräsentation domänenspezifischer Begriffe im technischen System, um eine verbesserte Nutzbarkeit von Daten und deren Austausch zu erzielen [13]. Dafür reicht die einfache Maschinenlesbarkeit der klassischen Modellierungssprachen nicht aus. Entscheidend ist die Interpretierbarkeit durch Maschinen, eine Funktionalität, die von semantischen Technologien, insbesondere von Ontologien angeboten wird. Ontologien bieten die Möglichkeit, Begriffe und ihre Beziehungen formal und explizit anzugeben und Inferenz- und Integritätsregeln abzuleiten, die automatisiert überprüfbar sind [14]. In den letzten Jahren wurden verschiedene neuere Modellierungssprachen von Wissenschaftlern entwickelt, um Ontologien zu spezifizieren und sie darzustellen [7]. Eine weit verbreitete Darstellung ist die Web Ontology Language (OWL) [15], die innerhalb der VPI für die Darstellung von Planungsinformationen verwendet wird.

Basierend auf den beschriebenen Techniken der Wissensrepräsentation besteht ein Ziel der aktuellen Forschung im Bereich der Fabrikplanung darin, das Wissen der Experten zu systematisieren. Dabei ist die Informationsmodellierung wiederum der grundlegende Ansatz. Innerhalb des EU-Projektes 'Virtual Factory Framework' wurde ein Informationsmodell mittels verschiedener Ontologien aufgestellt, um ein virtuelles Funktionsmodell zu realisieren [16]. Das entwickelte 'Virtual Factory Data Model' bietet ein Framework zur einheitlichen Definition von Daten, die zwischen den betrachteten Softwareanwendungen ausgetauscht werden. Darüber hinaus wurden in den vergangenen Jahren zahlreiche Informationsmodelle entwickelt, die sich auf verschiedene Aspekte der Fabrikplanung in unterschiedlichen Detaillierungsgraden fokussiert haben. Beispielsweise hat REZAEI ein Fabrikinformationsmodell für die integrierte Prozessgestaltung auf Basis von postrelationalen Datenbanken entwickelt [17], während WEIMER abstrakte Informationen für die Integration von Fabrikplanungen und betrieb modelliert hat [18]. Diese Ansätze bieten Konzepte, die größtenteils klassische Modellierungssprachen verwenden. Anstatt das Informationsmodell als zentrale Komponente in das technische System zu integrieren, dient das Informationsmodell lediglich der Definition einer gemeinsamen Sprache zwischen Ingenieuren und Anwendern sowie der Ableitung des logischen Datenmodells. Daher sind die vorgestellten Ansätze in Bezug auf die semantische Modellierung innerhalb eines integrativen Informationssystems nicht für die gewünschte Unterstützung von Fabrikplanungsprojekten anwendbar.

3 Semantische Informationsmodellierung der Virtual Production Intelligence

3.1 Anwendungsszenario "Aachener Fabrikplanungsvorgehen"

Virtual Production Intelligence bezeichnet das zugrundeliegende Konzept, das es Produkt-, Fabrik- und Maschinenplaner ermöglicht, Produkte und deren Produktion kollaborativ und ganzheitlich zu planen [9]. Das Konzept umfasst Methoden zur Konsolidierung und Verbreitung von Daten aus der Domäne der Virtuellen Produktion. Zudem beinhaltet die VPI Visualisierungs- und Interaktionstechniken, um die gewonnenen Informationen zu analysieren und eine ganzheitliche Betrachtung einzunehmen.

Um das Konzept der Virtual Production Intelligence und das Informationsprodukt – die VPI-Plattform – im Anwendungsfeld der Fabrikplanung zu demonstrieren, setzen wir ein auf der VPI basierendes Informationssystem zur Unterstützung des Aachener Fabrikplanungsvorgehens um [10]. Die Grundidee des CBFP ist die Modularisierung unterschiedlicher Planungsaufgaben nicht hinsichtlich des üblichen zeitlichen Vorgehens sondern hinsichtlich deren Inhalte [19]. Ein Planungsmodul umfasst dazu eine abgeschlossene Planungsaufgabe mit definierten Ein- und Ausgangsinformationen bzw. -parametern (Abbildung 2). Diese Parameter beinhalten die tatsächlichen Planungsinformationen. Somit ist das CBFP objektorientiert und deutlich flexibler als etablierte jedoch zumeist starre klassische Planungsvorgehen.

Bislang bietet der CBFP-Ansatz ein Konstrukt für den Informationsfluss. Darüber hinaus stellt er die Grundlage für ein Informationsmodell, das die notwendigen und optionalen Informationen ebenso wie deren Beziehung entlang des Planungsprozesses beschreibt. Dabei werden verschiedene Informationsarten unterschieden [19]:

- *Eingangsinformation* eines Moduls, die das Planungsergebnis (Ausgangsinformation) eines vorherigen Moduls ist bzw. die im Vorfeld der Planungsaufgabe des

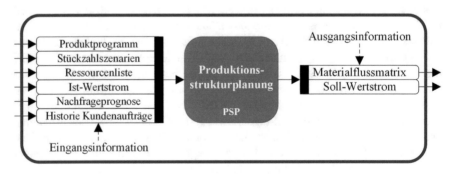

Abbildung 2 CBFP Modul 'Produktionsstrukturplanung' nach [19]

Moduls ermittelt werden muss (z. B. durch Datenexport aus einem ERP-System, durch Experteninterviews oder Workshops),

• *Ausgangsinformation* als Ergebnis eines bestimmten Planungsmoduls, das am Ende des Planungsschrittes erzeugt wird und in weiteren Modulen oder als endgültiges Planungsergebnis verwendet wird.

Die sich daraus ergebende Modullandkarte bildet die Grundlage für eine IT-basierte Analyse des Planungsprozesses. Die Handhabung der Daten und einzelner Informationen innerhalb eines konkreten Planungsprojektes ist dennoch sehr komplex. Es liegt eine verteilte Datenhaltung innerhalb der einzelnen Planungs- und Simulationsanwendung vor. CBFP bietet dafür kein Informationssystem, das den Planer auf einer Datenebene unterstützt. Folglich ist der für eine Gesamtbewertung erforderliche Datenaustausch zwischen den Quellen zeitaufwendig und fehleranfällig. Daher wird die bereits erwähnte Kopplung des CBFP mit dem VPI-Ansatz realisiert [10]. Das Ziel ist eine durchgehende Informationsmodellierung entlang des gesamten Planungsprojektes, um Fabrikplaner mit Methoden zur automatisierten Informationsanalyse und -auswertung zu unterstützen. Die größte Herausforderung liegt darin, die Interoperabilität der heterogenen Datenquellen mit Methoden der Informationsintegration zu erreichen. Die wesentlichen Prozesse der Informationsmodellierung und -integration werden in den folgenden Abschnitten vorgestellt.

3.2 *Prozess der Informationsmodellierung*

Der allgemeine, der VPI zugrundeliegende Prozess der Informationsmodellierung umfasst die folgenden vier Schritte:

1. Definition des domänenspezifischen Informationsmodells
2. Explizite Formalisierung der Begriffe, Beziehungen und Einschränkungen als Domänenontologie
3. Ableitung des logischen Datenmodells
4. Ableitung von Validierungsregeln zur Konsistenzprüfung

Der erste Schritt des Modellierungsprozesses beinhaltet die Definition des Informationsmodells der jeweiligen Domäne, hier der Fabrikplanung. Auf der Grundlage einer Analyse der Planungsdaten und Experteninterviews besteht das Ziel des Informationsmodells in einem gemeinsamen Verständnis der in der Domäne verwendeten Begriffe. Dazu wird das Modell zunächst mit klassischen Modellierungssprachen erstellt und visualisiert.

Um die gewünschte Maschineninterpretierbarkeit der domänenspezifischen Logik zu erreichen, werden die Begriffe, Beziehungen und Einschränkungen im nächsten Schritt mit Hilfe von Ontologien explizit formalisiert. Neben Ontologien zur Beschreibung von Strukturen liegt der Fokus auf Domänenontologien. Somit wird ein Schlussfolgern basierend auf einer Beschreibungslogik möglich, um eine automatisierte Identifizierung von Einschränkungsverletzungen und die Extraktion von nicht spezifizierten aber impliziten Informationen im technischen System umzusetzen.

Dieses generische Informationsmodell ist die Grundlage des logischen Datenmodells für eine konkrete Anwendung, das aus dem Informationsmodell abgeleitet werden kann. Als nächster Schritt wird die Abbildung des logischen Datenmodells auf die Ontologie definiert, um konkrete Daten unmittelbar den entsprechenden Begriffen des Informationsmodells zuzuordnen. Das Datenmodell beinhaltet dabei die objektrelationale Abbildung, z. B. durch die Konkretisierung von N:M-Beziehungen des Informationsmodells. Innerhalb dieses Schritts gehen die impliziten Informationen der Domänenontologie verloren, da sie nicht im Datenmodell abgebildet werden können.

Allerdings ermöglichen die Einschränkungen und Beziehungen innerhalb des Informationsmodells die Extraktion von Validierungs- und Konsistenzregeln, die während der Integration für eine Konsistenzprüfung der Planungsdaten genutzt werden. Dieser Prozess ist komplett automatisiert, sodass sich der Integrationsprozess automatisch an Veränderungen im Informationsmodell (und den entsprechenden Anpassungen im zugrundeliegenden Datenmodell) anpasst. Dieses Vorgehen gewährleistet eine hohe Datenqualität innerhalb der Informationsbasis der VPI-Plattform, da der Integrationsprozess nur dann abgeschlossen wird, wenn keine Konsistenzverletzungen auftreten.

Basierend auf diesem Prozess der Informationsmodellierung wird der anwendungsfallspezifische Prozess der Informationsintegration von Planungsdaten aus den Quellen im folgenden Abschnitt beschrieben.

3.3 *Prozess der Informationsintegration*

Mit Beendigung des Modellierungsprozesses ist die Grundlage für die Realisierung der Integration der verschiedenen Datenquellen gelegt. Dieser Prozess umfasst wiederum drei Schritte:

1. Identifikation bzw. Definition von Daten- bzw. Informationsquellen für alle Parameter
2. Integration dieser Daten- bzw. Informationsquellen
3. Persistieren und Visualisierung von Daten innerhalb der VPI-Plattform

Der erste Schritt umfasst die Identifizierung und Festlegung von Datenquellen für alle Parameter und die zugehörigen Informationen. Von besonderer Bedeutung sind dabei Kriterien wie die Korrektheit und Vollständigkeit der Daten, um bereits hier die Grundlage für eine hohe Informationsqualität zu legen. Basierend auf diesen Quellen wird die Methodik der adaptiven Informationsintegration für die tatsächlichen Integrationsprozesse angewendet. Diese bietet mehrere Integrationsdienste mit zwei Hauptfunktionalitäten [8]: Zunächst ermöglichen die Dienste das automatisierte Extrahieren, Transformieren und Laden der Daten aus den Quellen in die

Informationsbasis. Zweitens bilden sie die geladenen Daten auf die entsprechenden Begriffe der Informationsmodelle ab und reichern die Daten mit den impliziten Informationen der Ontologie an. Dieser Prozess der semantischen Annotation wird gefolgt von den erwähnten Konsistenzprüfungen. Im Fall inkonsistenter Daten oder anderen unerwarteten Problemen, wird der Nutzer über die Fehler informiert. Erst nach erfolgreicher Integration werden die integrierten Daten persistiert und automatisch in die Benutzerschnittstelle geladen, die Visualisierungen und Interaktionen für weitere Analysen bietet.

Der dargestellte Ansatz bietet die Möglichkeit, verschiedene Arten von Datenquellen an die VPI-Plattform anzubinden, ohne auf standardisierte Datenaustauschformaten zurückgreifen zu müssen. Somit ist eine konsolidierte Informationsbasis entlang des gesamten Planungsprozesses realisierbar. Dies ermöglicht schließlich ganzheitliche und nutzerspezifische Auswertungen und Visualisierungen, die in diesem Beitrag nicht weiter vertieft werden.

4 Ergebnisse

Im ersten Schritt des Modellierungsprozesses wurde das Informationsmodell für die Fabrikplanung definiert. Dazu wurden auf der Grundlage des abstrakten Informationsmodells des CBFP die unterschiedlichen Datenquellen analysiert und Experteninterviews durchgeführt. Die identifizierten Begrifflichkeiten und geltenden Einschränkungen wurden zu einem allgemein anerkannten Informationsmodell generalisiert. Die zunächst mittels eines UML-Klassendiagramms modellierten Hauptklassen einschließlich ihrer Beziehungen sind in Abbildung 3 dargestellt. Darüber hinaus wurde jeder dieser Begriffe durch zusätzliche Attribute ergänzt und das Informationsmodell weiter ausdetailliert.

Die Formalisierung des Informationsmodells erfolgt darauf aufbauend in Form von Ontologien mittels OWL. Dazu wurde eine Basisontologie als Vorlage spezi-

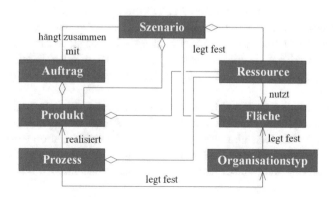

Abbildung 3 Hauptklassen des Informationsmodells für die Fabrikplanung

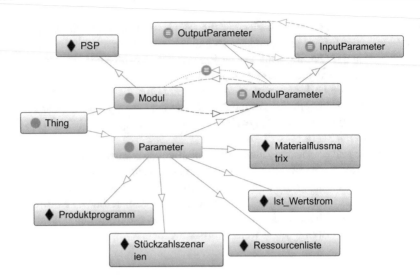

Abbildung 4 CBFP-Ontologie mit ausgewählten Instanzen

fiziert, um verschiedene Domänenontologien aufzustellen, die auf denselben Begriffs-, Beziehungs- und Attributstypen beruhen. In einem weiteren Schritt wurde das zuvor angedeutete Informationsmodell der Fabrikplanung in eine Ontologie überführt. Das abstrakte Informationsmodell des CBFP wurde als weitere Ontologie mit den beiden Grundkonzepten Modul und Parameter formalisiert (Abbildung 4).

Die Abbildung verdeutlicht die Beziehung zwischen Modulen und Parametern sowie einzelne Instanzen des in Abbildung 2 dargestellten Moduls Produktionsstrukturplanung. Mittels dieser Ontologie kann die Struktur eines Planungsprojektes während der Erstellung in der VPI-Plattform validiert werden. Das gesamte Vorgehen hat den Vorteil, dass die domänenspezifische Logik von der technischen Umsetzung des Planungssystems innerhalb eines Anwendungsszenarios entkoppelt wird. Somit wird eine Allgemeingültigkeit und Übertragbarkeit der Geschäftslogik sowie eine flexible Anpassung der Planungsmodule und Datenquellen erreicht.

Die Integrations- und Validierungsprozesse der Planungsdaten werden über die webbasierte VPI-Plattform ausgelöst. Für jedes Modul und jeden Parameter kann der Fabrikplaner zunächst die entsprechenden Datenquellen mit den definierten Datenvorlagen auf die Plattform hochladen und den Integrationsprozess starten (Abbildung 5).

Die Liste im linken Bereich enthält die aktivierten Module und Parameter des Planungsprojektes. Über die Auswahl eines Moduls kann der Nutzer auf der rechten Seite die Modulstruktur betrachten und die Datenintegration verwalten. Der entscheidende Mehrwert der VPI-Plattform liegt darin, dass dem Nutzer nach erfolgreicher Integration unmittelbar verschiedene Datenansichten wie die dargestellte

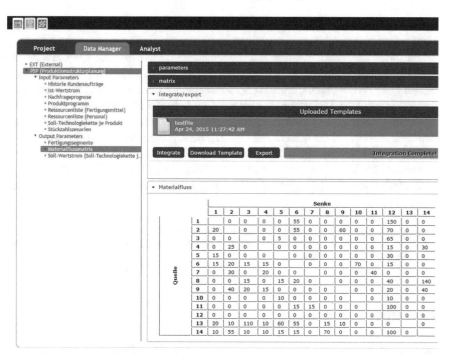

Abbildung 5 VPI-Plattform für Fabrikplanung

Materialflussmatrix präsentiert werden, die Daten aus unterschiedlichen Quellen zusammen führen.

5 Zusammenfassung und Ausblick

In diesem Beitrag wurden die grundlegenden Prozesse der Informationsmodellierung und -integration der Virtual Production Intelligence in Hinblick auf die Umsetzung eines kontinuierlichen und konsistenten Informationsmanagement für das Aachener Fabrikplanungsvorgehen vorgestellt. Der Fokus liegt dabei auf der semantischen Annotation von Planungsdaten, die mittels einer expliziten Formalisierung der Begriffe, Beziehungen und Einschränkungen in Form von Domänenontologien realisiert werden. Für einen einheitlichen Zugriff auf sämtliche Daten, die innerhalb eines Planungsprojektes generiert werden, bietet die VPI-Plattform einen Mehrwert, indem die relevanten Informationen aus den Quellen automatisiert integriert und visualisiert werden. Auf dem semantischen Informationsmodell basierend werden die Daten während der Integration automatisch validiert, um eine hohe Datenqualität zu gewährleisten. Die sich ergebende konsolidierte Informationsbasis ermöglicht es Fabrikplanern, ganzheitliche Bewertungen durchzuführen, die bisher nicht möglich waren.

Im nächsten Schritt werden das Informationsmodell und die Ontologien des genannten Anwendungsszenarios vervollständigt und validiert. Ziel ist es, alle relevanten Daten eines Planungsprojektes zu integrieren, um eine vollständige Informationsbasis zu bieten. Dafür werden weitere Module des CBFP integriert, um alle Arten von Fabrikplanungsprojekten zu modellieren. Gleichzeitig werden die bestehenden Analysealgorithmen verbessert und ergänzt, neue Kennzahlen zur Informationsbewertung entwickelt und die Visualisierung mit Hilfe eines Kennzahlen-Cockpits und Methoden der Visual Analytics verbessert. Schließlich wird die VPI-Plattform in realen Fabrikplanungsprojekten zur Bewertung der verbesserten Entscheidungsunterstützung eingesetzt.

Danksagung Die hier vorgestellten Ansätze werden von der Deutschen Forschungsgemeinschaft (DFG) im Rahmen des Exzellenzclusters Integrative Produktionstechnik für Hochlohnländer an der RWTH Aachen University unterstützt.

Literaturverzeichnis

1. C. Brecher, S. Jeschke, G. Schuh, S. Aghassi, J. Arnoscht, F. Bauhoff, S. Fuchs, C. Jooß, W.O. Karmann, S. Kozielski, S. Orilski, A. Richert, A. Roderburg, M. Schiffer, J. Schubert, S. Stiller, S. Tönissen, F. Welter, Integrative production technology for high-wage countries. In: *Integrative production technology for high-wage countries*, ed. by C. Brecher, Springer, Berlin and Heidelberg, 2012, pp. 17–76
2. P. Burggräf, Wertorientierte Fabrikplanung. Phd thesis, RWTH Aachen University, RWTH Aachen University, 2012
3. J.A. Tompkins, J. White, Y. Bozer, J. Tanchoco, *Facilities planning*, 4th edn. Wiley, Hoboken and NJ, 2010
4. U. Bracht, D. Geckler, S. Wenzel, *Digitale Fabrik. Methoden und Praxisbeispiele*. Springer, Berlin and New York, 2011
5. VDI Richtlinie. Digitale Fabrik – Grundlagen, 2008
6. F. Himmler, M. Amberg, Die Digitale Fabrik - eine Literaturanalyse. 11th International Conference on Wirtschaftsinformatik, 27th February – 01st March 2013, Leipzig, Germany, 2013, pp. 165–179
7. U. Leser, F. Naumann, *Informationsintegration. Architekturen und Methoden zur Integration verteilter und heterogener Datenquellen*, 1st edn. Dpunkt, Heidelberg, 2007
8. T. Meisen, P. Meisen, D. Schilberg, S. Jeschke, Adaptive information integration: Bridging the semantic gap between numerical simulations. Proceedings of the 13th International Conference on Enterprise Information Systems, 2011, pp. 51–65
9. R. Reinhard, C. Büscher, T. Meisen, D. Schilberg, S. Jeschke, Virtual production intelligence – a contribution to the digital factory. Proceedings of the 5th International Conference on Intelligent Robotics and Applications (ICIRA 2012) (1), 2012, pp. 706–715
10. C. Büscher, H. Voet, T. Meisen, M. Krunke, K. Kreisköther, A. Kampker, D. Schilberg, S. Jeschke, Improving factory planning by analyzing process dependencies. Procedia CIRP **17**, 2014, pp. 38–43
11. H. Krcmar, *Einführung in das Informationsmanagement*. Springer, Berlin, Heidelberg, 2011
12. K. Siau, Information modeling and method engineering. Journal of Database Management **10** (4), 1999, pp. 44–50

13. T. Berners-Lee, J. Hendler, O. Lassila, The semantic web: A new form of web content that is meaningful to computers will unleash a revolution of new possibilities. Scientific American **284** (5), 2001, pp. 34–43

14. T.R. Gruber, A translation approach to portable ontology specifications. Knowledge Acquisition **5** (2), 1993, pp. 199–220

15. w3c owl Working group, Owl 2 web ontology language document overview, 11 December 2012. http://www.w3.org/TR/owl2-overview/

16. W. Terkaj, G. Pedrielli, M. Sacco, Virtual factory data model. In: *Proceedings of the Workshop on Ontology and Semantic Web for Manufacturing (OSEMA 2012)*, ed. by D. Anastasiou, L. Ramos, S. Krima, Y.J. Chen, 2012, pp. 29–43

17. M. Rezaei, Konzeption eines fabrikinformationsmanagements auf basis von postrelationalen datenbanken: Ein beitrag zur unterstützung von concurrent und simultaneous engineering in der integrierten prozessgestaltung. Dissertation, Technische Universität Darmstadt, 2010

18. T. Weimer, Informationsmodell für die durchgängige datennutzung in fabrikplanung und -betrieb. Phd thesis, University of Stuttgart, University of Stuttgart, 2010

19. G. Schuh, A. Kampker, C. Wesch-Potente, Condition based factory planning. Production Engineering **5** (1), 2011, pp. 89–94

20. T. Meisen, Framework zur Kopplung numerischer Simulationen für die Fertigung von Stahlerzeugnissen. Dissertation. RWTH Aachen University, 2012

Implementing a Volunteer Notification System into a Scalable, Analytical Realtime Data Processing Environment

Jesko Elsner, Tomas Sivicki, Philipp Meisen, Tobias Meisen and Sabina Jeschke

Abstract The pace at which next-generation Internet of Things networks, consisting of wirelessly distributed sensors and devices, are being developed is speeding up. More and more devices produce data in automated manners and the demand of smartphones and wearable devices is continuously increasing. With respect to volunteer notification systems (VNS), the resulting vast amounts of data can be utilized for profiling and predicting the whereabouts of people that, combined with machine learning algorithms, complement artificial intelligence (AI)-based decision systems. Hence, VNS benefit from keeping pace with the current developments by using the corresponding data streams in order to improve decision making during the volunteer selection process. In emergency scenarios, the velocity, low latency and reaction times of the system are essential, which results in the need of online stream-processing and real-time computational solutions. This paper will focus on a basic concept for implementing a VNS approach into a scalable, fault-tolerant environment that uses state-of-the-art analytical tools to process information streams in real-time as well as on demand, and applies machine learning algorithms for an AI-based volunteer selection. This work concentrates on leveraging open source Big Data technologies with the aim to deliver a robust, secure and highly available enterprise-class Big Data platform. Within the given context, this work will furthermore give an insight on state-of-the-art proprietary solutions for Big Data processing that are currently available.

Keywords Volunteer Notification System · Internet of Things · Big Data · Stream Processing · Machine Learning

J. Elsner (✉) · T. Sivicki · P. Meisen · T. Meisen · S. Jeschke
IMA/ZLW & IfU, RWTH Aachen University, Dennewartstr. 27, 52068 Aachen, Germany
e-mail: jesko.elsner@ima-zlw-ifu.rwth-aachen.de

Originally published in "eTELEMED 2015 The Seventh International Conference on eHealth, Telemedicine, and Social Medicine", © International Academy, Research, and Industry Association (IARA) 2015. Reprint by Springer International Publishing Switzerland 2016, DOI 10.1007/978-3-319-42620-4_64

1 Introduction

As we are moving towards the Internet of Things (IoT), the number of sensors that are deployed around the world, and devices supporting various different sensory technologies, is growing at a rapid pace [1]. These sensors and devices continuously (and automated) generate high amounts of data. However, in order to add value to the collected raw data, further processing is required that will help understanding the meaning and correlations within.

Bundling the accumulated data into a so called real-time information pipeline does enable scalable real-time query/in-stream processing technologies [2] and regular batch processing, which is currently supported by various state-of-the-art Big Data analytical environments, as will be discussed later. To a given problem (query), the introduced approach will process both persisted as well as real-time data to generate results, which can be further processed instantaneously or stored for subsequent processing. Various machine learning extensions on top of the basic environment do furthermore provide possibilities for extensive profiling and learning approaches that are based on the collected data, whereas the resulting decisions are generated near real-time, enabling a scalable volunteer selection architecture within the application scenario of a Volunteer Notification System (VNS), as primary introduced in [3].

Hence, this paper is going to provide an insight of the various technologies that can be efficiently used in order to create a scalable, reliable and fault tolerant environment as architectural base for a reasonable VNS implementation.

1.1 Structure

Section 1 will continue by introducing the various terminologies that are used throughout this work, whilst Section 2 will discuss the state-of-the-art with respect to the (Big Data) domain specific technologies and analytical frameworks. Section 3 will give detailed insights on the basic implementation approach and the corresponding concepts and methods, discussing the scalability effects (of the most problematic system components) of the underlying technologies in comparison. The last section, Section 4, will present a brief conclusion on the elaborated approach and shortly discuss those proprietary solutions and standards that are currently well established in the industry.

1.2 Volunteer Notification System

A VNS is an approach to integrate laypersons and medically trained volunteers into emergency medical services (EMS). By tracking the users' location, and in case of a medical emergency, a VNS aims to alarm those potential voluntary first-aiders who can arrive on scene fast enough to provide the most urgent measures until professional EMS arrive at the victims location.

Whilst the volunteer selection process can be efficiently enhanced by an AI-driven selection system [4], rather than merely using the last known location of a volunteer, this general approach is greatly limited by the input data stream and the available processing power. Thus, in order to provide a technical solution for the basic research questions in regards to an intelligent VNS, the scope of this work will focus on providing a solution in which the supported input data – that is generated by a multitude of devices – ideally is limitless and the computational power will be matter of theoretically seamless scalability.

1.3   The Internet of Things Paradigm

The IoT paradigm proposes that everyday objects will be globally accessible over the Internet or other adequate network structures. Opposite to the Internet world, things with a physical shape usually belong to resource-challenged environments where energy, data throughput, and computing resources are scarce.

The focus of typical IoT activities lies on establishing connectivity at a certain protocol level to enable truly distributed machine-to-machine (M2M) applications. In the general protocol specification, the devices must communicate with each other (D2D). A device's data then must be collected and forwarded to the server infrastructure (D2S), whereas the server infrastructure will share the various device data (S2S), possibly providing it back to devices, analytical environments, people and any other subscriber for a specific type of data.

In regards to a VNS, the specific machines are handheld or wearable devices and corresponding servers. Hence, a device-to-server (D2S) infrastructure and a protocol that will secure this communication environment against data loss and eavesdropping, fulfills the basic requirement in the context of a VNS approach. A communication protocol of this type is the commonly used MQ Telemetry Protocol (MQTT) [5]. As device-to-device communication is not necessarily needed within a VNS approach, a pub/sub messaging system similar to a push notification system as lightweight as MQTT offers a suitable approach to fulfil the systems' communication requirements. A more in-depth view about MQTT and similar pub/sub systems will be discussed in Section 2.

1.4   Big Data in the Context of a VNS

In a data-driven society, massive amounts of data are being collected from people, sensors, algorithms and of course, the Web itself; storing it in conventional database systems (i.e., online transaction processing) or data warehouses (i.e., online analytical processing) that itself conform to an additional layer on top of single or multiple databases. The term Big Data describes the challenge for handling this continuously increasing data, whereas mainly three reasons posture the arising difficulties: the sheer volume, the velocity (how fast new data is continuously produced) and the variety of different data-types. For some time, an additional challenge has been observed; the so called veracity, which describes the challenge to exclude uncertainty and inconsistency within the collected data.

The VNS must handle these challenges gracefully and overcome the resulting difficulties with scalability and reliability in terms of the technologies that are being implemented. In general, the system approach that is to be illustrated in the upcoming sections of this work will be able to handle large amounts of continuously generated input data and will furthermore be able to detect faulty (i.e., inconsistent) information in an online matter.

1.5 Stream Processing

As computer systems are creating ever more data at increasing speeds, Hadoop-style batch processing has awakened engineers to the value of big data analysis, whereas the current trend is focusing on the demand for real-time processing. In essence, people do not only want all of their data analyzed, but they want it done as soon as possible, which is driving the current Big Data research trend towards so called high-velocity data [6]. Exemplary use cases within this context are real-time analytics, machine learning, and new generation of decision support and fraud detection systems [7].

The desire to extract real-time insight from high-velocity data led to the creation of so called Stream Processing Engines. These engines include open source projects, such as Twitter's Storm [8], Apache Spark [9] and LinkedIn's Samza [10] as well as proprietary solutions, such as Amazon Kinesis [11] or Google's BigQuery [12]. These engines provide functionalities for routing, transforming and analyzing streams of data at high-velocity for a specified time window or near real-time (depending of the velocity and volume of streamed data chunks). The classical approach in this context would instead store the real-time data in order to apply data warehousing techniques for batch-processing in a subsequent matter. Figure 1 illustrates the conceptual coherence of the IoT paradigm and real-time Big Data Analytics within the context of an intelligent volunteer selection system.

2 State of the Art

2.1 Pub/Sub Messaging Systems

Publish-subscribe is a messaging pattern in which occurring messages are not sent directly to a target receiver but rather published to a channel. Subscribers have the

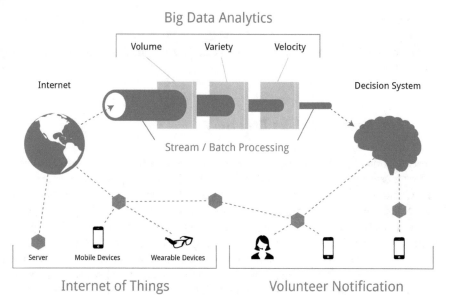

Figure 1 Big data analytics within a VNS

option to subscribe themselves on specific topics or channels and hence express their interest on receiving specific messages. The result is a lose coupling between publisher and subscriber, as they are unaware of each other.

In many pub/sub systems, publishers post messages to an intermediary message broker or event bus, and subscribers register subscriptions with that broker, letting the broker perform any type of necessary filtering. Pub/sub Messaging Systems allow implementation of a device-to-device, device-to-server and server-to-server interface, as have been introduced earlier.

The MQTT protocol on the other hand is a lightweight messaging protocol that uses a publish/subscribe architecture to deliver messages over low bandwidth or unreliable networks with a low footprint. Compared to a classical REST/HTTP implementation [45], MQTT imparts various advantages for the use within mobile applications, such as faster response times, higher throughput, higher messaging reliability, lower bandwidth usage and lower battery consumption [43].

In this context, Apache Kafka [13] is a publish/subscribe log for integrating data between applications, stream processing, and Hadoop data ingestion. The project aims to provide a unified, high-throughput, low-latency platform for handling real-time data feeds. The design is heavily influenced by transaction logs to prevent data corruption and/or loss. On the server side, Apache Kafka will be used to create a pipeline between the MQTT broker cluster and the Hadoop/Spark environment to persist and stream process data; it will be managed by Apache Zookeeper for scalability and reliability purposes.

An alternative to MQTT in a proprietary environment are Amazon SNS, Amazon SQS as well as Amazon Kinesis, which are all capable of real-time streaming/distributing data between applications merely within Amazon Web Services (AWS) [14].

2.2 The Apache Hadoop Ecosystem

Apache Hadoop [15] is an open source software project that enables the distributed batch processing of large data sets across clusters of commodity servers. It is designed to scale up from a single server to thousands of machines, with a very high degree of fault tolerance. Hadoop is supplemented by an ecosystem of Apache projects, such as Pig, Hive and Zookeeper and many more, which extend the value of Hadoop and improves its usability. The core part of Hadoop is the Hadoop file system (HDFS) which comprises two major components: namespaces and block storage service. The namespace service manages operations on files and directories, such as creating and modifying files and directories, whilst the block storage service implements the actual data node cluster management, resulting block operations and replication.

Hadoop was often criticized [16, 17] for its open-source implementation of the MapReduce model [18] based on so called JobTrackers, which due to its problematic structure have be resolved with the implementation of Apache YARN [19] and MapReduce 2 in the scope of Hadoop 2.x. YARN is a resource manager that is based on separating the processing engine and resource management capabilities of MapReduce as it was implemented in Hadoop's original approach. YARN is often called the operating system of Hadoop because it is responsible for managing and monitoring workloads, maintaining a multi-tenant environment, implementing security controls, and managing high availability features of Hadoop. One crucial advantage of YARN in the context of using the Hadoop ecosystem for the VNS implementation is that is allows multiple processing models to be implemented on top of HDFS, thereby allowing Apache Spark to fit into the Hadoop Ecosystem [20]. The resulting flexible architecture allowed companies as Amazon and Google to create cloud computing platforms (e.g., Amazon EMR and Google's Cloud Platform) which implement enterprise-features out of the box and give a transparent in-depth cost overview.

2.3 Apache Spark

Apache Spark is a cluster computing platform similar to Hadoop designed to be fast and of general-purpose. Spark extends the popular MapReduce model to efficiently support more types of computations, including interactive queries and stream processing. One of the main features that Spark offers, is the ability to run even huge computational queries fully in memory (split over various clusters), reaching performance gains of up to 100 times compared to general Hadoop MapReduce implementations under specific circumstances. However, the system itself is also faster than MapReduce when running merely on disc operations.

At its core, the Spark Engine itself is responsible for scheduling, distributing, and monitoring applications consisting of many computational tasks across many worker machines powered by a high-level structure of components. These components are designed to interoperate closely, supporting a library-like combination of the various data representations (graphs, matrices, SQL like queries). Spark revolves around the concept of a resilient distributed dataset (RDD), which is a fault-tolerant collection of elements that can be operated in parallel. There are currently two types of RDDs: firstly parallelized collections, parallelizing an existing collection in your driver program, and secondly by referencing a dataset in an external storage system supported by Hadoop (e.g., the local file system, HDFS, Cassandra, Amazon S3). This allows Spark to interoperate with various stable established solutions in order to efficiently focus on problems regarding the introduced big data challenges. A recent cloud service that is entirely based on Spark and runs on AWS has been introduced by Databricks (who also drove the adoption of the Apache Spark ecosystem) in 2014. It allows developers to create scalable computing clusters running on Apache Spark for data analysis, machine learning and similar use cases.

This work will incorporate Apache Spark and its core components as the main cluster computing platform to overcome weaknesses of classical Hadoop architectures and to support the incorporation of the various proprietary solutions, such as Amazon Web Services and the Databricks Cloud Platform.

2.4 Data Streaming and Processing

LinkedIn's Kafka was designed to support not merely the distribution of data, but also to provide the infrastructure primitives that will enable real-time data processing. Samza on the other hand provides elastic, fault-tolerant processing as being layered on top of real-time feeds. A simple analogy in respect to the batch domain is described by Kafka taking the role of HDFS while Samza relates to MapReduce.

While this architecture scales horizontally due to its MapReduce nature, speed is an important factor which needs to be considered. A combination of Apache Kafka with various Spark components (i.e., Spark SQL, MLlib and Streaming Processor) will result in a more reliable, vertically and horizontally scalable high-velocity architecture. The lack of security options within Kafka and Samza are an important criteria for using Spark's Security implementation and an integrated secure tunnel between Kafka and the corresponding MQTT brokers.

In terms of security, scalability and reliability a commercial solution with Amazon Kinesis and Amazon Elastic MapReduce provides leverage to these problems, including the high-velocity implementation of Spark components, which replicates the scenario in a more enterprise-ready fashion.

As the fault tolerance plays an additional key role for a successful scalable VNS implementation, Apache Cassandra [21] is the state-of-the-art database system in combination with Spark technologies; highly robust and fault tolerant. It protects against data loss or corruption by replicating blocks of data to multiple nodes and supporting replication between geographically distributed nodes. Amazon and Google offer similar enterprise ready data stores, such as Amazon Redshift [22], Amazon DynamoDB [23] and Google Cloud Datastore [44], whilst a general comparison between the Cassandra File System (CFS) and HDFS is given in [24].

2.5   Webinterfaces and API

Responsive web design architecture and supporting the HTML5 specification, esp. Websocket support [25], is efficiently incorporated by implementing Nginx [26] as a high-performance HTTP server for both, static web data as well as proxy requests to an underlying Node.js [27] runtime environment running server-side applications. Node.js applications are entirely written in JavaScript, whereas Express.js constitutes an adaptable MVC framework [28]. Node.js is characterized to be fast (due to event based architecture), offer high throughput, support high amounts of concurrent connections, support clustering and generally has a very low resource footprint. Offering advanced scalability, load balancing, health checks and some additional features, the Nginx Inc. released an enterprise version under the label: Nginx+ [29] Node.js in this context enables the implementation of simple server applications as well as the requirements in respect to APIs.

3   System Architecture

This section will illustrate the main strategy that incorporates the introduced technologies into a general system architecture that conforms to the requirements of an enterprise application.

3.1   Realtime Data Pipeline

Within a VNS, the data that is to be analyzed is generated by individual mobile or wearable devices. As illustrated in Figure 2, clients publish their data to a server which is connected to a message broker, which is responsible for broadcasting the received messages to the corresponding subscribers. Whilst a standard MQTT broker solution is lightweight and performant for a limited amount of connected clients (due to limits in the port range), a horizontally scalable approach will have to balance the various connections between multiple instances (load balancing) residing

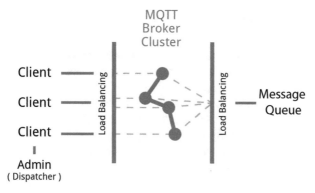

Figure 2 Realtime data pipeline

on different machines. As clients generally subscribe to specific topics in order to achieve push-like notifications, horizontal scaling will result in brokers having different information and topic structures.

To solve this problem, the various brokers (i.e., nodes) need to be connected with each other and share their message structure and permissions, forming a cluster of machines that can be scaled at will. Modern systems, such as RabbitMQ [30] and Apache's ActiveMQ [31], support the application of efficient clustering. Mirroring the message queues between all machines will allow the subscribers to connect to any existing node while still having access to the whole cluster. Established commercial projects that support scalable messaging systems and efficient load balancing for MQTT connections are: HiveMQ [32], CloudAMQP [33] and CloudMQTT [34].

3.2 Load Balancing

Since most standard load balancing approaches, such as Amazon's Elastic Load Balancer, only support Round Robin (RR) and Session Sticky Algorithms, they are not sufficient for balancing MQTT clients or applications between brokers. The already introduced commercial Nginx+ solution supports various advanced load balancing strategies [35], but even the open-source standard Nginx version can be extended with additional functionalities by incorporating the programming language LUA and a TCP-proxy module to support the programmatic injection of algorithms that can filter requests of clients and balance connections between brokers with high performance. This added functionality enables a distinguished consideration of the various active brokers in order to terminate obsolete sessions, run additional scripts for scaling the cluster, and perform regular health checks on running instances.

3.3   Ad-Hoc/Online Computation

As described in [36], ad-hoc computation on message brokers is efficiently achieved by combining Apache Kafka with Apache Sparks infrastructure; since Kafka efficiently persists the message queue on a data store (e.g., Cassandra or HDFS) while Apache Spark handles workloads both in real-time as well as by batch processing. Kafka is guaranteed to deliver reliable message durability and a fault-tolerant near real-time computation with Spark Streaming [37]. At this point, one might argue about missing security measures within Apache Kafka [38].

Whereas various other messaging platforms (e.g., RabbitMQ) support the persistence of incoming data on data stores, they are usually not performant enough or simply not optimized for processing environments such as Kafka, which itself is very robust in throughput of messages and during read/write operations [39]. Whilst the Apache Spark libraries provide methods for connecting to MQTT brokers and streaming data, the underlying communication has to be implemented manually. In contrast, Kafka can be implemented as a complementing stream processing layer between the MQTT cluster and Apache Spark [40].

Within a VNS, the streamed data will mainly consist of location data of individual volunteers and case update data. Thus, stream processing will be applied to regulate updates concerning a specific case in real-time; deriving decisional, predictive or anomaly detection results. However, an efficient volunteer selection, based on accumulated profile data, will mostly be computed in batches, as discussed in the upcoming section.

As data store, Apache Cassandra constitutes a high performance scalable database with linear scaling that secures an enterprise-ready solution for this work. Similar, proprietary options are Amazon DynamoDB and Amazon RedShift, whereas HDFS would partly limit the performance of Spark and other NoSQL data stores [41].

3.4   Batch/Offline Computation

Batch processing on big amounts of accumulated data is commonly implemented based on Hadoop clusters. Within a VNS, finding the most reasonable candidates for an ongoing medical emergency – within a minimum time interval – hereby constitutes a batch processing problem with an increasing (raw) data size over time. Location based data will be analyzed in order to compute behavioral patterns of volunteers; this can be done on a regular basis (iterative) based on batch processing of the acquired location data and in combination with various machine learning algorithms. The results will be available for additional real-time computations, whereas details for an AI-driven volunteer selection discussed in [42].

Apache Spark can accomplish both tasks of on- and offline computation quite reliable and fast allowing the results to be stored in data stores or be directly accessible via API or MQTT subscribed topics. While real-time per-case computation and live

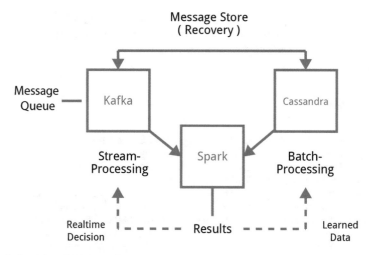

Figure 3 Real time data processing

updates would amass resource consumption it would be possible but unfeasible and unnecessary. With a proper modelling of data stores direct API access allows fast updates of a case without the need of costly computations. Behavioral patterns can be learned after an emergency scenario, as well due to the systems structure.

Figure 3 illustrates the general architecture for a real-time data processing environment, as has been discussed within this section. Nowadays, Amazon EMR, Google Cloud Platform and Databricks deliver the technologies needed for a successful computation environment for similar use cases and allow different services like data stores or real-time computation ecosystems to be fully implemented on commodity hardware.

4 Conclusion

This work illustrated details on how to implement a VNS into a distributed analytical environment with high velocity data support. Scalability and reliability is hereby achieved by utilizing merely open-source software solutions without relying on any commercially driven software or proprietary cloud solutions. While security and special solutions for load balancing and regulating the corresponding environments cannot be guaranteed by open-source Apache software alone, new Big Data challenges arise continuously and more open-source projects are being incubated or upgraded; hopefully solving both, newer as well as older challenges that were formally limited to enterprise solutions.

Acknowledgments This paper is based on work done in the INTERREG IVa project EMuRgency (www.emurgency.eu). The project is partially financed through the European Regional Development Fund (ERDF) and co-financed by several regions and partners of the EMuRgency consortium.

References

1. T. Danova. The Internet of Everything. http://uk.businessinsider.com/the-internet-of-everything-2014-slide-deck-sai-2014-2. October 8, 2014
2. M. Stonebraker, U. Çetintemel, S. Zdonik, The 8 Requirements of Real-time Stream Processing. ACM SIGMOD Record **34** (4), 2005, pp. 42–47
3. J. Elsner, M.T. Schneiders, M. Haberstroh, D. Schilberg, S. Jeschke, An Introduction to a Transnational Volunteer Notification System Providing Cardiopulmonary Resuscitation for Victims Suffering a Sudden Cardiac Arrest. In: *eTelemed2013*. 2013, pp. 59–64
4. J. Elsner, P. Meisen, S. Thelen, D. Schilberg, S. Jeschke, EMuRgency – A Basic Concept for an AI Driven Volunteer Notification System for Integrating Laypersons into Emergency Medical Services. International Journal On Advances in Life Sciences **5** (3 and 4), 2013, pp. 223–236
5. J. Elsner, M.T. Schneiders, D. Schilberg, S. Jeschke, Determination of the Relevant First Aiders within a Volunteer Notification System. In: *MedTel 2013, Luxembourg*. 2013, pp. 245–249
6. C.C. Aggarwal, ed., *Data Streams: Models and Algorithms, Advances in Database Systems*, vol. 31. Kluwer, 2007
7. J. Taylor, *Real-Time Responses with Big Data*. 2014
8. Apache Storm. http://storm.apache.org/. October 12, 2014
9. Apache Spark. https://spark.apache.org/. October 14, 2014
10. Apache Samza. http://samza.apache.org/. October 14, 2014
11. Amazon Kinesis. http://aws.amazon.com/de/kinesis/. October 14, 2014
12. Google BigQuery. https://cloud.google.com/bigquery/. October 14, 2014
13. Apache Kafka. http://kafka.apache.org/. October 15, 2014
14. Amazon Webservices such as SNS or SQS. http://aws.amazon.com/de/sns/. October 15, 2014
15. Apache Hadoop. http://hadoop.apache.org/. October 14, 2014
16. J. Polo, Big Data Processing with MapReduce. In: *Big Data Computing*, 2013, pp. 295–313
17. M. Stonebraker, D. Abadi, D.J. DeWitt, S. Madden, E. Paulson, A. Pavlo, A. Rasin, MapReduce and Parallel DBMSs: Friends or Foes? Communications of the ACM **53** (1), 2010, pp. 64–71
18. R. Lämmel, Google's MapReduce programming model – Revisited. Science of Computer Programming **70** (1), 2008, pp. 1–30
19. V.K. Vavilapalli, A.C. Murthy, C. Douglas, S. Agarwal, M. Konar, R. Evans, T. Graves, J. Lowe, H. Shah, S. Seth, B. Saha, C. Curino, O. O'Malley, S. Radia, B. Reed, E. Baldeschwieler, Apache Hadoop YARN: Yet Another Resource Negotiator. In: *Proceedings of the 4th Annual Symposium on Cloud Computing*. ACM, New York, NY, USA, 2013, SOCC '13, pp. 1–16
20. M. Zaharia, M. Chowdhury, M.J. Franklin, S. Shenker, I. Stoica, Spark: Cluster Computing with Working Sets. In: *Proceedings of the 2Nd USENIX Conference on Hot Topics in Cloud Computing*. USENIX Association, Berkeley, CA, USA, 2010, HotCloud'10, pp. 10–10
21. Apache Cassandra. http://cassandra.apache.org/. October 15, 2014
22. Amazon Redshift. http://aws.amazon.com/de/redshift/. October 15, 2014
23. Amazon DynamoDB. http://aws.amazon.com/de/dynamodb/. October 15, 2014
24. Datastax Corporation, Comparing the Hadoop File System (HDFS) with the Cassandra File System (CFS). Tech. rep., 2013. http://www.datastax.com/resources/whitepapers/hdfs-vs-cfs. April 16, 2015
25. I. Fette, A. Melnikov, The WebSocket Protocol. IETF, 2011
26. W. Reese, Nginx: the high-performance web server and reverse proxy. Linux Journal **173**, 2008
27. R. Fielding, U. Irvine, J. Gettys, J. Mogul, H. Frystyk, L. Masinter, P. Leach, T. Berners-Lee, Hypertext transfer protocol – HTTP/1.1. RFC Editor, 1999

28. C. Le, X. Yang, Research of applying MVC pattern in distributed environment. Computer Engineering **32** (19), 2006, pp. 62–64
29. Nginx+. http://nginx.com/products/technical-specs/. October 16, 2014
30. RabbitMQ. http://www.rabbitmq.com/. October 16, 2014
31. ActiveMQ. http://activemq.apache.org/. October 16, 2014
32. HiveMQ. http://www.hivemq.com/. October 16, 2014
33. CloudAMQP. https://www.cloudamqp.com/. October 17, 2014
34. CloudMQTT. http://www.cloudmqtt.com/. October 17, 2014
35. A. Piórkowski, A. Kempny, A. Hajduk, J. Strzelczyk, Load Balancing for Heterogeneous Web Servers. In: *Computer Networks*, Springer, 2010, pp. 189–198
36. Real time Analytics with Apache Kafka and Apache Spark. http://de.slideshare.net/rahuldausa/real-time-analytics-with-apache-kafka-and-apache-spark. October 17, 2014
37. M. Zaharia, T. Das, H. Li, T. Hunter, S. Shenker, I. Stoica, Discretized Streams: Fault-tolerant Streaming Computation at Scale. In: *Proceedings of the Twenty-Fourth ACM Symposium on Operating Systems Principles*. New York, NY, USA, 2013, SOSP '13, pp. 423–438
38. Samza / Kafka Security. http://bit.ly/1pl550b. October 17, 2014
39. J. Kreps, N. Narkhede, J. Rao, Kafka: A distributed messaging system for log processing. NetDB, 2011
40. R. Ranjan, Streaming Big Data Processing in Datacenter Clouds. IEEE Cloud Computing **1** (1), 2014, pp. 78–83
41. J. Zollmann, NoSQL Databases. Proceedings of the NetDB, 2011. http://sewiki.iai.uni-bonn.de/_media/teaching/labs/xp/2012b/seminar/10-nosql.pdf
42. J. Elsner, An AI Driven Volunteer Selection System. Ph.D. thesis, Aachen, 2015
43. MQTT Protocol 3.1.1. Spec. http://docs.oasis-open.org/mqtt/mqtt/v3.1.1/mqtt-v3.1.1.html. February 10, 2014
44. Google Cloud Datastore. https://cloud.google.com/datastore/. October 15, 2014
45. E. Wilde, Putting Things to REST. School of Information, 2007. Series: Recent Work

Continuous Integration of Field Level Production Data into Top-Level Information Systems Using the OPC Interface Standard

Max Hoffmann, Christian Büscher, Tobias Meisen and Sabina Jeschke

Abstract On the way to the fourth industrial revolution, one major requirement lies in reaching interoperability between hardware and software systems. Especially real-time propagation of shop floor information in top-level production planning and control systems as well as the consolidation of distributed information into a consistent data basis for comprehensive data analysis are still missing in most production environments. Existing approaches to serve interoperability through standardized interfaces are limited by proprietary data exchange protocols and information models. Within industrial manufacturing and automation, standardization attempts between these systems are primarily focused on industrial interfaces like OPC/OPC-UA. However, the aggregation of data created by devices like sensors or machinery control units into useful information has not been satisfactorily solved yet as their underlying models are carried out using different modeling paradigms and programming languages, thus inter-communication is difficult to implement and to maintain. In this work, an integration chain for data from field level to top-level information systems is presented. As Manufacturing Execution Systems or Enterprise Resource Planning tools are implemented in higher programming languages, the modeling of field level information has to be adapted in terms of a semantic interpretation. The approach provides integration capabilities for OPC-conform data generated on the field level. The information is extracted from low level information systems, transformed according to object-oriented programming paradigms and object-relational standards and finally integrated into databases that allow full semantic annotation and interpretation compatible to a common information model. Hence, users on management levels of the enterprise are able to perform holistic data treatment and data exploration along with personalized information views based on this central data storage by means of a reliable and comfortable data acquisition. This increases the quality of data and of the decision support itself, as more time remains for the actual task of data evaluation.

Keywords Interoperability · Information Integration · Ontology · Semantic Data · OPC · OPC UA

M. Hoffmann (✉) · C. Büscher · T. Meisen · S. Jeschke
IMA/ZLW & IfU, RWTH Aachen University, Dennewartstr. 27, 52068 Aachen, Germany
e-mail: max.hoffmann@ima-zlw-ifu.rwth-aachen.de

Originally published in "Procedia CIRP",
© Elsevier 2016. Reprint by Springer International
Publishing Switzerland 2016, DOI 10.1007/978-3-319-42620-4_65

855

1 Introduction

In terms of the fourth industrial revolution – also referred to as Industry 4.0 – it is the aim of the manufacturing industry to bring together all information from distributed production facilities and levels and in a holistic and continuous way to provide decision support on various levels and throughout the entire process chain. This includes not only the gathering of information from different levels and areas of the production. One major focus also consists in creating flexible automation systems and vertical control flows to enable reconfiguration capabilities of the production in real-time.

However, most automation systems in existing factories are characterized by static hierarchies and rigid control/automation flows. In most cases, control flows are organized top-down, whereas information flows and the gathering of data are bottom-up. Traditionally, the planning of processes and information exchange in industrial environments is inspired by the automation pyramid [1], which serves as a general design pattern for creating information and communication infrastructures (ICT) for the industry. A simple representation that focuses on the main layers of the automation pyramid is depicted in Figure 1. On top of the automation pyramid the Enterprise Resource Planning (ERP) systems are located.

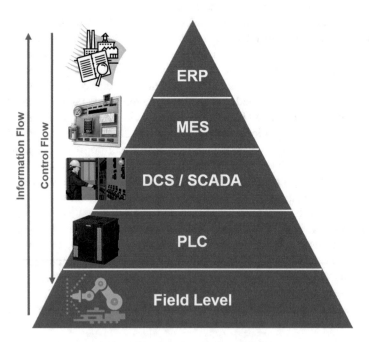

Figure 1 Automation pyramid for static control and information flows

ERP systems perform the resource planning of a company in terms of human and material resources. They are primarily used for long-term planning, e.g. in terms of process chains [2]. Manufacturing Execution Systems (MES) are responsible for mid-term production planning and execution. During the manufacturing, Process Distributed Control Systems (DCS) as well as Supervisory Control and Data Acquisition (SCADA) systems control conditions and system states during operation to avoid critical problems or serious failures in the production flow. Programmable Logic Controllers (PLC) are digital computers that are in charge of controlling signals of field devices like the machinery or other components of industrial environments. The field level is characterized by low level devices that are represented by the machines or sensors. These devices generate data that is needed to perform process optimization or to detect problems in the production flow.

Although there is a connection especially between the field level devices and PLC as well as SCADA systems, the information flow is mostly limited to rigid interfaces that cannot dynamically interpret and aggregate information in an appropriate way. In most cases, the data flow is limited to direct control of the production, i.e. MES are mostly not provided with real-time information from the field level, e.g. to perform rescheduling, real-time or even self-optimization [3]. This issue is due to the lack of powerful interfaces between low level and high level production systems. Another reason for preferring rigid interfaces over flexible automation systems is the aim for a complexity reduction on the field level, which is highly connected to robust and well-determined operational processes in the production to make processes more secure. However, whereas low level networks are primarily based on the communication over proprietary bus systems, high level information systems require networking connections that are based on (Industrial or office) Ethernet technologies.

In industrial practice, most of low level production systems are based on traditional data acquisition technologies as the Object Linking and Embedding (OLE) for Process Control – also known as OPC – that is performant and reliable on the one hand, however, on the other hand, very limited in terms of intelligent classification or annotation of data. Another drawback is lacking connectivity to networks of higher information systems. As a result, data from field level devices can only be manually integrated. This limits the usage of production information to batch-wise processing. This issue constitutes a major issue in providing real-time functionalities and optimization capabilities for production environments, which are required for addressing the ideas of the Industry 4.0.

In order to address this problem, production environments could be equipped with novel technologies like OPC Unified Architecture (OPC-UA), which involves an adaptation of the entire information and communication infrastructure (ICT) of a manufacturing enterprise. As these major changes of ICT environments within companies are usually connected to a high amount of time and money, most enterprises refuse a radical modernization of their running systems.

As a result, the true challenge of enabling scenarios of the Industry 4.0 in existing production sites is to extend capabilities of the hardware infrastructure that is in use to implement modern ways of information management in currently running environments without designing an entirely new process.

It is the aim of the present paper to introduce an approach to interconnect low level production networks that are based on traditional interface technologies like OPC with higher level information systems like MES or ERP by bridging OPC-based production information over an OPC-UA interface. By enabling full interoperability between these technologies, the data from field devices can be appropriately enriched with semantic information of the production environment or with additional meta-data that is important for data treatment steps at a later point or for real-time optimization purposes.

A real-time simulation of a combined OPC/OPC-UA production network is provided in order to proof the added value of creating generic interoperability between the technologies. While addressing these demands for interoperability, the paper attempts to answer the following research questions:

- Is it possible to couple OPC and OPC-UA based system within a production environment in a generic way?
- How can OPC components or sensors be appropriately simulated in order to investigate a holistic usage of data from OPC-based devices in higher information systems?
- Is it possible to simulate the behavior of such production networks under real conditions, i.e. in real-time?

To find resilient answers to these questions, the state of the art of automation systems with a focus on OPC technologies is summarized in Section 2. In Section 3, an approach for simulating a combined OPC/OPC-UA production network with dynamic data exchange and enrichment capabilities is described in detail. Section 4 validates the modelling approach by simulating a manufacturing environment based on a use case that has been derived from information of a real production environment. Section 5 concludes the results of the present paper and investigates further research opportunities in order to expand our research results into industrial practice.

2 State of the Art

In traditional automation, low level production facilities are integrated into the ICT in a manual and rarely standardized way. To connect field devices with higher systems of the automation pyramid, interfaces/drivers have to be carried out specifically for each different device in the machine layer.

To facilitate the workflow of embedding devices into the ICT, standard interfaces or data protocol conventions are used to enable plug-and-play or – in terms of the manufacturing area – plug-and-produce capabilities, which work similar to the connection of devices in computational environments.

Thus, in the beginning of the 1990s, manufacturing enterprises and members of the automation industry attempted to carry out a standard interface that was based on the Windows NT standard as most widespread operating system throughout all companies. This Operating System (OS) provides the OLE technology to interconnect multiple applications on the OS. The aim of the manufacturing companies was to establish similar approaches for connecting field devices with control systems of the automation layer. In 1995, a task force of big companies and automation providers like Siemens, Rockwell Automation, General Electric and ABB came up with the OLE for Process Control (OPC) standard, which uses the Distributed Component Object Model (DCOM) [4] for the linking of production facilities [5]. The central specification of this approach is the OPC Data Access (OPC DA) specification and has been published in 1996 [6]. By standardizing the access to data in automation systems, it is possible to embed driver or interface specific information directly into the field devices. This approach enables plug-and-produce capabilities without manually configuring each device.

The basic communication functionalities of OPC DA are based on the server-client principle. The OPC Server reads and propagates data from field level or from other data gene-rating devices. In terms of building an OPC network, the OPC Client creates an instance of the server. This server represents a single device or a group of data sources (Figure 2).

The OPCItems represent concrete objects, e.g. sensors in an automation system. These can be grouped to OPCGroups for similar items. The OPCServer represents these items and propagates single or aggregated data into automation systems. OPC Servers can store information, whereas OPC Clients read and redistribute these information. Besides, every OPC Client can also function as an OPC Server. Due to the success of this approach, the industrial users demanded for additional functionalities for the OPC standard, necessary for a full migration of the automation environment. These demands resulted in two additional OPC specifications, the OPC Alarms & Events (OPC A&E) and the OPC Historical Data Access (OPC HDA) specification in 1999 and 2001. The OPC A&E specification provides services to trigger real-time

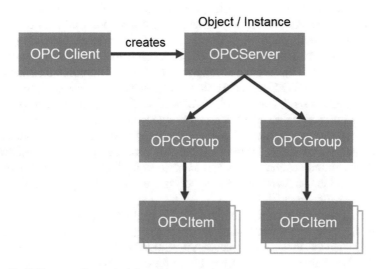

Figure 2 OPC server-client principle and item organization

actions based on events or critical system states that could be harmful to the process or to the automation systems. The OPC HDA specification provides functionalities to access data from previous processes for data acquisition purposes.

The success of OPC in the 1990s was mainly due to the high performance of the DCOM technology and due to the robustness of the resulting automation system [6]. However, the composition of OPC-based automation systems is rather rigid and hierarchical. Another drawback of the OPC standard is the lack of communication capabilities in networks as the DCOM standard cannot be properly configured to work with firewalls. As a result, the OPC consortium figured out ways to communicate over internet based systems, accordingly publishing the OPC XML-DA standard, which enables the propagation of OPC-based information via web services [5].

However, OPC XML-DA turned out to be comparatively slow, which is mainly due to the high overhead of information that has to be integrated in each XML message for each piece of information sent between OPC servers and clients.

Due to the resulting interoperability problems of classical OPC solutions, and due to the dependency on Windows-based systems and bad configurability characteristics in terms of industrial networks, the demand for a modernized version of OPC rose up from the industry. As a consequence, the OPC Unified Architecture (OPC-UA) standard had been carried out, which attempts to combine all specifications of the OPC Classic standard, and to extend their functionalities while enabling platform independence and interoperability [7]. One major functionality of OPC-UA is the possibility to configure lightweight and quick connections over the internet. As OPC-UA is not solely based on DCOM, the communication is performed over network borders and firewalls. This new approach solves major security problems and allows data exchange between distributed systems, as an encapsulation of system devices into a dedicated network, which is – unlike to OPC Classic – not necessary when using OPC-UA [8].

By providing abstract services, OPC-UA delivers basic functionalities to create a service-oriented architecture (SOA) of the automation system [8, 9]. OPC-UA integrates different specifications of OPC into a single set of services. Thus, components and automation networks based on OPC Classic can theoretically be integrated into OPC-UA networks [10]. Another advantage of OPC-UA is the configurability of the new standard. The OPC-UA specification only specifies the message format of the information that is send. Unlike to OPC Classic, OPC-UA does not specify an API [8]. Hence, the user of the OPC-UA server is able to use or implement an API of his personal preference. All communication in the OPC-UA standard is performed using the Communication Stack. There is a client-side and a server-side communication stack. Both communication stack APIs can be developed in individual programming languages, as long as their concepts support the technology mapping given by the OPC-UA specification [8].

The described configurability of the OPC-UA application programming interface provides the user with a freedom of choice in terms of the technology or the programming language that is used to access the information in OPC-UA networks. This way, OPC-UA delivers a flexible usability in different environments and for different purposes. This availability is linked to several advantages:

- The application or the purpose of an information system determines the properties of the OPC-UA communication stack API, e.g. whether the application is complex or lightweight for the usage in small embedded devices.
- The communication with OPC-UA network can be integrated into existing enterprise communication systems.
- Tool interoperability with other systems or components of the factory is guaranteed as the programming language is of free choice. This enables the embedding of OPC-UA systems in business process, e.g. data integration chains.

The advantages of OPC-UA are able to address challenges that are introduced by the Industry 4.0. Thus, many scenarios that are in the scope of the fourth industrial revolution can be realized, if automation systems are fully equipped with UA capable devices and according infrastructures. The embedding of OPC-UA into industrial environments is the first step towards the goals of the Industry 4.0, however there are many more challenges to meet besides technical and syntactical heterogeneity in industrial production. The next steps would consist in creating structural and semantic standards. However, existing production and automation systems are still in use of traditional automation systems, e.g. based on OPC or other rigid interface standards. Most manufacturers do not want to change their entire automation system or simply cannot take the risk of quitting a running system.

This results in a demand of embedding devices like sensors or automation components that are still based on traditional interfaces into an infrastructure that is designed according to the OPC-UA specification. The migration of DCOM-based OPC servers and clients into an OPC-UA infrastructure can be performed in several ways. The first way is to wrap the OPC components using the OPC-UA web service standards, the second approach is to directly implement the OPC Classic protocol according to the OPC-UA specification [8]. The wrapping approach is performed by storing the COM/ DCOM-based information into a web service envelope. This solution is not only slow due to the recombination of two different data formats, there are also security-related issues as the number of exploits to consider increases significantly.

In contrast to the wrapping of OPC DA components, the direct implementation provides several advantages in terms of homogeneity of the resulting application environment. The arising systems are easier to maintain and easier to update as there is only one technology to consider, which is OPC-UA.

However, there are no sufficiently generic approaches to migrate and examine the implementation of OPC/OPC-UA infrastructure as existing solutions require a manual configu-ration. The different paradigms of OPC and OPC-UA require a consistent mapping into one programming language and application environment to examine the cooperation of OPC and OPC-UA based networks under real conditions. The scope of this work is to combine the wrapping approach with direct implementation of OPC devices in terms of an object-oriented programming concept. These devices are consequently embedded into an OPC-UA network.

3 Continuous Integration of OPC Classic Components into an OPC-UA Production Network

The generic integration of OPC Classic components into an ICT based on the OPC-UA interface standard is performed in several steps that are based on a virtual representation of the components and the network for the embedding of a combined OPC/OPC-UA infrastructure:

1. Virtual representation of general OPC components through object-oriented programming concepts (OPC DA Clients),
2. Detailed mapping of OPC Classic networks by embedding of their components into a suitable server environment,
3. Configuration of an OPC-UA production network by using the same object-oriented programming approach as well as by creating an interoperable API,
4. Mapping of the OPC Data Access data model on OPC-UA by considering requirements of the OPC specification, and
5. Distribution of simulated data over OPC-UA networks and OPC Classic based consolidation of consistently annotated information into a database system for further processing.

In Figure 3 the abstract idea of integrating OPC Classic components into an OPC-UA network is visualized. As pointed out, the OPC Classic production environment, which may be present in many industrial applications, is located within the system limits of the OPC-UA production network.

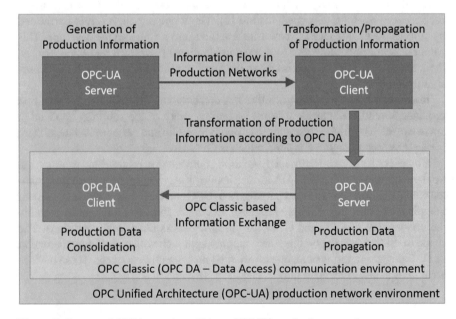

Figure 3 Concept of OPC integration within an OPC-UA production network

For our scenario, the Java programming language has been identified as suitable object-oriented approach to realize the combination of objects from various architectural approaches. In the first step, the OPC Classic components (OPC DA) are represented by Java classes based on the openSCADA [11] implementation of OPC. This openSCADA library provides basic functionalities of SCADA and delivers a full implementation of the OPC Classic standard. In terms of our scenario, the library is used to simulate an OPC DA Client that receives simulated production information from an OPC server. Data values are generated by simulated sensors and annotated with meta-data to meet format specifications of an OPC DA client.

The mapping of the OPC Classic network that comprises of multiple OPC DA clients and sensors is realized by the creation of a communication context, in which the different components of OPC Classic can interact with each other. This environment is encapsulated and is part of a higher OPC-UA information system context. The connection between the OPC Classic context and the OPC-UA environment is performed by a communication of the OPC DA server with an OPC-UA client that delivers functionalities of OPC Classic.

The configuration of the OPC-UA production network is also performed by the Java language. In terms of OPC-UA, a library of the Prosys OPC-UA implementation [12] is used to map OPC-UA specifications on Java objects. As mentioned, the API of OPC-UA is not standardized for an access by object-oriented programming languages. Thus, in terms of the OPC-UA server representation and control, we developed an API, which maps the Address Space information model of OPC-UA on Java classes that represent machines and devices of the production environment. This implementation contains all nodes, variables and references that are required to map the production information on suitable Java objects that are designed according to the OPC-UA standard. The technologies and implementations described have the capability of properly simulating a combined OPC/OPC-UA production network. Details of the simulation of such network are explained in the following section.

4　Simulation of an OPC/OPC-UA Production Network

In order to create a combined simulation of a production network that contains both, an emulation of traditional OPC components as well as novel OPC-UA servers and clients, an appropriate modeling environment and language syntax has to be selected. In the first place, the modeling environment needs to integrate object oriented approaches as concrete instances or objects of real devices need to be created for the simulation. Secondly, the modeling environment needs to provide an integration of database systems in order to enable a real-time propagation of production data in the course of the simulation. The entire ICT of the simulated OPC/OPC-UA production network is visualized in Figure 4.

The simulation of the production environment is initiated by a local repository that contains historical data of an enterprise. In our case, the data storage contains discrete sensor data of a machine that is used in the manufacturing industry for the

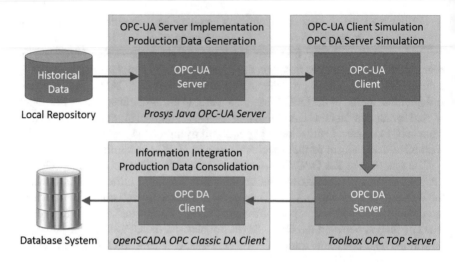

Figure 4 Realization of an OPC/OPC-UA production environment simulation

production of structural components. The repository contains signals of eight sensors that are directly associated to the machine or to its close environmental conditions.

The historical data consists of thousands of data sets that have been collected over several weeks. Each of these data sets contains sensor data that were recorded for a determined duration. Between each pair of data sets, a fixed waiting period is included. Each data set has a unique identifier (Id) and can be selected separately, whereby each set consists of 30000 values which represent 30 s of the process – one value per millisecond. The goal of the simulation is to pick data sets randomly from the local repository and to propagate the values in real-time. This approach guarantees a realistic system behavior as the data sets are executed without a fixed sequence, however the data sets itself contain data in chronological order. The implementation and execution of the Java-based OPC-UA server application is described as follows.

The application of the OPC-UA server and the according API has been implemented using the Prosys Java OPC-UA Client SDK. Based on our API, a server simulator class was developed that analyses the structure of the historical data and extracts data from the repository in discrete time steps. The data and the structure of the data sets are configured by Java classes that map production information on data streams that are conform to OPC-UA specifications. The implementation of the server application is depicted in Figure 5.

The simulator concept contains the DbHandler class that utilizes a method to randomly pick data sets from the local repository. The data mapping is performed by a Data class whereas the DataStructure class determines the according data format of the data sets that is transferred to the simulator. The simulator creates the Address Space Model of the OPC-UA server and loads the values of the current data set into the cache of the server. The single data points of the actual data set are then propagated into the simulated production system.

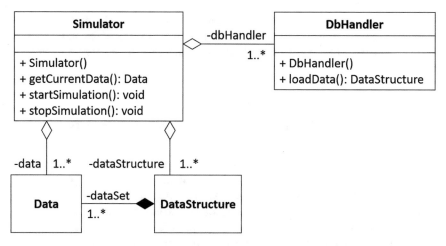

Figure 5 Simulator for randomized selection and propagation of data sets

The data that is propagated into the production network can be accessed by any OPC-UA client located within this network. In our case, the Toolbox TOP Server [13] is used to emulate a fully functional OPC-UA client that is able to communicate via web services in a Microsoft Windows architecture or between two computers in the same network. The OPC-UA client is able to browse for OPC-UA servers that are located within the same network domain and is accordingly able to find the running emulated OPC-UA server.

Besides OPC-UA clients, TOP Server also support functions of OPC Classic DA servers. Thus, through the server implementation, it is possible to utilize the native downward compatibilities of OPC-UA. The OPC-UA data stream from web services and initiated by an OPC-UA server environment is accordingly transferred into an OPC Classic based signal.

In order to browse for the OPC DA server that is emulated by the TOP Server, the openSCADA based OPC Classic DA client application is used. By using the Utgard module [14] of the openSCADA framework, a fully functional OPC DA client was implemented that can browse for OPC DA servers and nodes within an OPC Classic production network. The data stream between the emulated OPC DA server and the OPC DA client is exchanged via COM/DCOM technologies as the communication is based on pure OPC Classic.

As highlighted before, the communication in OPC Classic networks is difficult to establish, as the firewall setup in terms of DCOM is a very complicated task. Although we accomp-lished to browse for OPC DA servers, which are on a different computer in the same network, via DCOM, we conclude that a configuration of the OPC DA client on the same machine as the TOP Server OPC DA server is easier to perform.

After setting up the simulated production environment in the described way, the combined OPC/OPC-UA production simulator is started based on data in the local repository. The data stream is propagated as OPC-UA signal and transformed through the TOP Server. The data is then accessed by the OPC DA client. The client implementation that is also carried out in pure Java uses an object-relational database mapping library that enables a consolidation of incoming data stream into multi-dimensional database systems.

5 Validation and Evaluation of the Integration Process

The data consolidation enables an annotation of production data with meta-information. This information can e.g. contain sensor or component types, the installation place of sensors or additional information about the data, which is useful for later analysis steps. In order to validate the data integration process based on the implemented OPC DA client, we conducted data performance measurements concerning the data integration process as well as analysis and data mining operations on the consolidated production data. The data analysis itself is not in the scope of the current paper, however we have proven that a target-oriented analysis of the simulation production data is possible by using the OPC-UA based data generation.

The performance measurements aim at a validation of the real-time capabilities of the solution that have been carried out. In terms of the performance penetration tests, the consolidation step of the annotated information has been determined as the major bottleneck of the integration chain. As the streaming of the production data can be executed on a high-frequent basis, the critical point of the integration consists in the annotation of production data together with the subsequent consolidation into the multi-dimensional database system.

The according tests have been carried out using different data stream frequencies to determine the limits of the integration process. As the data sets contain 1000 values for each second, frequencies of the simulation can be adjusted to an extent that is able to challenge the real-time capabilities of the data integration chain. Hence, the frequency rate of the data propagation was adapted gradually until the limits of the data base integration of the OPC DA client were determined. The results have shown that data stream processing is smoothly possible to an integration frequency of 0.5 s. This result implies that about 500 values (for 500 ms) have to be buffered prior to each data integration operation. Further detailed performance measurement tests have been performed in order to determine the real-time capabilities in terms of an integration of multiple values and data streams. The results of these tests are beyond the scope of this architectural paper. In terms of an overall evaluation of these test scenarios, it can be concluded that the system architecture carried out in terms of this work is principally capable of real-time data integration. These satisfactory results are due to the fact that we carried out an integration chain that is based on a consistent usage of tools and technologies. The implementation of the simulation environments and the APIs for the utilized application ensures the integrity and interaction of the used solutions. Thus, the entire production environment turns out to be interoperable, although entirely different underlying technologies are in use.

6 Conclusion and Outlook

The aim of this work is to demonstrate generic interoperability of OPC-UA production networks by showing a simulated combined OPC/OPC-UA environment for the usage in real-time fashion. An integrated tool chain was developed based on pure Java applications that is capable of integrating data from OPC-UA sensors or components. The data that is integrated using an OPC Classic DA client contains all necessary production and meta-information to perform extensive data analysis in terms of the production optimization.

In the next steps, the focus will be on further integration of OPC components in OPC-UA environments. Extensible web services will be enabled that are enriched by domain or vendor specific meta-information. The domain specific knowledge that is integrated by expanding the communication stack of OPC-UA is realized by integrating ontologies into intelligent objects or smart devices in the production environment.

Together with full interoperability of OPC components with OPC-UA described in this work the capabilities of future smart OPC-UA devices can be fully exploited and implemented into older manufacturing environments. The described interoperability is a major step towards the realization of Industry 4.0 in existing production sites by creating the basis for the realization of Cyber Physical Production Systems.

Acknowledgments This work is support by the German Research Foundation in the Excellence Cluster "Integrative Production Technology for High-Wage Countries" at the RWTH Aachen University.

References

1. International Electrotechnical Commission. Iec 62264 - enterprise-control system integration
2. M. Hoffmann, T. Meisen, D. Schilberg, S. Jeschke, Multi-dimensional production planning using a vertical data integration approach: A contribution to modular factory design: Emerging technologies for a smarter world (cewit), 2013 10th international conference and expo on, 2013. doi:10.1109/CEWIT.2013.6713754
3. C. Brecher, S. Müller, T. Breitbach, W. Lohse, Viable system model for manufacturing execution systems. 46th CIRP Conference on Manufacturing Systems 2013 (Volume 7), 2013, pp. 461–466
4. N. Brown, C. Kindel, *Distributed Component Object Model Protocol - DCOM/1.0*. Microsoft Corp., 1998
5. J. Lange, F. Iwanitz, T.J. Burke, *OPC: From Data Access to Unified Architecture*, 4th edn. VDE-Verl., Berlin and Offenbach, 2010
6. W. Mahnke, M. Damm, S.H. Leitner, *OPC unified architecture*. Springer, Berlin [u.a.], 2009
7. U. Enste, W. Mahnke, OPC Unified Architecture: Die nächste Stufe der Interoperabilität. at - Automatisierungstechnik **59** (7), 2011, pp. 397–404
8. S.H. Leitner, W. Mahnke, Opc ua - service-oriented architecture for industrial applications. ABB Corporate Research Center, 2006
9. R. Hensel. Industrie-4.0-Konzepte rütteln an der Automatisierungspyramide, 2012. http://www.ingenieur.de/Themen/Produktion/Industrie-4.0-revolutioniert-Produktion

10. T. Hannelius, M. Salmenpera, S. Kuikka, Roadmap to adopting OPC UA. 6th IEEE International Conference on Industrial Informatics, 2008, pp. 756–761. doi:10.1109/INDIN.2008.4618203
11. openSCADA. open source Supervisory Control and Data Acquisition System. http://openscada.org/
12. Prosys. Prosys OCP UA Java SDK. http://www.prosysopc.com/products/opc-ua-java-sdk/
13. Toolbox. TOP Server OPC & I/O Server. http://www.toolboxopc.com/index.html
14. openSCADA. Utgard - vendor-independent, 100% pure JAVA OPC Client API. http://openscada.org/projects/utgard/

Assessment of Risks in Manufacturing Using Discrete-Event Simulation

Renaud De Landtsheer, Gustavo Ospina, Philippe Massonet,
Christophe Ponsard, Stephan Printz, Sabina Jeschke, Lasse Härtel,
Johann Philipp von Cube and Robert Schmitt

Abstract Due to globalisation, supply chains face an increasing number of risks that impact the procurement process. Even though there are tools that help companies address these risks, most companies, even larger ones, still have problems adequately quantifying the risks on their current process as well as on alternative process. The aim of our work is to provide companies with a software supported method for quantifying procurement risks and establishing adequate strategies for risk mitigation at an optimal cost. Based on the results of a survey on risk management practices and industrial needs, we developed a tool that enables them quantifying these risks. The tool makes it easier to express key risks via a process model that offers an adequate granularity for expressing them. A simulator incorporated in our tool can efficiently evaluate these risks through Monte-Carlo simulation techniques. Our main technical contribution lies in the development of an efficient Discrete Event Simulation (DES) engine, together with a Query Language that can be used to measure business risks from the simulation results. We show the expressiveness and performance of our approach by benchmarking it on a set of cases that are taken from industry and cover a large set of risk categories.

Keywords Discrete Event Simulation · Manufacturing · Supply Chain · Procurement Risks · Risk Management

R. De Landtsheer (✉) · G. Ospina · P. Massonet · C. Ponsard
CETIC Research Centre, Charleroi, Belgium
e-mail: rdl@cetic.be

S. Printz · S. Jeschke
IMA/ZLW & IfU, RWTH Aachen University, Dennewartstr. 27,
52068 Aachen, Germany

L. Härtel · J.P. von Cube · R. Schmitt
Fraunhofer Institute for Production Technology (IPT), Aachen, Germany

Originally published in "ICORES 2016",   869
© Springer 2016. Reprint by Springer International Publishing Switzerland 2016,
DOI 10.1007/978-3-319-42620-4_66

1 Introduction

Companies are faced with increasing procurement risks within the context of a global economy. These risks can be related to many different factors, such as, for example, the geographic location and the political and economic situation of the involved parties (suppliers, warehouses and factories). Assessing these risks alone is a difficult task as the risks may only reveal themselves at the end of the production chain and it also requires consideration be given to the impact of internal risks, as well as thought be given to the complexity of the manufacturing process (which could decrease the capacity for adaptation in the case of supply failure) or the level of optimisation in place (which could increase the impact in case of disruption).

Helping company managers make the right decision in the face of risks is not an easy task. Whilst analytical reasoning is impractical, model-based simulation has proved to be a competent approach [1]. Procurement risks present extra challenges as they occur at one end of the process (output), but can sometimes only be measured at the other end (input), therefore they require thought throughout the whole manufacturing process. The scope of our work is to address this challenge, focusing on small and medium enterprises in the field of mechanical engineering.

Our ultimate goal is to produce a user-friendly, tooled methodology that will guide the user through the whole process of risk assessment. In order to reach this goal, our work is structured as follows:

- A taxonomy of supplier and internal risks is identified, starting with the simplest risk - a shortage of raw materials, which can eventually lead the whole process chain to more elaborate risks, depending on the kind of order policy used.
- A survey was conducted on the practice of risk evaluation in an industrial context [2]. The results of this survey showed that nearly 66 % of the companies perform risk evaluations, although only 10 % rely on dedicated software tooling. This means that in practice risks are evaluated by an individual estimation of the cost factor and the probability of occurrence. In general, and by including historical data in the estimation, the quality of the estimation improves. However, relying on historical data and estimating the impact of factors, like delivery times, means that material quality is impossible to estimate, even in the case of changing suppliers or adding parallel processes to the chain. Based on the requirements identified in the survey, a software based risk management framework was defined.
- We developed a modelling and simulation toolset for identifying risks, quantifying them and deciding on design alternatives that can help mitigate risks. The main technical scope of this paper is to detail our toolset framework and show how it helps focus be retained on the risks during modelling, so as to stay efficient during the modelling time, simulation time and result analysis time.
- Finally, we are validating our work through a group of companies that are already trying out our tool via an easy to use web interface. Although this validation is not yet complete, we can already benchmark our approach on a number of industrial cases and in doing so assess the expressiveness and performance of our approach.

Our modelling framework includes concepts such as *storages*, where items can be stored or retrieved with a maximum capacity, and several types of production *processes*, each with different timing and failure behaviours. In addition, we defined a query framework on models that are fully declarative and include arithmetic, temporal and logic operators, as well as basic probes for the elements of our factory model (contents of a storage, whether a process is running or not, etc.). Based on this query language, the software tool is able to calculate the probabilities of different scenarios (e.g. delay in deliveries, defective parts or poor quality) and their impact, which is all based on a timed model of the relevant factory process.

This approach to monetary risk quantification is based on an approach developed in the Q-Risk project [3]. The simulation toolkit relies on the discrete event simulation module of the OscaR framework, particularly its simulation layer, and adds dedicated abstractions that are dedicated to the timed modelling of factories, and the modelling of risk-related queries [4].

Our main contribution lies not only in the risk-driven dimension of our framework, but also in regards to the usability factors. Its design is based on a number of trade-offs between the expressiveness and simplicity of the modelling language, as well as efficiency of the simulation engine.

The paper is structured as follows: Section 2 presents the context of our work; Section 3 presents our modelling language for representing factories; Section 4 presents our query language that can serve to evaluate risks; Section 5 illustrates how complex risks can be included in our query language; Section 6 describes our DES prototype and more specifically the Query Engine. Section 7 shows the benchmarking of our simulation tool both on the expressiveness and performance dimensions; Section 8 discusses some related work; Section 9 concludes the paper.

2 Background

In order to assess and quantify different kind of risks in manufacturing processes, we model the manufacturing process as a flow graph. This model captures the key procurement steps and the production process itself. Resource storage (in warehouses or stockrooms) and the flow of raw materials in basic processes will be explained in Section 3.1. The main graphical notations implemented by the graphical part of framework are shown in Figure 1, which is a model used later in the benchmarking process. Notations are quite self explanatory: a supplier is a little truck, storage types are represented by different variations of cylinders (the one with vertical bars can overflow) and processes are depicted with the industry icon (also with some variants: multiple horizontal lines means parallel batches, the cross means possible failure, the rounded, the rounded box depict a conveyor belt).

The operation of the whole manufacturing process can be described as a sequence of timed events. For instance, the first event involved with simulating a factory is fetching some materials for storage. This action can trigger a new order being sent to

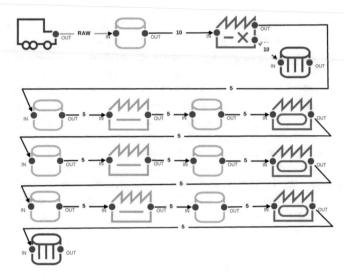

Figure 1 Beer game model

a supplier if the storage level reaches a certain threshold - in accordance with supply chain policy.

In the rest of this section we recall the nature of risks and the goal of risk management, then we give some details about Discrete Event Simulation and why it provides an adequate framework for modelling the operations of manufacturing processes and quantifying risks [5].

2.1 Risks and Risk Management

In order to develop a Discrete Event Simulation (DES) approach for quantifying the impact of risks in manufacturing enterprises, the nature of risk and the underlying process of risk management needs to be understood in detail.

2.2 Risk

Risks strongly affect an enterprise's business success and are directly related to costs, effort and yield [6]. Thereby, risk is understood as an event likely to occur with an undesired consequence. The most common and, for the approach, most convenient categories of risks are the cause and impact-oriented type. The root-cause-oriented approach considers information uncertainty and validity as risk [7]. The chance of not meeting a planned target is understood as an impact-oriented risk.

However, only combining both categories of risk leads to the necessary scope of information needed to properly manage risks. Hence, risks need to be understood as having a certain likelihood of missing a defined target. Hence, the concept of risk is defined through three components: the *hazards*, or potential dangers, the *consequences* of those hazards, and their predicted frequency, or *likelihood* [8]. A "natural" quantification of the hazards associated with a risk is the quantification of product consequences in order of likelihood. A cybernetic model of procurement-based hazards and their management is presented in [9].

Risk likelihood can be modelled with probability distributions [10], as the occurrence of a risk hazard in a process or system is *uncertain*. In [11], a theory of probabilistic risk analysis is developed, which is associated with the concept of system reliability. As a risk is defined as the deviation from a planned value, statistical measures can thus be applied to operationalise and compare the possible magnitude of such deviations [12]. Evaluation of the risk analysis and the reliability of a system can be done with the Monte-Carlo method [1].

2.3 *Risk Management*

The main objective of risk management lies in the assessment of major corporate goals in regards to risk policy strategies. Hence, risks affecting long lasting business success need to be controlled. However, enterprises will never be able to totally eliminate risks and will always have to consider a certain degree of residual risk [13]. One key task of risk management is to identify and analyse risks as early as possible in order to take cost optimal risk treating actions [6].

The basic process of risk management (Figure 2) is described in standards ISO 31000 and ONR 49000 ff. IEC 31010 provides an overview of corresponding risk management methods and techniques for a specific process.

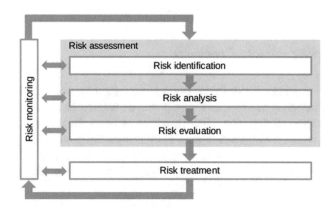

Figure 2 Risk management process

2.4 Discrete Event Simulation

There are two main approaches for representing time if we want to simulate the behaviour of a system: the first approach is to use a *continuous time*, in which the events affecting the system occur as time "ticks", all of which is proportional to the actual expected time of system operation. The other approach is to have a *discrete time*, and concentrate the simulation on only the operational events, rather than the time. This is the basis of Discrete Event Systems.

In the literature [14, 15], the main components of a DES model are described as: *entities* (which are the items that are flowing and transformed throughout the simulation), *queues* (representing storage devices or other areas in which entities wait to be used), *activities* (that actually perform some work on the entities), and *resources* (a special kind of entity that is required to operate activities).

DES models define *events* as discrete points in time in which the system state changes. The simulation of the model is simply a queue of different events triggered by the previous event, in other words the "next-in-time". Checking an event can trigger other events in the queue. For instance, checking the event starting an activity will trigger the events of fetching the corresponding entities needed to perform the activity, and thus ending the activity. The event of activity failure can also be triggered by a given probability.

Several software solutions exist to support DES based modelling for a variety of applications. Among the commercial software, we can cite AnyLogic [16], Arena [17] and Plant Simulation [18].

3 A Simulation Meta-Model for Factories

All the main elements of manufacturing processes are represented in our simulation meta-model, which allows us to define concrete models that are simulated in a Discrete Event Simulation engine. In addition to this, we designed a Query Language over concrete simulations in order to collect and analyse data.

3.1 Modeling Factory Processes

This section introduces the basic blocks for representing factories. In our approach, factories are modelled as a flow of items, processes and stocks (Figure 3).

Storages represent any kind of stock device or place for raw materials, like a warehouse, a barrel, a silo or a dumpster. They have a maximum capacity. When this capacity is reached, they either overflow, or create a blockage in the process, all of which is dependent on the settings of the storage. If a full stock storage overflows, any unloading material of that stock is lost.

Batch processes are factory processes that work in a batch fashion; supplies are collected from various stocks, then the process runs for some time, and finally the

Figure 3 Concepts of our process modelling languages

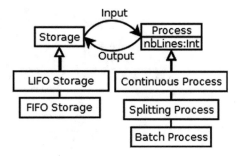

produced outputs are dispatched to their respective stocks before this whole cycle starts again.

Continuous processes are factory processes that typically run on a conveyor belt. Items are continuously picked from input stocks and undergo the process immediately at the physical end of the machine, they then pass through the machine in a queue, and when they reach the other end of the machine, the resulting items are dispatched to their respective stocks. A simple example is a conveyor belt that passes through a bakery oven; raw pastries are set on one end of the conveyor belt; they go through the oven and are cooked when they reach the other end of conveyor belt, from there they are dispatched to their output storage.

Splitting processes are similar to batch processes, except that they have several sets of outputs and when completed, one set of outputs is selected and the produced items are dispatched to the stocks associated with the selected output. This represents a quality assurance process, whose item flow is split into two (or more) separate flows based on the result of the quality assurance analysis.

Parallel processes are variants of the processes above, where several lines of the same process are running in parallel. Basically, all processes introduced here have a parameter specifying the number of process lines running in parallel.

Items flowing in processes and stocks are indistinguishable at any given point of the factory process as they all share the same part number. Yet, they have some intrinsic features: some items might come from a given process, others might be made out of poor quality supplies, etc. These intrinsic features can influence the behavior of some processes, such as the splitting process representing a quality assurance process. This notion of intrinsic features leads us to distinguish between two different types of storage, namely: First In-First Out (FIFO) storage and Last In-First Out (LIFO) storage.

3.2 Process Activation and Supply Chain Policies

Supply chain policies are also integrated in our model of the factory, together with activation policies that are able to turn a process on or off, depending on the demand

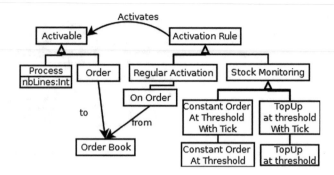

Figure 4 Concept for modelling activation rules

for the output stock. To model these two concepts, we introduce the notion of *activable* and *activation*. An activable is something that can be enabled through an activation. We also associate a magnitude with the activation, or in other words, an integer. An activable can be a process or a supply order. In the case of a process, the activation represents the number of batches that the process is allowed to execute. In the case of an order, the magnitude represents the number of ordered items.

In our model, an order is a stationary, activable object that represents a class of order that can be passed. The order is passed when the modelled order object is activated (Figure 4).

Activables can be activated based on various rules that are also part of our modelling framework. There are three types of activation rules, namely: regular activations (performs the activation on a regular basis over a period of time); order-based activations (performs the activation when an order is received); and stock monitoring activation (performs the activation when the stock level gets below the threshold).

3.3 Modelling Intrinsic Item Features

To represent these intrinsic features, we introduce the notion of *item class*, representing the set of intrinsically identical items. Item classes are characterised by a set of boolean *attributes*. A global set of attributes is defined on the whole simulation model; each item has its own attributes, which is a subset of the global set of attributes. This set of attribute attached to the item defines the item class to which it belongs.

When an item flows through a process, the process can update the attribute of the item to reflect the process that was applied to this item. Similarly, when an item flows through a splitting process, the selected output can be specified according to the attributes of the item. At this point, we had to set a trade-off between the expressiveness of the modelling language, its simplicity, and the efficiency of the simulation. Processes can update the class of items through three basic operations: setting an attribute, clearing an attribute, or loading a constant set of attributes.

Another restriction that we have implemented regards how the class of items produced by a process is linked to the classes of several potential inputs of this process. We consider the union of all attributes with all inputs performed in order to start a batch of attributes in the process, and set this union as the start class for the whole batch. The class transformation function of the process is then applied to this class, and every item output in the process from this batch share the same output class computed by this class transform function.

Since items are distinguishable through their item classes, we introduced two models of storages, namely FIFO and LIFO storages, representing queues and stacks of items, respectively.

4 Performing Queries Over Simulations

The goal of our approach is to perform risk-related queries on factory simulations. These queries are meant to be performed on single runs of simulations occurring inside the Monte-Carlo engine, which aggregates the query results over the runs. It can then be queried afterwards e.g. for mean, median, extremes, variance of these queries over the runs.

Our query language can roughly be split into three sets of operations, namely: probes on processes, probes on storages, and operators. Operators are split into five sets, namely logic operators, temporal logic operators, arithmetic operators, temporal arithmetic operators, and history recording operators. Arithmetic and logic operators differ by their their return types; they return numeric and boolean values respectively.

Since this query language runs over simulated time, we assume that the value of the queries are computed at the end of the trace on which they are evaluated. We define the operator of our language together with their semantics by using the $\models$ notation: $t \models P$ is the value of expression P when evaluated at position t of the current trace.

Some temporal operators refer to the previous position in time, denoted as $\mathrm{prev}(t)$, notably to compute deltas or assess changes. These should be used with care since we are in an event-based model of time, so adding such operators in the query will add extra time events in the simulation.

4.1 Probes for Processes

The probes on processes are atomic operators that extract basic metrics from processes of the simulation model. Suppose that p is such a process, the following probes are supported:

- $t \models \mathrm{running}(p)$ true if the process is running at time t, false otherwise.

- $t \models$ completedBatchCount(p) the total number of batches performed by the process between the beginning of the trace, and time t.
- $t \models$ startedBatchCount(p) the number of batches started by the process between the beginning of the trace, and time t. For a process with multiple lines, it sums up the started batches of each line.
- $t \models$ totalWaitDuration(p) the total duration where the process was not running between the start of the trace, and time t. for a process with multiple lines, it sums up the waiting time of each line.
- $t \models$ anyBatchStarted(p) true if a batch as started by the process at time t.

4.2 Probes for Storages

The probes on storages are atomic operators that extract basic metrics from storages of the simulation model. Suppose that s is such a storage:

- $t \models$ empty(s) true if the storage s is empty at time t, false otherwise.
- $t \models$ content(s) the number of items in the storage s at time t.
- $t \models$ capacity(s) the maximal capacity of s. This is invariant in time.
- $t \models$ relativeCapacity(s) the relative content of storage s at time t, that is: the content of the stock divided by the capacity of the storage.
- $t \models$ totalPut(s) the number of items that have been put into s between the beginning of the simulation and time t, not counting the initial ones.
- $t \models$ totalFetch(s) the number of items that have been fetched from s between the beginning of the simulation and time t.
- $t \models$ totalLostByOverflow(s) the number of items that have been lost by overflow from s between the beginning of the trace, and time t. If s is a blocking storage, this number will always be zero.

4.3 Operators

Logical operators

- $t \models$ true the constant true.
- $t \models$ false the constant false.
- $t \models\ !l$ the negation operator.
- $t \models l_1$ op l_2 where op is one of $\{\&, \|\}$ represent conjunction, and disjunction operators, respectively, returning their conventional results.
- $t \models a_1$ comp a_2 where $comp$ is one of $\{<, >, \leq, \geq, =, \neq\}$ represent comparison operators over numerical values, returning their standard results.

Temporal Logic Operators

- $t \models$ hasAlwaysBeen l true if for each t' in $[0; t]$, $t' \models l$

- $t \models$ hasBeen l true if there is a t' in $[0; t]$ such that $t' \models l$
- $t \models l_1$ since l_2 true if there is a position t' in $[0; t]$ such that $t' \models l_2$ and for each position t'' in $[t', t]$, $t'' \models l_1$
- $t \models @l$ true if both $t \models l$ and prev$(t) \models !l$.
- $t \models$ changed(e) e might be a logic or arithmetic expression; this evaluate to true when $t \models e$ and prev$(t) \models e$ have different values.

Arithmetic Operators

- $t \models n$ where n is a numerical literal represents a literal constant value
- $t \models a_1$ op a_2 where op is one of $\{+, -, *, /\}$ represent the classical arithmetic operators over numerical values, returning their conventional results.
- $t \models -a$ represents the unary negation.

Temporal Arithmetic Operators

- $t \models$ delta$(a1)$ is a shorthand for $t \models a - $prev$(t) \models a$
- $t \models$ cumulatedDuration(b) let be $T = (t_1, t_2) \| t_1 = prev(t_2)\&t_1 \models b\&t_2 \models b$ the accumulated duration of b is the sum over the couples (t_1, t_2) in T of $t_2 - t_1$
- $t \models$ time evaluates to t.
- $t \models$ min(a) the minimum over all the values of $t' \models a$ with t' in in $[0; t]$
- $t \models$ max(a) the maximum over all the values of $t' \models a$ with t' in in $[0; t]$
- $t \models$ avg(a) the average of all the values of $t' \models a$ with t' in in $[0; t]$
- $t \models$ integral(a) the integral of $t' \models a\, dt'$ with t' in $[0; t]$. The integral is computed through the trapezoidal rule taking the events as discretisation base.

History Recording

Two operators are available for recording the evolution of a query throughout the simulation run. The type of these operators is a history of value. The value itself is Boolean, or arithmetic, respectively. These operators cannot be composed with any other operator. Semantically, they are only evaluated at the end of the run, so that there is no position in time in their definition.

- Record(a) is the history of the value of the arithmetic query a over the simulation run.
- Record(b) is the history of the value of the boolean query b over the simulation run.

5   Modelling Risks as Queries

Our query language allows for computation of any metric measurement given in the simulated factory model, and metrics associated to risks in particular. Thus, it is necessary to identify the risks we want to assess. The risk identification process is partly generic because there are well-known categories of risks. We considered here the classical delay/quality/quantity triangle, which can be defined as follows:

- **Delay risks** are related to the possibility of taking more time than expected to produce goods, for instance due to process malfunction, or problems in stocking materials due to supply and storage failures.
- **Quantity risks** are related to the possibility of producing less goods than expected in a process or losing materials in the storage stage.
- **Quality risks** are related to the possibility of a process producing "bad" goods, or the degradation of materials whilst in storage.

The characteristics of these risks, such as their likelihood and their impact, may vary a lot. We quantify these risks by expressing them in our query language, generally as a cost value. Risks can occur at different levels of granularity. *Elementary risks* occur at component level (such as a provider, a storage type, a process) but not necessarily result in a system level risk if the supply chain is designed to cope with the risk. *System-level risks* occur at system level, for example, the global production latency, or the critical path of a supply chain.

In the rest of this section, we present how typical risks can be expressed in our query language. Queries can reference other queries; this makes it possible to refer fine grained risks in the definition of higher level risks.

5.1 Risks Specific to Suppliers

For a single supplier, the risk model is directly encoded in terms of quality, quantity and delay, so it makes little sense to measure it. We can consider more complex configurations, such as a combination of suppliers, reactive suppliers with less quality or quantity, and a slower supplier with high quality or quantity. We can assess the adequacy of the design of a supplier combination by measuring the probability of underflow in the common input storage they are supposed to replenish, e.g. using this query cumulatedDuration(content(sto) > MIN) where MIN is some minimal "safety" stock.

5.2 Risks Specific to Storages

As a delay risk for storage elements, we can compute the amount of time it takes before a storage overflow occurs ovst with the probe `cumulatedDuration` (`hasAlwaysBeen(ovst) < 1`). This allows us to know whether the storage has been losing pieces for a long time and be able to estimate the extent to which the storage should be enlarged.

The most basic quantity risk for an overflowing storage is the number of lost pieces. This is measured by the probe `totalLostByOverflow(ovst)`. Together with the previous probe, useful adaptations to the size of the storage container can be decided.

With respect to the quality risks, it is possible to find that a storage is over-sized, that is that the maximum contents of the stock throughout the simulation are a lot lower than the storage's capacity. This can be measured with the probe max(relativeCapacity(stock)) and it can be used to verify whether that value is higher than an acceptable percentage number.

5.3 Risks Specific to Processes

For quantity risks, we are interested in processes that do not work enough in the simulation. The percentage of idle time for a process p is measured by the probe (totalWaitDuration(p)/currentTime) ∗ 100. To detect whether the process p operated at all in the simulation, we can use the probe: hasAlwaysBeen (!anyBatchStarted(p)).

About quality risks, a basic query is used to look at the ratio of failed part from a failing batch process: totalPut(p)/completedBatchCount(p). This is however of limited interest as it will return a value close to the encoded probability of failure. More general attributes are however being defined on processes with more complex transformation functions – this includes more randomness, which would make such a query informative at that level.

5.4 System Level Risks

Quantifying risks at system level is highly dependent of each considered case. Let us illustrate some typical queries that can be used. A general consideration is that at system level risk are systematically turned into costs so they can be compared. In order to do this specific attributes are available: for example the cost of all parts going through stock, the cost to repair a damage machine, penalties for supplier being late, etc.

For quality risks, considering losses occur from failing processes FP1 and overflowing storage ST1. Considering there is no rework, the query LOSS is:

totalFailed(FP1) ∗ partCost(FP1) + totalLostByOverflow(ST1) ∗ partCost(ST).

Assuming there is also a similar query to define the VALUE_ADDED (based on the cost of incoming parts versus produced parts), the profitability is then defined as VALUE_ADDED/(LOSS + VALUE_ADDED). These queries could be made more general using a component selector to select all relevant parts to use during specified computation.

For quantity risks, an OUTPUT storage can be compared with an ORDER book: (content(OUTPUT) − content(ORDER)) ∗ partCost(ORDER). We use the agreed order cost which may differ from the cost of the produced parts.

Global delay risks are more difficult to express because the DES engine does simulate the flow of parts but only reports transformation events. It is however possible to monitor the time required to fulfill some order and compare it with the agreed delivery time and take into account possible penalties of being late.

6 Implementation

In this section, we first give an overview of the global architecture of the software tool before giving more details about the execution of queries in the simulation, which is the main body of our work.

6.1 Global Simulator Architecture

The simulator is implemented using the OscaR DES module [4] and is written in Scala [19]. A modelling web front-end was developed with JavaScript technologies, mainly Bootstrap [20], JQuery [21] and JointJS [22]. The lightweight Scalatra [23] web framework was used to wrap up the simulator as a set of web services. The web server contains a Monte-Carlo simulation engine that is able to aggregate the results (especially the queries) of several simulations over the same model. Figure 5 illustrates the architecture.

6.2 Supporting Risk Identification

The software tool includes a wizard helping the user to express risk using our query language. A succession of screenshot of this wizard is shown in Figure 6. It starts from the main risk categories (quantity/quality/delay) then guides the user into specifying how the risk can be measured both at system level and then at component level. A number of predefined queries are available in each context. Queries can also be edited

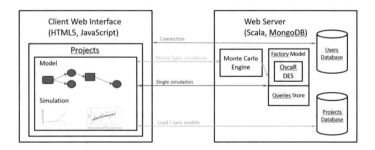

Figure 5 Global Architecture of the Software Tool

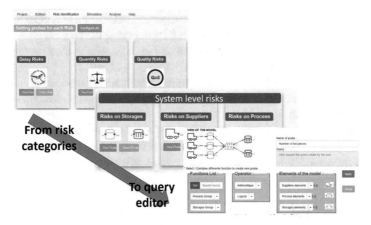

Figure 6 Encoding a query for a quantity risk

and designed from scratch using a plain text editor. The figure shows the encoding of a quantity risk measuring the number of lost pieces in a factory.

6.3   Efficient Evaluation of the Queries

All the elements of the factory feature optimal $O(1)$ complexity for their update operations. Attribute update operations performed by processes are collapsed into a constant number of bit-wise operation. Any combination of the operations on attributes can be aggregated into two efficient bit-wise operations performed using bit masks that represent attributes.

Queries must be evaluated efficiently during simulation runs. To this end, we have incorporated three mechanisms and the way they are evaluated on traces, namely:

- Incremental evaluation throughout the simulation run.
- Minimal updates that allows updating only the relevant fragment of queries.
- Bottom-up updates to allow the sharing of sub-queries.

6.3.1   Incremental Evaluation

Queries are evaluated incrementally throughout the simulation, so that no simulation trace is actually generated. At each step of the simulation, queries are notified about the new step, and they update their value accordingly.

A very simple expression illustrating this is a+b. At each step t, the value of this expression is the sum of the values of a and b: `sum.Value = a.value+b.value`.

A slightly more complex one is the max(a) expression. The `max` operator updates its output at each step t of the simulation run by applying the code: `if(a.value > max) max.value = a.value` where `max` is the output value of the operator, and `a.value` is the value of `a` at time t.

Some operators require more intricate update mechanisms with auxiliary internal variables. Consider the expression avg(a); the `avg` operator maintains two variables: the number of steps so far, and the sum of all the values of `a` collected at these step so far. The output value of the expression is the ratio between these two internal variables: `avg.value = sum_of_a / number_of_time_Steps-`.

All operators are able to perform such updates of their outputs.

6.3.2 Minimal Updates

To further increase the efficiency of the query evaluation, we distinguish two type of operators: non accumulating ones and accumulating ones. *Non accumulating operators*, such as the $+$ operator do not actually need to perform updates at every step, they just need to be updated at the end of the simulation trace in order to deliver a correct final value. *Accumulating operators* include all the temporal operators, and need to perform some sort of update at each step of the simulation run in order to deliver a correct value at the end of the run.

Some non-accumulating operators might also need to be updated at each step, for instance if their output value is needed by an accumulating operator. We therefore introduce the notion of accumulating expressions and non accumulating expressions. An *accumulating expression* is rooted at an accumulating operator, or is the direct sub-expression of an accumulating expression. emphNon-accumulating expressions are all the remaining sub-expressions.

At each step of the simulation, all operators rooted at accumulating expressions are updated. The non-accumulating expressions are only updated at the end of the simulation run.

6.3.3 Bottom-Up Update

A third mechanism that we have incorporated in our query evaluation engine is bottom-up update of the queries expressions. Queries are not structured into trees. They might actually be directed acyclic graphs because some fragment of queries can be shared by several queries. Bottom-up evaluation enables us to update shared fragment of the queries only once, thus gaining efficiency.

6.3.4 Overall Algorithm

When the simulation model and all the queries are instantiated into the engine, a one-shot analysis is performed on the structure of the queries in order to label each sub-expression as accumulating or non-accumulating. All accumulating expressions are then put on an evaluation list, which also captures the order to follow when updating them at each step to comply with the bottom-up order. At each step of the simulation, accumulating expressions are updated following this evaluation list. At the end of the simulation run, non-accumulating expressions are updated once, following a similar bottom-up process.

7 Benchmarking

In order to assess the approach, we took a benchmarking approach based on a set of representative models that were run on the implementations described in the previous section. Our benchmarking addresses our two main contributions:

1. **Expressiveness**: shows that all the risks identified in the cases can easily be captured by the modelling primitives and measurement probes, either by being based on the set of generic probes identified, or by writing case specific probes.
2. **Performance**: shows that running probes does not degrade the performance of the simulation engine a significant amount.

7.1 Benchmark Models

We selected four representative models out of a set of about 20 examples inspired by classic academic cases (these have specific, complex aspects) and other anonymous cases collected in industry. The cases also vary in the level of use of random variables. We describe relevant modelling aspects of each supply chain, together with specific risk issues associated with each model.

First Case: a Simple Assembler Factory

This case, illustrated in Figure 7, is a simple factory that builds industrial produce using two kinds of parts, Parts A and B. Each part has its supplier which feeds the stocks when they become lower than a given threshold. For part A, the supplier policy is to refill the stock to its maximum capacity. For part B, the policy is the delivery of a fixed amount of material. Part B must be preprocessed before assembly. The factory combines two units of part A with one unit of preprocessed part B, resulting in 80 % of products passing the quality tests. So the assembly process can be represented by a failing single batch process. The goal is to assess if the input stocks are kept within safe limits and are able to cope with production demand.

Second Case: a Beer Game Model

Our second case model is a classical problem called "beer game" [24]. It is a long linear supply chain going from the beer factory to the final retailer, passing by distributors and wholesalers. The beer factory is considered here as a supplier, and single

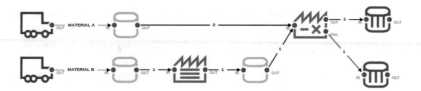

Figure 7 Model 1: a simple assembler factory

Figure 8 Model 3: multiple suppliers

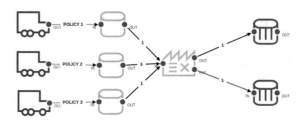

batch processes are used to represent external sources of delay in transport. Intermediary stocks also add extra delays. The continuity of retailing, distribution and wholesaler processes is modelled by conveyor belts. The resulting model is shown in Figure 1. The goal is to assess where potential bottleneck can occur.

Third Case: Multiple Suppliers

This case is inspired by a real industrial case, where manufacturing involved three different materials having their own supplier and refill policy, with random delays belonging to a Gaussian probability distribution. 90 % of produce built by a batch in the factory fulfilled the quality requirements. We want to evaluate the effects of different supplying policies in order to ensure the supply chains operate at optimal capacity, whilst minimising the frequency of orders (Figure 8).

Fourth Case: a Complex Assembly Process

Our last case is inspired by an industrial case in a factory where complex parts are assembled from 3 different materials following a complex process. Two of the parts are preprocessed on factory units that can fail (10 % of failures for the first one, 40 % for the second one). The process is shown in Figure 9.

7.2 Expressiveness Analysis

We identified a number of basic probes relating to risks directly related to model elements. Such probes are automatically generated. So, for each stock, we generated three different probes for measuring the average and maximum contents of the stock and for verifying whether the stock is full or overflowing. For each process, we

Figure 9 Model 4: complex assembly process

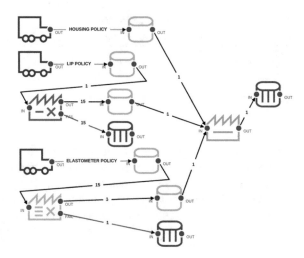

Table 1 Benchmarking table for expressiveness

Name	Size	Risk types	#probes	Comments
M1	9 (2/2/5)	Full stock Process failure	20	Simple manufacturing
M2	17 (1/7/9)	Blocked process Process failure	36	Beer game
M3	8 (3/1/5)	Full stock Process failure Supplier failure	20	Multiple suppliers
M4	14 (3/3/8)	Stock losses	31	Complex part assembly

generated a probe for measuring the amount of time in which the process was idle or blocked in the simulation.

In addition, the user can specify extra probes for expressing business specific risks that are typically more complex queries in the model. Table 1 summarises some model characteristics like size (suppliers/processes/storages), risks and number of probes. To assess expressiveness, we considered a single probe which actually proved enough to cover the targeted risks when dealing with basic probes. In the final two models we also explored risk mitigation strategies.

- In the first case, assessing whether the stocks of raw materials were full could be achieved with the probe $cumulatedDuration(relativeContent(stockA) = 1)$. For most of the time in the simulation, the stocks were full and the assembly process worked at full capacity.
- In the second case, we looked at the relative idle times in the process chain. We noticed that the distribution process, just after the fabrication, is the only one that blocks goods.

Table 2 Benchmarking table for performance

Name	No probes (ms)	Std probes (ms)	All probes (ms)	Overhead (%)
M1	7,3	11,9	12,5	71,2
M2	11,3	17,6	17,8	57,5
M3	25,8	29,2	34,2	32,6
M4	6,8	13,3	13,4	97,1

- In the third case, we both looked at process idle time (basic probe) and the average contents of stocks using the probe $avg(content(st))$. This helped us discover the best threshold to trigger an order, whilst minimising idle time.
- In the fourth case, a full stock was blocking the production. We mitigated the problem by experimenting with overflowing storage to estimate the right sized storage in order to avoid overflow by using the probe $totalLostByOverflow(lipStorage)$.

7.3 Performance Analysis

A Monte Carlo simulation was run for each case with a time limit of 10000 units and 2000 iterations to be more precise. Table 2 shows the computed average. We performed the benchmarks on an Intel Core i7-4600U CPU at 2.10 GHz with 8 GB of RAM. A single core is currently used. The simulation was triggered from the web interface on the same machine as the server.

The overhead in the simulation with probes varies from 32,6 % in Model 3 to 97,1 % in Model 4. Model 3 has the longest run time because of the randomness of the supply delays induced by probability distributions associated with suppliers. Model 4 has the shortest run time because the simulation stops at an earlier stage due to a full intermediary stock. In this case the relative overhead is bigger as a result of the DES engine - the load is more efficiently calculated due to the probe's evaluation in the modelling stage.

Globally, overheads are quite acceptable. Some improvements are still possible in the context of integration with a web application, especially in optimising the network requests between the web interface and the simulator, thus making that interface more responsive. The total simulation time allows thousands of simulations to run in a only a few minutes and explore risk mitigation alternatives within an hour.

8 Related Work

A typical risk assessment conducted on a given factory plan is reported in [25], which is based on the Arena simulation tool, featuring DES and Monte Carlo methods as is the case in our work. It stresses the importance of conducting stress-tests using such simulation platforms. Its focus is mainly on the disruption risk, unlike our work which can cope with other classes of risks, like quality for example. Our framework provides an added abstraction layer that can cut down the cost of performing these important stress tests, thus making them achievable by smaller industries.

A similar analysis has been performed on a beer supply chain in [24], whose model was presented in Section 7.1. This analysis leads to an evaluation of excessive accumulation in the inventory or back ordering sections. Again, no dedicated tooling was used for representing factories at a higher level, which would lead to higher costs for conducting such an evaluation in an industrial setting. Our tool could cope with using the available primitives.

Another simulation-based risk assessment is reported in [13]. It features an aerospace company with very low production volumes, and leads to the elaboration of a dedicated simulation engine. The engine was first developed with purely deterministic behavior, and then enriched with failure models and stochastic aspects. It was shown to be of great value to the company, despite mainly focusing on disruption risks, it helped the company develop a risk mitigation procedure. Our framework has a similar purpose and tries to propose a compromise between genericity and efficiency.

Almeder [26] presents a general framework that combines optimisation and the DES for supporting operational decisions in supply chain networks. Their idea is to iterate between a simulation phase, in which some parameters are estimated, and an optimisation phase that adapts the decision rules for the simulation. Our current work does not cover the minimisation of risk. The tool is rather designed to ease the identification of risk controls by the risk manager. We plan to address optimisation in a later phase, based on the optimisation engines also present in the OscaR framework [4].

9 Conclusions

We presented a Discrete Event Simulation Approach that is supported by a software tool for modelling the supply chain of manufacturing processes with the goal of assessing several kinds of risks on those processes, with a specific focus being placed on procurement risks. This assessment is performed through a simulation engine which uses Monte-Carlo techniques that can also be used to further explore risk mitigation strategies.

The strength of our approach is to support a declarative and easy to use graphical model for representing factory processes and stocks, together with a declarative query language for defining metrics to be measured whilst simulating the behaviour of the modelled system. We could successfully benchmark our approach both from the expressiveness and performance perspectives shown in several typical examples of factories, together with their supply policies.

Further work is still required in order to fully align our approach with the needs of industry. One of our current steps is to validate the tool by putting the tool in the hands of risk managers in a pilot case. We have already identified a number of requests regarding:

- Extending the modelling language, e.g. to support the notion of shared resources among processes and have a statistic model of process failures and breakdowns. More specialised processes allowing controlled fork/joins are also required.
- Identification of model parameters making easier the manual (and later optimised) exploration of risk mitigation strategies.
- Availability of a companion library of specific risks and related probes.
- Producing specific reports (e.g. business continuity plans). We have already explored some work in this direction [27].
- Possibly creating model refinements and better granularity of the simulation. However our aim is not to capture the full reality but what will help in assessing identified risks.
- Finally, parallelisation of the Monte-Carlo simulation engine to obtain better execution times, is needed.

Our framework combining usability, expressiveness and efficiency is an important milestone in our work regarding raising the company awareness, especially in smaller companies, of the need to evaluate their procurement risks and elaborate their supply policies in the most optimal manner. We believe it can be used to manage more general risks. Our design ideas can also be used to improve other risk management tools. Our framework is available online [28] and we plan to make it available open source.

Acknowledgments This research was conducted under the SimQRi research project (ERA-NET CORNET, Grant No. 1318172). The CORNET promotion plan of the Research Community for Management Cybernetics e.V. (IfU) has been funded by the German Federation of Industrial Research Associations (AiF), based on an enactment of the German Bundestag.

References

1. L. Deleris, F. Erhun, Risk management in supply networks using Monte-Carlo simulation. In: *2005 Winter Simulation Conference*. Orlando, USA, 2005
2. S. Printz, J.P. von Cube, C. Ponsard. Management of procurement risks on manufacturing processes - survey results. http://simqri.com/uploads/media/Survey_Results.pdf, 2015
3. J.P. von Cube, B. Abbas, R. Schmitt, S. Jeschke, A monetary approach of risk management in procurement. In: *7th Int. Conf. on Production Research Americas' 2014*. Lima, Peru, 2014, pp. 35–40
4. OscaR. OscaR: Scala in OR. https://bitbucket.org/oscarlib/oscar, 2012
5. F. Romeike, Der prozess der risikosteuerung und kontrolle. In: *Erfolgsfaktor Risiko-Management*, ed. by F. Romeike, Gabler, Wiesbaden, 2004, pp. 236–243
6. G.A. Zsidisin, B. Ritchie, *Supply Chain Risk: A Handbook of Assessment, Management, and Performance*. Springer, 2009
7. M. Siepermann, *Risikokostenrechnung: Erfolgreiche Informationsversorgung und Risikoprävention*. Erich Schmidt, Berlin, 2008
8. I. Sutton, *Process Risk and Reliability Management*, 2nd edn. Elsevier, 2015

9. S. Printz, J.P. von Cube, R. Vossen, R. Schmitt, S. Jeschke, Ein kybernetisches modell beschaf-fungsinduzierter störgößen. In: *Exploring Cybernetics - Kybernetik im interdisziplinären Diskurs*. Springer Spektrum, 2015
10. C. Artikis, P. Artikis, *Probability Distributions in Risk Management Operations*. Springer, London, 2015
11. E. Zio, *The Monte Carlo Simulation Method for System Reliability and Risk Analysis*. Springer, London, 2013
12. W. Gleißner, Quantitative methods for risk management in the real estate development industry. In: *Journal of Property Investment & Finance*, vol. 30(6). 2012, vol. 30(6), pp. 612–630
13. G.R. Finke, A. Schmitt, M. Singh, Modeling and simulating supply chain schedule risk. In: *2010 Winter Simulation Conference*. Baltimore, USA, 2010
14. S. Brailsford, L. Churilov, B. Dangerfield, *Discrete-Event Simulation and Systems Dynamics for Management Decision Making*. Wiley, 2014
15. C. Byong-Kyu, K. Donghun, *Modeling and Simulation of Discrete-Event Systems*. Wiley, 2013
16. AnyLogic. AnyLogic Multimethod Simulation Software. http://www.anylogic.com, 2015
17. Rockwell Automation. Arena Simulation Software. https://www.arenasimulation.com, 2015
18. Siemens. Plant Simulator. http://goo.gl/gH63jw, 2015
19. D. Wampler, A. Payne, *Programming Scala*, 2nd edn. O'Reilly media, 2015
20. Boostrap. Bootstrap website. http://getbootstrap.com, 2016
21. The jQuery Foundation. jQuery website. https://jquery.com, 2016
22. ClientIO. JointJS website. http://jointjs.com, 2016
23. Scalatra. Scalatra website. http://scalatra.org, 2016
24. R.A. Klimov, Y.A. Merkuyev, Simulation-based risk measurement in supply chains. In: *20th European Conference on Modelling and Simulation (ECMS 2006)*. Bonn, Germany, 2006
25. A. Schmitt, M. Singh, Quantifying supply chain disruption risk using Monte Carlo and discrete-event simulation. In: *2009 Winter Simulation Conference*. Austin, USA, 2009
26. C. Almeder, M. Preusser, R.F. Hartl, Simulation and optimization of supply chains: alternative or complementary approaches? In: *Supply Chain Planning*, ed. by H.O. Günther, H. Meyr. Springer-Verlag, 2009
27. A.E. Arenas, P. Massonet, C. Ponsard, Goal-oriented requirement engineering support for business continuity planning. In: *Proceedings of MReBA'15*. Stockholm, Sweden, 2015
28. SimQRi. Online SimQRi tool. https://simqri.cetic.be, 2015

A Framework for Semantic Integration and Analysis of Measurement Data in Modern Industrial Machinery

Tobias Meisen, Michael Rix, Max Hoffmann, Daniel Schilberg and Sabina Jeschke

Abstract The reliability of quality management in industrial processes mainly depends on information about the traceability, precision and accuracy of a measurement system as well as on its systematic bias. The progressive development of networking and sensing in industrial machinery facilitates a quality-related process monitoring regarding information of measurement systems and singular sensor nodes. Hereby, integration on the information level is mandatory. Furthermore, information from the shop floor and from enterprise applications is needed to provide a consistent and integrated quality analysis. Thereby, these systems use different standards and technologies for exportation and propagation of data. Besides, integrative quality management and data analysis require enriched data that does not only comprise, for example, the measured value and its standard-dependent unit on the sensor level; rather, additional information is needed (e.g. the production process or the time and place of measurement). In this paper, a framework is presented that facilitates the semantic integration and analysis of measurement and enterprise data according to real-time requirements. Semantic technologies are used to encode the meaning of the data from the application code. Herewith, the data is automatically annotated using terms and concepts taken from the application domain. Furthermore, a semantic integration and transformation process is facilitated. Thus, subsequent integration and, most importantly, analysis processes can take advantage of these terms and concepts using specialized analysis algorithms. Besides, the conceptual application of the presented framework and processes in a high-pressure-die-casting scenario is presented.

Keywords Real-time Capability in Data and Signal Processing · Semantic Integration · Sensor Data Analysis · Condition Monitoring

T. Meisen (✉) · M. Rix · M. Hoffmann · D. Schilberg · S. Jeschke
IMA/ZLW & IfU, RWTH Aachen University,
Dennewartstr. 27, 52068 Aachen, Germany
e-mail: tobias.meisen@ima-zlw-ifu.rwth-aachen.de

Originally published in "ISMTII 2013 - The 11th International Symposium
on Measurement Technology and Intelligent Instruments",
© Apprimus Verlag 2014. Reprint by Springer International
Publishing Switzerland 2016, DOI 10.1007/978-3-319-42620-4_67

1 Introduction

Due to the increasing complexity of production and manufacturing processes and, associated therewith, the increased challenge to facilitate a holistic process monitoring, measurement systems and sensors have become one of the main attention in process design. Nowadays, process parameters in manufacturing processes, like high-pressure-die-casting, are monitored and recorded by making use of bespoke measurement systems and sensor networks. These systems facilitate an early prediction of the process and the component quality as well as of the efficiency of the monitored subprocesses. The underlying physical cause-effect relationships and the related mathematical and stochastic models as well as their required process parameters are researched to a certain extent.

However, the measurement technique and the used measurement systems are mostly so-called insular solutions. As a consequence, the interoperability between the multiple systems used to monitor the whole process is not guaranteed due to a missing compatibility. This problem is bypassed by developing bespoke integration solutions that use adjusted interfaces and bundle the monitored data in silo solutions, which are afterwards used for quality evaluation. These silo solutions are neither applicable in other domains nor are they easy to maintain or to extend. As a consequence, an integrative evaluation of the entire process regarding aspects, like overall efficiency or quality, and correlations between data of different systems, especially solutions of different system manufacturers, is often not realizable or requires a high investment in the (information) technical infrastructure. Eventually, an integration of such monitored data and other business data recorded and analyzed at the enterprise or the operations command level requires additional adjustments in software applications and infrastructure, especially regarding the real-time requirements of sensor data monitoring and analysis.

Beyond this, the advances of the last years, especially in micro-sensor technology, have increased the attention on sensor networks and open up new applications of sensors within the production environment. Modern sensors are small, inexpensive, energy-efficient and reliable with wireless network capabilities. This leads to new possibilities of process monitoring, which technically can be performed in nearly every stage of the production [1].

These possibilities and further innovations in computing and computer science caused the evolution of the so-called "Cyber Physical Systems (CPS)" and the idea of an integrated industry. Thereby, CPS means networks of intelligent units that are connected within the "Internet of Things (IoT)" to exchange information among each other. This evolution towards an integrated industry facilitates manifold applications in multiple industrial domains [2]. However, the dire necessity of flexible and cost efficient solutions for information integration, propagation and analysis has further increased and bespoke infrastructural solutions like data silos are insufficient with respect to the requirements on the aforementioned topics. Hereby, CPS in industrial applications is referred to as "Cyber Physical Production Systems" (CPPS) [2].

This paper particularly focusses on the discussed integration and propagation problem of heterogeneous sensor and measurement systems in CPPS. Other challenges like hardware issues (e.g. constrains because of limited physical size), technical network issues (e.g. packet errors) or network topology issues (e.g. link and sensor-node failures) are not addressed. The application of the framework is presented in the basis of a real high-pressure-die-casting scenario.

In summary, the contributions in this paper include:

- The architecture of a framework that facilitates the integration and propagation of heterogeneous sensor and application data (Section 3.1),
- A multi-step monitoring and analysis processes with real-time capabilities (Section 3.2) and
- A conceptual application of the presented framework and processes in a high-pressure-die-casting scenario (Section 4).

2 Related Work

Measurement systems in industrial environments are sensor networks that are bespoken for a specific monitoring task. They often provide additional monitoring and reporting capabilities like condition monitoring of the machinery. Thereby, a sensor network, as defined by Trigoni, is "a collection of nodes with processing, communication and sensing capabilities deployed in an area of interest to perform a monitoring task" [3].

In the past, much attention has been devoted to the network topology in industrial application scenarios, for example wireless sensor networks (WSN) and energy efficiency, but still, as stated by Gungor, multiple challenges regarding these topics have not been met [4]. With the evolution of CPPS, the complexity of (wireless) sensor networks in industrial applications increases. As a consequence, a data management for integration, propagation and analysis becomes necessary not only for temporal and spatial sensor data, but also for data of other business applications. For specific domains like industrial manipulation (e.g. sensor-based robot control), data management and sensor integration have already been examined and suitable solutions have been proposed [5]. Nevertheless, more general solutions supporting CPPS scenarios and addressing temporal and spatial aspects of sensor data and process characteristics are required for the application in modern industrial scenarios. A comparison between the approaches that have been published during the last years points out that a middleware approach is a suitable architectural design principle for data management in sensor networks [6–9].

Hadim has evaluated different approaches for middleware-based sensor networks, but none of the evaluated WSN middleware solutions meet the requirements related to the identified heterogeneity issues [8]. Further evaluation and research according to this topic has been carried out by Yick and Gurgen [10, 11]. The latter has proposed the service oriented middleware SStreaMWare that is based on the concept of service orientation concerning sensor networks. This approach, previously published by Golatowski [12], has been picked up and put into practice in multiple middleware-based sensor networks. An essential benefit of this approach is the possibility to hide the heterogeneity of sensor nodes with the help of services, which can dynamically be discovered and executed [13–16].

As a consequence, several service-oriented middleware (SOM) for WSN, which represent a further development of middleware-based solutions, have been developed in order to address the manifold requirements and challenges of such systems. With respect to the field of manufacturing systems, Groba has proposed a SOM, which is heavily based on hard-coded adaptors and gateways and which therefore lacks flexibility and interoperability [17]. Mohamed has surveyed this and other approaches and, after having compared them to each other, has identified several necessity requirements for modern sensor networks [18]. Thereby, he pointed out that none of these approaches fulfilled all requirements concerning real-time constraints as well as interoperability and heterogeneity issues.

The latter has been researched further in particular in connection with the so-called Sensor Web [19] and, in general, with respect to the environmental monitoring via sensors. In contradiction to present industrial scenarios, the plug & play integration of heterogeneous sensors with existing infrastructures was and is regarded as highly promising, for instance in disaster scenarios [20]. However, as previously outlined, with the evolution of CPPS, interoperability, flexibility, integrative data management and horizontal integration will become more and more important [2, 21]. In connection with the Sensor Web Enablement initiative of the Open Geospatial Consortium, Bröring has published a semantically enabled service-oriented architecture [22]. A similar approach is applied in the ITA sensor fabric for fielded sensor networks [23] and in the semantic sensor service network proposed by Wang [24]. These concepts follow up the suggestions of Martinez, who alongside others [25] proposes the use of semantic web technologies to manage and to query data collected within sensor networks [26].

In industrial automation, the OPC Unified Architecture (OPC UA) is a similar approach, which extends the predecessor OPC by platform independent applications, information security and context-sensitive information models regarding reference types [27]. In contrast to the previously mentioned approaches, the OPC UA does not use semantic web technologies like OWL and SAWSDL. Instead, standardized information models for different application areas have been developed and have to be used. Hence, interoperability between different sensors and applications will only be ensured, if adaptors to and from this information model are available or if data transformation is already included in the driver model of the sensor. In addition, data has to be processed in a centralized way via OPC servers.

With emerge of semantic technologies in the architecture of sensor networks, research focused on the semantic description of sensors. In result, several sensor ontologies have been presented to semantically annotate hardware (e.g. linking method), software (e.g. data type of measured data) and domain aspects (e.g. context of measurement) of sensor networks [28]. These efforts have been merged in the SSN ontology of the W3C, which provides a conceptualization of hardware and software aspects and which is extendable by domain ontologies [29]. This approach leads to the possibility to describe sensor characteristics consistently at an abstract level using predefined vocabularies and concepts. Furthermore, new questions are raised, above all how to handle such linked data [30] and how to realize its full potential to facilitate a network of context-aware sensors and applications. Eventually, new problems such as the high response time for reasoning and semantic feature extraction have to be solved.

3   Real-Time Operating Architecture and Basic Concepts

CPPS require solutions that can handle both data of well-engineered and maintained enterprise applications as well as measured data gathered by sensors that operate in a highly dynamic physical environment. Furthermore, information integration, propagation, consolidation and analysis have to be highly flexible and adoptable to handle adaptations of production and manufacturing processes without the need of maintenance tasks or extensive reconfiguration. Thereby, the required information has to be ready for use when it is needed.

In the following, the architecture of a framework that is oriented on existing approaches is presented. The presentation is followed by a description of the applied multi-step data handling.

3.1   Framework Architecture

The architecture of the framework is based upon a service-oriented approach. As already suggested by Bröring [22], in the framework, sensor data is propagated by applying the service bus pattern. In addition, application information is distributed along this bus. Therefore, a resource-oriented approach has been implemented following the enterprise service bus pattern [31]. Figure 1 illustrates the architectural basics of the framework and its main components.

Sensor nodes are connected via *sensor node gateways*. As suggested by Wang, the framework supports sensor nodes with limited processing power or low power consumption using 6LowPan or other low-level protocols by providing specialized gateways [24]. Support for other, currently not supported protocols can be added programmatically. The linking of measurement systems is realized via *measurement system gateways* supporting access via application programming interfaces (API), databases and screen scraping. Further, data can either be pushed by the measurement system or pulled by the gateway via scheduling. Similar gateways, which are not depicted in the figure, exist in order to connect application data. The integration of such data is realized by the adaptive information integration technique [32].

The real-time capability, at least for raw sensor data, is ensured by introducing *fast path gateways* that provide data handling for raw data (streams). Instead of using

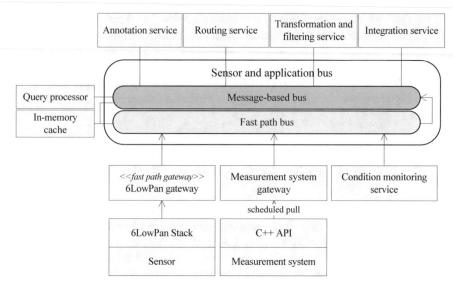

Figure 1 Framework architecture

semantic web protocols for data publishing, fast path data is directly routed to the connected fast path aware services. Thereby, each data package or stream is uniquely identifiable by a resource identifier assigned when the sensor is registered. Additional condition monitoring that requires for example raw data of multiple sensor nodes or high processing capabilities and that therefore cannot be processed on the node itself is supported by *condition monitoring services* configurable via an extensive rule language. This approach facilitates a flexible and adaptive real-time condition monitoring of data observed by multiple sensor nodes or measurement systems.

In parallel to the fast path data handling, data is semantically annotated by *annotation services* and routed to further services, either by an event-based publish/subscribe mechanism or by triggering configurable processes. Similar to the blackboard pattern, semantically annotated data is further processed by *transformation and filtering services* and integrated by *integration services*. Comparable to the aforementioned condition monitoring services, these services can be adopted to specific requirements by using rules. Thereby, each service can access processed data – raw data as well as semantically enriched data – via an in *memory data cache* that is automatically synchronized and accessible for all services. Routing is controlled and executed by *routing services*. The fast path data handling and semantic annotation is further discussed in Section 3.2.

Finally, the framework implements a semantically enabled data consolidation for further analysis, simulation or reporting purposes. The *query processor* realizes the access to the underlying *data storage*. Required data is automatically provided by using the adaptive information integration technique previously mentioned.

Figure 2 Multi-step
monitoring and analysis

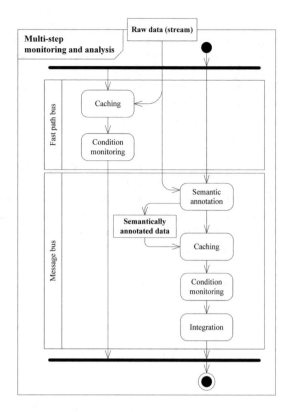

3.2   Real-Time Monitoring and Analysis

As previously mentioned, within the framework, the multi-bus approach is used to
facilitate a multi-step data monitoring and analysis. Figure 2 depicts the basic sensor
data handling processes and sketches the idea taken as a basis for the approach.

Sensor data or data of measurement systems is normally published to the fast path
bus that supports real-time Ethernet for data exchange and therefore provides the
required response time. The data is cached using the in-memory cache and is further routed to the configured condition monitoring services. These services can be
used to evaluate and to compare data of heterogeneous sensors directly. Thereby, the
in-memory cache facilitates stateful evaluations that, for example, can access data
of previous time frames. In parallel, the data is semantically annotated by mapping
the meta-data (the id and type of the sensor) as well as the value and the timing
information to the corresponding concept using reasoning. The semantically annotated data set is also cached and condition monitoring is applied. Further, the data
is integrated into a transactional database. In contrast to the in-memory cache, data
within this database is persisted long-term, so that it can be used for further analysis
and reporting.

Evaluations in both steps are defined by rules that are configurable via a rule expression language. For raw data, the evaluation is restricted to sensor metadata like the sensor id and group as well as the measured value and the timestamp. If a stateful evaluation is configured, information about previously observed data can be accessed via timestamp or index. In case of semantically annotated data, further information like the type of data or process information probably gathered from an ERP or MES can be used in the evaluation.

A rule is defined by three parts, the name, the condition and the effect. Conditions are formulated as Boolean expressions. Currently, data is bundled into data package objects. A package can be flagged as invalid, so that no further processing is done. In addition, a state can be assigned to a package in order to control the further process flow. Alarms can be raised by triggering events. For example, in order to raise an alarm event whenever the measured temperature of two independent measuring systems (following identified by "ms1" and "ms2") exceeds a threshold value of 750.0 Celsius, the corresponding rule has to be set to:

```
rule "Temperatur monitor"
when
     $d1 : DataPackage(sensorId=="ms1"&& value[2]>750.0)
     $d2 : DataPackage(sensorId=="ms2"&& value>750.0)
then
     raiseEvent(newAlarmEvent("Temperature"));
end
```

Because the measurement system represented by "ms1" records multiple values, the temperature value is accessed via the corresponding index of the supplied vector. The raised event is published to the registered services that can initiate appropriate measures. Currently, the evaluation of a stateful rule as described takes in average less than 50 microseconds (on commercially available PCs).

4 Application in High-Pressure-Die-Casting Scenario

The following description provides an overview of the scenario, to which an application of the framework will be applied in the foreseeable future (Section 4.1). Subsequently, the conceptual application of the framework is presented (Section 4.2).

4.1 Process and Data Acquisition

Regarding the rising product and process requirements in the high-pressure-die-casting machinery (HPDC), the growing need of new possibilities to appraise and to locate innovative approaches becomes obvious. Therefore, a high quantity of direct and indirect measurement equipment has been established and will be enlarged by

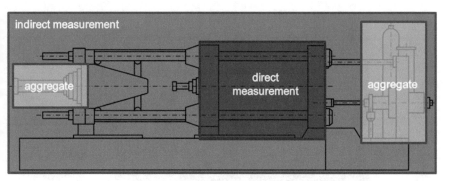

Figure 3 Measurement areas in the HPDC-machine

new systems. Due to the complexity of the HPDC-process, it is necessary to map all the peripheral equipment like the melting furnace or the robots to handle the finished parts. In the following, the HPDC-machine and its measurement system are illustrated (cf. Figure 3).

The actual situation comprises direct measurement equipment like thermocouples and pressure sensors inside the mold to rate directly the behavior of the melt. The indirect measurement equipment, such as strain gauges, which control the clamping forces inside the columns, is placed outside the mold. Another type of measurement data acquisition is preceded in peripheral aggregates like pumps in the hydraulic system. In case of defects or upcoming maintenance cycles, a signal is sent to the programmable logic controller (PLC). For this purpose, the signal can be generated in two ways: by controlling the aggregate or by monitoring the different process parameters of the aggregate, which are interpreted to gain a warning signal.

Beside the acquisition of measurement- and quality-data in the HPDC-machine as such, the same amount of various data is generated by other different peripheral systems. Initial melting of the raw material is an inherent part of the process and has to be controlled concerning the weight, the temperature and the composition of the alloy. According to the HPDC-machine, in the furnace, process-parameters are detected directly and indirectly by additional systems. Thus, temperature and weight are measured in real-time, but the composition of the alloy is quantified in batch mode as well as in a separate system. Consequently, each system generates dissimilar unsynchronized types of data.

Directly after being casted, the part-geometry is checked for completeness and inspected for faults by the operator. These faults are documented in a paper record. Moreover, casting parts are tested about internal defects by X-ray and computer tomography. These quality tests are managed by another operator and the documentation is stored as pictures. The last operation during the entire process is the machining operation that is, for example, drilling, milling, grinding or turning. In the same way as already described above, defects caused by an incorrect casting of the part have to be recorded in the machining process. The generated data of the different systems is either stored in associated databases (DB) that are part of each measurement system or in engineered solutions (cf. Figure 4).

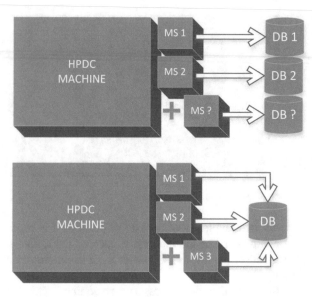

Figure 4   Existing measurement-system architecture and target architecture in the HPDC-foundry

4.2   Conceptual Application of the Framework

The existing system architecture is not able to handle the data volume and the various types of data. An integrated analysis of a production process is not practicable because of a missing data consistency and a missing interoperability. In addition, the communication between the different databases requires high efforts to realize data propagation and an open access. The aspired goal of the project is that, on the one hand, condition monitoring can be applied to sensor and measurement data independently from its source and that, on the other hand, process data (regardless of its granularity) can be stored into one central database for analysis and reporting purpose. Consequently, the amount of data rises within every installation of new sensors and measurement systems. Challenging the data volume, also known as Big Data and getting over restricted interfaces are objectives to be achieved.

5   Summary and Outlook

The presented framework is a lightweight solution to implement applications for semantic integration and analysis of measurement data in modern industrial machinery regarding the real-time requirements of such scenarios. It provides the basis for a modular, extensible architectural basis, which is able to consider constraints from various application domains. In contrast to existing MES solutions, the framework applies the idea of CPPS by distributed data acquisition, propagation and monitoring as well as real-time data processing and monitoring. Besides, further analysis is supported by providing data consolidation using adaptive information integration. Through this data integration technique a holistic information basis for production process monitoring is utilized.

The framework applies modern concepts and technology like semantic technology and service oriented architecture to facilitate the required loose coupling and distributed condition monitoring of CPPS scenarios. It facilitates the aggregation and analysis of data from distributed, heterogeneous sensor networks and therefore is a significant contribution to achieve vertical data integration among SPS and MES systems.

The application and adaption to a real industrial scenario (high-pressure-die-casting) has been discussed and the next steps of the implementation have been outlined. Hereby, the framework will be continuously developed, according to the sensor data capabilities of new sensor technology. In the near future, the analysis capabilities will be further extended. For instance, cameras should be plugged onto the machine and the data transfer as well as the analysis should be managed automatically. The latter addresses various research challenges regarding on the one hand the analysis capability of the raw data streams using the rule-based fast path bus and, on the other hand, the automatic extraction of semantic information from these data streams.

Acknowledgments The approaches presented in this paper are supported by the German Research Association (DFG) within the Cluster of Excellence "Integrative Production Technology for High-Wage Countries". The authors would also like to thank the Audi AG for their support and contributions to the presented research.

References

1. I.F. Akyildiz, T. Melodia, K.R. Chowdhury, A survey on wireless multimedia sensor networks. Computer Networks **51** (4), 2007, pp. 921–960
2. H. Kagermann, W. Wahlster, J. Helbig. Umsetzungsempfehlungen für das zukunftsprojekt industrie 4.0, abschlussbericht des arbeitskreises industrie 4.0, 2012. Vorabversion
3. N. Trigoni, B. Krishnamachari, Sensor network algorithms and applications. Philosophical Transactions of the Royal Society A: Mathematical, Physical and Engineering Sciences **370** (1958), 2012, pp. 5–10
4. V.C. Gungor, G.P. Hancke, Industrial wireless sensor networks: Challenges, design principles, and technical approaches. IEEE Transactions on Industrial Electronics **56** (10), 2009, pp. 4258–4265
5. T. Kröger, F.M. Wahl, Multi-sensor integration and sensor fusion in industrial manipulation: Hybrid switched control, trajectory generation, and software development. In: *Proceedings of IEEE International Conference on Multisensor Fusion and Integration for Intelligent Systems.* 2008, pp. 411–418
6. K. Römer, O. Kasten, F. Mattern, Middleware challenges for wireless sensor networks. ACM SIGMOBILE Mobile Computing and Communications Review **6** (4), 2002, pp. 59–61
7. K. Aberer, M. Hauswirth, A. Salehi, Infrastructure for data processing in large-scale interconnected sensor networks. In: *International Conference on Mobile Data Management, 2007.* IEEE, Piscataway, NJ, 2007, pp. 198–205

8. S. Hadim, N. Mohamed, Middleware: Middleware challenges and approaches for wireless sensor networks. Distributed Systems Online, IEEE **7** (3), 2006, p. 1
9. L. Mottola, G.P. Picco, Programming wireless sensor networks: Fundamental concepts and state of the art. ACM Comput. Surv. **43** (3), 2011, pp. 19:1–19:51
10. J. Yick, B. Mukherjee, D. Ghosal, Wireless sensor network survey. Computer Networks **52** (12), 2008, pp. 2292–2330
11. L. Gurgen, C. Roncancio, C. Labbé, A. Bottaro, V. Olive, Sstreamware: a service oriented middleware for heterogeneous sensor data management. In: *Proceedings of the 5th international conference on Pervasive services (ICPS '08), ACM, New York, USA,* 2008, pp. 121–130
12. F. Golatowski, J. Blumenthal, M. H, M. Haase, H. Burchardt, D. Timmermann, Service-oriented software architecture for sensor networks. In: *In Proc. Int. Workshop on Mobile Computing.* 2003, pp. 93–98
13. E. Avilés-López, J. García-Macías, Tinysoa: a service-oriented architecture for wireless sensor networks. Service Oriented Computing and Applications **3** (2), 2009, pp. 99–108
14. K.K. Khedo, R.K. Subramanian, A service-oriented component-based middleware architecture for wireless sensor networks. IJCSNS International Journal of Computer Science and Network Security **9** (3), 2009, pp. 174–182
15. H. Abangar, P. Barnaghi, K. Moessner, A. Nnaemego, K. Balaskandan, R. Tafazolli, A service oriented middleware architecture for wireless sensor networks. In: *2010 Future Network & Mobile Summit.* IIMC International Information Management Corp., Dublin, Ireland, 2010
16. J. Ibbotson, C. Gibson, J. Wright, P. Waggett, P. Zerfos, B. Szymanski, D.J. Thornley, Sensors as a service oriented architecture: Middleware for sensor networks. In: *Proceedings of the 2010 Sixth International Conference on Intelligent Environments,* ed. by V. Callaghan. IEEE Computer Society and IEEE, Washington, DC, USA, 2010, pp. 209–214
17. C. Groba, I. Braun, T. Springer, M. Wollschlaeger, A service-oriented approach for increasing flexibility in manufacturing. In: *IEEE International Workshop on Factory Communication Systems, 2008,* ed. by G. Cena, F. Simonot-Lion. IEEE, Piscataway, NJ, 2008, pp. 415–422
18. N. Mohamed, J. Al-Jaroodi, A survey on service-oriented middleware for wireless sensor networks. Serv. Oriented Comput. Appl. **5** (2), 2001, pp. 71–85
19. K.A. Delin, S.P. Jackson, Sensor web: a new instrument concept. Proc. SPIE Functional Integration of Opto-Electro-Mechanical Devices and Systems **4284**, 2001
20. V. Vescoukis, N. Doulamis, S. Karagiorgou, A service oriented architecture for decision support systems in environmental crisis management. Future Gener. Comput. Syst. **28** (3), 2012, pp. 593–604
21. R. Daisenroth, MES unterstützt Industrie 4.0: Wegweisendes Zukunftskonzept. VDI-Z **2013** (Nr 4), April, pp. 20–22
22. A. Bröring, P. Mauè, K. Janowicz, D. Nüst, C. Malewski, P. Maué, Semantically-enabled sensor plug & play for the sensor web. Sensors 2011 **11** (12), 2011, pp. 7568–7605
23. J. Wright, C. Gibson, F. Bergamaschi, K. Marcus, R. Pressley, G. Verma, G. Whipps, A dynamic infrastructure for interconnecting disparate isr/istar assets (the ita sensor fabric). In: *Proceedings of the 12th International Conference on Information Fusion, 2009.* IEEE, Piscataway, NJ, 2009, pp. 1393–1400
24. W. Wang, P. Barnaghi, G. Cassar, F. Ganz, P. Navaratnam, Semantic sensor service networks. In: *IEEE Sensors 2012.* IEEE, 2012, pp. 1–4
25. M. Lewis, D. Cameron, S. Xie, B. Arpinar, Es3n: A semantic approach to data management in sensor networks. In: *Semantic Sensor Networks Workshop.* 2006
26. K. Martinez, J.K. Hart, R. Ong, K. Martinez, J.K. Hart, R. Ong, Environmental sensor networks. Computer **37** (8), 2004, pp. 50–56
27. U. Enste, W. Mahnke, OPC Unified Architecture: Die nächste Stufe der Interoperabilität. at - Automatisierungstechnik **59** (7), 2011, pp. 397–404
28. M. Compton, C.A. Henson, H. Neuhaus, L. Lefort, A.P. Sheth, A survey of the semantic specification of sensors. In: *Proceedings of the 2nd International Workshop on Semantic Sensor Networks,* ed. by K. Taylor, D.D. Roure. CEUR-WS.org, 2009, CEUR Workshop Proceedings, pp. 17–32

29. M. Compton, P.M. Barnaghi, L. Bermudez, R. Garcia-Castro, Ó. Corcho, S. Cox, J. Graybeal, M. Hauswirth, C.A. Henson, A. Herzog, V.A. Huang, K. Janowicz, W.D. Kelsey, D. Le Phuoc, L. Lefort, M. Leggieri, H. Neuhaus, A. Nikolov, K.R. Page, A. Passant, A.P. Sheth, K. Taylor, The ssn ontology of the w3c semantic sensor network incubator group. Journal of Web Semantics **17**, 2012, pp. 25–32

30. T. Berners-Lee, Linked data – the story so far. International Journal on Semantic Web and Information Systems **5** (3), 2011, pp. 1–22

31. T. Meisen, *Framework zur Kopplung numerischer Simulationen für die Fertigung von Stahlerzeugnissen*, vol. Fortschritt-Berichte VDI. Reihe 10: Informatik / Kommunikation; 823, 1st edn. VDI-Verlag, Düsseldorf, 2012

32. T. Meisen, P. Meisen, D. Schilberg, S. Jeschke, Adaptive information integration: Bridging the semantic gap between numerical simulations. In: *Enterprise Information Systems, Lecture Notes in Business Information Processing*, vol. 102, ed. by R. Zhang, J. Zhang, Z. Zhang, J. Filipe, J. Cordeiro, Springer, Berlin / Heidelberg, 2012, pp. 51–65

33. D. Guinard, V. Trifa, F. Mattern, E. Wilde, From the internet of things to the web of things: Resource-oriented architecture and best practices. In: *Architecting the Internet of Things*, ed. by D. Uckelmann, M. Harrison, F. Michahelles, Springer Berlin / Heidelberg, Berlin, Heidelberg, 2011, pp. 97–129

Bitmap-Based On-Line Analytical Processing of Time Interval Data

Philipp Meisen, Tobias Meisen, Diane Keng, Marco Recchioni and Sabina Jeschke

Abstract On-line analytical processing is in the focus of research over the last couple decades. Several papers dealing with summarizability problems, cube computations, query languages, fact-dimension relationships or different types of hierarchies have been published. Nowadays, analyzing time interval data became ubiquitous. Nevertheless, the use of established, reliable, and proven technologies like OLAP is desirable in this respect. In this paper, we present an OLAP system capable to process time interval data. The system is based on bitmaps, enabling performant selection and fast aggregation. Moreover, we introduce a two-step aggregation technique, which enables the calculation of relevant measures in the context of time interval data. We evaluate the performance of our system using different bitmap implementations and a real-world data set. To our knowledge, there are no other systems available enabling OLAP and providing correct results considering the summarizability of time interval data.

Keywords Time Interval Data · Time Series · Bitmap · On-Line Analytical Processing · Two-Step Aggregation

1 Introduction

Time interval data are recorded in various situations, e.g. task executions, gesture tracking, or behavioral observations. Several disciplines like artificial intelligence, music, medicine, ergonomics, and cognitive science analyze time interval data to detect patterns (frequent sequences) or association-rules [1]. Lately, some attention on assessing the similarity between sets of time interval data (so-called event

P. Meisen (✉) · T. Meisen · S. Jeschke
IMA/ZLW & IfU, RWTH Aachen University, Dennewartstr. 27, 52068 Aachen, Germany
e-mail: philipp.meisen@ima-zlw-ifu.rwth-aachen.de

D. Keng
School of Engineering, Santa Clara University, Santa Clara, USA

M. Recchioni
Airport Division, Inform GmbH Aachen, Aachen, Germany

Originally published in "International Conference on Information Technology - New Generations (ITNG)", © IEEE 2015. Reprint by Springer International Publishing Switzerland 2016, DOI 10.1007/978-3-319-42620-4_68

sequences) arose [2]. Nevertheless, very limited attention has been given on the aggregation of time interval data in the context of on-line analytical processing (OLAP).

OLAP is generally described as a user-driven method to analyze multidimensional data. An important technique used within such applications is the aggregation of data along defined members of different hierarchies belonging to different dimensions [3]. In addition to the aggregating performance, the filtering capabilities (i.e. handling high selectivity) of the underlying database system are of great importance regarding the performance of an OLAP system. Since the beginning of multidimensional models (MDM), the research community has introduced three different approaches used to handle data within a MDM; multidimensional (MOLAP), relational (ROLAP), and hybrid approaches (HOLAP). The different advantages and disadvantages are frequently discussed in numerous publications. Likewise, other issues concerning MDM have been addressed over the last years; non-strict, non-onto or non-covering hierarchies [4, 5], many-to-many relationships and summarizability problems [6, 7], low granularities leading to high cardinality [3], computation of data cubes [8], and query languages [9].

When dealing with time interval data we are faced with several of these issues, e.g.

- Every interval defines a many-to-many relationship between the facts of the interval and the time dimension which leads to summarizability problems,
- Real-world data sets contain non-onto and non-covering hierarchies,
- The time dimension may use a lowest granularity leading to millions of granules (e.g. using milliseconds), and
- The querying of data for aggregated values needs temporal operators, e.g. as defined by Allen [10].

Our contributions: In this paper, we present an OLAP system for time interval data. The heart of our system is a bitmap-based implementation, enabling fast selection and aggregation. Additionally, we will introduce our two-step aggregation technique, which enables the calculation of relevant measures in the context of time interval data. To our knowledge, there are no other OLAP systems available providing sufficient solutions for the raised issues.

This paper is organized as follows: First, we provide a short recap of the time interval data analysis model (TIDAMODEL) presented in detail in [11] Next, we present the problems occurring while aggregating time interval data from a semantic and temporal point of view and we introduce our two-step aggregation technique. In the following section, we describe our bitmap-based implementation used to enable OLAP of time interval data supporting the presented aggregation methods. Finally, we provide performance measures of the system before the paper is concluded.

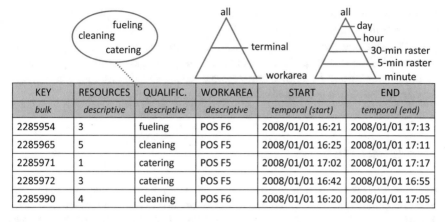

KEY	RESOURCES	QUALIFIC.	WORKAREA	START	END
bulk	*descriptive*	*descriptive*	*descriptive*	*temporal (start)*	*temporal (end)*
2285954	3	fueling	POS F6	2008/01/01 16:21	2008/01/01 17:13
2285965	5	cleaning	POS F5	2008/01/01 16:25	2008/01/01 17:11
2285971	1	catering	POS F5	2008/01/01 17:02	2008/01/01 17:17
2285972	3	catering	POS F5	2008/01/01 16:42	2008/01/01 16:55
2285990	4	cleaning	POS F6	2008/01/01 16:20	2008/01/01 17:05

Figure 1 Illustration of our running example TIDAMODEL, the specified interval is read to be half-open, i.e. [start, end)

2 Background: Modeling Time Interval Data

A characteristic of time interval data is the start and end time-value specifying the validity range of other descriptive values associated to the interval (cf. Figure 1). Sets of time interval data can be represented in many ways. When modeling time interval data in the context of analytics, it is necessary to define not only the schema of a time interval data record but also the MDM. A MDM is defined by the different dimensions, its hierarchies, levels and members. In the case of time interval data, the time axis and the time dimension are of special interest. In this paper, we use the TIDAMODEL introduced in [11]. The model is defined by a 5 tuple (P, Σ, τ, M, Δ) in which P denotes the time interval database, Σ the set of descriptors, τ the time axis, M the set of measures, and Δ the set of dimensions. The time interval database P contains the raw time interval data records and a schema definition of the contained data. The schema associates each field of the record (which might contain complex data structures) to one of the following categories; temporal, descriptive, or bulk. The model explicitly allows the definition of non-strict, non-onto, and non-covering hierarchies.

Our running example within this paper is illustrated in Figure 1. The figure shows five raw time interval records of a database P. Each record of the database is defined to provide values[1] for the fields KEY, RESOURCES, QUALIFICATION, WORKAREA, START, and END which are categorized in the specified groups. The example does not define the set of values, dimensions or mapping functions for every descriptor, which is normally necessary. Instead, it exemplifies the valid values for the QUALIFICATION descriptor (i.e. fueling, cleaning, and catering) as well as two hierarchies - one for the WORKAREA descriptor and the other one

---

[1]The values do not have to be provided, but if not the system uses a fallback strategy to determine a value, which can be defined by the mapping function (e.g. unknown, or not set).

for the time dimension. The time granularity within our example is defined to be minutes. Without introducing the term descriptor any further, it should be mentioned that a descriptive value of the raw record is mapped to one or many descriptor values. This enables the system to support generally many-to-many relationships between facts and dimensions. In our running example we use the identity function to map a descriptive value to a descriptor value.

3 Aggregating Facts

Aggregation is the predominant operation in OLAP systems [3]. The facts of the systems are aggregated along the defined members of the different hierarchies defined by the MDM. Typically, the OLAP system stores the collected facts in a cube (MOLAP) or retrieves facts from a relational database (ROLAP). The former aggregates the facts of each measure, while the data is loaded into the cube. The latter applies the aggregation of the measure when retrieving the data (e.g. using group by) or is loaded with pre-aggregated data generated during integration [12]. In Figure 2, a two-dimensional MOLAP cube and a ROLAP star schema are illustrated while taking into consideration that the raw records from our running example are used, excluding the START and END fields.

Whenever a user selects data (e.g. by roll-up, slice or dice operations), the collected facts are aggregated by the system according to the selected pre-defined measures or the aggregation specified in the multidimensional expression (MDX). The latter aggregation method is independent of the used technology. The use of aggregations within an MDX is mostly complicated, it is not applicable to end-users and quite error-prone. Considering the running example and the models shown in Figure 2: if a user were to select all qualifications, the work-area *POS F5* and a measure defined as SUM(RESOURCES), the calculation of the system would result in 9. More complex

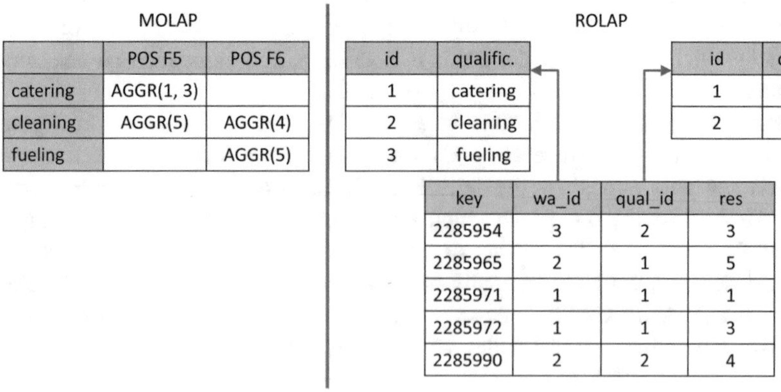

MOLAP

	POS F5	POS F6
catering	AGGR(1, 3)	
cleaning	AGGR(5)	AGGR(4)
fueling		AGGR(5)

ROLAP

id	qualific.
1	catering
2	cleaning
3	fueling

id	qualific.
1	POS F5
2	POS F6

key	wa_id	qual_id	res
2285954	3	2	3
2285965	2	1	5
2285971	1	1	1
2285972	1	1	3
2285990	2	2	4

Figure 2 MOLAP and ROLAP schemas of the running example excluding the START and END fields

queries, like retrieving the average amount of resources per work-area,[2] are not provided easily via a user-interface and therefore have to be formalized using MDX queries directly.

3.1 Aggregation and Many-To-Many Relationships

A many-to-many relationship occurs if a fact is associated to multiple members instead of exactly one. In the case of time interval data, each interval defines facts which are related to multiple members of the time dimension. Generally, modern OLAP systems like Microsoft Analysis Services, Oracle Database OLAP, IBM Cognos, or iccube support many-to-many relationships. Nevertheless, the (implicitly) adapted MDM increases the integration effort dramatically, especially when a low granularity (e.g. minutes) is used for the time dimension [7]. Additionally, the performance of the OLAP system decreases tremendously [13] and some systems are not capable of handling the amount of data necessary to be created in memory when querying time interval data [11].

The main difference between the aggregations of facts being related to dimensions by a many-to-many relationship is the selection of the facts to be aggregated. In the case of a many-to-many relationship, the system ensures that facts of the same record are not counted multiple times which would lead to summarizability problems. To achieve this, the OLAP system typically uses a unique identifier for each record to identify and remove duplicates. After the selection of facts is performed, the aggregation is done as described for the default case of one-to-many-relationships.

3.2 Aggregation of Time Interval Data

Figure 3 illustrates our running example in a Gantt-chart. The WORKAREAs are illustrated as swim-lanes, whereby different QUALIFICATIONs are color-coded. The figure shows the already mentioned many-to-many relationship, i.e. each fact of an interval is associated to multiple members of the time dimension (e.g. the descriptor value 1 of the time interval starting at 17:02 is associated to 15 granules 17:02, 17:03, …, 17:16). Additionally, the figure shows the values of three measures; *needed resources, time spent (min)* and *tasks finished*. The values are aggregated to the *5-min raster* members of the time hierarchy (i.e. [16:20, 16:25), …, [17:15, 17:20)). Defining these three measures within a modern OLAP system is currently not possible while only using multidimensional modeling. Heavily adapting and adding temporal logic to the integration process are some of the required steps to perform, accepting the negative impacts mentioned previously. To resolve these problems, we suggest

---

[2]This could be achieved by calculating the SUM per work-area and calculate the AVG of the retrieved summed values.

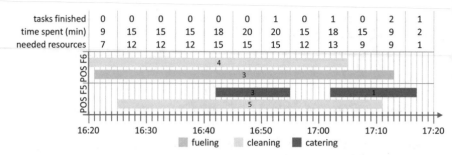

tasks finished	0	0	0	0	0	0	1	0	1	0	2	1
time spent (min)	9	15	15	15	18	20	20	15	18	15	9	2
needed resources	7	12	12	12	15	15	15	12	13	9	9	1

Figure 3 Running example illustrated in a Gantt-chart with three measures

a new aggregation technique, as well as the definition of new temporal aggregation operators like *count started*, *count finished* and *count running*. We introduce our aggregation technique initially by explaining each of the three measures. Thereafter, we define our technique in detail.

To calculate the measure *needed resources* for a specific member of a level of a time hierarchy, it is necessary to calculate an intermediate result for each member of the lowest granularity. The intermediate result of the measure of a granule is calculated by summing up all the RESOURCES facts. Looking at our running example shown in Figure 3 and the granule 16:42, the intermediate result is calculated by summing 5, 3, 3 and 4, i.e. 15. The intermediate result for the granule 16:41 is 12, calculated by 5 + 3 + 4. In a second phase, the maximum of all the retrieved intermediate results is picked as a result of the measure for the member of the higher level. In our example, the result of the measure for the member [16:40, 16:44] of the 5 min raster is thereby given by the maximum of 12 (16:40), 12 (16:41), 15 (16:42), 15 (16:43) and 15 (16:44).

The measure *time spent (min)* is also calculated using two steps. In the first step, the number of associated facts to a granule is counted. In the second step, the calculated counts are summed up. Looking at the example and the granule, 16:20 results in a count of 1, whereby 16:21 results in 2. Adding up all values for the member [16:20, 16:24] of the *5 min raster* results in 9, i.e. 1 (16:20) + 2 (16:21) + 2 (16:22) + 2 (16:23) + 2 (16:24).

The third measure *tasks finished* uses an aggregator based on temporal logic. In the first step, the temporal aggregation method *count finished* is applied on the lowest granularity counting the intervals finishing at the specified granule. In the second step, the retrieved values are summed up.

3.3   Two-Step-Aggregation Technique

As exemplified in the previous section, the calculations for measures in the context of time interval data are achieved by a two-step aggregation. In the first step, the value is calculated for the granules of the time-dimension by applying a default or a temporal aggregation method. In the second step, these calculated values are aggregated by applying a default aggregation method such as *count, sum, average*, or *mean*.

Before applying the two-step aggregation technique, the system has to select the data according to the defined filter and grouping criteria. To overcome summarizability problems arising by the mentioned reasons pertaining to many-to-many relationships, it is necessary that the system is capable of associating each fact to its interval. We present our bitmap-based solution considering the filtering in Section 4. We denote the set of selected intervals by Θ and the set of selected intervals for a specific group g by Θ_g. We denote the fact of a record ρ for the granule t of the time-dimension by $\phi(\rho, t)$. The set of all time granules selected by the query using a specified hierarchy level of the time-dimension or a time-window is denoted by T. Let the first aggregation method be denoted by AGGR1 and the second by AGGR2. The result of the first aggregation for a specific interval set Θ_g and a granule $t \in T$ is defined by

$$i_t := \mathrm{AGGR1}(\{\phi(\rho, t) \mid \rho \in \Theta_g\}). \tag{1}$$

The result of a measure mg for a specific group g and set of granules T is defined by

$$m_g := \mathrm{AGGR2}(\{i_t \mid t \in T\}). \tag{2}$$

Figure 4 illustrates the defined entities used when calculating a measure applying the two-step aggregation technique. The illustration is based on our running example, showing the application of the technique for Query Table 1.

As shown in the figure, a fact function may return *null* values. This indicates that no interval covers the specified granule. The *null* values might be important for the first aggregation function used. In some cases (e.g. *sum*), a *null* value can be understood as 0, in others (e.g. *average*) it might provide important information (e.g. *NaN* as result).

Table 1 Example query used in Figure 4

select	MAX(SUM(RESOURCES)) as 'needed Res.'
on	RASTER.5MIN
in	[17:00, 17:10]
where	QUALIFICATION in {catering,cleaning}
group by	WORKAREA

4 Bitmap-Based OLAP of Time Interval Data

As mentioned in the previous sections, aggregation is the predominant operation in an OLAP system. Nevertheless, selecting the needed data for aggregation and supporting incremental calculations are important aspects to increase aggregation performance. Additionally, the need to identify each interval individually is a significant requirement to ensure correct summarizability.

Bitmaps have been used in several implementations and publications to achieve fast data selection in mostly read-only environments. Depending on the application context of the bitmaps, it has been shown that the selection of the encoding and compression scheme of a bitmap is crucial, considering the performance gained and storage needed. Important criteria to select the best encoding and compression scheme are the types of queries [14], the order of the data [15], and the complexity considering the logic operations used within queries [16]. Furthermore, in specific scenarios implementations using bitmaps outperform popular commercial database management systems (DBMS) by a factor larger than ten [17].

Figure 5 illustrates the components of our bitmap-based OLAP system enabling fast selection and aggregation of time interval data. The illustration shows the different components focusing on the Data Repository and the Cache Persistence components. Before we introduce the application of the two-step aggregation technique presented in Section 3 and the selection of data needed for the aggregation based on the bitmap-based implementation, we briefly introduce the different components of the Data Repository.[3]

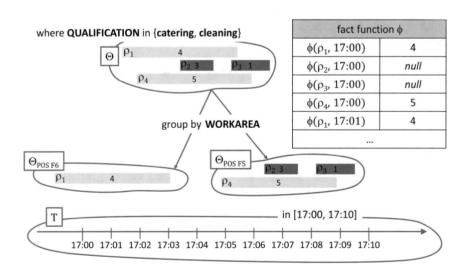

Figure 4 Entities defined for two-step aggregation technique

---

[3]The Insertion Process Is Introduced in Detail in [11]. We Refer the Interested Reader to that Paper for More Insights.

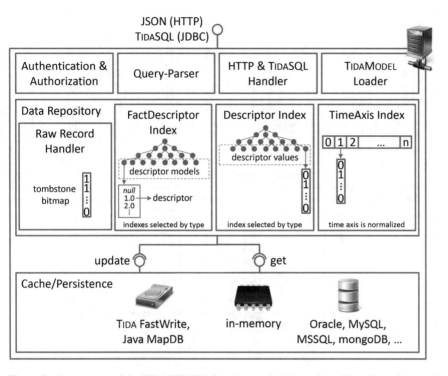

Figure 5 Components of the TIDASERVER focusing on the bitmap-based Data Repository and Cache/Persistence components

4.1 Raw Record Handler: Inserts, Deletes and Updates

Inserts, deletes and updates of time interval data records are handled by the *Raw Record Handler* to receive newly or updated raw time interval data records. The handler is responsible of managing the so called tombstone bitmap which identifies if a record is still valid (i.e. value of 1) or not (i.e. value of 0).

When a new record is received by the handler, it creates a unique identifier for that record. Additionally, the record is processed using the mapping functions defined by the TIDAMODEL to which the record belongs. Afterwards, the different bitmaps using the *Descriptor* and *TimeAxis Indexes* are updated. Finally, the tombstone bitmap is updated and set to 1 for the record.

The deletion of a record is achieved by setting the value within the tombstone bitmap to 0. A clean-up process picks up the dead stored records and performs a clean-up in the background. An update of a time interval record is performed by deleting the record and inserting the updated record as a new instance.

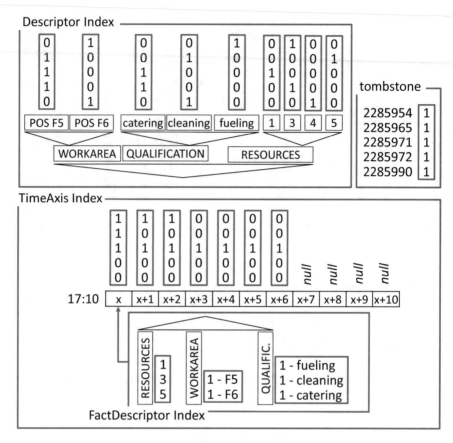

Figure 6 The indexes of the running example: TimeAxis Index is only illustrated for [17:10, 17:20] and only one FactDescriptor Index is shown

4.2 Descriptor, TimeAxis and FactDescriptor Index

The different indexes used within our running example are illustrated in Figure 6. In the example, it is assumed that each descriptor uses a fact-function invariant of the record (RESOURCES uses the identity function, WORKAREA and QUALIFICA-TION use a fact-function returning always 1).

The *Descriptor Index* is used to select the different bitmaps for the different descriptors. The system selects the best fitting index (from the ones known by the system) for the identifiers of the descriptors considering performance (e.g. TroveInt[4] is used for int values).

The *TimeAxis Index* is implemented using an array-like structure. Each granule of the time-axis is represented by an integer. The bitmaps for a specific granule can

---

[4]http://trove.starlight-systems.com/.

be reached by using the integer as index. Besides the bitmap, the index provides access to a *FactDescriptor Index* containing information about the facts associated to the granule of the time-axis. The *FactDescriptor Index* is used to retrieve the different facts for a granule. Facts are retrieved from a descriptor by its fact-function. The fact-function can thereby be defined to be invariant of the record or not. As defined in our example, the RESOURCES descriptor is an invariant descriptor, i.e. the value of the descriptor is provided by the descriptor itself. An example for a variant descriptor could be a temperature value measured over a time interval, assuming that the descriptive value (e.g. 120.54 °F) is mapped to a descriptor value categorizing the descriptive value (e.g. 120.54 °F is mapped to high'). When aggregating the facts of the temperature descriptor, the facts have to be retrieved directly from the record. The descriptive value high' does not define the exact temperature needed for aggregation. The *FactDescriptor Index* sorts all the facts of the different descriptors in ascending order. Variant descriptors are thereby moved to the top of the list (with a fact-value of *null*). Whenever the system has to retrieve a fact for a specific descriptor, it uses the index to retrieve a sorted list. If the list contains *null* values indicating variant facts, the fact of each record is retrieved. Otherwise the values of the list are used directly (cf. Figure 7).

4.3 Bitmap-Based Time Interval Data Selection

Whenever a (new) query is fired against the system, the *Query Parser* processes the query and receives the filtering, grouping and time dependent criteria. In addition, the parser determines the measures to be calculated and which level of the time hierarchy is to be used. Using this information, the system creates a set of bitmaps containing

- The resulting bitmap of the filtering criteria (determined by applying bitwise and $*$, or $+$, not $\neg$, or xor $\bigoplus$ operations)
- The bitmaps for each defined group, and
- The bitmaps for each selected time-granule.

The resulting set of bitmaps of the previously introduced Query 1 is shown in Formula (3), in which bit_x denotes the bitmap of the descriptor value x or the bitmap of the time-granule x.

$$r_1 = \{\text{tombstone} * (bit_{catering} + bit_{cleaning}),$$
$$\{bit_{17:00}, \ldots, bit_{17:10}\}, \{bit_{POS\ F5}, bit_{POS\ F6}\}\} \qquad (3)$$

To create the different time-series for each time-granule within each group, the algorithm iterates over each bitmap of a time-granule and each bitmap of a group. Within each iteration the algorithm combines the filtered bitmap with both the current time-granule bitmap and the group bitmap using the and operation. For our example, the resulting bitmap for time-granule 17:10 and the group POS F5 is shown in Formula (4).

$$\mathrm{bit}_{\mathrm{POS\ F5,\ 17:10}} = (0, 1, 1, 0, 0)^T \tag{4}$$

If a level of a hierarchy is used within the grouping or the filter criteria of the query, the algorithm resolves the granules for the specific descriptor and combines the bitmaps according to the hierarchy definition (using the or-operation).

4.4 Bitmap-Based Aggregation

After the selection is performed the aggregation algorithm is applied. The input for the aggregation is a set of bitmaps (cf. Formula 3) created by the selection algorithm. The following aggregation algorithm is based on the two-step aggregation technique presented in Section 3.

The first aggregation is performed on every bitmap of the sequence. The implementation, and thereby the performance, of the aggregation method depends on the type of aggregation. A *count* aggregation can be performed most efficiently by almost any bitmap implementation [16]. Temporal aggregations need additional information about the next or previous time-granule to be performed.[5] Other aggregations such as *sum*, *max* or *min* typically need the fact-values to be applied. These values are retrieved using the *FactDescriptor Index* to determine the sorted descriptors associated with the time-granule. If the facts of the descriptors are variant, the algorithm creates an array of facts retrieved from the records and applies the aggregation technique to the array. Otherwise, if the facts are invariant, the records are not needed and the result of the aggregation can be determined using bitmaps only. The algorithm iterates over the descriptors (descending if the aggregation is *max*, otherwise ascending) and combines the bitmap of the descriptor with the bitmap of the time-granule using the *and*-operation. For a *min* or *max* aggregation, the count of the resulting bitmap is used to determine the value (a value larger than 0 indicates that the algorithm can stop and return the invariant fact, otherwise the algorithm moves on to the next descriptor). In the case of a *sum* aggregation, the algorithm multiplies the count with the invariant value and sums all the results. Figure 7 illustrates selected aggregation methods useful as a first aggregation and the bitmap based calculation.

The second step aggregation is performed on the set of numeric values returned during the first step and mathematically straight forward. The aggregation collects the calculated values from the first step and applies the mathematical function to the set of numbers, e.g. *sum*, *min*, *max* or *count*.

---

[5]The aggregation combines the previous (count started) or following (count finished) bitmap with the actual one using the xor-operation and counts the result. We refer to the example shown in Figure 7.

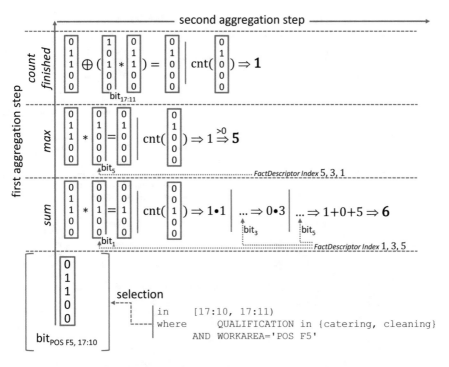

Figure 7 Example of the first aggregation step of *sum*, *max* and *count finished* based on bitmaps

5 Performance

To assess the performance of our implementation, we ran several tests on an Intel Core i7-3632QM with a CPU clock rate of 2.20 GHz, 8 GB of main memory, a SSD, and running 64-bit Windows 7 Enterprise. For our Java implementation, we used a 64-bit JRE 1.6.45, with XMX 4,096 MB and XMS 512 MB. We tested the runtime performance of our implementation using two different bitmap implementations, i.e. word-align (EWAH)[6] [16] and roaring bitmap [19]. We used a real-world data set containing 1,122,097 records collected over one year. The records have an average interval length of 48 minutes and three descriptive values (person (cardinality: 713), task-type (cardinality: 4), and work area (cardinality: 31)). The used time-granule was minutes (i.e. time cardinality: 525,600). Each test ran 100 times and the average CPU time is used as result.

The results are shown in Figure 8. As illustrated both implementations achieve almost the same performance when counting. Roaring bitmaps perform way better considering logical operations, which are used excessively when aggregating (e.g. *min* and *sum*). Additionally, the setter method of roaring bitmaps executes almost

---

[6]We use the JavaEWAH library, which is considered to offer the best query time for all Java world-align distributions [18].

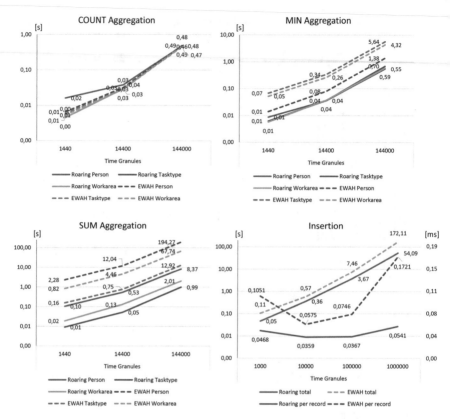

Figure 8 Results of the performance tests using our implementation based on word-align and roaring bitmaps using a real-world data set

constantly per record compared to word-aligned bitmaps when the size of the bitmap increases. The tests show that using roaring bitmaps can increase the performance by a factor of 25. With respect to results based on JavaEWAH shown in [11] using roaring bitmaps is promising.

6 Conclusions and Future Work

In this paper, we presented a bitmap-based OLAP system for time interval data. The system is capable of creating aggregations by loading the raw time interval data into the system, persisting the raw data (if required), and calculating measures needed to analyze time interval data using our introduced two-step aggregation technique. Additionally, we presented temporal aggregation methods like count finished, which are supported by the bitmap-based implementation as well.

In future work, we plan to use the implementation presented here in order to enable on-line analytical mining (OLAM) on time interval data. One of the challenges in that field is the remapping from aggregated time-series patterns back to the time-intervals. We also plan on defining a simplified query language to replace the mostly complex MDX queries. Additionally, we plan to present in detail the capabilities of distributed calculations and load balancing available through our implementation.

Acknowledgments The approaches presented in this paper are supported by the German Research Foundation (DFG) within the Cluster of Excellence "Integrative Production Technologies for High-Wage Countries" and the project "ELLI Excellent Teaching and Learning in Engineering Sciences" as part of the Excellence Initiative at the RWTH Aachen University.

References

1. P. Papapetrou, G. Kollios, S. Sclaroff, D. Gunopulos, Mining frequent arrangements of temporal intervals. Knowledge and Information Systems **21** (2), 2009, pp. 133–171
2. A. Kotsifakos, P. Papapetrou, V. Athitsos, Ibsm: Interval-based sequence matching. In: *Proceedings of the 13th SIAM International Conference on Data Mining, Austin, Texas, USA*. 2013
3. S. Agarwal, R. Agrawal, P. Deshpande, A. Gupta, J. Naughton, R. Ramakrishnan, S. Sarawagim, On the computation of multidimensional aggregates. In: *Proceedings of the 22nd VLDB*. 1996, pp. 506–521
4. T. Pedersen, Aspects of data modeling and query processing for complex multidimensional data. Ph.D. thesis, 2000. Ph.D. thesis
5. S. Banerjee, K. Davis, Modeling data warehouse schema evolution over extended heirarchy semantics. Journal on Data Semantics XIII, Lecture Notes in Computer Science **5530**, 2009, pp. 72–96
6. J. Mazón, J. Lichtenbörger, J. Trujillo, Solving summarizability problems in fact-dimension relationships for multidimensional models. In: *Proceedings of DOLAP '08*. 2008, pp. 57–64
7. J. Mazón, J. Lichtenbörger, J. Trujillo, A survey on summarizability issues in multidimensional modeling. Data & Knowledge Engineering **68**, 2009, pp. 1452–1469
8. A. Vaisman, R. A. Mendelzon, A temporal query language for olap: Implementation and a case study. In: *Proceedings of the 8th Biennial Workshop on Data Bases and Programming Languages (DBPL 2001), Frascati, Italy, September 8-10, 2001*
9. Y. Yuan, X. Lin, Q. Liu, W. Wang, J. Xu Yu, Q. Zhang, Efficient computation of the skyline cube. In: *Proceedings of the VLDB '05 Proceedings of the 31st international conference on Very large data bases*. 2005, pp. 241–252
10. J. Allen, Maintaining knowledge about temporal intervals. Commun. ACM **26** (11), 1983, pp. 832–843
11. P. Meisen, T. Meisen, M. Recchioni, D. Schilberg, S. Jeschke, Modeling and processing of time interval data for data-driven decision support. In: *Proceedings of the SMC2014, San Diego, USA, 2014*
12. S. Chaudhuri, U. Dayal, An overview of data warehousing and olap technology. SIGMOD Rec. **26** (1), 1997, pp. 65–74
13. M. Russo, A. Ferrrari. The many-to-many revolution 2.0: Advanced dimensional modeling with microsoft sql server analysis service. http://www.sqlbi.com/wp-content/uploads/The_Many-to-Many_Revolution_2.0.pdf. Last Checked: October, 2011
14. C. Chan, Y. Ioannidis, An efficient bitmap encoding scheme for selection queries. In: *SIGMOD 1999, Proceedings ACM SIGMOD International Conference on Management of Data, Philadephia, Pennsylvania, USA, June 1999*

15. K. Wu, A. Shoshani, K. Stockinger, Analyses of multi-level and multi-component compressed bitmap indexes. ACM Trans. Database Syst. **35** (1), 2008. Article 2
16. O. Kaser, D. Lemire. Compressed bitmap indexes: beyond unions and intersections, 2014. Eprint arXiv:1402.4466
17. K. Wu, S. Ahern, E. Bethel, J. Chen, H. Childs, E. Cormier-Michel, Fastbit: interactively searching massive data. Journal of Physics Conference Series 08/2009 **180** (1), 2009
18. G. Guzun, C. Guadalupe, D. Chiu, J. Sawin, A tunable compression framework for bitmap indices. In: *Proceedings of the 30th International Conference on Data Engineering (ICDE 2014)*
19. S. Chambi, D. Lemire, O. Kaser, R. Godin. Better bitmap performance with roaring bitmaps, 2014. Eprint arXiv:1402.6407

Modeling and Processing of Time Interval Data for Data-Driven Decision Support

Philipp Meisen, Tobias Meisen, Marco Recchioni, Daniel Schilberg and Sabina Jeschke

Abstract Over the past decades, several disciplines like artificial intelligence, music, medicine, ergonomics or cognitive science dealt with problems concerning analyses of data associated with time intervals. Topics like pattern recognition, comparison, quality, or visualization are in focus of current research. Using these techniques in the context of data-driven decision support is quite rare even though the importance of data to support better decision making can be enormous. Reasons lie above all in limited insufficient tooling support, expensive data processing, and inapplicable requirements. In this paper, we discuss the use of time interval data and name difficulties arising when processing such data for data-driven decision support. We discuss and present solutions for overcoming the identified problems and enabling the usage of time interval data for data-driven decision support. We introduce a time interval data analysis model that provides fast access to the raw time interval data but especially to aggregated time series, mostly needed when making meaningful decisions.

Keywords Time Interval · Data-Driven Decision Support · Multidimensional Modeling · Summarizability · Bitmap Index · Time Series

1 Introduction

In simple terms, a time interval (also referred to as temporal interval, event-interval, time segment, time range, or time period) is defined by two time points on a time axis, whereby the smaller (or first if equal) time point is called the start and the larger (or second if equal) time point is called the end. Using this generally understandable definition of a time interval, we define time interval data as data associated with a time interval. These associated data describe e. g. a state observed, items counted, and/or resources used within the defined time interval.

P. Meisen (✉) · T. Meisen · M. Recchioni · D. Schilberg · S. Jeschke
IMA/ZLW & IfU, RWTH Aachen University, Dennewartstr. 27, 52068 Aachen, Germany
e-mail: philipp.meisen@ima-zlw-ifu.rwth-aachen.de

Originally published in ?Proceedings of the IEEE International   923
Conference on Systems, Man, and Cybernetics (SMC2014)?, © IEEE 2014.
Reprint by Springer International Publishing Switzerland 2016,
DOI 10.1007/978-3-319-42620-4_69

Time interval data are recorded, collected, or generated in various situations (e. g. manufacturing processes, behavioral observations, or resource scheduling). Within each of these situations, time intervals are associated with data describing e. g. the results of a sub-process, the state observed, or the booking of a room. Generally, a time interval data record consists of descriptive data used to describe the important aspects and a time interval that bounds the description to a time window, defined by two time points.

In the area of decision-making, situations involving time interval data often arise. Nevertheless, the usage of time interval data in data-driven decision support systems (DSS) is quite rare. Answers to questions like "How many resources do I need to endure the next hours?", "Which process bottlenecks exist?", or "Is the observed behavior common?" are important to support decision-making dealing with resource-usage, process-optimization, or standardization. Common reasons[1] for the lack of usage of time interval data for decision support are:

- **Insufficient tooling support**: unsupported many-to-many relations, inadequate lowest granularity, poor query performance
- **Expensive data integration processes**: enormous redundant data creation, costly discretization of intervals
- **Inapplicable requirements**: shortage of integration of context specific aggregations, unsatisfying raw data and analytical result linkage

These reasons address one of the main problems arising when time interval data are used within data-driven DSS. Data-driven DSS are mostly based on multidimensional data models (MDM). An MDM assumes that each fact (i. e. descriptive data usable as measurement) can be associated with a single member of a dimension's lowest granularity. This reasonable assumption ensures that the model is free from summarizability problems.

However, while dealing with time interval data the assumption cannot achieve fulfilment. Figure 1 illustrates a simple transformation of a time interval data record into a valid multidimensional representation. The MDM consists of two dimensions (i. e. time and location) as well as the fact resources. Asking the model "How many resources are needed between 09:03 and 09:05?" demonstrates the summarizability problem. The answer would be six, which is incorrect. Instead, the correct answer is two.

Several approaches enabling the usage of many-to-many relationships in MDMs were introduced in the last years (cf. Section 3). Generally, these solutions can be applied as workarounds, but do not overcome the limitation inherit within the MDM and mostly do not solve the upcoming summarizability problems.

We introduce a model for time interval data enabling data-driven decision support. Additionally, we present an implementation based on our model. In Section 2, we

---

[1]These are the subsumed results considering the usage of time interval data retrieved from a survey of the business intelligence subdivision of the Inform GmbH. The survey "Business Intelligence: How do you use your temporal data?" was carried out during the "Inform Users Conference 2012". IT specialist of 64 international companies participated (aviation industry, logistics providers, and groundhandling service providers).

Figure 1 A simple transformation of a time interval data record into a MDM

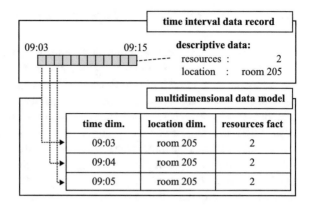

define needed terms and motivate the importance of time interval data in the context of data-driven DSS. In addition, we present requirements to be fulfilled by a model and an implementation. We discuss the related work done in the fields of data-driven DSS in context of time interval data in Section 3. In Section 4, we introduce our time interval data analysis model and present our implementation. We empirically evaluate the query performance of our implementation in Section 5. We conclude with a summary and directions for future work in Section 6.

2 Time Interval Data

2.1 Defition

A time interval data record ρ is a data record in which each field fits into one of the following categories: temporal, descriptive, or bulk. A time interval data record has exactly two fields categorized as temporal, i.e. one start and one end time point. We denote the value of the temporal start field of ρ by ρ_{start} and the end by ρ_{end}. We denote the set of possible values for ρ_{start} and ρ_{end} by Φ_{time}. Without loss of generality, we assume that the descriptive fields are numbered. We denote the value of the ith descriptive field by ρ_i. In addition, we define the set of all possible values of the ith descriptive field as Φ_i.

A time interval database P is a set of time interval data records. Without loss of generality, we assume that each record of P has the same temporal and descriptive fields. Additionally, we assume that the sets Φ_i of each descriptive field and Φ_{time} are equal. Thus, in the context of a time interval database Φ_i and Φ_{time} denote the sets generally, i. e. without any reference to a specific record.

2.2   Data-Driven Decision Support Systems

Data-driven DSS provide operational and strategic business intelligence using internal company data and sometimes integrated external data. A main feature of such a system is the user's possibility to interact with the data. A user can view a slice of data, drill-down to retrieve a more detailed understanding, or role-up to get a more generalized impression. The key to a successful data-driven DSS is having easy and rapid access to a large amount of accurate, well-organized data [1].

An MDM defines the dimensions and facts within the data. Such a model enables the use of online analytical processing (OLAP) [2] and facilitates rapid data access [3]. Data integration techniques [4] map the data to the MDM. Further activities are applied within an integration process, e.g. data cleansing to ensure data quality, or data enrichment to add missing data. Generally, it can be stated that the fast aggregation and separation of facts along defined dimensions' hierarchies in combination with fast filtering capabilities are mandatory to enable data-driven DSS [2, 5]. Thus, the question "How to speed up aggregation and filtering" is the central question to ask from a data perspective. Nevertheless, the question "How can summarizability problems be avoided while dealing with many-to-many relationships" cannot be put aside.

The aggregation of multidimensional data is achieved by application of an aggregation function to facts, which are selected by slicing, or dicing. The result of an OLAP query (also referred to as multidimensional expression (MDX)) is a multidimensional matrix or cube, containing the calculated value of the applied function for each selected dimensional member.

When calculating the result of an aggregation function the involved hierarchies of the different dimensions are not distinguished (e.g. by type) and not known by the function. Furthermore, the raw data records or their unique identifiers are typically unknown when the function is applied. Thus, it is impossible to differ if a fact is retrieved from the same or a different raw data record. This missing information is the reason for summarizability problems with non-strict fact-dimension relationships. Adapting the MDM and adding the information theoretically solves the problem [6]. The costs are an enormous amount of redundantly stored data, expensive integration processes, performance decreases, and an increased model complexity [7].

In the case of time interval data these costs are dramatically increased when choosing a low granularity (e.g. minutes) for the time dimension. This increase mostly leads to an MDM, which OLAP-application (e.g. Microsoft Analysis Services, Oracle Database OLAP, IBM Cognos, or icCube) can mostly interpret, but cannot handle. Additionally, the result of a query is a time series, i.e. a totally ordered set of calculated values, each associated with a member of a level of a hierarchy of the time dimension instead of a cube. The results have to be calculated using a two-step calculation for each member. At first, the facts of the lowest granularity are aggregated. Secondly, the calculated values of the first step are combined to determine the value of the hierarchy's level. This emphasizes the importance and the need for a special handling of the time dimension when handling time interval data within data-driven

(a) Gantt chart of the sample time interval data

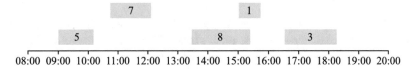

(b) minute based time series of the sample time interval data

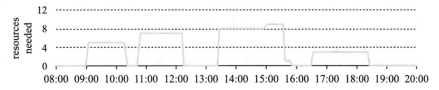

(c) hour based time series of the sample time interval data

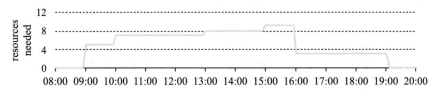

Figure 2 The time interval data as Gantt chart (**a**), aggregated as minute based time series (**b**), and hour based time series (**c**)

DSS. For example, reusing results of aggregations within the two-step calculation is a possibility to increase performance.

Figure 2 shows the time interval data as a Gantt chart (a) and two time series – retrieved from aggregation – illustrated as line charts. The minute (b) and hour based (c) time series are the result of queries summing up the resources of all data of the sample database. The time series is usable to find answers to questions about the amount of resources needed within e. g. the next hour or the next minute. Depending on the scope of the question, the granularity of the time series has to be adapted. For example, time series (c) states that in between [16:00, 17:00) three resources are needed, whereby a more detailed granularity as used in (b) shows that at 16:31 no resources, but at 16:32 three resources are needed.

In summary, the reasons for the lack of usage of time interval data in the context of data-driven DSS mentioned in Section 1 lie in the following problems: non-strict fact-dimension relationships and non-trivial aggregations.

2.3 Requirements

In this section, we outline the requirements that a MDM has to satisfy to support the integration and analysis of time interval data. These requirements are direct results of different industrial and research projects the authors carried out.

Due to already mentioned reasons and the associated complex integration process, most MDMs for time interval data only support batch updates. However, in multiple real-world scenarios, decision makers require the DSS to enable a decision-making on up-to-date data. This leads to the first requirement:

1. **Real-time integration**: The MDM has to support an integration (extraction, transformation and loading) of time related data directly after the creation to provide a real-time decision support by specialist.

Thereby, the integration has to consider the time characteristics and semantics as well as the complexity of data, which leads to additional requirements:

2. **Time semantics**: The processes of integration and analysis have to consider the underlying time semantic, e. g. multiplicity of hierarchies and different time zones. Furthermore, a query language implicitly has to support time characteristics like synchronicity and duration [8].
3. **Aggregation semantics**: The complexity of real-world data leads to complex data structures. This is particularly necessary for the aggregation of such data that has to consider the semantics (e. g., splitting a resource within an interval is impossible, otherwise splitting workload is).

Another special characteristic of time related data is the granularity of observed or measured intervals. In production scenarios, the granularity of measured time can differ between minutes or milliseconds. Other domains however, observe intervals that extend into hours. Such observation results in two more requirements:

4. **Time granularity**: The MDM has to support lowest granularities (i. e. above 10,000,000 granules) that lead to a time dimension with high cardinality.
5. **Raster time**: Besides the granularity, the DSS has to enable the user to raster time data on scales of days, weeks or years.

In times of Big Data and Big Data Analytics, some could argue that the size of data and therefore the granularity is an insignificant requirement. Even though this argument is valid for freely available data (e. g. social media data), data-driven decision support is related to business data and therefore has to consider legal and company related requirements. Data policies often do not allow the storage or distribution of data within external infrastructures. Hence, an MDM cannot relate on the usage of mass parallelization in multi-node infrastructures leasable from different companies. Rather the MDM has to support the computer infrastructure currently available to related businesses and end-users. This leads to the final requirement:

6. **Load balancing**: The MDM has to support parallelization regarding the processing of integration and analysis. However, optimization of performance should focus on the support of small computer networks.

3 Related Work

Over the past decades, several topics like pattern recognition [9, 10], data quality [11], comparison [12, 13], or visualization [14] were in focus of research dealing with time interval data. These topics were and still are addressed in various areas like music [9], medicine [15, 16], ergonomics [17] or cognitive science [18]. Within the field of data-driven DSS, research is focused on aspects of MDM: data modeling [19, 20], performance [21], solving summarizability problems [6, 22], and analytics, i. e. OLAP [2] or online analytical mining [22]. Furthermore, several research results were published addressing real-world problems considering MDMs [7, 23].

Our focus lies on an MDM for time interval data to enable the usage of such data in the context of data-driven DSS. Instead of solving a domain specific problem or generally focusing on problems occurring in the context of MDMs, we focus on an MDM for a specific type of data, i. e. time interval data. Our motivation for the selected related work are the problems and requirements stated in Section 2.

In [24], Pedersen presented an MDM for medical/clinical data. He formalized eleven advanced requirements that an MDM should satisfy. He evaluated the MDMs, current at that time, against the requirements. None of the MDMs satisfied the "many-to-many relationships between facts and dimensions" and the "support for aggregation semantics" requirements that are equivalents to our presented problems. Furthermore, Pedersen examined clinical data warehouse systems. He stated that "clinical data warehousing is still in its infancy" and "that most of the work has been done in industry, rather than in scientific environments". None of the systems fulfills the requirements formalized. According to our research, the limitations named by Pedersen are still present, even though additional features help to fulfill some mentioned requirements in specific cases (e. g. hidden placeholders are automatically added to support non-onto or non-covering hierarchies [24]). Additionally, formalized strategies and best practices on how to manage complex multidimensional data (e. g. decreasing the granularity) help reduce the requirements demanded by real-time application and enable data-driven DSS [20, 23].

Mazón et al. solve the problem of non-strict fact-dimension relationships generally on a conceptual level [6]. As shown earlier by Song et al. and mentioned in Section 2, the concrete application (i. e. the logical or physical model [25]) of these solutions lead to an enormous amount of redundantly stored data, expensive integration processes, performance decreases, and an increased model complexity [7]. Especially in the case of time intervals, the amount of stored data for each interval is increased by a factor of at least n (assuming the length of the interval is n granules).[2] Nevertheless, the presented conceptual view helps to reduce and solve summarizability problems on a logical and physical level in the case of non-strict relationships between facts and dimensions with low cardinality.

Querying dimensions with high cardinality or non-discrete values occurs when analyzing scientific data. Stockinger introduced the GenericRangeEval algorithm used to discretize scientific data and index it using a range encoded bitmap index [26]. Generally, bitmap indices are very efficient data structures for querying read-only [27] and indexing multidimensional data [3]. The used encoding and compression schemes differentiate implementations of bitmap indices.

- Three different types (and combinations of these, i. e. hybrids) categorize the different encoding schemes: equality, range, and interval [28].
- Compression schemes typically use run-length encoding (cf. PLWAH [29], CONCISE [30], EWAH [31], and PWAH [32]). However Chambi et al. recently presented a promising compression scheme based on packed arrays [33].

The selection of the encoding and compression scheme of a bitmap index is crucial, considering the performance gained and storage needed. Important criteria to select the best encoding and compression scheme are the used types of queries (i. e. range or equal) [28], the order of the data [34], and the complexity considering the logic operations used within queries [35]. Wu et al. implemented FastBit, a software tool used to query scientific data efficiently using bitmap indices, outperforming popular commercial database management systems (DBMS) in selected scenarios by a factor higher than ten [36].

4 Time Interval Data Analysis Model

4.1 Definition

We start to introduce our model by defining the term descriptor. A descriptor enables the system to map the descriptive values of a record to meaningful information, i. e. by putting it into context, associate it with facts, and cleaning noisy data. Given a time interval data record ρ of a time interval data base P with the defined sets $\Phi_1, \ldots, \Phi_n$ and Φ_{time}, we define a descriptor d_i as 3-tuple $(\varpi_i, \vartheta_i, \eta_i)$ that consists of

- The values of the descriptors (i. e. a finite set ϑ_i),
- The mapping function $\varpi_i \colon \Phi_i \to \mathbb{P}(\vartheta_i)$, and
- The fact function $\eta_i \colon \vartheta_i \to \mathbb{R}$.

---

[2]Assuming the storage needed by a field of a record is d, the time interval [14:00, 15:50] needs a capacity of 2d. Using a minute granularity (i. e. n = 110) the capacity to store the interval within a multidimensional model using Kimball's Bridge (cf. [6, 20]) would be increased to 2dn.

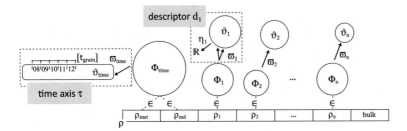

Figure 3 Illustration of descriptors and time axis

A descriptor's mapping function performs value related quality assurance, i.e. discretization, conceptualization, or cleaning. The specification of a fact function is optional, if defined it determines a number value (i.e. a fact) for the specified descriptor. We denote the set of all descriptors by $\sum$.

As the vast majority of research, we assume a discrete, bounded, and linear model of time. We define the time axis τ as 3-tuple (ϖ_{time}, ϑ_{time}, t_{grain}). The finite totally ordered set ϑ_{time} contains all the valid time points in the specified granularity train. The function $\varpi_{time}\colon \Phi_{time} \to \vartheta_{time}$ maps the ρ_{start} and ρ_{end} values to the bounded and discrete time axis used by the model. It ensures that the used time values are both values of the selected granularity t_{grain}.

Figure 3 illustrates the definitions of descriptors and a time axis. The cardinality of the domain and codomain (illustrated by the size of the circles) of a function ϖ_{time} or ϖ_i do not have to be equal. Instead, reducing (surjective) or increasing (injective) the cardinality can be necessary to ensure meaningful data-driven DSS. The former might be obvious (e.g. marking values of a descriptive field as unknown, combine synonyms, or reduce the resolution of time data), whereby the latter is used to create placeholders (e.g. from a business perspective it might be necessary to add a department or subdivision, even though time interval data is not collected).

Based on the definitions, we introduce three types of measures *implicit time measures*, *descriptor bound measures*, and *complex measures*. An *implicit time measure* is a measure defined by the temporal logic [8] (e.g. duration, count, or number of synchronal intervals). A *descriptor bound measure* defines the calculation based on a single descriptor (e.g. needed resources). A *complex measure* definition depends on other measures calculated (e.g. median or average) or defines a domain specific calculation. We denote the set of all defined measures as M.

The set of all defined dimensions Δ contains at most one dimension δ_i for each descriptor d_i and exactly one time dimension δ_{time}.

A dimension of descriptor d_i is a non-empty finite set of hierarchies, denoted as δ_i. We define a hierarchy of δ_i as a 2 tupel (G, L_G), whereby G denotes the relations among the members of the hierarchy and L_G the hierarchy's levels. We define G as a directed acyclic graph G := (V, V $\times$ V), whereby V denotes the members of the hierarchy. We define that $\vartheta_i \subseteq$ V and denote the other members by $V' := V \setminus \vartheta_i$. Each graph has a root r_G $\exists!r_G$ $v \in V'$: $\deg^+(v) = 0$. Additionally, G satisfies $\exists$ $v \in \vartheta_i$: $\deg^-(v) = 0$ and $\forall v \in V'$: $\deg^-(v) > 0$. These assumptions ensure that

- Exactly one sink (a.k.a. root) exists,
- The root is reachable from every member, and
- Every source (a.k.a. leaf) is a descriptor's value.

We define L_G as a partially ordered partition of V with binary relation $\preceq_G$ and $\{r_G\} \in L_G$. Additionally, L_G satisfies

$$\forall l_1, l_2 \in L_G, l_1 \prec G\, l_2 :$$
$$(\forall n_1 \in l_1, n_2 \in l_2 : \text{dist}(r_G, n_1) > \text{dist}(r_G, n_2)).$$

This assumption guarantees that the descendant of a level (according to the partial order $\prec_G$) increases the distance to the root and there exists at least one node of a level, which has a path to a precedent level.

Figure 4 illustrates three different types of hierarchies imaginable for a descriptor. Our definition covers all those types. The classical hierarchy ensures that the distance between root and leaves are equal. Additionally, the set of levels is totally ordered. The second hierarchy shows all the requirements demanded by [24] considering hierarchies. The third image shows a parent-child hierarchy. Each member of such a hierarchy is a descriptor's value, i. e. has facts directly associated with it. Additionally, our definition covers non-strict fact-dimension relationships as well, because of ϖ_i. The mapping function maps the record's value (which might be a list) explicitly to one or many values of the descriptor.

The time dimension δ_{time} is a non-empty set of hierarchies. We define a time hierarchy of δ_{time} as a 2-tuple (T, L_T), whereby T denotes the relations among the members of the hierarchy and L_T the set of the hierarchy's levels. We define T as a rooted plane tree with same depth for all leaves. To be consistent, we assume that T is directed towards the root. We denote the depth of the leaves as T_{depth} and the set of all nodes of depth k as N_k. The leaves $N_{T_{\text{depth}}}$ are the values of the time axis, i. e. $N_{T_{\text{depth}}} = \vartheta_{\text{time}}$. The set of levels $L_T := \{N_k \mid 0 \leq k \leq T_{\text{depth}}\}$ is a totally ordered partition of N, whereby the order $\prec_T$ is defined as $N_{T_{\text{depth}}} \prec_T \cdots \prec_T N_1 \prec_T N_0$.

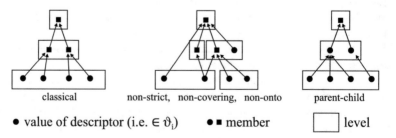

| classical | non-strict, non-covering, non-onto | parent-child |

- value of descriptor (i.e. $\in \vartheta_i$)   ●■ member   ☐ level

Figure 4 Three different types of descriptor dimensions

Additionally, we assume a total order for each set N_k with $0 \le k < T_{depth}$ and use the total order defined for ϑ_{time} as total order of $N_{T_{depth}}$. These assumptions ensure that

- The root is reachable from every member of the hierarchy by exactly one path,
- The length of the path to the root for all members of a level is equal,
- Each level (besides the root's level, it has no successor) has exactly one successor level and each level (besides the leave's level, it has no predecessor) has exactly one predecessor level,
- Each level's set of members is totally ordered, and
- Every leaf is a valid time point on the time axis.

The definition of a time hierarchy is stricter than the one of a descriptor's dimension, even though it is often referred to as a complex dimension [11, 20]. The reasons for that lie above all in the multiple hierarchies and the complex linkage between the levels of the different hierarchies. From a single hierarchy's point of view, this complexity is not present.

Figure 5 shows a sample time dimension with three totally ordered levels (hours $\prec_T$ parts $\prec_T$ all). Each leaf (i. e. each point in time) has exactly one path to the root and all those paths have the same length (i. e. each time point references exactly an hour, a part of the day and all).

To allow linkage between levels of different hierarchies of a dimension (independent if it is a time or descriptor dimension), we define shared levels as levels defined once and referred to in several hierarchies. A linkage enables an analyst to change the point of view at the data easily using roll-up and drill-down operations, instead of completely requerying the data. From a modeling perspective, shared levels are defined for a dimension and can be reused within a hierarchy. A slight modification of our presented dimension definition adds the support of shared levels. A dimension δ is a 2-tuple (H, L_{shared}), whereby H denotes the set of hierarchies of the dimension δ and L_{shared} the set of shared levels. Depending on the type of dimension, the sets L_T or L_G of a hierarchy can contain shared levels of the dimension. Within a time dimension, L_{shared} typically contains ϑ_{time}, because based on the definition of a time dimension each hierarchy must contain ϑ_{time} as leaves.

Based on the presented definitions of a time interval data-base P, the set of all descriptors $\sum$, the time axis τ, the set of all measures M, and the set of all dimensions Δ, we define our time interval data analysis model $tida_{Model}$ as 5-tuple (P, $\sum$, τ, M, Δ).

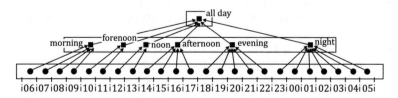

Figure 5 A sample time dimension

4.2 Data Structure

We present our data structure used to enable analytical processing of time interval data, based on a tida$_{Model}$. We work the data structure out by means of the insertion of a record.

Unique identifier for each record: For each record added to our implementation, a unique identifier is created.[3] This unique identifier is a consecutive integer. After a successful insertion, the implementation persists the records using a partitioned (based on nested array) file-based approach.

Map record to descriptor value: The mapping functions of each modeled descriptor are applied to the record (i. e. $\varpi_i(\rho_i)$). Our implementation creates a bitmap for each value of ϑ_i. A hash-map collects the different bitmaps. Thus, we implement a descriptor as bitmap index indexing records by the unique consecutive integer among the set of values determined by ϑ_i. Reference [31] describes the used bitmap implementation. We evaluated different high performance collections for Java and decided to use Trove[4] because of its performance considering insert and select. We optimized the indices by using the optimal implementation for different primitive data types (i. e. byte, short, int, or long).

The model implies that the codomain (i. e. ϑ_i) of the function is known. This is a valid assumption but in real-world applications, this is mostly not realizable. For example, assuming we modeled a marital status descriptor and business users defined the values for the descriptor as {married, single}. As mapping function, the users decide to use the identity function. In the future, time interval data with a value such as divorced might occur. In our implementation, we added three default mapping functions that define how to handle new data: *identity-*, *null-*, and *create-mapping function*. The null-mapping function maps each undefined value to a specified null-descriptor (i. e. marking the value as unknown), whereby the create-mapping function adds the new value to ϑ_i.

Map record to time axis: The time axis mapping function ϖ_{time} maps the interval values of the record (i. e. start and end value). As defined, the codomain ϑ_{time} is a totally ordered set. The total order allows us to map each element to an integer. We implemented the time axis as an array. Depending on the cardinality of ϑ_{time}, we use an optimized implementation based on the primitive data type byte, short, int or long. The array itself contains a bitmap for each granule of the time axis. The array allows fast selection (i. e. =(1) to select a specific bitmap), especially if a range query is fired. An optimized range bitmap index [27] is not reasonable because the calculation of measures needs single bitmaps, instead of a range-based aggregated bitmap.

In real-world applications and especially in the case of real-time data insertion, records are often incomplete considering the end of the time interval. Thus, we

---

[3]If the number of integers is exceeded the unique identifier is transformed to a nested array. A double nested array can contain $4.6 \cdot 10^{18}$ identifiers.

[4]We evaluated the implementations fastutil (http://fastutil.di.unimi.it), hppc (http://labs. carrotsearch.com), and Trove (http://trove.starlight-systems.com).

implemented three default strategies on how to handle unknown interval data: *boundaries, fail*, and *useknown*. The *boundaries* strategy uses the start (if start is unknown) or end value (if end is unknown) of the time axis. The *useknown* strategy uses the start value of the interval (if the end is unknown) or the end value (if the start is unknown).

Update dimensions: The dimensions of the descriptors and the time dimension are updated. The different hierarchies are implemented as defined by the model, i.e. as graph or rooted plane tree. Optionally, a node contains a bitmap that defines the records associated with the member. If the bitmap is not associated, the calculation is performed online, i.e. when the result is needed. If no members have associated bitmaps, the update dimensions step is skipped completely.

4.3 Querying the Data Structure

A query to retrieve time series or records from our implementation consists of

- The time range to retrieve data for (e.g. this year),
- The granularity of the time series (e.g. day),
- The measures to be calculated (e.g. count),
- The filter conditions (e.g. dim.hierarchy = member), and
- The grouping criteria (e.g. dim.hierarchy).

The result of a query is determined in five steps.

1. A filter bitmap is created through a combination of the descriptors' bitmaps specified by the filter conditions.
2. The bitmaps for the specified groups are determined.
3. The bitmaps of the time axis are selected using the defined time range.
4. Each bitmap of the time axis is combined with the filter bitmap and iterative with the group bitmaps. The result of the first aggregation (depending on the measure to be calculated) is determined.
5. Depending on the granularity defined, the second aggregation is applied using the results of the first.

Instead, the retrieval of records (i.e. the raw data) for a specific time range and filter conditions is performed in three steps. The first and second step is equal to the time series query processing. Within the third step, the records are retrieved from the file system (or the memory if cached).

4.4 Calculating Measures

As mentioned in this section, we distinguish three types of measures. Measures are of interest if the query selects time series, i.e. aggregated records. The execution costs define the distinction between the mentioned three types. An implicit time measure is calculated by the use of the bitmap determined within the third step. A descriptor bound measure is determined by retrieving the facts of the descriptor (using η_i) and applying the defined aggregation function. The most expensive execution is the determination of complex measures. Those are calculated after all other measures are known using all available information, i. e. the other measures results, the combined bitmaps and if needed, the raw records.

5 Performance

To assess the performance of our implementation, we run several tests on an Intel Core i7-3632QM with a CPU clock rate of 2.20 GHz, 8 GB of main memory, a SSD, and running 64-bit Windows 7 Enterprise. For our Java implementation, we used a 64-bit JRE 1.6.45, with XMX 1024 MB and XMS 512 MB. We tested our implementation against a ROLAP (Oracle DBMS 12c) and MOLAP system (icCube 4.1.1 Community Edition).[5] We tested the performance of our implementation using a real-world data set[6] containing 1,122,097 records collected over one year. The records have an average interval length of 48 min. We modeled the data in all three systems with four dimensions (i. e. the time dimension, as well as three flat dimensions: person (cardinality: 713), task-type (cardinality: 4), and work area (cardinality: 31)) and a measure to count the needed resources. For the ROLAP system, we integrated the data into an MDM following [6] to solve the summarizability problems (resulting in 52 M. records). Additionally, we added *Bitmap Join Indexes* and partitions to increase performance. The used MOLAP system supports dimensions *indexed by range*. Therefore, an additional integration was not necessary. Unfortunately, the lowest supported granularity of the MOLAP system is day. Using other granularities (i. e. hour or half-hour, as well as the integrated data) led to memory shortage or loading was not possible because of errors.

Figure 6 illustrates the result of the tests performed. Our implementation outperforms the MOLAP based technology. This result is not surprising, because our implementation is a, considering time interval data, optimized MOLAP solution. The ROLAP based technology outperforms our implementation in the case that the selectivity of a row-based index is high, compared to the dimensions cardinality. As seen from the *random days* test, this applies only as long as the number of used filter-criteria is small. If the number increases, a slower optimized full table scan is performed.

---

[5]We used those systems, because icCube supports so called range dimensions, the best implementation so far to enable the usage of interval data and Oracle has shown the best performances considering ROLAP.

[6]The data set, integration processes and all schemes used are available under https://github.com/pmeisen/public.

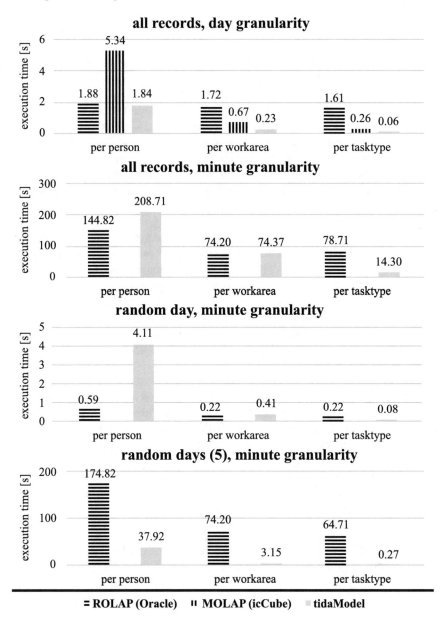

Figure 6 Results of the performance tests, showing the different results using our tida$_{Model}$ in comparison to a ROLAP (Oracle) and a MOLAP (icCube) based system

6 Conclusions and Future Work

The processing of time interval data in data-driven DSS is an important research topic to answer time interval dependent questions using well-known processing techniques. Most MDMs used to handle time interval data are too heavy for OLAP-applications and/or need an expensive integration process. We introduced our $tida_{Model}$ useful when handling time interval data in the context of data-driven DSS. A $tida_{Model}$ describes time interval data by descriptors, a time axis, as well as dimensions and measures. Additionally, we presented our implementation, which always performs better than a MOLAP and, considering the important queries for time interval based data-driven DSS, a ROLAP based technology. We have successfully integrated the presented prototypal implementation into different data-driven DSS that are in use by different companies.

Considering the requirements mentioned in Section 2, we conclude that our presented model and our implementation cover all stated. Nevertheless, some are covered better than others. As presented, requirement ii is covered using implicit time measure, requirement iii via mapping functions and complex measures, requirement iv by using a time based model, partitioning as well as bitmap indices, and requirement v through time hierarchies. The requirements i (we presented only insertion) and vi (not presented, but available through modularization) are not discussed in detail. Especially the deletion of records (as needed to fulfill requirement i) needs additional attention in the near future.

In future work, we will address the problem of deletion, e. g. by tombstone markers. We also plan to define a simplified query language to replace the, in the case of time interval data, mostly complex MDX queries. Additionally, we plan to present the persistence and load balancing available through our implementation in detail, as well as the handling of time interval data across different time zones. Another topic for future research is the evaluation of different visualizations. Currently, aggregated time interval data is mostly visualized using line-charts or heat-maps. Considering work done in the field of visual analytics, as well as the combination of visualizations, our model, and mining techniques are of interest.

Acknowledgments The approaches presented in this paper are supported by the German Research Foundation (DFG) within the Cluster of Excellence "Integrative Production Technologies for High-Wage Countries" at RWTH Aachen University.

References

1. D.J. Power, Understanding Data-Driven Decision Support Systems. Information Systems Management **25** (2), 2008, pp. 149–154
2. E.F. Codd, S.B. Codd, C.T. Salley, *Providing OLAP (On-line Analytical Processing) to User-analysts: An IT Mandate*. Codd & Associates, 1993
3. E.A.A. Abdelouarit, M. El Merouani, A. Medouri, Data Warehouse Tuning: The Supremacy of Bitmap Index. International Journal of Computer Applications **79** (7), 2013, pp. 7–10

4. C. White, Data Integration: Using ETL, EAI, and EII Tools to Create an Integrated Enterprise. Tech. Rep. tech. rep, The Data Warehousing Institute, 2005
5. N. Pendse. What is OLAP, 2002
6. J.N. Mazón, J. Lechtenbörger, J. Trujillo, Solving Summarizability Problems in Fact-dimension Relationships for Multidimensional Models. In: *Proceedings of the ACM 11th International Workshop on Data Warehousing and OLAP*. ACM, New York, NY, USA, 2008, DOLAP '08, pp. 57–64
7. I.Y. Song, I.y. Song, C. Medsker, E. Ewen, W. Rowen, An Analysis of Many-to-Many Relationships Between Fact and Dimension Tables in Dimensional Modeling. In: *Proceedings of the Int'l Workshop on Design and Management of Data Warehouses*. 2001, pp. 6.1–6.13
8. J.F. Allen, Maintaining Knowledge About Temporal Intervals. Commun. ACM **26** (11), 1983, pp. 832–843
9. P. Papapetrou, G. Kollios, S. Sclaroff, D. Gunopulos, Mining frequent arrangements of temporal intervals. Knowledge and Information Systems **21** (2), 2009, pp. 133–171
10. Sadasivam, Efficient approach to discvoer interval-based sequential patterns. Journal of Computer Science **9** (2), 2013, pp. 225–234
11. T. Gschwandtner, J. Gärtner, W. Aigner, S. Miksch, A Taxonomy of Dirty Time-Oriented Data. In: *proceedings of the CD-ARES 2012, 2012, Prague, Czech Republic*. 2012
12. O. Kostakis, P. Papapetrou, J. Hollmén, ARTEMIS: Assessing the Similarity of Event-Interval Sequences. In: *proceedings of the Conference on Machine Learning and Knowledge Discovery in Databases (ECML/PKDD), Athens, Greece*. 2011, Lecture Notes in Computer Science
13. A. Kotsifakos, P. Papapetrou, V. Athitsos, IBSM: Interval-Based Sequence Matching. In: *proceedings of the 13th SIAM International Conference on Data Mining, Austin, Texas, USA*. 2013
14. W. Aigner, S. Miksch, H. Schumann, C. Tominski, *Visualization of Time-Oriented Data*. Human-Computer Interaction Series. Springer London, London, 2011
15. C. Combi, M. Gozzi, J.M. Juarez, R. Marin, B. Oliboni, Querying Clinical Workflows by Temporal Similarity. In: *proceedings of the 11th conference on Artificial Intelligence in Medicine, Amsterdam, The Netherlands*. 2007
16. W. Aigner, P. Federico, T. Gschwandtner, S. Miksch, A. Rind, Challenges of Time-oriented Data in Visual Analytics for Healthcare. In: *IEEE VisWeek Workshop on Visual Analytics in Healthcare (VAHC), Seattle, USA*. 2012
17. K. Boonstra-Hörwein, D. Punzengruber, J. Gärtner, Reducing understaffing and shift work with Temporal Profile Optimization (TPO). Applied Ergonomics **42** (2), 2011, pp. 233–237
18. B. Berendt, Explaining preferred mental models in Allen inferences with a Metrical Model of imagery. In: *proceedings of the Eighteenth Conference of the Cognitive Science Society (COGSCI-96), USA*. 1996
19. R. Agrawal, A. Gupta, S. Sarawagi, Modeling multidimensional databases. In: *proceedings of the Thirteenth International Conference on Data Engineering*. 1997, pp. 232–243
20. R. Kimball, M. Ross, *The data warehouse toolkit: The definitive guide to dimensional modeling*, 3rd edn. Wiley Computer Publishing, 2013
21. S. Agarwal, R. Agrawal, P. Deshpande, A. Gupta, J.F. Naughton, R. Ramakrishnan, S. Sarawagi, On the Computation of Multidimensional Aggregates. In: *proceedings of the 22th International Conference on Very Large Data Bases*. 1996, pp. 506–521
22. H.J. Lenz, A. Shoshani, Summarizability in OLAP and statistical data bases. In: *SSDBM'97, IEEE Computer Society, Washington, DC, USA*. 1997, pp. 132–143
23. T.B. Pedersen, Managing Complex Multidimensional Data. In: *Business Intelligence*, ed. by M.A. Aufaure, E. Zimányi, no. 138 In: Lecture Notes in Business Information Processing, 2013, pp. 1–28
24. T.B. Pedersen, *Aspects of Data Modeling and Query Processing for Complex Multidimensional Data*. No. 4 In: Ph.D. thesis, Aalborg Universitetsforlag, Aalborg. Publication: Department of Computer Science, Aalborg University. 2000
25. D. Hay, What exactly is a data model? DM Review **13** (2), 2003

26. K. Stockinger, Design and implementation of bitmap indices for scientific data. In: *Applications Symposium, July 2001, IEEE Computer Society Press, Grenoble, France.* 2001, pp. 47–57

27. P. O'Neil, D. Quass, Improved Query Performance with Variant Indexes. In: *Proceedings ACM SIGMOD International Conference on Management of Data, Tucson, Arizona, USA.* 1997

28. C.Y. Chan, Y.E. Ioannidis, An Efficient Bitmap Encoding Scheme for Selection Queries. In: *SIGMOD 1999, Proceedings ACM SIGMOD International Conference on Management of Data, Philadephia, Pennsylvania, USA.* 1999

29. F. Delièbe, T.B. Pedersen, Position List Word Aligned Hybrid: Optimizing Space and Performance for Compressed Bitmaps. In: *proceedings of the 13th International Conference on Extending Database Technology (EDBT '10).* 2010, pp. 228–239

30. A. Colantonio, R. Di Pietro, CONCISE: Compressed 'n' Composable Integer Set. Information Processing Letters **110** (16), 2010, pp. 644–650

31. D. Lemire, O. Kaser, K. Aouiche, Sorting Improves Word-aligned Bitmap Indexes. Data & Knowledge Engineering **69** (1), 2010

32. S.J. van Schaik, O. de Moor, A Memory Efficient Reachability Data Structure Through Bit Vector Compression. In: *proceedings of the 2011 ACM SIGMOD International Conference on Management of data (SIGMOD '11).* 2011, pp. 913–924

33. S. Chambi, D. Lemire, O. Kaser, R. Godin, Better bitmap performance with Roaring bitmaps. eprint, 2014. arXiv:1402.6407

34. K. Wu, A. Shoshani, K. Stockinger, Analyses of Multi-level and Multi-component Compressed Bitmap Indexes. ACM Trans. Database Syst. **35** (1), 2008, pp. 2:1–2:52

35. O. Kaser, D. Lemire, Compressed bitmap indexes: beyond unions and intersections. eprint, 2014. arXiv:1402.4466

36. K. Wu, S. Ahern, E.W. Bethel, J. Chen, H. Childs, E. Cormier-Michel, C. Geddes, J. Gu, H. Hagen, B. Hamann, W. Koegler, J. Lauret, J. Meredith, P. Messmer, E. Otoo, V. Perevoztchikov, A. Poskanzer, Prabhat, O. Rübel, A. Shoshani, A. Sim, K. Stockinger, G. Weber, W.M. Zhang, FastBit: interactively searching massive data. Journal of Physics: Conference Series **180** (1), 2009

37. M. Bergeron, D. Conklin, Subsumption of Vertical Viewpoint Patterns. In: *proceedings of the Third international conference on Mathematics and Computation in Music, Paris, France.* 2011

38. J. Han, Towards On-line Analytical Mining in Large Databases. SIGMOD Record **27** (1), 1998, pp. 97–107

How Virtual Production Intelligence Can Improve Laser-Cutting Planning Processes

Rudolf Reinhard, Urs Eppelt, Toufik Al-Khawly, Tobias Meisen, Daniel Schilberg, Wolfgang Schulz and Sabina Jeschke

Abstract The complexity of modern production conditions demands integrative approaches in the fields of simulation and analysis to improve product quality and production efficiency. Existing concepts of virtual production meet this need to some extent. However, problems of application interoperability and data compatibility remain. One approach is the definition of a standardized file format, which is costly to create and to maintain. Other approaches avoid the need for a uniform standard by mapping data structures onto a canonical data model. Although these methods allow for simulation and examination of individual elements, the analysis of the integrated process remains a challenge. Here, the data analysis solutions from the field of the so-called intelligence-solutions can prove useful. Within this paper, a use case scenario taken from the field of laser cutting is presented. Herein, the planning for laser cutting is conducted in a modular format. A new concept is presented that addresses the requirements aforementioned and that conforms to the principles of the integration and examination of data. The new concept, called Virtual Production Intelligence, is formed by combining the concept of virtual production with "intelligence solutions" or the goal of gaining knowledge through the analysis of already completed processes.

Keywords Application Integration · Data Analysis · Decision Support · Digital Factory

1 Introduction

Production companies in highwage countries are facing a growing complexity in their production conditions due to the global price competition, the fast changes in customer requirements and the short product lifecycles [1]. In order to address this complexity, it would be necessary not only to obtain the consultancy of experts for each single available manufacturing process, but also to establish a communication

R. Reinhard (✉) · U. Eppelt · T. Al-Khawly · T. Meisen ·
D. Schilberg · W. Schulz · S. Jeschke
IMA/ZLW & IfU, RWTH Aachen University, Dennewartstr. 27, 52068 Aachen, Germany
e-mail: rudolf.reinhard@ima-zlw-ifu.rwth-aachen.de

Originally published in ?Proceedings of the 22nd International Conference
on Production Research (ICPR 22)?, © IFPR 2013. Reprint by Springer International
Publishing Switzerland 2016, DOI 10.1007/978-3-319-42620-4_70

between them to answer specific questions on how to optimize a complete production line.

A modern approach to reduce this complexity, which has gained importance in recent years, is the use of computational techniques and simulation applications in the field of production technology [2].

Due to the increase of the computational performance with respect to speed and storage, these simulation applications have positively influenced research and development (R&D) in the industrial environment. They turned out to be useful for production planning as they enable the user to cope with the complexity mentioned above.

Although successful stories have been presented in [3], where the use and benefit of computational technology for simulation was demonstrated, they still seem to be sporadic due to the fact that it was not obvious to everybody so far that the simulation can provide a nice overview of a specific manufacturing process and its process domains.

In fact, the industrial developer or machine operator faces essentially the same problem as a navigator in real space, that is: to find points-of-interest in the configuration space of a single machine or of a complete production line. Crucial differences can be subsumed as follows: The required map (in case of the manufacturing of a process map) has more than two or three dimensions and there are few to no "cartographers" who describe the relevant process domain space.

Exploratory analysis of manufacturing processes resp. its configuration spaces may be a helpful tool for mastering the variety of dimensions which have to be taken into consideration. This leads to the idea of a navigating cockpit, which can navigate developers of manufacturing technologies or even machine operators through the high-dimensional design resp. configuration space. This cockpit is an interactive visualization area that can be used for R&D purposes as well as for production planning purposes. A prototypic implementation of this kind of cockpit in combination with selected methods of analysis, such as a multidimensional, multi-objective optimization, is currently in progress.

The challenges just depicted do not only occur in laser cutting but in many other manufacturing processes.

Instead of engineering a concrete prototype at an early stage of product design, a digital model of this prototype is now being drafted containing a description of its essential characteristics. In a further step, this model is passed to a simulation application to predict characteristics of the prototype that may change after the manufacturing step. The usage of these digital models is subsumed under the notion of virtual production, which "is the simulated networked planning and control of production processes with the aid of digital models. It serves to optimize production systems and allows a flexible adaptation of the process design prior to prototype realization" [4, 5].

Nowadays, various simulation applications exist within the field of virtual production. They allow for the simulated execution of manufacturing processes, such as heating and rolling. Herein, different file formats and file structures are independently developed to describe digital models. Through this, the simulation of single aspects of production can be examined more easily. Nevertheless, the integrative simulation of complex production processes cannot be executed without large costs and time efforts as the interoperability between heterogeneous simulation applications is commonly not given.

One approach to overcome this challenge is the creation of a new standard file format, which supports the representation of all considered digital models. However, regarding the variety of possible processes and application domains, such an approach leads to a highly complex data format whose potential and expressiveness is not required in most of its use cases. Its comprehension, maintenance and usage, again, require large costs and time efforts. Furthermore, necessary adaptations and extensions take time until their implementation is finished [6, 7].

Another approach considers the usage of concepts from data and application integration. It avoids the definition of a uniform standard. Within this approach, the interoperability between the simulation applications is guaranteed by mapping the aspects of different data formats and structures onto a so called canonical data model [8, 9]. This approach is called Adaptive Application and Data Integration [10]. Newer approaches extend these concepts with regard to semantic technologies by implementing intelligent behavior into such an integrative system.

Consequently, new possibilities concerning the simulation of whole production processes emerge, which allow the examination of different characteristics of the simulated process, e. g. material and machine behavior. With regard to the analysis of the integrated processes, new questions arise as methods for the analysis of the material or machine behavior mentioned above cannot be transferred to the analysis of the corresponding integrated process. A further challenge comes up as soon as suitable user interfaces are added. These are necessary for the handling of the integrated process and its traceability.

Thus, with regard to the preparation of the building of data analyzing applications for manufacturing purposes, questions arise concerning in particular the methods of data analysis and the presentation of analysis results.

Similar questions emerge during the analysis of enterprise data. Applications giving answers to such questions are, in many cases, subsumed under the notion of Business Intelligence. These applications have in common that they identify, aggregate, extract and analyze data within enterprise applications [11, 12].

This paper introduces an integrative concept that applies solutions inspired by those taken from the field of Business Intelligence to the field of virtual production. This concept provides for the integration, analysis and visualization of data that are generated during the simulation of single manufacturing processes. Because of its application to and its goal of contributing to the knowledge of integrative virtual production processes, it is called Virtual Production Intelligence (VPI). This paper illustrates the development of the concept of VPI and its exemplary application in the field of laser cutting technology. Section 2 presents a general introduction into VPI, whereas Section 3 shows the application of VPI in the domain of laser cutting R&D. In Section 4, data handling methods and data analysis tools are presented. A conclusion and an outlook are given in Section 5.

2   Virtual Production Intelligence: Definition and Objective

Virtual production aims at an entire mapping of the product as well as of the production processes within a numerical model. Thereby, the mapping should comprise the whole lifecycle of the product and the production system [13]. Within an enterprise, nowadays, virtual production is established by employees, software tools such as Product-Lifecycle-Management applications (PLM applications) and organizational processes [13].

The requested analysis possibilities serve the purpose of gaining knowledge about various production processes. This is achieved by examining already completed processes or such ones, which have been planned but not implemented. As already mentioned, the term "intelligence" is commonly used to describe activities that are linked to those analyses. Software tools, which support the analysis and the interpretation of business data, are subsumed under the term "Business Intelligence".

As this term can be defined in different ways, at this point, the basic idea of "Business Intelligence" will be pointed out [14]: A common feature of the various definitions consists in the aggregation of relevant data from different data sources, which are applications within a company, into a data storage, which is available whenever needed. The transmission of data taken from the application data bases into this central data storage is realized by the well-known Extracting, Transforming and Loading process (ETL). Common concepts of intelligence solutions like those employed for reporting or customer relationships management (CRM) are not considered in this paper. Instead, the integration and analysis of data will be focused.

Requirements for a system, which in particular supports the planning process described in Section 3.1, especially the data and application integration, and which follows the idea of Business Intelligence, can be subsumed as below:

- Interoperability: Facilitating the interoperability between applications in use.
- Analytical abilities: Systematic analyses providing the recognition of potentials towards optimization and delivering fundamental facts for decision support.
- Alternative representation models: Taylor-made visualization for the addressed target group, which provides appropriate analyzing facilities based on a uniform data model.

In order to find a solution, which fulfills the requirements mentioned above, a concept formation is needed that addresses the field of application, that is, in this case, the virtual production as well as the aim of gaining knowledge. This aim is also addressed by the term "Intelligence". The concept formation will take into account approaches and methods which will contribute to the gaining of knowledge concerning the processes considered within the field of virtual production. Therefore, the concept formation results in the term "Virtual Production Intelligence".

In [15], an exemplary use case taken from the field of "factory planning" can be found, in which data integration methods belonging to the idea of Virtual Production Intelligence are applied.

3 Virtual Production Intelligence Concepts Applied in the Field of Laser Cutting

3.1 The Laser Cutting Process

Laser cutting is a thermal separation process widely used in shaping and contour cutting applications. The industrially most relevant laser cutting process is the fusion metal cutting process, since the cutting of large metal sheets into smaller pieces with specified contours is addressed in many branches of manufacturing industry. Laser cutting has the advantage over conventional cutting techniques that it is a fast and accurate technology with the optical tool laser not being exposed to any wear.

There are, however, gaps in understanding the dynamics of the process, especially issues related to cut quality (cf. Figure 1). It was found that the modeling and simulation of the laser process provide that understanding without the need to execute a lot of experiments [16].

The ablation process in fusion metal cutting is mainly based on thermodynamics and hydrodynamics, as the absorbed laser energy is converted to heat, which melts the material. In the end, this melt is driven out of the cut kerf by a gas jet coming out of a cutting nozzle coaxially aligned with the laser beam.

There are some criteria that are of major interest in the context of this manufacturing technology, such as cut quality, adherent dross and maximum cutting speed.

The modeling of a laser cutting process requires the modeling of at least three entities at the same time. The three elements involved in cutting are the gas jet, the laser beam and the material to be cut. Therefore, it is evident that the modeling of the cutting gas flow, the radiation propagation and the ablation of the material (in fusion cutting: removal by melt ejection) have to be accomplished as well as the numerical implementation of these models. As these models are already accomplished, nowadays, there are appropriate models available [16].

3.2   Exploration of Parameter Domains in Laser Cutting Supported by Virtual Production Intelligence

The current state of the art in modeling and simulation of manufacturing processes consists in executing single individual simulations. Each individual simulation is characterized by a high-dimensional input parameter set and is executed on interesting points in this parameter space called seed points. In order to understand and to optimize the process, the simulation has to be performed on different seed points within a wide design space.

However, having simulation results on single un-connected seed points is not enough to gain knowledge and to understand the process. Therefore, the results from separate simulation runs have to be logically connected to form a knowledge base that users can operate on. This knowledge base is a map of process criteria along process parameters. It allows the user to have a complete overview of the solution properties that contribute well to any design optimization process or visual exploration of specific parameter domains either for R&D, planning or educational purposes.

The technique for logically connecting single data points representing the relationship between input parameters and output criteria in the first place is a function interpolation in high-dimensional spaces. The mathematical interpolation can be executed with the help of various techniques, such as Radial Basis Function Network (RBFN) [17, 18], Taylor series expansion (Kriging) [19, 20] or artificial neural networks (ANN) [21]. The process of making a discretely sampled space operative via a continuous mapping is called metamodeling. The metamodeling process considered in this paper consists of several steps as shown in Figure 2:

(a) The choice of a suitable (initial) distribution of samples in the parameter space, which is commonly called design-of-experiments-method (DOE). This initial population of the parameter space can be random and sparse and could be ini-

Figure 1 The laser cutting process (**a**) and the resulting cutting kerf (**b**)

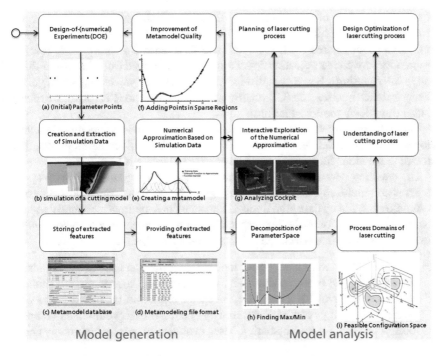

Figure 2 Illustration of the VPI workflow for the laser cutting process

tialized by any design of experiment (DOE) method [22], e. g. Factorial Designs (FD), Latin Hypercube Design (LHD), Box-Behnken Design (BBD) etc.

(b) The creation and extraction of simulation data with appropriate physical numerical models: Spatially and time resolved simulations provide whole distributions of physical properties (e. g. temperature distributions), which need to be reduced to criteria significant for the process (i. e. scalar quantities). In laser cutting, these characteristic criteria are extracted from a cutting model such as [23, 24] and are then potentially transferred to a modeling database.

(c) The saved parameters and criteria from the single simulations are opposed to each other to give a discrete function between the input (process parameters) and the output (process criteria). The numerical approximation of these discrete relations results in a continuous mapping between the input and the output, which is essentially what metamodeling is about. The interactive exploration (cf. step e) of this continuous mapping (metamodel) is done with the aim of gaining insights into the considered process (such as laser cutting) for the purpose of optimizing it (e. g. aiming at gas savings or energy savings without a loss of quality of the cut product).

The steps (a–c) discussed above may be extended via additional steps of processing. This either results in an iterative improvement of the metamodel function or in a decomposition of the whole high-dimensional parameter space in so-called process domains. Within these domains, significantly different properties of the solution can be encountered, which should be distinguished from each other (e. g. the evaporation-driven melt-ejection domain should be separated from the gas-jet-driven one in laser cutting). For accomplishing those additional processing steps, general Computational-Intelligence (CI)-algorithms will be implemented as future models made available by the VPI-Platform.

Computational intelligence (CI) is a set of nature-inspired computational methodologies and approaches which addresses complex real-world problems, especially in providing solutions for complicated and/or inverse problems [25].

The resulting new evolutionary metamodeling process considered in this paper focusses on a multi-objective evolutionary optimization. It uses the concept of evolution control to build up the optimal and, at the same time, minimal metamodel training data base necessary for the best approximation of the parameter-criteria-relationship possible. The process is described below as well as in Figure 2:

(d) Starting with an initial, very poor model approximation (due to a small size of the training data base), in each step, the quality of a metamodel is optimized iteratively by invoking points from the sparse regions of the covered parameter space. These points can be discovered by making use of evolutionary computation, which mimics the population-based evolution of species commonly known as genetic algorithm [26].

(e) After having found the best approximation of the parameter-criteria-relationship, the analysis part of VPI can be started. One vital ingredient of the analysis is the decomposition of the underlying parameter space, as this introduces the process domains from which a typical machine configuration is chosen. The process is described in Figure 2.

(f) The domain decomposition of the metamodel is approached, for example, by the computation of the Morse-Smale complex from unorganized scattered data [27]. The Morse-Smale complex provides a topologically meaningful decomposition of the domain by finding all the global and local maxima and minima at the beginning of the process.

(g) Decomposing and visualizing the domain will help to acquire a qualitative understanding of the laser cutting process. Moreover, it will help to find a feasible configuration space, which is in accordance with potentially pre-defined quality criteria.

All the methods introduced above play an important role in the analysis of complex systems. They serve as an effective way of mapping input-output relationships for assessing the impact of input parameters on the output criteria. In addition, the method is applied to solve numerous types of optimization problems that involve

	PL.W.	focus.m.	rbd.mm.	wkerf.mum.
1	1000.00	-10.000000	0.0150000	562.264
2	5655.17	-10.000000	0.0150000	769.176
3	5137.93	-9.310340	0.0150000	733.526
4	4620.69	-8.620690	0.0150000	699.018
5	4103.45	-7.931030	0.0150000	664.268
6	3586.21	-7.241380	0.0150000	629.733

Figure 3 Exemplary input quantities of an examined process (here: laser cutting)

computationally demanding functions like the process simulations meanwhile used to model laser cutting processes. This means that the concrete manufacturing process, which is simulated and, in a further step, examined, is not focused. Instead, the applied methods for the examination are in the focus of interest.

4 Data Integration and Analysis

4.1 Integration Process and the Data Model

The simulation results are written out into output files whose syntax is kept in a table form to keep the output in a human readable form. The columns' names describe both, input and output quantities as well as the unit to measure the corresponding quantity. Figure 3 illustrates the first rows of such a file describing an examined cutting process. Four columns have been selected to show exemplary quantities. Their heads describe the name of the quantity and its unit, that is:

- PL.W: Power of laser given in Watt,
- Focus.m: Laser focus position in meters,
- rdb.mm; Raw beam diameter (of ablating gas) in millimeters,
- Wkerf.mum: Width of (cutting) kerf within the manufacturing component in micrometers.

To ensure the integrity of data, the user must be aware of the units used within the simulation application. In the case this is not given, appropriate computation steps have to be performed before integrating the data containing the application results into the data storage. A simple example for such a computation is the unit conversion. This behavior can be varied by the application's operator. If a varying unit for the

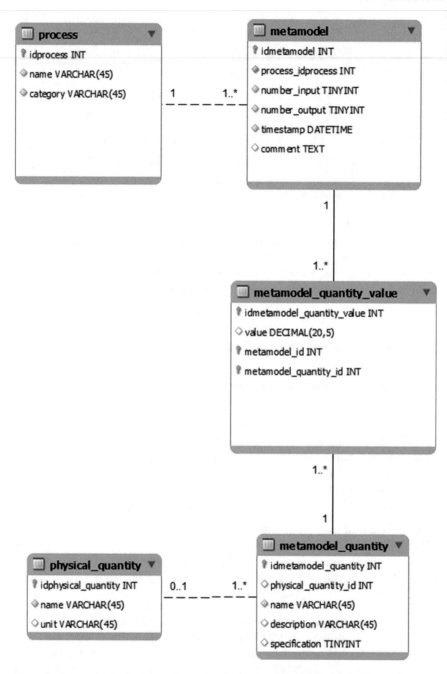

Figure 4 Data model for the integration and analysis of laser cutting data in UML-notation

focus of the laser lens is selected during the examination of one laser cutting process, the recalculation of extracted data will be caused, so that the given quantity will fit to the new unit.

The data for analysis is prepared by making use of five tables depicted in Figure 4. Each manufacturing process, which was already analyzed in the presented framework, determines a row in the table **process**. For the later purpose to build up a history of all examined processes, each process is named properly. Naming policies are not considered in the data model and have to be implemented by the use of appropriate ontologies [28]. With this approach, the separation of content and data is ensured. The table also contains an attribute category to ensure the generality for manufacturing processes, which are examined by making use of the method of metamodeling as defined in Section 3.1.

Within the examination of a laser cutting process, several discrete simulation data are considered. To ensure this relationship within the data model, the table **metamodel** contains a foreign key with the ID of the associated process. This table manages the inputs and the outputs for a discrete set of input and output data as already described in Section 3.2. To facilitate the versioning within the considered simulation data for examining a single process, each row within the table **metamodel** has a timestamp. It contains information about the point in time at which the simulation data was created as well as a comment for the process analyst.

The rows in the table **metamodel-quantity-value** represent each value (a parameter or a criterion value) involved in an examined process. A row in this table always corresponds to a single row in the table **metamodel**.

Each quantity involved within a numerical simulation generates a row in the table **metamodel-quantity**. It can either serve as input (parameter) or as output for the numerical approximation (criterion). To distinguish these two cases, the descriptor "specification" is used by the following convention: specification = 0 means input, specification = 1 means output. By following this approach, future specifications such as boundary conditions can also be considered for a numerical approximation and can be mapped into the data model without the need of creating a new attribute for this table.

Each input or output quantity has its abstracted physical quantity, which is managed in the table physical_quantity. The fact that the diameter of a laser beam corresponds to the **physical_quantity** "length" as well as to the cutting length of the product might serve as an example.

The presented approach for the modeling of data within an examined laser cutting process facilitates the implementation of requirements for other manufacturing processes explored by the method of metamodeling.

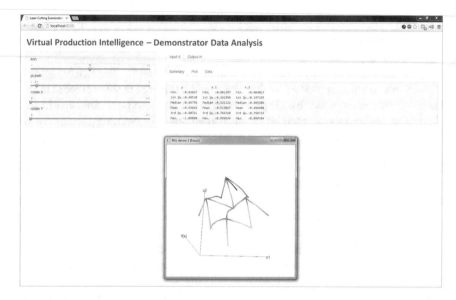

Figure 5 Screenshot of the current version of the data analysis tool

4.2   Data Analysis: Methods and Tools

As already described in Section 3.2, the aim of the analysis is to determine the relationships between input variables and output values in a high-dimensional parameter space.

The analysis goal can be described as follows: Assuming a specified target criterion. What are the sensitive parameters involved to obtain this criterion? Due to the high-dimensionality, which is characteristic for the examined data sets, dimension reduction is a requirement, which also has to be met.

Since the relationships between input and output cannot be described in an analytical manner, the exploration of data in an interactive way is the most promising approach to investigate these relationships. To facilitate the exploration, an online tool is used. The user interface consists of various manipulations and variation tools that enable the user to obtain insights into the data resp. the relationships manifested by the data (cf. Figure 5).

Since the philosophy of Virtual Production Intelligence contains in particular integration and collaboration, an online tool is built to provide the usage of implemented and stable data analysis methods. Within the last years, the programming language R [29] is used more frequently; one of the reasons is the fact that this programming language is not proprietary.

The supply of a fast communication with little latency is a crucial requirement. This requirement is met by servers using the web socket protocol [30], which allows for the bidirectional communication between client and server.

In order to build a fast and stable prototype web application that also provides the integration of data analysis methods written in the programming language R [29], which is used more frequently within the last years, the web application framework "Shiny" [31] is used. It is based on the web socket specification.

The implementation of promising statistical methods, such as clustering, dimension reduction and singular value decomposition, is initiated and first results encourage the implementation of further methods. Most of these methods are available via packages written in R. One of those packages offering many benefits is the Morse-Smale-Package [27], which is already available at the VPI-platform (cf. Figure 5).

5 Summary and Outlook

Within this paper, the general idea of Virtual Production Intelligence and its application in the planning of laser cutting processes were presented. The aim of VPI is to support the processes within the Virtual Production by providing the integration and the analysis of data generated along the planning or analysis of production processes. In particular, the analysis requires the usage of appropriate methods and tools. These ideas were pointed out in Section 2.

One concrete application of this idea consists in the use of the computational technologies mentioned above for the R&D of the laser cutting process as depicted in Section 3.2. In Section 4, the requirements for data integration were described and it was pointed out how these requirements can be met. Furthermore, an online analysis tool was described, which is suitable for the analysis of high dimensional data sets.

Further progress will contain the elaboration of additional methods for the refinement of the numerical approximation and the identification of suitable stochastic methods. In both cases, an implementation of the validation of the identified methods is intended to show how the additional methods have improved compared to previous methods.

Acknowledgments The approaches presented in this paper are supported by the German Research Association (DFG) within the Cluster of Excellence "Integrative Production Technology for High-Wage Countries".

References

1. G. Schuh, S. Aghassi, S. Orilski, J. Schubert, M. Bambach, R. Freudenberg, C. Hinke, M. Schiffer, Technology roadmapping for the production in high-wage countries. In: *Prod. Eng. Res. Devel. (Production Engineering)*, vol. 5, 2011, pp. 463–473
2. B. Ilschner, R. Singer, *Werkstoffwissenschaften und Fertigungstechnik: Eigenschaften, Vorgänge, Technologien.* Springer, 2010

3. G.J. Schmitz, U. Prahl, *Integrative computational materials engineering: Concepts and applications of a modular simulation platform.* Wiley-VCH; John Wiley [distributor], Weinheim and Chichester, 2012. P. 1 online resource
4. Verein Deutscher Ingenieure. Digital factory - fundamentals, 2008
5. Verein Deutscher Ingenieure. Digital factory - digital factory operations, 2011
6. M. Nagl, B. Westfechtel, *Modelle, Werkzeuge und Infrastrukturen zur Unterstützung von Entwicklungsprozessen: Symposium.* Forschungsberichte (DFG). John Wiley & Sons, 2003
7. C. Horstmann, *Integration und Flexibilität der Organisation durch Informationstechnologie.* Gabler Verlag, 2011
8. S. Hoberman. Canonical data model, 2008. URL http://www.information-management.com/issues/2007_50/10001733-1.html. Zugriff am 25.02.2013
9. D. Schilberg, *Architektur eines Datenintegrators zur durchgängigen Kopplung von verteilten numerischen Simulationen.* VDI-Verlag, Aachen, 2010. P. 156
10. T. Meisen, P. Meisen, D. Schilberg, S. Jeschke, Application integration of simulation tools considering domain specific knowledge. In: *Automation, Communication and Cybernetics in Science and Engineering 2011/2012*, Springer, 2013, pp. 1067–1089
11. B. Byrne, J. KLING, D. McCarty, G. Sauter, P. Worcester, *The value of applying the canonical modeling pattern in SOA, The information perspective of SOA design*, vol. 4. 2008
12. M. West, *Developing High Quality Data Models.* Elsevier Science, 2011
13. U. Bracht, D. Geckler, S. Wenzel, *Digitale Fabrik.* Springer Berlin Heidelberg, Berlin and Heidelberg, 2011. 10.1007/978-3-540-88973-1
14. H. Luhn, A business intelligence system. IBM Journal, 1958, pp. 314–319
15. R. Reinhard, C. Büscher, T. Meisen, D. Schilberg, S. Jeschke, Virtual production intelligence – a contribution to the digital factory. In: *Intelligent Robotics and Applications*, vol. 7506, ed. by C.Y. Su, S. Rakheja, H. Liu, Springer Berlin Heidelberg, 2012, pp. 706–715
16. R. Poprawe, W. Schulz, R. Schmitt, Hydrodynamics of material removal by melt expulsion: Perspectives of laser cutting and drilling. Physics Procedia **5**, 2010, pp. 1–18
17. M. Orr. Introduction to radial basis function networks, 1996. URL http://dns2.icar.cnr.it/manco/Teaching/2005/datamining/articoli/RBFNetworks.pdf
18. S. Rippa, An algorithm for selecting a good value for the parameter c in radial basis function interpolation. Advances in Computational Mathematics (11), 1999, pp. 193–210
19. D. Jones, M. Schonlau, W. Welch, Efficient global optimization of expensive black-box functions (13), 1998, pp. 455–492
20. J. Martin, T. Simpson, Use of kriging models to approximate deterministic computer models. Aiaa Journal (Vol. 43, No. 4), 2005, pp. 853–863
21. S. Haykin, *Neural Networks and Learning Machines*, 3rd edn. Prentice Hall. Pearson Education, 2009
22. F. Jurecka, *Robust design optimization based on metamodeling techniques.* Shaker Verlag, 2007
23. W. Schulz, *Die Dynamik des thermischen Abtrags mit Grenzschichtcharakter.* Shaker, 2003
24. W. Schulz, M. Niessen, U. Eppelt, K. Kowalick, Simulation of laser cutting. In: *The theory of laser materials processing*, vol. 119, ed. by J.M. Dowden, Springer, Dordrecht, 2009, pp. 21–69
25. T. Yoel, Chi-Keong G., *Computational Intelligence in Optimization*, vol. 7. Springer, Berlin, 2010
26. D.E. Goldberg, *Genetic Algorithms in Search, Optimization and Machine Learning.* Addison-Wesley Longman Publishing Co, Boston, 1989
27. S. Gerber, K. Potter, Data analysis with the morse-smale complex: The msr package for r. Journal of Statistical Software **50** (2), 2012, pp. 1–22
28. M. Gagnon, M. Gagnon, Ontology-based integration of data sources. IEEE Xplore, [Piscataway and N.J.], 2007, pp. 1–8

29. The R Foundation for Statistical Computing. The r project for statistical computing. URL http://
www.r-project.org/. Zugriff am 26.02.2013
30. I. Fetter, A. Melnikov. The websocket protocol: Rfc 6455, 2011. URL http://tools.ietf.org/
html/rfc6455. Zugriff am 26.02.2013
31. I. RStudio. Easy web applications in r, 2013. URL http://www.rstudio.com/shiny/

An Agile Information Processing Framework for High Pressure Die Casting Applications in Modern Manufacturing Systems

Michael Rix, Bernd Kujat, Tobias Meisen and Sabina Jeschke

Abstract Modern production of high pressure die casting parts raise new challenges regarding planning, scheduling and analyzing of the underlying manufacturing process. The smart factory approach and the research and development pursuit of the fourth industrial revolution necessitate the refurbishment and upgrade of already existing manufacturing systems and the introduction of new information and communication technologies (ICT) in automation systems in order to achieve a holistic, company-wide information exchange and a pervasive traceability of product and manufacturing data. According to this approach, previous programmable logical controls (PLC), established business intelligence solutions and existing manufacturing execution systems (MES) with mutually lacking interfaces are integrated into a new ecosystem for planning, executing and analysis applications. Due to the fact that each system persists on its own user interface, the implementation has to be strongly coupled to a user centered design of innovative human machine interfaces, joined into one distributed, networked application. In this paper, an agile information processing framework for foundry purposes is presented. Every underlying application is accessible via web-based user interfaces providing control of each single system. This leads into a service oriented architecture triggering the individual underlying systems as services, which are connected using web communication technology to exchange data along a shared information model. The data storage is modular to ensure scalability and interoperability with other manufacturing departments. During the actual manufacturing process, different services like inline data mining analysis are deployed and the results are visualized in user demanded dashboards and reports. For new requirements in business intelligence and MES the developed interfaces are provided in a unique library and a content management system. The described architecture enhances the development of new information applications, accelerates the planning and execution process and is completely orientated to the demands of users, as fast planning procedures and analysis driven user interfaces.

M. Rix (✉) · T. Meisen · S. Jeschke
IMA/ZLW & IfU, RWTH Aachen University, Dennewartstr. 27, 52068 Aachen, Germany
e-mail: michael.rix@ima-zlw-ifu.rwth-aachen.de

B. Kujat
AUDI AG, Ettinger Str., 85045 Ingolstadt, Germany

Originally published in "Research and Innovation in Manufacturing: Key Enabling Technologies for the Factories of the Future - Procedia CIRP", © Elsevier 2016. Reprint by Springer International Publishing Switzerland 2016, DOI 10.1007/978-3-319-42620-4_71

Keywords High Pressure Die Casting · Agile Architecture · Smart Factory · User Centered Design

1 Introduction

High class car design requires emission reduction concurrent to low fuel consumption and the increase of driving dynamics [1]. One innovative and expedient approach to reach these objectives is the introduction of lightweight design. Realizing such innovative lightweight concepts like the space frame technology leads to the intensive utilization and application of high pressure die casting (HPDC) parts for structural purposes in car bodywork [2]. The production of these high tech parts requires a high invest on manufacturing means providing stable and controllable process conditions. To secure these circumstances, it is necessary to monitor every fault during the process, to support the worker with useful information on changing process conditions and to detect possible material flow bottlenecks in advance. Process-accompanying quality measurements has to assure the feedback of information to the central production process and avoid inaccurate machine settings. Therefore, a wide range of information systems has been introduced in the past which led to a heterogeneous and complex system of systems. Especially in research and development plants, like HPDC testing faculties, the high variety of changing projects necessitate a modular and agile architecture of information systems, which current implemented system does not provide. Accompanying new manufacturing technologies, a high variety and quantity of measurement systems is immutable. In addition to this heterogeneous situation, the "Smart factory" approach [3] as well as the research and development pursuit of the fourth industrial revolution [4] demand for vertical and horizontal integration solutions in information and communication technologies (ICT). Contrary to these requirements, automation and information systems are in general segregated into separate assignments performing encapsulated tasks, which is a stark contrast to the distributed, networked information concept behind smart factories. In result, the information and data flow between these different layers is fixed and often hardcoded by making use of proprietary interfaces and information exchange protocols. This actual static and hierarchical automation model cannot accomplish the mentioned requirements due to the fact, that interoperable interfaces are restricted or encrypted. Furthermore, the business model of commercial software avoids downwards compatibility and the independency of the operation system in order to introduce a stringent license sale model.

On top of this, the fast developing progress in information and communication technologies (ICT) and the long term life cycle of production facilities and manufacturing information systems are diverging, so that the incurrence of a technological gap appears inevitable. Therefore, agile approaches are needed to solve this discrepancy by including existing information and automation systems into an agile framework, which provides open interfaces and operating system independency as well as downward compatibility. In addition, modern information architectures must be easily portable on other manufacturing and production departments within the company.

In this paper the introduction of a distributed, networked information processing framework for high pressure die casting applications is illustrated on the basis of a use case in the Audi testing foundry. In order to meet the mentioned challenges, this paper addresses the following research questions:

- How can the dilemma of horizontal and vertical integration of modern information systems be solved by suitable approaches?
- How can heterogeneous data be transferred into pervasive and traceable information, regarding the high pressure die casting use case?
- How can real time data streams be analyzed with data mining methods and the related results are visualized?

Section 2 gives an overview of technologies and methods suitable to realize an integration of different information systems. The use case of the high pressure die casting manufacturing process from the ICT point of view is specified in Section 3. Based on a newly developed agile information system, the central service for real time analysis by means of data mining technologies is discussed in Section 4. Section 5 concludes the described approaches and shows future prospects or manufacturing information applications, due to agile approaches.

Nomenclature	
DWH	Data warehouse
ERP	Enterprise resource planning
HMI	Human machine interface
HPDC	High pressure die casting
ICT	Information and communication technology
IoT	Internet of Things
KPI	Key performance indicator
MES	Manufacturing execution system
MQTT	Message queue telemetry transport
OPC-UA	OLE for process control unified architecture
PLC	Programmable logic control
SCADA	Supervisory control and data acquisition
SOA	Service oriented architecture

2 Related Work

Control and data acquisition systems along automated production processes are in general implemented as static and hierarchical structures. One prominent infrastructural architecture is the so called automation pyramid (Figure 1a). The following layers are described as top-down formulation by task and highlight the static dilemma of the standard automation approach.

Enterprise resource planning (ERP) acts as global system for scheduling the distribution of human and material resources as well as storing, purchase and financial controlling system. On the shop floor level manufacturing execution systems (MES) were deployed for planning and controlling production processes directly on machines or production lines. Sensors and actuators within the field level are operated by programmable logic controls (PLC). The human machine interfaces (HMI) are effectuated in the layer of supervisory control and data acquisition (SCADA) on decentralized devices located near to the field level.

Each sublayer is used for different planning or controlling tasks. A different approach is presented in Figure 1b [4]. Compared to the automation pyramid, one single information layer connects the different subsystems to the field layer. The information and data flow is managed directly underneath the single applications and systems. For example, reporting engines directly access data warehouses (DWH) or MES datasets to create reports or dashboards. The information exchange becomes possible by negotiating between each system on using an extensible information layer as middleware.

Hence, a testing and development department with high variety of measurement systems and sensors needs dynamic solutions. As mentioned before, the life cycle of mechanical facilities, automation systems, and software differ blatantly [5], which results in the necessity to deploy new software solutions, which follow a model driven development procedure and a modularization of each subsystem. The terms of "Industrie 4.0" or "Smart factory" comprise the integration and consolidation of information flow in horizontal direction from ERP to field level and in vertical direction between information devices and machines within the field level [5]. As a consequence, existing communication interfaces between these systems and the

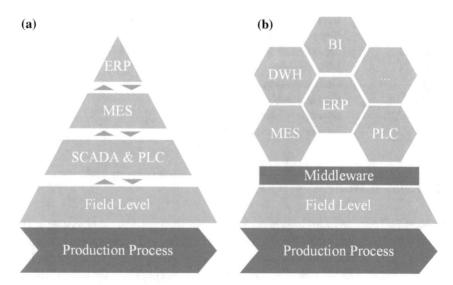

Figure 1 **a** Automation pyramide paradigm; **b** Agile information system approach according to [5]

encapsulated implementation must be broken up and adapted to the level of a middleware. These adapters respectively integrators are provided by a service repository to secure the interoperability between each application and to enable the communication to each other. The smart factory approach includes the communication between terminal devices among themselves and real-time-compliant software that can be interlinked spontaneously [6]. Therefore, existing distributed object architectures (DOA), running on different devices with static interfaces are transformed to service oriented architectures (SOA) [7]. Regarding the holistic approach of scalable, company-wide information systems, SOAs are a proper solution to achieve this goal. Concerning existing DOAs in the financial driven point of view, it has to be balanced out if a reengineering solution or a new implementation is appropriated [8]. A lifetime extension of stock software and modernization to SOA is realized by wrapping these applications into services [9].

In order to use historical data sets in unconsolidated states, the extract transform load (ETL) process according to [10] provides a practical solution. During the extraction, raw data sets from files or different databases are preprocessed and stored into the so-called staging area. Within the staging area data cleansing is processed and the data sets are transformed into information, persisted in a DWH. Based on this DWH, information can be used via object relational mapping (ORM) by service oriented architectures. An approach for stream processing purposes is the lambda architecture according to [11], where new data are segregated into three different data streaming layers (Figure 2):

- Batch layer: This layer comprises the master data set and defines views of different entities as well as data mining analysis models. The training of these models runs batch driven in this layer.
- Speed layer: New data streaming is processed in real time by static views, which were created before or are imported cyclically from the batch layer.

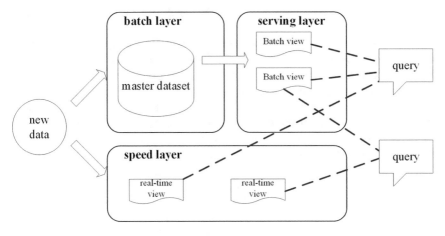

Figure 2 Stream processing in the lambda architecture [12]

- Serving layer: Queries from services to the agile framework are handled in this layer, who acts as handler for the real-time demands in the speed layer or more complex queries for analysis in the batch layer.

Regarding the field level, the integration of different PLCs, sensors or serial signals into a holistic framework needs an adaptable and flexible standard. Therefore, different protocols and standards like OPC-UA and MQTT are upcoming solutions from the level of machine to machine communication (horizontal) up to the communication of World Wide Web (vertical), the so called Internet of Things (IoT).

OPC stands for OLE for process control, where OLE means object linking and embedding. It was developed as standard protocol to provide a communication between PLCs and SCADA. OPC relies on the Microsoft DCOM Standard [13] and is bound to this slow communication protocol. Because of that, OPC unified architecture (UA) was introduced, which uses the TCP/IP protocol for its communication and as a consequence it is independent from the operating system [14]. The major advantage of OPC-UA besides the information flow from sensors into higher systems is the domain driven structure, which describes the setup and sensor configuration of the automated systems. OPC-UA follows an inverted client-server paradigm, i.e. that many servers are deployed directly on the field level and only less clients from upper levels are sending request. Machine to machine communication is implemented by the installation of thin clients on the field level devices [15]. Another protocol for horizontal communication is standardized in the message queue telemetry transport protocol (MQTT). MQTT is easier to implement on fieldbus systems with higher latency in comparison to OPC-UA, which consist a hierarchical structure, albeit several domain specific implementations are distributed [16]. Afterwards in this paper, the usage of OPC-UA is chosen. On wrapping serial signals or PLC interfaces by OPC-UA servers, it makes it possible to connect these data sources directly into an information processing framework.

Generally spoken, OPC-UA implies the functionality to run UA Servers as Services and to define a XML-based Protocol. In conclusion OPC-UA can be used as SOA itself.

Concerning research and development of data analysis in high pressure die casting, first steps were initiated by [17]. Using artificial neural networks, this work aimed on detecting optimized process parameter properties including four central measurement values of temperature of melt, vacuum in the mold, velocity of the piston and the clamping force. Another approach comprises the implementation of a non-generic application showing the capabilities of data mining methods in manufacturing, which resulted in a reduction of production rejects of nearly eighty percent [18]. An actual research project focusses on developing and testing a real time cognitive control system especially for high pressure die casting applications [19].

3　Agile Information Processing Framework for HPDC

3.1　Use Case Testing Foundry

The implementation of our agile information processing system is evaluated by a use case in the production domain of high pressure die casting. Due to the high variety of process parameters, the different communication standards in this manufacturing domain, and the lack of information systems, the use case in HPDC is ideal for refurbishing older PLCs and the deployment of modular ICT. The initial situation is characterized by isolated solutions for information systems with lacking interfaces.

The manufacturing process and the different steps are listed in Figure 3. The initial step is the melting of lightweight metal, like aluminum. One of the characteristic attributes of liquid aluminum is the high affinity to hydrogen entrainment. To avoid this, the liquid melt is treated by an impeller flushing inert gas into the metal in order

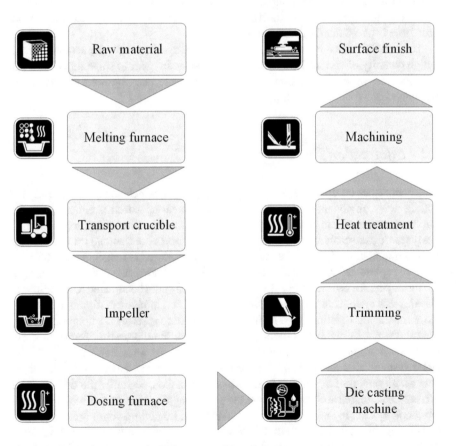

Figure 3 Production process for high pressure die casting parts

to substitute the hydrogen. Otherwise, the major die casting defects, the so-called gas porosity is very probable. Due to the fact that this continuous production process changes to a batch process for producing single parts on the die casting machine, it is necessary to have enough melt as reservoir. Dosing furnaces accomplish this task during the production process. Quality determining analytics like spectral analysis are taken out after each step until the meld reaches the die casting machine. The main production step takes place on the die casting machine and is accompanied by several measurement and automation systems. HPDC is characterized by the injection of liquid nonferrous metals under high pressure into a mold. The injection process is achieved by a piston, pressing the melt by high velocities into the mold, the so-called die during 20–50 ms. The main values for accompanying measurement systems during this manufacturing step are different temperatures, pressures, forces and piston velocities. Process-related to HPDC, dispensable areas of the part are removed by trimming and are melted as recycled material in the melting furnace. Rejected parts are detected by optic user examination and are documented in a quality system. The last three steps of machining, heat treatment and surface finished are performed by second tier partners. The information flow during the external manufacturing processes is transmitted batch wise by reports.

The initial situation for this use case comprised paper based quality acquisition and a high quantity of variant PLC and MES interfaces. In order to solve this challenge, the focus lies on the integration of these different interfaces into one service based framework for foundry purposes.

3.2   Information Framework

This framework comprises a repository of integrators, which can be used easily for adding new components and devices. This adaptable approach is extended by the implementation of data driven services, e.g. real time data streaming for data mining analyses. A user-centered design aspect quickly solves the demand on new visualization applications by using actual web technologies of HTML5. Thus, different types of reports and dash boards are implemented and provided as libraries. Hereby, the time spent for user interface development cycles could be reduced to a minimum. Besides this agile frontend development, the backend implementation is adapted in the same manor. Existing MES solutions and database systems are not rebuild, they are transferred by wrapping them into a service in the die casting framework. By following this approach, a modular an agile solution for the transformation of elder static systems is implemented. There is no necessity of reengineering or the totally new development of adapted information systems. Consequently, the development of a service oriented architecture is achieved by wrapping old systems into the framework and by creating new applications as services. Regarding communication abilities on field level, like sensors or different PLCs, the usage of the OPC-UA standard enables a fast integration methodology, linking serial signals via an OPC-UA server on an embedded device into the network. Java based OPC-UA clients are used to communicate with the embedded OPC-UA servers via Ethernet and TCP/IP. Thus, different field level data sources can be easily integrated into the agile information processing framework for HPDC.

As mentioned before, the framework is based on open source code and it was implemented in Java, which guarantees an operating system independent runtime environment. The development and implementation uses the Spring framework, principally out of three main reasons:

- Object relational mapping (ORM): Spring allows an easy integration of ORM frameworks like Hibernate, which enables the ORM functionalities, i.e. elder relational database models can be converted and manipulated in an object oriented language by mapping entities to objects.
- Web functionalities: User interfaces are developed consistently web based for the HPDC requirements. Spring provides therefore the needed Web-Servlet functionalities.
- Test driven development (TDD): In order to ensure a stable, nearly bug-free software, every service is implemented with adapted tests, before delivering the new services in the production runtime environment.

Based on this development environment, new requirements from users in the HPDC foundry can be realized in short term developing cycles. Close to process planning and mechanical engineering, this approach can easily implement requirements of users. For example, aggregated data from real time processes like the measurement of temperatures in the melting furnace can be quickly visualized in the control station. Process planning properties like linking alarms and events to maintenance plans or the key performance indicator (KPI) of the overall equipment efficiency (OEE) became possible, without setting up a new bought application. Real-time and aggregated data, as the mentioned temperature and the OEE are handled oriented to the lambda architecture paradigm. It is not necessary to run every data stream on the same service.

Based on the described architecture a wide range of services are deployed for different requirements, like process planning or data-mining analyses and the related visualizations.

4 Data Mining and Annotation Service

In this section an approach of stream processing on the focus of inline data mining analysis is presented. This evaluates the development of an agile information processing framework for HPDC. The cycle time for one part on a die casting machine has a duration from 80 s up to 130 s. During this period the inline analysis must predict the decision, if the part can be delivered to the next manufacturing processes or if the part has to be rejected and melted back in the furnace. Figure 4 illustrates the data flow into the inline prediction process and into the batch process, which runs in daily periods.

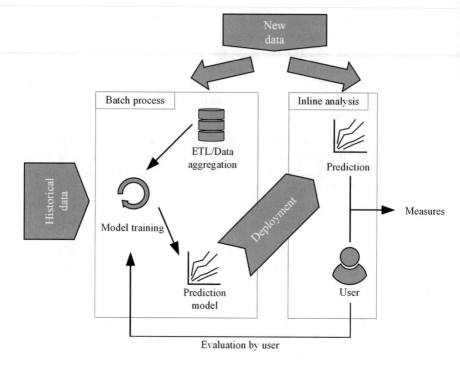

Figure 4 Stream processing for inline data mining analysis

This prediction is based on a data mining model, which is trained every night in the batch processing layer. The collected data from every part of the last production day is used to train the data mining model and evaluates the prediction results. The information input to the batch process is accomplished by new data from measurement systems and quality information systems, by historical data from former casting projects and from a web based user interface for the worker, who evaluates the displayed prediction results by entering the real result. After training the data mining model, a prediction model is derived, which is deployed to the inline analysis stream. The described stream processing is inspired by the lambda architecture in big data technologies and deployed as a service.

Thus, the composition of different services is implemented in a task scheduler, who runs the services by distinct types of triggers (Figure 5). Signals from PLCs, tolerance limits of specific measurement values or simply periodically used time intervals are triggering the specific services for handling the data streams.

At last, the interoperability between different manufacturing departments or information systems is accomplished by an annotation service. A master data management (MDM) repository is designed for each manufacturing line and provides the most important information about each data source or product. This information is annotated as additional metadata to the information storage, where the actual data is persisted. Databases of measurement systems are extended by metadata entities

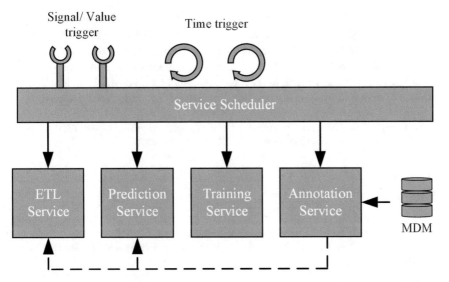

Figure 5 Service composition in the HPDC framework

via the mentioned ORM in Hibernate. On field level, the usage of OPC-UA enables labeling the gathered information into the UA specific information model. Providing services ensure via open interfaces to other information systems. Therefore, it is inevitable to specify dynamically the exchange format, for example in XML-files as OPC-UA or in another descriptive technology.

5 Conclusion and Outlook

In this work, requirements for modern information processing systems for the production process of high pressure die casting were examined. The introduction of an agile approach was highlighted by different technologies, as OPC-UA on every level of the automation systems and the implementation of a service oriented architecture. Due to that, the dilemma of holistic vertically and horizontally integrated information system can be solved.

Hence, the heterogeneous information structure in the foundry use case is presented and a solution for a pervasive traceability approach is presented. The initial situation concerning information systems in HPDC is solved by an implementation of a modular and agile information processing framework. The complex requirements on measurement and quality recording systems are fulfilled by the described framework. By annotating every gathered data, information about the origin of data regarding the process chain is not lost and can be aggregated easily by dynamic ORMs. For batch driven stream processing, the new type of inline analyses stream process was introduced and enables fast visualizations and web based technologies provide an independent deployment on every device. The next step in research of agile information processing systems could by the implementation of domain specific ontologies with the goal to match it on a variety of differing companywide ontologies. As a result of that, every information in one company is included in one holistic system and leads to accomplish the goals of the smart factory approach.

Acknowledgments The approaches presented in this paper are supported by the AUDI AG within a Ph.D. Research cooperation project with RWTH Aachen University.

References

1. A. Lakeit. Resource efficient body-in-white production at audi, 2011
2. Peter Wanke. Key factors when producing highly integrative light-weight structural parts, 05.03.2015
3. D. Zuehlke, Smartfactory—towards a factory-of-things. Annual Reviews in Control **34** (1), 2010, pp. 129–138. doi:10.1016/j.arcontrol.2010.02.008
4. T. Bauernhansl, M. ten Hompel, B. Vogel-Heuser, *Industrie 4.0 in Produktion, Automatisierung und Logistik: Anwendung - Technologien - Migration*
5. B. Vogel-Heuser, C. Diedrich, A. Fay, S. Jeschke, S. Kowalewski, M. Wollschlaeger, P. Göhner, Challenges for software engineering in automation. Journal of Software Engineering and Applications **07** (05), 2014, pp. 440–451. doi:10.4236/jsea.2014.75041
6. O. Sauer, Information technology for the factory of the future – state of the art and need for action. Procedia CIRP **25**, 2014, pp. 293–296. doi:10.1016/j.procir.2014.10.041
7. S. Baker, S. Dobson, Comparing service-oriented and distributed object architectures. In: *On the Move to Meaningful Internet Systems 2005: CoopIS, DOA, and ODBASE, Lecture Notes in Computer Science*, vol. 3760, ed. by D. Hutchison, T. Kanade, J. Kittler, J.M. Kleinberg, F. Mattern, J.C. Mitchell, M. Naor, O. Nierstrasz, C. Pandu Rangan, B. Steffen, M. Sudan, D. Terzopoulos, D. Tygar, M.Y. Vardi, G. Weikum, R. Meersman, Z. Tari, Springer Berlin Heidelberg, Berlin, Heidelberg, 2005, pp. 631–645. doi:10.1007/11575771_40
8. A. Umar, A. Zordan, Reengineering for service oriented architectures: A strategic decision model for integration versus migration. Journal of Systems and Software **82** (3), 2009, pp. 448–462. doi:10.1016/j.jss.2008.07.047
9. G. Canfora, A.R. Fasolino, G. Frattolillo, P. Tramontana, A wrapping approach for migrating legacy system interactive functionalities to service oriented architectures. Journal of Systems and Software **81** (4), 2008, pp. 463–480. doi:10.1016/j.jss.2007.06.006
10. R. Kimball, J. Caserta, *The data warehouse ETL toolkit: Practical techniques for extracting, cleaning, conforming, and delivering data*. Wiley, Indianapolis, IN, 2004
11. N. Marz and J. Warren, *Big Data: Principles and Best Practices of Scalable Realtime Data Systems*. Manning Publications, New York, 2014
12. Michael Hausenblas & Nathan Bijnens. Lambda architecture, 2014. http://lambda-architecture.net/
13. W. Mahnke, S.H. Leitner, M. Damm, *OPC unified architecture*. Springer-Verlag, Berlin and Heidelberg, 2009
14. J. Lange, T.J. Burke, F. Iwanitz, *OPC: Von Data Access bis Unified Architecture*, 5th edn. VDE Verl., Berlin, 2014
15. M.S. Mahmoud, M. Sabih, M. Elshafei, Using opc technology to support the study of advanced process control. ISA transactions **55**, 2015, pp. 155–167. doi:10.1016/j.isatra.2014.07.013
16. P. Papageorgas, D. Piromalis, T. Iliopoulou, K. Agavanakis, M. Barbarosou, K. Prekas, K. Antonakoglou, Wireless sensor networking architecture of polytropon: An open source

scalable platform for the smart grid. Energy Procedia **50**, 2014, pp. 270–276. doi:10.1016/j. egypro.2014.06.033

17. J.K. Rai, A.M. Lajimi, P. Xirouchakis, An intelligent system for predicting hpdc process variables in interactive environment. Journal of Materials Processing Technology **203** (1-3), 2008, pp. 72–79. doi:10.1016/j.jmatprotec.2007.10.011

18. Dörmann Osuna, Hans W, *Ansatz für ein prozessintegriertes Qualitätsregelungssystem für nicht stabile Prozesse*. Univ.-Bibliothek and Univ.-Verl. Ilmenau, Ilmenau and Ilmenau, 2009

19. Franco Bonollo and Nicola Gramegna, *The MUSIC guide to key-parameters in High Pressure Die Casting*. Assomet srl and Enginsoft SpA, 2014

Virtual Production Intelligence – Process Analysis in the Production Planning Phase

Daniel Schilberg, Tobias Meisen and Rudolf Reinhard

Abstract To gain a better and deeper understanding of cause and effect dependencies in complex production processes it is necessary to represent these processes for analysis as good and complete as possible. Virtual production is a main contribution to reach this objective. To use the Virtual Production effectively in this context, a base that allows a holistic, integrated view of information that is provided by IT tools along the production process has to be created. The goal of such an analysis is the possibility to identify optimization potentials in order to increase product quality and production efficiency. The presented work will focus on a simulation based planning phase of a production process as core part of the Virtual Production. An integrative approach which represents the integration, analysis and visualization of data generated along such a simulated production process is introduced. This introduced system is called Virtual Production Intelligence and in addition to the integration possibilities it provides a context-sensitive information analysis to gain more detailed knowledge of production processes.

Keywords Analysis · Digital Factory · Laser Cutting · Production Technology · Virtual Production · Virtual Production Intelligence · VPI

1 Introduction

Considering the individualization and increasing performance of products the complexity of products and production processes in mechanical and automatic processing is constantly growing. This, in turn, results in new challenges concerning the designing as well as the production itself. In order to face these challenges, measures are required to meet the demands which are based on higher complexity. One measure to face this challenge is a more detailed planning of the design and manufacturing of the products by the massive use of simulations and other IT tools which enable the user to fulfill the various demands on the product and its manufacturing. To a

D. Schilberg (✉) · T. Meisen · R. Reinhard
IMA/ZLW & IfU, RWTH Aachen University, Dennewartstr. 27, 52068 Aachen, Germany
e-mail: daniel.schilberg@ima-zlw-ifu.rwth-aachen.de

Originally published in "Transactions on Engineering Technologies",
© Springer 2013. Reprint by Springer International Publishing Switzerland 2016,
DOI 10.1007/978-3-319-42620-4_72

further improvement of simulations and IT tools it is important not to evaluate them separately but in their usage context: which tool is used to which planning or manufacturing process. It has to be fathomed which information on which effort between the tools are exchanged.

To formulate and execute an appropriate measure, it is necessary to create a basis which allows a holistic, integrative examination of deployed tools in the process. Aim of such an examination is an increasing product quality, efficiency and performance. Due to the rapid development of high-performance computers, the use of simulations in product design and manufacturing processes has already been well-established and enables users to map relations more and more detailed virtually. This has led to a change concerning the way to perform preparatory and manufacturing activities. Instead of an early development of physical existent prototypes, the object of observation is developed as a digital model which represents an abstraction of essential characteristics or practices. The subsequent simulation the digital model is used to derive statements concerning practices and properties of systems to be examined. The use of digital models in production processes is described by the term of virtual production which specifies a "mainstreaming, experimental planning, evaluation and controlling of production processes and plant by means of digital models" [1, 2].

This paper will show an integrative concept which describes an important component to achieve the objective of a virtual production by the integration and visualization of data, produced on simulated processes within the production technology. Taking account of the application domain of production technology and used context-sensitive information analysis, with the aim of an increasing improvement of knowledge concerning the examined processes, this concept is called Virtual Production Intelligence (VPI). The aim of this paper is to present how the VPI contributes to optimize manufacturing processes like laser cutting. The usage of the VPI in a factory planning process is shown in [3].

2 Problem

As a central issue of the virtual production the heterogeneous IT landscape can be identified. As indicated in the introduction, a variety of software tools to support various processes are used. Within these software tools data cannot be exchanged without effort. The automation pyramid offers a good possibility to demonstrate this difficulty. The automation pyramid is depicted in Figure 1. It shows the different levels of the automation pyramid with the corresponding IT tools and the flow of information between the levels. The level related processes are supported by the mentioned IT tools very good or at least sufficient. At the top level command and control decisions for the company management are supported by Enterprise Resource Planning (ERP) systems. Therefore, these systems allow the decision-makers in the management to monitor any enterprise-wide resources like employees, machinery or materials.

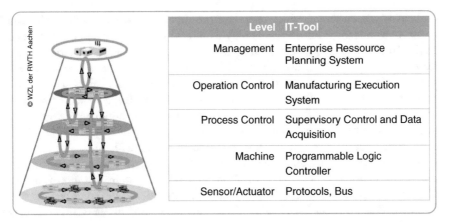

	Level	IT-Tool
	Management	Enterprise Ressource Planning System
	Operation Control	Manufacturing Execution System
	Process Control	Supervisory Control and Data Acquisition
	Machine	Programmable Logic Controller
	Sensor/Actuator	Protocols, Bus

Figure 1 Automation pyramid [4]

At the lower levels, the Manufacturing Execution Systems (MES), the data acquisition (Supervisory Control and Data Acquisition SCADA) and programmable logic controllers (PLC) are arranged according to the increasing complexity. The field level is the lowest level. Corresponding to protocols the data exchange is organized on this level. The used software tools are developed very well to support the corresponding processes on the appropriate level. The Association of German Engineers (VDI) addressed for the virtual production a unified data management as a way to use data and information across all levels, but this is not realized yet. Without a unified data management the data exchange from a PLC via the SCADA and MES up to the ERP system requires a great effort for conversions and aggregating. The goal is that the ERP can support decisions on the base down to the PLC Data and changes in the ERP system will change the input for the PLC. Currently most companies only exchange data between different levels instead of a flow of information across all levels. This is why a holistic picture of production and manufacturing process is not possible [5].

At present, a continuous flow of information is available only with the application of customized architectures and adapters to overcome the problem of heterogeneity. This involves high costs, why usually small and medium-sized enterprises (SMEs) have no integration of all available data into a system.

There are a high number of different IT tools for virtual production. These enable the simulation of various processes, such as in manufacturing technology, the realistic simulation of heat treatment and rolling process or the digital viewing of complex machinery such as laser cutting machines. At this juncture various independent data formats and structures have developed for a representation of the digital models. Whereas an independent simulation of certain aspects of product and manufacturing planning is possible, the integrative simulation of complex manufacturing processes involves high costs as well as an high expenditure of time because in general an interoperability between heterogeneous IT tools along the automation pyramid is not given. One approach to overcome the heterogeneity is the homogenization with the help of a definition of unified data standards. In this context a transfer of the data formats into a standardization of data by the use of specific adapters as mentioned above. However, this approach is not practical for the considered scenario for two reasons. Firstly, the diversity of possible IT tools that are used lead to a complex data standard. This is why its understanding, care and use are time and cost intensive. Secondly, the compatibility issues for individual versions of the standard are to be addressed (see STEP [6]). Therefore the standard must be compatible with older versions and enhanced constantly to reflect current developments of IT tools and to correspond to the progressive development through research [7, 8].

Another approach, which is chosen as basic in this paper, includes the use of concepts of the data and application integration, which do not require a unified standard. The interoperability of IT applications must be ensured in a different way so that no standard data format is necessary. This is done by mapping the various aspects of the data formats and structures on a so-called integrated data model or canonical data model [8, 9]. In current approaches to these concepts are extended to the use of semantic technologies. The semantic technologies enable a context-sensitive behavior of the integration system. The continuation of this approach enables the so-called adaptive application and data integration [10, 11].

The integration of all data collected in the process in a consolidated data management is only the first step to solving the problem. The major challenge that must be overcome is the further processing of the integrated data along a production process to achieve a combination of IT tools across all levels of the automation pyramid. The question of the analysis of data from heterogeneous sources is addressed in the analysis of corporate data for some time. The applications that enable integration and analysis of data are grouped under the term "Business Intelligence" (BI). BI applications have in common that they provide the identification and collection of data that arise in business processes, as well as their extraction and analysis [12, 13].

The problem in the application of BI on virtual production is that the implementation of the BI integration challenges of heterogeneous data and information conceptually solves in the first place which causes significant problems in the implementation of functional systems. Thus, in concept, for example, a translation of the data into a common data format and context-sensitive annotation is provided. A translation may not be achieved because it is proprietary information which meaning is not known to the annotation. This is also the reason why so many BI integrations have failed so far [14].

The following shows that the previously addressed problems should be solved by the vision of the digital factory. Because this vision is not realized yet, the section heterogeneity of simulations and solution: Virtual Production Intelligence will outline next steps towards the realization of a digital factory. The term "Virtual Production Intelligence" was selected in reference to the problem introduced in the term "business intelligence", which has become popular in early to mid-1990s. It called "business intelligence" methods and processes for a systematic analysis (collection, analysis and presentation) of a company's data in electronic form. Based on gained findings, it aims at improved operative or strategic decisions with respect to various business goals "Intelligence". In this context "Intelligence" does not refer to intelligence in terms of a cognitive size but describes the insights which are provided by collecting and preparing information.

3 Digital Factory

The digital factory is defined by the working group VDI in the VDI guideline [1] as:

"the generic term for a comprehensive network of digital models, methods and tools – including simulation and 3D visualization – integrated by a continuous data management system. Its aim is the holistic planning, evaluation and ongoing improvement of all the main structures, processes and resources of the real factory in conjunction with the product."

According to the VDI guideline 4499 the concept of the digital factory does not include individual aspects of the planning or production but the entire product life cycle (PLC) (Figure 2). All processes from the onset to the point of decommissioning shall be modeled. Therefore the observation starts with the collection of market requirement, the design stages including all the required documents, project management, prototypes (digital mock-ups), the necessary internal and external logistic processes, planning the assembly and manufacturing, the planning of appropriate manufacturing facilities, installation and commissioning of production facilities, the start-up management (ramp up), series production, sales to maintenance and ends with the recycling or disposal of the product all these points should be part of the

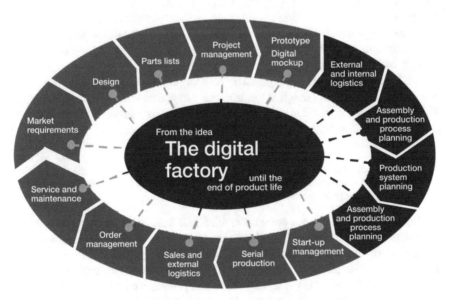

Figure 2 Product Life Cycle (VDI 4499) and localization of virtual production within the product life cycle in accordance with VDI Directive 4499

Digital Factory. Currently there is no platform which complies with this integration task. But there are already implemented some elements of the digital factory at different levels of the automation pyramid or in phases of the PLC.

Existing PLC Software products help companies to plan, monitor and control the product life cycle in parts. However, these applications are usually only isolated solutions and enable the integration of IT tools that have the same interfaces for data exchange and are provided by the same manufacturer. The detail of the images of individual phases of the product life cycle does not reach this high spatial resolution of special applications to the description of individual phases of the product life cycle or of IT tools that focus on aspects of individual phases. Therefore the recommendation of the VDI to design data management and exchange as homogeneous as possible can only be considered for new developments. Besides there is still no approach about how to implement a standard for such a homogeneous data exchange and how to prevent or avoid the known issues of a standardization process. Therefore even a project that wants to realize the homogenization of the flow of information cannot succeed, because it is not defined what such a condition has to look like. Moreover there is no standard or efforts to standardize as for example the Standard for the Exchange of Product Model Data (STEP) compete with proprietary formats. It must be considered that the proprietary formats were also used to protect the knowledge and skills of the software provider.

With view to a visualization of the digital factory there are tools of Virtual and Augmented Reality which enable users to realize 3D models of factories with or without people as well as to interact with it and to annotate information. A real time-control of a physical existent plant via virtual representation, at which data from the operation in virtual installation are illustrated and further processed for analysis, is right now not possible. The running times of individual simulations do not meet the real-time requirement. With the present techniques, its developments and innovations the goal of digital manufacturing is to be achieved.

The Virtual Production Intelligence serves as a basic building block for the digital factory. To achieve this goal, it is not necessary to address the overall vision of the digital factory, but rather it is sufficient to focus the area of simulation-based virtual production (see Figure 2). Again, the VDI guideline 4499 is cited to the definition of virtual production:

"is the simulated networked planning and control of production processes with the aid of digital models. It serves to optimize production systems and allows a flexible adaptation of the process design prior to prototype realization."

The production processes are here divided into individual process steps, which are described by simulations. The simulation of the individual process steps is done using modern simulation tools which can represent complex production processes accurately. Despite the high accuracy of individual simulations the central challenge in virtual manufacturing is the sum of individual process steps in a value chain.

The VPI is developed to set the interoperability of IT tools in a first step with distinctly less effort than using tailored solutions mentioned above. In a second step the integrated data is consolidated, analyzed and processed. The VPI is a holistic, integrative approach to support the implementation of collaborative technology and product development. Thereby enabling optimization potentials are identified and made available for the purpose of early identification and elimination of errors in processes. To better understand the terms holistic, integrative and collaborative will be defined as follows:

- Holistic: all parts of the addressed processes will be taken into consideration.
- Integrative: use and integration of existing solutions.
- Collaborative: consideration of all processes addressed in involved roles as well as their communication

In the next section, the above-mentioned heterogeneities that should be overcome by the use of the VPI, a closer look.

4 Heterogenity

Regarding ISO/IEC 2382-01 [15] interoperability between software applications is realized when the ability exists to communicate, to run programs, or to transfer data between functional units is possible in such a way that the user need no information about the properties of the application. Figure 3 summarizes the heterogeneities, which contribute significantly to the fact that no interoperability is achieved without using customized adapters [16–18].

The syntactic heterogeneity describes the differences in the technical description of data, for example different coding standards such as ASCII or binary encoding, or the use of floating-point numbers as float or double. These two types of heterogeneity can be overcome relatively easy by the use of adapters. Therefore a generic approach should be applied, so that the implemented adapters are reusable. Existing libraries and solutions are available to address the problem of technical heterogeneity. Most modern programming concepts contain methods for implicit type adjustments and controlling explicit conversion of data [16–18].

Overcoming the structural and semantic heterogeneity is the much greater challenge. Structural heterogeneity differences specify the representation of information.

Kind of Heterogeneity	Description/Examples
Syntactical	Presentation of data; e.g. format of numbers, encoding.
Structural	Order, in which data attributes are exported.
Semantical	Meaning of attribute denominations; t = *time* or *temperature* ?

Figure 3 Types of heterogeneity of simulations

Semantic heterogeneity describes the differences in the importance of domain specific entities and concepts used for their award. E. g. the concept of ambient temperature is used by two simulations, simulation A, uses the concept to define the room temperature of the site where the heating furnace is located. Simulation B uses the concept to define the temperature inside the heating furnace so the temperature in the immediate vicinity of the object to be heated is specified.

In the following section, the VPI is presented, which provides methods to overcome of the mentioned heterogeneity and to facilitate interoperability between applications [16–18].

5 Virtual Production Intelligence

The main objective for the use of the "Virtual Production Intelligence" is to gather results of a simulation process, to analyze and visualize them in order to generate insights that enable a holistic assessment of the individual simulation results and aggregated simulation results. The analysis is based on experts know how and physical and mathematical models. Through an immersive visualization requirements for a "Virtual Production Intelligence" are completely covered.

The integration of result data from a simulation process in a canonical data model (Figure 4) is the first step to gain knowledge from these data and to realize the extraction of hidden, valid, useful and actionable information. This information includes, for example, the quality of the results of a simulation process or in concrete cases, the causes for the emergence of inconsistencies.

Right now the user who has to identify such aspects has currently limited options to do so. With the realization of an integration solution a uniform view to the data gets possible. This includes on the one hand the visualization of the entire simulation

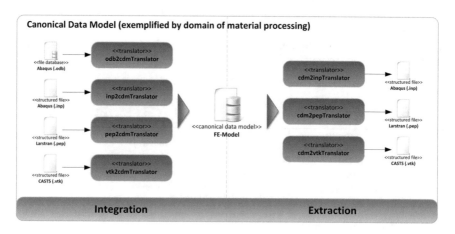

Figure 4 Canonical data model VPI

process in a visualization component and on the other hand, the analysis of the data over the entire simulation process. For this purpose, different exploration methods can be used.

First, the data along the simulation process is integrated into a canonical data model. This is implemented as a relational data model, so that a consistent and consolidated view of the data is possible. Subsequently, the data is analyzed on the analysis level by the user. The user can interact in an immersive environment to explore and analyze the data. With the ability to provide feedback to the analysis component, the user can selectively influence the exploration process and make parameterizations during runtime.

In addition to a retrospective analysis by experts, it is also useful to monitor the data during the simulation process. Such a process monitoring assures compliance with parameter corridors or other boundary conditions. Therefore, if a simulation provides parameter values outside the defined parameter corridors the simulation process will be terminated. Then experts can analyze the current results in order to subsequently perform a specific adaptation of the simulation parameters. A process monitoring could also enable the extraction of point-of-interests (POI) on the basis of features that would be highlighted by the visualization (Figure 5). The components of the "Virtual Production Intelligence" are shown in [3].

An effective optimization of different structures of production, such as determining the number of process chains and production segment is made possible only by mapping the interdependencies of different planning modules.

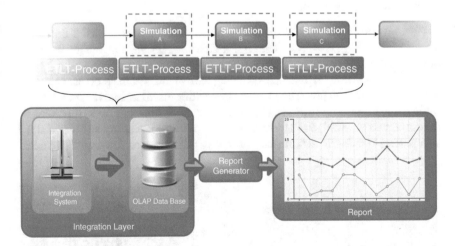

Figure 5 Extraction of point-of-interests

6 Application Domain Laser Cutting

The VPI is used to identify relevant machine parameters to optimize laser cutting processes concerning different goals like quality or speed. At first a brief look on the process itself is given. Laser cutting is a thermal separation process widely used in shaping and contour cutting applications. Therefore the laser cutting process has some advantages over conventional cutting techniques. The cutting process is very fast and accurate. Because an optical tool is used there is no risk of additional wear.

The ablation process in fusion metal cutting is based on thermodynamics and hydrodynamics. The absorbed laser energy is converted to heat which melts the material. This melt is driven out of the cut by a gas jet that is coming out of the nozzle, coaxially aligned to the laser beam. The VPI is a simulation based tool; therefore not the real process is used for optimization but the simulated. Hence, the analysis results of the VPI strongly correlate to the quality of the simulated laser cutting process. The core of each simulation is the simulation model that is used. The modelling of a laser cutting process requires the modelling of at least three entities at the same time. The optical tool – the laser beam – must be included, the material that should be cut by the laser beam, and the gas jet separating the melt. To gain a good model it is evident that the modeling of the following quantities has to be accomplished as well as their numerical implementation:

- The cutting gas flow
- The radiation propagation
- The ablation of the material

Figure 6 shows the simulation results based on a numerical model developed by the NLD at RWTH Aachen University, Germany, for the ablation ant the beam propagation into the cut kerf.

There are, however, gaps in understanding the dynamics of the process, especially issues related to cut quality. The user of a laser cutting machine needs to know how the surface roughness on cut edge can be better influenced. How can dross formation on the cut bottom be avoided and the inclination of cut edge be controlled? Especially it is important to understand the influence of laser parameters on those quality criteria.

Figure 6 Ablation simulation for laser cutting and simulated beam propagation into cut kerf

The most important parameters may be the wave length and the modelling of the wave length. Is a gas laser better than a solid state laser? The shape of the beam must be analysed: What is the influence of a Gaussian or a tophat shaped beam? Should the polarization circular or radial? Hence, the goal of the VPI is the reduction of numbers of parameters that are relevant for reaching a certain cutting quality. For that the correlation between chosen output or criterion and parameters or inputs must be determined. The VPI uses different methods to find the correlations (Figure 7). The VPI uses a sensitivity analysis:

"The study of how uncertainty in the output of a model (numerical or otherwise) can be apportioned to different sources of uncertainty in the model input." [19]

The results of qualitative methods can be visualized by scatter plots and quantitative methods will be used like the computation of rank correlation coefficients between various criteria and parameters.

The VPI is used for the planning of the laser cutting process. It supports the user in three ways. At first by using the VPIs data integration possibilities the user can gather data from various sources and get a consolidated view on these data. In the

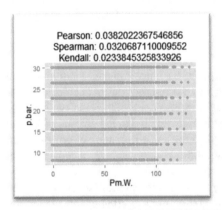
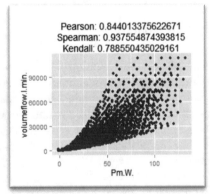

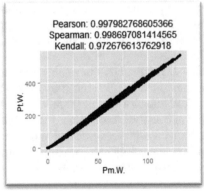

Figure 7 Methods used by the VPI for reduction of number of parameters

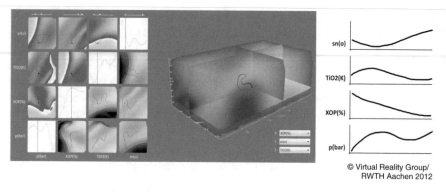

Figure 8 VPIs explorative visualization

second way the VPI provides an explorative visualization (see Figure 8) to present the data and to facilitate interaction. The last way is the data analysis to determine where and how to optimize process outcome.

7 Conclusion

In this chapter the Definition of the VPI was given as a holistic and integrative concept for the support of the collaborative planning, monitoring and control of core processes within production and product development in various fields of application. As a role model the idea of the Business Intelligence is used, applied to the domain of Virtual Production. The aim of the VPI is the identification and elimination of error sources in planning processes as well as detecting and taking advantage of enhancement potentials.

With the VPI an essential contribution to the realization of the vision of the digital factory can be achieved. The VPI is an integration platform that enables heterogeneous IT tools in the phase of product and production planning to interoperate with each other. Based on information processing concepts it supports the analysis and evaluation of cause-effect relationships. As product and production planning is the core area of Virtual Production, as part of the digital factory, the contribution is focused on this part. The VPI is the basis to establish interoperability. The functionality of the VPI was presented and illustrated by using the example of factory planning. The use of the VPI allows a significant reduction in engineering effort to create tailored integration and analysis tools, since the VPI is an adaptive solution. Now it is possible to start with a process-oriented and so contextual information processing. Information is now not only based on a single process step, it is related to the overall process, so that the importance and validity of information can be considered.

The future work concerning the VPI in the domain laser cutting will be the determination of machine parameters depending on desired machine states. That will be an optimization problem for a multidimensional function. The solution could be an explorative visualization based on the concept of hyperslices linked with 3D volume visualization. It is important to evaluate what cause-effect relationships can be identified through the exploration process. Furthermore, it must be examined how this information can be presented to the user in an immersive environment, and how can context information understandable and comprehensible be presented. For this purpose, there are various feedback-based techniques in which experts assess results of analysis and optimization. A bidirectional communication is needed, the user gives feedback and this feedback will be used to correct the displayed information. The system will store this feedback to avoid imprecise or erroneous statements.

References

1. VDI Richtlinie 4499, Blatt 1, Digital factory. Tech. rep., 2008
2. VDI Richtlinie 4499, Blatt 2, Digital factory. Tech. rep., 2011
3. D. Schilberg, T. Meisen, R. Reinhard, Virtual production – the connection of the modules through the virtual production intelligence. In: *Lecture Notes in Engineering and Computer Science: Proceedings of The World Congress on Engineering and Computer Science 2013, WCECS 2013, 23-25 October, 2013, San Francisco, USA*. pp. 1047–1052
4. R. Lauber, P. Göhner, *Prozessautomatisierung 1*, 3rd edn. Springer, Berlin, 1999
5. H. Kagermann, W. Wahlster, J. Helbig, *Umsetzungsempfehlungen für das Zukunftsprojekt Industrie 4.0 – Ab-schlussbericht des Arbeitskreises Industrie 4.0*. Forschungsunion im Stifterverband für die Deutsche Wissenschaft, Berlin, 2012
6. DIN EN ISO 10303
7. M. Nagl, B. Westfechtel, *Modelle, Werkzeuge und Infrastrukturen zur Unterstützung von Entwicklungsprozessen*, 1st edn. Symposium (Forschungsbericht (DFG)). Wiley-VCH, 2003
8. C. Horstmann, *Integration und Flexibilitat der Organisation Durch Informationstechnologie*, 1st edn. Gabler Verlag, 2011
9. D. Schilberg, *Architektur eines Datenintegrators zur durchgängigen Kopplung von verteilten numerischen Simulationen*. VDI-Verlag, Aachen, 2010
10. T. Meisen, P. Meisen, D. Schilberg, S. Jeschke, Application integration of simulation tools considering domain specific knowledge. In: *Proceedings of the 13th International Conference on Enterprise Information Systems*. 2011
11. R. Reinhard, T. Meisen, T. Beer, D. Schilberg, S. Jeschke, A framework enabling data integration for virtual production. In: *Enabling Manufacturing Competitiveness and Economic Sustainability – Proceedings of the 4th International Conference on Changeable, Agile, Reconfigurable and Virtual production (CARV2011), Montreal, Canada, 2-5 October 2011*, ed. by A.E. v. Hoda. Berlin Heidelberg, 2012, pp. 275–280
12. B. Byrne, J. Kling, D. McCarty, G. Sauter, P. Worcester, *The Value of Applying the Canonical Modeling Pattern in SOA. IBM (The information perspective of SOA design, 4)*. 2008
13. M. West, *Developing High Quality Data Models*, 1st edn. Morgan Kaufmann, Burlington, MA, 2011
14. W. Yeoh, A. Koronios, Critical success factors for business intelligence systems. Journal of computer information systems **50** (3), 2010, p. 23
15. ISO/IEC 2382-01
16. M. Daconta, L. Obrst, K. Smith, *The Semantic Web: The Future of XML, Web Services, and Knowledge Management*. 2003

17. D. Schilberg, A. Gramatke, K. Henning, Semantic interconnection of distributed numerical simulations via soa. In: *Proceedings World Congress on Engineering and Computer Science 2008*, ed. by I.A. of Engineers. Newswood Limited, Hong Kong, 2008, pp. 894–897
18. D. Schilberg, T. Meisen, R. Reinhard, S. Jeschke, Simulation and interoperability in the planning phase of production processes. In: *ASME 2011 International Mechanical Engineering Congress & Exposition*, ed. by ASME. Denver, 2011
19. A. Saltelli, S. Tarantola, F. Campolongo, M. Ratto, *Sensitivity Analysis in Practice: A Guide to Assessing Scientific Models*. John Wiley & Sons, Ltd., 2004

Text Mining Analytics as a Method of Benchmarking Interdisciplinary Research Collaboration

Stefan Schröder, Thomas Thiele, Claudia Jooß, René Vossen, Anja Richert, Ingrid Isenhardt and Sabina Jeschke

Abstract This paper introduces the process of adopting and implementing modern text mining approaches of analysis within the Cluster of Excellence (CoE) Tailor-Made Fuels from Biomass (TMFB) at RWTH Aachen University and presents initial results of the analysis of research output by use of common clustering algorithms, namely Principal Component Analysis and k-means. As one main part of this paper the data driven approach is classified into benchmarking efforts, which are part of the research work of the so called Supplementary Cluster Activities. The SCA, supporting the cluster management, are initiated in order to promote interdisciplinary collaboration of CoE researchers with different disciplinary backgrounds. This cross-linking is aided by means of knowledge engineering and knowledge transfer strategies, such as the exploration of synergies and benchmarking of research results as well as progress. In this course an adoption of current benchmarking efforts to the specific cluster research framework conditions is described. At this, in case of differing data sources according to those used in widespread business organisational benchmarking, possible TMFB data sources are outlined and a selection for analysis is reasoned. While benchmarking is usually differentiated in internal and external benchmarking, in this case focus lies on internal analysis of publications in order to reflect research work. Benchmarking of publications is used and implemented to identify (best) methods, practices and processes of CoE to improve the research organization. Second major part and central question within the scope of this paper is in which way text mining respectively clustering algorithms are sensitive applicable to TMFB publications and are able to be used as benchmark for research clusters. Thus thematically priorities of TMFB researchers will be investigated in order to create an overview according to research topics, keywords and methods. In case of an outlook further steps, e.g. dealing with generated results, data visualisation or further acquisition of data corpora, are formulated.

Keywords Benchmarking · Interdisciplinarity · Text Mining · Clustering · K-Means · Principal Component Analysis

S. Schröder (✉) · T. Thiele · C. Jooß · R. Vossen · A. Richert · I. Isenhardt · S. Jeschke
IMA/ZLW & IfU, RWTH Aachen University, Dennewartstr. 27, 52068 Aachen, Germany
e-mail: stefan.schroeder@ima-zlw-ifu.rwth-aachen.de

Originally published in "12th International Conference on Intellectual Capital, Knowledge Management & Organisational Learning (ICICKM 2015)", © Academic Conferences and Publishing International Limited 2015. Reprint by Springer International Publishing Switzerland 2016, DOI 10.1007/978-3-319-42620-4_73

1 Introduction: Cluster of Excellence 'Tailor-Made Fuels from Biomass' and 'Supplementary Cluster Activities' at the 'RWTH Aachen University'

In 2012 the further funding of the research Cluster of Excellence [1] "Tailor-Made Fuels from Biomass" (TMFB) at RWTH Aachen University was announced by the German Research Foundation (German abbreviation: DFG) and the German Council of Science and Humanities (German abbreviation: WR). Since 2007, TMFB has been working on a solution for one of the major challenges that society is facing today: a rising energy demand and the limited availability of fossil energy resources. With this challenge in mind, researchers from the fields of chemistry, biology, process engineering and mechanical engineering have joined forces in this CoE to develop new alternative fuels from biomass which will not be competing with the food chain [2]. In an interdisciplinary approach, more than 70 scientists are engaged in the synthesis and combustion of such fuels.

Over the course of the past decade, the number of interdisciplinary CoE at German Universities has been increasing, primarily because of the emergence of research issues that cannot be met by only disciplinary approaches [3, 4]. Thus, German CoE are initiated to focus on current issues and research questions which require the input of more than one discipline and their researchers. With the initiation of CoE the complexity of cooperation increases. Hence, as part of TMFB, the Supplementary Cluster Activities (SCA) were initiated to promote a cross-linking and collaboration of CoE members. The SCA pursues a two-fold strategy (cf. Figure 1). One working area mainly focuses on the promotion of early career researchers, promotion of gender equality and performance measurement as well as knowledge engineering. These tasks are summarized under the so called Collaboration-Enhancing Services, which comprise e. g. strategy workshops, colloquia, trainings and public relations. The second working area, in order to strengthen interdisciplinary collaboration, is called Collaboration-Enhancing Research. These tasks include e. g. identification of key performance indicators, exploration of synergies, benchmarking and support of knowledge transfer within and outside of the CoE. Hereafter measures according to the research work package, especially to benchmarking task, are outlined.

This paper points out an approach to investigating and adopting data driven analytical methods in order to achieve objectives regarding tasks like establishing (internal) benchmarks, utilizing synergies and supporting knowledge transfer. Therefore Section 2 deals with the relevance and connection of benchmarking, data-mining analytics and supplementary cluster activities tasks. On this occasion the necessity of benchmarking and differences to common used business benchmarking approaches are lined out. In Section 3 existing and allegedly useful data bases for analysis are described. The text mining driven approach by itself, its application for logical reasons and its targets are described in Section 4. Afterwards, first initial generated results by applying two frequently exploited clustering algorithms are presented to proof if the mentioned approach may lead to further process able results. In form of an outlook, in Section 5, the idea of adopting clustering algorithms in order to

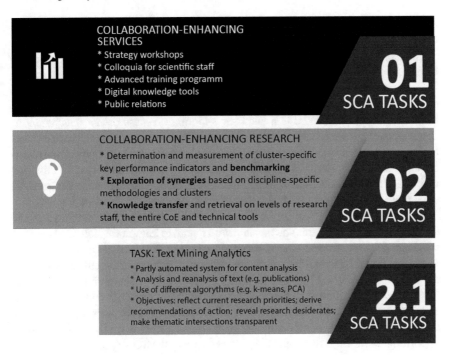

Figure 1   SCA tasks within the Cluster of Excellence TMFB (own diagram)

benchmark scientific research are discussed as well as the procedure of comparing TMFB research priorities to relevant topics from the general scientific community is briefly sketched. Likewise further steps, e.g. in form of visualisation and data acquisition, are pointed out.

2   Supplementary Cluster Activities in the Context of Benchmarking

Benchmarking is one task within the SCA portfolio. Following the interpretation and necessity to adopt available approaches is explained. Originated from the 'business world', benchmarking is traditionally thought of as a management tool that improves performance by identifying and applying best documented practices. Modern commercial benchmarking has now come to refer to the process of identifying the best methods, practices, and processes of an organization and implementing them to improve one's own organization [5, 6]. In order to do so, the best entities in a sector have to be identified and specific indicators like unit costs have to be compared [6–8]. The objective of benchmarking is to help the management of an organisation to improve performance. Over the last years many governmental and public organisations adopted a range of approaches to research benchmarking [9, 10, 32–34].

In an effort to create comparable and compatible quality assurance and academic degree standards, universities increasingly employ benchmarking strategies, too [5]. Originated from the DFG and WR, who interpret benchmarking as "evaluating and steering [...] research performance", "increasing self-reflection" and "avoiding [...] ton technology" [11], there is a need for adopting business driven benchmarking approaches to research clusters and its differing frameworks. Particularly in the context of acquiring university data, the CHE reveals common challenges for university benchmarking [12]. Especially, points regarding to the data acquisition (quality, format and effort) as well as assignment and interpretation problems are outlined. To meet these challenges applying data analytics is described in Sections 3 and 4.

Jackson and Lund [9] define benchmarking as collection of approaches for self-evaluation activities. On this account and with regard to Levy and Valcik [13] the SCA are dealing with benchmarking in a functional primarily internal and even external way. Functional benchmarking examines similar functions in organization that are not direct competitors [5]. Levy and Valcik's [13] outline benchmarking in a manner, that it is well transferable to the SCA understanding: "Benchmarking is a strategic and structured approach whereby an organization compares aspects of its processes and/or outcomes to those of another organization or set of organizations to identify opportunities for improvement." [13] Transferred to the SCA approach:

- Organization stands for the Cluster of Excellence TMFB
- Another organization or set of organizations stands for the relevant
- Scientific community (applies for other CoE as well as for other similar specialist disciplines)
- Improvement in this case includes: revealing synergies and intersections as well as uncovering research methods, practices and processes

Whatever it's (benchmarking) origins, implicit in the concept of benchmarking is the use of references by which other objects can be measured, compared, or judged [35]. Which 'references' can be considered for analysis is discussed in Section 3.

At this point the question may arise, what objectives are pursued by implementing and adopting a benchmarking approach to TMFB. Objective is to ...

- ...create an overview according to research topics, keywords and methods
- ...compare actual and desired values respectively results
- ...reflect on research focus and alignment
- ...identify synergies and intersections between TMFB researchers
- ...identify synergies with external entities
- ...analyse analogies and differences of TMFB research priorities in comparison to general scientific community research topics
- ...adapt user-specific visualization of results

This mentioned objectives are not only motivated theoretically or based on research funding sources like the WR. As Jooß [14] inter alia worked out in a CoE questioning, the success of interdisciplinary cooperation depends on methodological exchange, identification of key persons, reflection of research results, identification and visualization of interfaces, cross-linking and reflection [14]. In order to handle these demands, the following SCA approach of analyzing data is superimposed. As this paper outlines the whole SCA approach and holistic objectives are introduced, the results presented in Section 4 regard to the above mentioned objective 'create an overview onto research topics, keywords and methods', which are used within TMFB.

Section 3 briefly explains benchmarking differences between CoE and commercial operating companies and lists a selection of possible information bases which can be analyzed.

3 Benchmarking and Data Sources

In the context of benchmarking, there is a need for measuring, comparing and judging standards or references [5]. University frameworks and CoE funding objectives are unlike companies. While CoE are driven and judged by scientific output and knowledge production, companies are sales and product output driven [6]. Epistemic interests (WR) are paramount to CoE, contrary to companies' products or service output. On this account, the basis of valuation differs. If the quality of CoE performance is to be judged, scientific output (e. g. publications) than (none existing) sales revenues are being consulted (sure, CoE are managed on financial basis, but no direct revenues of CoE are expected from political or funding perspective. Return on investments are desired for the business location Germany and therefore are hardly quantifiable). Therefore, contrary to business units, the success and progress of CoEs must be evaluated on another information basis. That's why possible access to CoE information from an external point of view is identified. In case, as part of the management, the SCA have further internal access, the external access resulting from endeavour of transferability is preferred. Table 1 lists possible sources (without any claim to completeness) and corresponding content:

It is obvious to deal with this existing information. Noticeable, nearly almost all types of data sources contain to a greater or lesser extent vast text amounts. In order to develop a benchmarking approach to performant analyse this text amounts to achieve results regarding to content, method and intersections of TMFB work, the SCA decided to adopt data driven analytical methods. In a first step, publications are chosen as information source because of their informative value, accessibility and comparability. This paper outlines the processing of a TMFB publication sample in order to reflect current research foci and together used keywords (e. g. methods or materials). Future analysis will consider general scientific publication databases, to achieve the SCA objectives mentioned in Section 2 (e. g. analyze analogies/differences of TMFB research priorities in comparison to general scientific community research topics). In the case of analyzing publications text mining analytics are sensitive. Later on, in accordance with the holistic analysis of data mentioned in

Table 1 Free access to CoE data sources (own diagram)

Source	(derivable) Content
Online presence CoE	• General information onto project structure and objectives • Background information according to funding volume and runtime • News and announcements • Contact persons and responsibilities
Online presence involved partners	• Partner objectives • Researcher assignment and expertise • Number of employees • Dissertation projects
Publication databases TMFB	• *Publications (e. g. conferences, journals)* • *Research results (e. g. research questions, methodology, findings, further research desiderates)* • *Cooperation structure (who cooperates with whom – internal and external)* • *Quality of research output (e. g. journal impact factor)*
Publication databases general	• *Research results (e. g. research questions, methodology) according to the scientific community* • *See above*
Newsletter	• Current information and news
Social Media (Xing, LinkedIn, facebook, twitter, YouTube)	• Employee information (professional and personal) • CoE/project news • Announcements (e. g. Call for Papers, new Professorships) • Images/photos/illustrations • Videos • Project visits
Brochures, flyer and demonstrator	• General information • Research questions • Specific advertisements • Lego©demonstrator (process description)
Research funding	• General information regarding to e. g. funding criteria • General annual reports (e. g. funding volume)
Online Stores (e. g. Amazon, Springer)	• Book publications
Proposals, research reports, project reviews, advisory board advises etc.	• No external access
...	• ...

Table 1, web-mining and further analytics have to be applied as well (e. g. in case of analyzing social media content).

At this point the question may arise, why (semi-)automatized approaches should be used? There are several reasons for the implementation of an automated system for the content analysis of text. Krcmar [15] also describes the necessity of adopted information management systems for the management of organizations. He outlies the efficient usage to supply researcher and stakeholders with relevant information [15]. The benefit for TMFB is the automation of this process to reduce the manually workload and to allow more efficient analysis and reanalysis of text. Such a system will be applicable to extremely large quantities of text where there is little possibility of intense human analysis [16]. In case of the TMFB publication sample (described in Section 4) and especially in view of analysing holistic TMFB publications as well as community relevant publications, the process automation is essential. Otherwise the workload in manually analysing is disproportionate and it could be possible that patterns cannot be revealed by classical text analysis.

Section 4 clarifies functionalities of text mining analytics and assesses existing text mining respectively clustering algorithms and presents first results coming from a data sample of 28 TMFB publications.

4 Applying Text Mining Analytics

"While technology enables us to capture and store ever larger quantities of data, finding relevant information like underlying patterns, trends, anomalies, and outliers in the data and summarizing them with simple understandable and robust quantitative and qualitative models is a grand challenge. Data mining helps to discover underlying structures in the data, to turn data into information, and information into knowledge." [17]. Text mining is the extraction of implicit, previously unknown, and potentially useful information from data [17]. In view of the provided data, a quantity of text (28 publications) is analyzed, in order to investigate applicability of clustering algorithms to prepare existing data for further analysis regarding to the earlier mentioned objectives.

Within the help of text mining, patterns can be discovered and it is possible to predict future values, derive recommendations, optimize the choice of methods and identify possible or even new research paths [17] (Miner 2009). Text mining extends the applicability to un- or semi-structured data like texts from documents, news, customer feedback, e-mails, web pages, internet discussion groups and social media [17]. The following subsections elaborate further on pre-processing the data and applying clustering algorithms. In the end of Section 4 results according to TMFB data are described.

4.1   Clustering and Pre-processing Publications

An access to deal with TMFB publications by the use of text mining algorithms in order to create an overview onto research topics, keywords and methods is clustering, i.e., the task of automatically grouping objects into groups of similar objects. Cluster analysis divides a heterogeneous group of records into several more homogeneous classes or clusters [18]. Below, the choice of a cluster algorithm is roughly sketched. According to Miner [19] Figure 2 underlines the selection of clustering data. To summarize, clustering data is sensible because in this case we exploratory analyze sorted documents with no given categories. Whether existing algorithms are applicable to the TMFB application will be determined in the following paragraph.

"Clustering is a technique that is useful for exploring data. It is particularly useful where there are many cases and no obvious natural groupings." [17]. As mentioned by Kempf and Siebert [20], benefits of using cluster analysis are inherent in grouping none clearly assignable, poor structured and assorted data [20]. Thus patterns within heterogeneous publications can be revealed [6]. By clustering groups of publications with mainly related properties are built [17]. Affiliations are expressed by homogeneity of objects within the clusters and heterogeneity between them [21, 22]. A good clustering method produces high-quality clusters to ensure that the inter-cluster similarity is low and the intra-cluster similarity is high; members of a cluster are more like each other than they are like members of a different cluster.

Specifically, functionally related properties can be grouped together to i. e. novel research pathways or identify patterns based on statistical similarity [7]. This depends on a distance metric, such as Euclidean distance (squared root sum of distances along each feature) or Manhattan distance (sum of absolute feature differences) [18]. The aim of clustering is twofold. Firstly, it seeks to separate data items into a number of groups so that items within one group are more similar to other items in the same group. Secondly, it aims to arrange so that items in one group are different from items in other groups. Visualizing how the clusters are formed and which data points are in which cluster is difficult [17]. Clustering techniques can be divided into different approaches i.e. agglomerative and divisive approaches as well as portioning approaches (e. g. [23]). [7, 17].

Before clustering can be proceeded, pre-processing the data is an important step in text mining processes, because the given data has to be transformed in a machine-readable format. In the case of analyzing publications the original text has to be

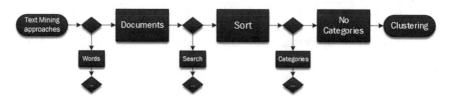

Figure 2   Choice of clustering algorithm (own diagram following [19])

word	total	in class (TMFB_Literatur)
production	1175,0	1175,0
reaction	1077,0	1077,0
model	789,0	789,0
temperature	775,0	775,0
pressure	713,0	713,0
conditions	706,0	706,0
process	672,0	672,0
glucose	599,0	599,0
reactions	585,0	585,0
cellulose	576,0	576,0
experiments	520,0	520,0
ignition	520,0	520,0
oxygen	511,0	511,0
energy	501,0	501,0
yield	491,0	491,0
concentration	462,0	462,0
liquid	444,0	444,0
biomass	439,0	439,0
lignin	437,0	437,0
phase	430,0	430,0
hydrogen	399,0	399,0
method	395,0	395,0
formation	383,0	383,0
species	380,0	380,0
product	374,0	374,0
water	370,0	370,0

Figure 3 Word occurrences after pre-processing sorted by frequency

reduced to only relevant words, which are accessible for statistical analysis. Therefore steps like tokenizing, filtering stopwords and tokens by length as well as POS-Tagging are conducted.

Figure 3 shows first results with regard to the data pre-processing. Occurences of recurrent words respectively attributes are investigated after decompositing words from the text, reducing of articles, conjunctions, präpositions and negations, as well as assigning words and punctuations to verbosities. Further investigation according to the comparison of word occurences and document belongings have to be done and may give insights in terms of specific word usage or common cluster topics. For instance at first view the identifcation of common used substances (e. g. glucose, cellulose, oxygen, lignin, hydrogen etc.) is possible. By applying clustering algorithms word occurences influence the assignment. Whether clustering techniques can be applied sensitive is worked out in the following.

4.2   Principal Component Analysis (PCA)

Often in the context of cluster analysis k-means and principal component analysis (PCA) are used to reduce dimensions of parameter space. PCA is one of the most widely employed and useful tools in the field of exploratory analysis [24]. PCA is used if the structure of the data is no obvious. It turns the orthogonal coordinate system so that the covariance matrix is diagonalized. The basic assumption of PCA is that the direction with the biggest variance contains most of the information [25]. The axis of the greatest variance is called the first principal component. Another axis, which is orthogonal to the previous one and positioned to represent the next greatest variance, is called the second principal component [26].

PCA is often used to identify the strongest predictor variables in a data set. PCA is a technique for revealing the relationships between variables in a data set by identifying and quantifying a group of principal components. [18]. Principal components have been used frequently in studies as a means to reduce the number of raw variables in a data set [27, 28]. PCA is interesting because it serves as a starting point for many modern algorithms [29].

Applying PCA to TMFB research papers, the following components can be investigated. Table 2 gives exemplary insights to principal components and their attributes. It becomes clear, that attributes are successfully assigned by its eigenvectors. Concerning the contents, intersections between different TMFB authors can be exposed as every principal component is determined by not less than two different publications.

Looking up the principal component assignment of papers according to Table 2, the following distribution of publications within the principal components (PC) is assigned: PC3 is mainly described (highest impact) by nearly 5 publications, PC6 by 2 publications and PC 8 by 2. In this case, validating the results by looking up the eigenvectors in the documents shows a successful application of the PCA. This

Table 2   Exemplary Eigenvectors PCA TMFB publications

Attribute	PC 3	Attribute	PC 6	Attribute	PC 8	Attribute	PC 12
Cellulose	0,278	Fuel	0,394	Lignin	0,497	Descriptors	0,235
Ignition	0,234	Bio fuel	0,290	Organosolv	0,219	Xylan	0,177
Flame	0,190	Methylfuran	0,115	Liquids	0,188	Lignin	0,159
Laminar	0,186	Toxicity	0,098	Conductivity	0,183	Catalysts	0,131
Rules	0,156	Bio mass	0,097	Viscosity	0,146	Xylose	0,110
Delay	0,144	Ignition	0,071	Alcohol	0,140	Silica	0,107
Radicals	0,135	Motor	0,065	Solutions	0,119	Conformer	0,097
Hydrolysis	0,132	Bio diesel	0,062	Catalytic	0,094	Organosolv	0,092
Glucose	0,131	Butanol	0,047	Acetate	0,087	Catalysis	0,088
…	…	…	…	…	…	…	…

approaches allows inter alia the identification of intersections between researchers as well as revealing of commonly together used keywords. Concluding at this point leads to an assumption, that clustering TMFB publications by PCA will lead to the desired objectives in case of analysing publications and reflecting research work. Further investigation by applying the algorithm to the whole TMFB data has to be done, especially in case of extracting and separating methods from research topics or substances and approaches. Likewise mapping researchers to keywords has to be integrated in a next step.

4.3 K-Means Algorithm

The application of the most common clustering algorithm k-means and its results are presented in the following paragraph. The classical k-means algorithm was introduced by [36]. Its basic operation is explained hereafter: given a fixed number (k) of clusters, assign observations to those clusters so that the means across clusters are as different from each other as possible. The points are iteratively adjusted so that each of the N points is assigned to one of the K clusters, and each of the K clusters is the mean of its assigned points [18, 30]. The underlying mathematical function is

$$J = \sum_{i=1}^{k} \sum_{x_j \ni S_i} ||x_j - \mu_i||^2| \tag{1}$$

whereby x_j is defined as data points and y_i as main focus of clusters S_i. The target function is based upon the method of smallest squares and is described as clustering by variance minimization. Because $||x_j - \mu_i||^2$ is the squared Euclidean distance, k-means arranges every object to the ones close by.

The k-means algorithm finds the best location for k centroids, each corresponding to the center of a cluster. Starting with k random, examines the algorithm data points closest to the centroid. For each cluster, the points within it are considered to determine a new centroid. This new centroid is then used to determine the closest points potentially causing points to move into different clusters [17, 18]. This process is repeated until the centroid stops moving or until an error function is minimized. It is best suited for clusters that are spherical in shape and are sufficiently distant from one another to be distinct. Its weaknesses are that it is limited to spherical clusters of similar density and it sometimes gets stuck in a local minimum where the centroids do not represent the best clustering. Another aspect is that the number of clusters k, is not determined automatically, however this does allow different runs to be performed in order for cluster validity measures to be compared [17].

Applying k-means to TMFB data, by setting k to 25 (according to the PCA 25 clusters with sufficient impact were built) the publications were clustered. Exemplary results are displayed in Table 3.

Table 3 Exemplary attributes per cluster (k-means)

Attribute	Cluster_0	Attribute	Cluster_4	Attribute	Cluster_6
Hexanol	0,456	Toxic	0,357	Flame	0,505
Phase	0,309	Mutagen	0,278	Stencil	0,302
Analyst	0,267	Bio diesel	0,194	Isotherm	0,193
...	...	...	...	...	...

Validating the results as done by the PCA leads to similar results. Building content-related clusters upon TMFB publications by use of k-means algorithms works. Compared with the PCA algorithm, the performance of k-means algorithm is worse. In case of greater numbers of publications at this point PCA seems to be the performant alternative. To sum up, applying text mining algorithms to research publications is suitable. Relevant key words and their allocation can be described by using k-means as well as PCA algorithms. This is the starting point for further investigations in order to create an overview onto research topics, keywords and methods. Further steps and thoughts are pointed out in Section 5.

5 Outlook: Clustering, Web Mining and Visualization

To sum up the considerations made in the introduction of this paper: The introduced concept of benchmarking and reflecting of scientific research output by applying data-mining approaches will be possible. Clustering data with common algorithms reveals first suitable and promising results. Correlations concerning the contents have been investigated. Based on the findings in Section 4, results can be derived in order to satisfy objectives like identifying current research topics, keywords and methods. Nevertheless, the used algorithms and maybe even further algorithms must be adopted and refined to get more detailed and more performant information on specific content generated by the CoE. Thus, in case of the PCA, reducing the threshold by excluding of attributes may be one possible way which will be further investigated. Additional attention lies on the validation of clustering approaches. For example, the given examples (cf. Section 4: Tables 2 and 3) must be reflected against the backdrop of the generated clusters and their attribute composition.

Besides, the acquisition of a second data corpus is initiated. Hereby publications from associated fields (e. g. mechanical engineering, biology and chemistry) are collected in order to apply k-means and PCA algorithms to provide replies onto the SCA objectives mentioned in Section 2. Thus the reflection of research focus and alignment, the identification of synergies with external entities and the analysis of analogies and differences of TMFB research priorities in comparison to general scientific community research topics should be facilitated. Thus the benchmarking approach, as described in Section 2, can be applied appropriate.

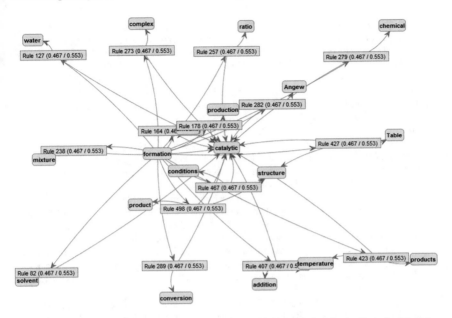

Figure 4 Association Rules- graphic view onto intersections related to the attribute "catalytic"

So far not notified, dealing with this topic also includes questions regarding to the visualisation of results in order to interpret and present results and thus produce benefits for researcher in an easy accessible and understandable way. For this, beside the usage of semantic mapping, association rules algorithms can be investigated as well. First studies show promising results (cf. Figure 4 and [31]. Simultaneously adjustments get obvious in case of reducing generic attributes, i. e. complex, angew or table. Possibilities will be figured out within the scope of SCA research activities.

Acknowledgments This research has been funded by the German Research Foundation (DFG) as part of the Clusters of Excellences 'Tailor-Made Fuels from Biomass' and 'Integrative Production Technology for High-Wage Countries' at RWTH Aachen University.

References

1. CoE Förderantrag, Exzellenzcluster Maßgeschneiderte Kraftstoffe aus Biomasse, *Proposal for the Establishment of Clusters of Excellence: Unveröffentlichter Förderantrag, eingereicht bei der Deutschen Forschungsgemeinschaft (DFG) und dem Wissenschaftsrat (WR)*. 2007
2. Cluster of Excellence. Renewal proposal for the cluster of excellence "tailor-made fuels from biomass": Excellence initiative by the german federal and state governments to promote science and research at german universities: unpublished, 2011
3. E.M.K.a.B. CoE Förderantrag, *Renewal Proposal for the Cluster of Excellence: Tailor-Made Fuels from Biomass: Unveröffentlichter Förderantrag, eingereicht bei der Deutschen Forschungsgemeinschaft (DFG) und dem Wissenschaftsrat (WR)*. 2011

4. Deutsche Forschungsgemeinschaft (DFG), *Verwaltungsvereinbarung zwischen Bund und Ländern gemäß Artikel 91 b Abs. 1 Nr. 2 des Grundgesetztes über die Fortsetzung der Exzellenzinitiative des Bundes und der Länder zur Förderung von Wissenschaft und Forschung an deutschen Hochschulen - Exzellenzvereinbarung II (ExV II)*. Bonn, 2009

5. G.D. Levy, S.L. Ronco, How benchmarking and higher education came together. New Directions for Institutional Research **2012**(156), 2012, pp. 5–13. doi:10.1002/ir.20026

6. J. Rehäuser, Prozeßorientiertes informationsmanagement-benchmarking. In: *IV-Controlling auf dem Prüfstand*, ed. by H. Krcmar, A. Buresch, M. Reb, Gabler Verlag, Wiesbaden, 2000, pp. 189–229. doi:10.1007/978-3-322-90245-0_10

7. X. Dai, T. Kuosmanen, Best-practice benchmarking using clustering methods: Application to energy regulation. Omega **42**(1), 2014, pp. 179–188. doi:10.1016/j.omega.2013.05.007

8. P. Bogetoft, *Performance benchmarking: Measuring and managing performance*. Management for Professionals. Springer, New York, 2012

9. N. Jackson, H. Lund, *Benchmarking for higher education*. Buckingham, 2000

10. H. Lund, P. Vestergaard-Poulsen, I. Kanstrup, P. Sejrsen, The effect of passive streching on delayed onset muscle soreness, and other detrimental effects following ecccentric exercise. Scandinavian Journal of Medicine & Science in Sports 8(4), 1998, pp. 216–221

11. D. Dzwonnek. Statement zum forum 1 „informationsbedarf der hochschulen und relevanz von leistungsvergleichen" im rahmen der „nationalen tagung zur bedeutung des forschungsratings als instrument der strategischen steuerung und kommunikation" des wissenschaftsrats, 21. september 2012, bonn., 2012

12. Y. Hener, P. Giebisch, I. Roessler, Entwicklung geeigneter Indikatoren und Kennzahlen für die Steuerung der University Leipzig – Benchmarking von Fakultäten, CHE Centrum für Hochschulentwicklung gGmbH. Workingpaper (103), 2008

13. G.D. Levy, N.A. Valcik, eds., *Benchmarking in Institutional Research*. New Directions of Instituional Research. San Francisco, 2012

14. C. Jooß, *Gestaltung von Kooperationsprozessen interdisziplinärer Forschungsnetzwerke*. Aachen, 2014

15. H. Krcmar, *Einführung in das Informationsmanagement*, 2nd edn. Springer-Lehrbuch. Springer Gabler, Berlin, Heidelberg, 2015

16. A.S. Smith, M.S. Humphreys, Evaluation of unsupervised semantic mapping of natural language with leximancer concept mapping. Behavior Research Methods (38), 2006, pp. 262–279

17. M. Hofmann, R. Klinkenberg, Rapidminer: Data mining use cases and business analytics applications, 2014

18. R. Nisbet, J. Elder, G. Miner, *Handbook of Statistical Analysis & Data Mining Applications*. Elsevier, 2009

19. G. Miner, D. Delen, J. Elder, A. Fast, T. Hill, R. Nisbet, eds., *Practical Text Mining and Statistical Analysis for Non-structured Text Data Applications*. Elsevier, 2012

20. K. Mertins, G. Siebert, S. Kempf, *Benchmarking: Praxis in deutschen Unternehmen*. Springer-Verlag, Berlin and New York, 1995

21. K. Backhaus, *Multivariate Analysemethoden – Eine anwendungsorientierte Einführung*, 6th edn. Berlin, 1990

22. J. Hartung, B. Elpelt, *Multivariate Statistik - Lehr- und Handbuch der angewandten Statistik*. München Wien, 1984

23. C. Ding, X. He, K-means clustering via principal component analysis. Proceedings of the 21st International Conference on Machine Learning. Canada., 2004

24. M. Monfreda, Principal component analysis: A powerful interpretative tool at the service of analytical methodology. In: *Principal Component Analysis*, ed. by P. Sanguansat, 2012, pp. 49–66

25. I.T. Jolliffe, *Principal Component Analysis*, 2nd edn. Springer, 2010

26. P. Sanguansat, Two-dimensional principal component analysis and its extensions. In: *Principal Component Analysis*, ed. by P. Sanguansat, 2012, pp. 1–22

27. I. Fodor. A survey of dimension reduction techniques, 2002. http://e-reports-ext.llnl.gov/pdf/240921.pdf

28. M. Hall, G. Holmes, Benchmarking attribute selection techniques for discrete class data mining. In: *Proceedings of IEEE Transactions on Knowledge and Data Engineering*, vol. 15. 2003, vol. 15, pp. 1437–1447

29. O. Maimon, L. Rokach, *Data mining and knowledge discovery handbook*, 2nd edn. Springer, New York and London, 2010

30. C.M. Bishop, *Pattern Recognition and Machine Learning*. Springer, 2007

31. S. Schröder, T. Thiele, C. Jooß, R. Vossen, A. Richert, I. Isenhardt, S. Jeschke, Text mining analytics as a method of benchmarking interdisciplinary research collaboration. 12th International Conference on Intellectual Capital, Knowledge Management & Organisational Learning, ICICKM 2015, 05.11.2015-06.11.2015, Bangkok, Thailand, 2015, pp. 408–417

32. R. Farquhar, Higher education benchmarking in Canada and the United States of America. In: A. Schofield (ed.), *Benchmarking in Higher Education: An International Review*. London: United Nations Educational, Scientific and Cultural Organization, CHEMS; Paris, 1998

33. V. Massaro, Benchmarking in australian higher education. In: A. Schofield (ed.), *Benchmarking in Higher Education*. United Nations Educational, Scientific and Cultural Organization, London: CHEMS; Paris, 1998

34. U. Schreiterer, Benchmarking in european higher education. In: A. Schofield (ed.), *Benchmarking in Higher Education: An International Review*. United Nations Educational, Scientific and Cultural Organization, London: CHEMS; Paris, 1998

35. S.L. Ronco, *Internal benchmarking for institutional effectiveness*, In: New Directions for Institutional Research, Issue 156, 2012, pp. 15–23

36. J.A. Hartigan, *Clustering Algorithms (Probability & Mathematical Statistics)*. John Wiley & Sons Inc., 1975

Printed in the United States
By Bookmasters